THE
HUMAN CAREER

HUMAN CAREER

Human Biological and Cultural Origins

SECOND EDITION

Richard G. Klein

The University of Chicago Press • Chicago and London

Richard G. Klein is professor of anthropology and human biology at Stanford University. His books include *Ice Age Hunters of the Ukraine* and (with Kathryn Cruz-Uribe), *The Analysis of Animal Bones from Archeological Sites*, both published by the University of Chicago Press.

The University of Chicago Press, Chicago 60637
The University of Chicago Press, Ltd., London
© 1989, 1999 by The University of Chicago
All rights reserved. Published 1999

08 07 06 05 04 03 02 01 00 99   1 2 3 4 5

ISBN: 0-226-43963-1 (cloth)

Library of Congress Cataloging-in-Publication Data
Klein, Richard G.
    The human career : human biological and cultural origins / Richard
G. Klein. — 2nd ed.
        p.   cm.
    Includes bibliographical references and index.
    ISBN 0-226-43963-1 (cloth : alk. paper)
    1. Human beings—Origin.   2. Fossil hominids.   3. Human evolution.
I. Title.
GN281.K55   1999
599.93′8—dc21   98-31387
                                                                              CIP

♾The paper used in this publication meets the minimum requirements of the American National Standard for Information Sciences—Permanence of Paper for Printed Library Materials, ANSI Z39.48-1992.

To Gail, for her support and patience

CONTENTS

ILLUSTRATIONS

TABLES

PREFACE TO THE
SECOND EDITION

Paleoanthropology—the parent discipline for human paleontology and paleolithic archeology—has never been more vigorous or productive. In the ten years since the first edition of this book was published, paleoanthropologists have discovered human fossils that antedate 4 million years ago and that show ever more clearly the ape origins of the human family; they have found new fossils and archeological sites that imply a complex history of branching events within the genus *Homo*; and they have uncovered a wealth of new fossil, archeological, and genetic evidence that the last shared ancestor of all living humans existed in Africa no more than 200,000 years ago.

The purpose of this edition is to show how the new discoveries and analyses supplement previous ones to reveal the basic course of human evolution. Two conclusions are especially clear. The first is that the australopithecines and other very early people who lived between 5 and 1.8 million years ago were morphologically intermediate in appearance and behavior between apes and unquestionable humans, and that the first "true" humans, who appeared 1.8 to 1.7 million years ago, were morphologically and behaviorally intermediate between the australopithecines and living people. An ironic corollary is that the human family now provides one of the most persuasive fossil cases for the occurrence of macroevolution.

The second major conclusion is that fresh fossil, archeological, genetic, and geochronological findings confirm earlier ones suggesting that modern humans originated in Africa and later replaced other kinds of people in Eurasia. The first edition espoused this "Out of Africa" scenario only cautiously, and it retained an earlier view that human evolution comprised a series of grades or stages, from the relatively apelike australopithecines through *Homo habilis*, *Homo erectus*, and "early" (or "archaic") *Homo sapiens* to modern *Homo sapiens*.

The richer fossil, archeological, and genetic records that bolster "Out of Africa" now indicate that the stage system must be discarded in favor of a branching scheme. Thus the text argues that after an initial human dispersal from Africa by 1 million years ago, at least three geographically distinct human lineages emerged. These culminated in three separate species: *Homo sapiens* in Africa, *Homo neanderthalensis* in Europe, and *Homo erectus* in eastern Asia. *Homo sapiens* then spread from Africa, beginning perhaps 50,000 years ago to extinguish or swamp its archaic Eurasian contemporaries. The spread was prompted by the development of the uniquely modern ability to innovate and to manipulate culture in adaptation. This ability may have followed on a neural transformation or on social and technological changes among Africans who already had modern brains. Whichever alternative is favored, the fossil, archeological, and genetic data now show that African *H. sapiens* largely or wholly replaced European *H. neanderthalensis.* The situation in eastern Asia is more obscure, because the fossil, archeological, and geochronological data are much sparser, but fresh data will probably show that *H. sapiens* replaced *H. erectus* in the same way.

The change from a synthesis centered on stages to one based on branches required a new chapter titled "Evolution of the Genus *Homo*" in place of the first edition's separate chapters titled "*Homo erectus*" and "Early *Homo sapiens.*" The new chapter stresses the fossil and archeological evidence for multiple, contemporaneous species within fossil *Homo*, but it also acknowledges the continuing meagerness of the record, particularly before the advent of unequivocal *Homo sapiens* and *Homo neanderthalensis* by 130,000 years ago. To ensure that readers appreciate just what the evidence is (and what it is not), the new chapter illustrates many of the key fossils and artifacts.

Most of the remaining chapters retain their original titles, but every one has been updated and expanded, and they all have many new illustrations. The total number of figures for the book has been increased from 151 to 223, and I could not have produced them without the willing, skilled assistance of my archeological colleague and friend Kathryn Cruz-Uribe. Even a cursory examination of the following pages will reveal her enviable ability to render informative, tasteful line drawings of complex fossils and artifacts. I used the computer programs Adobe Streamline and Macromedia Freehand to enhance her drawings with labels for key landmarks or features.

Like the first edition, this one differs from most commercial texts in emphasizing not only the broad pattern of human evolution but also the fossil and archeological data behind it. The book includes far more fossil and archeological detail than standard introductions to human evolution, and it is intended to be as much a sourcebook as a text. Nonetheless, the first edition was often used as a text, and with this in mind I have tried to make the second edition more accessible to undergraduate students. The most significant change is the addition of an introductory chapter that outlines how species evolve and how evolutionary biologists (including paleoanthropologists) classify and name them.

My own students have often complained about the standard academic author-date referencing system used in the first edition, because it obstructs text flow and intelligibility, especially when many citations are chained together. In this edition I have therefore adopted a system in which numbers in parentheses in the text refer to numbered items in a summary reference list. A secondary benefit of this system is that it shortens the text by more than five thousand words. The reference list is alphabetized, so works by a particular author are easy to locate. There is also a separate "reference index" that lists the text pages on which individual numbered works are cited. I used the Endnote computer program to ensure that the numbers cited in the text matched the numbered works in the reference list.

The reference list includes more than 2,500 separate items, yet this is only a tiny fraction of the literature on human evolution. It is in fact only a small fraction of the items that have appeared since the first edition was published, and it may neglect some significant discovery or idea in a reference I have missed. I hope that anyone who sees a glaring omission will notify me so I will not overlook it in the future.

In preparing the second edition, I have benefited from the research and writing of innumerable colleagues, but I particularly want to thank James Bischoff, Frank Brown, Kathryn Cruz-Uribe, Robert Franciscus, Clark Howell, and Henry McHenry for critical comments on portions of the text. I am also grateful to the many wonderful students at the University of Chicago and Stanford University who endured the lectures and text drafts that underlie both editions.

PREFACE TO THE
FIRST EDITION

Superficially, the study of human evolution seems remarkably abstruse and impractical. Yet each year the popular press covers major finds, large prime-time audiences watch televised documentaries, and thousands of students enroll in pertinent university courses. *Australopithecus* and Olduvai Gorge feature in nationally syndicated cartoons, and some discoverers of important fossils are better known than their counterparts in more practical fields such as physics and medicine. Clearly, in spite of its apparent irrelevance to everyday affairs, the origin of the human species is intensely interesting to most modern humans, who are fascinated by the growing number of fossils, artifacts, and related facts that scientists have amassed. They want to know what these data tell us about the appearance and behavior of our remote ancestors. This book is a summary of what I think the data say.

There are many perspectives on how the data should be interpreted, and this book, of necessity, reflects just one. In writing it, I have tried to steer a middle course between what I see as two extreme approaches—one in which the data are simply a springboard for stimulating speculation about what might have happened in the past, and another in which they are meaningless except to test and eliminate all but one competing explanation of what really happened. The difficulty with the first perspective is that it emphasizes imagination over validity. The difficulty with the second, whose roots lie in a perception of how the physical sciences have advanced, is that it assumes an unrealistic degree of control over data quantity and quality. In fact, substantial control is unusual in human evolutionary studies, where carefully planned experiments are rare and most data are obtained through excavations and field surveys whose success often depends more on chance than on design. Under these circumstances, I think that the physical sciences provide a less suitable role model than the judicial system, in which often limited evidence is weighed to determine which of two or more competing explanations or interpretations seems most reasonable. In most instances the possible alternatives are not pared unequivocally to one, but one is selected because it seems more justifiable, given the evidence on hand.

Of course in human evolutionary studies, as in the judicial system, both laypeople and specialists may disagree about what is reasonable or justifiable and also about the soundness of the supporting evidence. All too often the evidence is incomplete, ambiguous, or even contradictory, and it cannot be used to bolster any particular theory or explanation very strongly. In this book I have tried hard to present the major competing opinions on prominent unresolved issues, and whenever possible I have explained why I think one view is more reasonable than others. More often than I would like, I have had to say that a firmer choice will require more data. I know that not everyone will agree with the positions I have taken or even with my decision to abstain on some matters. However, I think that such differences of opinion are unavoidable, given the imperfect nature of the evidence; and the point is that this book is inevitably just one of many possible summaries of what we know about human evolution. Its success will depend on the extent to which the readers, experts and laypeople alike, think that the presentation and argumentation are sensible.

Philosophical approaches aside, there are several possible ways to organize the evidence for human evolution. The way I have chosen is perhaps the most conventional, focusing on a series of chronologically successive stages—beginning with the earliest primates, dating from perhaps 80 million years ago, and ending with the emergence of anatomically modern people within the past 200,000 years. The presentation does depart from the norm, however, in that I have given roughly equal weight to the fossil record and to the accompanying archeological evidence over the 2.5 million year period for which this is available. Most summaries focus largely on the fossils or on the archeology, thereby forgoing one of the major points I have sought to make—namely, that the human form and human behavior have evolved together and that neither can be fully understood or appreciated without a full understanding of the other. At the same time, however, it remains true that the fossils are far easier to arrange into a set of chronologically successive, interrelatable units, and since the fossil record is also far longer than the archeological one, I have relied on the fossils to define the chronologically successive stages that structure the text. This is not to say there are no problems in defining the fossil units, but these pale beside the difficulties in defining and interrelating corresponding archeological categories.

The difference stems from our much weaker understanding of the mechanisms underlying artifactual (cultural) change and differentiation.

A second way this survey differs from many others is that it includes more information on specific sites, fossils, artifacts, and so forth—in short, it is more detailed, with more concern for the factual evidence that underlies our understanding of human origins. It is a formal rendering of the lectures I give in an upper-level undergraduate course at the University of Chicago, and it has been written with upper-level undergraduates, graduate students, and professionals in mind. In my experience, the audience for whom it is primarily intended will already have a basic understanding of how evolution occurs (through natural selection, mutation, gene flow, and gene drift), and therefore this latter topic is not explicitly addressed. However, at least some members of the audience will lack essential background information on skeletal anatomy, zoological classification and nomenclature, stone-tool typology and technology, and especially the geologic time frame for human evolution, and so these subjects are covered. In general I think the book is too detailed to be a central text in lower-level courses, especially ones that also deal with modern human variation, genetics, and the like; but I hope it will find use there as one of the sources the instructor consults or recommends to those students who are especially curious about the fossil and archeological records.

In keeping with the comparatively technical orientation of the book, I have employed an in-text citation system that is common in professional scientific publications. I rejected the usual textbook system of grouping sources at the end of each major section or chapter because I felt the target audience for the book would prefer to know precisely where to look for further information on a particular topic. I also wanted to give credit directly where it was due. I rejected a system of linking references to numbers because I thought the risk of serious error would be too great when so many references are involved. The large number of references was unavoidable, given the broad theme, but I have tried to keep the list manageable by stressing recent sources that can serve as guides to older ones, and I have also excluded many non-English primary sources in favor of secondary English ones with their own extensive bibliographies of important non-English publications.

No synthesis of human evolution would be successful without good illustrations, but these can be very expensive and time consuming to produce. As a result, even many commercially produced texts are underillustrated. I have attempted to compensate for the limitations of time and expense that were important here by adapting many illustrations from published sources, which are gratefully acknowledged. Thanks mainly to the efforts of Kathryn Cruz-Uribe, most of the illustrations have been substantially modified to support pertinent points in the text and to provide stylistic consistency. In addition, whenever possible I have labeled important features directly on fossils, artifacts, site plans, and so forth, and I have attempted to make the captions freestanding supplements to the text, to emulate the useful sidebars that are common in commercially produced texts. My goal was to make the illustrations especially helpful to those with little or no prior knowledge of skeletal anatomy, stone artifacts, stratigraphies, and such.

It was not easy to choose a title for the book, because the most obvious ones, such as "Human Origins" and "Human Evolution," have been used many times before. The final choice—*The Human Career*—is the name of a graduate course I took at the University of Chicago in 1962, in which F. Clark Howell introduced me to the concept of human evolutionary studies as an amalgam of human paleontology and paleolithic archeology. Howell's alternative name for the subject matter, "paleoanthropology," would do equally well—though it too has been used before and has often been applied to human paleontology alone rather than to the broader paleontological/archeological field that Howell had in mind. In both the title and the text, I intend the vernacular term *human* (and its complement, *people*) to refer to all members of the zoological family Hominidae, as conventionally defined, and not simply to living humans.

My own research on human evolution has focused mostly on behavioral (archeological) evidence from middle and late Quaternary sites in southern Africa and parts of Europe, and my acquaintance with the remainder of the record comes mainly from published sources. In synthesizing these, I have tried hard to make the text as accurate, comprehensive, and up-to-date as possible, and I have been greatly helped by comments and criticisms from Peter Andrews, Kathryn Cruz-Uribe, Janette Deacon, Leslie Freeman, Fred Grine, Clark Howell, Philip Rightmire, Chris Stringer, Russell Tuttle, and Tom Volman. I hope they find that I have employed their suggestions productively and that I have not introduced any new errors in the process. Inevitably, however, some defects remain, and I would be grateful to hear from anyone who finds a specific problem or who has suggestions on how the interpretations or overall organization can be improved.

EVOLUTION, CLASSIFICATION, AND NOMENCLATURE

1

This chapter introduces basic evolutionary terms and concepts that are essential if we are to synthesize sites, fossils, artifacts, dates, and other "facts" into a coherent overview of human evolution. The discussion draws on much more extensive treatments of process and principle in works listed in the references (411, 586, 670, 681–683, 1364, 1480, 1482, 1759, 1760, 1974, 1976, 2036, 2037).

The Biological Species Concept

The species is the least arbitrary and most fundamental evolutionary unit, and no discussion of evolution, whether general or focused on a unique species like *Homo sapiens*, can proceed very far without a clear understanding of what a species is. In modern evolutionary biology, a *species* is defined as a group (or population) of organisms that look more or less alike and that can interbreed to produce fertile offspring. Practically speaking, individuals are usually assigned to a species based on their appearance, but it is their ability to interbreed that ultimately validates (or invalidates) the assignment. Thus, no matter how similar two populations may look, if they are incapable of interbreeding, they must be assigned to different species.

Species are commonly composed of smaller units called *breeding populations* (or sometimes demes or Mendelian populations) in which most individuals find their mates. In modern human terms, major political units such as cities, states, or countries usually correspond to increasingly more inclusive breeding populations. The ultimate breeding population is always the species, outside which no individuals may find their mates. Topographic or other barriers between breeding populations may prevent them from interbreeding, but as long as interbreeding is possible and fertile offspring would result, the breeding populations still belong to the same species. Breeding populations tend to fluctuate in size, membership, and total number over relatively short periods, which means they are harder to delineate than species. However, as discussed below, they have great evolutionary significance as potential species in the making.

The modern, population-oriented definition of the species is clearly not the only one possible, and it is worthwhile to consider why it has replaced a definition that was popular before evolution became widely accepted in the latter half of the nineteenth century. This earlier concept defined each species according to the characters of a type specimen, usually the first one to be found or described. Specimens that were not identical to the type specimen were allowed in the same species, but they were regarded as "sports" or deviants, and their anomalous features did not contribute to the species description.

The typological definition of the species was problematic even at the time it was most popular. To begin with, it was clearly arbitrary, and it floundered when the original type specimen turned out to be less typical of a species than later discovered deviants. Also, there was the nagging problem of deciding just how different two specimens had to be before they could be assigned to different species. From a modern perspective, typological thinking led to an unfortunate tendency to exaggerate small differences and thus to define an unnecessarily large of number of species. Most important, however, the typological definition literally fixed species in time. They were immutable, for if they could change then there could be no type specimen, only an endless series of deviants. The typological definition was therefore incompatible with evolution as promulgated by Darwin and others, and the very intraspecific variability the typological definition sought to suppress was fundamental to Darwin's novel idea of how species evolved, via natural selection of the fittest (discussed below).

Although only the population definition of the species is consistent with the idea of evolution, applying it to the fossil record can be a problem. This difficulty arises partly because interbreeding cannot be observed among fossils, and paleontologists must therefore rely solely on morphology. More precisely, they must decide whether a group of fossils represent one species or more, based on how much morphological variability tends to characterize the living species most closely related to the fossils. Making firm decisions is obviously complicated by the highly incomplete or fragmentary nature of most fossils, which is perhaps the major reason fossil species are often so hotly debated.

A potentially more serious problem is that the identification of fossil species inevitably requires dividing the evolutionary continuum into arbitrary units. If the record is complete enough, some dividing lines must eventually fall between succeeding generations that could have interbred and that therefore could not belong to different species. This problem seems insuperable, except for the discovery that whereas evolution is continuous, the fossil record generally is not. As discussed below, the fossil history of most species suggests long periods of morphological stability or stasis, followed either by extinction or by brief bursts of rapid change during which new species emerge. Since the bursts appear to have been very short compared with the periods of stasis, and since they may have occurred primarily in small, peripheral breeding populations, truly intermediate fossils are rarely if ever found. Most fossil species therefore have distinct beginning and end points that probably correspond at least broadly to the reproductive boundaries of living species.

Natural Selection

It is sometimes thought that Charles Darwin (1809–1882) invented the idea of evolution in *The Origin of Species* (published in 1859). In fact the idea had been proposed long before, and Darwin's monumental contribution was the discovery of a natural mechanism to explain how evolution had occurred. Before Darwin, the notion that species had evolved was hardly less mystical than the more common traditional idea that they had been created supernaturally.

Darwin's great discovery was the principle of natural selection, which he synthesized from his own natural history observations and from the writings of other eminent nineteenth-century thinkers, perhaps above all Thomas Malthus (1766–1834) on population growth. As framed by Darwin and still employed by evolutionary biologists today, natural selection flows from the observation that individuals within a breeding population vary in morphology and behavior, that offspring tend to inherit features of morphology and behavior from their parents, and that not all individuals contribute equal numbers of offspring to the next generation. If we assume that differential capacity to survive and reproduce is linked to morphology and behavior, then any new traits that enhance survival and reproduction will tend to increase in frequency from generation to generation. Such traits are commonly said to be *adaptive* or, in the parlance of evolutionary biology, to make their bearers more *fit*, and *natural selection* can be defined as the sum total of environmental forces or pressures that determine fitness.

Fitness is defined totally, if somewhat tautologically, as a measure of individual ability to produce offspring who themselves survive and reproduce.

As Darwin noted, natural selection is in fact exemplified by the well-known process that people have used to produce breeds of domestic animals. In this case the selective force is human preference or desire, and an individual animal's fitness (ability to survive and reproduce) depends on the extent to which it exhibits desired characteristics. New breeds can be produced relatively quickly, even within historical memory, if selection is stringent enough. Darwin recognized that far more time would be required to transform one species into another or to produce two species from one, and he probably would not have proposed natural selection as the underlying mechanism if geologists had not simultaneously been demonstrating the great antiquity of the planet.

Darwin was intensely aware of contemporary geological research, and he was especially influenced by the work of Charles Lyell (1797–1875), often known as the founder of modern geology and formulator of the *uniformitarian* principle that guides it. In its simplest form, uniformitarianism may be defined as the idea that "the present is the key to the past." Less cryptically, it is the idea that erosion, sedimentation, volcanism, crustal movements, and other detectable geologic processes have operated throughout the earth's history and are sufficient to explain every aspect of the geologic record. From a uniformitarian perspective, for example, the occurrence of shelly marine limestone far above modern sea level may be explained by crustal uplift, and to accept this an observer must realize only that the effects of uplift and other geologic processes have had eons to accumulate. In a sense, Darwin's idea that natural selection explains the origin of species simply extended the uniformitarian principle to the biological realm.

Genetics and Evolution

To Darwin, natural selection operated on individuals, and fitness in the classical Darwinian sense applies exclusively to individuals. But modern evolutionary biologists often think of selection as favoring or discouraging particular traits. The difference is profound, because it reflects an understanding of how traits are inherited—in short, of genetics—that was totally lacking in Darwin's day. Strictly speaking, the fundamental nature of inheritance had been established, but the discovery was published in an obscure journal by an obscure Austrian priest named Gregor Mendel (1822–1884), whose pioneering research was recognized only in 1900, long after both he and Darwin had died.

When Darwin wrote *The Origin of Species*, it was generally believed that inheritance involved the averaging or blending of parental traits in offspring and that it was a blend of blends that offspring transmitted to their own offspring. This view of inheritance presented an immediate problem for the idea of natural selection, for it followed that a selectively advantageous novelty (or mutation) could not become fixed in a population. Instead, it would become diluted or lost through progressive blending with the much larger number of less advantageous alternatives. In essence, Mendel showed that inheritance was not a blending process, but involved discrete units or particles, inherited half from one parent and half from the other. The particles, which we now call *genes,* retain their integrity from generation to generation, even though their expression may vary depending on the presence of other genes or on particular environmental conditions. Thus an adaptive or advantageous novelty will not be diluted out of existence but will spread because it confers greater fitness.

Today a particular gene may be defined by its position or locus on a chromosome or by its singular chemical structure, and evolutionary divergence may be investigated directly at the gene level by comparing genetic material (DNA) or its low-level products (proteins) among related organisms. For those interested in the origin of species, however, it is useful to define a gene simply as a Mendelian particle or a discrete unit of inheritance. The complete set of genes an individual possesses is known as his or her *genotype.* The genotype interacts with the environment to produce the *phenotype,* comprising all observable traits of individual appearance and behavior.

The complete complement of genes that characterizes a species or a breeding population within a species is known as its *gene pool,* and this can be described by the frequencies of different genes (or gene variants) it contains. In a way totally consistent with Darwin's definition of evolution as "descent with modification," evolution can then be alternatively defined as change through time in gene frequencies. In the fusion of Darwinian thought and genetics that informs modern evolutionary biology, four forces are commonly said to alter gene frequencies through time. The most important is natural selection, formulated essentially in Darwin's terms. The others are mutation, gene flow, and random gene drift.

1. *Mutation* is spontaneous, accidental change in the chemical structure of a gene. Mutation is thought to be unusual (generally affecting fewer than one in 100,000 genes), and in

complex organisms like people, the mutations that do occur are much more likely to be harmful than helpful. This means that mutation has much less potential to alter gene frequencies than does natural selection. Mutation is important primarily as the ultimate source of all genetic variability, but if mutation were to cease immediately, natural selection could continue to operate indefinitely on the genetic variability that already exists. This is particularly true since the variability in most species includes genes whose effects are largely suppressed or masked by those of others. In changed circumstances, natural selection could reduce the frequency of the masking genes, resulting in significant, even profound, morphological change totally in the absence of mutation.

2. *Gene flow* is simply the exchange of genes between populations. By definition, the populations must belong to the same species (or gene flow is impossible). In a population that is newly formed from two others, the new gene frequencies will be intermediate, and in the absence of natural selection, their precise values will depend on the relative sizes of the two original populations. Whatever the exact outcome, however, in a strictly technical sense evolution has occurred.

3. *Random gene drift* is chance change in gene frequencies. It is intuitively obvious that random drift is most likely to occur in small populations and to affect genes that are represented in very low frequencies to begin with. In a special instance called the founder principle or effect, chance alone may explain why gene frequencies differ between a parent population and the descendants of a small colonial party that splintered from it. The founder principle is often thought to account for the some of the genetic differences between Native Americans and the east Asian ancestral population they are derived from.

The genetic definition of evolution allows insights that were previously not possible, but its application to the fossil record is not straightforward. This is because fossils reveal only the phenotype, which may change independently of the genotype and vice versa. The genotype in fact includes "noncoding" genes that have no apparent function and that by definition have no phenotypic expression. They are thus not subject to natural selection, and they seem to accumulate changes (via mutation) at a relatively rapid and constant rate. Similarities and differences among them may be especially useful for determining the degree of relationship among recently evolved species, but their pattern of change stands in sharp contrast to the pattern that appears to characterize the fossil record. As discussed below, change here appears to have been mainly epi-sodic, that is, to have occurred in rapid bursts, separated by long intervals of little or no change. The genetic basis for this remains unclear. On the one hand, it is possible that the episodes of pro-found phenotypic change that produced new species were usu-ally accompanied by wholesale alteration of associated geno-types. Alternatively, it is conceivable that major phenotypic change was often promoted by changes in only one gene or a handful of genes that regulate individual development. For the moment, all one can say is that the relation between genetic and phenotypic change in the fossil record is poorly understood and may be very complicated.

Speciation

Many evolutionary biologists make a distinction between *microevolution,* in which genetic change is relatively minor and occurs within separate breeding populations that can still exchange genes, and *macroevolution,* in which genetic change is far more dramatic and ultimately produces populations that can no longer exchange genes—that is, new species. In the case of macroevolution, the genetically isolated populations could be an ancestor and a descendant separated by many generations, or they could be contemporaneous descendants of what was once a single breeding population. Faced with a relatively sparse database and long time spans, paleontologists generally focus on macroevolution, and they must therefore be especially con-cerned with *speciation,* the process by which species form.

　　Darwin believed that natural selection caused species to change more or less continuously and that the slow accumula-tion of gradual change eventually produced new species. He was not particularly concerned with how one species could evolve into two or more, but his notion of gradual change can be ex-tended to fit the modern concept of *adaptive radiation,* whereby a species spreads into a variety of environments to which local populations then adapt (meaning they respond to differing nat-ural selection pressures). Assuming that a topographic or cli-matic barrier then arises or that sheer distance prevents adapt-ing populations from exchanging genes, their gene pools will begin to diverge. Differently adapted populations may later come to overlap in environments that provide niches (oppor-tunities) for both, but by this time they may no longer be capable of exchanging genes; if they are, the hybrids will be less successful than nonhybrids in exploiting either paren-tal niche. Thus, if hybridization does occur, selection will be against hybrids (and individuals who hybridize), and the parent populations will continue to diverge until they become true species that no longer exchange genes. For present purposes the important point is that, in common with Darwin, the traditional model of adaptive radiation emphasizes what has been called *anagenesis,* or gradual change along separate branches of the evolutionary tree. The fundamental concept involved is now often called *phyletic gradualism.*

Phyletic gradualism is plausible in perhaps every respect but one—it fails to explain the way evolutionary change appears to occur in the fossil record. The rule here is for species to appear and disappear relatively abruptly and to change very little in the interim. In Darwin's time one could argue, as Darwin did, that the record was too incomplete to bear on how evolution proceeds, but this is no longer the case. Many paleontologists have therefore abandoned gradualism in favor of what Eldredge and Gould (685) have termed *punctuated equilibrium.* Evolution by punctuated equilibrium is basically the idea that speciation occurs mainly in rapid bursts during brief branching episodes and that thereafter species generally change very slowly if at all. The emphasis in punctuated equilibrium is strongly on branching or *cladogenesis* (the formation of branches or clades) as an event, and anagenesis is largely denied (or replaced by stasis). Figure 1.1 shows how, unlike gradualism, the punctuationalist model postulates significant evolutionary change only at branching time and not along branches.

Advocates of punctuated equilibrium argue that the innovations that mark new species usually become fixed in small, peripheral breeding populations that paleontologists are unlikely to uncover. It is only after speciation, when a new species expands to overlap and sometimes replace its parent, that it crosses the threshold of paleontological visibility. This is why the fossil record rarely reveals a series of graduated, transitional forms or "missing links" between parent species and their descendants. In the punctuationalist model, the driving force behind speciation remains natural selection, but the selection tends to be pulsed rather than continuous. Thus, as discussed in chapter 4, African antelopes and primates (including people) apparently speciated in bursts in response to episodes of profound climatic and environmental change about 5 million years (my) ago and again about 2.5 my ago (2350). In combination with mountain building, sea level changes, and other major geographic alterations, climatic change could fragment a widespread species into isolated populations and thus hasten their tendency to diverge genetically (2352).

The fossil record clearly makes much better sense if cladogenesis is the rule and anagenesis the exception, but it is

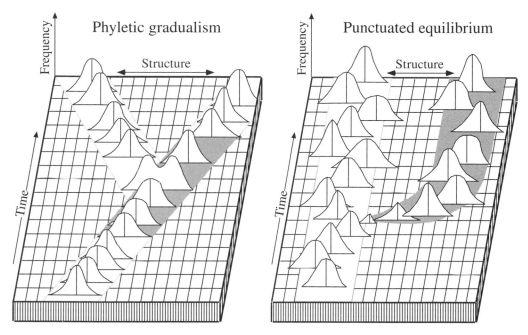

Figure 1.1. The process of speciation as visualized by gradualists (*left*) and by advocates of punctuated equilibrium (*right*) (redrawn after 2343, p. 62). The bell-shaped curves reflect the distribution of one or more morphological traits at different times. Distinct species are indicated by the blank and shaded paths the curves rest on. The gradualist model postulates a steady, continuous shift in modal morphology along branches. When the shift is extensive enough, a new species is born. The punctuationalist model postulates essentially random, noncumulative morphological change through time. A new species emerges only when innovations become fixed in a small peripheral population. This population then diverges rapidly before settling into a conservative mode where change once again tends to be random and noncumulative.

important to emphasize that it is the nature of the record, not logic or experiment, that most strongly supports punctuated equilibrium. In effect, punctuated equilibrium is more a thoughtful generalizing of data than a theory, and there is no claim, even by strong advocates, that punctuated equilibrium should be assumed to characterize the evolution of people or any other species in the absence of empirical (fossil) support.

Phylogeny and Classification

The noun *phylogeny* and its adjective *phylogenetic* are often used as synonyms for evolution and evolutionary. However, strictly speaking a phylogeny is a tree diagram illustrating the evolutionary history and relationships of a species or species group. It is thus to a species what a genealogy or family tree is to an individual. Figure 1.2 illustrates a currently popular phylogeny of the Hominidae, the zoological family that includes

Figure 1.2. A currently popular phylogeny of the hominids, illustrating proposed ancestor-descendant relationships between known fossil and living species (redrawn after 900, fig. 13). The question mark in the phylogeny reflects continuing debate about the origins of the *Homo* lineage. As discussed in chapters 4–6, some authorities believe the *Homo* lineage also comprises several branches.

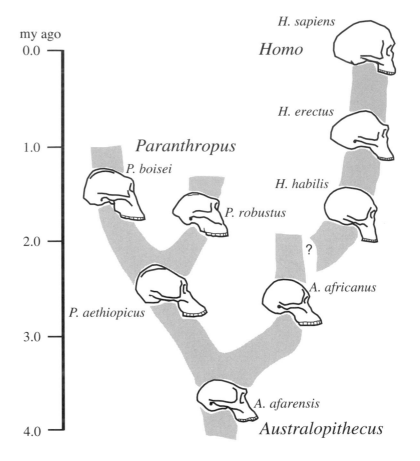

all people, living and extinct. Like many phylogenies, it includes branches that do not reach to the present, because species often become extinct without issue. In fact, final extinction (or termination) is the eventual fate of all species and should be conceptually separated from the situation, also reflected in figure 1.2, where "extinction" results from the evolution of one species into another.

Phylogeny is intimately tied to *classification,* the assignment of species to higher categories based on their presumed evolutionary relationships. A phylogeny is in fact largely a diagrammatic representation of a classification scheme, and insofar as phylogenetic reconstruction cannot occur without classification, classification is more fundamental. The modern system of biological classification was devised in the seventeenth and eighteenth centuries and achieved essentially its present form in the tenth edition of Carolus Linnaeus's *Systema Naturae,* published in 1758. Its most basic unit is the species, which, as already discussed, is defined today as a population of organisms that look more or less alike and that can interbreed and produce fertile offspring. Based on their (presumed) degree

Table 1.1. A Classification of Living People Involving Twenty-one Potential Levels in the Linnaean Hierarchy

---

*KINGDOM: Animalia
  *PHYLUM: Chordata
    SUBPHYLUM: Vertebrata
      SUPERCLASS: Tetrapoda
      *CLASS: Mammalia
        SUBCLASS: Theria
          INFRACLASS: Eutheria
            COHORT: Unguiculata
              SUPERORDER: ———
                *ORDER: Primates
                  SUBORDER: Anthropoidea
                    INFRAORDER: Catarrhini
                      SUPERFAMILY: Hominoidea
                        *FAMILY: Hominidae
                          SUBFAMILY: Homininae
                            TRIBE: Hominini
                              SUBTRIBE: ———
                              *GENUS: *Homo*
                                SUBGENUS: (*Homo*)
                                  *SPECIES: *sapiens*
                                    SUBSPECIES: *sapiens*

Note: A dash follows a level for which no taxon is in common use. Asterisks designate the seven obligatory and most basic levels in the Linnaean system.

of evolutionary relationship, species are then classified into progressively higher categories, minimally including (from bottom to top) genus, family, order, class, phylum, and kingdom. Thus species that are presumed to share a very recent common ancestor are generally placed in the same genus; species that are more distantly related are placed in different *genera* (the plural of *genus*); genera that share a relatively recent common ancestor are placed in a common family; more distantly related genera are placed in separate families; and so forth, up to the level of the kingdom.

Since Linnaeus's time, many new levels have been inserted among the seven principal ones, mainly to accommodate the great proliferation of known species. Depending on the species being classified, twenty or more levels may be recognized today; twenty-one are illustrated in table 1.1, which presents a classification of *Homo sapiens* according to the Linnaean system. Whatever number of levels are used, the principle remains the same. A category at any given level contains a group of species whose overall degree of relationship is reflected by the level at which they are grouped in the hierarchy. Thus, although *H. sapiens* is the only surviving human species, as discussed in chapters 4 and 5 below, it is commonly said to have at least two very close extinct relatives (*H. habilis* and *H. erectus*), which are therefore included in the same genus. Other extinct species

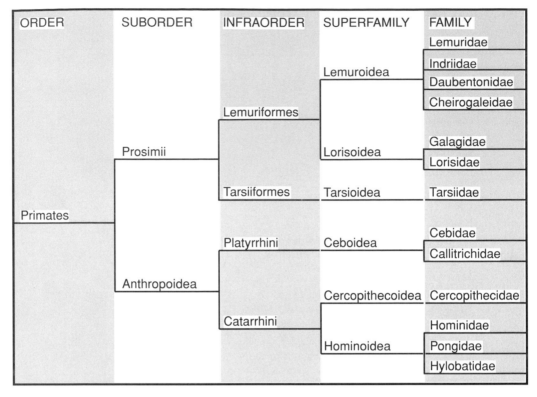

ORDER	SUBORDER	INFRAORDER	SUPERFAMILY	FAMILY
				Lemuridae
			Lemuroidea	Indriidae
				Daubentonidae
		Lemuriformes		Cheirogaleidae
	Prosimii		Lorisoidea	Galagidae
				Lorisidae
		Tarsiiformes	Tarsioidea	Tarsiidae
Primates				
		Platyrrhini	Ceboidea	Cebidae
				Callitrichidae
	Anthropoidea		Cercopithecoidea	Cercopithecidae
		Catarrhini		Hominidae
			Hominoidea	Pongidae
				Hylobatidae

Figure 1.3. A popular hierarchical classification of the living Primates down to the family level. Chapter 2 discusses alternatives.

that are more distantly related and that are human only in the broad sense are usually placed in the separate genus *Australopithecus* and are united with species of *Homo* only at the family level, in the Hominidae. The Hominidae in turn are usually joined with the apes (several families living and extinct) in a common superfamily, the Hominoidea, the Hominoidea with the Old World monkeys (constituting a single superfamily, the Cercopithecoidea) in the infraorder Catarrhini, the Catarrhini with the New World monkeys (Platyrrhini) in a common suborder, the Anthropoidea (higher primates), and finally, the Anthropoidea with the lower primates (Prosimii) in the order Primates. Figure 1.3 graphically summarizes the hierarchy just described to include all living primate taxa down to the family level. As discussed in chapter 3, the details of this particular hierarchy are subject to debate and revision, but the broad outline is firm.

From what has been said, it follows that each category above the species comprises a group of related species (except in those rare instances where a species has no close relatives living or extinct, in which case a higher category may include only one species). For convenience, each species in the Linnaean hierarchy and each group of related species at any level is called a *taxon* (plural *taxa*). Thus the species *Homo sapiens* is a taxon, as are the genus *Homo*, the family Hominidae, the superfamily Hominoidea, the infraorder Catarrhini, the suborder Anthropoidea, the order Primates, and even the kingdom Animalia. Species and genera are commonly referred to as *lower taxa*, as opposed to categories above the genus, which are known as *higher taxa*. Since it is taxa that are classified, the term *taxonomy* is often used as a synonym for classification, though it more narrowly refers to the system of rules for constructing a classification.

Although evolutionary relationships are the basis for classification, they cannot be observed directly and must be inferred from the degree of similarity among taxa. In general, the greater the similarity, the closer the assumed relationship. This explains how the modern classification system could develop before evolution became popular and also how it can be used today even by antievolutionists. More important, the unavoidable reliance on degree of resemblance can distort evolutionary classifications, since common descent is not the only cause of similarities among taxa. Among other causes, undoubtedly the most important is similar adaptation to shared environmental conditions. Resemblances due to similar adaptations are often called *analogies* (or *convergences*) in distinction from *homologies*, which are similarities due to common descent. A very conspicuous and often-cited analogy is the streamlined, finned body form shared by fish and whales, which, despite this morphological similarity, are not very closely related. A less famous but equally clear analogy is the independent, parallel development in apes and some New World monkeys of an upper limb structure that permits hanging below branches. In contrast, the numerous obvious, detailed similarities in head and body form between people and chimpanzees exemplify homologies, inherited from a relatively recent common ancestor.

In many instances (fish and whales, for example) careful scrutiny of multiple characters allows the unambiguous separation of homologies from analogies. In other cases the distinction can be difficult, particularly in fossil taxa, in which the number of assessable characters is limited. In addition, even when homologies have been isolated, they are not in themselves sufficient to establish the degree of evolutionary relationship among taxa. A further distinction must be made between homologies that are widely shared and ones that have a narrower distribution within a taxonomic group. In general, widely shared characters have only limited taxonomic value,

even if they are prominent and numerous. Characters with a
limited distribution are more likely to reveal the basic evolutionary links among taxa in the group. The realization that different characters must be assigned different weights in evolutionary studies is central to the increasingly popular perspective
on classification known as *cladistics* (684, 2469). Following
Hennig (1014), cladists usually distinguish between two fundamental kinds of homologous characters:

1. *Primitive* (or generalized) characters (plesiomorphies or sym-
   plesiomorphies), which arose early in the evolutionary history of a taxonomic group. These characters will be very
   widespread and will therefore not help in dividing the group
   into lower-level taxa, that is, in determining their genealogical relationships. A well-known primitive feature in mammals, for example, is the occurrence of five digits on the end
   of each limb. This feature has been retained in Primates, but
   its ubiquity renders it useless for subdividing the order. It
   cannot even be used to distinguish Primates from non-
   Primates, many of whom also retain five digits as a result of
   distant, shared descent.
2. *Derived* (or advanced) characters (apomorphies), which arose
   relatively late in members of a group and will differ among
   them. By definition, in contrast to primitive characters, derived characters are useful in assessing genealogical links
   among taxa. For example, structural modification of the rear
   limb to permit habitual bipedal (two-legged) locomotion is
   a key derived feature of the hominids (people broadly understood) that ultimately underlies their separation from
   other closely related taxa within the larger hominoid group
   (apes and people). Derived features can be further subdivided
   into two basic types: shared, derived characters (synapomor-
   phies), which demonstrate a special evolutionary tie among
   taxa that have them, and unique, derived characters (au-
   tapomorphies) or novelties, which distinguish a taxon from
   all others. Unique, derived characters are not useful for inferring evolutionary connections, though they may eliminate one taxon from the ancestry of another that lacks them.

Applying cladistics to a group of related taxa produces a
cladogram or tree diagram, which organizes taxa according to
the number of derived features they have in common. The
more derived features two taxa share, the more likely it is that
they will reside on branches that connect to each other before
connecting to other branches farther down the tree. In form,
cladograms resemble traditional phylogenies; but unlike phylogenies, which place taxa at different points or branching nodes

within a tree, cladograms range them at the ends of terminal branches. Also unlike phylogenies, cladograms do not take time relationships into account, and they generally do not include a time scale. The essential difference is that cladograms illustrate perceived degrees of derived similarity among taxa, not ancestor-descendant relationships. Figure 1.4 illustrates a cladogram for some living higher primates based on the presumed degree of derived similarity in the structure of the beta-type globin gene.

Cladists usually regard all derived features as equally important, and this can produce classifications that radically revise traditional ones. Thus, as discussed in chapter 3, the derived biomolecular similarities that the chimpanzees, gorilla, and people share imply that their last common ancestor lived too recently to justify their continued inclusion in separate families (the chimpanzees and gorilla in the Pongidae and people in the Hominidae). However, it can be argued that people should be assigned to a distinct family because they have departed further from the last common ancestor in gross morphology and behavior or, perhaps more precisely, because their unique derived features—habitual bipedalism and especially the large and complex human brain—have produced a novel adaptation or grade of life. This argument provides a plausible rationale for maintaining the status quo and for avoiding the confusion that inevitably follows terminological change, but from a strictly logical standpoint the cladistic reclassification is less arbitrary and easier to defend. Cladistic procedure does not preclude the identification of *grades* (similar levels of structural organization or of adaptation); it only excludes them from consideration in determining the level at which related species are classified together. This is because grades need not bear on the degree of evolutionary relationship that classification is supposed to reflect.

Moreover, in focusing exclusively on degree of similarity or difference, cladograms are less abstract than phylogenies, which must be more conjectural or hypothetical. If the principles behind cladistics are straightforward, however, the practice is often problematic. Cladistic analysis may stumble if shared, derived features that are actually analogies (or parallelisms) are mistaken for homologies. In addition, it is not always easy to determine whether a character is primitive or derived within a group. In fact it can be both, since character reversals are an occasional feature of the evolutionary process. Skull bone thickness, for example, changed from relatively thin in the earliest hominids (the australopithecines and *Homo habilis*) to relatively thick later on (in *H. erectus* and early

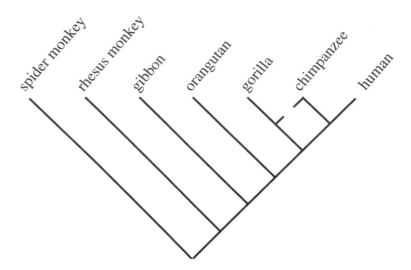

Figure 1.4. A cladogram illustrating the evolutionary relationships among humans (*Homo*), chimpanzees (*Pan*), gorillas (*Gorilla*), orangutans (*Pongo*), gibbons (*Hylobates*), rhesus monkeys (*Macaca*), and spider monkeys (*Ateles*), based on inferred derived similarities in the structure of the beta-globin gene (data in 858). Other biomolecular and morphological analyses produce similar cladograms, except that morphological data often suggest that the chimpanzees and gorilla are more closely related to each other than either is to people. If the chimpanzees and gorilla share a closer common ancestor with each other than either does with people, then the cladogram should be altered, and the broken line should replace the solid line leading to the chimpanzees.

H. sapiens), then back to thin (in later *H. sapiens*, including living people). In a situation like this, functional understanding of a character or character state would obviously be useful for determining its cladistic value. A functional understanding is also crucial for determining whether two (or more) shared, derived characters are truly independent. If not, they may constitute only one character, and their implications for an especially close taxonomic relationship are diminished.

Finally, cladistic methodology is clearly best when it is applied to discrete characters, that is, characters that are either present or absent. It produces more equivocal results when it is applied to characters that differ among taxa mainly in their degree of expression; yet such characters are the rule when the taxa are very closely related, as they are, for example, within the genus *Homo*. Some of the problems, together with the substantial fruits of cladistic analysis, are illustrated in the text, though the emphasis there is on potential phylogenies and on broad evolutionary trends, independent of classification or phylogeny.

Nomenclature

In theory, the names that are assigned to zoological taxa could be formulated without rules, but long experience has shown that a regulated system reduces potential ambiguity and improves communication. Thus, all biologists adhere to a single naming system that essentially follows the practice of Linnaeus. In this system taxa are always given Latin names, or at least names whose form has been latinized. The Latin name of a taxon may be intended as a substantive characterization (*Homo sapiens*, for example, means "wise man"), but in general a name should be regarded simply as a label, not as a definition. The name of a species always consists of two words—the genus (generic) name followed by the species (specific) designation in the narrow sense. Grammatically, the genus name is a Latin noun and the species designation is a Latin modifier (either an adjective or a noun in the genitive singular) or another Latin noun in apposition. The name of a genus is always capitalized and italicized (or underlined in typescript), whether it stands alone or is modified by a species designation. The second part of the species name (the species designation narrowly understood) is italicized but not capitalized. If a genus has already been cited, a closely following citation of a species in the same genus may abbreviate the genus name to its first letter and a period: for example, *H. sapiens* or *H. erectus* closely following the citation of *Homo sapiens*.

In those relatively rare instances when a subgenus name is used, it is capitalized, italicized, and placed in parentheses between the genus and species terms: for example, *Australopithecus (Paranthropus) robustus*. If a subspecies name is used, it is italicized but not capitalized and is appended to the full species name (e.g., *Homo sapiens sapiens*). Subspecies are large-scale breeding populations within a species that are morphologically and geographically distinct, but strictly considered they have no place in the Linnaean hierarchy except where there is evidence that such populations are in the process of becoming separate species (that is, are undergoing progressive genetic divergence because of differing natural selective pressures in different environments). There is no evidence that any populations of living *H. sapiens* are such incipient species, but a case can be made for some extinct populations, such as the Neanderthals, whom some specialists thus place in the subspecies *H. sapiens neanderthalensis*.

Categories above the genus level are capitalized but not italicized. Higher taxa, up to the level of the superfamily, always derive their names from one of the genera they include. By convention, they also have endings that designate their level: (in vertebrates) -ini for tribe, -inae for subfamily, -idae for family, and -oidea for superfamily. Above the superfamily level, the only rule is that the name must be a Latin (or latinized) noun. The Latin names are usually anglicized by dropping the Latin ending (for example, hominid is used instead of Hominidae). In their anglicized form, they are generally not capitalized and can be used as either adjectives or nouns.

Technically, the full name of a taxon at any level includes the name of its inventor and the date of its invention (for example, *Homo sapiens* Linnaeus, 1758). In practice this requirement is frequently ignored, particularly when, as in this book, taxonomy itself is not a primary concern.

Ideally, all specialists would refer to the same taxon by the same name, but problems arise when one taxon has been given two or more names and there is disagreement about which has priority (which was suggested or legitimized first). Even if one name clearly has priority, many specialists may retain a second name that has a long history of use. However, the most important nomenclatural disagreements do not involve only names but are concerned with more basic differences of opinion about the evolutionary relationships among taxa. Thus, as discussed in chapter 6, many specialists believe the Neanderthals and modern humans should be assigned to the same species, *Homo sapiens,* while others place the Neanderthals in a separate species, *Homo neanderthalensis.* The difference is not trivial, since the single-species alternative implies that the Neanderthals could have participated in modern human origins, whereas the separate-species alternative rules this out. Nomenclatural disagreements that stem from taxonomic (classificatory) disputes are particularly common among paleontologists, as is illustrated in this book.

THE GEOLOGIC TIME FRAME

2

Among all the variables behind human evolution, none is more crucial than time, and a discussion of the geologic time frame must precede any consideration of fossils and artifacts. In the best of all possible worlds, the ages of all ancient sites or fossils would be known in calendar years, and a discussion of time could focus entirely on how calendric or numerical ages are obtained. In fact, methods for determining numerical ages are treated here, but not exclusively, because they have not been or cannot be applied to many sites. The most important and widely useful methods were developed only in the late 1940s and 1950s, nearly a century after the first real scientific inquiry into human origins. In the early decades of serious study, dating was almost entirely relative, based mainly on the obvious or inferred stratigraphic position of one site with respect to others. Even today many sites can be dated only in this way, either by their stratigraphic location in a sedimentary sequence or, more often, by the stratigraphic implications of their fossils or artifacts. Additionally, even at those sites where numerical (or "absolute") dating is possible, stratigraphic context furnishes a vital cross-check on the results. Finally, there is a genuine sense in which numerical dates are only unusually precise indicators of stratigraphic position. In sum, the concept of stratigraphy remains fundamental to research on human evolution.

Stratigraphic Units and the Geologic Time Scale

In general, any geological objects that can be arranged in a sequence from younger to older can be used to construct a stratigraphy. The most obvious and most basic items are successive layers of rock whose distinctive qualities can be used to define *rock-stratigraphic* (lithostratigraphic) *units*. Fossils extracted from rock layers provide the basis for *biostratigraphic units,* which can be subdivided into faunal and floral stratigraphic units or into particular kinds of faunal or floral units. Thus, as illustrated below, biostratigraphies can be founded on fossils of rodents, pigs, horses, or elephants or on some combination of the fossils of these creatures and others. Even human (or pre-human) fossils and artifacts can serve, though entities based on artifacts or other behavioral debris are probably best placed in a separate class of *culture-stratigraphic units.*

Although stratigraphic units are always based on objects found in rock layers, different kinds of units need not correlate one-to-one, and a single biostratigraphic unit, for example, may incorporate several rock units. Units that are correlated are assumed to have formed or existed during the same time period, and the successive periods that correspond to particular rock units, biostratigraphic units, and so forth are generally called *time-stratigraphic* or *chronostratigraphic units.* Whereas units defined by rock type (lithology), fauna, flora, or other properties tend to be geographically localized, chronostratigraphic units have no spatial bounds, and different rock or biostratigraphic units from many different areas can correspond to a single chronostratigraphic unit. Chronostratigraphic units in turn correspond to periods of real geologic time, which are at once the most abstract and the most continuous components of the system. They are not themselves stratigraphic units but simply are named time spans to which geologists ascribe particular chronostratigraphic units.

In both theory and practice, the various kinds of stratigraphic units can usually be recognized at different scales, for which sets of hierarchical terms are available (995). Thus, with respect to rock-stratigraphic units, the smallest definable ("mappable") unit is generally a *bed*, spatially and lithologically related beds are grouped into *members*, and related members are grouped into *formations*. The australopithecine site descriptions in chapter 4 illustrate the principle. The hierarchy of words for biostratigraphic units is less formalized, but many specialists use terms such as *faunal complex, stage, zone,* or *biozone,* sometimes prefixed to indicate hierarchical rank or status. The smallest chronostratigraphic unit in common usage is the *stage.* A related group of successive stages constitutes a *series,* related series constitute a *system,* and related systems constitute an *erathem.* The corresponding time terms (from least to most comprehensive) are *age, epoch, period,* and *era.* Table 2.1 presents a list of named eras and a partial list of included periods and epochs, spanning the entire history of the earth, from roughly 4.6 billion years ago to the present. The scheme is rooted mainly in biostratigraphy, while the boundary dates have been estimated mainly by the absolute dating methods discussed below. The table is relatively undetailed for eras before the Cenozoic because they are largely irrelevant here, since this book focuses on the evolution of the primates, which began at earliest in the latest part of the Mesozoic era.

Into the 1970s, treatments of human evolution accentuated the Quaternary (or last) period of the geologic time scale, partly because it seemed to coincide with the appearance and evolution of people and partly because it was considered climatically or faunally unique. However, it is now clear that the base of the Quaternary, formally defined by local changes in marine fossils in Italy, dates to at most 2.1 million years (my) ago (1283) and perhaps to only 1.7–1.6 my ago (tentatively accepted here) (2146), whereas, as shown below, human evolution began more than 4.5 my ago. In addition, there is no sharp faunal or climatic break between the Quaternary, formally defined, and the immediately preceding Pliocene. The Quaternary will therefore not be treated separately here, except to note that it is now commonly subdivided into three parts: (1) the early Quaternary, between 1.7 my ago and the beginning of the Brunhes Normal Polarity Chron, roughly 780 thousand years (ky) ago; (2) the middle Quaternary, between roughly 780 ky ago and the beginning of the Last Interglacial about 127 ky ago; and (3) the late Quaternary, spanning the past 127 ky. The Brunhes Normal Chron and the Last Interglacial are discussed below.

Relative Dating Methods

As indicated previously, the techniques for placing sites, fossils, artifacts, and other objects in geologic time may be divided between absolute methods that determine ages *in years* and relative methods that say only whether one item is older than another. Relative methods were developed first and are more widely applicable. In addition, all absolute methods are obviously also relative ones, and their credibility depends in part on their consistency with other, standard relative methods. In short, relative methods remain basic to all historical geologic studies, including research on human evolution.

Among relative methods narrowly defined, the most fundamental is unquestionably the principle of stratigraphic superposition. This underlies the concept of stratigraphic units introduced in the previous section and in essence states that, all other things being equal, objects found in higher rock layers postdate ones found in deeper layers. The qualification "all other things being equal" is necessary because burrowing animals, invading roots, and the like can displace objects into lower or higher layers, while crustal movements, landslides, and other geomorphic events can even reverse a stratigraphic sequence, placing older layers on top of younger ones. Where such distur-

Table 2.1. The Geologic Time Scale

Era	Period	Epoch	Millions of Years Ago	Some Firsts
			0	
		Holocene or Recent		
			0.006	First cities
			0.0009	First farmers
	Quaternary		0.01	
			0.012	First people in the Americas
			0.05	First fully modern humans
		Pleistocene		
			0.5	Oldest (demonstrated) use of fire
			1.4	First people in Eurasia
			1.75	
		Pliocene		
CENOZOIC			2.5	Oldest (known) stone artifacts and *Homo*
			4.4	Oldest (known) bipedal hominids
			5.2	
		Miocene		
			23.5 23	First monkeys and apes
	Tertiary	Oligocene		
			34	
			35	First higher primates
		Eocene		
			50	First lemurs and tarsiers
			56.5	
		Paleocene		
			65 65	First primates
	Cretaceous			
			120	First placental mammals
			146	
			160	First birds

bances occur, however, their effects are often minor or detectable, or both, and the principle of superposition has been fruitfully applied at countless archeological and fossil sites. Its main limitation is that, in the most literal sense, it cannot be used to date objects in layers that do not physically overlap, that is, ones that occur at physically separate sites.

In situations where physical overlap does not occur, stratigraphic dating depends on perceived similarities or differences in the properties of two (or more) layers. The properties can be lithological, physical, chemical, or fossil, taken separately or in combination. In a sense, absolute dating methods simply illustrate the stratigraphic principle applied to the physicochemical composition of rock layers. Absolute methods are special not only because they indicate stratigraphic ages in years but also because they usually do not require a special knowledge of local geological history and can be applied in the same way anywhere

Era	Period	Epoch	Millions of Years Ago	Some Firsts
MESOZOIC	Jurassic		208	
			220	First mammals and dinosaurs
	Triassic		245	
	Permian		290	
			300	First reptiles
	Pennsylvanian		323	
	Mississippian		363	
			370	First amphibians
PALEOZOIC	Devonian		409	
	Silurian		439	
	Ordovician			
			500	First vertebrates (fish)
			510	
	Cambrian			
			550	First chordates
			570	
			800	First multicellular life (sponges, algae)
Precambrian or Proterozoic			1,400	First nucleated cells (eularyotes)
			3,500	First unicellular life (prokaryotes)
			4,000	First complex organic molecules
			4,600 4,600	Origin of the solar system and Earth
			15,000	Origin of the universe

Source: Modified after 951, p. 12, and 1876, p. 26.

in the world. In contrast, relative dating methods require detailed local information, which restricts their application to limited geographic areas. The size of the area can vary from the immediate neighborhood of a site to a large region or even a continent. As illustrated below, relative methods that employ physicochemical analysis tend to be most tightly restricted, while ones that rely on fossils tend to have the broadest geographic application.

Relative Dating by Chemical Content: The Fluorine Method

Among relative dating methods that depend on chemical analysis, the most influential in human evolutionary studies has certainly been the fluorine method (1627). It is based on the observation that buried bones adsorb fluorine from groundwater.

Bones that were buried in the same site at the same time should contain the same amount of fluorine, and gross differences in fluorine content thus suggest noncontemporaneity. The fluorine method was instrumental in unraveling the notorious Piltdown hoax, named for a site in Sussex, southern England, where a seemingly ancient skull and mandible were found in 1911–1912 (466, 2030, 2200, 2201). The skull was thick walled but otherwise remarkably modern in appearance, while the jaw was very apelike in basic structure, including a bony (simian) shelf behind the mandibular symphysis (chin region). The combination became increasingly incongruous as new fossil finds from elsewhere failed to replicate it, and in 1953 it was shown to be a forgery (2405, 2406). A major part of the evidence was that the skull and mandible contained different concentrations of fluorine and that both contained much less fluorine than most of the associated animal bones that were the main evidence for great antiquity. It is now known that the skull came from a relatively recent human and the mandible came from an orangutan (1398).

In the Piltdown case, and in others where the fluorine method has been used to check the contemporaneity of bones from the same site, the value of this method is plain. But it cannot be used to determine the relative ages of bones from different sites, because there is great geographic variation both in the amount of groundwater and in the amount of fluorine it contains. The same limitation affects virtually all other relative dating methods based on the accretion or deletion of chemicals (for example, nitrogen or uranium) in buried objects.

Biostratigraphic Dating

In most instances, sites that do not physically overlap in some way and that are not amenable to absolute dating can be dated relative to one another only by their contents. There is an obvious danger of circularity here, if dating is based on objects whose relative age has not been independently determined. Thus it would be unwise to conclude that one site is older than another simply because it contains simpler artifacts, because irrefutable examples have shown that artifact change is not always directional. Artifacts can be used for dating only when the cultural stratigraphy of a region has been established on independent empirical grounds. Likewise, fossils can be used only when the basic biostratigraphy of a region has been worked out in advance, but fossils have at least two clear advantages over artifacts. First, of course, they can be used to date sites where artifacts do not occur or that formed before artifacts were made. Second, at least with regard to the earlier stages of human evolution, they often define stratigraphic units that cover larger areas and shorter time spans than do units defined by artifacts.

For biostratigraphic purposes, the most useful fossils come either from taxa (species, genera, etc.) that spread very quickly and very widely, from taxa that appear to have died out over large areas at about the same time, or from taxa that were evolving rapidly, so that their stage of evolutionary development is itself a clue to their relative age (1385). The aim is to define biostratigraphic units that are as fine as possible and that transgress time as little as possible (that is, that correspond to the same time period in different regions). Some time transgression is inevitable, since no species can appear or disappear everywhere simultaneously and since evolutionary changes cannot occur at exactly the same time in all populations of a widespread species. But both theory and empirical research suggest that some taxa provide a basis for defining units whose time transgressiveness is negligible, at least compared with the antiquity of the human fossils or artifacts they can be used to date. In this context, in Eurasia the most productive biostratigraphy developed so far involves the microtine rodents, but large mammals are also useful for relative dating. In Africa, elephants, pigs, and horses have proved most helpful.

Microtine Biostratigraphy in Europe. The microtine rodents, including the voles, the lemmings, and their relatives, are a branch of the cricetids (hamsters) characterized by hypsodont (high-crowned) molars whose occlusal surfaces comprise a series of alternating, triangularly prismatic cusps (fig. 2.1). The heartland of microtine evolution has been in northern Asia, and over the past 5 million years, at times when climatic conditions were right, various microtine species have spread both eastward to North America and westward to Europe (1752, 1753). They are currently the most successful rodent group in the Northern Hemisphere, and their past success is reflected in numerous fossil occurrences, particularly where bones were accumulated partly or wholly by owls or other birds of prey.

The biostratigraphy of the microtines has been studied most thoroughly in Europe, especially in central and north-central Europe (719, 1002, 1003, 1132, 1258, 1259), France (442, 443), and Britain (2096). In each place, microtine biostratigraphic units can be defined by first appearances, by evolutionary changes within established lineages, or by both. Important first appearances include the immigration (from northern Asia) of the Norway lemming, *Lemmus*, in the mid-Pliocene, perhaps 3.5 my ago, and of the collared lemming, *Dicrostonyx*, at or shortly after the beginning of the Pleistocene, 1.7–1.6 my

Figure 2.1. Stages in the evolution of the voles *Mimomys, Arvicola,* and *Microtus,* based on morphological changes in the lower first molar (modified after 1003, p. 402). The horizontal arrows indicate approximate times of first appearance for species in the *Mimomys-Arvicola* lineage. Broken lines indicate two major transitions in this lineage, the first approximately 3.2–3.3 my ago, when cementum first appeared between molar prisms, and the second between about 800 and 600 ky ago, when ever-growing, rootless molars evolved.

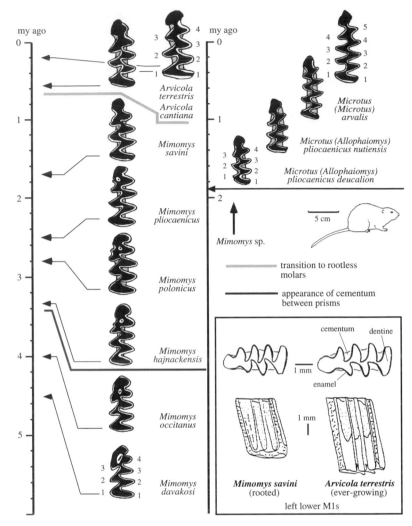

ago. (This date and others presented below depend mainly on correlations between microtine evolutionary events and the global paleomagnetic stratigraphy outlined later in this chapter.) Based on their historic distribution and climatic preferences, *Lemmus* and *Dicrostonyx* probably both migrated to Europe during periods of very cold climate. Their episodic expansion counts among the clearest biological evidence for repeated climatic deterioration (glaciation) later on, as discussed in the final section of this chapter.

In nearly all microtine lineages, evolutionary change involved the development of increasingly high-crowned molars. In some this was accompanied by a tendency for cementum to pack between the triangular prisms of the molar crowns. Separately and together, these trends made for more efficient feeding on grasses, which are an important but highly abrasive component of most microtine diets. Increases in crown height and cementum packing are particularly useful for characterizing successive stages in the evolution of the modern water vole, *Arvicola*, from its immediate ancestor, *Mimomys*. Based on changes in the first lower molar (M_1), figure 2.1 illustrates the major stages and shows two especially striking transitions, one about 3.25 my ago when cementum first appeared between prisms and a second roughly 600–500 ky ago when the molars became ever-growing and ceased to develop roots, even in elderly individuals. As the figure suggests, the evolution of rootless molars is used to define the appearance of *Arvicola*. Subsequent stages in the evolution of *Arvicola* can be defined especially by a tendency for the enamel to become thinner on the rear (convex) surfaces and thicker on the fore (concave) surfaces of the molar prisms (1002, 1003, 2096).

Figure 2.1 also illustrates stages in the evolution of the common (or field) vole, *Microtus*. The sequence is founded mainly on a tendency for the first molars to increase in length and on an associated increase in the number of triangular prisms or angles. *Microtus* perhaps evolved in Asia from a species of *Mimomys* and then migrated to Europe roughly 1.7–1.6 my ago. Like *Arvicola*, it was initially distinguished from *Mimomys* by the development of ever-growing, rootless molars. Together, *Mimomys*, *Arvicola*, and *Microtus* can be used to construct biostratigraphies that are plainly relevant to human evolution and prehistory. Figure 2.2 shows one that has wide application in central and southeastern Europe over the entire time span (middle Pleistocene to Holocene) that people have occupied the area.

The Biostratigraphic Implications of Large Eurasian Mammals.
Large Eurasian mammals are less helpful than small ones, because their remains are less abundant and they do not permit such fine stratigraphic subdivision (1259). The microtines, for example, underlie three well-defined biostratigraphic stages within the past 1.7–1.6 million years: the Villányan, Biharian, and Toringian. Figure 2.2 shows only the Biharian and Toringian, because no unequivocal artifacts or human remains have been found with Villányan species. Over the same long period, large mammals define only two stages: the late Villafranchian and the succeeding Galerian. The boundary between the two is often placed near 1 my ago (97), where it would coincide broadly with the boundary between the early and late Biharian. The distinction of the Galerian from the late Villafranchian is fuzzy, however and Villafranchian faunas may have given way to Galerian ones only gradually between 1 and 0.5 my (1259).

Time Units	Regional Climatic Stratigraphy	Central European Biostratigraphy (Arvicolidae)		Important Fossil Sites, with Values for an Index of Molar Enamel Thickness in *Mimomys* and *Arvicola*	
				Central Europe	Southeastern Europe (Pannonian Basin)
Quaternary — Pleistocene / Holocene	Holocene	*Arvicola terrestris* Zone		*Arvicola*	*Arvicola*
				Euerwanger Buhl H: 83.03	Pilisszánto : 84.48
	Weichsel Glaciation	*Arvicola-Microtus* Stage (Toringian)		Krockstein/Rübeld : 89.08	Peskö : 89.31
				Kremathenhöhle : 89.23	**Istállóskö** : 89.54
				Dzerová Skala : 92.04	
				Roter Berg/Saalfeld: 97.25	**Subalyuk:** (96.43)
				Burgtonna 2 : 98.44	
	Eem Interglacial			Untertürkheim: 100.81	**Tata:** 99.22
				Adlerberg/Nördlingen: 100.83	
				Taubach: 105.15	
				Schönfeld : 106.02	
	Saale Complex	*Arvicola cantiana* Zone			Hórvölgy: 101.91
				Weimar-Ehringsdorf (lower travertine) : 112.30	Solymár: 108.32
					Budapest-Várbarlang 2: (116.83)
	Holstein Complex			Dobrkovice 2: (123.21)	Budapest-Várbarlang 1: 123.08
				Bilzingsleben: (132:52)	
				Mosbach: (133.54)	Tarkö 4: (129.00)
				Hundsheim: 135.15	**Vértesszöllös:** (136.14)
	Elster & Cromer Complex	*Microtus-Mimomys* Stage (Biharian)	*Mimomys savini* Zone	*Mimomys*	
				Prezletice: 132.98	
				Voigtstedt: 139.09	
				Koneprusy C 718: 141.69	

Figure 2.2. Middle and late Pleistocene biostratigraphy of central and southeastern Europe, based on the voles *Mimomys, Arvicola,* and *Microtus* (modified after 1003, p. 395). The far right column lists key sites with vole fossils, followed by average values for a biostratigraphically significant index of first lower molar enamel thickness (thickness on the concave margin of the molar prisms/thickness on the convex margin) × (100). Sites shown in boldface type are those with important human fossils or artifacts.

This does not mean the Eurasian large mammal fauna remained static. On the contrary, the alternation of glacial and interglacial episodes discussed below, coupled with the tendency for later glacial periods to be more intense, produced dramatic changes, including the progressive appearance after 500 ky ago of distinctly cold-adapted or cold-resistant species like the arctic hare (*Lepus timidus*), arctic fox (*Alopex lagopus*), woolly rhinoceros (*Coelodonta antiquitatis*), woolly mammoth (*Mammuthus primigenius*), and reindeer (*Rangifer tarandus*). The recurrent appearance and disappearance of these in mid-latitude Europe in fact counts as some of the best evidence for glacial/interglacial alternation after 300 ky ago. They may not define a clear biostratigraphic stage, but individually they and other taxa can be used for broad temporal placement. The giant deer (*Megaloceros giganteus*), the red deer (*Cervus elaphus*), and the mammoth lineage (*Mammuthus*) provide particularly good cases in point (1385).

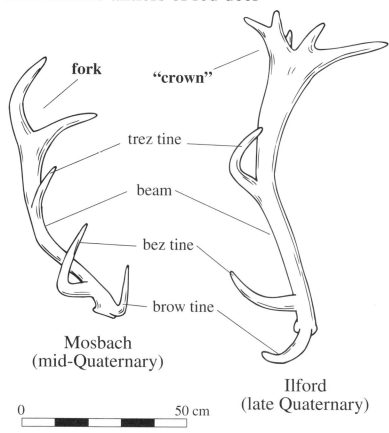

shed mature antlers of red deer

fork

"crown"

trez tine

beam

bez tine

brow tine

Mosbach
(mid-Quaternary)

Ilford
(late Quaternary)

0   50 cm

Figure 2.3. Shed mature antlers of red deer (*Cervus elaphus*) from (*left*) the Mosbach mid-Quaternary fossil site, Germany, and (*right*) the Ilford late Quaternary site, England (redrawn after 1385, p. 335). Mid-Quaternary antlers typically terminate in a simple two-pronged fork, whereas late Quaternary specimens exhibit a multipronged "crown." The difference can be used to arrange European Quaternary sites in time.

The giant deer (or "Irish elk") is useful mainly for its abrupt disappearance. It is unknown at any site after 10,600 radiocarbon years ago (2097), although there are many later sites where it might be expected if it had survived. Its presence in an otherwise undated context thus implies a time before 10 ky ago. The red deer (American elk) is useful especially for a change in antler form between roughly 400 and 250 ky ago, in which a simple double-pointed terminal fork evolved into a multipronged crown (1384) (fig. 2.3). Crowned antler tips support an age near 250 ky ago for the Swanscombe site in England, with its famous fossil human skull, while simple two-point (acoronate) antlers imply a time before 250 ky ago for the well-known Ambrona and Torralba hand ax (Acheulean) sites in Spain, discussed in chapter 5. Finally, the mammoth lineage originated in Africa and appeared in Eurasia about 2.5 my ago. Shortly thereafter, mammoths became extinct in Africa and new species or subspecies arose primarily in Siberia, from which they subsequently spread to Europe. Like the fossil African elephants considered in the next section, successive species or subspecies of mammoths were distinguished by ever more complex molars, in which the enamel that surrounded individual molar plates became thinner, the plates became more tightly packed, and the number of plates per molar increased through time. Mean enamel thickness and the mean number of plates can thus date mammoth-rich sites relative to each other (1386, 1387) (fig. 2.4).

The Biostratigraphy of Elephants, Pigs, and Horses in Africa. Since the microtines are restricted to Eurasia and North America, they can play no role in African biostratigraphy. Many African groups can be used, however; from a paleoanthropological perspective, the most noteworthy are clearly the elephants, pigs, and horses (550). During the time span of human evolution, all three groups were characterized by important first and last appearances and also by lineages for which directional evolutionary trends have been established. Fossil elephants, pigs, and horses provide the main basis not only for dating many important African sites relative to each other but also for cross-checking the validity of absolute dates when they are available.

Figure 2.5 summarizes the biostratigraphic aspects of the elephants. Until the late Quaternary, the most common genus in Africa was *Elephas*, which appeared shortly before 4 my ago. An early species, *Elephas ekorensis*, was ancestral to the especially common *Elephas recki*, which evolved somewhat before 3 my ago and then underwent progressive change through time. This change is particularly obvious in the molars, which became increasingly high crowned while the enamel surrounding individual molar plates became thinner and more tightly folded (165, 556, 1418). From molar change, several stages of *Elephas recki* have been defined, and where appropriate fossils occur, these can be used to bracket sites within an average range of about 500 ky.

The pigs and horses provide a broadly comparable picture (figs. 2.6 and 2.7). There is minor disagreement on details (552, 958) with respect to the pigs, but both first appearances and directional changes in molar dimensions and morphology can be used for biostratigraphy. The horses are most useful with regard to first appearances, especially since most major events in horse evolution took place outside Africa and since immigrant taxa then dispersed rapidly through much of the continent. Both the spread of the three-toed horse, *Hipparion*, and that of the modern one-toed form, *Equus*, are key time markers for relative dating. From a north African entry, three-toed horses

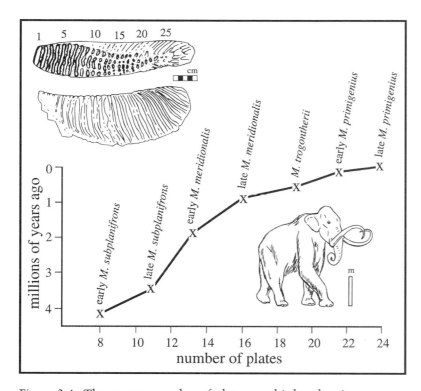

Figure 2.4. The average number of plates on third molars in successive species of *Mammuthus*, from its first appearance in Africa before 4 my ago until its extinction on the Eurasian mainland, roughly 1.5 my ago (redrawn after 1387, p. 150). The earliest species— *M. subplanifrons*—was exclusively African, but later species— *M. meridionalis, M. trogontherii,* and *M. primigenius*—were Eurasian. *M. primigenius* is known popularly as the woolly mammoth from its hairy coat, which Paleolithic people depicted in their art and which is sometimes partially preserved in permanently frozen ground. The mammoths can be distinguished from other elephants by their domed braincases and their inwardly curved tusks, shown on a reconstructed woolly mammoth in the figure (redrawn after 2096, p. 44). The molars of mammoths and other elephants comprise a series of subparallel enamel plates that are held together by cementum. Each plate has an enamel shell surrounding dentine. Occlusal and lateral views (*top left*) illustrate the basic structure on an upper third molar of *Mammuthus primigenius* from Last Glacial deposits in Britain (redrawn after 2096, p. 47). The alternation of enamel and dentine produced a rough occlusal surface that helped in grinding vegetal foods. In *Mammuthus* (and *Elephas,* as discussed in the text), natural selection for a more abrasive diet favored an increase in the number of plates through time. In Last Glacial *M. primigenius*, the third molars sometimes had more than twenty-five plates, as illustrated on the occlusal view in the figure.

Figure 2.5. Time ranges of the Elephantinae in Africa (modified after 550, fig. 2). The successive stages of *Elephas recki* are particularly useful for establishing the chronological relations among otherwise undatable early and middle Pleistocene sites.

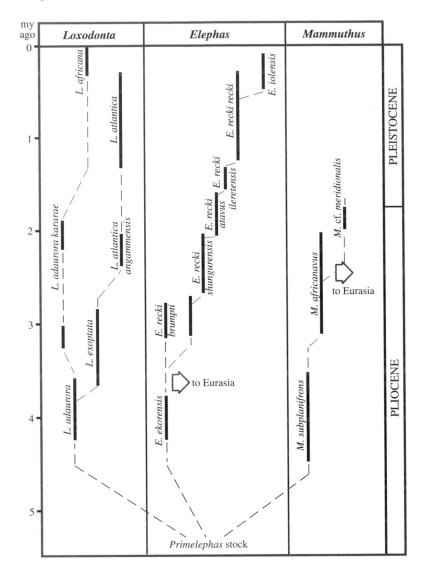

spread between 12 and 10.5 my ago, perhaps mostly (if not entirely) around 10.5 my ago. One-toed horses dispersed shortly before 2 my. For perhaps 1.5 my they overlapped with three-toed species, which became extinct about 500 ky ago. Together with pig taxa, the one-toed horse was especially crucial in demonstrating a large numerical (radiopotassium) dating discrepancy between the Koobi Fora and the Lower Omo sites in eastern Africa. At first it seemed that a fauna including the one-toed horse had appeared some 600 ky earlier at Koobi Fora than in the Lower Omo, only 100 km north. As discussed in chapter 4, this led to further checking of the absolute dates, and it was ultimately shown that the original Koobi Fora estimates were in fact roughly 700–600 ky too old.

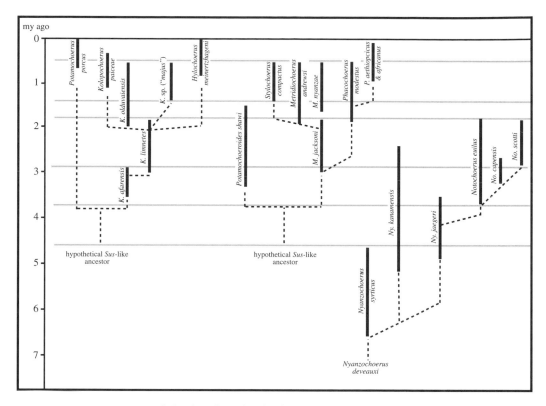

Figure 2.6. Time ranges of the fossil suids of Africa. The dotted lines separate proposed biostratigraphic zones or stages (redrawn after 550, fig. 3). The chart shows that the suids mushroomed between roughly 5 and 1.5 my ago but then suffered severe losses (extinctions), probably mainly between 1 my ago and 400 ky ago. The reasons for the losses remain conjectural, but the ongoing parallel success of the bovids was probably at least partly responsible.

Numerical Dating Methods

Although biostratigraphy and other relative dating methods remain basic to human evolutionary studies, the impact of numerical ("absolute") dating methods cannot be overstated. This is especially true of isotopic (radiometric) methods, which have revolutionized paleoanthropology since their initial development in the 1940s and 1950s. Until 1960, the best guesstimate for the time span of human evolution was perhaps the past 1 million years or so. This notion changed radically in the early 1960s, when the publication of radiopotassium dates from Olduvai Gorge, Tanzania, showed that people in the broad sense (hominids) already existed at least 1.75–1.85 my ago (696, 1324). More recently, as detailed in chapter 4, the radiopotassium method has been applied at many other east African sites, where, in conjunction with relevant evidence from biostratigraphy and other sources, it shows that human evolution began substantially earlier, certainly before 4.4 my ago. The radiocarbon method has had a comparable impact at the recent end of

Figure 2.7. Time ranges of the later Miocene to Holocene equids of Africa, with some important sites where equid fossils occur (modified after 550, fig. 1). From a biostratigraphic point of view, the two most important events were the arrival and dispersal of three-toed horses (*Hipparion*) sometime between 12 and 10.5 my ago and of one-toed horses (*Equus*) shortly before 2 my ago.

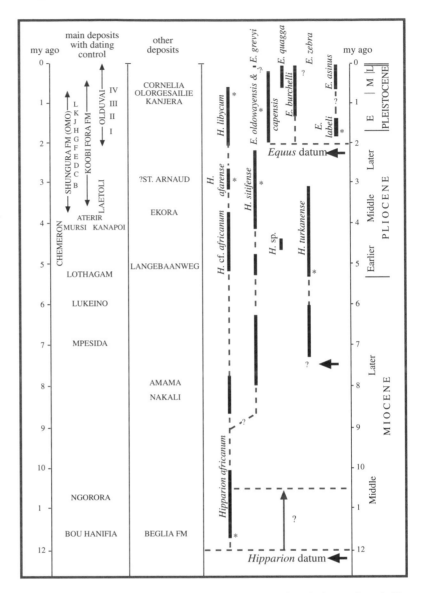

the time scale, where, for example, it has helped show that fully modern people did not appear everywhere at the same time. The implications of this are considered in chapters 6 and 7.

The main purpose of this section is to summarize the principles behind radiopotassium dating, the radiocarbon technique, and other isotopic methods that have become so vital to human evolutionary studies. It also touches briefly on non-isotopic numerical methods, of which the most important is clearly paleomagnetism (paleomagnetic stratigraphy). Inevitably, it also points up some important limitations of numerical dating, above all the nagging difficulty of obtaining reliable dates between the minimum limit of the conventional radiopotas-

Table 2.2. Some Basic Parameters of Isotopes Used in Numerical Dating

Isotope	Material Dated	Potential Range (years)	Half-Life (years)
Carbon-14	Wood, charcoal, shell	40,000 using conventional technology; 100,000 using linear accelerators	5,730
Protactinium-231	Deep-sea sediment, shell, coral, travertine	150,000	34,300
Thorium-230	Deep-sea sediment, shell, coral, travertine	350,000	75,200
Uranium-234	Coral	1 million	250,000
Chlorine-36	Groundwater	500,000	310,000
Beryllium-10	Deep-sea sediment, polar ice	8 million	1,600,000
Potassium-40	Volcanic ash, lava	(No practical limit)	1.25 billion
Uranium-238	Igneous and metamorphic rocks	(No practical limit)	4.5 billion
Rubidium-87	Igneous and metamorphic rocks	(No practical limit)	48.8 billion

sium method at perhaps 500–300 ky ago and the maximum limit of conventional radiocarbon dating at 40–30 ky ago.

Some General Features of Isotopic Dating

All isotopic dating methods rely on the decay of naturally occurring radioactive elements or isotopes (varieties of elements). In the process of decay, radioactive atoms are literally transformed from one isotope into another (not necessarily of the same element). For example, the justly famous radioactive isotope of carbon, carbon-14 (^{14}C), decays to nitrogen-14 (^{14}N). For each isotope, the rate of decay is a constant that is unaffected by ordinary environmental variables such as temperature and humidity or by the chemical compound in which the isotope occurs. Each isotope's decay rate is ordinarily expressed as its half-life, which is the average amount of time necessary for half the radioactive atoms in a sample to decay. Table 2.2 presents the half-lives of several radioactive isotopes that are important in isotopic dating.

As an example, the table lists the half-life of ^{14}C as 5,730 years, which means that roughly half the ^{14}C atoms in a sample will disintegrate within 5,730 years, half of the remainder at the end of 11,460 (5,730 × 2) years, half of what is then left at the end of 17,190 (5730 × 3) years, and so forth. Technically, some ^{14}C will always remain in a sample, though with conventional technology it will be very difficult to measure the amount after a lapse of 25–30 ky.

To obtain an isotopic date, several conditions must be met. First, of course, the object to be dated must contain a radioactive isotope with a known half-life. Second, at the time of formation, the object should in general have contained only the radioactive "parent" and none of the "daughter" into which it decays. In mathematical terms, at time of formation the daughter/parent ratio should have been 0:1 (zero). Finally, there must be some means for measuring the amount of parent and daughter in the object today, that is, for establishing the modern daughter/parent ratio. Then it is possible to use the half-life of the parent (the known rate at which it decays to the daughter) to calculate the time that has elapsed since the daughter/parent ratio was zero, that is, since the object formed. In practice, the condition that there be no initial daughter (that the daughter/parent ratio be zero) is frequently violated, but fortunately it is often possible to separate original daughter from "radiogenic" daughter.

From a paleoanthropological perspective, the most important isotopic dating techniques are the radiopotassium (K/Ar) methods, the fission-track method, the uranium-series method(s), the radiocarbon method, the luminescence method(s) (TL and OSL), and electron spin resonance (ESR). Figure 2.9 (after 1886) shows the approximate time ranges each method covers and the percentage of error or uncertainty associated with the dates it produces.

Radiopotassium (Potassium/Argon) Dating

This method relies on radiopotassium (^{40}K), which makes up about 0.01% of all naturally occurring potassium (K). The ^{40}K decays to argon-40 (^{40}Ar) and calcium-40 (^{40}Ca) in a known ratio, and theoretically either the ^{40}Ca/^{40}K or the ^{40}Ar/^{40}K ratio could be used for dating a geological sample. However, in practice the ^{40}Ca/^{40}K ratio is not useful because in most samples radiogenic ^{40}Ca cannot be distinguished from original ^{40}Ca, which is very abundant in nature.

In contrast, the ^{40}Ar/^{40}K ratio is very useful because rocks heated to a very high temperature (generally above 300°C) tend to lose any original argon (an inert gas) they contain. When they cool, radiogenic argon begins to accumulate again. Since the rate of accumulation is known (it is the rate at which ^{40}K decays into ^{40}Ar), the ^{40}Ar/^{40}K ratio in a rock is a direct function of the time elapsed since the rock cooled.

Radiopotassium dating is more complicated in practice than in principle (572, 797, 945). It works best on rocks—or more precisely on minerals within rocks—that are rich in potassium. Some minerals tend to lose their argon under physical or chemical stress, independent of heating, and secondary heating may cause some but not all radiogenic argon to escape. A date for a rock that has been secondarily heated may thus reflect a mixture of two (or more) heating events.

Because radiopotassium dating generally requires high temperatures to set the clock (daughter/parent ratio) to zero, its use is restricted mainly to volcanic extrusives, such as lavas and ash falls. Meteorites that were heated during passage through the atmosphere may also be dated, but they are much rarer than volcanic rocks. The very long half-life of ^{40}K means that the radiopotassium method has no practical maximum limit (it can be used to estimate the age of the earth) but that in most cases it cannot be used to date rocks younger than a few hundred thousand years old. This limitation occurs because they contain too little ^{40}Ar for accurate measurement and because the statistical error associated with the age estimate therefore may be as large as the estimate itself.

Since the late 1980s, the conventional radiopotassium (^{40}K/^{40}Ar) method has been largely replaced by a variant that often allows more accurate and precise dates and that can be applied to otherwise undatable contaminated or mixed samples. In this variant, known as the ^{40}Ar/^{39}Ar method (632, 1497), the target sample is irradiated to convert its natural ^{39}K content to ^{39}Ar. The amount of resulting ^{39}Ar is directly proportional to the content of ^{40}K, and it can be measured more precisely than ^{40}K content. The biggest advantage of the ^{40}Ar/^{39}Ar variant, however, is that it can be used to estimate the ^{40}K/^{40}Ar ratio in extremely small samples, including single grains of volcanic origin. It thus allows age estimates where grains from more than one volcanic event may be mixed in a single sample. A prominent example is the GATC tuff (volcanic ash layer) that immediately underlies the oldest known hominid fossils at the Aramis locality in the Middle Awash Valley of Ethiopia. Twenty-six individually dated feldspar crystals from a sample of this tuff divide between nine that formed an average of 23.6 my ago and seventeen that formed about 4.4 my ago (2476). The 23.6 my old grains represent contaminants (xenocrysts) from a much older volcanic event, and only the 4.4 my old grains truly bear on the age of the hominid fossils. In this instance a conventional K/Ar result from the mixed sample is a seemingly precise, but obviously erroneous, (mixed) age of about 15.5 my ago.

The technology that underlies the ^{40}Ar/^{39}Ar method is complex, and it often relies on lasers for the intense heat that is required to release argon from an irradiated sample. Like the conventional K/Ar method, the ^{40}Ar/^{39}Ar technique usually cannot produce dates that overlap the radiocarbon time range within the past 50 ky, because very young samples usually contain too little radiogenic ^{40}Ar. In unusual circumstances, however, when a large or particularly potassium-rich volcanic crystal is available, ^{40}Ar/^{39}Ar can produce remarkably young dates. The most striking example is a date of about 1925 B.P. (years before the present) on large sanidine grains from the eruption of Mount Vesuvius that buried Pompeii. This closely approximates the known historical date of 1918 B.P. (A.D. 79) (1751).

The radiopotassium method (broadly understood to include ^{40}Ar/^{39}Ar) has been extensively applied in eastern Africa, where vulcanism has been nearly continuous since the early Miocene. Dates on volcanic extrusives, such as ash layers, that lie stratigraphically above or below a fossil or archeological site can be used to estimate the latter's age, as, for example, in the Middle Awash or at Olduvai Gorge, where a series of dates indicate that the oldest deposits, with their artifacts and bones, accumulated about 1.8 my ago. Another paleoanthropologically important application of the radiopotassium method is the dating of paleomagnetic reversals recorded in volcanic extrusives (see below). Since paleomagnetism is also recorded in nonvolcanic sediments, it can often be used in turn to provide rough dates at fossil or archeological sites where direct radiopotassium dating is impossible.

Fission-Track Dating

This method utilizes the "tracks" formed by spontaneous fission of uranium-238 (^{238}U) in naturally occurring glasses and minerals (744, 887, 1606, 2359). The tracks are annealed (erased) when a substance is heated to a high temperature, and they reform when it cools. Since the rate of track formation is directly proportional to the half-life of ^{238}U, the last time a sample cooled can be determined from the density of tracks relative to the amount of ^{238}U it contains. The tracks, usually enlarged by acid etching in the laboratory, are counted under high magnification. A potential complication is that, depending on the substance, some annealing (track loss) can occur at low temperatures. This is particularly likely in glasses, which therefore tend to produce ages that are "too young." Like the radiopotassium technique, fission-track dating is applicable mainly to volcanic extrusives. It requires minerals that are moderately rich in ^{238}U (if there is too much, the tracks are too closely packed for counting). It has been used much less often than radiopotassium, largely because it is more tedious. From a paleoanthropological perspective, its main importance has been to check radiopotassium dates at the important east African sites of Olduvai Gorge (743), Koobi Fora (846, 1102), Middle Awash (944), and Hadar (2379).

Uranium-Series Dating

Uranium-series (U-series) dating is actually a set of methods based on the parent isotopes ^{238}U, ^{235}U, and thorium-232 (^{232}Th) (1278, 1885–1887, 1890). All three decay, through a series of intermediate radioisotopes, to stable isotopes of lead. Some of the intermediate products decay too rapidly to be useful for dating geologic events, while those with longer half-lives tend to decay at rates about equal to the rates at which they are produced. This means they can be used for dating only when they are transferred from the system in which they were produced to another, where their removal or introduction sets the clock to zero.

In general, U-series dating depends on the high solubility of uranium in water, whereas its decay products, such as ^{230}Th (from ^{238}U via ^{234}U) and protactinium-231 (^{231}Pa) (from ^{235}U) tend to precipitate out as they form. Uranium in seawater or lake water thus tends to remain in solution, while ^{230}Th and ^{231}Pa tend to become incorporated in sediments that accumulate on the floor below. Assuming that the ^{230}Th and ^{231}Pa content was always the same as it is at the sediment surface today, the quantity of ^{230}Th or ^{231}Pa in a buried layer can be used to calculate its age. This method is obviously applicable only to sea- or lake-bottom sediments, where it can, however, date climatic events that may bear on human evolution. There are several sources of potential error, including the possibility that the ^{230}Th and ^{231}Pa in ancient sediments came partly from uranium in particles of detritus washed in from a terrestrial source. In addition, the sedimentation rate (the rate at which ^{230}Th and ^{231}Pa settle out) is not necessarily constant over time.

In theory, the principles behind U-series dating of sea- or lake-floor deposits should also permit U-series dating of aquatic shells, which are usually built of carbonates that have been dissolved in the surrounding water. With the carbonates, there is also some dissolved uranium but virtually no (insoluble) ^{230}Th or ^{231}Pa. The $^{230}Th/^{234}U$ and $^{231}Pa/^{235}U$ ratios in a shell should therefore reflect the time elapsed since the shell formed. U-series analysis of shells has dated ancient (raised) beaches in the Mediterranean that are stratigraphically related to archeological or fossil sites (2040), but in general shell dates are questionable because shells tend to adsorb uranium from their burial environment. This is also commonly the case with bones, since fossil bones typically contain 10 to 1,000 ppm (parts per million) of uranium compared with only 0.1 ppm in fresh bone (548). Single large bones may also adsorb uranium unevenly, and the date will then depend on which part of the bone is sampled (227). Differential adsorption among bones must also explain why different bones within the same layer often provide discordant U-series ages (227). The bottom line is that U-series dates on bone should be regarded critically, and they are presented in succeeding chapters mainly when alternative dating is unavailable.

Among carbonates of organic origin, the most reliable ones for U-series dating are corals, partly because they tend to be relatively rich in uranium to begin with and partly because uranium intake after death is rarer and relatively easy to detect. Fossil corals from various parts of the world have provided concordant dates of 125 ± 10 ky ago for a high sea level generally correlated with the Last Interglacial (discussed in the section on Cenozoic climates below) (953, 2039). In addition, accurate, high-precision U-series dates on coral show that radiocarbon dates (discussed in the next section) often underestimate true calendar ages, and the U-series dates may thus be used to correct (or "calibrate") the radiocarbon time scale to 22 ky ago and beyond (136, 137).

Finally, the ^{230}Th/^{234}U and ^{231}Pa/^{235}U ratios can be used to date inorganic carbonates—limestones, speleothems (stalagmites and stalactites), travertines, and such—that precipitate out of solution in cave, spring, or lake deposits. Since thorium and protactinium are relatively insoluble, they tend to be absent from recently precipitated carbonates, which will, however, contain tiny amounts of coprecipitated uranium. After precipitation, ^{230}Th and ^{231}Pa accumulate in the carbonate as a result of uranium decay. Either the ^{230}Th/^{234}U ratio or the ^{231}Pa/^{235}U ratio can then be used to estimate the age of the carbonate.

From a paleoanthropological perspective, U-series dating of inorganic carbonates is certainly its most important application, because these materials occur in many archeological sites and because U-series dating covers a time range that is not well covered by other methods. The ^{231}Pa/^{235}U and ^{230}Th/^{234}U ratios permit dates back to about 150 and 350 ky ago, respectively, well beyond the lower limit of radiocarbon and near the practical upper limit of radiopotassium. Cave dripstones (speleothems) and spring limestones (travertines) form mainly under humid conditions, and the extensive application of U-series dating to these deposits has confirmed inferences from other sources that global climate was much moister during the Last Interglacial, between roughly 127 and 90 ky ago, than it was during the Last Glaciation, between about 90–80 and 15–10 ky ago (1012).

The U-series method has also been applied to inorganic carbonate horizons in some key archeological and human fossil sites, especially in Europe (548, and chapters 5 and 6 below), but the results remain controversial, partly because they are not always internally consistent and partly because they often seem to contradict dates implied by biostratigraphy or other methods. Sources of error include the possibility that the dated carbonate was contaminated by "detrital" uranium (in particles introduced by wind or flowing water) and the possibility that uranium or its daughter products were leached from the carbonate after precipitation. The method is obviously most successful when it can be applied to sealed carbonate layers that were minimally altered after they formed, and it is most compelling when samples from a single site provide easily replicable dates that are in proper stratigraphic order.

The 17 m thick sequence of travertines in the Abric (Rock-shelter) Romaní near Barcelona, Spain, illustrates the point. The travertines were generated by moss growth during wet periods when cascades of carbonate-charged waters often sprayed the floor of the shelter. During drier periods, Middle Paleolithic (or Mousterian) and early Upper Paleolithic (early Aurignacian) people occupied the shelter, and the travertines thus sandwich successive Middle and early Upper Paleolithic horizons. U-series analysis has produced more than twenty stratigraphically concordant dates that begin at 70 ky ago near the bottom of the travertine sequence and end at 40 ky ago near the top (228, 229). The dates near the top place the earliest Upper Paleolithic at about 43 ky ago, while accelerator (AMS) radiocarbon dates on charcoal from Abric Romaní and other Spanish sites place the earliest Upper Paleolithic between 39 and 37 ky ago. However, for reasons discussed in the next section, in the 40–35 ky range radiocarbon may underestimate true calendar ages by at least 3,000 years, and even if the radiocarbon dates reflected true calendar time, they need be only minima (40–35 ky old or older). Thus, in a manner that may always elude radiocarbon, the Abric Romaní U-series dates show that the Upper Paleolithic replaced the Middle Paleolithic in northeastern Spain about 43 ky ago.

Radiocarbon (Carbon-14) Dating

The radiocarbon or carbon-14 (^{14}C) method is the most celebrated of all isotopic techniques because of its widespread application in archeology. It relies on four major assumptions: (1) that the amount of ^{14}C, which results from a reaction between cosmic radiation and ^{14}N in the atmosphere, has remained constant in the atmosphere over the time period covered by the method; (2) that ^{14}C is as likely to be oxidized to carbon dioxide as is nonradioactive carbon (mainly ^{12}C); (3) that carbon dioxide mixing in the atmosphere is relatively rapid, so that the atmospheric ratio of radioactive and nonradioactive carbon is about the same everywhere; and (4) that most organisms do not discriminate between radioactive and nonradioactive carbon when they build their tissues. Since most plants obtain all their carbon from atmospheric carbon dioxide, and since most animals obtain their carbon directly or indirectly from plants, the $^{14}C/^{12}C$ ratio in plants and animals will closely reflect the ratio in the atmosphere. After an organism dies and carbon assimilation ceases, however, the amount of ^{14}C in its tissues decreases by half approximately every 5,730 years (the half-life of ^{14}C). The ^{14}C decays back to ^{14}N, but dating depends on the $^{14}C/^{12}C$ ratio in a sample, which is compared with the $^{14}C/^{12}C$ ratio in the atmosphere. The difference is a function of the time elapsed since the death of the organism providing the sample (893, 1370, 2148, 2161).

Some of the assumptions behind ^{14}C dating are clearly problematic. First, the $^{14}C/^{12}C$ ratio in the atmosphere has patently varied through time. The combustion of huge amounts of fossil fuel (coal, oil, and gas) following the Industrial Revolution in the nineteenth century has released large quantities of "old" carbon (deficient in ^{14}C) into the atmosphere, while nuclear explosions since 1945 have produced an opposite bias, increasing the proportion of ^{14}C. Fortunately, these factors can be circumvented by relying on the $^{14}C/^{12}C$ ratio in calendrically dated objects from before the Industrial Revolution, and they do not materially affect dates obtained since 1955, when a standard ratio was widely adopted.

More difficult to accommodate are much earlier fluctuations in the atmospheric $^{14}C/^{12}C$ ratio that are indicated by repeated discrepancies between ^{14}C dates and calendrically dated artifacts or between ^{14}C dates and dates from more secure methods, such as tree-ring dating (dendrochronology) (1917, 2150). Such fluctuations were probably caused mainly by changes in the strength of the earth's magnetic field, which acts as a shield against cosmic radiation, by variation in solar activity (the main source of cosmic radiation), and perhaps partly also by long-term changes in the amount of carbon dioxide dissolved in the world's oceans (1483, 2098). Paired ^{14}C and tree-ring dates allow the radiocarbon chronology to be corrected or

cal yr B.P. (x 1000)

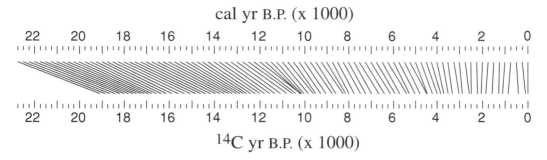

Figure 2.8. The relation between radiocarbon (^{14}C) and calendar (or calibrated) years from roughly 22,000 calendar years ago to the present (redrawn after 142, fig. 1). In general radiocarbon dating underestimates true calendar ages, and the discrepancy increases with time.

"calibrated" over the past 10,000 calendar years, while paired ^{14}C and high-precision U-series (^{234}U/^{230}Th) dates on marine coral extend the calibration to about 22,000 calendar years ago (fig. 2.8). And ^{14}C dates on organic debris in annually laminated sediments (varves) from the bottom of Lake Suigetsu, Japan, allow less secure calibration to 45,000 calendar years ago (1227).

In general, ^{14}C dates for the past 45 ky are younger than true (calendar) ages, and over the 22 ky period for which calibration is well established, the inconsistency tends to grow with time. The greatest discrepancy is at 22 ky ago, when ^{14}C dates come out about 18.5 ky, or about 3,500 years "too young" (136, 137). At 30 ky ago, ^{14}C dates may underestimate true ages by 4,300 to 3,500 years, and at roughly 43 ky ago, near the limit of the conventional radiocarbon method, they may be 5,000 years too young (229). The bottom line is that pending fuller, more secure calibration, radiocarbon dates that antedate 22 ky ago are more useful for arranging sites relative to one another than for placing them in absolute time. The most general point is that no radiocarbon date implies a true calendar age unless it has been calibrated, and this is currently impossible before 22 ky ago.

Another questionable assumption is that all the carbon in a sample originated from the atmosphere. Some living creatures (particularly some mollusks and water plants) are known to assimilate carbon from old rocks. They will thus contain more ^{12}C than they "should," and if they are dated they will seem to be "too old." An even more serious problem occurs when carbon enters the remains of a creature after its death. Most common and most serious is contamination by "young" carbon, for example, from humic acids penetrating a buried archeological level. If the young carbon is not detected and eliminated, the ^{14}C age will be "too young."

There are sample preparation procedures to detect and eliminate contaminants, but they are not foolproof, and the best course remains careful sample selection in the field. Some materials are also easier to cleanse or less likely to be contaminated to begin with. Charcoal (charred wood) is probably best in this regard. The inorganic components of shell and bone are generally not useful, because they tend to exchange carbon with their burial environment. The organic (protein) component of bone or shell is suitable but is rarely present in its original form. More commonly, only some amino acid constituents of protein remain, and these might come from the burial environment rather than from the bone or shell itself. The problem is particularly acute for ancient bones that retain little datable organic matter (2033, 2147). Since ^{14}C dates on such material are often younger than dates on associated materials like charcoal, the organic extract was probably contaminated by more recent, intrusive carbon. Among the constituents of bone protein, one—hydroxyproline—is especially sought for ^{14}C dating, because it is unlikely to have originated from anything but bone (823). Unfortunately, it is highly soluble and is therefore often leached from bones after burial.

Radiocarbon (and other isotopic) dates are properly stated with a plus-or-minus figure, for example, 14,000 ± 120 years ago. This figure is a measure of the statistical and analytical uncertainty associated with a particular age determination. The plus-or-minus figure is computed so that the chances are about 67% that the actual radiocarbon age of the sample has been bracketed. In the case of a date given as 14,000 ± 120 years ago, the bracketing period would be 13,880 to 14,120 years ago. The figure will be larger for a sample that contains relatively little carbon. If the figure is doubled, the probability that the actual age of the sample has been bracketed increases to 96%. If the figure is tripled, there is no practical chance that the age of the sample lies outside the bracketing period.

The time range covered by the ^{14}C method is limited by the short half-life of ^{14}C. In general, samples that are older than 30–40 ky contain too little ^{14}C for accurate measurement by conventional technology. This technology counts only radioactive emissions (decaying ^{14}C atoms), not all ^{14}C atoms in a sample. Linear accelerators functioning as sensitive mass spectrometers now allow direct counting of ^{14}C atoms before they decay, and since the overwhelming majority in a sample do not decay during the brief time when a sample is analyzed, the ^{14}C content of old samples can be measured much more precisely (193, 996, 997, 1383, 1842, 2149). In theory, accelerator mass spectrometer (AMS) dating could provide reliable ages up to

100 ky ago or even beyond. It also allows very small samples to be dated, and so far this has been its main application (1018). One reason is the problem of obtaining truly suitable samples for very old dates. Samples that are older than 35 ky will retain less than 2% of their original, natural ^{14}C, and even a tiny amount of recent contaminant will thus make them seem much younger. For example, a 1% increment of modern carbon in a sample whose true age is 67 ky ago will produce a ^{14}C age of 37 ky ago. This leads to the important point that, in general, a date of more than 35 ky ago should be regarded as only a minimum age (697). The true age may be significantly greater.

The Luminescence Methods and Electron Spin Resonance

The luminescence dating methods and electron spin resonance depend on the observation that irradiation by naturally occurring uranium, thorium, and radiopotassium causes electrons to accumulate in defects within crystalline substances. The aggregate number of trapped electrons can be measured, and the rate at which they accumulated can be estimated from the level of background radioactivity to which a substance was exposed. The number of trapped electrons divided by their annual accumulation rate (the annual radiation dose) then provides the last time the crystal traps were empty. The luminescence methods are applied primarily to objects where heat or light emptied (or zeroed) the traps before burial, while ESR is applied mainly to dental enamel, where formation (precipitation during life) produced initially empty traps. The time range covered by the luminescence methods and ESR depends on the time it takes for traps to fill (or saturate) so that no more electrons can be added. This varies from substance to substance and from site to site, but in general the methods are applicable to materials that are between a few thousand and a few hundred thousand years old (812) (fig. 2.9).

The luminescence methods comprise thermoluminescence (TL), which has been applied mostly to objects like potsherds and occasional flint artifacts in which the crystalline traps were emptied by intense heat before burial (17, 18, 2471, 2472), and optically stimulated luminescence (OSL), which has been applied mainly to quartz sand grains in which the traps were emptied by exposure to sunlight (16, 715, 1101). TL and OSL differ mainly in the way they release and measure trapped electrons. TL relies on intense heat and OSL on intense light. In either case, the object glows as the electrons escape, and the intensity of the glow (luminescence) is directly proportional to

Figure 2.9. Time ranges covered by the numerical dating methods discussed in this chapter (adapted from 1886, fig. 1). Note that the horizontal scale (years before the present) is logarithmic. In practice, dates near the limits of each method may be difficult to obtain, and their reliability may be especially questionable. This is perhaps particularly true of very "old" dates from the ^{14}C method and very young ones from the ^{40}K/^{40}Ar method.

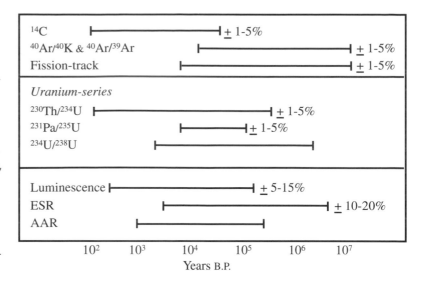

the number of trapped electrons. Traps typically are of two kinds—shallow ones from which the electrons are easily released and deeper ones from which they escape less readily. Either set of traps can provide a date, but the deep traps are less likely to have been fully emptied before burial, and they are thus more likely to provide a date that is "too old." OSL is superior to TL for measuring the glow from the shallow traps, and it is thus preferable whenever there is doubt about full zeroing. Such doubt is especially likely to affect sand grains as opposed to potsherds or burned flints. With regard to sand, TL is more problematic because it generally requires many thousands of grains for a single date, whereas OSL can be done on only a few dozen grains or even on single grains (1789). This advantage makes OSL the method of choice when there is some chance that older and younger grains have been geologically mixed in a single sample.

TL and OSL rely equally on a sound estimate of the annual radiation dose, which depends partly on radioactivity within the target object and partly on radioactivity in the burial environment. Radioactivity within the target (the internal dose rate) is readily estimated, but radioactivity in the burial environment (the external dose rate) is more difficult, since it may have varied over time. Variation may occur because groundwater subtracted (leached) or added uranium through time, because variation in soil moisture content variably buffered buried objects from irradiation, or because the texture of the surrounding deposit changed. Finer-grained deposits (for example, unconsolidated sand) inhibit irradiation less than lumpier ones (for example, undecomposed sandstone blocks).

In short, the luminescence methods depend on site-specific details that do not affect the radiocarbon, radiopotassium, or uranium-series methods, and it is for this reason that the error or uncertainty estimates associated with luminescence dates tend to be relatively large (fig. 2.9). Like U-series dating, however, the luminescence methods can provide dates in the critical gap between the practical lower limit of radiocarbon at about 50 ky ago and the practical upper limit of radiopotassium at about 300 ky ago. They can also, of course, be applied where suitable materials for radiocarbon or radiopotassium are absent. At some late Paleolithic European sites where TL and radiocarbon have been used together, they closely agree (17, 2275), and this buoys confidence in TL. Unquestionably, its most important application so far has been in Israel (1549, 2274), where dates on burned flints show that modern or near-modern people were present between 120 and 90 ky ago, when Neanderthals were the sole occupants of Europe. The Israeli TL dates and the associated fossils provide crucial support for the "Out of Africa" theory of modern human origins, as presented in chapters 6, 7 and 8.

Electron spin resonance (ESR) in its conventional application to dental enamel is more questionable. It requires the same estimate of background radioactivity as the luminescence methods (918, 1011, 1891), but unlike them it must confront the problem that teeth tend to adsorb large quantities of uranium after burial. Success in ESR dating thus depends on modeling the history of uranium uptake. Ordinarily two discrete models are considered, one in which the uranium is supposed to have accumulated shortly after burial (the early uptake or EU model) and a second in which it is supposed to have accumulated more continuously (the linear uptake or LU model). The models produce particularly discrepant dates for teeth that are rich in uranium, and in the extreme case a date based on the assumption of linear uptake can be twice as old as one based on early uptake. The true age is usually assumed to lie in between. ESR has provided dates at many of the same sites to which TL has been applied, but as discussed in chapter 6, the results are not always consistent. Most notably, ESR suggests a much shorter time span than TL for the Middle Paleolithic in Israel (1551, 1552). The problem may be largely that the history of uranium uptake is often more complicated than the theoretical models allow, and it may even include uranium subtraction (leaching). ESR dates are reported routinely below, and some are surely accurate. Their overall reliability remains to be established, however, and they have to be carefully assessed site by site.

Nonisotopic Absolute Dating Methods

The most important nonisotopic numerical dating methods are
tree-ring dating (dendrochronology) and varve dating. They are
highly accurate but can be applied only in certain restricted geographic regions, and they cover only the past 8,000 to 10,000
years. Only varve dating will be summarized here, but tree-ring
dating follows broadly similar principles. An additional numerical dating method—amino acid racemization—also de-
serves mention, because its application to ostrich eggshell has
provided useful dates at some African sites.

Varve Analysis

A varve is a distinctive band of sediment, often made up of two
subbands, that is laid down each year on the floor of a lake or
other relatively calm body of water. In general the floor must
also be anoxic (poorly oxygenated), so that the sediments are
not disturbed by burrowing organisms. For dating purposes, the
most useful varves are ones that form in quiet glacier-fed lakes.
Each year, after the spring thaw, material washed into such a
lake consists of both coarse and fine particles. The coarse ones
settle out first, forming the lower subband of a varve. The finer
particles settle out later, forming the upper subband. The overall characteristics of a varve, particularly its thickness, are determined by annually variable events, especially the intensity
of the thaw.

   In an area with several glacier-fed lakes, each year will
produce its own distinctive varve, alike from lake to lake. In
some regions, above all Fennoscandinavia, there are varved sediments that formed under lakes whose times of existence overlapped. The bottommost varves in one set thus correspond to
the uppermost varves in another. In Fennoscandinavia it has
thus been possible, when starting with varves of known age, to
establish a reliable varve sequence covering the past 10–12 ky
(2144). The varves can be used to date any materials they contain and to trace the retreat of the Last Glaciation ice sheet in
its waning phases. Varves of known age have also been used to
calibrate the ^{14}C method, though they are less useful for this
purpose than are tree rings.

Amino Acid Racemization

The amino acids that make up proteins can exist in two mirror-
image molecular forms: L-, or left-handed, and D-, or right-
handed. In general, only L-amino acids occur in the tissues of

living animals, but they are converted after death to D-amino acids at a rate that depends on ambient temperature, moisture, and pH (acidity) and that differs for each amino acid (686). The reaction that produces the conversion is known as racemization (or epimerization). Fossil shells and bones often still contain some amino acids, and if the postmortem temperature and moisture history of a specimen are known, the D-forms/L-forms ratio for a particular amino acid is a measure of the time since death (949, 950).

Like U-series, luminescence, and ESR dating, amino acid racemization (AAR) is appealing because it can provide dates in the period between 40 and 100 ky or more that is not well covered by either radiopotassium dating or radiocarbon dating (fig. 2.9). The lower limit is constrained by postdepositional temperature, but the method can extend to 200 ky ago in the tropics and to 1 my ago in cooler regions. Using AAR on bones was briefly popular in the 1970s (100), but it fell from favor, mainly after it was applied to prehistoric human skeletons from California. AAR suggested that some of these were between 40 and 70 ky old, but archeological context, together with U-series determinations (231) and ^{14}C dates (2147), sometimes on bones from the same skeletons, indicated that the skeletons were no more than 11 ky old. In this instance (and others) the problem was probably a grossly incorrect estimate of the racemization rate, which may proceed more rapidly in old bones that have lost most of their original amino acids (100). In other instances AAR, like ^{14}C dating, can provide mistaken ages when foreign amino acids contaminate a buried bone, and using AAR on bone has been largely abandoned.

AAR has been more fruitfully applied to mollusk shells (2398), and ostrich eggshell (OES) is also very suitable (1155). Like mollusk shell, OES retains its native organic material better than bone, it is less subject to postdepositional leaching and contamination, and it commonly provides radiocarbon dates that agree closely with dates on associated charcoal (362). It is particularly abundant in African sites, but it also occurs across southern and central Asia to Mongolia and northern China, wherever ostriches ranged in the past. AAR dating of OES still requires an estimate of average temperature after burial, but this can be obtained from the amino acid ratio in OES fragments that have been radiocarbon-dated between 30 and 20 ky ago. Such fragments will have experienced a rough average of glacial and interglacial temperature extremes.

The most important site to which AAR on OES has been applied is perhaps Border Cave, South Africa. Internally consistent AAR determinations bracket the Middle Stone Age sequence here between 145 and 70 ky ago (1558). Other methods provide similar ages for comparable Middle Stone Age artifacts elsewhere. The Border Cave AAR dates may pertain to some well-known early modern or near-modern human fossils, though the position of these within the stratigraphy is uncertain (chapter 6). OES is also associated with the even more famous modern or near-modern human remains at Qafzeh Cave, Israel. TL on associated burned flints suggests the Qafzeh human fossils are about 92 ky old, and AAR could provide a useful cross-check, since it relies on totally different assumptions. Alternatively, if the 92 ky age is accepted, AAR could be used to estimate the average temperature since burial. This could then be used for AAR dating at other Israeli Stone Age sites.

Paleomagnetism (Paleomagnetic Stratigraphy)

Paleomagnetic (or geomagnetic) dates are generally much less precise than those of other numerical dating methods, and paleomagnetism is perhaps best described as a cross between numerical and relative dating. In essence it provides the basis for a special kind of stratigraphy, with broadly the same kind of dating implications that biostratigraphy has. At its root are past fluctuations in the intensity and direction of the earth's magnetic field. From a paleoanthropological perspective, the most important changes are in polarity, from times when a compass would point north to times when it would point south and vice versa. Reversals in currents within the earth's fluid core almost certainly cause these shifts, though the currents are poorly understood and the timing of shifts appears irregular (563, 564, 951, 1179). Shifts do, however, clearly occur on two scales: very long periods characterized by essentially the same polarity are punctuated by much shorter ones of opposite polarity. The long periods, lasting hundreds of thousands or even millions of years, are called polarity chrons (formerly epochs). The shorter ones, lasting no more than a few tens of thousands of years, are known as polarity subchrons (formerly events).

Ancient polarity can be detected most readily in volcanic rocks and in fine-grained sediments that settled into place relatively slowly, for example, on the ocean floor. The alignment of ferromagnetic particles in a volcanic rock reflects the direction of the field when it cooled, while the alignment of particles in fine-grained sediments reflects the direction at settling time. Subsequent compaction of the sediments prevents realignment. Polarity changes are also recorded in magnetic lineations or stripes that are fixed in oceanic crust as it spreads laterally from midocean ridges. Magnetic profiling of the oceanic crust has extended the paleomagnetic stratigraphy to 80 my ago and before (418).

The sequence of polarity changes is especially well known for the past 5 my. It comprises four polarity chrons and ten or more polarity subchrons (fig. 2.10). The chrons and subchrons have been dated from paired radiopotassium and paleomagnetic readings in volcanic rocks (1496) and from paired climatic and paleomagnetic determinations in deep-sea sediments (1923). The deep-sea record reveals a long sequence of cold/warm oscillations whose ages can be established from an understanding of the astronomical variables that probably forced them (the Milankovic theory discussed briefly below). The radiopotassium and astronomical datings agree closely where they can be cross-checked (for example, on the placement of the boundary between the Matuyama and Brunhes Chrons at about 780 ky ago [112, 150]). The boundary dates between chrons remain subject to further empirical and statistical revision, but they are reasonably well fixed. The dates for subchrons are less secure because of their short duration, which may be less than the statistical uncertainty associated with radiopotassium dates. Even the existence of some subchrons remains controversial.

Accurate measurement of remanent magnetism in ancient rocks or sediments requires not only proper instrumentation but also care in field removal. Complications can be introduced by secondary heating of a volcanic rock and by postdepositional chemical processes that alter the behavior of ferromagnetic particles.

Unlike either the first appearance of a biological taxon (a species, genus, etc.), which may be time transgressive and regionally restricted, or global climatic change, whose impact may be harder to detect in some places than in others, a change in polarity affects the entire globe simultaneously, and given appropriate deposits it should be detectable everywhere. This makes the geomagnetic stratigraphy attractive for defining boundaries between time periods, including, for example, that between the Pliocene and the Pleistocene, now sometimes fixed at the top of the Olduvai Subchron, roughly 1.75 my ago, and that between the early Quaternary and the middle Quaternary, now commonly equated with the boundary between the Matuyama Reversed Chron and the Brunhes Normal Chron, about 780 ky ago.

The geomagnetic stratigraphy is also very useful for bracketing sites in time that cannot be dated more directly. Two prominent examples are the famous Peking man site of Zhoukoudian, northern China, and the Trinchera (Gran) Dolina site, Atapuerca, Spain, which is the oldest unequivocal human

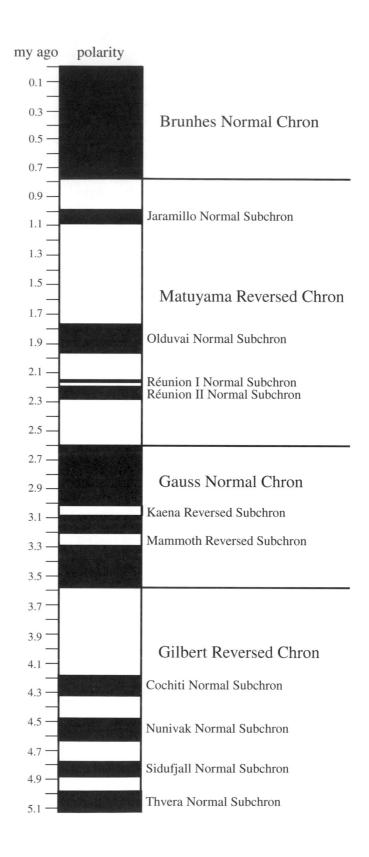

Figure 2.10. Global paleomagnetic stratigraphy for the past 5 million years (based on 418, 1496, 1923). By convention, normal polarity (compass needle pointing north) is indicated in black, reversed polarity in white. The existence and dating of subchrons is often controversial, especially within the Brunhes Normal Chron, and only widely accepted examples are shown.

occupation site in Europe. Neither Zhoukoudian nor the Trinchera Dolina has provided material that is ideally suited for isotopic age determination, but at Zhoukoudian, paleomagnetic determinations from fine-grained parts of the fill have isolated the Brunhes/Matuyama boundary at a level below that containing the earliest Peking man fossils and artifacts (1388). In combination with paleoclimatic evidence allowing broad correlation with the global marine record, the Zhoukoudian paleomagnetic data suggest that the deposits with human fossils and artifacts span the period from about 500 to 230 ky ago.

Paleomagnetic analysis at the Trinchera Dolina has identified the Brunhes/Matuyama boundary in sediments slightly above layers with the oldest artifacts and human remains (1650). The artifacts and fossils are thus estimated to be about 800 ky old. Elsewhere in Europe no site unequivocally demonstrates human occupation before 600–500 ky ago, but the Trinchera Dolina implies that people were present at least sporadically before then.

The Zhoukoudian and Atapuerca examples illustrate the most common application of paleomagnetism in archeology. This is to determine whether a site antedates or postdates 780 ky ago, based on two observations or assumptions: that the oldest known artifacts date from the very beginning of the Matuyama Reversed Chron, and that geomagnetic reversals (if any) within the Brunhes Normal Chron were probably too brief to be detected in archeological deposits. Artifacts in reversely magnetized sediments must thus date from within the Matuyama Chron, between 2.5 my and 780 ky ago. Artifacts in normally magnetized sediments are more ambiguously dated. They could date to the Brunhes Chron, after 780 ky ago or, where faunal remains suggest greater antiquity, to one of the normal subchrons inside the Matuyama Chron. Sequential magnetization change (for example, from normal to reversed and back to normal) can help to narrow the alternatives.

Cenozoic Climates

Beyond furnishing the chronological framework for human evolution, geologic studies can also illuminate the natural selective forces that drove it. Among these, none is potentially more important than climate or, more precisely, climatic change. For example, as discussed in the next chapter, the development of drier, more seasonal climates between 10 and 5 my ago (in the late Miocene) could at least partly account for the broadly contemporaneous emergence of the human family (Hominidae). The final section of this chapter aims to summarize Cenozoic climatic change as it may bear on human evolution, with particular reference to the later Cenozoic Ice Age, which, in one of its milder phases, still grips the planet. The evidence for Cenozoic climatic change is incomplete and sometimes ambiguous or even contradictory (322, 1547), but the broad trends are clear (1213, 1560, 1561); only the details remain subject to revision.

In the early Cenozoic, global climates were relatively warm and equable, and there was remarkably little difference in temperature between the equator and the poles. World temperatures rose slightly during the Paleocene to a peak in the early Eocene, before beginning a long decline. The descent occurred in steps (1213), including sharp ones in the early Oligocene roughly 34 my ago, the mid-Oligocene about 31 my ago, the mid-Miocene about 14.5–14 my ago, the late Miocene between 6.5 and 5 my ago, and the mid-Pliocene between 3.1 and 2.8 my ago (fig. 2.11). The main underlying cause was the changing configuration and topography of the continents as it altered oceanic and atmospheric circulation (1304). Glaciers were absent during the Paleocene and Eocene, but the Oligocene temperature minima probably initiated transient glaciation in eastern Antarctica. The sharp temperature plunge that occurred in the mid-Miocene, about 14.5–14 my ago, coincided with the formation of a permanent East Antarctic Ice Sheet that has retained roughly its present dimensions ever since.

The cold episode at the end of the Miocene, between 6.5 and 5 my ago, was particularly acute, and it probably added a permanent West Antarctic Ice Sheet. During this cold period, repeated glacier growth often drained so much water from the oceans that the Atlantic and the Mediterranean were disconnected, and the Mediterranean periodically dried up, leaving behind vast salt deposits (1028). In the period immediately following 5 my ago, global temperatures recovered, and the Mediterranean refilled. But in the mid-Pliocene, between about 3.2 and 2.8 my ago, temperatures plunged again, and the northern continents were glaciated for the first time. A strong increase in debris ice-rafted from polar glaciers to the surrounding seas about 2.8 my ago (1133) signals a continuing temperature fall that coincided with the emergence of a highly conspicuous cycle in which glacial peaks returned on average every 41 ky (2186). About 1 my ago, the average duration of glaciations expanded to about 100 ky, and their intensity increased sharply (1921). Peak ice volume after 1 my generally exceeded earlier volume by 50% or more.

The mechanisms that drove late Cenozoic cold/warm oscillations have been fiercely debated, but most specialists now assign a fundamental role to astronomical variables or, perhaps

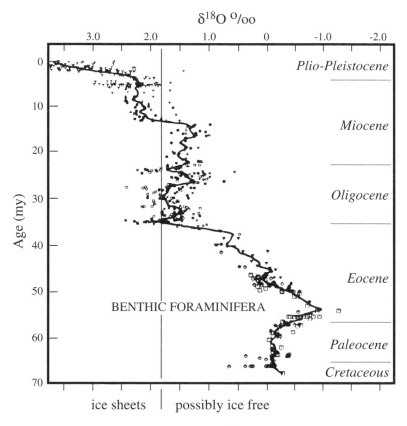

$\delta^{18}O$ %o

Figure 2.11. Composite δ oxygen-18 ($\delta^{18}O$) record for Atlantic Cenozoic localities (modified after 1561, fig. 1). The vertical scale is time
in millions of years before the present. The horizontal scale tracks
relative change in the concentration of ^{18}O in deepwater (benthic)
foraminifera. Small positive or negative $\delta^{18}O$ values (to the right) indicate a low likelihood for continental ice sheets. Large positive values (to the left) indicate a high probability. The continuous curve is
based on individual $\delta^{18}O$ values obtained in different deep-sea cores
that cover the entire Cenozoic era. A transient ice sheet probably
formed in east Antarctica during the Oligocene, but it became permanent only from the mid-Miocene, beginning about 14 my ago.
Ice sheets formed in the Northern Hemisphere only after the mid-
Pliocene, beginning about 3.2 my ago.

more precisely, to their interaction with global oceanic and atmospheric circulation (1645). The "astronomical theory" is
commonly credited to the Serbian astronomer Milutin Milankovic, who refined it between 1915 and 1940 (188). Milankovic
suggested that glaciations began at times when Northern
Hemisphere summers became relatively cool, and he noted
that insolation (solar heat) reaching high latitudes in summer
oscillates regularly, in correspondence with rhythmic change
in three factors: the wobble of the earth's spin axis (the preces-
sion of the equinoxes), which affects the time of year when the

earth is nearest to the sun; the obliquity (tilt) of the spin axis, which alters the angle at which sunlight hits the surface; and the shape (eccentricity) of the earth's orbit, which varies from nearly elliptical to nearly round. Insolation lows (times of reduced solar input) in response to the three factors occur in cycles of 19/23, 41, and 100 ky, respectively (993, 1106–1108, 1922). The durations of late Cenozoic glacial periods appear to match astronomical expectations, although the influence of each cycle (or astronomical variable) has varied through time. Only the 19/23 ky (or precessional) cycle is apparent before 2.8 my ago, while the 41 ky (or obliquity) cycle dominated from roughly 2.8 my ago, when large ice sheets first appeared in the Northern Hemisphere and highly conspicuous cold/warm oscillations began, until about 1 my ago. After 1–0.9 my ago, the 100 ky (or eccentricity) cycle became much more important, but the signal of the 41 ky cycle remained strong, especially in higher latitudes. The shift to dominance of the 100 ky cycle not only prolonged glaciations (to an average of 100 ky), it also initiated a sharp increase in their amplitude (intensity).

The late Cenozoic cold periods, with their greatly enlarged ice sheets, are commonly called glaciations or glacials, and the intervening warmer periods, with their reduced ice sheets, are commonly called interglacials or interglaciations. The Holocene of table 2.1 is simply the most recent interglacial and might better be known as the Present Interglacial. From a climatic perspective, it is not truly separate from the preceding Pleistocene, and some specialists prefer to extend the Pleistocene to the present, equating it with the Quaternary (1397). During the past 1–0.9 my, the cooling that initiated each glaciation occurred slowly and irregularly over 70–90 ky, whereas the terminal warming occurred much more abruptly, usually in less than 10 ky. Glaciations were generally interrupted by slightly milder intervals or "interstadials" that alternated with periods of greater cold or "stadials." Compared with interglacials, however, both stadials and interstadials were characterized by greatly expanded glaciers.

The existence of glaciers far beyond their present limits was documented first in Norwegian and Swiss mountain ranges in the early 1800s (1397). Evidence for greatly expanded continental ice sheets was developed somewhat later in the nineteenth century, initially by the great Swiss zoologist Louis Agassiz. As land-based evidence for ancient glaciations accumulated in both North America and Europe, it seemed to indicate four major late Cenozoic glaciations, separated by three main interglacials (excluding the last or present one). The names of the four main glaciations (from older to younger, Günz,

Mindel, Riss, and Würm) identified in a classic study by Penck and Bruckner (1909) in a small Alpine region south of Munich in Germany were widely extended to incorporate supposedly like-aged glacial phenomena elsewhere, and sites, fossils, artifacts, and such were commonly dated by reference to the time periods these four glaciations or the intervening interglacials were supposed to occupy. Beginning in the mid-1950s, however, studies of sediments from the deep-sea floor have shown that the number of major glaciations far exceeded the number proposed on land (1811). There were eight glaciations during the past 780 ky alone and perhaps fifty during the past 2.5 my, although those before 1–0.9 my ago were shorter and less intense (1921). The four-glaciation Alpine scheme survives in some textbooks, but it should now be totally abandoned.

The reason climatic inferences drawn from deep-sea sediments are more reliable is that deposition on the deep-sea floor tends to be more continuous. Major interruptions or discontinuities are more common in land-based sequences, especially in areas that were glaciated, since each succeeding glaciation tended to erase the evidence of its predecessors. In fact, the only land-based records that approach the deep-sea floor for continuity over long periods come from ice drilled out of permanent glaciers in Greenland (1153, 1154) and Antarctica (1163–1165) and from areas beyond the reach of the glaciers, such as central Europe and central China, with their deep covers of windblown dust (loess) (1280–1283), and the lake bottoms of Lake Bogotá in the Colombian Andes (1059, 2293), Owens Lake in California (230), Grande Pile in northwestern France (159, 934, 1916, 2474, 2475), Lake Phillipi in northern Greece (2294, 2468), Lake Biwa in Japan (771, 1184, 1556), and Lake Baikal in south-central Siberia (2470). For the periods these unusually long land-based records cover, they corroborate and fill out the basic history of late Cenozoic climatic change now standardized based on data from the deep-sea floor.

The deep-sea record for late Cenozoic climatic change has been established in sediment cores removed by sophisticated equipment that barely disturbs the sediment structure. Climatic change is inferred from the species of foraminifera and other sea-dwelling microorganisms represented in different parts of a core and also from the changing chemical composition of their tests (or shells). Variation in the content of oxygen-16 (^{16}O) and oxygen-18 (^{18}O) is particularly informative. These are naturally occurring, stable (nonradioactive) varieties (isotopes) of oxygen, which foraminifera and other marine organisms extract from seawater and build into their tests. The ^{18}O is heavier, and water molecules with ^{18}O evaporate less readily than those with ^{16}O. During glacial periods, when large amounts of water from the oceans became locked in ice sheets, seawater thus became enriched in ^{18}O relative to ^{16}O. The $^{18}O/^{16}O$ ratio in microorganismal tests partly depends on the temperature and salinity of the water in which the microorganisms lived, but it mainly reflects the original ratio in the water, and changes in the oxygen composition of tests from deep-sea cores therefore track the episodic growth and decline of the ice sheets (1918–1920).

The climatic stages detected from oxygen-isotope analysis are designated by arabic numerals, with odd numbers for the interglacial stages and even numbers for the glacial ones (150, 687, 688, 1919, 1924) (fig. 2.12). Thus the Holocene or Present Interglacial is known as isotope stage 1. There is the complication, however, that a long, milder period, generally agreed to fall within the Last Glaciation, has been designated stage 3 and that the Last Glaciation thus includes stages 4 through 2. Stage 5 corresponds to the Last Interglacial, though only its earliest substage, known as 5e, was comparable to stage 1 in terms of ice sheet reduction and global warmth (fig. 2.12). The later substages, 5d through 5a, reflect varying but generally cooler climate, and some authorities prefer to place them in the early part of the Last Glaciation. Almost certainly, they correspond to some land-based deposits that have long been assigned to the early part of the Last Glaciation, for example, to the Würm I in France (402).

The oxygen-isotope stages have been dated by a variety of methods, including ^{14}C used near the top and a combination of paleomagnetism and extrapolated sedimentation rates used farther down. One important conclusion from the dating is that, within the past 800–900 ky, periods that approximated the warmth of stage 1 or substage 5e—that is, interglacials in the narrowest sense—have probably lasted only about 10 ky each, whereas the intervening cold periods have lasted far longer, about 100 ky each (fig. 2.12). This means that during the later phases of human history—or more precisely prehistory—people lived mainly under glacial conditions.

The most obvious aspect of glaciations was the mushrooming of ice on the continents. At their maximum the ice sheets incorporated 50 million cubic kilometers of water, and they covered nearly a third of the earth's surface (fig. 2.13). In partial compensation, however, sea level dropped by 130–160 m, and fresh land was exposed on the continental shelves. The emergent shelves not only became available for occupation but, in some cases, connected previously isolated landmasses,

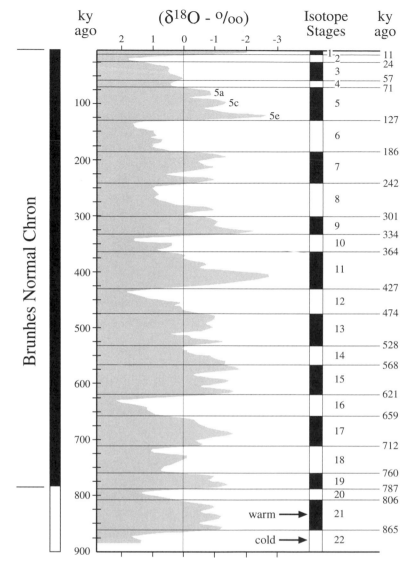

Figure 2.12. The δ oxygen-18 $(\delta^{18}O)$ record for the past 900,000 years, based on deep-sea sediment cores MD900963 in the tropical Indian Ocean and 677 in the equatorial Pacific Ocean (modified after 150, pp. 103, 106). Odd-numbered isotope stages indicate relatively warm periods, while even-numbered stages indicate colder ones. The boundary dates have been estimated primarily on the assumption that predictable variation in the relation between the earth and the sun forced the cold/warm oscillations.

allowing people and other animals to inhabit new areas. For example, it was the lowering of sea level during a glaciation between 2 and 1 my ago that allowed early people to occupy Java, and the broad land bridge that connected Siberia and Alaska during the Last Glaciation promoted or at least permitted the initial human colonization of the Americas.

The impact of glacial climate was worldwide (2072). From a human point of view, one of the most important effects was a large-scale redistribution of plants and animals. Tundra and steppe replaced forest in midlatitude Eurasia, and arctic animals such as reindeer penetrated far south of their interglacial range. Mean annual temperatures declined everywhere, by as

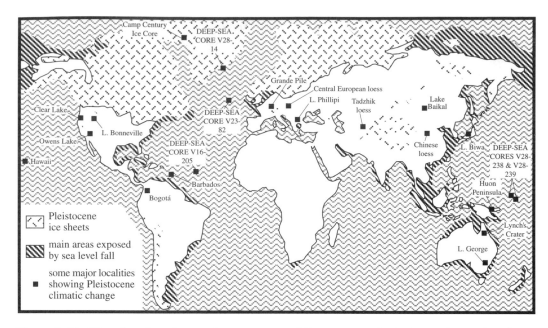

Figure 2.13. Map showing the maximum extent of Quaternary glaciation and associated sea level change (redrawn after 1788, fig. 2.1).

much as 16°C in higher latitudes and perhaps as much as 3°C near the equator. In most places precipitation also declined, probably mostly because of reduced evaporation from colder oceans. The effects of reduced precipitation are particularly obvious at lower latitudes (between roughly 30° S and 30° N), where large areas that are now rain forest became grassland or savanna, while grassland and savanna regions turned to desert. The reduced vegetation cover fostered wind erosion (deflation) at low latitudes, and some of the wind-borne dust accumulated in deep-sea cores adjacent to the continents. Analysis of the cores shows that dust volume increased in step with ice volume after 2.8 my ago, and like ice volume, it oscillated rhythmically in a 41 ky cycle between 2.8 and 1 my ago and in a 100 ky cycle afterward (647, 648).

The close correlation between glacial climate and significantly greater aridity contradicts an ill-founded earlier idea that wetter conditions (pluvials) characterized lower latitudes while glaciations affected higher ones. In fact, pluvial conditions were mainly an aspect of interglacials (398), though lakes did grow during glaciations in some midlatitude areas because of altered atmospheric circulation, lowered evaporation, or both. The recurrent and persistent aridity of the tropics during glaciations probably affected human populations as much as or more than increased cold did in higher latitudes, and the natural selection associated with recurrent aridity may explain some of the important evolutionary events discussed in later chapters. These events include the origin of the hominids between 6 and 5 my ago, the evolution of the genus *Homo* and of the robust australopithecines between 3 and 2.5 my ago, the emergence of first truly human species, *Homo ergaster*, 1.8–1.7 my ago, and perhaps even the origin and dispersal of fully modern people about 50 ky ago.

PRIMATE EVOLUTION: LATE CRETACEOUS TO LATE MIOCENE $\;3$

The Primates are a zoological order comprising roughly two hundred living species grouped informally as lemurs, lorises, tarsiers, monkeys, apes, and humans. The order also includes an equal or greater number of fossil species whose relation to living ones is hotly debated. This chapter outlines the main features of primate evolution before the emergence of humans (broadly understood) in the late Miocene. The presentation is relatively detailed, but the emphasis is on basic stages in primate history rather than on specific ancestor-descendant relationships. This reflects the growing realization among specialists that such relationships have often been wrongly inferred because the potentially ambiguous meaning of similarities among species has not been adequately appreciated. The crucial similarities are ones that indicate a shared common ancestor, but to begin with these can be very difficult to distinguish from similarities that reflect only adaptation to shared circumstances (analogy) or parallel evolution (homoplasy) in distantly related lines.

Equally important, it is increasingly clear that similarities due to common descent must be divided between "primitive" ones that developed early in the history of a species or species group and "advanced" or "derived" ones that developed much later. Only shared derived features indicate a closely shared origin or a possibly close ancestor-descendant relationship between time-successive species. Yet, as discussed in this and later chapters, shared derived characters are often hard to separate from shared primitive ones or even from shared characters developed independently in two species as a result of adaptation to similar natural conditions. In these circumstances, compelling ancestor-descendant relationships are far harder to establish than to disprove. (In general, for disproof it is necessary only to show that the supposed descendant lacks derived or advanced features found in its putative ancestor.) The key issue outlined here is addressed again below and is discussed in greater detail in chapter 1, which also summarizes some basic principles of zoological classification and nomenclature that are integral to this and subsequent chapters.

The rest of this chapter consists of three parts: a brief introduction to the skeleton, mainly to define basic anatomical terms relevant here and in later chapters; a longer discussion of the living primates, emphasizing those aspects that are essential for understanding primate (including human) evolution; and finally the central focus, a summary of the nonhuman primate fossil record organized by time periods, from the Cretaceous through the Miocene.

The Skeleton

For descriptive purposes, the skeleton is conventionally divided into two main segments—the *cranium* (skull) and the *postcranium* (trunk and limbs). The cranium itself is often divided into three principal sections—the braincase (the vault or neurocranium); the face (including the upper jaw or maxilla), which is attached to the braincase by bone; and the lower jaw or mandible, which is attached to the braincase and face only by soft tissue. The braincase and face are formed of several bones that grow separately and that fuse together only in adults, if at all. Figure 3.1 illustrates a hominid cranium with the conventional names for the principal parts.

Teeth are especially prominent in fossil studies, because they are the most durable skeletal parts and therefore dominate most fossil samples. Their specific morphology usually also reflects the dietary adaptation of a species, and since much evolution has involved dietary specialization and divergence, teeth play a key role in reconstructing evolutionary trends and relationships. Like other mammals, primates have four basic types of permanent teeth—known, from front to back, as incisors, canines, premolars, and molars (abbreviated as I, C, P, and M). The juvenile or deciduous dentition contains only incisors, canines, and premolars (abbreviated as dI, dC, and dP).

On each side of each jaw, the earliest mammals had three permanent incisors, one canine, four premolars, and three molars. By convention, the incisors are called (from front to back) I1, I2, and I3, the premolars P1, P2, P3, P4, and the molars M1, M2, and M3. Superscripts designate upper teeth (e.g., I^1), and subscripts designate lower ones (e.g., I_1). Most living mammals have lost teeth in the course of evolution, but the ones they retain are designated by the numbers of their primitive mammalian antecedents or homologues. Thus humans and their

Figure 3.1. Recon-
structed skull of
*Australopithecus
africanus* ("Mrs.
Ples"), showing the
principal anatomi-
cal parts or regions
(redrawn after 1317,
p. 130).

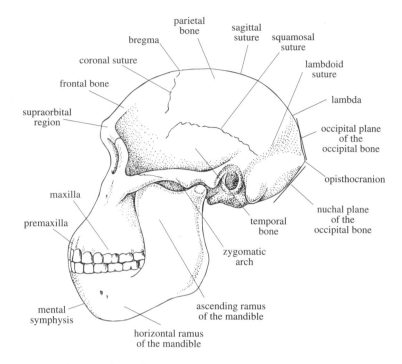

Australopithecus africanus ("Mrs. Ples")

closest living relatives within the Primates have lost the first two premolars of their remote ancestors, and the two premolars still found on each side of an adult human jaw are called P3 and P4. Figure 3.2 illustrates the adult dentitions of some representative primates, with the numeric designations of their teeth.

The description of teeth often requires an indication of their separate parts or orientation, and for this purpose four terms are in common use: *buccal*, to indicate the portion of a tooth nearer the cheek (it is replaced by *labial* for teeth that abut the lips); *lingual*, for the portion nearer the tongue; *mesial*, for the portion nearer the front of the mouth (or nearer the midline, depending on circumstances); and *distal*, for the portion nearer the rear of the mouth (or farther from the midline).

At fossil sites, postcranial bones are usually less common than teeth, because they tend to be softer and more fragile. They are often at least equally informative, however, revealing behavioral aspects, such as locomotor pattern, that are much harder to detect from teeth or skulls. Since much evolution has involved specialization and divergence in locomotor modes and other postcranial behaviors, postcranial bones supplement and complement teeth and skulls for the reconstruction of broad evolutionary patterns and relationships. All mammals have

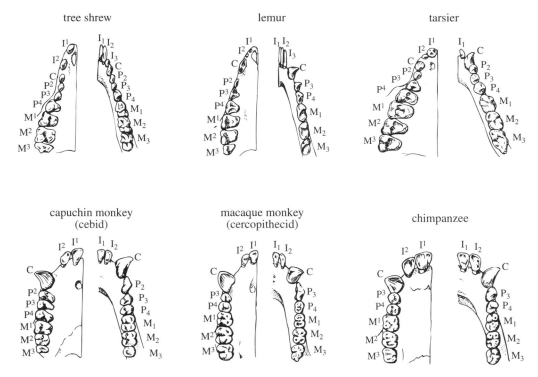

Figure 3.2. Right upper and lower dentitions of various primates (in each case, the upper is to the left and the lower is to the right) (redrawn after 1884, p. 102). The primitive mammalian dentition is thought to have comprised, on each side of each jaw, three incisors, one canine, four premolars, and three molars. In the course of evolution, all living primates have lost the first premolar (P1), while all catarrhine primates have also lost the second (P2). Both catarrhines and platyrrhines have lost the third incisor (I3), and the platyrrhines have either lost the third molar (M3) or retain it in reduced form, as in the capuchin monkey, whose dentition is illustrated here.

fundamentally similar postcranial skeletons, inherited from their common ancestor, and primates are distinguished mainly by a tendency to retain specific parts that many other mammals have lost during their evolution. Figure 3.3 illustrates a hominid skeleton, giving the names of the main postcranial bones.

Like teeth, individual postcranial bones can be described with respect to their orientation in a complete skeleton. The common terms are *proximal*, for the portion of a bone closer to the skull; *distal*, for the portion farther from the skull; *medial*, for the portion closer to the midline of the body; *lateral*, for the portion farther from the midline; *anterior* or *ventral*, for the portion closer to the front of the body; and *posterior* or *dorsal*, for the portion closer to the back. These terms can be combined, as, for example, in "anterior distal humerus" to mean the elbow end of the humerus when viewed from the front.

Figure 3.3. Reconstructed skeleton of *Australopithecus afarensis,* showing the names of the main skeletal parts.

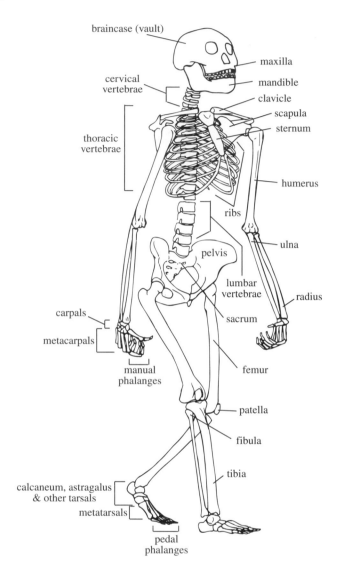

Where context requires, other anatomical parts and terms are introduced below, with illustrations when practical.

Definition of the Primates

The order Primates was first defined by the eighteenth-century inventor of modern biological classification, Carolus Linnaeus, who wanted the name to indicate a natural primacy for the order, since it included people. Linnaeus's original definition has long since been discarded, but no universally accepted replacement has emerged. The problem is that, compared with the Cetacea (whales and dolphins), Rodentia (rodents), Carnivora (carnivores), Artiodactyla (even-toed ungulates), Perissodactyla (odd-toed ungulates), Proboscidea (elephants), and most of the other eighteen living orders of mammals, the Primates are difficult to characterize based on unique traits they all share (1975, 2465). In distinction from other mammals, however, the living primates generally possess the following features (49, 430, 432, 1316, 1466, 1468):

1. A body structure preserving the clavicle (collarbone), pentadactyly (five digits on the end of each limb), and other primitive characters that have commonly been lost or modified in other mammalian orders.

2. Grasping extremities (hands and feet) with highly mobile digits, including a divergent hallux (big toe) and usually also an opposable pollex (thumb).

3. Flattened nails replacing primitive mammalian sharp, bilaterally compressed claws on the hallux and usually also on other digits, associated with highly sensitive tactile pads on the tips of the digits opposite the nails. Besides enhancing the sense of touch, the pads bear friction ridges (dermatoglyphs) that aid in grasping. Although nails themselves are not preserved in the fossil record, their presence is reflected in characteristic (dorsoventral) flattening of the underlying terminal phalanges.

4. Orbits (eye sockets) that tend to be convergent, that is, closely spaced and facing in the same direction, producing substantial overlap between the fields of vision and thus a high degree of stereoscopic, three-dimensional vision (depth perception). Also, in primates, unlike many other mammals, the orbits are completely surrounded by a bony ring (the post- or circumorbital bar), supplemented in higher primates by a bony wall (the postorbital plate or septum) separating the orbits from the skull behind. The high degree of stereoscopic vision reflected in the orbits is associated with a unique neural apparatus for processing visual signals and with enlargement of the visual centers in the occipital and temporal lobes of the brain.

5. A shortened muzzle or snout compared with that in most other mammals, generally associated with a more limited olfactory sense (sense of smell) and with a tendency for the olfactory bulbs in the brain to be relatively small.

6. A fully bony auditory bulla or middle ear in which the ventral floor is composed either of an extension of the petrosal bone that encloses the inner ear or of a combination of the petrosal bone and the ectotympanic bone or tympanic ring (the bone across which the eardrum is stretched) (fig. 3.4). In

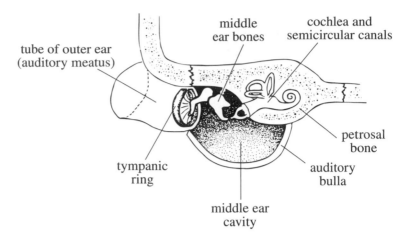

tube of outer ear
(auditory meatus)

middle
ear bones

cochlea and
semicircular canals

petrosal
bone

tympanic
ring

auditory
bulla

middle ear
cavity

Figure 3.4. Diagrammatic section through the right ear region of a therian mammal, seen from the front (redrawn after 430, fig. 3). In most nonprimates the auditory bulla either is cartilaginous or is formed by a separate entotympanic bone. In primates it is formed by an extension of the petrosal bone. The primate character of the bulla has been used to assign the Paleocene group known as plesiadapiforms to the Primates, but it is conceivable that the floor of the plesiadapiform bulla was formed by an entotympanic that fused seamlessly with the petrosal. This occurs in some nonprimates today. The true character of the plesiadapiform bulla could be determined only from fetuses, which are lacking in the fossil record.

other mammals the bulla is generally floored by an independent entotympanic bone, which has been lost in primates.

7. Reduction in the number of incisors and premolars compared with those in the earliest mammals and many living ones, combined with a relatively simple and primitive cusp pattern on the molars.

8. A unique sulcal (fissure) pattern on the surface of the cerebral cortex of the brain. Also, relative to body size, primates tend to have larger brains than other mammals.

Classification of the Primates

Modern biological classification is hierarchical, closely following the scheme devised by Linnaeus in the mid-eighteenth century. The species is the most basic unit and is usually defined today as a group of organisms that look more or less alike and that can interbreed to produce fertile offspring. Groups of related species are combined into *genera* (singular *genus*), genera into families, families into superfamilies, superfamilies into infraorders, infraorders into suborders, suborders into orders, and so forth. A species or group of related species at any level in the hierarchy is known as a *taxon* (plural *taxa*). Individual species and genera—for example, the species *Homo sapiens* and the genus *Homo*—are known as lower taxa. Groups of re lated species above the genus level—for example, the family Hominidae, the superfamily Hominoidea, the infraorder Catarrhini, or the suborder Anthropoidea—are known as higher taxa. Chapter 1 provides additional background for those unfamiliar with the basic system, together with information on how taxa are named.

In modern biological classification, the arrangement of species within genera and higher taxa is supposed to reflect their evolutionary relationships, such that species placed in the same higher-level taxon share a closer (more recent) common ancestor than ones placed in different taxa at the same level. Thus the inclusion of modern people, *Homo sapiens,* and the (extinct) "robust" australopithecine, *Paranthropus* (or *Australopithecus*) *robustus,* in the family Hominidae reflects the belief that they share a more recent common ancestor than either shares with the orangutan, *Pongo pygmaeus,* which is placed in a separate family, the Pongidae. In turn, modern humans, the robust australopithecines, and the orangutan are placed in a common superfamily, the Hominoidea, based on evidence that they are more closely related to each other than any are to the Old World (African and Eurasian) monkeys, which are placed in a separate superfamily, the Cercopithecoidea.

Until recently, zoological classification in general and primate classification in particular depended mainly on gross morphological similarities and differences among living species, on the assumption that detailed anatomical similarity implies close evolutionary relationship. However, especially since the 1960s, the results of gross anatomical studies have been significantly supplemented and altered by information on biomolecular differences and similarities. At the same time, new discoveries have shown that many fossil species cannot be readily accommodated in the same higher taxa as can living ones.

As a result of fresh biomolecular and fossil research, the long-standing traditional classification of the Primates has been abandoned. No new consensus is in sight, but at least for heuristic purposes most authorities would probably accept the breakdown into suborders, infraorders, and superfamilies presented in table 3.1. A major problem concerns the proper classification of the Tarsiiformes (tarsiers). They have been included here in the suborder Prosimii (lower primates), though many specialists believe they are more closely related to creatures in the suborder Anthropoidea (higher primates). In the

Table 3.1. Traditional Classification of the Primates

Classification	Known Temporal Distribution	Geographic Range
Order Primates		
Suborder Anthropoidea (or Simii)		
Infraorder Catarrhini		
Superfamily *Propliopithecoidea	Eocene-Oligocene	Africa
Superfamily Hominoidea	Early Miocene to Recent	Africa, Europe, Asia (worldwide today)
Superfamily *Parapithecoidea	Oligocene	Africa
Superfamily Cercopithecoidea	Early Miocene to Recent	Africa, Europe, Asia
Infraorder Platyrrhini		
Superfamily Ceboidea	Late Oligocene to Recent	South and Central America
Suborder Prosimii		
Infraorder Tarsiiformes		
Superfamily Tarsioidea	Miocene to Recent	Asia
Superfamily *Omomyoidea	Paleocene to Oligocene	Europe, North America, Asia, Africa
Infraorder Lemuriformes		
Superfamily *Adapoidea	Early Eocene to Late Moocene	Europe, North America, Asia, Africa
Superfamily Lorisoidea	Early Miocene to Recent	Africa, Asia
Superfamily Lemuroidea	Pleistocene to Recent	Madagascar
Suborder Praesimii		
Infraorder Tupaiiformes		
Superfamily Tupaioidea	Late Miocene to Recent	Asia
Infraorder *Plesiadapiformes		
Superfamily *Plesiadapoidea	Middle Paleocene to late Europe	North America, Europe
Superfamily *Microsyopoidea	Early Paleocene to late Eocene	North America, Europe

Source: Modified after 829, p. 60, to separate the Tarsioidea from the Omomyoidea and the Propliopithecoidea and Parapithecoidea from the Cercopithecoidea/Hominoidea and to correct the dates for the earliest Propliopithecoidea, Ceboidea, and Omomyoidea.
Note: Asterisks designate extinct taxa.

latter view, which is perhaps gaining momentum, the names Prosimii and Anthropoidea are commonly abandoned in favor of Strepsirhini (the Prosimii without the Tarsiiformes) and Haplorhini (the Anthropoidea with the Tarsiiformes). The issue involved here is addressed again below, where some of the evidence on both sides is introduced.

Another problem, which is more central to this book, concerns families, above all those within the Hominoidea. Traditionally, specialists have recognized three—the Hominidae, for people; the Pongidae, for the "great apes"; and the Hylobatidae, for the "lesser apes" (gibbons and siamangs). But it is increasingly clear that most fossil apes cannot be comfortably included in the Pongidae or Hylobatidae, and they will thus have to be placed in their own extinct families. Perhaps even more important, biomolecular studies demonstrate unequivocally that the living African pongids—the gorilla (*Gorilla*) and the chimpanzee (*Pan*)—are actually more closely related to people than either is to the living Asian pongid, the orangutan (*Pongo*). Minimally, this means that the African apes should be lumped with people into the Hominidae, leaving the orangutan as the sole living member of the Pongidae. Much more radical revisions are plausible, and the relatively limited degree of DNA and protein divergence among the living apes and people has been used to lump them all into the Hominidae, as follows (110, 858, 1569):

Superfamily: Hominoidea
Family: Hominidae
Subfamily: Hylobatinae
Genus: *Hylobates*
Subfamily: Homininae
Tribe: Pongini
Genus: *Pongo*
Tribe: Hominini
Subtribe: Gorillina
Genus: *Gorilla*
Subtribe Hominina
Genus: *Pan*
Genus: *Homo*

Insofar as this scheme may reflect recency of common ancestry and actual degree of genetic relationships more closely than any other, it (or something like it) may someday be widely accepted. However, adopting it here would surely introduce more confusion than enlightenment, and in the interests of communication I have continued to employ a more traditional classification in which Hominidae (and its anglicized form, hominid) refers exclusively to people, living and extinct.

Fossil primates and thus primate (including human) evolution can be understood only by reference to the living primates, and any overview of fossil forms must include a survey of living ones. The following brief survey is designed only to provide essential background information. It begins with those taxa that are most closely related to people and proceeds downward through table 3.1 to progressively more distant relatives, which it treats in progressively less detail. It is based mainly on information from broad syntheses in references 732; 1468, chap. 1; 1608; 1690, chap. 3; 1758; 1884; and 2265. It should be read in conjunction with figure 3.5, which illustrates the geographic distribution of various living nonhuman primates. The figure shows that (excepting people) surviving primates are restricted almost entirely to the tropics and subtropics. This is particularly true of the most primitive forms (the prosimians or lower primates), and the sum suggests that primates originated under relatively warm conditions.

Figure 3.5.
Geographic dis-
tribution of the
living nonhuman
primates (modified
after 1884, pp. 39,
41, and 1608, fig. 4).

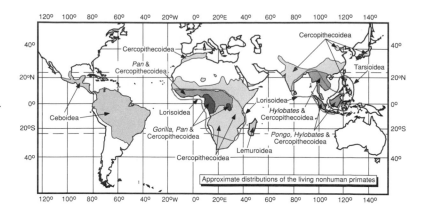

The Living Primates

The Anthropoidea

The living Anthropoidea or higher primates belong to three superfamilies—the Hominoidea, including people and apes; the
Cercopithecoidea, including the African and Eurasian or Old
World monkeys; and the Ceboidea, including the South and
Central American or New World monkeys. The Old World
monkeys share many primitive features with their New World
counterparts, but numerous derived anatomical and biomolecular characters show that the Old World monkeys are actually
more closely related to apes and people. Among the derived features that Old World monkeys, apes, and people share are a common dental formula, comprising two incisors, one canine, two
premolars, and three molars on each side of each jaw; an external auditory meatus (bony tube) projecting outward from the
middle ear; and complete closure of the bony wall behind the orbit. In New World monkeys the dental formula involves the
same number of incisors and canines, but there are three premolars on each side of each jaw, and in some species (the marmosets
and tamarins discussed below) there are only two molars. New
World monkeys also lack the external auditory meatus, and
their postorbital closure is less complete.

There are many other differences, among which the most
commonly cited is perhaps the orientation of the nostrils
(fig. 3.6). In general, in the New World monkeys the nostrils tend
to be widely spaced and to face sideways, whereas in the Old
World monkeys and Hominoidea they tend to be closer together
and to face more forward or downward. This has led to the use
of the terms *platyrrhine* ("flat-nosed") and *catarrhine* ("down-
ward-nosed") to distinguish the two groups. More formally, the

Old World higher primates, living and extinct, are commonly placed in the infraorder Catarrhini, while the New World forms are assigned to the infraorder Platyrrhini.

Within the catarrhines, the apes are clearly closest to people. In distinction from catarrhine monkeys, apes and people share numerous derived, gross anatomical features such as a short, flat, broad trunk; a shoulder structure that permits free rotation of the arms around the shoulder joint; and the absence of an external tail. The apes and people inherited their shared trunk form and shoulder structure from a common ancestor in which it was an adaptation for underbranch suspension and for forelimb-based tree climbing with the body in an orthograde (upright) position. Some apes use it in exactly this way today, while for various reasons other apes and people have largely abandoned living in the trees. In contrast to apes and people, but like most other land mammals, the catarrhine monkeys have long, narrow, deep trunks and a shoulder structure that restricts arm movement largely to a plane paralleling the body (fig. 3.7). This reflects the retention in monkeys and most other mammals of primitive quadrupedalism. Unlike apes (and people), the monkeys are quadrupedal even in the trees, where they walk along the tops of branches. They commonly use their tails as balance organs to keep their center of gravity over a branch, and it is possible that the common ancestor of the living hominoids lost its tail when it evolved an underbranch (vs. overbranch) posture or locomotor mode.

As already noted, the apes have been traditionally divided between two families—the Pongidae and the Hylobatidae—both readily distinguished from people (Hominidae) in numerous characters. Of these characters, perhaps the most notable are a relatively smaller brain, lower limbs that are not constructed for efficient bipedalism, and a dentition in which the anterior teeth (incisors and canines) are large relative to the posterior ones (premolars and molars). The canines are especially large and, unlike human ones, tend to wear along the mesial and distal (fore and rear) surfaces rather than at the tips. Associated with this, the lower third premolar (P_3) just behind the lower canine is sectorial (elongated mesiodistally and essentially unicuspid). When the mouth is shut, the mesial surface of the upper canine shears between the distal surface of the lower canine and the mesial edge of the sectorial premolar (fig. 3.8). The lower canine fits into a gap (diastema) between the upper canine and the upper lateral incisor (I^2). In the Pongidae, not only are the canines large overall, but, in further distinction from people, males have much larger ones than females.

New World Monkey
Cebus

Old World Monkey
Macaca

Figure 3.6. Representative New World and Old World monkeys, illustrating the sideways orientation of the nostrils typical in the New World monkeys and the downward or forward orientation typical in the Old World monkeys (redrawn after 1884, p. 22).

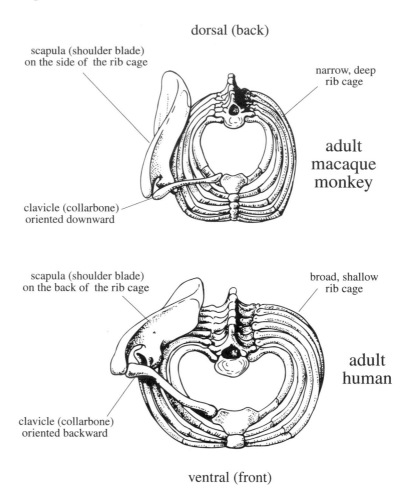

dorsal (back)

scapula (shoulder blade)
on the side of the rib cage

narrow, deep
rib cage

adult
macaque
monkey

clavicle (collarbone)
oriented downward

scapula (shoulder blade)
on the back of the rib cage

broad, shallow
rib cage

adult
human

clavicle (collarbone)
oriented backward

ventral (front)

Figure 3.7. Cephalic (top down) views of the rib cage and right
shoulder girdle of an adult macaque monkey and of an adult human, showing the deeper, narrower chest of the monkey and the
different arrangement of the scapula and clavicle, which limits the
monkey's ability to rotate its arm around the shoulder (redrawn after 1884, p. 81).

Both gross anatomical and biomolecular studies show that
the Pongidae resemble people far more than the Hylobatidae
do. There are three extant pongid genera—*Pongo*, the orangutan (and nominate genus); *Pan*, the chimpanzees; and *Gorilla*,
the gorilla. As indicated above, on biomolecular grounds the
chimpanzees and the gorilla are more closely related to people
than either is to the orangutan. Detailed similarities in the wrist
and hand linked to a shared mode of locomotion ("knuckle
walking," described below) and impressive, recently discovered
correspondences in dental enamel development and structure
suggest that chimpanzees and gorillas share a closer common

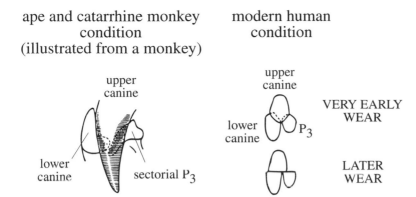

ape and catarrhine monkey
condition
(illustrated from a monkey)

modern human
condition

upper
canine

lower
canine   sectorial P₃

upper
canine

lower
canine   P₃   VERY EARLY
WEAR

LATER
WEAR

Figure 3.8. Occlusal relation between the upper canine, lower canine, and lower third premolar (P₃) in a nonhuman catarrhine primate (*left*) and a human being (*right*) (early wear above, later wear below) (redrawn after 1317, p. 182). In catarrhine monkeys and apes, the upper canine is part of a shearing complex involving the lower canine and the elongated, unicuspid (sectorial or "cutting") lower third premolar. When the jaws close the canines interlock, and wear occurs mainly on their fore and rear surfaces. In people the canines do not interlock, and wear occurs primarily at the tips. In addition, in people the lower third premolar is bicuspid and more rounded in outline.

ancestor with each other than either does with people (47, 48, 55, 616).

However, most analyses of proteins and of nuclear and mitochondrial DNA suggest that chimpanzees and people are more closely related to each other than either is to the gorilla (233, 409, 857, 858, 1058, 1062, 1856, 1857, 1942–1944, as opposed, for example, to 668 and 1452). It is difficult for biomolecular analyses to resolve the issue conclusively (109), because the number of phylogenetically informative genes may be very small if, as the genetic data themselves suggest, the branching events that produced the gorilla, chimpanzees, and people occurred in rapid succession (probably within 1 to 3 million years). Conceivably the issue is moot and there was only a single event that produced all three lineages more or less simultaneously (1451, 1816). Biomolecular methodology might also fail to resolve the human-chimpanzees-gorilla trichotomy if the last shared ancestor was genetically highly variable (polymorphic). In this instance different genes or different individual gorillas, chimpanzees, and humans might always suggest different, contradictory branching sequences (different phylogenetic relationships between species pairs) (1816). If continuing biomolecular studies confirm a special chimpanzee-human relationship, the extent of parallel morphological evolution between the chimpanzees and gorilla is truly remarkable.

Chimpanzees. Chimpanzees comprise two species, known popularly as the common chimpanzee (*Pan troglodytes*) and the pygmy chimpanzee or bonobo (*P. paniscus*). Historically, common chimpanzees occurred in suitable forest and woodland across equatorial Africa, from Gambia on the west to the shores of Lake Victoria and Lake Tanganyika on the east. Bonobos were restricted to the forests of Zaire (now the Democratic Republic of the Congo), south of the Zaire River. Common chimpanzees vary somewhat in size from place to place, but on average, free-ranging adult males probably weigh about 54 kg and females about 40 kg. Despite the "pygmy" appellation, bonobos are not significantly smaller, and they differ from common chimpanzees mostly in their relatively longer trunks, longer legs, and shorter arms. They are also less dimorphic, particularly in major skeletal dimensions.

Both species of chimpanzees feed mainly on ripe fruits, climbing trees to obtain them. They usually ascend with the trunk upright and depend mainly on their long arms for lift and stability. Young individuals often brachiate (arm swing) from branch to branch, but adults are too heavy to do this routinely. They travel between trees mainly on the ground, where they commonly assume a four-legged posture, with the feet flat and the hands curled, so that the weight of the forequarters rests on the knuckles (fig. 3.9). Their habitual locomotion is thus *quadrupedal* (four-footed), though it could equally be called *quadrumanous* (four-handed), since chimpanzee feet are very handlike from a human perspective. Quadrumanism is perhaps a particularly apt description for chimpanzee climbing, in which both the hands and feet are used to grasp branches.

Free-ranging common chimpanzees form loosely knit communities ranging from fewer than twenty to more than one hundred individuals (2507). Males usually remain in their natal groups for life, whereas adult females tend to disperse to other groups. Each female forages through a small territory either by herself or with her dependent young, while males form bands that patrol the territories of several females, fending off neighboring males. Female and mixed-sex aggregations occur at particularly fruitful trees, but individuals scatter when the food is exhausted. Bonobos have not been as thoroughly studied, but their social behavior seems broadly similar. The most conspicuous differences are that they forage more often in small, mixed-sex groups and their communities tend to more stable or cohesive (2100, 2435, 2436). Compared with common chimpanzees, bonobos are also marked by stronger bonds between females, weaker ones among males, markedly fewer aggressive

adult pig-tailed macaque

human child

adult chimpanzee

juvenile gorilla

Figure 3.9. Quadrupedal postures in (clockwise) an Old World monkey, a human, a gorilla, and a chimpanzee (redrawn after 1884, p. 55). Note that the monkey and human are standing with their palms flat, whereas the chimpanzee and gorilla are standing with their knuckles curled. Note also that the chimpanzee and gorilla have much longer arms relative to their legs.

interactions, and a greater inclination to share food. The differences might occur in part because, unlike common chimpanzees, bonobos nowhere overlap with gorillas (2508). When the fruits they favor become seasonally scarce, they thus can more readily fall back on the leaves and shoots that gorillas otherwise preempt. Bonobos' reduced susceptibility to food stress may partly explain why they are less aggressive toward each other, and reduced aggression among males could explain why they are less dimorphic.

Beginning with Jane Goodall in Gombe Stream National Park, western Tanzania (856), observers have now repeatedly recorded two unexpected, humanlike aspects of common chimpanzee behavior: they routinely hunt and eat other animals, such as monkeys or small antelope with which they share their range, and they make and use tools. In general, chimpanzees hunt much less than historic human hunter-gatherers, but some chimpanzee communities still consume several hundred kilograms of meat per year (2034). In addition, like hunter-gatherers

(978, 979), chimpanzees do not hunt only to eat. Males are the principal hunters, and they trade meat for sexual access (2034, 2035). They may also share meat with other males to cement political alliances.

Although chimpanzees hunt and use tools in many different places, the specifics vary significantly from region to region (253, 254, 1500). For example, common chimpanzees in the Taï National Park, Ivory Coast, commonly initiate hunts before the prey are in range, often stalk in groups, and regularly share the spoils. In contrast, chimpanzees at Gombe hunt much more opportunistically, more rarely cooperate in stalking, and share more reluctantly. Similarly, in western Africa many chimpanzee groups routinely use stone or wooden hammers and anvils to crack nuts, at least some modify twigs to pluck marrow from the bones of hunted prey, and most if not all probe insect nests with stems, twigs, or pieces of vine (the insects that cling to the invading probes are extracted and eaten). In contrast, no east-central African chimpanzees use tools to crack nuts or sticks to extract marrow, and they vary in how much they use vegetal probes to obtain insects.

Since nuts, marrow bones, and insect nests are more or less ubiquitous, some observers argue that regional differences in hunting, tool use, and other behaviors at least partly reflect chimpanzee "cultures" that anticipate human ones (1500, 1992). Unlike humans, however, chimpanzees transmit their "cultures" exclusively through direct observation or imitation, which is to say that only behavior is passed between generations, not ideas or meanings. In this regard chimpanzees are like lions or birds, whose offspring also learn specific behaviors by watching their elders. Thus, like all other animals, chimpanzees lack both Culture and cultures in the true anthropological sense. Nonetheless, their behavior strongly suggests that propensities to meat eating and tool use already existed in the last shared ancestor of chimpanzees and people.

Gorilla. Gorillas overlap chimpanzees in distribution, but gorillas require much denser forest, and they are far less widespread and abundant. They are concentrated in two distinct enclaves—a larger one in the lowland forests around the Gulf of Guinea in western Africa and a smaller one in the mountainous areas of eastern Zaire, western Uganda, and western Rwanda in east-central Africa. The separate populations are known, respectively, as the lowland and mountain gorillas, and they are usually regarded as subspecies of a single species, *Gorilla gorilla*. Some authorities further divide the lowland gorilla between a more western and a more eastern subspecies (910).

Gorillas are the largest living apes, but the sexes differ markedly in size. Free-ranging adult males reach 180 kg or more, while females average closer to 90 kg. Young gorillas are adept brachiators, but the adults are so large that they rarely enter trees, except occasionally to sleep. Like chimpanzees and bonobos, gorillas are quadrupedal on the ground, and they move from place to place with the feet flat and the hands resting on the knuckles. They feed mostly on leaves and shoots that they find at ground level. Their characteristic social groups average perhaps twelve individuals including a single dominant mature male (often designated a "silverback" from the mat of gray white hairs found on the back), some subadult males, and several unrelated females with immature young. "Surplus" adult males live alone or in small bachelor groups. Males compete fiercely for females, but gorilla communities are more stable and coherent than their chimpanzee counterparts, perhaps because preferred gorilla foods are more abundant and evenly distributed. Studies of free-ranging gorillas (such as 748 and 1872) indicate that, unlike chimpanzees, gorillas rarely if ever eat meat or use tools.

Orangutan. Unlike chimpanzees and gorillas, which are exclusively African, orangutans are exclusively Asian. There is only one living species, *Pongo pygmaeus,* which was narrowly restricted in historic times to the Southeast Asian islands of Borneo and Sumatra. Fossils, however, show that during the Pleistocene the same species was spread through mainland Southeast Asia to southern China. Orangutans are like chimpanzees in size but like gorillas in the degree of sexual dimorphism. Free-ranging adult males probably average about 65 kg, while females weigh only about half as much. In sharp contrast to both gorillas and chimpanzees, orangutans are almost exclusively arboreal and come to the ground relatively rarely. Young orangutans actively brachiate, but adults climb through trees very cautiously, so as to avoid potentially fatal falls. They are mostly quadrupedal on the ground, but unlike chimpanzees or gorillas, they tend to walk on the sides of their fists rather than on their knuckles. They feed mainly on fruits, supplemented to some extent by insects. Many of their preferred fruits are hard coated, which may explain why they have thicker dental enamel than either chimpanzees or gorillas. They seem to be essentially solitary (nongregarious), and the only coherent social unit is a female and her dependent young. Adult males apparently defend small territories containing a handful of adult females with whom they mate. Free-ranging orangutans rarely if ever eat meat, but like chimpanzees, they sometimes use twigs or other natural objects as tools (2299). Also like chimpanzees, orangutans may vary from place to place in the specifics of tool use and in other learned, socially transmitted behaviors.

Lesser Apes. As their vernacular name implies, the Hylobatidae or lesser apes are relatively small. Depending on the species, adults weigh between 4 and 13 kg, and males and females are generally very similar in overall size and appearance. In their near total lack of sexual dimorphism, the lesser apes contrast strongly with the great apes, from which they are also readily distinguished by the extraordinary length of their arms relative to their trunks and by hardened pads of skin (ischial callosities) that overlie broadened, roughened margins (ischial tuberosities) on the ischial bones of the pelvis. The pads permit prolonged sitting on the haunches and are also found in the Old World monkeys. They may therefore represent a primitive catarrhine feature that was lost in the great apes only after the line leading to the lesser apes had diverged.

Historically, the lesser apes were distributed from southern China through the forests of Southeast Asia, including the offshore islands of the Indonesian Archipelago, where they overlapped the orangutan. As many as eight species are recognized, divided between two subgenera of the genus *Hylobates.* The single species placed in the subgenus *H. (Symphalangus)* comprises the largest of the lesser apes, often known as the siamangs. The several species placed in the subgenus *H. (Hylobates)* include a variety of smaller apes commonly known as the gibbons. Like the term lesser apes, however, the term gibbons is often used to embrace both the gibbons and siamangs.

Gibbons and siamangs rarely leave the trees, where they use their powerful, elongated arms to brachiate from branch to branch and tree to tree. The gibbons proper eat mainly ripe fruits, supplemented with insects and birds' eggs, whereas the siamangs concentrate more on fresh leaves and shoots, supplemented with fruits and insects. Unlike other apes, both gibbons and siamangs have a social organization involving male-female pairs that bond for life. Each pair and its immature offspring inhabit a territory that they defend against neighboring pairs. Adult males do not regularly compete for females, which perhaps explains the limited amount of sexual dimorphism.

Old World Monkeys. The cercopithecoid or Old World monkeys are far more widespread and diverse than the apes, though this is a relatively recent development, and the fossil record shows that apes were once much more diverse than monkeys. Historically, cercopithecoid monkeys were found more or less throughout Africa and southern Asia, except in extreme deserts. During parts of the Pleistocene they also occurred in Europe, and an isolated population persists today on the Rock of Gibraltar.

The taxonomy of the cercopithecoids is disputed, but there are perhaps seventy-five living species in seven genera usually divided into two subfamilies—the Cercopithecinae and the Colobinae—within a single family, the Cercopithecidae. The cercopithecines are the common monkeys of sub-Saharan Africa (*Papio, Theropithecus, Cercocebus, Erythrocebus,* and *Cercopithecus*), with only a single genus (*Macaca*) in northern Africa and Eurasia. In contrast, the colobines are more abundant in Asia (*Nasalis, Presbytis, Pygathrix,* and *Rhinopithecus*), though one genus (*Colobus*) is restricted to Africa, where it is locally common.

The difference between cercopithecines and colobines is primarily dietary: whereas cercopithecines tend to focus on fruits (both ripe ones and ones not ripe enough for ape consumption), colobines emphasize leaves. To subsist on leaves, which are relatively high in nonnutritive bulk (fiber), colobines have evolved elaborate, sacculated stomachs. They also have relatively long, sharp crests on their molars, which are thus well suited for shearing leaves—unlike the shorter, blunter crests on cercopithecine molars, which are better suited for crushing fruits.

All cercopithecoids are primarily quadrupedal and arboreal, though some appear equally at home on the ground and a few cercopithecines are largely or wholly terrestrial. Knuckle walking is unknown; most arboreal species tend to rest and move on their palms and soles, while largely terrestrial species commonly stand and walk on the tips of their digits. Some colobines routinely hang from branches by their arms, but they are anatomically incapable of the arm-swinging locomotion (brachiation) observed in gibbons and immature great apes. Terrestrial species tend to be larger than arboreal ones, and body size varies from the highly arboreal, 1–2 kg talapoin monkey (*Cercopithecus talapoin*) to the largely or wholly terrestrial baboons (*Papio* and *Theropithecus*), in which adult males reach 40 kg. Most species are organized in mixed-sex troops, numerically dominated by females. Adult males compete vigorously for females, which perhaps explains why they usually have much larger bodies and canines (fig. 3.10).

New World Monkeys. The ceboid or New World monkeys are commonly divided into two families: the Callitrichidae or marmosets and tamarins, and the Cebidae or New World monkeys

Papio mandrillus

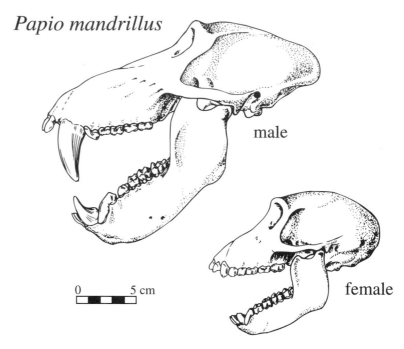

male

female

0 ⬛⬛⬛⬛ 5 cm

Figure 3.10. Skulls of male and female adult mandrills (*Papio mandrillus*), illustrating the extraordinary degree of sexual dimorphism, in both overall size and canine development, that characterizes many cercopithecoid monkeys (redrawn after 1884, p. 202).

in the narrow sense. The Cebidae are more diverse, with a minimum of eight genera (*Saimiri, Cebus, Aotus, Callicebus, Pithecia, Alouatta, Lagothrix,* and *Ateles*), versus five (*Callimico, Callithrix, Cebuella, Leontopithecus,* and *Saguinus*) for the marmosets and tamarins. The cebids or New World monkeys proper tend to resemble the Old World monkeys in overall form, diet, and sociality, but there are no large terrestrial species (comparable to the baboons) or dedicated leaf eaters (comparable to the colobines). Instead, all cebid species are more or less strictly arboreal, and they feed mainly on fruits, variably supplemented by leaves and insects. They are all primarily quadrupedal, but like the colobines, some larger cebid species have developed shoulder and arm specializations that allow them to hang from branches. The same larger species also commonly have a prehensile tail that functions as a fifth limb in underbranch suspension. Like most colobines, they occupy forests where their suspensory abilities allow them to feed near the ends of slender branches that could not be reached by exclusively quadrupedal forms.

On average, the marmosets and tamarins are much smaller than the cebids, and in size and some morphological attributes they recall lower primates. Perhaps most notable, like most lower primates but unlike any higher primates, they have a mix of nails and claws. Bilaterally compressed claws occur on every digit but the big toe, and they are used to secure the body on branches that small hands and feet could not grasp. In addition, like most lower primates, some marmoset and tamarin species feed mainly on insects or gum, as opposed to fruits or leaves, which are the main foods of cebids and other higher primates. True lower primates are absent in the New World, and to a degree the marmosets and tamarins fill their niche. In key features of head and body, however, the marmosets and tamarins closely resemble the cebids, and like virtually all other higher primates, they are also diurnal and highly social. In Africa and Asia, where lower primates overlap higher primates in the same way that marmosets and tamarins overlap cebids, the lower primates are nocturnal and solitary.

The Prosimii

The living prosimians or lower primates include three superfamilies in two infraorders—the Tarsioidea in the Tarsiiformes and the Lemuroidea and Lorisoidea in the Lemuriformes. A fourth superfamily—the Tupaioidea—is sometimes placed in the Lemuriformes, when it is included in the Primates at all. In accordance with the classification in table 3.1, it has been subsumed here in its own infraorder—the Tupaiiformes—within a third primate suborder, the Praesimii.

Tarsiers. The Tarsioidea or tarsiers are tiny (100–200 g) animals represented by a single genus (*Tarsius*) with three or four species. They are restricted to Sumatra, the southern Philippines, and the Celebes Islands off Southeast Asia, where they inhabit bush and forest. They are almost totally nocturnal in their activity pattern, and they have enormous eyes for night vision. They feed primarily on insects, supplemented with small vertebrates and perhaps with vegetal material. They take their name from their highly elongated tarsus (ankle), which they use to propel themselves in long leaps between near-vertical branches to which they cling in an upright position. This distinctive mode of locomotion has been called vertical clinging and leaping, and it is aided by a long nonprehensile tail that can be shifted in midair to ensure an upright landing. Like the lorises and some lemurs described below, tarsiers appear to be mainly solitary, although semipermanent male-female pairs have sometimes been reported.

As already indicated, the taxonomic and phylogenetic status of the tarsiers has been a matter of intense disagreement (10, 1899) because, unlike other so-called lower primates, they share a suite of important derived features with the higher primates (734, p. 18; 1468, pp. 663–670). For example, they lack both a naked rhinarium (a spot of hairless, glandular moist skin around the nose) and an accompanying cleft in the upper lip, their orbits are partly separated by bone (a partial postorbital plate) from the jaw musculature just behind, and their right and left frontal bones are completely fused. They also resemble the higher primates in key characters of reproductive biology, including the presence of an advanced hemochorial type of placenta. Many authorities (such as 156, 1041, 1468, 2126, 2128) believe the features tarsiers share with higher primates, together with the arguable results of biomolecular studies (99, 599, 1261, 1569, 1697, 1868), indicate that tarsiers should be classified with the higher primates in a shared primate suborder, the Haplorhini, as distinct from the suborder Strepsirhini for the Lemuriformes.

Lemurs and Lorises. The Lemuroidea or lemurs are a diverse group of monkey-sized animals confined entirely to Madagascar and the nearby Comoro Islands. In keeping with their limited geographic distribution, they are often known as the Malagasy lemurs. The Lorisoidea or lorises are broadly similar creatures that live in Africa and southern Asia. Among the anatomical features that imply the lemurs and lorises share a single (monophyletic) origin, the most notable are the alignment and forward protrusion of the lower canines and incisors (fig. 3.2) to form a "dental comb" used in grooming, and the presence of a claw on the second toe, also used for grooming.

In contrast to higher primates, both lemurs and lorises lack a postorbital plate, and they tend to have faces like those of nonprimate mammals, with a relatively long snout, a rhinarium, a cleft upper lip bound down to the gum, relatively large and free-moving ears, and an immobile facial expression. Like the tarsiers, they further differ from higher primates in having more laterally placed incisors; relatively sharp, unmolarized premolars; and in most species a mandible in which the two halves remain unfused at the symphysis (midline) (fig. 3.11). Also like the tarsiers, they are much less sexually dimorphic than most higher primates. Among external features, those that link them most clearly to the higher primates are their hands and feet. These are functionally adapted for grasping, with mobile digits tipped by flattened nails except, as already mentioned, for the second toe, which bears a grooming claw.

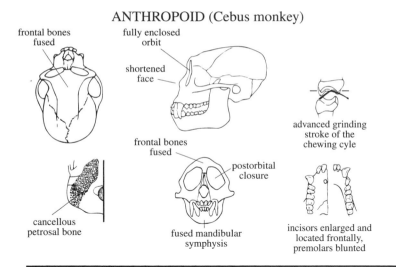

ANTHROPOID (Cebus monkey)

frontal bones fused

fully enclosed orbit

shortened face

advanced grinding stroke of the chewing cyle

frontal bones fused

postorbital closure

cancellous petrosal bone

fused mandibular symphysis

incisors enlarged and located frontally, premolars blunted

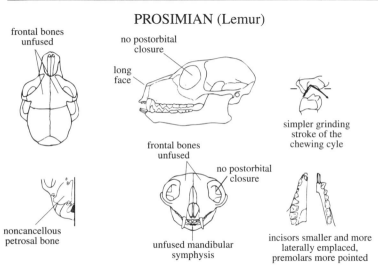

PROSIMIAN (Lemur)

frontal bones unfused

no postorbital closure

long face

simpler grinding stroke of the chewing cyle

frontal bones unfused

no postorbital closure

noncancellous petrosal bone

unfused mandibular symphysis

incisors smaller and more laterally emplaced, premolars more pointed

Figure 3.11. Comparison between the skulls of a Malagasy lemur (*Lemur*) and a New World monkey (*Cebus*) to illustrate some of the features that commonly distinguish prosimians from anthropoids (redrawn after 1834, fig. 4).

The lorises comprise two major types—the Lorisidae or lorises in the narrow sense and the Galagidae or galagos. The lorises are both south Asian (two species in two genera) and equatorial African (two species in two genera). The galagos or "bushbabies" (four to six species in one genus) live exclusively in Africa, where they have a broader distribution than the lorises, extending far to the south in savanna and bush. In sharp distinction from the exclusively diurnal catarrhine monkeys and apes that share their range, the lorises and galagos are all nocturnal. They are also relatively unsociable, and individuals usually forage alone. All lorises are almost entirely arboreal, but the lorises proper tend to be slow-moving climbers, while the galagos are more energetic runners and jumpers. Like the tarsiers and some lemurs, the galagos have powerfully built hind limbs that aid in vertical clinging and leaping, but galagos vary in how much they cling and leap, and some species are pre-dominantly quadrupedal. Both lorises and galagos include spe-cies that feed primarily on insects and others that focus more on gum or fruits.

The lemurs are a much more diverse group of animals; this reflects the minicontinental size of Madagascar, its envi-ronmental diversity, and its separation from Africa beginning in the Cretaceous. By the middle or late Paleocene, when the first lemurlike animals may have appeared elsewhere, Mada-gascar was already in roughly its present position, 400 km east of Africa (fig. 3.12). Incredible as it seems now, the ancestors of the lemurs must therefore have reached the island on floating rafts of vegetation. The stochastic (or "sweepstakes") nature of rafting events precluded the arrival of most other modern mammals, and the ancestral lemurs thus evolved in isolation from monkeys, apes, and many other potential competitors. This isolation allowed them to radiate into niches that lower primates could not occupy elsewhere.

At present there are twenty-two lemur species in perhaps ten genera, and at least fourteen additional species and seven genera probably became extinct only within the past 1,500 years, after the initial human colonization of Madagascar (390, 654). Most lemurs are arboreal, but some are more terrestrial, and to judge from their morphology and large size, some recently extinct forms may have been totally terrestrial. In feeding, some species focus on insects or gum while others emphasize fruit or leaves. They also vary greatly in locomotor pattern, preference for nighttime versus daytime activity, social behav-ior, and other features; and though many species recall lorises in their overall level of organization, others are more reminis-cent of monkeys. In keeping with contrasts between monkeys and prosimians elsewhere, the more monkeylike lemurs tend to be larger bodied, more diurnal, more dependent on leaves or fruit (vs. insects or gum), and more social (vs. solitary) than their less monkeylike relatives (1468). The monkeys and mon-keylike lemurs are often cited as an example of parallel evolu-tion, whereby creatures with similar genetic backgrounds ex-posed to similar environmental conditions have evolved in similar or parallel fashion.

The Praesimii

The only extant superfamily of Praesimii is the Tupaioidea or tree shrews, widely distributed throughout southern Asia and Southeast Asia, including the offshore islands. In total, there

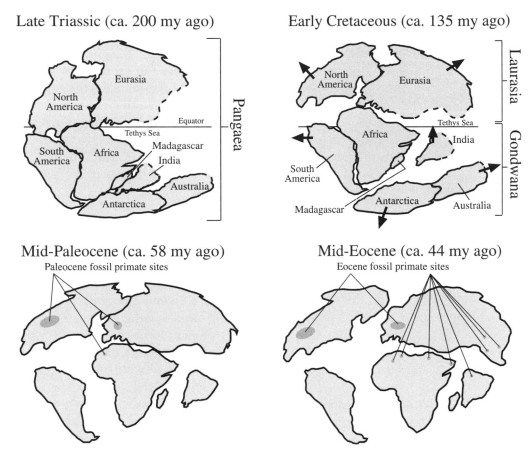

Figure 3.12. Changing positions of the continents from the late Triassic, roughly 200 my ago, to the middle Eocene, roughly 44 my ago (modeled after 2534). In the late Triassic, the modern continents were essentially joined in a single supercontinent known to geophysicists as Pangaea. Subsequent fragmentation (drift) divided Pangaea into a northern hemisphere landmass known as Laurasia and a southern hemisphere mass known as Gondwana. Yet further fragmentation divided Laurasia and Gondwana into separate parts, foreshadowing the modern continents. The conjunction of North America and Eurasia in the Paleocene and early Eocene accounts for the close similarity of their early primate faunas, as discussed in the text.

are perhaps five genera, all resembling true shrews (order Insectivora) or tropical squirrels (order Rodentia) in overall appearance. All are highly active diurnal inhabitants of forest undergrowth, where they feed mainly on insects, supplemented with vegetal material and small vertebrates.

Like the undoubted primates already discussed, tree shrews have a complete post- or circumorbital bar, and they possess a few minor features of the teeth, eye, brain, and limbs that suggest some relationship to lemurs (1193, 1316). However, they lack the bony ear structure (petrosal bulla, discussed above) that all other primates share, their orbits are not notably convergent, they have claws on all the digits, and their hands and feet are not especially well adapted for grasping. These and other nonprimate features may be used to exclude them from the Primates; but even authorities who place them in their own distinct order (Scandentia) generally agree that the tree shrews share a very close common ancestry with the Primates and that they are probably more like the common ancestor than any living primate. Thus, taxonomic disagreements aside, virtually all authorities continue to include the tree shrews in discussions of primate evolution.

Primate Evolution and the Biomolecular Clock

The genes and proteins of extant primates not only inform on their evolutionary relationships, they can also be used estimate the time(s) when extant taxa last shared a common ancestor. The method is often called the biomolecular clock, since it relies on the assumption that DNA or protein differences between taxa accumulate at a more or less constant (linear) rate. The rate can be determined from genetic or protein differences between taxa whose divergence time has been established in the fossil record, and it can then be applied to estimate divergence times between other taxa whose fossil record is less clear (46). The result is a set of working hypotheses that must ultimately be checked against new fossils.

Thus, if we assume a fossil (geological) date of 25 my ago for the divergence of Old World monkeys from apes, DNA hybridization data imply that the human and chimpanzee lines split about 5.5 my ago and that the gorilla lineage became distinct about 7.7 my ago, the orangutan lineage about 12.2 my ago, the gibbon lineage about 16.4 my ago, and (by definition) the line leading to Old World monkeys about 25 my ago (1943). These age estimates agree closely with ones developed from an earlier set of the same data by using a fossil date of 12 my ago for the emergence of the orangutan line (1689) (fig. 3.13). If we assume that the orangutan line actually diverged about 17 my ago, the full data set places the separation of humans and chimpanzees at 7.7 my ago and the emergence of the gorilla, orangutan, gibbon, and Old World monkey lines at 11, 17, 23, and 34 my ago, respectively.

The biomolecular clock has been controversial since it was first applied to human evolution (by 1866, 1867, 1869), mainly because it often contravenes dates that fossil evidence appears to support. One obvious objection is that proteins or DNA need

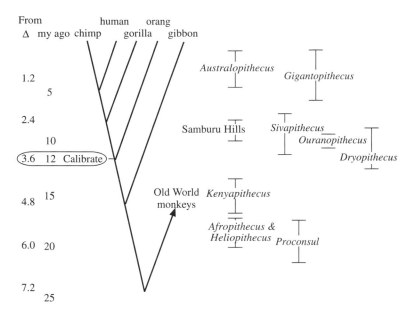

Figure 3.13. Framework of catarrhine evolution as deduced from DNA hybridization data (modified after 1689, fig. 8). The branching times are based on the assumption that DNA differences accumulate at a linear rate, which can be calculated from the branching time of the orangutan "clade," estimated here from the appearance of *Sivapithecus* roughly 12 my ago. The known or probable time ranges of some important fossil hominoids are shown to the right. Dates of divergence based on most other molecular data are broadly similar, and they are broadly consistent with the fossil record.

not diverge at a constant rate (1365). DNA-DNA hybridization experiments suggest that whole genomes evolve at a more or less constant rate, but particular parts may not (409). There is in fact evidence that the mutation rate in protein-coding DNA has slowed progressively from the most primitive to the most advanced primates, perhaps reflecting an increase in generation span and thus a decrease in the number of germ-line DNA replications and in the number of possible mutations per unit time (110). If the mutation rate in protein-coding DNA has slowed, then a rate based on the initial (fossil) separation of lower and higher primates will underestimate true divergence times within the higher primates, while a rate based on (fossil) divergence times within higher primates will overestimate the time when higher and lower primates split. The implication is that a biomolecular clock based on mutations in protein-coding DNA must be calibrated differently for different taxa or for particular portions of a phylogenetic tree to which it is being applied.

Dates based on differentiation of one of the globin genes illustrate the point. If we assume (from fossils) that Old World (catarrhine) and New World (platyrrhine) higher primates separated 30 my ago and that the mutation rate slowed progressively in ever more advanced higher primates, then globin gene differences suggest that humans and chimpanzees diverged about 6.5 my ago and that the lines leading to the gorilla, orangutan, and Old World monkeys split off at roughly 6.8, 12.3, and 21.5 my ago (975). These estimates differ somewhat from the DNA-DNA hybridization times summarized above, but they are close enough to suggest that each set approximates reality. The most significant similarity is in the timing of the human/chimpanzee divergence, which both sets place between 8 and 5 my ago. The 8–5 my estimate is further consistent with dates based on differentiation in mitochondrial DNA (1062) and, more important, with the known fossil record, as discussed in this chapter and the next.

Biomolecular estimates are bound to be imprecise, if only because the rate of genetic divergence may not be accurately established from the fossil record, where few if any times of evolutionary splitting are precisely fixed (829, 832). There is also the possibility that viral infections have occasionally transferred genetic material between closely related species, with the result that they appear to share a much more recent common ancestor than they really do (2124). Nonetheless, the congruence between current biomolecular and fossil dates for the last shared ancestor of people and apes is striking. This is especially so since the biomolecular estimates came first, and they were instrumental in provoking a reevaluation of a fossil date that was popular in the 1960s and 1970s. At that time, fossil jaws from the Indian subcontinent suggested that people had split from apes by 12–14 my ago (1957). Additional, more complete fossils and fresh analyses suggested that the jaws actually documented the divergence of the orangutan lineage from a joint African ape and human line (1689). Virtually all paleontologists now agree that the human/African ape split occurred between 8 and 5 my ago as indicated by the molecular studies (1686).

The Ancestors of the Primates

The great zoological class Mammalia, to which the Primates belong, evolved from mammal-like (therapsid) reptiles during the Triassic period of the Mesozoic era, roughly 220 my ago. The early mammals soon diversified into several stocks, including one leading to the therians, a branch or subclass of the mammals that includes the marsupials (metatherians), placentals (eutherians), and some other, extinct infraclasses. The marsupial and placental lineages diverged during the early Cretaceous period, between 130 and 100 my ago (1616). The earliest placentals were diverse, but all were variously similar to shrews, moles, tenrecs, or other living creatures that have traditionally been lumped into the order Insectivora.

It is from Cretaceous insectivore-like creatures that all later placental orders sprang. A tiny lower molar (1.85 mm long) from late Cretaceous (70–80 my old) deposits in eastern Montana suggests that the Primates were among the first to differentiate. The tooth has been assigned to the genus *Purgatorius*, which is better known from succeeding early Paleocene deposits in the same region (2126, 2300) The higher taxonomic assignment of *Purgatorius* is debatable, but even if it represents only an insectivore in the narrow sense, a Cretaceous genesis for Primates is still suggested by their diversity and abundance in the Middle and Upper Paleocene. Some background on the Cretaceous world is thus pertinent to an understanding of primate origins.

Cretaceous geography was very strange by modern standards (fig. 3.12). In the very early Cretaceous, South America was still joined to Africa, but the two continents began to drift apart roughly 125 my ago, and by 70–80 my ago, when the Primates perhaps emerged, the continents were separated by a narrow but expanding South Atlantic. South America and North America were separated throughout the Cretaceous, but North America was connected to Europe via Greenland. Also, global climate was remarkably mild more or less throughout, and there was relatively little temperature difference between the equator and the poles. In both the Southern and Northern Hemispheres, temperate forests thrived at high latitudes.

Biologically, the Cretaceous is often known as the "age of dinosaurs," because they were its most conspicuous, if not its most numerous, vertebrates. But it also witnessed important evolutionary developments and diversification in other kinds of reptiles, in birds, in early mammals, and not least in plants. Before the Cretaceous, the principal plants were gymnosperms (nonflowering plants such as conifers, palms, and cycads). Angiosperms (flowering plants, including trees, grasses, herbs, etc.) arose in the early Cretaceous and subsequently radiated to become the dominant plant forms by the early Paleocene (2109, table 1). Their success created niches for creatures that could feed on the nectar, nuts, berries, or fruits that flowering plants produce. One result was an explosion in insects, particularly ones that promote plant pollination. Birds and mammals diversified to exploit the increase in both edible plant parts and insects. Virtually all Cretaceous mammals had sharp cutting ridges on their molar teeth that were well suited for slicing through insect tissue. The earliest primates probably continued to eat insects, but their molars had lower and less pointed cusps, blunter ridges, and other distinctive features that suggest they took more fruits, seeds, and other vegetal matter than did their more insectivore-like ancestors (1827, 2125).

In order to specialize on insects or fruit, early primates were probably at least partly arboreal, and it has long been assumed that a primeval adaptation to life in the trees accounts for such characteristic primate features as grasping extremities (hands and feet), relatively sophisticated vision, a diminished sense of smell, and a relatively large brain (1316). Grasping extremities assist in movement over branches, and vision is more useful than smell for locating food or for moving safely and rapidly in a three-dimensional, arboreal habitat. Much of the brain in all species is dedicated to muscular control and coordination, and arboreal life places a special premium on the ability to coordinate various muscle groups with each other and with the eyes. However, squirrels provide abundant proof that distinctive primate grasping extremities, visual specializations, and expanded brains are not essential for active arboreal life, and the Cretaceous ancestors of the primates probably had extremities, vision, and brains more like those of squirrels than like those of later primates. It is thus necessary to find a more specific explanation for why primate specializations evolved. Perhaps the most plausible alternatives are that they helped in visual predation on insects among slender branches in forest undergrowth (429, 430, 432), that they permitted access to fruits growing at the ends of slender branches (2109), or that they assisted in feeding on both insects and fruits (1734). The position favoring enhanced visual predation is incorporated in figure 3.14, which illustrates two similar but competing views of early primate evolution linked to major dietary shifts.

Paleocene Primates

After dominating the earth's fauna for 150 million years, the dinosaurs became extinct at the end of the Cretaceous. The reasons are hotly debated (847), but the resulting environmental opportunities help explain why mammals burgeoned in the Paleocene. Among the mammals that profited were the early primates.

Geographically, the Paleocene differed little from the late Cretaceous. North and South America were separated by ocean,

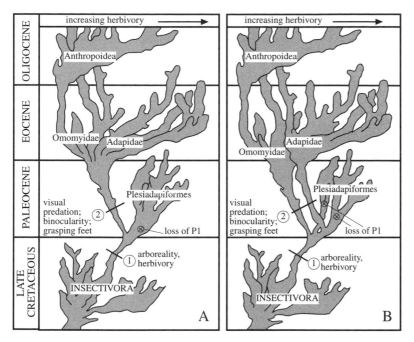

Figure 3.14. Two interpretations of early primate phylogeny (adapted from 430, fig. 8). In both, the earliest primates diverged from their insectivore (or insectivore-like) ancestors by a greater emphasis on arboreality and herbivory. In both, the evolution of the omomyids and adapids (Eocene prosimians) involved a subsequent shift back toward insectivory, based now on binocular vision and grasping extremities. Finally, in both, increasing emphasis on herbivory is a crucial element in the evolution of the anthropoids. The main difference between the two interpretations is that in A (*left*), the omomyids and adapids share a single Paleocene ancestor, whereas in B (*right*) they have separate ones in which binocular vision and grasping extremities evolved independently. On present evidence, A is more likely. Both interpretations suggest that the omomyids and not the adapids were ancestral to the anthropoids, but this issue remains debatable. The circled xs mark the loss of the first premolar (P1) in the plesiadapiforms. This is one of the specializations that makes them unlikely ancestors for Eocene prosimians.

while North America and Europe were still connected (fig. 3.12). Climate was generally very mild, and tropical or subtropical forests flourished in middle latitudes.

So far, Paleocene primate fossils are best known from western North America and western Europe, which were connected by a broad, forested land bridge stretching over what is now the subpolar ocean and islands in between. Environmental conditions were apparently very uniform across the land bridge, and the primate faunas on both sides were very similar (824). They shared at least two genera, *Chiromyoides* and the better-known *Plesiadapis*, which, with two or three Eocene genera and *Homo*, were the only primate genera to become naturally distributed in both the Old World and the New World.

At most Euramerican sites where Paleocene primates occur, they are very abundant and very diverse (314, 536, 562, 732, 1827, 1959, 2125, 2126), suggesting they were highly successful. There is incomplete agreement on which taxa are truly primates, but many specialists would accept at least sixteen genera in four families: Paromomyidae, Picrodontidae, Carpolestidae, and Plesiadapidae. Sometimes the Paleocene Microsyopidae and, less often, the Apatemyidae are also regarded as primates (831). The primate status of the Paromomyidae has recently been questioned (1187, 1193) because they share derived cranial features with *Cynocephalus*, the colugo or "flying lemur," the single surviving genus within the order Dermoptera. Like the colugo, the paromomyids may have had a web or fold of skin (patagium) that could be stretched between the limbs for gliding (154), but the presence of this fold in paromomyids depends on uncertain associations between first and second phalanges and between phalanges and diagnostic cranial remains (1274). Yet even if the number of generally accepted Paleocene primate families is limited to three, it is clear that Paleocene primates were much more abundant and diverse than living prosimians. Unfortunately, however, in spite of their abundance, most Paleocene primate fossils are fragmentary jaws and isolated teeth. Limb bones and skulls are much rarer, and the available skulls tend to be badly damaged.

The known Paleocene primates were all relatively small, from shrew size (100 g or less) to house cat size (roughly 5 kg). All were perhaps omnivorous to a degree, but differences in body size, dental morphology, or both suggest that some concentrated more on insects, others more on seeds or fruits (562). Insofar as their skulls are known, mainly for *Plesiadapis*, they were very primitive, with large snouts and laterally placed (nonconvergent) orbits that indicate a well-developed sense of smell and limited overlap between the fields of vision (fig. 3.15). The orbits were open to the sides, as in primitive mammals generally, rather than surrounded by a post- or circumorbital bar, as in later primates. The postcranium was equally primitive, at least as deduced from *Plesiadapis* fossils. The elbows and ankles were relatively mobile and probably encouraged climbing (2128, 2129), but bilaterally compressed, pointed terminal phalanges indicate that the digits retained sharp claws (vs. nails) (831) (fig. 3.15), and there is no evidence for the grasping hands and feet of later primates.

In overall form, most Paleocene primates probably resembled modern tropical squirrels, and we may reasonably ask why

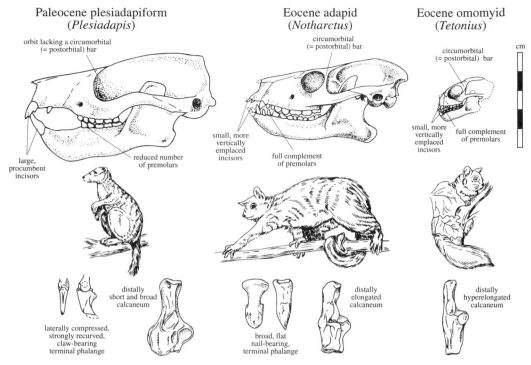

Figure 3.15. *Top:* Skulls of the Paleocene plesiadapiform *Plesiadapis,* the Eocene lemurlike adapid *Notharctus,* and the Eocene tarsierlike omomyid *Tetonius* (redrawn to the same scale after 1827). *Middle:* Reconstructions of *Plesiadapis, Notharctus,* and *Tetonius* (redrawn to different scales after 1825, p. 162). *Bottom:* Terminal (third) phalanges and calcanea (heel bones) of the same taxa (redrawn to different scales after 1825, p. 162). Note the large snouts, laterally placed, nonconvergent orbits open to the side, large procumbent incisors, and reduced number of premolars in *Plesiadapis.* Terminal phalanges show that *Plesiadapis* had claws whereas *Notharctus* had nails. *Notharctus* and especially *Tetonius* also had elongated calcanea, indicating that they were well adapted for leaping from branch to branch. In its enhanced ability to leap, *Tetonius* probably resembled the living tarsier or the African galago.

they should be classified as primates. The answer is in fact equivocal. Perhaps most important, their cheek teeth (premolars and molars) closely resembled those of later, undoubted primates (1825). It has also been suggested that they shared the uniquely primate petrosal bulla (or middle ear) discussed above. However, it is now known that the bulla in the Paromomyidae was floored mainly or entirely by a separate entotympanic bone (versus an extension of the petrosal) (1193, 1194), and the precise configuration of the bulla is one of the key features that suggest the paromomyids should be transferred from the Primates to the order Dermoptera (colugos), as noted above. In *Plesiadapis,* the only other Paleocene primate for which the middle ear is known, the floor has been diagnosed as petrosal (2126), but it could in fact be made of an entotympanic bone that fused imperceptibly with the petrosal (1416). Seamless fusion of the entotympanic and petrosal occurs in some extant nonprimates, and in these cases only fetuses show that the bulla is not truly petrosal. Since fetuses of Paleocene primates are unknown and are unlikely to be found, the true character of the bulla is indeterminate, and this feature does not clearly link any Paleocene primates to later ones. Their primate status thus depends primarily on their cheek teeth and on the assumption that primates existed in the Paleocene and must be represented by those Paleocene mammals that are most like later, undoubted primates.

Curiously, in spite of their abundance and diversity, the presumed Euramerican Paleocene primates provide few clues to the origins of later, Eocene to Recent forms. They are unlikely ancestors themselves, because they evolved dental specializations that later forms lack (429, 430, 749, 1827, 1959). These specializations include large, procumbent central incisors, perhaps used to grasp food, and a reduced number of lateral incisors, anterior premolars, or both. Except for some basically Paleocene taxa that survived into the early Eocene before becoming extinct, Eocene primates tended to have smaller, more generalized incisors, and they commonly retained incisors or premolars that the Paleocene forms had lost (fig. 3.15). Among known Paleocene taxa, only the earliest, *Purgatorius*, was sufficiently generalized to be ancestral to Eocene to Recent forms, but there is no reason to suppose it evolved in their direction. More likely it gave rise to later, more specialized Paleocene primates.

Because the Euramerican Paleocene primates combined very primitive features with specializations lacking in all later primates, they are now commonly placed in their own infraorder, the Plesiadapiformes, named for *Plesiadapis*, the best-known genus (1959). In vernacular terms, they might equally well be called *archaic primates* (732, 2126) as opposed to the primates of modern aspect, or *euprimates*, that succeeded them (1039). Alternatively, they may simply be the most primatelike of known Paleocene mammals (734), in which case they could be removed from the Primates altogether and placed in a separate order that shared a close Cretaceous ancestor with true Primates. One authority (833, 835) has proposed such an order, the Proprimates, whose content would be essentially the same as his previously proposed primate suborder, Praesimii, listed here in table 3.1.

None of the known Plesiadapiformes survived the Eocene, and their extinction could have resulted at least in part from unsuccessful competition with evolving rodents, bats, and euprimates (1411). The origins of euprimates remain obscure for lack of fossil evidence. A plesiadapiform root is unlikely for reasons given above, and North America or the combined North American–European landmass thus becomes an unlikely birthplace. South America can probably also be excluded, since its relatively well known late Cretaceous to Eocene fossil record contains no early primates or likely primate ancestors. Asia remains possible, and a case can be made from three jaws recovered in mid-Paleocene deposits of the Wanghudun Formation, Anhui Province, southern China (2127). These have been assigned to the genus *Decoredon*, whose teeth have been likened to those of Eocene tarsiiform euprimates discussed below. The specimens are poorly preserved, however, and the primate status of *Decoredon* is questionable (1946).

That leaves only Africa, which is arguably most plausible a priori (830, 831, 1038), since it hosted so many later major events in primate evolution. Unfortunately, African Paleocene and Eocene fossil sites formed mainly on the continental margin, and they tend to be poor in terrestrial mammals (549, 1871). An important exception is the site of Adrar Mgorn 1 at the foot of the High Atlas Mountains in southern Morocco (834, 1946). Here relatively abundant fossils of terrestrial mammals occur in association with sharks' teeth indicating a late Paleocene age, roughly 60 my ago. The mammalian fossils include ten isolated teeth that share several derived features with teeth of Eocene tarsiiforms in the family Omomyidae. The Adrar Mgorn specimens have been assigned to a previously unknown omomyid genus and species, *Altiatlasius koulchii* (1946), and if this diagnosis is correct it supports an African origin for more advanced primates in the Paleocene or late Cretaceous, followed by their spread to northern continents in the very late Paleocene or earliest Eocene.

Eocene Primates

Climatically, the Eocene was mainly similar to the Paleocene. Subtropical forests covered much of the western United States and also western Europe, as far north as the modern English Channel. As in the Paleocene, North and South America were separated by ocean, but in the early Eocene, perhaps about 50 my ago, the previous land connection between North America and Europe was broken (fig. 3.12).

Like Paleocene primate fossils, Eocene ones come mainly from western North America and western Europe. The Euramerican forms include some specialized plesiadapiforms that survived into the Eocene before becoming extinct, but most were broadly similar to living prosimians. Thus, excepting only the poorly known late Paleocene *Altiatlasius koulchii* that was just mentioned, the Eocene primates were the first euprimates or "primates of modern aspect" (1959), and if they were alive today they would undoubtedly be called lemurs and tarsiers. Technically, they can be readily accommodated in the extant infraorders Lemuriformes and Tarsiiformes within the suborder Prosimii. Arguably, some may even be placed in extant prosimian superfamilies or families.

Eocene Prosimians. The origins of the Eocene prosimians remain to be resolved, but by the very early Eocene, *Cantius*, a primitive lemuriform, and *Teilhardina*, a primitive tarsiiform, had appeared on both sides of the Euramerican land bridge. Together with *Donrussellia*, a like-aged (very early Eocene) lemuriform so far known only from Europe, *Cantius* and *Teilhardina* share numerous dental characters that suggest a relatively recent common ancestry (828–830, 1825, 1826). At least tentatively, *Altiatlasius koulchii* indicates that the shared ancestor evolved in Africa during the Paleocene or even the late Cretaceous. Whatever their origin, *Cantius*, *Donrussellia*, and *Teilhardina* are likely stem forms for the separate Eocene prosimian radiations that occurred in Europe and North America when the continents separated in the early to mid-Eocene.

Specialists disagree on the precise taxonomy of later Eocene prosimians, but the Euramerican forms are usually divided among forty to fifty genera in two families: the Adapidae, which were more lemurlike, and the Omomyidae, which were more tarsierlike (536, 562, 732, 1827, 1959, 2126). Less abundant fossils demonstrate that one or both families also existed in southern and eastern Asia (157, 158, 1851) and northern Africa (1972). Their presence on the Indian subcontinent is particularly striking, since in the Eocene the area was a large island, separated from mainland Asia and Africa by substantial stretches of ocean (fig. 3.12). The implication is that like the Malagasy lemurs discussed above and the New World monkeys discussed below, the Eocene Indian prosimians must descend from forms that reached their homeland by chance, on floating rafts of vegetation.

Like their plesiadapiform predecessors, the Eocene prosimians tend to be very abundant at sites where they occur, and they are also represented primarily by jaws and teeth. However, there are more skulls and limb bones and even some nearly complete skeletons. The lemurlike adapids tended to be larger than their tarsierlike omomyid contemporaries, and in overall body size some approached the larger historic Malagasy lemurs.

The relatively large size of the adapids, combined with the morphology of their cheek teeth and jaws, suggests that most fed mainly on fruits or leaves (562). In contrast, the smaller size of the omomyids, together with their tooth and jaw morphology, implies that they focused more on insects. Relatively small orbit size indicates that the adapids were mainly diurnal, while much larger orbits suggest that the omomyids tended to be nocturnal like the living tarsier or the African galago.

Unlike *Plesiadapis* and probably other plesiadapiforms, those Eocene prosimians for which pertinent postcranial bones are known had hands and feet well adapted for grasping. The digits were long and mobile, with nails instead of claws on most, if not all, digits. The postcranial bones of the adapids indicate that some were hind-limb-dominated, highly active arboreal quadrupeds like living lemurs (fig. 3.15), while others were relatively slow arboreal quadrupeds like living lorises (562, 1734, 1825, 1828). In contrast, the omomyids had elongated tarsal bones that probably aided in tarsierlike or galagolike vertical clinging and leaping (fig. 3.15) (562, 827).

The relatively numerous Eocene prosimian skulls suggest evolutionary advances comparable to those in the hands and feet. In contrast to the plesiadapiforms, in Eocene prosimians the orbits were completely surrounded by a post- or circumorbital bar, as in later primates (fig. 3.15). Probably even more significant, by comparison with plesiadapiform skulls, Eocene prosimian skulls indicate less reliance on the sense of smell and more on vision. The snout was reduced, and the orbits faced more forward and less laterally, so that there was greater overlap between the fields of vision. Endocasts show that Eocene prosimians had correspondingly smaller olfactory bulbs and an expanded visual cortex (larger occipital and temporal lobes) (1722). More generally, after correction is made for body size, they had much larger brains than the plesiadapiforms did, though by the same measure most probably had smaller brains than modern prosimians (1722).

The known Eocene adapids may not include the ancestors of the living Malagasy lemurs, but they were certainly close to them. Thus they shared a different, unique derived ear structure with living lemurs—the tympanic (or ectotympanic) ring floated more or less freely within the auditory bulla instead of fusing to the opening of the outer ear as it does in all living primates but lemurs (1468). In addition, at least some Eocene adapids closely resembled living lemurs and lorises in derived features of the wrist and ankle (155, 1467, 1468). Derived similarities in ear morphology also imply that the Eocene omomyids were closely related to living tarsiers (156, 1469), but only the very earliest ones are potential ancestors, since isolated teeth suggest that the tarsier itself (genus *Tarsius* in the family Tarsiidae) had already appeared in China by the mid-Eocene, about 45 my ago (157).

The overall similarity of Eocene adapids to living lemurs has even led some authorities to place the Eocene forms in the extant superfamily Lemuroidea. The known Eocene adapids were more primitive than living ones in several important characters, however, such as the retention of four premolars on each side of each jaw (vs. three in living lemurs) and the possession of generalized lower incisors and canines (vs. the protruding, elongated ones that form a specialized dental comb in living forms). Thus the adapids are probably best assigned to a separate (extinct) superfamily, the Adapoidea (838, 1038). But it is possible that a so far unknown adapid, transported to Madagascar accidentally on a natural raft of vegetation sometime in the Paleocene or Eocene, was the direct ancestor of the Malagasy lemurs. During the Eocene, Madagascar was actually closer to India than to Africa (fig. 3.12), and an Asian ancestry may therefore be more likely than an African one. The same adapid or a similar one may also have been ancestral to the modern African and Asian lorises (838).

Further, an adapid may be ancestral to the higher primates (826, 1736), though this role has been more commonly assigned to one of omomyids, in the superfamily Omomyoidea (645, 1038, 1520, 1838, 2126, 2128) (fig. 3.14B). Omomyid teeth and skulls are generally similar to anthropoid ones in some important respects, and biochemical and morphological similarities suggesting a link between living tarsiers and anthropoids have already been noted. Yet not all biochemical studies agree that anthropoids are closer to tarsiers than to lemurs (1868), and the morphological resemblances between the tarsiers and anthropoids could be parallelisms (analogies). This might apply, for example, to the shared presence of a postorbital plate and to the shared loss of the rhinarium. Omomyids lacked a postorbital plate, and a pronounced gap between their upper incisors indicates they had a rhinarium (10, 1463). If the last common ancestor of the tarsiers and the anthropoids was an omomyid, then the living tarsiers and anthropoids must have lost their rhinarium and developed their postorbital plates in parallel (independently). Parallel evolution is perhaps particularly indicated for the postorbital plate, which is less complete and much differently constructed in tarsiers than in anthropoids (1970).

Equally important, the teeth of some adapids also generally resemble those of anthropoids, so much so that specialists disagree about the anthropoid versus advanced adapid identification of some late Eocene and early Oligocene dentitions from Asia and Africa. In addition, adapids and early anthropoids share several features that neither taxon shares with omomyids (826, 838, 1734, 1736, 1970). These include a relatively long snout with nearly parallel tooth rows (vs. a short snout and more divergent tooth rows in omomyids); symphyseal fusion of the two halves of the mandible in several adapids and all anthropoids but in no omomyids; short, vertically placed, spatulate incisors in adapids and anthropoids (vs. more procumbent, more sharply pointed ones in omomyids); large, interlocking, sexually dimorphic canines in some adapids and in anthropoids (vs. small, nondimorphic, premolarlike canines in omomyids); and the lack of an external auditory meatus (tubular ectotympanic) in adapids and early anthropoids (vs. its presence in omomyids).

The controversy concerning an omomyid versus adapid ancestry for the anthropoids illustrates a common problem in the construction of phylogenetic (evolutionary) trees—disagreement over which similarities and differences among taxa are truly relevant to establishing their evolutionary relationships and, especially, uncertainty among specialists whether shared specializations genuinely reflect close relationship (recent common descent). As indicated above, the alternative possibility is that the specializations evolved independently (in parallel or convergently) in distantly related lineages that were adapting to similar conditions. Often there is the additional problem of deciding whether shared similarities are unique specializations inherited from a recent common ancestor or primitive retentions from a much more distant one. Like other phylogenetic controversies, the one surrounding omomyids versus adapids can be resolved only by a denser, more complete fossil record.

Eocene Anthropoids. A fuller fossil record may show that the ancestor of the anthropoids was neither an adapid nor an omomyid. This possibility is increasingly implied by a small but growing number of anthropoid or possible anthropoid fossils from Eocene sites in Asia and Africa. The Asian fossils come from Burma and China, and their higher primate status is arguably more dubious. The Chinese examples comprise jaws from the Shanghuang Fissure fills near Shanghai (157) (fig. 3.16). Associated mammal species imply an age near 45 my ago (mid-Eocene). The same deposits have provided adapid and omomyid fossils, together with the oldest proposed specimens of *Tarsius* mentioned above. At least four possible higher primate species

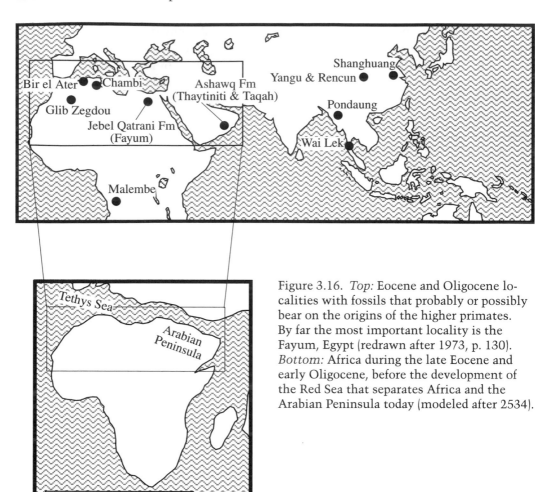

Figure 3.16. *Top:* Eocene and Oligocene localities with fossils that probably or possibly bear on the origins of the higher primates. By far the most important locality is the Fayum, Egypt (redrawn after 1973, p. 130). *Bottom:* Africa during the late Eocene and early Oligocene, before the development of the Red Sea that separates Africa and the Arabian Peninsula today (modeled after 2534).

have been identified, but only one, *Eosimias sinensis,* has been described. Its anthropoid status depends on derived features it shared with undoubted anthropoids, including a 2-1-3-3 dental formula, a single-rooted anterior lower premolar (P_2), shortening and crowding of the third and fourth premolars (P_3 and P_4), labial (lip- or cheekwise) expansion of their crowns, and subtle features of molar occlusal morphology. *E. sinensis* differed from most, if not all, known anthropoids in its diminutive size (its body weight was probably between 67 and 137 g), in its unfused mandibular symphysis, and in various, presumably primitive dental details. A larger number of more complete fossils will be necessary to show that its derived similarities to anthropoids truly reflect shared descent and not parallel (or convergent) evolution.

The Burmese fossils are similarly problematic for their incompleteness. They comprise fragmentary jaws from alluvial

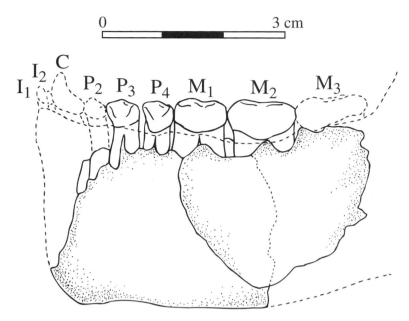

Figure 3.17. Partially reconstructed mandible of *Amphipithecus mogaungensis* (redrawn after 471, p. 30). Dated to about 40 my ago, it is arguably the oldest higher primate fossil yet found. An assignment to the higher (vs. lower) primates is suggested by several features, including the great depth of the jaw relative to the molar crown height and the nearly equal depth of the jaw from back to front. Typically, in lower primates the mandible is shallower overall, and it becomes especially shallow toward the front.

sandstones of the Pondaung Formation, where they are associated with other species that suggest an age between 44 and 40 my ago (late-middle Eocene) (471, 474, 1479). At least two possible anthropoid species are represented—*Amphipithecus mogaungensis* and *Pondaungia cotteri*. *A. mogaungensis*, which was apparently about gibbon size, is somewhat better known. It differed from prosimians and resembled anthropoids in at least three ways (474):

1. Its mandible was deep compared with its molar crown height, and the depth remained nearly the same all along the jaw (fig. 3.17). In prosimians, the mandible is shallower overall, and it becomes especially shallow toward the front.
2. Its second molar was parallel sided, as in anthropoids, and did not narrow toward the front, as in prosimians.
3. The two halves of its mandible were fused at the symphysis, as in all known anthropoids, and the symphysis was reinforced internally by two distinct horizontal shelves, the inferior and superior transverse tori. Those prosimians that have fused symphyses (some adapids and Malagasy lemurs) have only an inferior transverse torus.

In addition, the molar crowns of *Amphipithecus mogaungensis* were relatively flat with low, blunt cusps, suggesting that
it fed less on insects and more on leaves and fruit than prosimians commonly do. In this respect also it may have been more
like known anthropoids. In its premolar morphology, *Amphipithecus* was similar enough to *Eosimias* to be its descendant,
and in its overall form it was generalized enough to be ancestral
to both the New World anthropoids (with which it shared three
premolars on each side of the mandible) and the Old World ones
(which retain only two). Fresh, more complete fossils may show
that *Amphipithecus* and *Pondaungia* were advanced adapids
rather than primitive anthropoids (826), but they could still lie
close to the ancestry of later anthropoids.

The African fossils come from the base of the Jebel Qatrani
Formation in the Fayum Depression of Egypt and from the site
of Glib Zegdou in Algeria (fig. 3.16). The Jebel Qatrani Formation has long been attributed to the early (or early and middle)
Oligocene, but a late Eocene age has been proposed on the assumption that an erosional nonsequence (unconformity) at the
top of the formation reflects the rapid drop in global sea level
that occurred at the Eocene/Oligocene boundary (2290). An
alternative correlation between Fayum sedimentary nonse-
quences and global sea level changes places the Jebel Qatrani
Formation entirely within the Oligocene (836), while the Fayum geomagnetic stratigraphy may mean that it straddles the
Eocene/Oligocene boundary (recently revised upward from 36
to 34 my ago) (1178, 1179, 1181). The differences in estimated
age might be trivial if they involved only dating. But the Eo-
cene/Oligocene boundary coincided with a sharp drop in global
temperatures (fig. 2.11 in the previous chapter), and if the
boundary lay at the very base of the Jebel Qatrani Formation,
climate change could be invoked to explain the initial diversification (adaptive radiation) of the higher primates as a punctuational event (836).

If the geomagnetic dating is provisionally accepted, the
large, well-known sample of higher primate fossils from the
middle part of the formation (discussed below) would still belong to the early Oligocene, but a smaller, much more recently
discovered sample from near the base would date from the very
late Eocene, about 36 my ago (fig. 3.18). This sample comprises
partial jaws, skull fragments, and limb bones that have been divided among four new higher primate genera—*Catopithecus,
Proteopithecus, Serapia,* and *Arsinoea* (1738, 1966, 1967, 1970,
1973). A fifth genus, *Plesiopithecus*, was also originally assigned
to the higher primates when only its lower molar morphology was well known. Additional, more complete specimens,

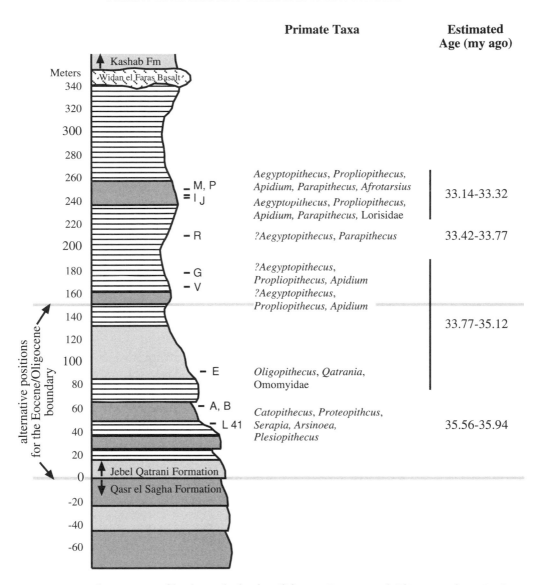

Figure 3.18. Schematic profile through the fossiliferous Eocene and Oligocene deposits in the Fayum Depression, Egypt (redrawn after 1178, fig. 4 and table 1). Gray tones indicate relatively fine sediments; horizontal lines designate coarser ones. The letters immediately to the right of the profile denote collecting localities ("quarries"). The middle column lists the primate taxa recovered at various localities, and the right-hand column shows the ages inferred from correlation between the geomagnetic reversal stratigraphy of the Fayum section and the global geomagnetic reversal time scale. The Eocene/Oligocene boundary may lie within the Jebel Qatrani Formation (1178, 1181), or it may lie at its very base (836).

including a nearly complete skull, now show that it lacked the postorbital closure that marks all widely accepted anthropoids. It was equally primitive in its retention of four premolars, and it has been removed to a new advanced prosimian family with possible links to the living lorisoids (1972). It illustrates the difficulty of distinguishing derived prosimians and primitive anthropoids when only partial jaws or teeth are available.

Judging from tooth size, adults of all the Fayum, late Eocene higher primates weighed less than 900 g, and among living anthropoids, only marmosets and New World squirrel monkeys are as small. As a group, the Fayum taxa are assigned to the higher primates primarily because their cheek teeth (premolars and molars) resemble those of their longer-known early Oligocene successors. The higher (vs. lower, especially adapid) primate status of *Catopithecus* is further confirmed by a nearly complete skull with totally fused frontal bones, a postorbital plate or septum (separating the orbit from the braincase behind), and an ectotympanic bone (the circular bony support for the eardrum) that is fused to the margin of the auditory bulla (rather than lying freely within it). A skull fragment of *Arsinoea* also shows total midline fusion of the frontals. If the Fayum taxa had been represented only by teeth, they might have been mistaken for advanced adapids, and their molar occlusal morphology suggests they focused less on fruits and more on insects than most later anthropoids (1738). This implies that the emergence of the anthropoids involved a gradual rather than a sudden shift in diet from insects to fruits (or leaves). Postcranial evidence may eventually show that the transition centered largely on a change in locomotion, involving a decrease in climbing and leaping between vertical supports and an increase in quadrupedal walking over horizontal ones (790, 791). Conceivably it also involved a reorganization of the special senses. However, the interorbital distance in *Catopithecus* suggests it may have relied on smell to the same extent as many prosimians, and its brain was demonstrably smaller than that of like-sized living anthropoids.

In detail, the dentitions of the Fayum genera are remarkably diverse, and several families may be represented. Based on dental morphology, *Catopithecus* and *Proteopithecus* have been tentatively assigned to the family Propliopithecidae, which also contains their two most prominent Oligocene successors discussed below. But it is now clear that, like *Amphipithecus*, *Proteopithecus* retained the relatively primitive 2-1-3-3 dental formula still found in the platyrrhines, whereas *Catopithecus* possessed the more derived 2-1-2-3 formula of the Oligocene propliopithecids and all later catarrhines, Arguably the difference implies that *Catopithecus* and *Proteopithecus* should be assigned to different families, and barring an unlikely evolutionary reversal in premolar numbers, only *Proteopithecus* could lie near the ancestry of both platyrrhines and catarrhines.

For the moment, however, the taxonomy and precise affinities of the Fayum Eocene anthropoids are probably less important than their sheer diversity. Supplemented by a single lower first molar from Nementcha (Bir el Ater), Algeria, that has been assigned to yet another genus, *Biretia* (261), and that probably also dates from near the Eocene/Oligocene boundary (1735), the plethora of Fayum taxa suggests that the anthropoids could have originated in Africa as early as or even earlier than their putative adapid or omomyid ancestor.

The fossils from Glib Zegdou, Algeria, provide more direct, if limited, support for an early African emergence of the anthropoids. They comprise perhaps fifteen isolated teeth that are estimated to be about 50–46 my old (late early Eocene), based on the kinds of algal cysts (charophytes) that occur in the same deposits. They have been divided among four or five proposed higher primate species, of which only *Algeripithecus minutus* and *Tabelia hammadae* have been described (849, 850). In basic tooth shape, both species resembled adapids, but they anticipated later higher primates in the inflation and blunting of the molar cusps. The difference suggests that they shared a higher primate tendency to feed more on fruits and less on insects. Two cusps—the protocone and paracone—on the upper second molar (M^2) of *A. minutus* illustrate the point most clearly. In adapids, the paracone exhibits a steep buccal (cheekside) edge and the protocone is high in side view, whereas in *A. minutus* the paracone is buccally expanded and the protocone is relatively low (fig. 3.19). In the details, *A. minutus* is particularly similar to the undoubted Oligocene higher primate *Aegyptopithecus zeuxis*, discussed below.

Based on dental dimensions (fig. 3.19), adults of *Algeripithecus minutus* probably weighed between 150 and 300 g and were thus much smaller than *Aegyptopithecus zeuxis*. Among living primates, only some prosimians and callitrichids (marmosets) are as small or smaller. However, *Catopithecus, Proteopithecus*, and the other Fayum late Eocene higher primates were only marginally larger, and the fossil record shows clearly that large size is a derived feature in higher primates. On present evidence, the first species to approximate the largest living higher primates appeared only in the Miocene.

In overall dental morphology, *Algeripithecus* and the Fayum late Eocene higher primates resemble adapids more closely than they do omomyids (1965, 1966), but this need mean only that adapids and anthropoids share a more recent common ancestor than either shares with omomyids. Fresh discoveries of Paleocene and Eocene fossils in Africa may eventually show

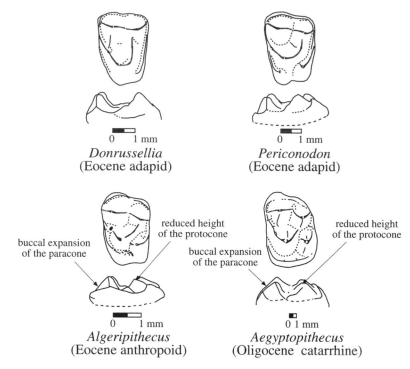

Figure 3.19. Occlusal morphology of the upper second molar (M^2) in two Eocene adapids (*Donrussellia* and *Periconodon*), in *Algeripithecus*, and in *Aegyptopithecus*, illustrating two key derived features— buccal expansion of the paracone and reduced height of the protocone—that *Algeripithecus* and *Aegyptopithecus* share (redrawn after 849). Derived dental features like this suggest that *Algeripithecus* was a higher primate, even though so far, unlike *Aegyptopithecus*, it is not represented by diagnostic skulls. Note that, though all the M^{2}s are drawn to the same length, the *Aegyptopithecus* specimen is much larger than the others.

that the anthropoids originated in Africa as early as or even earlier than either adapids or omomyids. Unfortunately, additional fossils may appear very slowly, since Africa is so poor in terrestrial Paleocene and Eocene fossil sites.

Oligocene Primates

In the very late Eocene or early Oligocene prosimians disappeared from Europe, and by the very late Oligocene or early Miocene they were also gone from North America. The Oligocene prosimians of North America were all omomyids, and to judge by their limited numbers and diversity, they were much less successful than their Eocene forebears. Adapids survived to the late Miocene (until approximately 9 my ago) in southern and eastern Asia (837, 1648), and evolution toward living prosimians obviously continued during and after the Oligocene in

Africa, Asia, or both. Except in Madagascar, however, beginning with the Oligocene the numbers and diversity of prosimians were greatly reduced, and they became largely confined to nocturnal niches.

In both Europe and North America, the Oligocene extinction of the prosimians coincides broadly with a trend toward cooler, drier climatic conditions, which eventually eliminated suitable prosimian habitat (subtropical forest) from middle latitudes. Suitable habitat persisted in Africa and Asia, but here the prosimians suffered from the highly successful radiation of anthropoids (higher primates) into diurnal forest niches. As discussed in the previous section, fossils from the base of the Jebel Qatrani Formation in the Fayum Depression, Egypt, demonstrate that anthropoids had already radiated extensively by the late Eocene, 36–35 my ago. A rich fossil assemblage recovered from higher levels within the same formation, a smaller one from two localities in the Ashawq Formation, Sultanate of Oman, and a canine from Malembe, Cabinda, northern Angola (fig. 3.16), record the emergence and early radiation of the catarrhines by the very early Oligocene, roughly 34–33 my ago. By the late Oligocene, roughly 25 my ago, the platyrrhines were also established in South America.

African Oligocene Anthropoids. The early Oligocene portion of the Jebel Qatrani Formation has been bracketed between 35 and 33 my ago by correlation of the Fayum geomagnetic reversal stratigraphy with the global reversal time scale (1178, 1181). This age is constrained by radiopotassium dates on an overlying basalt, and it is broadly supported by estimated sedimentation rates and by faunal correlations with Eurasia (735, 736, 1735, 1737, 1958, 1961, 1965) (fig. 3.18). Plant and animal fossils, together with sedimentologic-geomorphologic analyses, indicate that the Jebel Qatrani sediments accumulated on a swampy, heavily vegetated plain crossed by large, meandering streams (311, 313, 1637). The streams reached the nearby sea and were flanked by large trees, whose silicified trunks are preserved in the deposits. Together, the sediments and fossils imply warm and relatively nonseasonal climatic conditions.

Like the Jebel Qatrani Formation, the roughly contemporaneous Ashawq sediments apparently accumulated in a warm, moist, nearshore setting (2177, 2178). During the Oligocene the Red Sea did not exist, and the Arabian Peninsula on which Oman is situated was the northeastern corner of Africa. Northeastern Africa was well watered, and forest probably stretched more or less continuously between the present-day Fayum and Oman. The early Oligocene primate faunas of the Fayum and

Oman appear to have been similar, and together with the very late Eocene primates of the Fayum, they reveal not only the early diversification of the anthropoids but also the corresponding decline of the prosimians. The early Oligocene primate sample from the Fayum comprises roughly one thousand specimens, of which only a handful derive from lower primates, including two tarsiiforms and a possible adapoid or lorisoid (731, 736, 1968, 1969). The Ashawq sample is smaller and less completely described, but among several dozen isolated teeth and some fragmentary jaws, higher primate specimens predominate heavily. They can be assigned to two or more of the same genera represented in the Fayum (1738).

The single canine from Malembe, Angola, may also represent one of the Fayum taxa. It is significant mainly because it demonstrates that higher primates were widely distributed in Africa by the early Oligocene (1682).

As now most commonly described (736, 739, 1969, 1971), the Fayum anthropoids comprise six genera: *Qatrania, Apidium, Parapithecus* (including *Simonsius*), *Propliopithecus* (including *Moeripithecus* and *Aeolopithecus*), *Aegyptopithecus*, and *Oligopithecus. Propliopithecus* is represented by four species; *Qatrania, Apidium,* and *Parapithecus* include two each; and *Oligopithecus* and *Aegyptopithecus* have one each, for a total of twelve species. Not all were coeval, and some may have evolved from others. Probably most important, an early species of *Propliopithecus* could have been ancestral both to later species of *Propliopithecus* and to *Aegyptopithecus.* The Ashawq higher primate fossils come mainly from a species of *Propliopithecus* that is also represented in the Fayum, where it was originally assigned to a separate genus (*Moeripithecus*). Arguably, the Ashawq material suggests that this genus should be resurrected (2179).

Fossil teeth and jaws predominate for all Fayum taxa, but there are postcranial bones for *Apidium, Parapithecus, Propliopithecus,* and *Aegyptopithecus,* while *Apidium, Parapithecus,* and above all *Aegyptopithecus* are represented by partial skulls, the second oldest known for any anthropoids (the oldest is from *Catopithecus,* discussed above). By analogy with living primates, the teeth of the Fayum species can be used to infer their body size and ecology (727, 739, 1191). The Fayum taxa were all at or below the low end of the modern anthropoid range, varying from mouse size (300 g) (*Qatrania*) to very small monkey size (900–1,700 g) (*Apidium,* perhaps some *Parapithecus,* and *Oligopithecus*) to medium monkey size (3,000–4,000 g) (most *Parapithecus* and *Propliopithecus*) to perhaps gibbon size (5,900 g)

(*Aegyptopithecus* and at least one species of *Propliopithecus*). Except for *Qatrania*, however, they were all larger than living (mainly prosimian) primates that are primarily insectivorous, and they fit comfortably within the range of those that are mainly fruit eaters (frugivores) or leaf eaters (folivores). To judge by their relatively short molar shearing crests (shorter than those of living folivores but comparable to those of living frugivores), most of the Fayum taxa, including *Qatrania*, were probably frugivorous. Frugivory in *Aegyptopithecus* is further implied by its relatively large incisors. Only (some?) *Parapithecus* possessed sufficient molar shearing to suggest some folivory.

By analogy with living primates of similar size, most if not all the Fayum species were probably arboreal. The morphology of the available limb bones implies they were either arboreal quadruped-climbers (*Propliopithecus* and *Aegyptopithecus*) or arboreal quadruped-leapers (*Apidium*) (791, 793). There is no evidence that they had the suspensory abilities of the living apes. The available faces (from *Apidium*, *Parapithecus*, and *Aegyptopithecus*) all have small orbits, indicating an essentially diurnal lifestyle, as in all known later Old World anthropoids (739).

In early descriptions, the Fayum taxa were divided into probable or certain early hominoids (*Oligopithecus*, *Propliopithecus*, and *Aegyptopithecus*) and possible early cercopithecoids (*Apidium* and *Parapithecus*) (1958, 1959). This interpretation is still arguable (793, 1964), but a more recent, more conservative view is that they are all simply early anthropoids, antedating the divergence between hominoids and cercopithecoids (45, 731, 734, 736, 738, 968, 1188, 1961, 1973).

The question of their phylogenetic relationships to hominoids and cercopithecoids hinges largely on their teeth. The molars of undoubted cercopithecoids are uniquely derived in form, with two pairs of relatively high cusps, each pair linked buccolingually by a crest or loph (fig. 3.20). The full pattern is usually called *bilophodont*, and *Apidium* and especially *Parapithecus* show a configuration that might anticipate it. However, other dental features suggest they were only distantly related to cercopithecoids or to any other catarrhines. Both genera retained three premolars on each side of each jaw, versus two in cercopithecoids and all known hominoids, including *Propliopithecus* and *Aegyptopithecus*. If they stand near the cercopithecoid stem, it follows that the reduced number of premolars that cercopithecoids and hominoids share (two on each side of each jaw) is a result of convergence (parallel evolution) rather than of shared (common) descent. This seems unlikely.

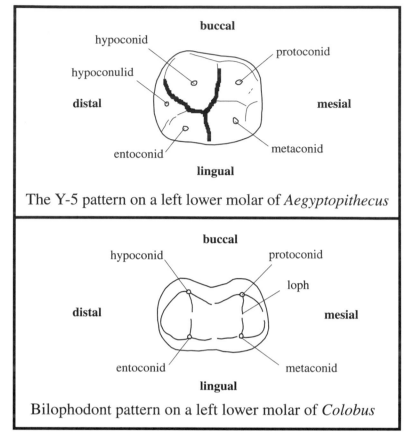

Figure 3.20. Lower molars of the early Oligocene catarrhine *Aegyptopithecus* and of the extant leaf-eating monkey *Colobus* (adapted from 391, figs. 2 and 8). The monkey molar exhibits the typical bilophodonty of all cercopithecoids, with two pairs of cusps linked by shearing crests (lophs). The *Aegyptopithecus* molar shows a pattern of five distinct cusps separated by a Y-shaped fissure system that is broadly characteristic of all Miocene to Recent hominoids. The Y-5 pattern is believed to be primitive in catarrhines, and the bilophodont condition probably evolved from it.

Furthermore, *Parapithecus* is clearly excluded from cercopithecoid origins by the loss of permanent lower incisors, which were reduced to either one or none on each side of the jaw in *Parapithecus*. (In the species of *Parapithecus* that some authors place in the separate genus, *Simonsius*, the canines met at the midline of the symphysis in a derived condition unique among the anthropoids [1192, 1963].) Only *Apidium* retained the right number of incisors (two in each quadrant; fig. 3.21) to be near the ancestry of the cercopithecoids.

The relatively well known dentitions of *Propliopithecus* and *Aegyptopithecus* exhibit no uniquely derived cercopithecoid features, and in virtually all important respects they recall

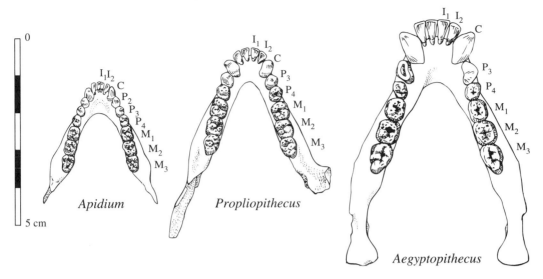

Figure 3.21. Mandibles of *Apidium, Propliopithecus,* and *Aegyptopithecus* (partially restored) (drawn by Kathryn Cruz-Uribe from photos in 2126, fig. 156, and 738, figs. 4, 9, 10; © 1999 by Kathryn Cruz-Uribe). Note the presence of three premolars on each side of the jaw in *Apidium,* versus two in *Propliopithecus* and *Aegyptopithecus.* The mandibles and teeth of *Propliopithecus* and *Aegyptopithecus* are strikingly apelike in overall form and differ from each other only in size and some basic proportions, particularly in the relatively larger size of M_1 in *Propliopithecus.* Based on mandibular and dental morphology, either *Aegyptopithecus* or *Propliopithecus* could be ancestral to all later catarrhines.

hominoid teeth far more than they do cercopithecoid ones. Especially notable are the crowns of the lower molars, which resemble those of all Miocene to Recent hominoids, with three rounded cusps on the lingual side and two slightly larger, shorter ones on the buccal side, separated by a Y-shaped groove or fissure (fig. 3.20). It is conceivable, however, that this represents the primitive catarrhine condition and that cercopithecoid bilophodonty evolved from it. The earliest known, unquestionable cercopithecoid molars, from the early Miocene of eastern and northern Africa, exhibit a morphology, including a fifth cusp and incomplete lophs, that is plausibly derived from the molar morphology of *Propliopithecus* and *Aegyptopithecus* (639). By comparison with the round-cusped early catarrhine-hominoid type of molar, which is well suited for crushing fruits, bilophodont molars are better suited for shearing leaves, and the development of bilophodonty may signal a shift from frugivory toward folivory among the earliest cercopithecoids (41, 2159).

Much new fossil evidence will probably be necessary to resolve the relation of the Fayum catarrhines to later ones, but for the moment *Qatrania, Apidium,* and *Parapithecus* (including *Simonsius*) are probably best placed in their own extinct superfamily, the Parapithecoidea. A parapithecoid may have been ancestral to the cercopithecoids, but at present it seems more likely that the parapithecoids were a specialized group of early anthropoids with no special link to any later catarrhines (968). Conceivably they represent a totally separate, extinct anthropoid branch, parallel to the catarrhines and platyrrhines (740).

Propliopithecus and *Aegyptopithecus* may also be placed in an extinct superfamily, the Propliopithecoidea, which perhaps includes the ancestors of both hominoids and cercopithecoids. Taxonomically, *Oligopithecus* remains the most problematic genus, partly because it is known only from a single mandible fragment and a handful of isolated teeth (1736) and partly because it exhibits a unique combination of features. It had only two premolars, like *Aegyptopithecus, Propliopithecus,* and all other known Miocene to Recent catarrhines, but it retained a more primitive molar occlusal morphology. It could represent a distinctive early anthropoid group, perhaps ancestral to all other Oligocene primates in the Fayum (1188). It could even be an advanced adapid (825, 826), though this seems unlikely, since its teeth closely resemble those of its late Eocene predecessor, *Catopithecus,* which is known not only by teeth, but also by a skull with undoubted anthropoid features.

Taxonomic and phylogenetic considerations aside, the relatively abundant jaws, limb bones, and partial skulls of *Aegyptopithecus* exhibit a remarkable mix of advanced and primitive features that could never have been predicted (1188, 1964). The most obvious advanced features are in the teeth, which were strikingly apelike, even including substantial sexual dimorphism in the size of the canines and lower anterior premolars. Other significant, typically anthropoid derived features include fusion of the two halves of the mandible at the symphysis, which was buttressed internally by both inferior and superior transverse tori; fusion of the right and left frontal bones; olfactory bulbs that were significantly smaller and a visual cortex that was significantly larger than in most prosimians (1723); and cuplike orbits that, unlike prosimian ones, were closed off from the skull behind by a postorbital plate or septum. This was somewhat less complete than in extant catarrhines, resembling the more primitive condition retained in platyrrhines. Arguably it developed as a structural response to static stresses created in the skull when the anterior teeth were used to open hard-coated fruits or large nuts (1834). The same or similar stresses could explain the fusion and internal buttressing of the mandibular symphysis.

Among strikingly primitive skull characters, *Aegyptopithecus* had a snout that was somewhat variable in size but

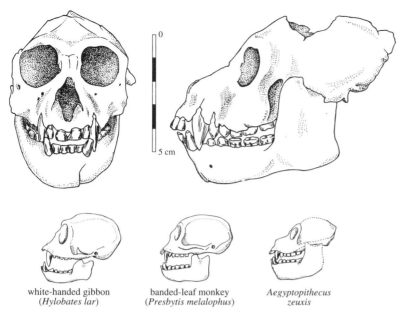

white-handed gibbon
(*Hylobates lar*)

banded-leaf monkey
(*Presbytis melalophus*)

*Aegyptopithecus
zeuxis*

Figure 3.22. *Top:* Facial and lateral views of an *Aegyptopithecus* skull (drawn by Kathryn Cruz-Uribe from photos in 1958; © 1999 by Kathryn Cruz-Uribe). The mandible has been partially reconstructed from pieces that were not directly associated with the skull. *Bottom:* Lateral view of the same skull compared with those of extant catarrhines of roughly similar size (redrawn after 313, fig. 2). Note that *Aegyptopithecus* had a significantly longer snout, more laterally placed orbits, and a braincase that was absolutely and relatively smaller than that in the other catarrhines.

that was always longer and more protruding than in later anthropoids (fig. 3.22); a brain that was bigger for body size than that in prosimians but was still below the lower limit for other anthropoids; more extensive postorbital constriction than in any other known anthropoid, recalling the condition in some Eocene adapids; orbits that faced more laterally than in other known anthropoids; and a bony ear that lacked an external auditory meatus (tubular extension of the ectotympanic bone).

The best-known postcranial bone of *Aegyptopithecus*, the humerus, also exhibits manifestly primitive features, including an entepicondylar foramen at the distal end (fig. 3.23). This occurs in prosimians and some platyrrhine monkeys but not in extant cercopithecoids or hominoids. The known postcranial bones show no specializations for suspensory postures or locomotion but instead suggest that *Aegyptopithecus* was a slow-moving, arboreal quadruped broadly similar to many living monkeys (Fleagle in 313; 728). Also like them, it may have had a tail. On present evidence, in the postcranium as in the skull, *Aegyptopithecus* was sufficiently primitive to be near the ancestry of both monkeys and apes.

The postcranium of *Propliopithecus* is more poorly known, but the only obvious difference from *Aegyptopithecus* is smaller size. *Propliopithecus* also had primitive apelike teeth and jaws, which differed from those of *Aegyptopithecus* mainly in the proportions of the molars (1188). In *Propliopithecus* the first molar, M_1, and the second molar, M_2, were about the same size, whereas in *Aegyptopithecus* M_1 was much smaller

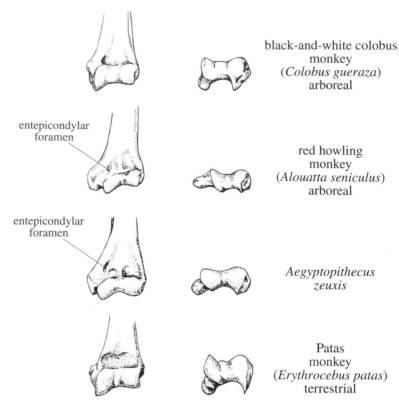

black-and-white colobus
monkey
(*Colobus gueraza*)
arboreal

entepicondylar
foramen

red howling
monkey
(*Alouatta seniculus*)
arboreal

entepicondylar
foramen

*Aegyptopithecus
zeuxis*

Patas
monkey
(*Erythrocebus patas*)
terrestrial

Figure 3.23. Anterior and distal views of the distal left humerus in a black-and-white colobus monkey, a red howling monkey, the primitive Fayum catarrhine *Aegyptopithecus*, and a patas monkey (redrawn after 313, fig. 3). Note that, like all extant catarrhines, the colobus and patas monkeys lack an entepicondylar foramen, which does occur in some platyrrhines and in *Aegyptopithecus*. Note also that in basic distal humerus structure *Aegyptopithecus* was more like the colobus and howling monkeys, which are arboreal, than like the patas monkey, which is terrestrial.

than M_2 (fig. 3.21). In this important respect and some others (1962), *Aegyptopithecus* was more like Miocene to Recent hominoids, for which it is therefore a more likely ancestor than is *Propliopithecus*.

If *Aegyptopithecus* and *Propliopithecus* had been represented only by postcranial bones or skulls without teeth, they might have been identified as monkeys (cercopithecoids). When their postcranial bones, skulls, and teeth are considered together, the conclusion is more complex. Whether *Aegyptopithecus* and *Propliopithecus* antedate or postdate the hominoid/cercopithecoid split, together with early Miocene fossils discussed below, they suggest that the split resulted from divergent dietary adaptations (39, 2159), not from locomotor ones as the contrasts between living monkeys and apes might imply.

South American Oligocene Anthropoids. The oldest known anthropoid fossils from South America are fourteen fragmentary dentitions, possibly from only two or three individuals, assigned to the species *Branisella boliviana* (1037, 1038, 1040, 1413, 2133, 2477). They come from the Salla Beds of northern Bolivia, which were originally thought to date from the early Oligocene, that is, broadly the same time as the Jebel Qatrani Formation. However, radiopotassium and fission-track dates now show that the Salla Beds postdate 27 my ago, and constrained by this age, the Salla geomagnetic stratigraphy suggests that *Branisella* dates from the very late Oligocene or early Miocene, sometime between 26 and 22.6 my ago (1179, 1412–1414, 1521, 1605). In its greatly reduced M^3 and other dental features, *Branisella* clearly anticipates extant platyrrhine (ceboid) monkeys, and more abundant fossils from localities in Argentina and Colombia show that platyrrhine monkeys remarkably similar to extant genera were already established by the early to middle Miocene, between roughly 19 and 13 my ago (312, 730, 732, 733, 826, 1040, 1187, 1190, 1413, 1836, 1837).

Since the early Tertiary mammals of South America included no primates or likely primate ancestors (1520, 1661), the ancestors of the platyrrhine monkeys almost certainly evolved on another continent and migrated to South America, probably sometime between the late Eocene and the late Oligocene. As the home continent, some authorities have favored North America or Africa or Asia via North America (471, 838, 1834), while others have preferred Asia via Africa or Africa directly (473, 973, 1038, 1040). Since the Central American land bridge formed only about 3 my ago and South America was thus an island continent from the middle Cretaceous until this time (fig. 3.12), the dispute depends partly on whether it was more accessible from North America or from Africa. A second vital question concerns the nature of the relations between platyrrhines and other primates.

Wherever the platyrrhines originated, their ancestors surely crossed to South America accidentally, probably on rafts of vegetation washed out to sea. The channels that separated South America from North America and Africa were hundreds of kilometers wide, but both probably contained intervening islands to serve as stepping stones in animal dispersal (838, 1038, 1520). The plausibility of a crossing from Africa to South America is enhanced by increasing evidence that African hystricomorph rodents reached South America in the Oligocene (731, 1413). For the moment, however, based strictly on the paleogeographic evidence, the early platyrrhines could have migrated from either North America or Africa.

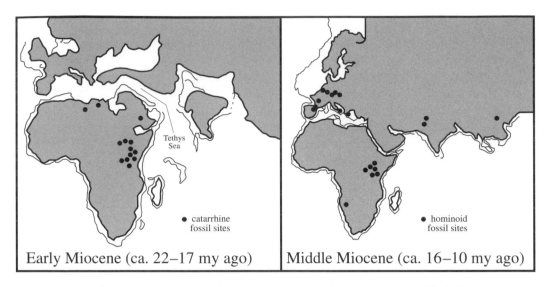

Figure 3.24. Relative positions of Africa and Eurasia in the early and middle Miocene (modified after 1304, figs. 3 and 4). The middle Miocene higher primates of Eurasia probably evolved from African forms that dispersed to Eurasia when the northward drift of the African plate promoted faunal interchange roughly 17–16 my ago.

The available phylogenetic and fossil evidence is less ambiguous. At least some early platyrrhines exhibit striking dental similarities to some Fayum Oligocene catarrhines (730), which supports biochemical studies that link platyrrhines more closely to catarrhines than to any extant prosimians (99, 1868). The implication is clear that platyrrhines and catarrhines are descended from a common anthropoid ancestor rather than from separate prosimian ones. The upwardly revised dating of *Branisella* suggests that the earliest platyrrhines may postdate the earliest African anthropoids (from the Fayum) by as much as 10 my. It thus seems increasingly possible that the platyrrhines evolved from late Eocene or early Oligocene African anthropoids, perhaps broadly similar to the Fayum parapithecoids. A firm test of this hypothesis will require so far unknown late Oligocene (30–25 my old) anthropoid fossils from Africa, South America, or both.

Miocene Primates

During the Miocene, the continents came to occupy basically the same positions they do today (fig. 3.24). In the early mid-Miocene, about 17–18 my ago, the northward drift of the African plate brought it into broad contact with Europe and western Asia, reducing both the lateral extent of the intervening Tethys Sea and its climatic influence on the adjacent continents.

As a result, southern Eurasia became generally cooler and drier, and the trend was accclerated by broadly simultaneous mountain building in the Mediterranean and Himalayan regions (1304). Equally important, the newly developed land connections promoted an unprecedented degree of faunal interchange between Africa and Eurasia. Among the African species that migrated to Eurasia were hominoids, which rapidly radiated through Eurasian forests, from Spain on the west to China on the east.

In most discussions of primate evolution, the acceleration of faunal exchange between Africa and Eurasia is used to divide the Miocene between an early part, before 17–16 my ago, when hominoids are known only from Africa, and a later part, after 16 my ago, when they are better known from Eurasia. The early Miocene hominoids, after 22 my ago, are plausibly derived from Fayum Oligocene predecessors, but 9–10 my separate the two groups (fig. 3.25). So far, the only specimens to fill the gap are three isolated teeth and two fragmentary jaws from Lothidok, northwestern Kenya, dated by radiopotassium between 27 and 24 my ago (late Oligocene) (284). The specimens have been assigned to the new genus *Kimoyapithecus*, but they are too sparse for higher taxonomic assignment (1342). Although they share some features with specimens of the early Miocene hominoid *Afropithecus,* discussed below, they may still antedate the cercopithecoid/hominoid divergence. Biomolecular estimates place this between 25 and 20 my ago (fig. 3.13) (1686, 1689), and a split by 20 my ago is confirmed by primitive but unequivocal cercopithecoid fossils from 19–17 my old deposits in eastern Africa. As indicated previously, the New World (ceboid) monkeys were also firmly established by this (early Miocene) time, but their further evolution will not be discussed here because it does not bear on human origins.

The Early Miocene. So far, early Miocene primates are known principally from western Kenya and neighboring Uganda, mainly from sites in shallow sedimentary basins near modern Lake Victoria (fig. 3.26). The fossiliferous layers are interstratified with alkaline volcanic ashes and lavas that helped to preserve the bones from acid dissolution and that also permit radiopotassium dates confirming an early Miocene age (234, 1682, 2291).

Associated plant and animal fossils indicate that early Miocene primates in eastern Africa inhabited tropical forest and woodland. Forest dominated and stretched along the equator right across the continent into areas that became woodland,

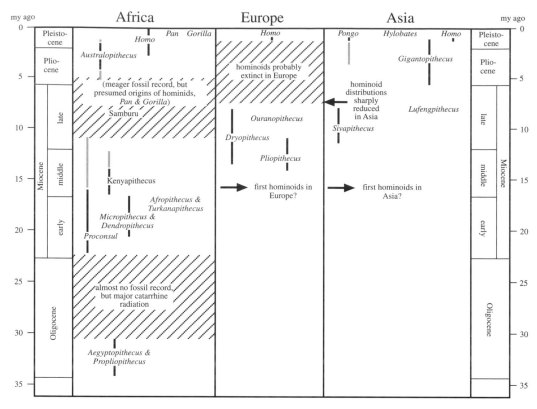

Figure 3.25. Temporal distribution of some important hominoid genera discussed in the text. Note the unfortunate gaps in the African fossil record between roughly 31 and 22 my ago and between roughly 12 and 5 my ago. The first gap was almost certainly a time of major catarrhine evolution when the hominoids differentiated from the cercopithecoids, and the second gap includes the time when the line leading to hominids diverged from the lines leading to the chimpanzee and the gorilla. Hominoids are unknown in Eurasia before the mid-Miocene, probably because they evolved in Africa and were unable to reach Eurasia until the contact between Africa and Eurasia was significantly broadened about 17–16 my ago. In the late Miocene, about 8 my ago, hominoids became extinct in Europe and their numbers and distribution were greatly reduced in Asia, probably because climatic change eliminated suitable habitat in middle latitudes.

bushland, or savanna in the succeeding mid-Miocene to Pleistocene (39, 40, 50, 59, 1680). Hominoids dominate the primate samples, but prosimians similar to modern lorises and galagos also occur (2366), and there are infrequent, primitive cercopithecoid monkey jaws and teeth, assigned to the genus *Prohylobates* (968, 2056). The rarity of cercopithecoids and the abundance of hominoids, which, it is stressed below, had a primitive, monkeylike postcranium, suggests that early Miocene hominoids were occupying many niches later occupied by monkeys. At least early on, cercopithecoids may have been more at home in less forested environments, such as those that existed in the early Miocene of northern Africa, where

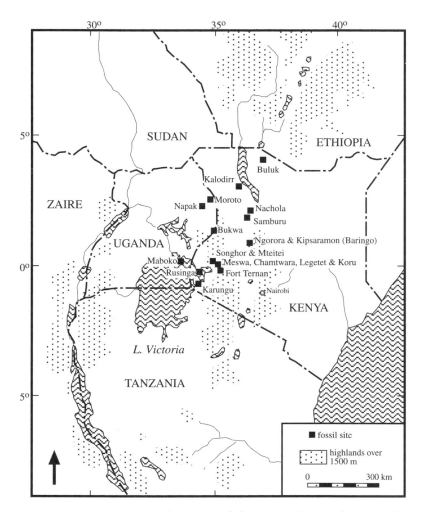

Figure 3.26. Approximate locations of the main Miocene hominoid fossil sites in eastern Africa.

cercopithecoids occur without hominoids (639). The major cercopithecoid adaptive radiation probably occurred only much later, in the late Miocene and early Pliocene, at a time when hominoid fortunes were in decline, owing at least in part to progressively cooler, drier global climate and to the accompanying growth of savannas at the expense of forests and woodlands.

Despite a growing fossil sample and more detailed, more sophisticated analyses, the taxonomy of the Miocene hominoids has undergone numerous revisions, and no final consensus is in sight (compare, for example, 46, 50, 169, 968, 1202; for overviews of disagreements see 470, 1686). Some disagreement is inevitable, since taxonomic judgments must often be based on small samples, dominated by partial jaws and isolated teeth. Skulls and postcranial bones are much rarer, and it is usually difficult to ascertain which postcranial bones belong with which teeth. In these circumstances the number of characters available for comparison tends to be small, and it can be very hard to separate interspecific variability from intraspecific variability, particularly intraspecific variability due to sexual dimorphism. The problem is exacerbated by the possibility that some fossil hominoid species may have been much more dimorphic than their living counterparts, which are thus misleading guides (1202). Finally, there is the perennial problem of determining whether some of the similarities observed among fossil taxa reflect close common descent (homology) versus common adaptation (analogy) or evolutionary parallelism (homoplasy).

With these difficulties in mind, however, specialists generally agree that most early Miocene (20–17 my old) hominoid fossils belong to a single genus, *Proconsul*. This genus comprised two subgenera, *P. (Proconsul)* with two or three species and *P. (Rangwapithecus)* with one or two. Dentitions and teeth dominate the samples, but there are also some partial skulls and postcranial bones. The postcranial bones show that the species varied greatly in adult weight, from 11 kg (siamang size) for the smallest to 87 kg (orangutan or female gorilla size) for the largest (1725). *P. heseloni,* which was siamang size, and *P. nyanzae,* which was pygmy chimpanzee size, have provided the largest number of fossils, including partial skeletons (fig. 3.27 top) (2365, 2374, 2375, 2385).

In the skull (fig. 3.27 bottom), *Proconsul* was far advanced over *Aegyptopithecus,* and it resembled the living apes in the full forward rotation of the orbits, the limited amount of postorbital constriction, the small size of the snout relative to the braincase, and the large size of the brain, both absolutely and relative to body size (2371). It was arguably less advanced over *Aegyptopithecus* in the external morphology of the cerebral cortex (700, 1722). Dentally, *Proconsul* resembled both *Aegyptopithecus* and the living apes, and in both cranial and dental features it is a plausible Miocene link between them. It lacked the secondarily enlarged incisors (for peeling fruit) of the chimpanzee and orangutan and the especially high-cusped molars (for masticating leaves) of the gorilla and was thus generalized enough to be near the ancestry of all three. Its primitive, low-cusped, thin-enameled molars suggest a basically frugivorous diet (739).

Postcranially, *Proconsul* was more complex. Its arms and legs were subequal in length, as in living monkeys (2374) (fig. 3.27 top), and the wrist was designed to bear the weight of the forequarters, probably on the palms rather than on the

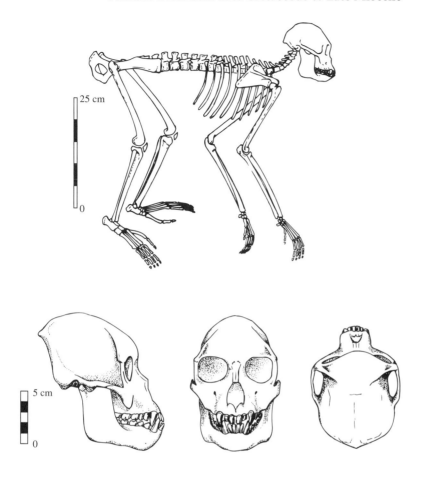

Proconsul heseloni (KNM-RU 2036)

Figure 3.27. *Above:* Reconstruction of *Proconsul heseloni* (Rusinga Island specimen no. 2036 in the collections of the National Museums of Kenya) (redrawn after 2374). *Below:* Reconstructed skull of female *P. heseloni* (redrawn after 2375). The specimens were recently removed from *P. africanus* to *P. heseloni* (2365).

knuckles (1514). The upper humerus shaft was curved to ac-commodate a deep, narrow monkeylike chest, and the lumbar (waist) portion of the vertebral column was monkeylike in its length and flexibility (2382, 2385). In these and other features, *Proconsul* appears to have been a generalized arboreal quadru-ped (50, 728, 1829, 2374, 2382), yet it had relatively mobile joints that probably permitted underbranch suspension, and it lacked a tail (2383). In its movements and positional behavior, it was probably more like a monkey than like an ape, but it was not identical to either. The nearest living analogues may be those New World monkeys that combine arboreal quadrupedalism with greater suspensory capabilities than are found in their Old World counterparts. Like the skull and dentition, the postcranium exhibits no specializations that exclude *Proconsul* from the ancestry of the living apes.

Besides *Proconsul,* several other hominoids are known from early Miocene deposits in eastern Africa. Their taxonomy and phylogenetic relationships are debatable, but at minimum there were five genera. Three—*Dendropithecus* (including "*Limnopithecus*" in part), *Micropithecus,* and *Morotopithecus*—come from the same area of western Kenya that has provided abundant fossils of *Proconsul,* while two others—*Afropithecus* and *Turkanapithecus*—come from early Miocene sites in northern Kenya where *Proconsul* is unknown. One of the north Kenyan sites has also provided fossils of a third hominoid genus, *Simiolus* (1348), which is broadly similar to *Dendropithecus* as described immediately below.

Dendropithecus is a genus of roughly gibbon-sized apes that are known from dentitions and a small number of postcranial bones. In its relatively elongated, slender limb bones, daggerlike canines, highly sectorial lower third premolars, and relatively simple lower molars, *Dendropithecus* resembled the gibbons and siamangs. However, the shared similarities may all be primitive catarrhine characters that the gibbons have simply retained (38, 729), and *Dendropithecus* possesses no clearly derived features that link it uniquely to the gibbons or to any other later hominoids. It need not be specially related to the gibbons, and it could have been on or near the line leading to all living hominoids.

Micropithecus was an extremely small ape, smaller than a gibbon, that is represented by jaws, isolated teeth, and a partial face (729, 742). Like the gibbons and siamangs, it had a short, broad face, narrow molars, and relatively large, high-crowned incisors and canines, but in other dental features it either is unique or resembles *Dendropithecus* (38). As with *Dendropithecus,* its small size can be used to postulate a special ancestral relationship to gibbons and siamangs, but the morphological case is not compelling.

Morotopithecus was a chimpanzee-sized ape that is known from a palate and partial face, two fragmentary mandibles, parts of two femurs, a nearly complete lumbar vertebra, fragments of four other vertebrae, and a proximal scapula. The dental elements have been variously likened to those of *Proconsul* and *Afropithecus,* and some specialists have kept them in informal limbo as the "Moroto hominoid," after the Moroto find site in Uganda. Their taxonomic assignment remains disputable, but the postcranial bones imply separate generic status (792). The intact lumbar vertebra hints at a short, stiff waist region like that of a modern ape (1861, 2382), while the scapula suggests sufficient shoulder mobility for hand over hand climbing or underbranch suspension (792). If the inferred morphology is confirmed by additional, more complete fossils, *Morotopithecus* would be the oldest known hominoid to anticipate living apes and people in postcranial form and function.

Afropithecus was an ape the size of a female gorilla that is known from a partial cranium, fragmentary jaws, isolated teeth, and a few postcranial bones (1343, 1346). It is readily differentiated from most early Miocene apes by its large size and from *Proconsul* by a series of cranial and dental features, including a longer, narrower snout that merges almost imperceptibly into a steeply inclined frontal bone, a mandible that is very deep and robust, upper cheek teeth that flare outward at the base so that the occlusal surfaces seem small by comparison, upper premolars that are enlarged relative to the molars, and thicker cheek tooth enamel. Postcranially it resembled *Proconsul*, and it apparently lacked the derived postcranial specializations that have been tentatively inferred for *Morotopithecus*.

When *Afropithecus* was known only from partial dentitions, their thick enamel and some other features suggested they came from *Sivapithecus*, which is best known from late Miocene (12–8 my old) deposits in India and Pakistan (1350). But this assignment was contested (641, 1689), and the fuller sample now on hand implies a distinct genus, possibly near the line that ultimately produced *Sivapithecus*.

Afropithecus may also be represented by a maxilla and four isolated teeth from the Ad Dabtiyah early Miocene (ca. 17 my old) site in Saudi Arabia. Pending additional discoveries, however, this material has been referred to the separate genus *Heliopithecus* (50, 55–57). In their enlarged premolars, thickened enamel, and some other features, *Afropithecus* + *Heliopithecus* are advanced in the direction of the later great apes and people. Since living gibbons maintain the primitive condition for these features (relatively small upper premolars, thin enamel, a relatively slender, shallow mandible, etc.), even in the absence of fossils with derived gibbon features, *Afropithecus* + *Heliopithecus* argue that the great apes and gibbons must have diverged before 17–18 my ago (55). This is broadly consistent with a divergence estimate of 22–18 my ago based on one molecular system and calibration date (1942, 1943), but not with others of 16 my ago (1689, 1943) and 15–12 my ago (565) based on other systems or calibration points.

Turkanapithecus was a gibbon-sized ape represented by a partial cranium, mandible, and some possibly associated post-cranial bones (1344, 1347). It is easily distinguished from *Afropithecus* by its much smaller size and by its shorter, less protruding snout, which was nonetheless more conspicuous than in *Proconsul*. It is differentiated from all other known hominoids by the presence of small accessory cusps (cuspules) on the upper molars and fourth premolars. Postcranially, it appears to have been primitive in the manner of *Proconsul* and *Afropithecus*. It has no obvious affinities with any contemporaneous or later hominoids, and, together with *Afropithecus*, it is significant mainly because it shows that early Miocene hominoids were probably much more diverse than discoveries in western Kenya and neighboring Uganda had previously implied. There is no reason to suppose that the discovery of *Afropithecus* and *Turkanapithecus* exhausts the actual diversity, and it is increasingly clear that early apes (broadly understood) were far more successful than their historic counterparts.

Middle and Late Miocene. During the middle and later Miocene, global climatic change produced generally cooler conditions in middle and upper latitudes and drier ones in lower latitudes, at least seasonally. In eastern Africa, drier and probably more seasonal climate combined with large-scale crustal movement (rifting) to reduce the extent of tropical forest, which was largely replaced by more open tropical woodland in the middle and later Miocene (39, 1304, 1684). In eastern Africa, middle and especially later Miocene fossils are less abundant than early Miocene ones, but they suggest that *Proconsul* or a descendant persisted until at least 14 my ago and perhaps until 8 or 9 my ago (1023, 1024). A second hominoid, *Kenyapithecus*, occurs at several Kenyan sites dated between roughly 16 and 14 my ago or possibly somewhat later (1681, 1683). Even more than *Afropithecus* + *Heliopithecus*, it was distinguished from *Proconsul* by a robust mandible, enlarged upper premolars, thick enamel, and other advanced features that ally it with later great apes and people (55). A proposed ancestor-descendant relationship between an older (16–15 my old) species, *K. africanus*, and a younger (ca. 14 my old or later) species, *K. wickeri*, is debatable, mainly because *K. wickeri* is represented by only a small number of highly fragmentary fossils. The better-known *K. africanus* is a plausible though not compelling ancestor both for several later Miocene hominoids in Eurasia (*Ouranopithecus*, *Sivapithecus*, and *Gigantopithecus*) and for the living African hominoids (1492, 1681, 1683).

As drier, more seasonal woodlands spread in eastern Africa during the mid-Miocene, the distribution of fruits probably became patchier and more seasonal, and leaves became more attractive as a dietary staple. Probably in keeping with this, mid-Miocene (ca. 15 my old) deposits contain more monkey fossils (968), and the molars are clearly advanced over early Miocene ones, with complete bilophodonty. So far only one genus, *Victoriapithecus*, has been identified. It was formerly thought to comprise two species (639, 644), but only one is recognized at present (187, 1493, 2056). Morphologically it shares no indisputably derived features with either cercopithecines or colobines (186), and it probably antedates their divergence. So far, some undoubted (derived) colobine fossils from later Miocene (ca. 10.5 to 8.5 my old) sites in eastern Africa and Namibia constitute the oldest evidence for the colobine/cercopithecine divergence (187, 543).

Eastern Africa figures prominently in reconstructions of Miocene higher primate evolution, partly because it contains so many relevant sites and partly because other parts of Africa have provided almost none. However, a right mandible and three postcranial bones from Berg Aukas, northern Namibia, show that by 12–14 my ago hominoids had spread to 20° S (537, 540–543) (fig. 3.28). The specimens have been assigned to the new species and genus *Otavipithecus namibiensis*. In most observable features, including the nonrobust mandibular body and the thinness of the molar enamel, *Otavipithecus* was more primitive than *Kenyapithecus* and allies, but its evolutionary relationships are indeterminate pending the recovery of additional fossils. At present it is significant mainly because it shows that mid-Miocene hominoids were not restricted to equatorial eastern Africa, and this in turn explains how they were able to extend their range to Eurasia, beginning about 17–18 my ago. It was already noted that broadened land connections encouraged dispersal from Africa to Eurasia at this time.

Once across the land bridge, hominoids flourished in a broad band of broad-leaved sclerophyllous woodland that stretched from southern Europe to southwestern China (39) (fig. 3.28). Thanks to the vagaries of preservation and discovery, Eurasian middle and late Miocene forms are actually better known than their African contemporaries. Their taxonomy and relationships are controversial, but at a minimum there were probably seven genera—*Dryopithecus, Pliopithecus, Ouranopithecus, Sivapithecus, Gigantopithecus, Lufengpithecus,* and *Oreopithecus*—considered separately below. Late Miocene deposits in southern Europe have also provided an undoubted

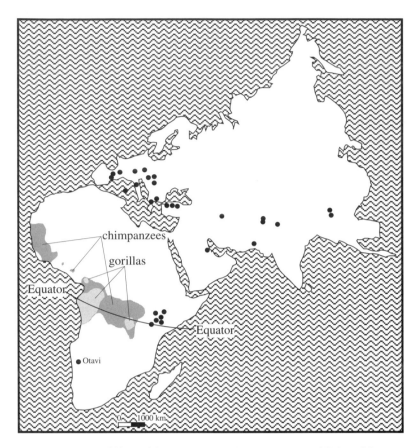

Figure 3.28. Middle and later Miocene (ca. 15–6 my old) fossil hominoid localities in relation to the historic distribution of the chimpanzees and the gorilla (redrawn after 542, p. 356). Note the concentration of fossil sites in equatorial eastern Africa and their spread across midlatitude Europe from Spain to China.

(derived) colobine monkey, *Mesopithecus* (54). Together with colobine fossils from late Miocene sites in southern Asia (140, 1004), this reconfirms African evidence for the late Miocene divergence of colobines and cercopithecines. By the very late Miocene or early Pliocene, both the colobines and cercopithecines had undergone extensive radiations (644), perhaps in response to the same general environmental change that reduced overall hominoid diversity and sparked hominid origins.

Dryopithecus includes two to four species of chimpanzee-sized apes, which are relatively well known from sites that date between roughly 14 and 8 my ago in central and western Europe. Climatic change, which eliminated its preferred forest habitat, probably forced its extinction at or shortly after 8 my ago. Dentally and postcranially, it was sufficiently similar to *Proconsul* to suggest a relatively close common ancestor,

perhaps *Proconsul* itself. With *Proconsul* and most other hominoids, it shared projecting, sexually dimorphic canines and a
Y-5 cusp and fissure pattern on the lower molars. This configuration was first recognized in *Dryopithecus*, and it is consequently usually known as the *dryopithecine Y-5 pattern*. The
postcranium, including a partial skeleton from the site of Can
Llobateres, Spain (1596), indicates, however, that *Dryopithecus*
was significantly advanced over *Proconsul* in the direction of
the living apes. Recall that *Proconsul* was primitive in its torso
shape, limb proportions, and other features, all of which show
that it was habitually pronograde (with the spinal column parallel to the ground) and quadrupedal in the manner of living
monkeys. In contrast, *Dryopithecus* had a derived, apelike body,
with a shortened, relatively inflexible lumbar vertebral (waist)
skeleton, a broad, flat thorax (chest), scapulae (shoulder blades)
that were situated behind the thorax (rather than alongside it),
and long powerful arms that could rotate around the shoulder
joint. In living apes these features promote upright (orthograde)
posture (with the spinal column perpendicular to the ground)
and an enhanced ability to hang below branches or to climb
hand over hand. *Dryopithecus* is in fact the oldest known hominoid for which such characteristically apelike posture and locomotion can be securely inferred.

Dryopithecus also diverged from *Proconsul* in several derived dental and craniofacial traits that suggest a sister relationship (closely shared common ancestor) with other late
Miocene hominoids (*Sivapithecus* and *Ouranopithecus*, discussed below); with the living African great apes, *Pan* and
Gorilla; and with the fossil hominid *Australopithecus* (170–
172). Similarities to the living African apes and *Australopithecus* may be particularly telling, because they are presumably
primitive for the group, and they include at least ten characters
that *Dryopithecus* shares with *Gorilla* (and other fossil hominoids) but not with *Pan* and *Australopithecus*. These traits include, for example, (1) the size of the incisive foramen (a perforation in the premaxilla behind the upper incisors for the passage
of nerves and a blood vessel), which is larger in *Dryopithecus*
and *Gorilla* and smaller in *Pan* and *Australopithecus* (fig. 3.29);
(2) the length of the premaxilla or portion of the upper jaw that
bears the incisors (shorter in *Dryopithecus* and *Gorilla*, longer
in *Pan* and *Australopithecus*) (fig. 3.29); (3) the shape of the lateral orbital margin along the frontal zygomatic process (thicker
in *Dryopithecus* and *Gorilla*); (4) the orientation of the ascending ramus of the mandible (more sloping in *Dryopithecus* and
Gorilla, more vertical in *Pan* and *Australopithecus*; (5) the

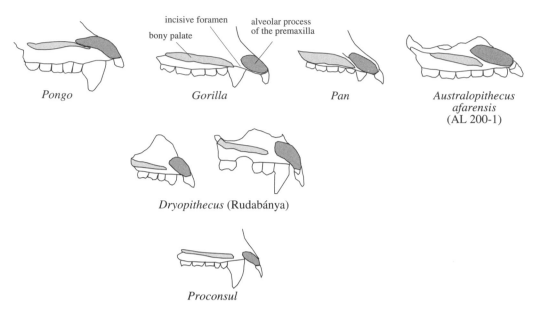

Figure 3.29. Midsagittal sections through the faces of *Pongo* (orangutan), *Gorilla*, *Pan* (chimpanzee), *Australopithecus afarensis* (from Hadar, Ethiopia), an early hominid, *Dryopithecus* (from Rudabánya, Hungary), a mid-Miocene ape, and *Proconsul* (from Kenya), an early Miocene ape (redrawn after 2389 and 172). Light shading outlines the bony (or hard) palate. Dark shading outlines the alveolar (or incisor-bearing) process of the premaxilla. The gap between the two is the incisive canal with the incisive foramen at its base. In *Pongo* the premaxilla tends to be longer and more horizontally oriented than in *Gorilla* and *Pan*, and the incisive canal is very small. It is especially wide in *Gorilla* and narrower in *Pan* and *Australopithecus*. Its great breadth in *Dryopithecus* suggests that *Gorilla* maintains the primitive condition, whereas *Pan* and *Australopithecus* are derived in the same way. If this is accepted, then palatal-premaxillary morphology joins other features suggesting that *Pan* and *Australopithecus* share a more recent common ancestor with each other than either does with *Gorilla*.

shape of the upper lateral incisors (narrower and more peg-shaped (vs. spatulate) in *Dryopithecus* and *Gorilla*); (6) the length of the P_4 relative to the M_1 (longer in *Dryopithecus* and *Gorilla*); and (7) relative proportions of the lower molars (longer and narrower in *Dryopithecus* and *Gorilla*). If the occurrence of these features in *Dryopithecus* means they are primitive for the African great apes, then they suggest that *Pan* and *Australopithecus* are similarly derived with respect to *Gorilla* and thus that *Pan* and *Australopithecus* (and by extension *Homo*) are more closely related to each other than they are to *Gorilla*.

In short, the primitive features that *Dryopithecus* and *Gorilla* share with each other but not with *Pan* and *Australopithecus* reinforce biomolecular evidence for an especially close relationship between chimpanzees and people. The anatomical argument is not completely compelling, however, partly because the chimpanzees and gorilla share derived features of locomotion and enamel ultrastructure that they do not share with hominids (55) and partly because the derived features that chimpanzees and *Australopithecus* share uniquely could reflect convergence or parallelism rather than common descent (616).

Pliopithecus includes perhaps eleven species of gibbon- and siamang-sized apes found at western and south-central European sites dated between roughly 16 and 11 my ago (839). The genus has also been identified in southern China (54). It appeared in Europe before *Dryopithecus* and, to judge by the associated fauna, was generally tied to moister, more heavily wooded areas (1607). It usually does not occur in the same sites as *Dryopithecus* and became extinct earlier, probably because the mid-Miocene trend to cooler, drier climate eliminated its habitat first. Its taxonomy is controversial, and based mainly on detailed differences in the dentition, some specialists prefer to divide *Pliopithecus* among at least three genera—*Pliopithecus* in the narrow sense, *Crouzelia*, and *Anapithecus* (54, 173, 839, 968).

Whatever the appropriate taxonomy, by far the best-known species is *Pliopithecus vindobonensis*, which is represented by several partial skeletons from mid-Miocene deposits at Neudorf an der March, Slovakia (fig. 3.30). *P. vindobonensis* shared with the gibbons and siamangs some cranial and dental features such as a short, broad face, projecting orbital margins, a gracile mandible that deepened toward the front, relatively narrow incisors, canines that were sexually dimorphic but large in both sexes, and some degree of intercusp cresting (644). Unlike gibbons and siamangs, however, it had third molars that were long both relatively and absolutely, orbits that were backed by bone only to the extent seen in *Aegyptopithecus* and the platyrrhine monkeys, and an external auditory meatus that was incompletely ossified. Like gibbons and siamangs, it had a gracile postcranial skeleton with relatively indistinct muscle markings, but it lacked their forelimb specializations for brachiation. Its limb proportions resembled those of quadrupedal monkeys, and its overall postcranial morphology suggests it was an arboreal quadruped with some suspensory capabilities (728). Like *Aegyptopithecus*, it maintained an entepicondylar foramen on the distal humerus, and it may have had a tail.

In retaining such primitive features as an entepicondylar foramen and an incompletely ossified auditory meatus, *Pliopithecus* recalls *Aegyptopithecus* and differs from all other known Miocene to Recent hominoids. It could be ancestral to gibbons and siamangs only if they independently developed

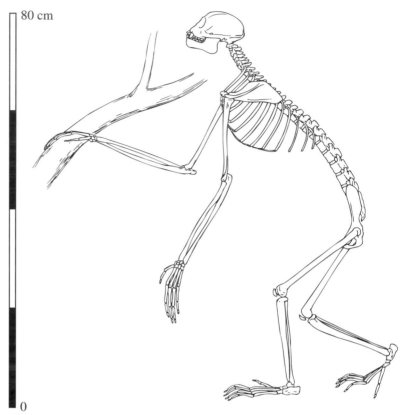

80 cm

0

Pliopithecus vindobonensis

Figure 3.30. Reconstructed skeleton of *Pliopithecus vindobonensis* from mid-Miocene (ca. 15 my old) deposits in Slovakia (redrawn after 2529, fig. 106). In its short, broad face, slender postcranial bones, and other features, *P. vindobonensis* resembled the living gibbons and siamangs, but it lacked their extraordinarily long arms and other morphological specializations for brachiation. It also retained some remarkably primitive features, such as an incompletely ossified external auditory meatus and an epicondylar foramen on the distal humerus. Most authorities now regard it as one of a group of closely related small hominoids that inhabited western and central European forests between roughly 16 and 11 my ago and that bear no relationto any later hominoids.

many of the derived features they share with the great apes. These include not only a completely ossified meatus and complete postorbital closure but also numerous detailed similarities in the structure of the shoulder, elbow, and trunk. More likely the similarities between *Pliopithecus* and the gibbons and siamangs reflect a combination of shared, retained primitive features and of convergence, relating, for example, to a shared gracile postcranium. Most probably, then, *Pliopithecus* had nothing to do with gibbon origins and was simply a unique Miocene catarrhine that became extinct without issue. Its peculiar mix of primitive and advanced features suggests it should be placed in its own family, the Pliopithecidae, within the Hominoidea—or perhaps within the Propliopithecoidea, separate from the Hominoidea. Placement within the Propliopithecoidea would imply descent from Oligocene relatives of *Aegyptopithecus* and *Propliopithecus* through an as yet unknown early Miocene form.

The removal of *Pliopithecus* from possible gibbon and siamang ancestry, and the failed case for a special relationship between the gibbons and siamangs on the one hand and early Miocene *Dendropithecus* and *Micropithecus* on the other, leaves the gibbons and siamangs with no fossil record until they appear in modern form in Pleistocene and Holocene deposits in their historic Southeast Asian range (729). By geographic location, estimated 16 my old (mid-Miocene) date, and observed morphology, a poorly known Chinese genus, *Dionysopithecus*, and an even more poorly known, like-aged, possibly congeneric form from Pakistan could represent early members of the gibbon-siamang lineage (141, 729), but the fossils are insufficient for a binding judgment.

Ouranopithecus (= *Graecopithecus* [50, 616]) was a chimpanzee-sized ape represented by jaws, teeth, and a partial face from late Miocene (ca. 11–9 my old) deposits in Greece (258, 262, 263). It is readily distinguished from most early Miocene hominoids and from *Dryopithecus* and *Pliopithecus* by its thick molar enamel, which it shared with *Sivapithecus*, *Gigantopithecus*, and *Kenyapithecus*. Thick enamel in *Ouranopithecus* and the other middle to late Miocene hominoids may have derived from a shared east African ancestor, in which it permitted feeding on nuts, seeds, subterranean tubers, and other hard or grit-encrusted foods (1186). Such foods probably increased relative to fruits as woodland replaced forest in the early mid-Miocene (40, 42). The animal species that accompany *Ouranopithecus* and other thick-enameled hominoids almost invariably imply an environment of open woodland or savanna (as opposed to forest).

Like modern woodlands, middle to late Miocene ones probably provided an incomplete tree canopy, and many nuts and other hard foods undoubtedly occurred at ground level. They were probably also distributed in patches that could be reached only by movement on the ground, and *Ouranopithecus* and other thick-enameled hominoids may thus have been more terrestrial than their thin-enameled relatives. This is especially interesting because thick enamel and some degree of terrestriality almost certainly marked the line leading to *Australopithecus,* the earliest known hominid. Besides thick enamel, *Ouranopithecus* and *Australopithecus* also shared round, swollen molar cusps, the absence of a honing facet (from contact with the upper canine) on the mesial (anterior) face of the lower third premolar (P_3), and substantial sexual dimorphism in cheek tooth size (especially notable in the oldest *Australopithecus*). The postcranium of *Ouranopithecus* remains undescribed, but on dental grounds, if *Ouranopithecus* occurred in Africa, it might be a plausible ancestor for *Australopithecus.* On combined geographic-morphologic grounds, however, *Kenyapithecus* or its immediate descendant is a better candidate.

Sivapithecus comprised several siamang- to orangutan-sized species represented in later mid-Miocene (ca. 12.5–7 my old) deposits in the Siwalik Hills on the India-Pakistan border and in Turkey. Arguably it also occurs in late middle Miocene sites in Europe (Neudorf an der March, Slovakia) and western Kenya (*Kenyapithecus wickeri* of Fort Ternan) (1202, 1689). Whether or not the European, Chinese, and African specimens are included, *Sivapithecus* is still best known from the Siwaliks, which have provided numerous dentitions, some limb bones, and a partial skull.

Sivapithecus can be distinguished from *Ouranopithecus* by several features, including a much more robust mandible; larger incisors and canines relative to the cheek teeth; a narrower, higher-crowned, and more sectorial lower third molar; and a distinctly derived nasal and subnasal morphology resembling that of the orangutan (see below) (58, 1202, 1960). In other respects, however, especially in the thickness of the molar enamel, the two taxa were very similar, and, like *Ouranopithecus*, *Sivapithecus* apparently exhibited considerable sexual dimorphism. It appears that specimens formerly assigned to a separate genus, *Ramapithecus,* may simply represent a small species of *Sivapithecus,* and many specialists now regard *Ramapithecus* as a "junior" synonym of *Sivapithecus.*

As already noted, *Sivapithecus* inhabited relatively open woodland where thick enamel would have been advantageous for feeding on hard or abrasive subterranean plant foods. These were probably more abundant and more reliable than soft-coated fruits. Circumstantially, then, it has been suggested that *Sivapithecus* at least partly fed on hard objects rather than being strictly frugivorous (42). This remains a reasonable hypothesis, though preliminary studies show that in dental microwear

Sivapithecus was more like the chimpanzee, known to eat soft fruits, than like other primates that eat hard foods (2158).

In those dental features that *Sivapithecus* shared with *Ouranopithecus*, it also resembled *Australopithecus*, with which it further shared molars that were large relative to the front teeth (and to estimated body size) and relatively short, stubby canines that tended to wear from the tips as well as from the mesial and distal surfaces. Because of obvious dental resemblances to *Australopithecus*, fossils assigned to *Ramapithecus* (− *Sivapithecus*) were once thought to represent the earliest known hominid. But in its face, known from both Pakistan and Turkey (58, 1688, 2390), *Sivapithecus* shares several derived features with the orangutan. These include a narrow interorbital region; high, narrow orbits; a subnasal plane that is continuous with the floor of the nasal cavity (not stepped); and upper central incisors that are much larger than the lateral ones (fig. 3.31) (52, 1464, 2388). The orangutan also has thick enamel on its molars, and there is now wide agreement that *Sivapithecus* was on or near the line leading to the orangutan.

One might argue from enamel thickness that *Sivapithecus* (including *Ramapithecus*) was near the ancestry of both the orangutan and hominids, but this hypothesis is highly questionable. The biggest objection is that biomolecular evidence shows that living people are much closer to the chimpanzee and the gorilla than to the orangutan (52); yet unlike hominids and the orangutan, the chimpanzee and the gorilla have thin enamel. The implication is that enamel thickness by itself may not be taxonomically relevant. In fact, scanning electron microscopy indicates that hominid, chimpanzee, and gorilla enamel coatings are basically similar in ultrastructure and development and that the thin enamel of the chimpanzee and gorilla is secondarily derived from thick enamel (1462). The chimpanzee, the gorilla, and hominids could thus share a thick-enameled ancestor, perhaps *Kenyapithecus* or its immediate descendant in Africa.

In this light *Sivapithecus* is significant not because it was ancestral to hominids but because it shows that the line leading to the orangutan had probably split from the line leading to the African apes and hominids by 12 my ago (fig. 3.13). Unfortunately the fossil record sheds very little light on the probable common ancestor of chimpanzees, gorillas, and people or on the actual history of chimpanzees and gorillas. The main problem is the extreme rarity of hominoid fossils from African late Miocene deposits (between 14 and 5.5 my ago). The most

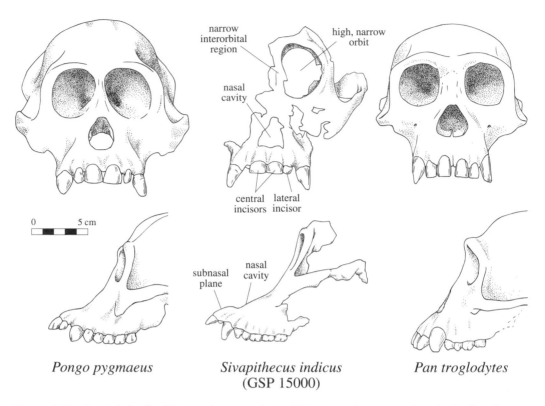

narrow
interorbital
region

high, narrow
orbit

nasal
cavity

central   lateral
incisors incisor

0   5 cm

subnasal   nasal
plane   cavity

Pongo pygmaeus   *Sivapithecus indicus*   *Pan troglodytes*
                          (GSP 15000)

Figure 3.31. Partial skull of *Sivapithecus indicus* (GSP 15000) compared with skulls of an orangutan, *Pongo pygmaeus*, and of a chimpanzee, *Pan troglodytes* (drawn by Kathryn Cruz-Uribe from photographs and casts; © 1999 by Kathryn Cruz-Uribe). In its high, narrow orbits, narrow interorbital region, (nonstepped) continuity between the floor of the nasal cavity and the subnasal plane, and large size of the central incisors compared with the lateral ones, the *Sivapithecus* skull closely resembles that of the orangutan and differs from those of the African great apes. The implication is that *Sivapithecus* is near the ancestry of the orangutan. (GSP = Geological Survey of Pakistan.)

significant specimen is a fragmentary maxilla from the Namurungule Formation in the Samburu Hills (Baragoi region) of north-central Kenya (1122). Fission-track and radiopotassium dates, together with the associated fauna, indicate an age between 10.5 and 6.7 my (1478). The maxilla has gorillalike molars and very thick enamel, and it could represent the ancestor of the gorilla or even the common ancestor of the gorilla, chimpanzees, and people, but additional, more complete finds are necessary for confirmation (48, 55, 1689). The history of the gorilla and chimpanzees may always be obscured by a lack of Miocene and Pliocene localities in their historic west and central African habitats.

The postcranium of *Sivapithecus* is moderately well known, and it appears to have been predominantly a non-knuckle-walking, monkeylike quadruped, perhaps broadly like *Proconsul* and other earlier Miocene hominoids (1574, 1687,

1743, 1829–1831). Particularly telling is the curvature of the upper humerus shaft, which suggests that the humerus generally moved alongside a deep, narrow, monkeylike chest and that it was not habitually raised for underbranch suspension or hand over hand climbing. In its lack of derived (apelike) postcranial features, *Sivapithecus* poses an evolutionary conundrum. If *Dryopithecus*, which possessed a straight humerus shaft and other derived apelike postcranial features, was near the line leading to the African great apes and *Sivapithecus* was on or near the line leading to the orangutan, then the African great apes and the orangutan must have developed their shared postcranial specializations in parallel (50, 1687). The alternative is that the facial similarities of *Sivapithecus* and the orangutan represent parallelisms or shared primitive features (186, 1493, 1686, 1687) and that the ancestry of the orangutan must be sought in another mid-Miocene hominoid.

Gigantopithecus, as the name suggests, was perhaps the largest primate that ever lived, exceeding even the gorilla in size. It occurs in Siwalik deposits dated to approximately 6.3 my ago (141), and it may have survived in southern China and Vietnam until 0.5 my ago, where it apparently overlapped with *Homo erectus* (references in chapter 5). It is known only from mandibles and isolated teeth. In its dentition, *Gigantopithecus* dwarfed *Sivapithecus*, but it was otherwise generally similar: the cheek teeth were very large relative to the front teeth, the molars had thick enamel, and the canines were robust but low crowned, wearing from the tips (644). The dental similarities suggest a broadly similar diet, if not a closely shared ancestor. Since postcranial bones have not been found, the precise positional behavior of *Gigantopithecus* remains speculative, but given its size it was probably completely terrestrial. Its origins remain unclear, but together with *Ouranopithecus* and *Sivapithecus* it may ultimately descend from *Kenyapithecus*.

Lufengpithecus was a hominoid the size of a siamang to a small chimpanzee that existed in southern China roughly 8–7 my ago (102). So far it is known from only a single locality in Lufeng County, Yunnan Province, but the site is very rich and the sample comprises several partial skulls, fragmentary jaws, and roughly a thousand isolated teeth (1896, 1897). The specimens were originally divided between two species (each in a separate genus), but they probably represent a single highly dimorphic species (1200).

Lufengpithecus has often been likened to *Sivapithecus*, but it lacked the very narrow interorbital region, tall, oval orbits, nonstepped subnasal region, and other facial and dental features that distinguish *Sivapithecus* (and that liken it in turn to the orangutan). Except perhaps for relatively thick enamel, it shares no conspicuously derived features with any other hominoids, and it exhibits some unique characters, including, for example, the depression (or concavity) of the (glabellar) region between and just above the orbits and thick, raised temporal lines that course straight backward from the upper middle margin of the orbit (1896, 1897) (fig. 3.32). In other large-bodied Miocene to Recent hominoids, the glabellar region either is flat or bulges outward, and the temporal lines either are not as marked, follow a different course, or both. In short, *Lufengpithecus* has no obvious close relatives among either fossil or living hominoids (50, 1201), and it is perhaps significant mainly for further underscoring the remarkable diversification of Miocene hominoids in Eurasia and for showing that the degree of sexual dimorphism in living species need not be a guide to its past extent (1203). *Lufengpithecus* may have been even more dimorphic than the gorilla and the orangutan, in which males are sometimes twice the size of females.

Finally, *Oreopithecus* was a siamang-sized anthropoid that inhabited swampy, late Miocene (ca. 9–8 my old) forests in Italy (95, 96). Its relatively flat face, vertically implanted incisors, low-crowned canines, and bicuspid lower third premolars, together with features of its pelvis, once suggested it might be on the line leading to humans. But its relatively long, slender arms, short hind limbs, curved phalanges, and extremely mobile limb joints imply it was primarily arboreal and suspensory in its habits (739). After *Dryopithecus*, it is in fact the oldest known hominoid for which orthograde posture and ability to climb hand over hand have been inferred. Its lower molars possessed six major cusps, in distinction from all other known hominoids, and the cusps on both the upper and lower molars were partly linked by crests similar to those in cercopithecoids (644). Its affinities are highly controversial, and it is commonly placed in its own extinct family, the Oreopithecidae. It could be either an aberrant hominoid, to judge from significant postcranial specializations it shares with the living apes (967), or a strange cercopithecoid, to judge from its dentition (1835, 2126). If it was as unique as its dentition suggests, and if its postcranial specializations for underbranch suspension thus evolved independently, this would clearly enhance the possibility for comparable independent (parallel) postcranial evolution in the lines leading to the orangutan and the African great apes. More generally, postcranial suspensory specializations would be of limited value for establishing phylogenetic relationships within the hominoids.

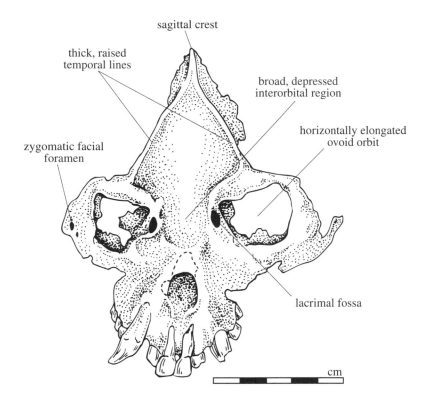

Lufengpithecus PA 644

Figure 3.32. Skull PA 644 of *Lufengpithecus* from Yunnan Province, southern China (redrawn after 1896, fig. 1). *Lufengpithecus* had thick dental enamel like *Sivapithecus* and most other mid- and late-Miocene hominoids, but it was unique in many features, including its thick, upraised temporal lines, its broad, depressed interorbital (glabellar region), and perhaps its extreme sexual dimorphism. It underscores the remarkable diversity of late- and mid-Miocene hominoids in Eurasia. (PA = Paleoanthropology specimen, Institute of Vertebrate Paleontology and Paleoanthropology, Beijing.)

From limited dental similarities, it has been suggested that *Oreopithecus* evolved from Oligocene *Apidium* of the Fayum (1959), but its most secure phyletic link is to *Nyanzapithecus,* an early mid-Miocene (19–16 my old) catarrhine from western Kenya (966, 967). Dentally, *Nyanzapithecus*—including specimens formerly assigned to *Proconsul (Rangwapithecus) vancouveringi*—is nicely intermediate between more generalized early Miocene east African anthropoids and *Oreopithecus.* It possessed some dental specializations that eliminate it from the direct ancestry of *Oreopithecus,* but it still implies an east African origin for the Oreopithecidae, and

Oreopithecus itself may be represented in the later mid-Miocene (ca. 14 my old) deposits of Fort Ternan and a neighboring site in western Kenya (966, 967).

Summary and Conclusion

From the preceding account, it is clear that the primate fossil record is flawed by major temporal and geographic gaps and that even where fossils occur they are mainly fragmentary jaws and isolated teeth. The spotty record with its incomplete fossils is the main reason informed specialists often disagree about the probable behavior and ecology of fossil species and especially about their phylogenetic relationships. Some disagreement also stems from different a priori expectations or theoretical perspectives, which themselves are partly determined by the state of the record (431, 737).

Much research on fossil primates has understandably been motivated by a desire to trace the origins of living ones, but as the fossil record has improved, especially since the 1960s, tracing the descent of living primates has become more of a problem, not less. This is because the fuller record shows that few known fossil primates possessed derived features or specializations that link them unequivocally and uniquely to any living form. Many fossil species, in fact, had their own unique specializations that effectively eliminate them from the ancestry of later species and that suggest they became extinct without issue. To the extent that fossil forms resemble living ones (or each other), it is often difficult to determine if the shared features are actually primitive characters inherited from a relatively distant common ancestor or, alternatively, are parallelisms that genetically similar taxa developed independently as they adapted to similar circumstances. Parallelism (convergence or homoplasy) is especially problematic and difficult to detect, yet it is the only reasonable explanation for why some specializations (for example, in body form and inferred locomotion) often imply one set of evolutionary relationships while others (for example, in dental form and inferred diet) suggest a contradictory set. On top of this, the fossil record presents many more potential ancestors than the living primates require, and this underscores the point that many fossil forms have no living descendants.

But even if it is not possible to draw a full and compelling phylogeny linking successively younger fossil primates to living species, it is possible to outline it, along with the probable times when major primate groups diverged from each other (fig. 3.33).

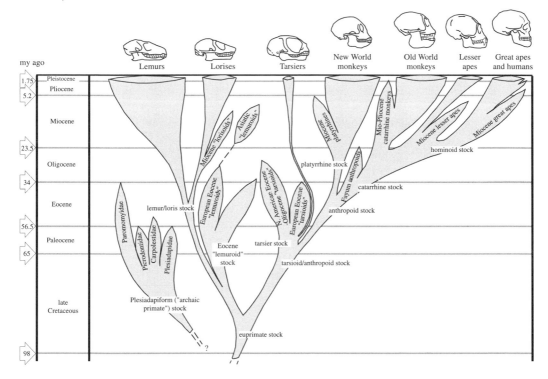

Figure 3.33. A provisional phylogeny of the Primates (modified after 1468, p. 46). The times when major groups diverged are subject to revision, but the order of divergence is reasonably firm.

Equally important, fresh discoveries and increasingly sophisticated analyses have revealed the main stages of primate evolution, which can be conceptualized as a series of adaptive radiations preceding the emergence of hominids:

1. An initial radiation beginning probably in the late Cretaceous, leading to the Paleocene Plesiadapiformes. These were distinguished from their insectivore (or insectivore-like) ancestors primarily by the morphology of their cheek teeth and perhaps by their ear structure. They barely differed from insectivores in their brains, special senses, or locomotion, and their main behavioral distinction was perhaps the significant addition of seeds, fruits, or other vegetal matter to a basically insectivorous diet.

2. A second radiation beginning in the Paleocene or even the late Cretaceous, producing a wide variety of much more advanced lemurlike and tarsierlike forms. These were distinguished from the Plesiadapiformes and from their own more primitive ancestors by the development of typically primate grasping hands and feet with nails instead of claws and by a reorganization of the brain and face to emphasize vision over

olfaction. The selective advantage of these morphological advances was perhaps an enhanced ability to hunt insects among slender branches in forest undergrowth.

3. A third radiation beginning in the early or middle Eocene, producing a variety of primitive higher primates. Like their lemurlike or tarsierlike ancestors, they were mainly arboreal and quadrupedal, but they may have spent more time walking on top of branches and less time leaping between them. They probably also focused less on insects and more on fruits and leaves, and their brains tended to be larger, with an even greater emphasis on vision over olfaction.

4. A fourth radiation or set of radiations during the late Oligocene or early Miocene, in which higher primates differentiated into forms clearly anticipating extant apes and monkeys. To begin with, the main difference between apes and monkeys was dental, and this suggests that the earliest monkeys differed from their apelike contemporaries mainly in their greater propensity to eat leaves. The early apes and monkeys were probably very similar in their palm-flat, quadrupedal, nonsuspensory habitual postures and locomotion. Almost certainly both were basically arboreal, though the early monkeys may have been relatively more terrestrial. To begin with, the apes were much more abundant and diverse than the monkeys, probably because forested, fruit-rich settings dominated the tropical and subtropical portion of Africa where monkeys and apes first diverged.

5. A fifth radiation in the middle and later Miocene (after 17–16 my ago) that produced the first apes in which torso shape and limb mobility imply a modern apelike proclivity to orthograde posture and to hand over hand climbing. Some apes also developed novel dentitions, in which the enamel was thickened, the molar surfaces were enlarged, and the canines were relatively small and noninterlocking. These dental innovations were probably a response to the increased abundance of hard or grit-encrusted foods relative to fruits. The change in food availability was probably sparked by a middle to late Miocene trend toward cooler, drier, more seasonal continental climates that promoted the spread of woodlands and savannas at the expense of forests, particularly in tropical and subtropical latitudes. The novel dental complex is especially noteworthy, because it anticipated the dental morphology of the early Pliocene australopithecines, after 5 my ago.

It is all but certain that the lines leading to people and chimpanzees diverged in equatorial Africa between 7 and 5 my ago and that the last shared ancestor inhabited woodlands that increasingly replaced forests during the late Miocene, after 10 my ago. To find food or to move between food sources in these woodlands, the common ancestor was probably at least partly terrestrial, and the hominid lineage emerged when one population adapted bipedalism as its habitual mode of terrestrial locomotion. Arguably, a particularly sharp reduction in tree cover near 6 my ago provided the selective stimulus for bipedalism. Unfortunately, the African fossil record between 10 and 5 my ago is stubbornly sparse, and the morphology of the protohominids remains conjectural. If the australopithecines, described in the next chapter, are a reasonable guide, the protohominids had chimpanzeelike bodies and skulls that they inherited from a middle to late Miocene predecessor. They probably also had thick enamel, enlarged molars, and reduced canines, but this inference raises questions because it implies that thin enamel, relatively small molars, and enlarged canines are novelties that chimpanzees evolved in parallel with gorillas. The alternative is that the australopithecines evolved thick enamel and related dental features independently as they adapted to broadly the same environmental circumstances that favored these characters in some of their Miocene predecessors. A much expanded fossil record will be necessary to make an informed choice.

THE AUSTRALOPITHECINES AND *HOMO HABILIS*

4

In 1939 Gregory and Hellman (892) proposed that a group of very early hominid fossils from South Africa should be placed in the subfamily Australopithecinae, to be distinguished from a second subfamily, the Homininae, containing later (more advanced) hominids. The formal term Australopithecinae has found little favor, but its informal, anglicized derivative is now widely used to designate not only the original South African fossils but also a growing number of similar fossils from eastern Africa.

The australopithecines are the oldest known hominids, with a fossil record beginning before 4.4 million years (my) ago and extending to sometime between 1.2 my ago and perhaps 700 thousand years (ky) ago. In the vernacular, they have sometimes been called "man-apes" because they combined bipedal locomotion—the defining characteristic of the hominid family—with small, ape-sized brains. It is becoming increasingly clear that they were also apelike in other respects, including the length and power of their arms, and they probably mixed human-style bipedal locomotion on the ground with ape-style agility in the trees. Males were much larger than females, suggesting an apelike social organization in which males competed vigorously for females and males and females did not cooperate economically. There is no archeological evidence for technology, and tool use was probably no more developed than among chimpanzees. In short, morphology and behavior together suggest that the australopithecines were bipedal apes rather than primitive humans. Humans, or more precisely the genus *Homo*—defined above all by brain expansion—appeared only about 2.5 my ago, and it is probably no coincidence that the oldest known stone tools date from about the same time.

The evolutionary origins of the australopithecines remain vague because of a sparse African fossil record between 8 and 4.5 my ago, and the relationships of australopithecine species to each other and to the earliest widely recognized species of

Homo, H. habilis, remain controversial. But continuing discoveries in Africa have narrowed the phylogenetic possibilities, and they have also revealed the basic morphological and behavioral course of early hominid evolution. It is on the known pattern that this chapter focuses.

History of Discovery: South Africa

The first australopithecine fossil was discovered by Raymond Dart, a British-trained anatomist who in 1922 was appointed professor of anatomy at the University of the Witwatersrand in Johannesburg, South Africa. There are minor discrepancies between the discovery as recounted by Dart (576) and as reconstructed by Tobias (2197, 2198). Dart's account is summarized here.

Dart planned to create a museum in his department, and he encouraged students to collect suitable fossils during their vacations. In 1924 a student brought him a fossil baboon skull from the Buxton lime quarry at Taung in the northern Cape Province of South Africa, about 320 km southwest of Johannesburg (fig. 4.1). Through a geologist colleague, Dart contacted the quarry manager, who then sent him two crates of fossiliferous breccia—rock-hard blocks of sandy sediment and fossils cemented together by limy glue—from a small cave exposed in the quarry. The fossils in the crates were mainly uninteresting, but one crate contained a natural endocast (mold of the inside of a skull) that Dart immediately realized came from a hominoid primate. Among the breccia blocks in the crates, Dart found one with a depression into which the endocast fitted. Inside the block he saw traces of bone that he hoped was the front of the skull belonging with the endocast.

The front was present (fig. 4.2), but it took Dart many weeks to expose it with a hammer, chisels, and a sharpened knitting needle. The dentition showed that the skull came from a juvenile whose first molars were just erupting, and various other features showed that in important respects the creature was intermediate between apes and people. For example, it was obviously apelike in the small size of its brain—or more precisely in its endocranial (internal skull) volume—which Dart estimated would have reached 525 cc at adulthood. Subsequent reestimates have reduced this to between 404 and 440 cc (703, 1044), but even Dart's figure was much closer to the average for common chimpanzees (ca. 400 cc) and gorillas (ca. 500 cc) than to the average for living people (ca. 1,350 cc). At the same time, the deciduous canines were much smaller and less projecting than those in apes, and the foramen magnum ("large hole"),

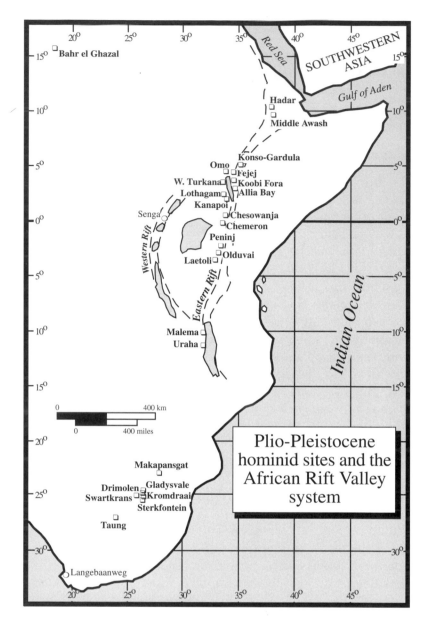

Figure 4.1. Approximate locations of African Plio-Pleistocene hominid fossil sites and of the Langebaanweg early Pliocene faunal site. Tectonic activity associated with the Rift Valley system in eastern Africa created numerous basins that trapped and preserved bones beneath lake and riverine sediments. Volcanoes also linked to rifting provided lavas and ashes that permit radiometric dating. Tectonic activity, often sparse vegetation, and episodic and frequently violent rainfall have promoted gullying that now exposes fossils at the surface. The Rift system does not extend to southern Africa, where sedimentary basins tend to be rare and very shallow. The southern African Plio-Pleistocene hominid sites are all in limestone caves, and they lack volcanic materials for dating. The known caves are clustered mainly in an intensively explored area between Johannesburg and Pretoria, South Africa. The Langebaanweg site lacks any traces of hominids, although it dates from the time when they first emerged and has provided more animal bones than all the other sites combined. The implication is that the earliest hominids were confined to more equatorial latitudes.

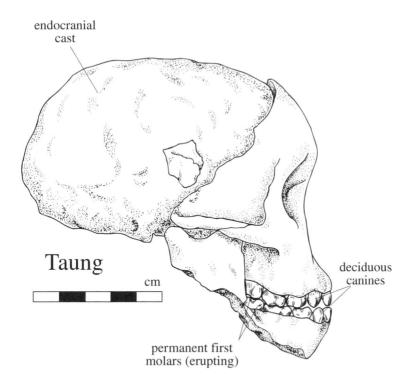

endocranial
cast

Taung

cm

deciduous
canines

permanent first
molars (erupting)

Figure 4.2. Facial skeleton and endo-cast of *Australo-pithecus africanus* from Taung, viewed from the right side (drawn by Kathryn Cruz-Uribe from photos and casts; © 1999 by Kathryn Cruz-Uribe).

through which connections passed from the spinal column to the brain, was more downwardly oriented on the base of the skull. This suggested that the skull was balanced on top of the spinal column, as in people, where this position is a natural concomitant of upright, bipedal locomotion (fig. 4.3). In short, the position of the foramen magnum implied that the Taung individual was bipedal like people rather than quadrupedal like apes.

Dart published a description of the skull in the February 7, 1925, issue of *Nature* and concluded that it came from a previously unknown species "intermediate between living anthropoids and man" (574). He named the species *Australopithecus africanus* ("African southern ape") and stressed both the skull's intermediate or transitional morphology and its discovery far outside the geographical ranges of the living apes. Taung lies in a nonforested region where living apes could not survive, and from geological studies Dart argued that the regional environment was probably similar when the fossils accumulated. The ancient vegetation was thus inappropriate for an ape in the narrow sense. It was, however, clearly suitable for an ape that was developing human traits, especially bipedal locomotion.

In the following issue of *Nature*, Dart's assessment was severely criticized by several anthropologists. A consensus soon developed that *Australopithecus* was simply a fossil ape,

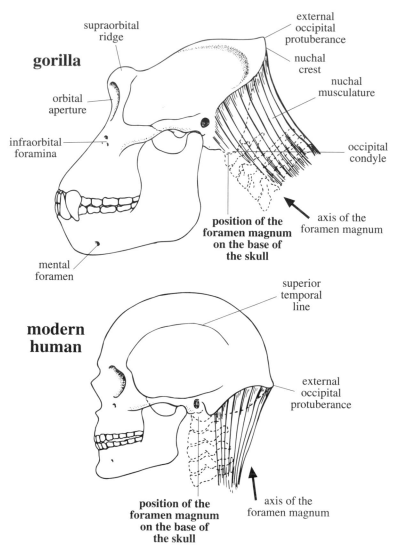

Figure 4.3. Gorilla and modern human skulls, showing the differences in the position of the skull with respect to the spinal column (adapted from 1318, p. 67). In people, the skull is balanced on top of the spinal column, and the foramen magnum consequently lies much farther forward on the base of the skull. The orientation of its axis is also more vertical. The more posterior position of the foramen magnum in the gorilla, together with its larger, more protruding face, also requires larger, more powerful nuchal (neck) muscles to stabilize the head.

with no special relevance to human evolution. According to a prominent eyewitness to the debate (1318), the reasons for doubting Dart included nonscientific considerations, such as Dart's flowery, rhetorical writing style, his reputation for coming to hasty conclusions, and his unbelievable good luck in finding such a spectacular fossil just two years after arriving in Africa.

However, there were fundamental scientific concerns as well. There was very little from which to estimate the antiquity of the Taung skull, but some authorities felt that the associated baboon fossils implied it was too recent to be a human ancestor. In addition, the discovery of primitive human fossils ("*Pithecanthropus*") in Java in 1891 had convinced many specialists that Asia, not Africa, was the cradle of humanity. Finally, the skull did not fit either of the current theories of human evolution (2394). The first held that human ancestors should be equally primitive in all traits, whereas Dart's find showed a mix of advanced and primitive features. The second theory, supported by the famous Piltdown skull (later exposed as an elaborate fake), proposed that the peculiarly human brain had evolved before other uniquely human traits. The Taung skull, as interpreted by Dart, suggested just the reverse—that the uniquely human mode of bipedal locomotion had evolved before the brain.

Beyond these theoretical obstacles, some specialists disputed Dart's claim that the endocranial cast indicated a humanlike (vs. apelike) brain. This point remains controversial (1051). Even more problematic was the very young age of the Taung individual at the time of death, which was at five to seven years if one applies the same maturation rate and dental eruption schedule as is seen in living people. The actual age may have been only three to four years (544), since growth increments in australopithecine enamel (204, 358, 359), together with the relative timing of crown and root development in various teeth (1995), suggest that the australopithecines matured at about the same rate as modern apes, perhaps 30% to 50% faster than modern people. The crucial point is that very young apes resemble very young humans much more than adult apes resemble adult humans, and some of Dart's critics suggested that as an adult the Taung individual would have been much more apelike in all features, including the orientation of the foramen magnum. Finally, there was the problem that Dart was inferring bipedal locomotion from a portion of the anatomy not directly involved in locomotion. Without actual bones from the locomotor skeleton—especially from the pelvis or leg—it could be argued that *Australopithecus* may not have been bipedal.

Unfortunately, Dart did not visit the Taung site to seek additional fossils, especially adult skulls or postcranial bones that could have won his case, and by the time qualified specialists did go, the cave that had contained the juvenile skull had been quarried away. Dart continued to remove breccia from the skull, and in 1929 he succeeded in separating the upper and lower jaws to permit a more complete assessment of the dentition. He distributed casts of the dentition to many specialists, and W. K. Gregory, an eminent American authority on mammalian teeth, concluded that it came from a hominid rather than from an ape.

Gregory's support aside, however, additional specimens were obviously necessary to convince many other authorities, and Dart was not personally prepared to make the search. This task fell to Robert Broom, a Scottish-born physician and authority on fossil reptiles who had settled in South Africa. He visited Dart's laboratory two weeks after the formal announcement of the Taung find in *Nature* and was an early convert to Dart's conclusions. But he had other work to complete, and it was not until the 1930s that he was able to initiate research on *Australopithecus*. In 1936, working from his base at the Transvaal Museum in Pretoria, about 90 km north of Johannesburg, Broom had his attention called to the fossiliferous breccia in a cave on the farm Sterkfontein, near the town of Krugersdorp, roughly 25 km northwest of Johannesburg (fig. 4.1). On visiting the site, he almost immediately recovered a partial adult australopithecine skull.

In 1938, during one of his periodic visits to Sterkfontein, Broom heard of another promising cave at Kromdraai, approximately 3 km to the east. Again he was quickly rewarded with an adult skull, but it differed in important respects from the earlier Sterkfontein skull. He concluded that it represented a different kind of australopithecine, a view that is still widely accepted.

Having found adult skulls, by 1939 Broom could show that adult australopithecines were no more apelike than the Taung child. Equally important, at Sterkfontein he had recovered a distal femur and at Kromdraai an astragalus (talus) that showed the australopithecines probably were bipedal, as Dart had argued.

Partly for economic reasons and partly because of World War II, research in the Sterkfontein region largely ceased between 1939 and 1945. In collaboration with G. W. H. Schepers, Broom spent the war years preparing a monograph on the Sterkfontein and Kromdraai finds. This appeared in 1946 (366) and was widely read and acclaimed. As a result, most authorities came to accept the australopithecines as hominids, though some were disturbed by Broom's tendency to divide the known fossils into more taxa than seemed warranted: he placed the Sterkfontein and Kromdraai finds in their own genera—*Plesianthropus* ("near man") and *Paranthropus* ("alongside man"), respectively—distinct from Dart's original *Australopithecus*. Most specialists today place all the fossils in the single genus *Australopithecus*, divisible into two species—*A. africanus* (Taung and Sterkfontein) and *A. robustus* (Kromdraai). A growing number have resurrected the category *Paranthropus* as originally defined by Broom, however, and *Paranthropus* is used here. In the vernacular, *A. africanus* is often called the "gracile (or slender) australopithecine," and *Paranthropus* (or *Australopithecus*) *robustus* is the "robust australopithecine," although they differed less in body mass or stature than the names imply.

In 1947–1948, assisted by John T. Robinson, Broom renewed work at Sterkfontein, with spectacular success. Among the crucial specimens they recovered was a partial australopithecine skeleton, including a pelvis that demonstrated beyond all doubt that the owner was bipedal. They also found a nearly complete adult skull that is probably the most famous South African australopithecine find after the original Taung find. It was regarded as a female *Plesianthropus* and became popularly known as Mrs. Ples. (Fig. 3.1 shows a reconstruction.)

In 1947 James Kitching, participating in an expedition organized by Dart, found an australopithecine fossil at the Makapansgat Limeworks Cave in the northern Transvaal, approximately 300 km north of Johannesburg, and in 1948 Broom and Robinson made the first discoveries in a cave on the farm Swartkrans, less than 2 km from Sterkfontein. Research in the late 1940s at Sterkfontein, Makapansgat, and Swartkrans greatly enlarged the available australopithecine sample, producing many skulls or partial skulls, dentitions, and isolated teeth plus a small but diagnostic sample of postcranial bones that pointed consistently to bipedalism. By 1950 the mounting evidence had convinced nearly all skeptics that the australopithecines were early members of the human family. Debate continued regarding their relation to later people, however, and the issue is still not settled.

Sterkfontein, Kromdraai, Makapansgat, and Swartkrans retain fossiliferous breccias, and each is being investigated or has recently been investigated. They were supplemented in 1992 by the discovery of two australopithecine teeth from the long-known Gladysvale Cave site, about 13 km northeast of Sterkfontein (191), and again in 1994 by the recovery of more complete fossils at the previously unstudied Drimolen Cave, roughly 7 km north of Sterkfontein (1215). Drimolen promises to be particularly productive, and yet other sites undoubtedly exist, particularly in the dolomitic limestones of the Krugersdorp region. Together the southern African sites have provided a very large sample of australopithecine bones, comprising more than thirty-two skulls or partial skulls, roughly one hundred jaws or partial jaws, hundreds of isolated teeth, and more than thirty postcranial bones (594, 901, 1073, 2103). Since 1959, however, southern Africa has been largely eclipsed by spectacular discoveries of australopithecine fossils in eastern Africa.

History of Discovery: Eastern Africa

The east African sites now occupy the limelight for two important reasons. First, in eastern Africa the fossils often occur in relatively friable ancient stream or lake-edge deposits rather than in rock-hard cave fills. At some sites, such as Olduvai Gorge, australopithecine fossils can even be excavated using dental picks and brushes in the accepted archeological manner. At the southern African sites, dental picks and brushes have been second to dynamite, pneumatic drills, hammers, and chisels. Second, and at least equally important, many of the east African sites are stratigraphically related to volcanic extrusives (mainly ancient volcanic ash layers or tephra) that can be dated in years by the radiopotassium and fission-track methods. So far no reliable material for absolute dating has been found in the southern African sites, which have been dated mainly by comparing their animal fossils with fossils from dated layers in eastern Africa.

Although a partial australopithecine maxilla was found in 1939 at Laetoli (Garusi) near Olduvai Gorge, northern Tanzania, its significance was obscured by its singularity and by the onset of World War II. It was only in 1959 that Louis and Mary Leakey found the first east African australopithecine fossil that was widely recognized for what it was. The fossil was a skull from Olduvai Gorge, and it was important not only because it extended the known range of the australopithecines but also because it occurred in deposits that were soon dated to about 1.75 my ago by a pioneering application of the radiopotassium method (1324). The 1.75 my estimate was the first indication of the true antiquity of the australopithecines, and it nearly doubled the total time span that most specialists had allowed for human evolution.

Following their 1959 discovery, the Leakeys obtained funds for much more extensive research at Olduvai Gorge, leading to the recovery of many new early hominid fossils, including the first remains of *Homo habilis*, the oldest known species of *Homo* (1325). Equally important, the Olduvai finds stimulated successful searches for australopithecines and very early *Homo* at other sites in eastern Africa (fig. 4.1), beginning with Peninj

(Lake Natron) (1964) and continuing with Chemeron (1965), the lower Omo Valley (1966), East Turkana (Koobi Fora) (1968), Chesowanja (Chemoigut) (1970), Hadar (1973), Laetoli (1974), the Middle Awash (1981), West Turkana (1984), Fejej (1989), and Konso (originally Konso-Gardula) (1991). Several of these sites, including Olduvai, continue to produce key fossils and early artifacts, and many new sites undoubtedly await discovery. So far, eastern Africa has provided only about half as many australopithecine fossils as southern Africa, but many of the east African specimens are remarkably complete, and the associated data, especially the absolute dates, have no parallel in southern Africa.

Geology of the Southern African Australopithecine Sites

The southern African australopithecine sites are all limestone caves in which sands, silts, natural stones, fossils, and sometimes artifacts accumulated in unsorted jumbles or breccias. Calcareous glue precipitated from groundwater commonly hardened the breccias to a cementlike consistency. The sites were discovered mainly after mining for bands of calcitic flowstone (travertine) that occur in and around the breccias. Since the breccias tend to be very hard, the excavators could not employ standard archeological techniques, which are applicable only to much more friable deposits. The cave fills cannot be securely dated by radiometric methods, and their antiquity has been estimated mostly by their faunal contents (compared with faunal assemblages from radiometrically dated sites in eastern Africa). In addition, the fossils, including those of the australopithecines, were introduced to the caves mainly by carnivores or raptors (at Taung) (190), by carnivores or natural pitfall trapping (at Kromdraai) (320, 2345, 2353), by hyenas (at Makapansgat) (1420, 1421), and by large cats (at Sterkfontein and Swartkrans) (320, 329). Strictly speaking, then, the caves can tell us little about australopithecine behavior. However, the abundance of australopithecine or cercopithecoid monkey fossils at three sites (Sterkfontein, Kromdraai, and especially Swartkrans) suggests that the sites offered rocky ledges or other inaccessible sleeping places. Living baboons often seek out such places to reduce the risk of predation, but even if the strategy was successful, it would still produce a bone assemblage in which the sleepers were relatively abundant.

The taxonomy of the South African australopithecines is not completely settled, but most authorities recognize just two species—*Australopithecus africanus* (at Taung, Makapansgat, and Sterkfontein) and *Paranthropus* (or *Australopithecus*) *robustus* (at Swartkrans, Kromdraai, and the new, largely undescribed site of Drimolen). *Australopithecus africanus* is unquestionably more ancient, with an inferred age range of 3 to 2.5 my ago. No South African site unquestionably samples the period between 2.5 and 2.0 my ago, when *Paranthropus robustus* probably appeared. Its documented range is between 1.8 and perhaps 1.2 my ago. Animal fossils and other paleoenvironmental indicators show that woodland contracted and grassland expanded dramatically between 2.5 and 1.8 my ago. The concluding section of this chapter considers what this environmental shift might mean for evolutionary change in the australopithecines.

The essential features of the following site-by-site presentation are summarized in figures 4.4, 4.5, and 4.7 and in the following section on the geologic antiquity and geographic range of the earliest hominids. The site descriptions refer to artifact assemblages using the terms Oldowan and Acheulean, which are defined fully only in the following sections. Readers who do not need or want a detailed introduction to the sites may wish to simply skim the individual sketches.

Taung

The child's skull that is the type specimen of *Australopithecus africanus* remains the only australopithecine fossil ever found at Taung. The cave it lay in was largely quarried away before it could be studied by interested geologists, but it was probably part of an extensive system formed within the oldest of four limestone aprons (tufas) mantling the Gaap Escarpment near Taung (395, 399, 1657, 1665, 2198). Reddish sands, silts, and occasional bones entered the cave from the surface through fissures. The materials were subsequently cemented together by calcite (crystallized calcium carbonate) precipitated from water that periodically permeated the deposits. The origins of the bones remain obscure, but they may have been introduced by leopards or other large cats denning in the cave. Carnivore-damaged fossil baboon skulls have been found in a nearby cave that may have been part of the same system (324). Alternatively or jointly, the accumulator could have been a large bird of prey. Nearly all the Taung animals, including the child, fall within the prey-size range of an eagle, and some of the animal bones exhibit damage that may be from eagle talons or beaks (190).

The geologic age of the Taung skull is difficult to establish, in part because its geologic provenance is so poorly documented. An age of 1 my ago or less has been suggested by calculating the time when geomorphic change first opened the cave (1654, 1657) and by uranium-series dates of 942 ky ago on the limestone apron in which the cave occurred and of 764 ky ago on the next youngest apron (2333). But the assumptions behind the geomorphic dating are highly questionable, and uranium is now known to have entered the aprons after they formed (2202). The dates thus underestimate the true age, perhaps by a substantial amount. The Taung fauna contains relatively few species that have been dated elsewhere, and its stratigraphic association with the skull is unclear, but it includes cercopithecoid monkey fossils that resemble east African ones dated to about 2.3 my ago (640, 642). The Taung skull itself has figured in the dating controversy, since at an age of 1 my ago or less, it would be the latest known *Australopithecus africanus*, perhaps by more than 1 my. This raises the possibility that it actually derives from the robust australopithecine species *Paranthropus robustus*, which, on evidence from Swartkrans (presented below), may have survived to 1 my ago or later. However, a detailed analysis of the Taung dentition (896) unequivocally allies it with Sterkfontein and Makapansgat fossils of *A. africanus* that surely antedate 2 my ago. The age of the Taung skull will always be speculative, but applying a model of how caves formed within the limestone apron to the known or probable historical positions of the child's skull and key cercopithecoid fossils suggests that the skull actually antedates the cercopithecoids (1517, 1518). In this case, the skull could fall within the range of 2.8 to 2.3 my ago that associated fauna implies for *A. africanus* at the Sterkfontein site, discussed immediately below.

Sterkfontein

At Sterkfontein, silts, sands, and other materials fell or were washed through shafts descending from the surface into a solution cavern within the local dolomitic limestone. Once inside, the foreign debris, together with fragments of the cavern roof and walls, were cemented to a rock-hard breccia by calcite, which also occurs as pure interbeds and lenses within the breccia. Subsequently, erosion removed most of the cavern roof, exposing the breccia at the surface. In places, subsequent decalcification has transformed the breccia into a more friable mix of sand, rocks, and bones.

On average, the Sterkfontein cave fill (or Sterkfontein Formation) comprises about 20 m of deposit, divided among six successive members (1655). These are numbered 1–6 from oldest to youngest, but they are not neatly stacked. The contacts are often complex, partly because deposition often occurred at an angle on talus slopes, partly because sediments sometimes subsided into younger, lower-lying solution cavities, and partly because solution pockets and cracks, filled with later material, sometimes penetrated the members (506, 1660). The result is that younger sediments can lie immediately alongside or even under older ones.

Dating has proceeded mainly by faunal comparisons with eastern Africa. Only Members 4–6 have provided adequate numbers of fossils, and unfortunately many of these were collected before the members were defined. Their original stratigraphic provenance is thus uncertain. This complication aside, however, faunal dating suggests that Member 4 (the "type site") probably accumulated sometime between 2.8 and 2.3 my ago, perhaps mainly about 2.5 my ago; that Member 5 (the "extension site") probably formed between 2 and 1.4 my ago, perhaps in discontinuous bursts; and that Member 6 accumulated much later, after 200 ky ago (fig. 4.4) (506, 642, 1657, 1659, 2346, 2347). The faunal estimate for Member 4 is broadly corroborated by an average ESR age of 2.1 ± 0.5 my ago on enamel from eight bovid teeth (1893) and by paleomagnetism (1156). Members 2–4 are characterized by normal or intermediate polarity, which suggests that they formed during the Gauss Normal Polarity Chron, between 3.4 and 2.48 my ago.

Hominid fossils have been found only in Members 2, 4, and 5 (506, 509, 2196), and stone artifacts occur only in Member 5 (1284, 1285, 1796, 2050). The hominid specimens from Member 2 comprise four articulating bones of the ankle and big toe (designated Little Foot) that are said to exhibit both human and apelike features (509, 678). The most notable apelike feature is the divergence of the big toe. The specimens represent *Australopithecus africanus* or its immediate ancestor, and they could be more than 3 my old (1519).

Member 4 has provided approximately 650 specimens that come either exclusively from *Australopithecus africanus* or mostly from *A. africanus* and secondarily from another, larger-toothed species that was ancestral to *Paranthropus* (504, 506). Member 5 has produced roughly 20 specimens from *Homo habilis, H. ergaster/erectus,* or both and three teeth that are probably from *Paranthropus robustus.* The Member 5 artifacts have been divided between an Oldowan assemblage (2,800 pieces with no bifaces), putatively dated between 2 and 1.7 my ago, and an Early Acheulean assemblage (635 pieces including bifaces) that postdates 1.7 my ago (1284–1286). Much later Middle Stone Age artifacts rest on top of Member 5 but have never been found in place within it.

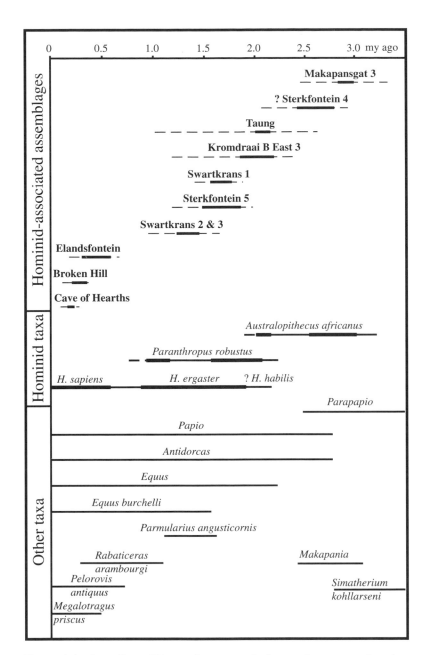

Figure 4.4. Ages (in millions of years ago) of some important South African hominid fossil sites, based on their faunal contents (modified after 2346, p. 741). The numbers following site names are stratigraphic units (members). Also shown are the known time ranges of some important mammalian taxa used in the dating, as well as the inferred time ranges for fossil hominid species. (*Parapapio* and *Papio* are baboons, *Equus* and *Equus burchelli* are zebras, and the remaining species are buffalo or antelopes.)

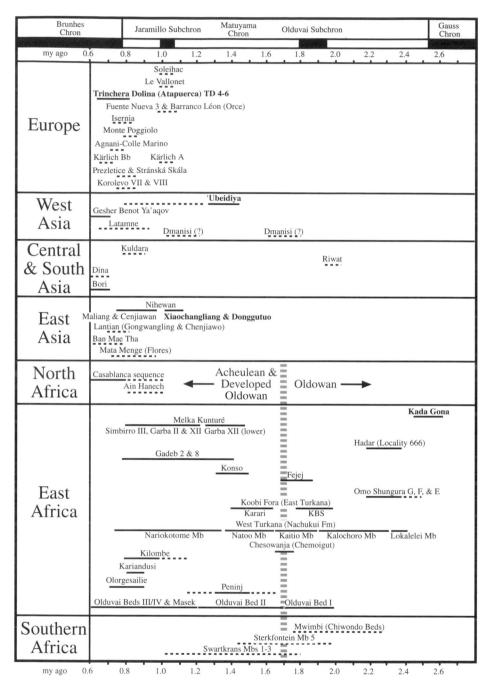

Figure 4.5. Dating of the earliest artifact industries in Africa and Eurasia (in a format suggested by 1119, fig. 13.2). Dotted lines indicate possible or probable dates based mainly on geologic inference or faunal correlations. Boldface marks the oldest secure sites on each continent. The best evidence for people in Eurasia before 1 my ago comes from ʿUbeidiya in Israel, only marginally outside Africa. On the known record, people colonized the farther reaches of Asia and Europe only after 1 my ago.

Antelope species indicate that the environs of Sterkfontein changed from relatively moist and wooded in Member 4 times to much grassier and drier afterward (1516, 2341, 2342, 2344, 2349). Fossil wood fragments imply that large tropical or subtropical trees grew nearby when Members 2 and 4 accumulated.

Swartkrans

The geology of Swartkrans is better understood than that of any other southern African australopithecine site, thanks to the insight and persistence of C. K. Brain (319–321, 323, 325, 327, 328, 330; also 396). Like nearby Sterkfontein, Swartkrans began as a subterranean dolomite cavern into which sediments fell or were washed through a shaft descending from the surface. The accumulating fill or breccia was cemented by calcite and later was exposed at the surface when erosion removed most of the cavern roof (fig. 4.6). The stratigraphy is complex, because later materials sometimes filled erosion or solution hollows within earlier breccia, and objects in close proximity horizontally may thus differ greatly in age.

The Swartkrans cave fill (or Swartkrans Formation) comprises five members, numbered 1 to 5 from bottom to top. Member 1 includes two distinct parts—the Lower Bank below and the Hanging Remnant above that were separated when erosion removed the intervening deposit. Member 2 subsequently filled the void, and its presence both below and above Member 1 illustrates the stratigraphic complexity that Brain had to unravel. Based on faunal comparisons with eastern Africa (2346), Member 1 probably accumulated sometime between 1.8 and 1.5 my ago (fig. 4.4). Member 2 is more difficult to date, because the available fauna almost certainly contains many elements inadvertently mixed in from Member 3, which fills a gully incised through Member 2 into the Lower Bank of Member 1. Both Members 2 and 3 appear broadly similar to Member 1 in fauna and artifacts, however, and they may have formed immediately afterward, between perhaps 1.5 and 1 my ago. Member 4 contains Middle Stone Age artifacts, suggesting that it postdates 250 ky ago. Bones from Member 5 have been radiocarbon-dated to about 11 ky ago.

Members 1–3 have provided craniodental fragments from more than one hundred individuals of a robust australopithecine (320, 901, 2397). The first specimens were assigned to the species *Paranthropus crassidens* (365), distinct from the previously known *Paranthropus robustus* at nearby Kromdraai. Some specialists continue to argue for two species (900, 1073),

Figure 4.6. Three stages in the evolution of the Swartkrans australopithecine cave (modified after 328, pp. 31–32). In the first stage, outside sediment has just begun to funnel down a recently formed shaft. In the second, erosion has partly removed earlier sediment, and fresh material has filled the erosional gap. In the third stage, erosion has totally removed the roof, and the complex fill is exposed to the elements.

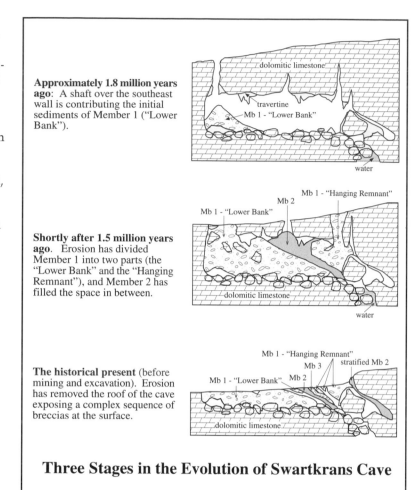

Approximately 1.8 million years ago: A shaft over the southeast wall is contributing the initial sediments of Member 1 ("Lower Bank").

Shortly after 1.5 million years ago. Erosion has divided Member 1 into two parts (the "Lower Bank" and the "Hanging Remnant"), and Member 2 has filled the space in between.

The historical present (before mining and excavation). Erosion has removed the roof of the cave exposing a complex sequence of breccias at the surface.

Three Stages in the Evolution of Swartkrans Cave

but most now lump the Swartkrans fossils into *Paranthropus* (or *Australopithecus*) *robustus*. Most come from Member 1, and erosion of Member 1 could have introduced some fragments into Members 2 and 3. The specimens that are in situ in Member 3 could represent the youngest robust australopithecines on record, if they were deposited near 1 my ago. Members 1 and 2 have also provided craniodental elements from more than seventeen individuals of *Homo ergaster* or *H. erectus*, including a partial skull (SK 847/45) originally assigned to *"Telanthropus capensis."* All three members have furnished postcranial elements that probably come mostly from *P. robustus* (2103). Assignment to *Homo* is especially improbable for Member 1, where 97% of the diagnostic craniodental fragments represented *P. robustus*.

Members 1–3 have yielded more than 875 flaked stone artifacts (490) and 77 bone fragments that appear polished or worn from use (331). The stone artifacts in situ do not include

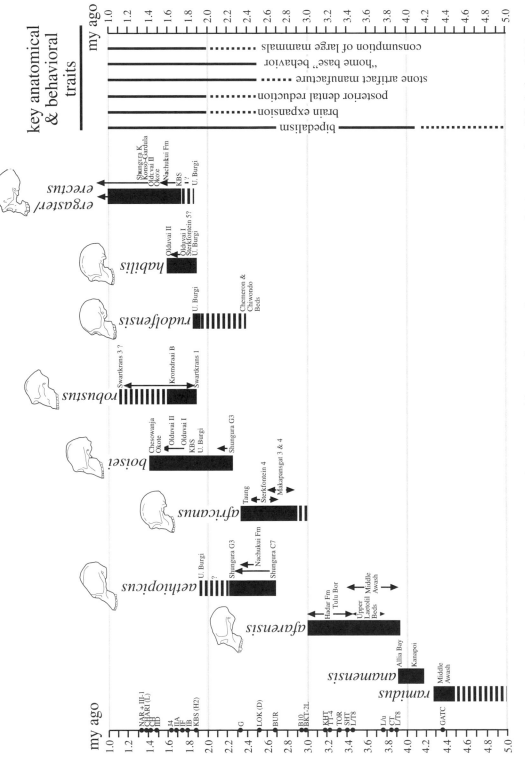

Figure 4.7. *Left*: Time spans of the most commonly recognized hominid species between 4.4 and 1.0 my ago (modified after 1219, p. 428). *Right*: Dating of some key behavioral and anatomical traits (adapted from 959, fig. 1). Broken lines imply uncertain or insecure records. Circles and associated abbreviations on the time scale indicate especially well-dated east African volcanic layers. Names alongside the species bars designate sites or units within sites that have provided key fossils.

any unequivocal bifaces, and they have been assigned to the Developed Oldowan "Culture." Member 4 contains Middle Stone Age artifacts, while Member 5 is artifactually sterile. Occasional burned bones first appear in Member 3, perhaps indicating incipient human control over fire (326, 332, 1951). The burned bones are not locally concentrated, however, and there is no evidence for hearths.

The antelopes, monkeys, and other mammals represented in Members 1–3 imply relatively dry, grassy conditions like those indicated for the broadly contemporaneous deposits of Sterkfontein Member 5.

Kromdraai

Kromdraai comprises two adjacent fossil caves, Kromdraai B, the "ape-man site," which has provided both australopithecine and animal bones, and Kromdraai A, the "faunal site," which has provided animal bones and artifacts but no human fossils. A and B may have been part of a single cavern system into which sediments fell or were washed through shafts descending from the surface, as at Sterkfontein and Swartkrans. Subsequent cementing of the cavern fills by calcite and removal of the roof(s) by erosion exposed the hardened cave fills (breccias) at the surface. The geology of Kromdraai B is better known, thanks to a systematic study (1657, 1658) associated with excavations between 1977 and 1980 (2345, 2353).

Kromdraai B comprises two spatially distinct areas, called East and West. To date, hominid fossils have been found only at Kromdraai B East, where they occur in Member 3 within a sequence of five members comprising the Kromdraai B East cave fill (or Kromdraai B East Formation). The fossils comprise the type (and arguably the oldest known) specimens of *Paranthropus robustus* (364) and represent a minimum of eight individuals. Member 3 has not yet provided artifacts, but these occur in one or both of the overlying members (4 and 5). The fauna is not especially useful in dating, but the morphology of the hominid fossils (895) suggests that Member 3 accumulated about 2 my ago (2346) (fig. 4.4). It could thus fall between Sterkfontein Member 4 and Swartkrans Member 1. An age of 2 my ago is consistent with the mainly reversed polarity of Kromdraai B sediments, which places them within the Matuyama Reversed Chron, between 2.48 and 0.78 my ago (1156). The Member 3 animal bones come mainly from cercopithecoid monkeys, including leaf-eating (colobine) forms whose presence suggests relatively wooded and moist surroundings.

The artifacts from Kromdraai A include seventy-nine flaked stones and twenty-one unmodified pebbles or rock fragments that only people could have introduced (1288). The associated animal bones come mainly from ungulates that indicate relatively dry, grassy conditions. Since the composite sequences from Sterkfontein and Swartkrans suggest increasing aridity over the period when the Kromdraai deposits formed, the fauna may imply that the Kromdraai A artifacts postdate the Kromdraai B *Paranthropus* remains. The Kromdraai A assemblage includes very large flakes (> 10 cm in maximum length) and other pieces that broadly recall early Acheulean artifacts from Sterkfontein Member 5, and it may date from the same period, between roughly 1.7 and 1 my ago.

Gladysvale and Drimolen

Gladysvale and Drimolen are remnants of cave systems with breccias like those at nearby Sterkfontein. Swartkrans, and Kromdraai, but few details are available. At Gladysvale, blocks of breccia from mining dumps have provided an extensive mammalian fauna and two isolated hominid teeth (189, 191, 1693). The teeth have been tentatively assigned to *Australopithecus africanus*, and the associated fauna suggests they may be between 2.5 and 1.7 my old. However, the fauna comprises a peculiar mix of species that elsewhere do not overlap in time, and if the teeth were in a breccia unit that contained only the older species, they could be even older than 2.5 my. Breccia at Drimolen has produced a skull, partial dentitions, and isolated teeth of *Paranthropus robustus* (1215). No stone artifacts have been reported, but there are possible bone artifacts like those at Swartkrans. The associated fauna suggests an age comparable to that of Swartkrans Member 1, between roughly 1.8 and 1.5 my ago.

Makapansgat

Whereas the Krugersdorp australopithecine caves (Sterkfontein, Swartkrans, Kromdraai, Gladysvale, and Drimolen) were probably subterranean receptacles linked to a gently undulating surface by strongly sloping or even near-vertical shafts, the Makapansgat Limeworks cave was probably a tunnel-like cavern with a nearly horizontal entrance from the flank of a steep-sided valley (1420). Relative to gravity, water flow most likely played a greater depositional role at Makapansgat than at the Krugersdorp sites, and some of the older deposits probably accumulated under standing water. It is also possible that animals,

especially hyenas, regularly penetrated deeper into Makapansgat than into the Krugersdorp caves. As in the Krugersdorp sites,
however, calcium carbonate, precipitated from groundwater, in-
durated the Makapansgat sediments after they were deposited.

The Makapansgat cave fill (or Makapansgat Formation) has
been divided among five members, numbered 1 to 5 from bottom
to top (1656, 1657). It is possible, however, that this scheme seriously oversimplifies much greater stratigraphic complexity
and that supposedly successive members consist in part of
broadly contemporaneous depositional variants or facies (1420).
Member 3 (the "gray breccia") has provided roughly twenty-
five hominid fossils, and Member 4 has produced at least four
(1746). Only *Australopithecus africanus* is represented. No
definite stone artifacts have been uncovered, but some believe
that *A. africanus* modified and used many of the animal bones
in Member 3 (575, 576).

Faunal comparisons with eastern Africa imply that Member 3 formed about 3 my ago, and Member 4 is probably only
slightly younger (fig. 4.4) (1420, 1657, 2346). Makapansgat is
the southern African australopithecine site most suitable for
paleomagnetic dating, because the lower members were laid
down partly underwater and the sediments were not heavily
disturbed afterward (1742). The available paleomagnetic readings support an age near 3 my ago for Member 3 (357, 1498,
1499, 1659). Makapansgat is thus the oldest known *Australopithecus africanus* site.

Geology of the East African Australopithecine Sites

In contrast to the southern African australopithecine sites, the
east African ones include no caves. Instead, they all occur in
ancient stream or lake deposits that have been exposed by recent gullying. They are closely associated with the eastern
branch of the Great Rift Valley (fig. 4.1), where tension between
massive continental plates that began more than 20 my ago
(1006) forced a strip of land more than 2,000 km long and 40–
80 km wide to fall with respect to its sides. Associated tectonic
activity has repeatedly created and destroyed natural dams, al-
ternately creating and draining lakes and redirecting streams.
The same tectonic activity frequently faulted or tilted preexisting lake and river deposits, and earth movement, often sparse
vegetation, and episodically violent rainfall have encouraged
erosion, which has favored fossil exposure and discovery. Vulcanism, also linked to earth movement, has persistently contributed lavas and ashes (tuffs) that can be used to date enclosing
lake and river deposits. In many places flowing water displaced

fossils from their original resting places even before burial, and most were discovered only after they had eroded from their burial sites. In some places, however, fossils and artifacts occur in near primary position, in ancient "living sites." This is especially true at Olduvai Gorge, Koobi Fora (East Turkana), and the Lokalalei site (West Turkana), as discussed below. The western branch of the Rift Valley has been much less productive, in large part because associated deposits are less richly fossiliferous and the fossils are less abundantly exposed.

As with the previous section on southern African sites, a reader who does not need or want detailed information on specific east African sites may prefer to skim the individual sketches, attending mostly to the accompanying figures. The principal conclusions are summarized in figures 4.5 and 4.7, as well as in the text sections that follow this one.

Olduvai Gorge

Olduvai Gorge is a usually dry valley on the western margin of the Eastern Rift Valley in northern Tanzania. It was first investigated in 1913 by Hans Reck (1744), who reported fossils that later attracted the attention of Louis Leakey. In 1931 Leakey, Reck, and Hopwood returned for further work, and in 1935 Louis and Mary Leakey initiated regular visits that recovered large samples of artifacts and animal bones and culminated in the discovery of the famous "*Zinjanthropus*" australopithecine skull in 1959. Between 1960 and 1973, Mary Leakey excavated many early archeological sites at Olduvai, some of which provided additional human fossils (1327, 1328, 1338). The Leakeys' research revolutionized paleoanthropology, not only because it provided key fossils and artifacts at Olduvai, but because it encouraged others to tap the great paleoanthropological potential of other east African sites. Olduvai itself is not exhausted, and specialists following in the Leakeys' footsteps have discovered important human fossils, animal fossils, and artifacts (246, 1149).

The gorge comprises two usually dry branches, known as the Main Gorge and the Side Gorge (fig. 4.8). They have a combined length of about 50 km, and they expose up to 100 m of lacustrine, fluvial, and eolian deposits within a shallow basin. Important sites often occur where gullies cut into the sidewalls of the gorge, and the sites are often named for a combination of their discoverer and the Swahili word for gully (Korongo), as in "Frida Leakey Korongo," abbreviated FLK. The deposits are divided among seven successive mappable stratigraphic units known from bottom to top as Bed I, Bed II, Bed III, Bed IV, the

Masek Beds, the Ndutu Beds, and the Naisiusiu Beds (fig. 4.9) (980, 982, 1330, 1331, 1336). Volcanic tuffs at various levels within the sequence aid in horizontal correlation and mapping, which has been complicated by extensive faulting. K/Ar dating, ^{40}Ar/^{39}Ar dating, paleomagnetism, and ^{14}C dating fix the Olduvai sequence between 2.03 my ago and 15 ky ago (696, 1324, 2135, 2380).

During the accumulation of Bed I, the Olduvai Basin contained a broad, shallow saline and alkaline lake that fluctuated up to 3 m in depth and varied between 7 and 15 km in diameter. Most archeological and paleontological sites occur along the southeastern lakeshore, where freshwater streams drained into the lake from nearby volcanic uplands. K/Ar and ^{40}Ar/^{39}Ar dates indicate that volcanic deposits at the base of Bed I formed about 2.03 my ago but that most of the sedimentary deposits, including all the fossiliferous sites, accumulated very rapidly between about 1.86 and 1.75 my ago.

Lava flows and associated sediments at the base of Bed I exhibit reverse magnetism, but the overlying deposits into the lower part of Bed II show normal magnetism. The normally magnetized deposits define the Olduvai Normal Subchron, which interrupted the Matuyama Reversed Chron between about 1.95 and 1.77 my ago.

The lake persisted during the deposition of Lower Bed II, roughly until the accumulation of a sequence of eolian tuffs known as the Lemuta Member. A major disconformity above the Lemuta Member marks the beginning of widespread faulting and folding in the central part of the basin, and as a result the lake progressively shrank, disappearing entirely shortly before the end of Bed II deposition. The disconformity also marks an abrupt faunal change at Olduvai, after which some archaic large mammals represented in Bed I and Lower Bed II no longer occur and several new species make their first appearance. Another disconformity separates Bed II from overlying Bed III and is believed to reflect a major phase of Rift Valley faulting that has been radiopotassium-dated elsewhere to 1.20–1.15 my ago. Thus Bed II is believed to have formed between about 1.75 and 1.20 my ago. The top of the Olduvai Normal Subchron occurs just below the Lemuta Member, and the disconformity above it that separates Lower Bed II from Upper (or Middle and Upper) Bed II must thus date to about 1.7–1.6 my ago.

The widespread faulting that produced the disconformity between Bed II and Bed III led to substantial erosion of Bed II. The Olduvai Basin was transformed into an alluvial plain, and Beds III and IV were primarily stream laid. They are readily distinguished only on the eastern side of the basin.

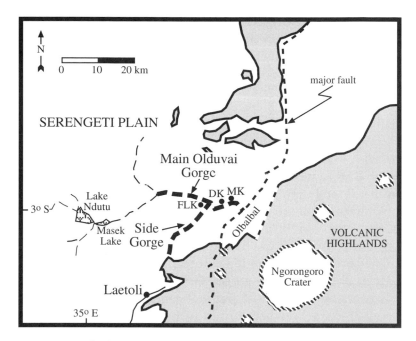

Figure 4.8. The location of Olduvai Gorge on the Serengeti Plain
(redrawn after 982, fig. 1).

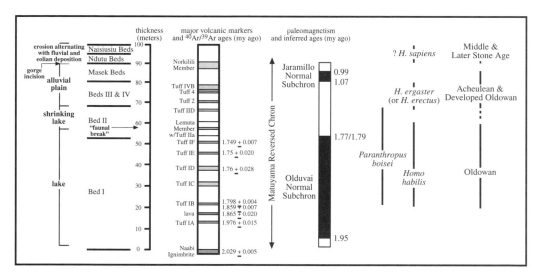

Figure 4.9. Schematic stratigraphy of Olduvai Gorge, showing the principal beds, their depositional environment, their paleomagnetism, the main volcanic stratigraphic markers, their
^{40}Ar/^{39}Ar ages, and the local stratigraphic extent of hominid taxa and artifact industries
(modified after 1336, fig. 2, and 2135, fig. 6).

The Masek Beds are mainly fluvial deposits similar to those of Beds III and IV and were the last sediments to accumulate before major incision of the gorge. An early paleomagnetic study suggested that Bed IV contained the boundary between the Matuyama Reversed Chron and the Brunhes Normal Chron at 780 ky ago. Fresh research suggests that Bed IV sediments are reversed more or less throughout, that the lower part of the Masek Beds is reversed, and that the upper part is again normal (2135). It seems most likely that the lower part of the Masek Beds formed during the Jaramillo Normal Subchron between 1.07 and 0.99 my ago, in which case the entire underlying Olduvai sequence formed within the Matuyama Reversed Chron.

The Ndutu Beds are fluvial and eolian deposits that accumulated over a lengthy period of intermittent faulting, erosion, and partial filling of the gorge. The upper unit of the Ndutu Beds contains eolian tuffs that are mineralogically similar to but much more weathered than tuffs in the Naisiusiu Beds, whose age has been established by the radiocarbon method. If a constant weathering rate is assumed, the upper unit of the Ndutu Beds seems to have a mean age of 75 ky, from which it has been suggested that the Lower Ndutu Beds accumulated between roughly 400 ky ago (the top of the Masek Beds) and 75 ky ago.

Finally, the Naisiusiu Beds are mainly eolian tuffs deposited after the Upper Ndutu Beds had been severely eroded and after the gorge had been cut to its present level. Radiocarbon dates fix the Naisiusiu Beds between about 22 ky and 15 ky ago.

Past climates at Olduvai can be inferred from the proportionate representation of windblown sediments (980), the stable oxygen- and carbon-isotope composition of soil carbonates (439), pollen (265), and vertebrate remains (1176). Overall, the vicinity of the gorge became drier from 2 my ago to the present, but the trend was interrupted by wet periods that may correspond to Northern Hemisphere interglaciations (1177). The upper part of Bed I (between Tuffs IB and IF in fig. 4.9), which probably accumulated in less than 20 ky, records a dramatic shift from wetter to drier and back to wetter (2380). The abruptness of the change recalls equally rapid shifts that are recorded in late Pleistocene and Holocene deposits in eastern Africa.

Fossils of a robust australopithecine (*Paranthropus boisei*) have been found in Bed I, Lower Bed II, and Upper Bed II, while fossils assigned to *Homo habilis* occur in Bed I and Lower Bed II. Fossils of *H. ergaster* (or *H. erectus*) have been found in Upper Bed II, Bed III, and Bed IV, while very fragmentary fossils that may derive from early *H. sapiens* are known from the Masek and Lower Ndutu Beds. A skull of early *H. sapiens* from Lake Ndutu, at the headwaters of Olduvai Gorge, came from deposits that are probably coeval with the Upper (Norkilili) Member of the Masek Beds (1336). Stone artifacts occur throughout the Olduvai sequence: Oldowan in Bed I and Lower Bed II; Developed Oldowan and Acheulean in Upper Bed II to Bed IV; Acheulean in the Masek Beds; Middle Stone Age near the top of the Lower Ndutu Beds and in the Upper Ndutu Beds; and Later Stone Age in the Naisiusiu Beds. The Lake Ndutu skull, from deposits that may correlate with the Upper Masek Beds, was associated with Acheulean artifacts.

Laetoli

The Laetoli locality is approximately 45 km south of Olduvai Gorge in northern Tanzania (fig. 4.8). Technically, it was the first site in eastern Africa to provide australopithecine fossils. These were a lower canine found in 1935 and a small fragment of a right maxilla, with premolars and an isolated third molar, found in 1939. However, the canine was initially misidentified as cercopithecoid (2448), and the significance of the maxillary fragment and molar was appreciated only much later, after more tightly controlled and abundant discoveries at the site between 1974 and 1979 (1332). These comprised teeth, jaws, and a very fragmentary immature skeleton from a maximum of twenty-three australopithecine individuals (1334). Most were found on the surface, but they clearly derive from the upper part of the Laetolil Beds (also known as the Garusi or Vogel River Series), which consist mainly of eolian and air-fall volcanic tuffs covering an area of about 70 square km in the Laetoli region (956, 981, 1331, 1336, 1337).

Based on their find spots, the individual Laetoli australopithecine fossils can be at least broadly related to eight widespread marker tuffs, numbered 1–8 from bottom to top. Most of the specimens came from between Tuff 3 and Tuff 8, and in age they are bracketed between radiopotassium estimates of 3.76 my ago from below Tuff 1 and 3.46 my ago for Tuff 8 (fig. 4.10) (673). One specimen (a mandible designated Laetoli Hominid 4) is the holotype (type fossil) for *Australopithecus afarensis* (1152), to which the remaining fossils have also been assigned. *A. afarensis* was probably responsible for some spectacular human footprints preserved on a paleosurface within Tuff 7 (the "footprint tuff") (983, 1333, 1335). Surfaces within the tuff also preserve abundant prints from a wide variety of animals.

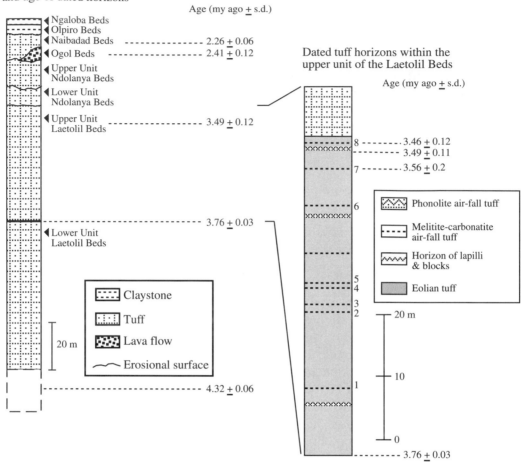

Figure 4.10. *Left:* Schematic stratigraphic column for the Laetoli area, showing the position and age of dated horizons. *Right:* Dated tuff horizons within the upper unit of the Laetolil Beds (redrawn after 673, figs. 2.10, 2.11).

In addition to australopithecine fossils, the Laetoli locality has provided an archaic *Homo sapiens* skull (Laetoli Hominid 18) and a massive, heavily rolled mandible (LH 29) tentatively assigned to *Homo erectus* (or *H. ergaster* as used here) (1334). The skull came from the Upper Ngaloba Beds, which cap the local stratigraphic sequence, and the mandible probably came from the somewhat older Lower Ngaloba Beds. A giraffe vertebra from the same level as the skull has been dated to 129 ± 4 ky ago by ^{230}Th and to 108 ± 30 ky ago by ^{231}Pa (981). These dates are tentative, but they support a previous age estimate for the skull. This was based on the skull's position just above a tuff that is mineralogically very similar to one near the top of the Lower Ndutu Beds at Olduvai. Based on both its stratigraphic position and the extrapolated sedimentation rates at Olduvai, the correlated Ndutu tuff appears to have an age of 120 ± 30 ky.

No stone artifacts have been found in the Laetolil Beds. However, Acheulean or Developed Oldowan bifaces and other artifacts occur in the Olpiro Beds (961), which unconformably overlie the Laetolil Beds and may have formed about the same time as Olduvai Bed II, after 1.7 my ago. The Ngaloba Beds, unconformably overlying the Olpiro Beds, contain Middle Stone Age artifacts, some directly associated with the Ngaloba skull.

Peninj

The Peninj site is west of Lake Natron, approximately 80 km northeast of Olduvai Gorge in northern Tanzania. The fossiliferous sediments at Peninj accumulated mostly in and around a lake within a broad, relatively shallow basin (1112). The deposits have been divided into two major units—the Humbu Formation (older) and the Moinik Formation (younger). Paleomagnetic readings and radiopotassium dates on a basalt near the base of the Moinik Formation and on a basalt within the Humbu Formation suggest that the Humbu Formation probably dates from between 1.6 and 1 my ago (1120). This age is consistent with the Humbu Formation fauna, which resembles that from Upper Bed II at Olduvai.

The mandible of a robust australopithecine was found in Humbu Formation deposits that are probably just older than 1.6 my, and Acheulean artifacts occur in deposits that are probably only slightly younger.

Tugen Hills, Baringo (Chemoigut and Chesowanja)

The Tugen Hills, west of Lake Baringo, central Kenya, contain more than 3,000 m of fluviatile and lacustrine sediments spanning the interval between roughly 16 my ago (mid-Miocene) and near present (Holocene) (234–236, 1023, 1024). The sequence has been divided among twelve successive geologic formations, including seven that have provided fossils, artifacts, or both. Of these, only the Chemeron and Chemoigut Formations are relevant here, but four others deserve mention. These are (1) the Muruyur Beds, dated to about 16–15 my ago, which have furnished nearly fifty specimens of large hominoids, including teeth like those of *Kenyapithecus*; (2) the Ngorora Formation, bracketed between roughly 13.2 and 8.5 my ago, which has produced four teeth from at least two large hominoid taxa, including what may be the youngest known representative of *Proconsul*; (3) the Lukeino Formation, bracketed between 6.3 and 5.6 my ago, which has delivered a lower molar that could represent the common ancestor of people and chimpanzees (1025); and (4) the Kapthurin Formation, which formed between 580 and perhaps 230 ky ago (557, 2134) and has provided two partial mandibles, an isolated incisor, and several postcranial bones that represent a robust, extinct variety of *Homo* (2013, 2503), perhaps early *Homo sapiens* as this term is used here. The Kapthurin sediments also contain numerous Acheulean artifacts.

The Chemeron Formation consists of fluviolacustrine deposits divided among five members, numbered 1 to 5 from older to younger. Member 1, or correlative deposits, has provided a hominoid proximal humerus, possibly from an australopithecine, and a fragmentary mandible at the Tabarin locality that has been referred to *Australopithecus afarensis* (1022, 1024, 2387). Radiopotassium dates and faunal correlations bracket the Tabarin site between 5.25 and 4.15 my ago, and the mandible is thus older than unequivocal specimens of *A. afarensis* discussed below. It is broadly coeval with specimens of *Ardipithecus ramidus*, the oldest well-documented hominid species, to which it may ultimately be reassigned. The proximal humerus may be equally old (1684), but it is not reliably dated (1024).

Deposits that probably belong to Member 4 of the Chemeron Formation at site JM85 have provided a temporal fragment of early *Homo* that is dated by ^{40}Ar/^{39}Ar and associated fauna to about 2.4 my ago (718, 1027). If the age has been correctly assessed, the temporal may be added to a mandible from the Chiwondo Beds, Malawi, a maxilla from the Kada Hadar Member, Ethiopia, and isolated teeth from Shungura Formation Members E–G, Ethiopia, which together suggest that *Homo* emerged 2.5 to 2.4 my ago. A significant gap exists between *Homo* at 2.4 to 2.3 my ago and the next oldest indisputable fossils of *Homo*, dated to 1.9 to 1.8 my ago at Koobi Fora, northern Kenya, and Olduvai Gorge, Tanzania.

The Chemoigut Formation, which overlies the Chemeron Formation, has provided two fragmentary skulls and some fragmentary teeth attributed to a robust australopithecine (*Paranthropus boisei*) (237, 238, 427, 876). Faunal contents bracket the Chemoigut Formation between 2 and 1.5 my ago, in keeping with a radiopotassium date of approximately 1.42 my ago for the Chesowanja Basalt that immediately overlies it (1060). Artifacts from several localities within the Chemoigut Formation have been assigned to the Chemoigut Industry, which is closely similar to the Oldowan of Bed I and Lower Bed II, Olduvai (876). At one site, artifacts were associated with baked clay fragments that may represent the oldest traces of humanly controlled or produced fire (495). However, natural baking—for example, below a smoldering tree stump—cannot be ruled out.

The Chesowanja Formation, which unconformably overlies the Chemoigut Formation, has provided no hominid remains, but it does contain numerous stone artifacts. These have been assigned to the Losokweta Industry, which may be regarded as a variant of the Developed Oldowan or of the Acheulean, though it lacks bifaces (876, 960). Bifaces that occur on the surface of the Chesowanja deposits apparently represent a post-Chesowanja Acheulean industry.

Lothagam and Kanapoi

Lothagam and Kanapoi are approximately 75 km apart, in the drainage of the Kerio River southwest of Lake Turkana in northern Kenya (fig. 4.11) (175). At both sites, the sequence comprises lacustrine sediments that originated when Lake Turkana was much larger and fluvial sediments that accumulated when the lake had substantially shrunk. At Lothagam, fluvial deposits have supplied a hominoid mandible fragment comprising a portion of the right horizontal ramus with a heavily worn M_1 and with the roots for P_4, M_2, and M_3 (1662). There are no tightly constraining radiometric dates, but the associated fauna suggests an age of about 5.6 my ago (1026). The mandible shares a molar root pattern and other features with *Australopithecus afarensis* (1023, 1025, 1026), but the relative thinness of the enamel recalls *Ardipithecus ramidus*, which is also closer in geologic age. The specimen is probably too incomplete for unambiguous taxonomic assignment (2452), but it may represent the oldest known hominid.

At Kanapoi, the sequence consists of a basal fluvial unit overlain by lacustrine deposits covered in turn by fluvial sediments (1340). A deltaic facies (variant) of the lake deposits provided a hominid distal humerus in 1965 (1663) and more than eight other fragmentary hominid specimens (mainly partial dentitions and isolated teeth) after 1994 (1340, 2384). Tuff units within the basal fluvial unit have been fixed by $^{40}Ar/^{39}Ar$ dating to about 4.2 my ago, and the Moiti Tuff that caps the same lake deposits elsewhere in the Lake Turkana Basin has been dated to about 3.9 my ago. The hominid fossils are thus bracketed between 4.2 and 3.9 my ago. They have been assigned to *Australopithecus anamensis*, which has also been identified from 3.9 my old deposits at Allia Bay on the eastern shore of Lake Turkana. At both Kanapoi and Allia Bay, the associated fauna implies an open woodland or bushland environment.

Figure 4.11. The Lake Turkana Basin showing the geological formations (in boldface) that have provided fossils of australopithecines, early *Homo*, or both (redrawn after 370, p. 287). Collectively, localities within the Koobi Fora Formation are often referred to as East Turkana, those within the Nachukui Formation as West Turkana, and those within the Usno and Shungura Formations as Omo or Lower Omo.

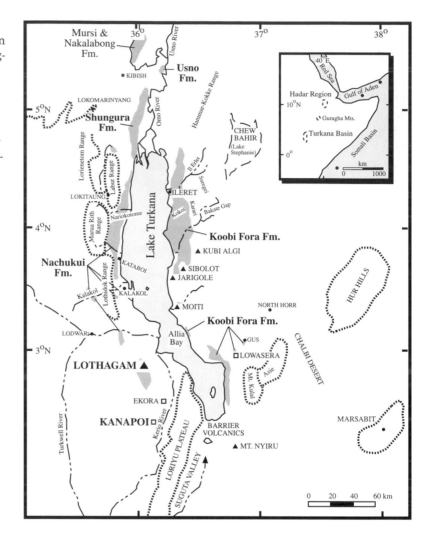

Figure 4.11. The Lake Turkana Basin showing the geological formations (in boldface) that have provided fossils of australopithecines, early *Homo*, or both (redrawn after 370, p. 287). Collectively, localities within the Koobi Fora Formation are often referred to as East Turkana, those within the Nachukui Formation as West Turkana, and those within the Usno and Shungura Formations as Omo or Lower Omo.

Koobi Fora (East Turkana, East Rudolf)

The localities collectively called Koobi Fora are on the east side of Lake Turkana (formerly Lake Rudolf) in northern Kenya (fig. 4.11). The fossiliferous sediments at Koobi Fora were laid down mainly by streams flowing toward Lake Turkana from uplands to the east (2340). Lacustrine sedimentation in the Turkana Basin began shortly before 4 my ago and was probably initiated by the eruption of basalts that disrupted previous drainage patterns (370, 371). Between 4 and 2.5–2 my ago, the basin was sometimes filled by a large lake, but it was more often dominated by an extended Omo (or "Turkana") River that flowed south and east to the Indian Ocean. A lake like the historic one was present more continuously after 2 my ago. Vulcanism and tectonic movement determined the alternation

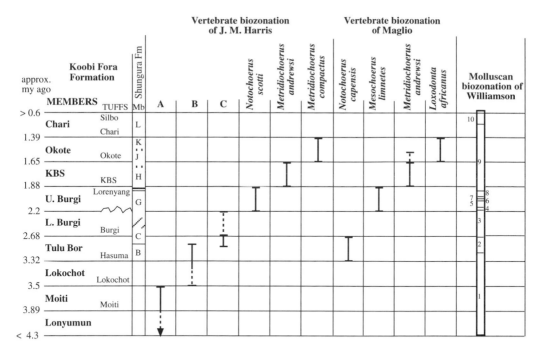

Figure 4.12. Schematic stratigraphy of the Koobi Fora Formation, with correlations to the Shungura Formation of the lower Omo River Valley and to biostratigraphies suggested by Harris, Maglio, and Williamson (modified after 372, fig. 9). *Notochoerus, Mesochoerus,* and *Metridiochoerus* are genera of extinct pigs. *Loxodonta africana* is the living African elephant.

between lake and river, but when a lake was present, climatic factors affected its volume.

More than 130 layers of volcanic tuff (or tephra) have been identified in the Lake Turkana Basin. These provide material for radiopotassium, $^{40}Ar/^{39}Ar$, and fission-track dating, and their geochemical signatures are often unique. They thus allow stratigraphic correlations within the Koobi Fora region and between Koobi Fora and other areas, including the lower Omo River Valley to the north, the West Turkana region to the west, the Kerio Valley to the southwest, the Hadar and Middle Awash regions of east-central Ethiopia, and even deep sea cores from the Gulf of Aden (370, 375, 717).

The Plio-Pleistocene sequence in the Koobi Fora region comprises over 560 m of fluvial, lacustrine, and deltaic deposits that are grouped into the Koobi Fora Formation (369, 372, 717). This has been divided in turn among eight successive members, each named for a tuff at its base (figs. 4.12, 4.13). A combination of K/Ar and $^{40}Ar/39Ar$ dates, paleomagnetic readings, and faunal correlations indicates that the deposits accumulated discontinuously between about 4.34 and 0.7 my ago. In historical perspective, the Koobi Fora Formation provides a highly

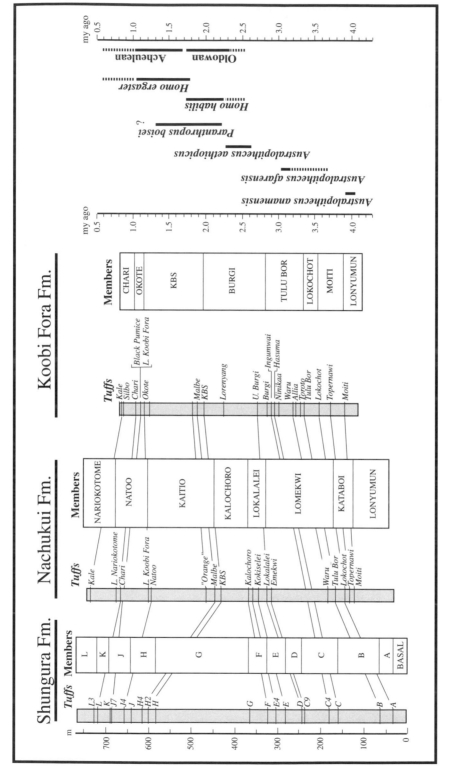

Figure 4.13. Correlation of the Plio-Pleistocene successions in the lower Omo River Valley (Shungura Formation), West Turkana region (Nachukui Formation), and Koobi Fora region (Koobi Fora Formation) (modified after 370, 717). The formations and their constituent members are scaled according to thickness, not according to age. Isotopic dates on tuffs determine the ages indicated on the time scale to the right of the Koobi Fora Formation. Tuffs can be identified in more than one region, based on their distinctive geochemical signatures.

instructive example of how radiometric dates may be checked by faunal remains. Initially, dates of approximately 2.6 my ago for the KBS Tuff implied that various mammal taxa (*Equus*, for example) had appeared at Koobi Fora at least 600 ky before they first appeared in the lower Omo River Basin (Shungura Formation) at the north end of Lake Turkana (374). This seemed unlikely and prompted redating, which eventually showed that the original 2.6 my estimate was about 700 ky too old.

At Koobi Fora proper, fossils of australopithecines and early *Homo* are bracketed between roughly 2.1 and 1.3 my ago. Most come from deposits overlying the KBS Tuff (in what was formerly known as the Upper Member of the Koobi Fora Formation or the Ileret Member at the Ileret Locality) (2368, 2372) (fig. 4.13). The KBS Tuff is now dated at about 1.88 my ago. Fossils of *Paranthropus boisei* have been found mainly in the KBS Member, but they also occur below it and above it, especially in the overlying Okote Member. In sum, they occur in deposits spanning the interval from somewhat before 2 my ago to perhaps 1.4 my ago. Fossils commonly attributed to early *Homo* (*Homo habilis, H. rudolfensis,* or both, including the famous KNM-ER [Kenya National Museum–East Rudolf] 1470 and 1813 skulls) have been recovered in the Upper Burgi Member just below the KBS Tuff, where they probably date to about 2 my ago. Fossils representing *Homo ergaster* or *H. erectus* have been found in the KBS Member just below and also above the Okote Tuff, where they probably range from slightly before 1.7 my ago to perhaps 1.3 my ago. Fossils possibly representing yet another taxon (*A. africanus* or a new, small-brained species of *Homo*) have been found both in and below the KBS Member, where they probably date from before 2 my ago to perhaps 1.7 my ago.

Finally, at Allia Bay, about 40 km south of Koobi Fora, fossils assigned to *Australopithecus anamensis* have been recovered about 5 m below the Moiti Tuff in the Lonyumun Member of the Koobi Fora Formation (519, 1340). The Moiti Tuff is dated to about 3.9 my ago, and the fossils are thought to be only marginally older. Elsewhere, *A. anamensis* has been recognized only at Kanapoi in the southwestern sector of the Lake Turkana Basin. It is the second oldest hominid species and the oldest for which bipedalism has been conclusively shown.

In addition to fossils of australopithecines, Koobi Fora has provided a cranium of near-modern *Homo sapiens* (341). It was a surface find, but it is presumed to have come either from the top of the Chari Member of the Koobi Fora Formation or from the bottom of the overlying (and probably much younger)

Galana Boi Formation. A very robust but still essentially modern human femur was found in similar circumstances nearby (1341, 2230). Uranium-series dates on both specimens suggest an age of up to 300 ky ago (348), but the reliability of such dates on bone depends on the timing of uranium uptake and loss, and this is unknown.

The oldest stone artifacts found at Koobi Fora occur just below and within the KBS Tuff, where they are dated to about 2–1.8 my ago. They have been assigned to the KBS Industry, and they resemble broadly contemporaneous Oldowan artifacts from Bed I at Olduvai Gorge (962, 1121). Somewhat younger artifacts from the Okote Member, dated to roughly 1.4 my ago, have been assigned to the Karari Industry, which is broadly similar to the Developed Oldowan at Olduvai. An early Acheulean site has been excavated in deposits that probably correlate with the Chari Member, slightly postdating 1.4 my ago. However, the early Acheulean is poorly represented at Koobi Fora, particularly compared with its abundance at Konso, roughly 200 km to the northeast (89). The difference may reflect the rarity of suitable raw material for hand ax manufacture at Koobi Fora.

The Lower Omo River Basin

The localities collectively referred to as Lower Omo occur in the lower valley of the Omo River, just north of where it enters Lake Turkana in southwestern Ethiopia (fig. 4.11). The Lower Omo preserves more than 1,000 m of Plio-Pleistocene sediments, deposited in a subsiding basin mainly by the meandering ancestral Omo River (456, 1074). The sequence also records four or five episodes of mainly lacustrine sedimentation under a greatly expanded ancient Lake Turkana. As at Koobi Fora, numerous interbedded tuffs provide material for radiometric dating and also serve as markers for stratigraphic correlations, not only within the Lower Omo region but also between the Lower Omo and other regions, especially Koobi Fora and West Turkana (figs. 4.12, 4.13) (375, 717, 957). As a result of extensive paleontological collecting, preceded by extremely careful and thorough geologic mapping and dating, the Lower Omo has provided the standard Plio-Pleistocene biostratigraphic sequence with which other east African records are routinely compared.

Radiometric dates and complementary paleomagnetic readings show that the Plio-Pleistocene deposits of the Lower Omo span the interval between roughly 4.1 and 0.8 my ago. The sequence includes four main, spatially discrete geologic formations: the Mursi Formation, dated to about 4.1 my ago; the Nkalabong Formation, dated to about 3.95 my ago; the

Usno Formation, bracketed between 4.1 and 2.97 my ago; and the Shungura Formation (the classic "Omo Beds"), with exposures spanning the interval from approximately 3.6 my ago to perhaps 1 my ago. These formations are now grouped with the Koobi Fora Formation and other correlative deposits in the Lake Turkana Basin into the Omo Group (1005). Paleontologically, the Shungura Formation is by far the most productive, and it is the principal source of information on dated Plio-Pleistocene faunal change in eastern Africa. It comprises about 760 m of fluvial, lacustrine, and deltaic sediments that have been divided among twelve members, labeled (from bottom to top) the Basal Member and Members A–L (excluding I, which was skipped). Except for the Basal Member, each member includes a tuff at its base with the same letter designation.

At most localities, Shungura Formation sediments have been deformed by tilting and subsequent faulting, so that their bedding planes intersect the surface at a distinct angle. They are overlain by much younger (late mid-Pleistocene to mid-Holocene) horizontal sediments assigned to the Kibish (or "Omo-Kibish") Formation (397).

Most early hominid fossils from the Lower Omo come from the Shungura Formation (236 specimens), but a small number have also been recovered in the Usno Formation (23 specimens) (fig. 4.14). In contrast to Koobi Fora, where the hominid fossils are mostly younger than 2 my ago, the Lower Omo specimens are mainly between 3 and 2 my ago. The reason for the difference is not clear, but the records are thus more complementary than supplementary. The Lower Omo fossils also tend to be much more fragmentary, because they accumulated mainly in higher-energy (fluvial) environments. The vast majority (225) are only isolated teeth (2112), and this hinders both sorting and diagnosis. However, at least five taxa are recognized at present: (1) *Australopithecus afarensis* from the Usno Formation and Shungura Formation Member B, in deposits that are probably about 3 my old; (2) *Australopithecus aethiopicus* from Shungura Formation Member B and perhaps Members D–G, dated between roughly 2.6 and 2.3 my ago; (3) *Paranthropus boisei* from Shungura Members G–L, spanning the interval from about 2.3 to perhaps 1.2 my ago; (4) *Homo* cf. *habilis* or simply *Homo* sp., based on a relatively small number of fragmentary specimens from Shungura Members E–H and perhaps L, that is, between roughly 2.3 and perhaps 1.3 my ago; and (4) *H. ergaster* (or *H. erectus*) in Shungura Member K, that is, about 1.4 my ago.

The oldest well-dated artifacts in the Lower Omo succession come from Shungura Members F and G and are therefore

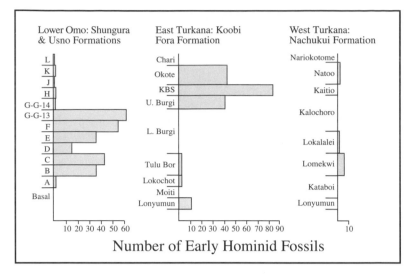

Figure 4.14. The numbers of hominid fossils by member within the principal formations of the Omo Group, Lake Turkana Basin (modified after 717, p. 615). The fossils are mainly fragmentary jaws, isolated teeth, or fragmentary limb bones, but some (especially in the Koobi Fora and Nachukui Formations) are more complete, including a partial skeleton (from the Nachukui Formation). Bars for members within the Koobi Fora and Nachukui Formations are scaled to correspond to like-aged members within the Shungura and Usno Formations. It is thus obvious that the majority of fossils from the Lower Omo are older than the majority from Koobi Fora.

probably 2.4 to 2.3 my old (1079). Potentially older (2.5 to 2.4 my old) artifacts may occur in Member E or even in Member C, but this has yet to be confirmed by excavation. All the artifacts fit within the Oldowan Industrial Complex, as originally defined at Olduvai Gorge.

In addition to fossils of australopithecines and early *Homo*, the Lower Omo has provided fossils of near-modern or modern *Homo sapiens*. These come from Member 1 of the Kibish Formation, which disconformably overlies the Omo Group Formations and is believed to date from the early part of the late Quaternary, after 127 ky ago.

West Turkana

West Turkana is the name informally applied to the area immediately west of Lake Turkana, northern Kenya (fig. 4.11), which contains up to 715 m of fossiliferous Plio-Pleistocene deposits that closely resemble those at Koobi Fora and in the lower Omo River Valley. These deposits have been designated the Nachukui Formation, which is divided among eight successive members that can be directly correlated with members of the Koobi Fora and Shungura Formations based on shared volcanic tuffs (fig. 4.13) (375, 717, 957). The correlations and direct dating show that the Nachukui Formation accumulated between about 4.3 and 0.7 my ago.

The rich mammalian fauna from the Nachukui Formation (more than a thousand mammalian fossils from ninety-three species) corroborates and supplements the basic biostratigraphy previously established in the lower Omo River Valley and at Koobi Fora (957). The Nachukui sequence is particularly notable for fossil assemblages dated between roughly 3 and 2 my ago, a period that elsewhere in Africa is well controlled only in the Lower Omo. Hominid fossils have been recovered from several horizons. They are not as numerous as those from the Lower Omo or Koobi Fora (fig. 4.14), but they include remarkably complete specimens, particularly the famous Black Skull (KNM-WT [Kenya National Museum–West Turkana] 17000), a cranium stained blue black by manganiferous minerals (1353, 2363), and the partial skeleton of the "Nariokotome boy" (KNM-WT 15000) (373, 2370). *Australopithecus afarensis* or a similar taxon occurs in the lower part of the Lomekwi Member with a probable age of about 3.3–3.2 my. *Australopithecus aethiopicus* is represented by the Black Skull and other fossils in the Lokalalei Member, where they are dated to 2.5–2.4 my ago, and *P. boisei* occurs in the Kaitio Member, where it is bracketed between about 2.3 and 1.6 my ago. A cranial fragment attributed to *Homo habilis* has been found in the Kalochoro Member, which is probably slightly more than 2 my old. Finally, *H. ergaster* or *H. erectus* (most notably the Nariokotome boy) is represented in the Natoo Member at a level dated to about 1.6 my ago.

Stone artifacts assigned to the Oldowan Industrial Tradition are known from the Kalochoro and Kaitio Members (1216, 1217), and Acheulean bifaces occur in the Natoo Member (1797). The Kalochoro artifacts are estimated to be about 2.35 my old, and they are equaled or exceeded in age only by artifacts from sites in the Shungura Formation (Omo Valley) (1079) and at Hadar, both in Ethiopia (1222, 1912). The Natoo bifaces are dated to roughly 1.65 my ago. They are the oldest known, and they suggest that *H. ergaster* (or early African *H. erectus*) and the Acheulean emerged more or less simultaneously.

Fejej

The Fejej Plain lies roughly halfway between the mouth of the Omo River and Chew Bahir (former Lake Stephanie) in extreme southern Ethiopia. The sequence comprises fluviolacustrine deposits and interbedded volcanics that span much of the later Cenozoic (87). Seven isolated hominid teeth have been found at a locality that is probably between 4.2 and 4 my old, based on ^{40}Ar/^{39}Ar dating and paleomagnetism (1182). The teeth closely resemble those of *Australopithecus afarensis* (741), but they could equally derive from *A. anamensis* or from another so far undefined species. At a second locality, Oldowan artifacts occur in sediments that probably accumulated about 1.9 my ago (87, 88).

Konso (formerly Konso-Gardula)

The Konso locality is in the Rift Valley, north of Fejej, southern Ethiopia. The sequence includes more than 50 m of richly fossiliferous fluviolacustrine deposits and interbedded tuffs (89, 2111). Direct dating, correlation of the tuffs to dated tuffs elsewhere, and faunal correlations indicate that the deposits span the interval between roughly 2 and 1 my ago. Most fossils come from two discrete time horizons, the first at about 1.9 my ago and the second at about 1.4 my ago. The assemblage at 1.4 my ago contains nine specimens of *Paranthropus boisei,* including a nearly complete skull (KGA 10-525), and two fragments of *Homo ergaster* (or *H. erectus*) (90, 2111). All nine fossils of *P. boisei* come from a single locality where the associated mammalian fossils imply a dry, grassy paleoenvironment. Deposits dated to about 1.4 my ago have also provided abundant Acheulean bifaces that are among the oldest known.

Hadar

The Hadar site is in the Awash River Valley within the Afar Depression, about 300 km northeast of Addis Ababa, Ethiopia. The fossiliferous sediments at Hadar have been grouped into the Hadar Formation, which has four members: (from bottom to top) the Basal Member, the Sidi Hakoma Member, the Denen Dora Member, and the Kada Hadar Member (fig. 4.15). The Hadar Formation sediments were accumulated mainly by streams, in a basin that was also periodically inundated by a large lake (456, 1150, 1222). Interbedded volcanic tuffs and a basalt layer provide materials for radiometric dating, and the tuffs may be used for regional and interregional correlations.

The lower part of the Kada Hadar Member and the three underlying members have furnished more than 320 fossils from the australopithecine species *Australopithecus afarensis,* also represented in the Middle Awash nearby and at Laetoli. High-precision single-crystal laser fusion ^{40}Ar/^{39}Ar dates now bracket the Hadar *A. afarensis* fossils between 3.4 and 2.9 my ago (1220)

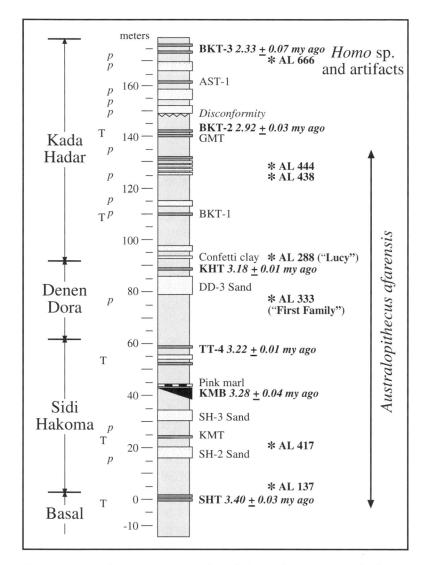

Figure 4.15. Schematic stratigraphy of the Hadar Formation (redrawn after 1222). The constituent members are listed on the extreme left. A T in the adjacent column marks deposits that accumulated in a large ("transgressive") lake, and a *p* marks the positions of paleosols, which formed when deposition was relatively quiescent. ^{40}Ar/^{39}Ar dating provided all the ages to the right of the stratigraphic column. Asterisks indicate the stratigraphic horizons of particularly important fossil sites (Afar localities, each abbreviated by AL and an appropriate serial number). Other abbreviations refer to key geological marker horizons.

(fig. 4.15). Most come from the Denen Dora Member, in which a single site (Afar Locality [AL] 333) produced more than two hundred specimens from at least nine adults and four juveniles (the "first family") (1150, 2458). Other particularly notable finds include 40% of a single skeleton ("Lucy" = AL 288-1) and a nearly complete skull (AL 444-2). The skeleton comes from a diminutive individual and the skull from a very large one, and comparable size variation marks the rest of the Hadar sample. The implication is that *A. afarensis* actually comprises two (or more) species or that it was characterized by a high level of sexual dimorphism, comparable to that in gorillas and orangutans. Careful morphometric analysis favors a single, highly dimorphic species.

Sediments near the top of the Kada Hadar Member have provided a maxilla of *Homo* that is firmly dated by ^{40}Ar/^{39}Ar dating and associated fauna to about 2.3 my ago (1222). The only older fossils of *Homo* may be a temporal bone from the Chemeron Formation at Lake Baringo, Kenya, dated to about 2.4 my ago (1023, 1027), and a mandible from the Chiwondo Beds at Uraha, Malawi, also thought to be about 2.4 my old (1883).

Artifacts from Hadar suggest that the emergence of *Homo* may have coincided with the first manufacture of flaked stone artifacts. The oldest artifacts from Hadar (narrowly understood) accompany the maxilla of *Homo* and are thus about 2.3 my old. Somewhat older artifacts, dated to 2.6–2.5 my ago by a combination of ^{40}Ar/^{39}Ar dating and paleomagnetism, occur near the top of the Kada Hadar Member along the Gona River at adjacent Kada Gona (959, 963, 1912). Both the Hadar and the Kada Gona artifacts resemble those from broadly contemporaneous or somewhat younger sites in the Lake Turkana Basin and at Olduvai Gorge, and they can be assigned to the same Oldowan Industrial Complex. The next youngest artifacts reported from the Hadar region are Acheulean bifaces and associated pieces found in fluvial deposits that unconformably overlie the Hadar Formation and that are clearly much younger (early or middle Quaternary) (559, 560).

Associated fauna suggests that the Hadar environment varied between predominantly dry bush/woodland (Sidi Hakoma and lower Kada Hadar Members) and riverine forest and wetland (Denen Dora Member) during the time span of *A. afarensis*. The vegetation may have been significantly grassier (less wooded) when the maxilla of *Homo* and associated artifacts were deposited (upper part of the Kada Hadar Member) (1222).

Middle Awash

The Middle Awash is the name assigned to a region straddling the Awash River in the Afar Depression, just south of Hadar. Between 1975 and 1978, J. E. Kalb led a team that demonstrated the fossiliferous and archeological potential of the Middle Awash region (1171, 1172). In 1981 and again from 1990 to the present, T. D. White and J. D. Clark led teams that produced many new fossils and artifacts and a wealth of detail on stratigraphic and paleoenvironmental context (492, 494, 2460, 2462). Geologic interpretation is complicated by faulting and uplift, by a tendency for contemporaneous deposits to differ strongly over short distances, and by the cyclical nature of deposition, which caused deposits of different ages to look very similar. However, the sequence contains numerous tuff horizons, which have been dated by the ^{40}Ar/^{39}Ar method and by geochemical correlation to dated tuffs at Hadar, Omo, and the Lake Turkana Basin (2462, 2476). A combination of dates, fauna, and artifacts shows that the deposits span the interval between the late Miocene (ca. 6 my ago) and the late Quaternary (less than 200 ky ago). The sediments older than about 3.9 my ago formed mainly under a large lake, whereas those that postdate 3.9 my ago were deposited mainly by the proto–Awash River or by streams that seasonally entered the Awash Basin from the east.

Erosion has exposed different portions of the Middle Awash sequence at discrete localities, each named for a local drainage feature (a wadi or mostly dry tributary of the Awash River). The oldest unequivocal hominid fossils discovered anywhere come from the Aramis locality, where they occur in lacustrine silts sandwiched between tuffs each dated by ^{40}Ar/^{39}Ar to about 4.4 my ago. The fossils exhibit numerous very primitive, apelike features, and they have been assigned to *Ardipithecus* (or *Australopithecus*) *ramidus,* so far known only from Aramis. Associated animal and plant fossils indicate that *A. ramidus* died in relatively wooded (as opposed to grassy) surroundings. Animal remains at other, mostly later sites imply less wooded conditions.

Australopithecus afarensis is represented by frontoparietal skull fragments dated to 3.8 my ago at Belohdelie and a nearly complete mandible, three mandible fragments, two isolated molars, a humerus, a fragmentary ulna, and a femur dated to roughly 3.4 my ago at the Maka locality (86, 2450, 2462). An archaic form of *Homo* is represented by a partial skull, a parietal bone, and a distal humerus fragment at the Bodo locality (85, 494, 538, 539), dated by ^{40}Ar/^{39}Ar to about 600 ky ago. Faunal similarities to Olorgesailie (Kenya) and Olduvai Gorge Bed IV (Tanzania) support this age. Finally, modern or near-modern *Homo sapiens* is represented by skull fragments at the Aduma locality, where associated (Middle Stone Age) artifacts imply an age of less than 200 ky.

The Middle Awash sediments contain numerous archeological sites, ranging from Oldowan or Developed Oldowan and Acheulean to Middle and Later Stone Age (492, 494, 1170). So far the oldest artifactual occurrences are associated with faunal elements implying an age of 1.5–1.3 my. Several Acheulean or Developed Oldowan occurrences are known at younger (early and middle Quaternary) sites. These include the Bodo locality, where Acheulean artifacts occur in the same sediments as the partial skull and other remains of archaic *Homo*. The Bodo artifacts broadly resemble like-aged Acheulean artifacts from the upper horizons of Olorgesailie and from Olduvai Gorge Bed IV. The Middle and Later Stone Age sites are not directly dated, but occurrences elsewhere imply that they postdate 200 ky ago.

Geological Antiquity and Geographic Range of the Earliest Hominids

From the site summaries that have been presented, it is clear that the australopithecines existed from before 4.4 my ago until at least 1.4–1.2 my ago. The gracile australopithecines probably became extinct before 2 my ago, either through evolution into *Homo* or through physical replacement by *Homo*. The oldest commonly accepted species of *Homo, H. habilis*, is known from roughly 2.4 my ago until perhaps 1.8–1.7 my ago, when it (or one of its constituents) was replaced by or evolved into *H. ergaster* (or African *H. erectus*). The robust australopithecines became extinct sometime between 1.2 and 0.7 my ago. A more precise determination is currently impossible because the fossiliferous portions of the long Plio-Pleistocene sequences in the Lake Turkana Basin essentially top out at 1.4–1.2 my ago, while most sites that probably monitor the immediately succeeding period (Olduvai Gorge and perhaps Melka Kunturé and the Middle Awash) have not yet provided large fossil samples. The upper limit of 0.7 my ago is based on the invariable absence of robust australopithecines in fossil assemblages that postdate this time. The absence is surely significant, because in regions where robust australopithecines lived before 1.4 my ago, their fossils usually outnumber those of *Homo*, and they may have been even more common on the ground.

So far the australopithecines and earliest *Homo* have been found only in Africa, between roughly 27° S (Taung) and 16° N (Bahr el Ghazal). More than a hundred years ago, Thomas Huxley (1104) and later Darwin (577) postulated that people originated in Africa, since it was the home of their closest living relatives, the chimpanzee and the gorilla. The accumulated data amply confirm Huxley's and Darwin's prescience.

The known fossil sites are concentrated in two areas—in the northeastern portion of South Africa and along the east African Rift Valley, from northern Tanzania in the south to east-central Ethiopia in the north (fig. 4.1). The site concentrations and specific site locations reflect the occurrence of sedimentary traps with good conditions for bone preservation, but the Bahr el Ghazal site in Chad demonstrates that early hominids ranged widely within tropical and subtropical Africa, into areas where fossil sites are rare or may not exist.

It may never be possible to pinpoint the geographic range of the australopithecines precisely, but at this stage it seems increasingly likely that they were limited to lower latitudes within Africa, especially early on. Beyond direct evidence for this at sites where australopithecine fossils occur, there is also evidence from sites where they are absent. The most significant of these nonaustralopithecine sites are probably Langebaanweg (Varswater Formation), at 33° S near Cape Town, South Africa (fig. 4.1) (1008–1010), Sahabi (Sahabi Formation), at 30° N near the Gulf of Sidra, in northeastern Libya (248), and Ahl al Oughlam, at roughly 32° N in Atlantic Morocco (799–802, 1741).

Langebaanweg and Sahabi have provided thousands of bones from species that indicate a geologic age of about 5 my ago (early Pliocene), when the australopithecines or their immediate hominid forebears had probably emerged farther north. The Langebaanweg fauna is the largest, most diverse Pliocene fauna yet found in Africa, and among the extant orders of African mammals, only the Sirenia (dugongs) is missing. At least fifty-six species of large mammals are present, yet in spite of careful searches not a single hominid fossil has been recovered. The Sahabi fauna is less rich, but it is as extensive as most early Pliocene east African faunas, and claims to the contrary notwithstanding, it has also produced no hominid specimens (1023, 2455, 2463). Ahl al Oughlam is estimated to be roughly 2.5 my old, and it approaches Langebaanweg in richness, yet hominid fossils and artifacts are again totally absent. Almost certainly, Langebaanweg, Sahabi, and Ahl al Oughlam lack hominids, because they lie in temperate regions that people colonized only after 2 my ago. The relatively complete fossil and archeological records of northern Africa suggest that the first colonists arrived only about 1 my ago (1740), when they were probably members of *Homo ergaster* (or early African *H. erectus*).

If the earliest hominids penetrated low middle latitudes in Africa only after 2 my ago, their absence from Eurasia is readily understandable, since they could not leave Africa before they reached its northeastern corner.

Taxonomy and Morphology of the Australopithecines

Most authorities today recognize at least seven species of australopithecines, divided among one to three genera. There is incomplete agreement on precisely which fossils characterize each species (in technical terms, on what their "hypodigms" comprise), but most specialists would probably accept the listing in table 4.1 (from 2055) as a reasonable starting point. With regard to genera, *Australopithecus* is accepted by everyone, but some experts place the so-called robust australopithecines in a separate genus, *Paranthropus*, originally proposed by Broom for robust fossils from Kromdraai and Swartkrans. This usage is accepted here and reflects growing evidence that the robust australopithecines represent a unique and highly specialized development within the hominid family. The oldest known australopithecines may also be sufficiently different from later ones to merit a distinct genus, for which *Ardipithecus* has been proposed. This usage is also followed here, although its final acceptance depends on full publication of the growing fossil sample to which *Ardipithecus* refers. The seven commonly recognized australopithecine species, with their known sites and probable time ranges in parentheses, are as follows.

1. *Ardipithecus* (or *Australopithecus*) *ramidus* (Middle Awash [Aramis] and possibly Lothagam, Baringo, and Tabarin [Chemeron Formation], Kenya; from before 5 my ago at Lothagam until at least 4.4 my ago at Aramis)
2. *Australopithecus anamensis* (East Turkana [Allia Bay] and Kanapoi; from perhaps 4.2 to 3.9 my ago)
3. *Australopithecus afarensis* (Hadar, Middle Awash [Belohdelie and Maka], Laetoli, and possibly also Fejej, Omo [Usno Formation and Shungura Formation Member B], Koobi Fora [Tulu Bor Member], West Turkana [lower Lomekwi Member]; from roughly 3.9 until roughly 2.8 my ago, or from perhaps 4.1 my ago if Fejej is included)
4. *Australopithecus africanus* (Taung, Sterkfontein [Member 3 and possibly Member 2], Makapansgat [Members 3 and 4], and probably Gladysvale, between roughly 2.8 and 2.3 my ago)
5. *Australopithecus aethiopicus* (West Turkana [Lokalalei Member of the Nachukui Formation] and Omo [Shungura Formation Member C and possibly Members D through lower G], between perhaps 2.7 and 2.3 my ago)
6. *Paranthropus* (or *Australopithecus*) *robustus* (Kromdraai B [Member 3], Swartkrans [Members 1–3], Drimolen, and possibly Sterkfontein [Member 5]; from roughly 1.8 my ago until perhaps 1 my ago or somewhat later)

Table 4.1. The Cranial and Mandibular Fossils That Strait et al. Used to Infer the Evolutionary Relationships among the Australopithecines and *Homo*

Australopithecus afarensis
Craniums:
Hadar (AL) 33-125, 58-22, 162-28, 199-1, 200-1, 288-1, 333-1, 333-2, 333-45, 333-105, 417-1, 444-2
Garusi 1
East Turkana (KNM-ER) 2602
Mandibles:
Hadar (AL) 128-23, 145-35, 188-1, 198-1, 207-12, 206-1, 277-1, 311-1, 333w-1, 333w-12, 333w-60, 400-1a, 417-1
Laetoli (LH) 4
Maka (Middle Awash) (MAK) VP-1/12

Australopithecus africanus
Craniums:
Sterkfontein "Type Site" (Sts) 5, 17, 20, 26, 67, 71, 52a, TM 1511, TM 1512
Sterkfontein "West Pit" (Stw) 13, 73, 252, 505
Taung 1
Makapansgat (MLD) 1, 6, 9, 37/38
Mandibles:
Sterkfontein "Type Site" (Sts) 7, 36, 52b
Sterkfontein "West Pit" (Stw) 384, 404, 498, 513
Makapansgat (MLD) 2, 12, 22, 29, 34, 40, 45

Australopithecus aethiopicus
Craniums:
West Turkana (KNM-WT) 17000
Omo L 338-y-6
Mandibles:
West Turkana (KNM-WT) 16005
Omo L 55-s-33, L 860-2, 18-1967-18, 44-1970-2466, 57-4-1968-41

Paranthropus robustus
Craniums:
Swartkrans (SK) 12, 13/14, 46, 47, 48, 49, 52, 55, 65, 79, 83, 848; (SKW) 8, 11, 29, 2581; (SKX) 265
Kromdraai B TM 1517
Mandibles:
Swartkrans (SK) 6, 12, 23, 34, 1586; (SKW) 5; (SKX) 4446, 5013
Kromdraai B TM 1517

Paranthropus boisei
Craniums:
Olduvai Gorge (OH) 5
East Turkana (KNM-ER) 405, 406, 407, 732, 733, 13750, 23000
West Turkana (KNM-WT) 17400
Chemeron (KNM-CH) 1
Omo 323-896
Mandibles:
East Turkana (KNM-ER) 403, 404, 725, 727, 729, 801, 805, 810, 818, 1468, 1469, 1483, 1803, 1806, 3229, 3230, 3729, 3954, 5429, 5877, 15930
West Turkana (KNM-WT) 16841
Omo L 7a-125, L 74a-21
Natron (Peninj)

Source: After 2055, p. 20.
Note: Parentheses enclose museum/locality abbreviations. Taxonomic and phylogenetic assessments depend mainly on these specimens; other specimens are too incomplete, too undiagnostic, or too weakly described to be included.

7. *Paranthropus* (or *Australopithecus*) *boisei* (Konso [KGA 10], Omo [Members G–L], West Turkana [Kaitio Member of the Nachukui Formation], Koobi Fora [Upper Burgi, KBS, and Okote Members of the Koobi Fora Formation], Chesowanja [Chemoigut Formation], Olduvai [Beds I and II], Peninj [Humbu Formation], and Malema [Chiwondo Beds, Malawi]; from roughly 2.3 my ago until sometime between 1.2 and 0.7 my ago)

To these could be added yet an eighth species, *Australopithecus bahrelghazali,* founded on a fragmentary upper third premolar and the anterior portion of a mandible retaining one incisor, the alveoli (sockets) for the remaining three, both canines, and all four premolars (379, 380). The specimens come from the Bahr el Ghazal region of central Chad, and the associated fauna implies an age between 3.4 and 3 my ago, coeval with *A. afarensis.* To the extent that comparisons are possible, the mandible resembles those of *A. afarensis,* but it is distinguished by a few features listed below. Pending additional discoveries, it is tentatively included here in *A. afarensis.* It is significant primarily for its point of origin (fig. 4.1), 2,500 km west of like-aged australopithecine sites in the Rift Valley of eastern Africa, and it raises the possibility that north-central Africa was equally important in human origins. If future research confirms that *A. bahrelghazali* was a distinct species, it would provide the best available evidence for a branching event in human evolution before 3 my ago.

The australopithecines were all bipedal, and it is this more than any other feature that guarantees their hominid status. As a group, they are distinguished from *Homo* by small cranial capacity (roughly 400–550 cc); relatively large, prognathic (projecting) faces with apelike noses rather than typically human external ones; large posterior teeth (premolars and molars), usually with very thick enamel; and aspects of the postcranium that suggest an apelike ability to climb. The cranium and dentition are particularly well known for each species, and they suggest that relative to the earlier australopithecines (*Ardipithecus ramidus, Australopithecus anamensis, A. afarensis,* and *A. aethiopicus*), the later ones (*A. africanus* and particularly *Paranthropus boisei* and *P. robustus*) anticipated *Homo* in many derived craniodental features, including somewhat enlarged brains, reduced canines, fully bicuspid lower third premolars (P_3s), less prognathic (flatter) faces, more strongly flexed cranial bases, and a deeper depression on the skull for articulation with the mandible. The postcranium is more poorly known, but it implies a more complicated picture in which, relative to the earlier australopithecines, one or more of the later

Table 4.2. Mean Body Weight and Stature in Modern *Pan troglodytes*, the Best-Known Australopithecine Species, *Homo habilis* (Broadly Understood), *Homo ergaster*, and Modern *Homo sapiens*

	Geologic Age (my ago)	Body Weight (kg)		Stature (cm)	
		Male	Female	Male	Female
P. troglodytes	Modern	54	40		
A. afarensis	3.8–2.9	45	29	151	105
A. africanus	3.0–2.4	41	30	138	115
P. robustus	1.8–1.4	40	32	132	110
P. boisei	2.3–1.4	49	34	137	124
H. habilis	2.4–1.6	52	32	157	125
H. ergaster	1.7–0.7	63	52	180	160
H. sapiens	Modern	65	54	175	161

Source: After 1509, p. 18.
Note: The body weight estimates for fossil species were calculated from leg joint size in a sample of modern humans. Values predicted from a composite human-ape sample are 4 to 23 kg higher. The stature estimates were calculated from femoral lengths via a human stature/femur ratio of 3.74 defended in 721.

ones and even earliest *Homo* may have been more apelike in some respects (1513). The cranial and postcranial bones unite in demonstrating that people evolved from apes, but they also show that much larger samples will be necessary to determine the exact evolutionary relation of the australopithecines to each other and to early *Homo*. This point is emphasized again below.

The australopithecines differed from later people not only in their morphology, but also in their pattern of growth and maturation and in their shared tendency to very small body size, particularly in females. Dental crown and root development in the Taung "child" and in other australopithecine juveniles implies that the first permanent molar erupted at roughly three years of age versus an average of six years in living people (1996, 1997). If it is fair to extrapolate from this to later life history events, individual australopithecines probably reached sexual maturity at a much younger age than later people, and they probably did not live beyond fifty years, the approximate maximum life span of the great apes. It would follow that australopithecine females, like great ape females, did not enjoy a significant postreproductive life span. A decade or more of female postreproductive life is a uniquely human trait that may have appeared first in *Homo ergaster* at roughly 1.7 my ago, in *H. sapiens* after 500 ky ago, or perhaps incrementally between the two. The crucial point here is that if the australopithecines were apelike in their life history, they were probably also apelike in related behaviors. Perhaps most important, older australopithecine females, unlike their much later human counterparts, could not have provisioned their grandchildren (or grandnieces and -nephews), since they would usually have had young children of their own. Arguably, long postreproductive life spans were selected for in *H. ergaster* or later people, when provisioning by postreproductive females allowed younger women to produce additional children more rapidly (977, 1621).

Small body size and strong sexual dimorphism also have behavioral implications. Estimates of australopithecine body weight and stature vary widely (1167, 1501, 1502, 1505, 1509, 1510, 1795, 2045), mainly because they must be based on the mathematical relation between skeletal dimensions and body size in living primates (including humans). The results will differ depending on which skeletal dimension(s) and which living primates are used. Body weight is particularly difficult to reconstruct reliably, because the relation between skeletal element size and body weight is relatively loose in living species; the australopithecines had unique body proportions, which are not replicated in any living species; and the supply of relevant fossil parts is very limited.

With these caveats in mind, table 4.2 (after 1506, 1509, 1510) nonetheless presents mean body weight and stature estimates that are surely in the ballpark. The table shows that, on average, australopithecine males probably weighed 40–45 kg (88–99 lbs) and females 30–35 kg (66–77 lbs). The difference between males and females is significantly greater than in living humans and recalls the difference in chimpanzees. Comparably small weights and perhaps even greater sexual dimorphism characterized *Homo habilis,* and mean weights and statures approaching modern ones were achieved only in *H. ergaster,* after roughly 1.7 my ago. *H. ergaster* was also the first species in which sexual dimorphism fell to the modern level, and the table shows that the reason was an especially large increase in female size. Arguably, like the apelike pattern of maturation, strong sexual dimorphism in the australopithecines implies an apelike social organization in which males competed vigorously for females and the sexes did not cooperate economically.

By definition, each australopithecine species was unique in detail, and each is therefore described separately below. The descriptions rely heavily on references 214, 1073, 1224, 1317, 1318, 1512, 2055, 2449, 2459. The most important similarities and differences are illustrated in figures 4.16–4.33. At the outset, particular attention should be paid to figure 4.16, which shows the location and attachment of the masseter and temporalis muscles involved in mastication (chewing). Many of the key cranial differences among the earliest hominids reflect a shift in the arrangement of these muscles and an associated

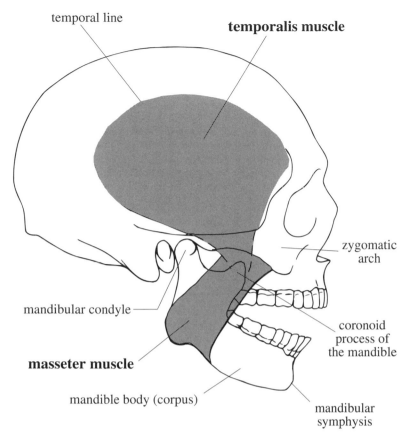

temporal line

temporalis muscle

zygomatic arch

mandibular condyle

coronoid process of the mandible

masseter muscle

mandible body (corpus)

mandibular symphysis

Figure 4.16. Representation of a modern human skull showing the insertion of the masseter and temporalis muscles involved in mastication (adapted from 415, p. 177). In the evolution of the later australopithecines from *Australopithecus afarensis*, the anterior (fore) part of the temporalis muscle became increasingly important relative to the posterior (rear) part. This change was accompanied by a forward shift in the root of the zygomatic arch to which the masseter muscle attaches. The result was an increase in the grinding power of the cheek teeth, also reflected in the teeth themselves.

shift in the nature of mastication, emphasizing grinding with the cheek teeth (premolars and molars).

Ardipithecus ramidus

Ardipithecus ramidus was originally described from thirteen partial dentitions or isolated teeth, a left temporal, associated fragments of a right temporal, a left temporal, and an occipital, an isolated humerus, and an associated humerus, radius, and ulna (2460, 2461). Subsequent fieldwork has recovered a mandible associated with a partial skeleton as well as other more fragmentary fossils, all of which are under study. The original

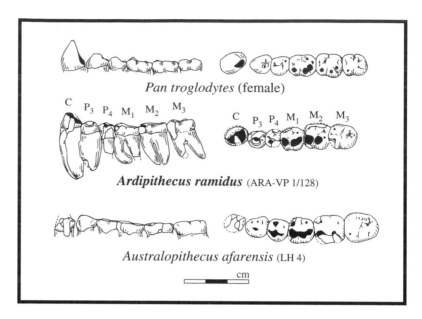

Figure 4.17. Left mandibular dentitions of a female chimpanzee (*top*), *Ardipithecus ramidus* (*center*), and *Australopithecus afarensis* (from Laetoli) (*bottom*) (redrawn after 2460, p. 311). The lower canine in *Ardipithecus ramidus* was as large as that in *Australopithecus afarensis* or larger, but the premolars and molars were significantly smaller. In this regard *Ardipithecus ramidus* more closely recalls the chimpanzee. (ARA-VP = Aramis; LH = Laetoli.)

A. ramidus fragments reveal a wealth of apelike features, including the largest canines relative to cheek teeth of any known hominid (fig. 4.17); a lower P_3 that is highly asymmetrical and unicuspid (fig. 4.18); a lower deciduous P_3 (= dP_3) that is very chimpanzeelike (fig. 4.19); dental enamel that is only slightly thicker than in chimpanzees; a very shallow mandibular (glenoid) fossa (hollow on the base of the temporal bone for articulation with a rounded bar of bone, or condyle, atop the mandible); an inconspicuous articular eminence (mound of bone) anterior to the mandibular fossa (figs. 4.21, 4.25); a temporal bone that is heavily pneumatized (inflated by air pockets), an external auditory meatus (the bony tube between the eardrum and the external ear) that is significantly smaller than in *Australopithecus afarensis* and all later hominids, and a strongly muscled arm in which the elbow could have been locked to aid in tree climbing. Derived features that distinguish *A. ramidus* from apes and link it to hominids include the incisiform morphology of the canines, the absence of a honing facet on the P_3 (indicating it did not occlude with the upper canine as it does in apes), and the forward position of the foramen magnum on the basioccipital. From the position of the foramen magnum, it

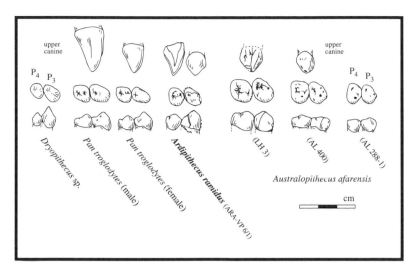

Figure 4.18. Upper canines and lower premolars of *Dryopithecus,* male and female chimpanzees, *Ardipithecus ramidus,* and *Australopithecus afarensis* (redrawn after 2460, p. 311). In *Ardipithecus ramidus* the upper (and lower) canines were more incisorlike than in apes but less incisorlike than in *Australopithecus afarensis.* The P₃ was remarkably apelike in the angle between its long axis and the axis of the lower dental row (indicated by the position of the P₄), in its conspicuous asymmetry, and in the lack of a cusp on the lingual (tongue) side. *A. afarensis* was also apelike in these respects, but somewhat less so. (ARA-VP = Aramis; LH = Laetoli; AL = Hadar.)

could be argued that *A. ramidus* was bipedal, but this can be established firmly only when the lower limb bones are described. Assuming that *A. ramidus* was a hominid, it is the most primitive found so far.

Australopithecus anamensis

A. anamensis is sparsely known from ten isolated teeth, a left maxilla, a left mandible fragment, and a nearly complete radius from Allia Bay, East Lake Turkana, Kenya, and by a nearly complete mandible and associated temporal bone, two associated mandible fragments, a nearly complete maxilla and associated lower incisor, four partial dentitions or isolated teeth, a distal humerus, and most of a tibia from Kanapoi, southwest of Lake Turkana, Kenya (1340). Some additional partial dentitions and postcranial bones from both sites support the original species diagnosis but have not yet been described in detail (2384). *A. anamensis* resembled both the chimpanzee and *A. ramidus* in its heavily pneumatized temporal bone, small external auditory meatus, shallow mandibular fossa coupled with an indistinct articular eminence, large canines, and asymmetrical,

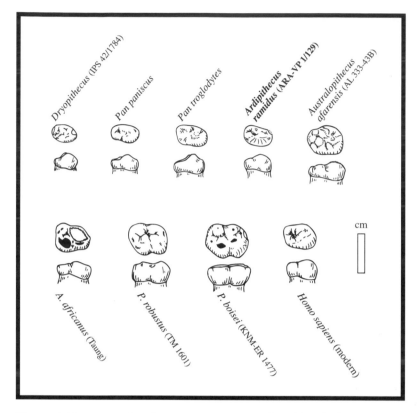

Figure 4.19. Lower deciduous anterior premolars (dP₃s) of *Dryo-pithecus*, pygmy chimpanzee, common chimpanzee, *Ardipithecus ramidus*, *Australopithecus afarensis*, *A. africanus*, *Paranthropus robustus*, *P. boisei*, and living humans (redrawn after 2460, p. 307). In the small size and (buccolingually) narrow crown of its dP₃, *Ardipithecus ramidus* was remarkably like a chimpanzee. (IPS = Institut Paleontologic Dr. M. Crusafont, Sabadell, Spain; ARA-VP = Aramis; AL = Hadar; KNM-ER = Kenya National Museum–East Rudolf; TM = Transvaal Museum.)

nearly unicuspid lower P₃. Unlike *A. ramidus*, however, *A. anamensis* shared more advanced features with *A. afarensis* (described below), including relatively thick dental enamel and large, relatively broad molars. Also like *A. afarensis*, it was unquestionably bipedal. This is shown by the form of the tibia (fig. 4.20), including the near right angle between the proximal shaft and the proximal articular surface, the large size of the lateral (outer) proximal condyle, and the uniquely human buttressing of the proximal and distal shaft. These features and others indicate the typically human transfer of weight from one leg to the other during bipedal locomotion. In contrast, in chimpanzees, where such weight transfer does not take place, the proximal articular surface forms a discernible angle with the

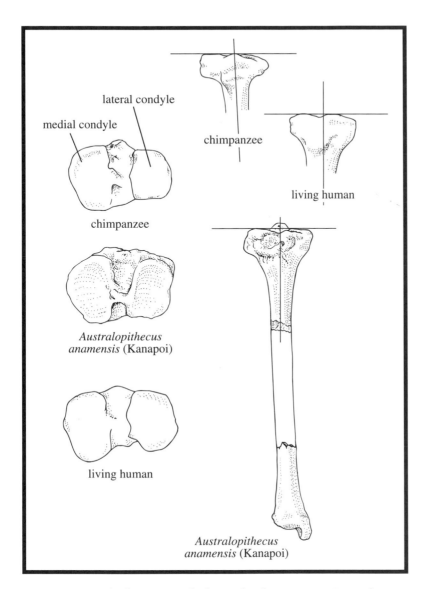

Figure 4.20. *Left:* the proximal tibiae of a chimpanzee, *Australo-pithecus anamensis* from Kanapoi, and a living human viewed from the surface that articulates with the distal femur. *Right:* The proximal tibiae of a chimpanzee and a living human viewed from the front and compared with the reconstructed tibia of Kanapoi *A. anamensis* (redrawn after 1339, p. 45). The Kanapoi tibia resembles the human one in the subequal size of the lateral and medial condyles and in the right angle between the proximal surface and a line bisecting the shaft. In chimpanzees the medial condyle is conspicuously larger than the lateral one, and the proximal surface meets the shaft at an oblique angle. The shared human and Kanapoi condition reflects the demands of habitual bipedal locomotion.

shaft, the lateral condyle is perceptibly smaller than the medial (internal) one, and the shaft lacks conspicuous buttressing.

The humerus and radius indicate that like both *A. ramidus* and *A. afarensis*, *A. anamensis* had powerful arms that would have aided in tree climbing (1001). Body weight, estimated from the tibia, was between 47 and 55 kg (103 and 121 lbs), which slightly exceeds estimates for *A. afarensis*. Sexual dimorphism may have been greater than in *A. afarensis*, judging by greater apparent variation in canine root size (2384). As *A. anamensis* and *A. afarensis* are defined at present, *A. anamensis* differed from typical *A. afarensis* in several important respects, including the form of the tympanic plate (the bone that partially encloses the middle ear and that supports the eardrum), longer and more robust canine roots (at least in presumed males), nearly parallel (vs. more splayed) left and right cheek tooth rows and supporting mandibular rami, less vertical (more backward sloping) mandibular symphyses ("chin" regions), and a tendency for the lower molar walls to slope inward on the buccal (cheek) side while the upper molar walls slope inward on the lingual (tongue) side. In *A. afarensis*, both lower and upper molar walls are more vertical. Arguably, however, *A. anamensis* differs less from the (older) Laetoli portion of the *A. afarensis* sample than from the (younger) Hadar/Middle Awash portion, and future discoveries may force a reevaluation of the present species distinctions.

Australopithecus afarensis

A. afarensis is far better known than *A. ramidus* or *A. anamensis*. Its cranium displays numerous primitive, apelike features, including very small endocranial capacity (the published range for three specimens of 380–485 cc, with a mean of 415 cc [1053], will be somewhat increased when an estimate appears for the most recently found skull [1220]); mastoid processes and other lateral structures of the cranial base heavily pneumatized (that is, filled with air cavities) (figs. 4.21, 4.24); cranial base only weakly flexed; tendency for the development of compound temporal-nuchal crests, reflecting relatively expanded posterior temporalis muscles; steeply inclined (vs. more horizontal) nuchal plane on the occipital bone; forward projection (prognathism) of the face, especially below the nose, greater than in any other known hominid; an abrupt transition or step between the floor of the nasal cavity and the premaxillary bone (nasoalveolar clivus) just below; prominent canine juga (bony bulges) over the canine roots, forming lateral pillars that do not reach the nasal (pyriform) aperture; a lightly built, relatively short

Small brains

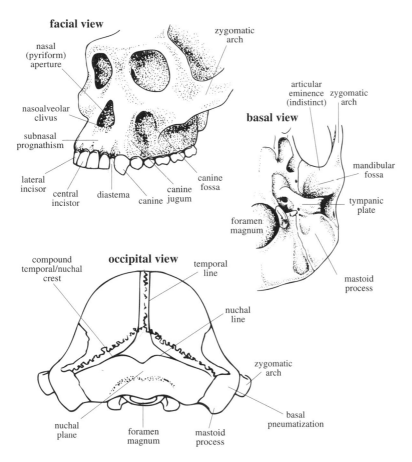

Figure 4.21 Three views of the skull of *Australopithecus afarensis*, showing important features mentioned in the text (redrawn after 2459, figs. 8, 10, 11). *A. afarensis* exhibits many primitive, apelike features, including extensive pneumatization of the cranial base, the presence of a compound temporal-nuchal crest, pronounced subnasal prognathism, the absence of a distinct articular eminence, and a diastema or gap in the upper tooth row between the lateral incisor and the canine.

zygomatic bone, arising relatively far back on the maxilla and separated from the canine juga by a deep depression (canine fossa); no distinct articular eminence anterior to the mandibular fossa; narrow, flat, and shallow palate, with limited premaxillary shelving; maxillary tooth rows converging posteriorly (fig. 4.26); frequent diastema (gap) between the upper second incisor and the upper canine; upper incisors relatively large and procumbent; upper central incisors notably broader than lateral ones; upper canine generally asymmetric and projecting, frequently with a contact facet from the lower canine on the fore (mesial) edge and one from P$_3$ on the rear (distal) edge (fig. 4.27); canines much larger in males than in females;

Figure 4.22. *Top:* Lateral view of a mandible from Swartkrans articulated with a reconstructed *Paranthropus boisei* skull fromOlduvai Gorge. *Bottom:* Occlusal view of the mandible (redrawn after 415, fig. 7.8). The occlusal outline has been slightly distorted by pressure in the ground that compressed the horizontal branches inward. Note such characteristic robust australopithecine characters as the anteriorly placed sagittal crest, dish-shaped face, thick, deep mandibular body, and molarized premolars.

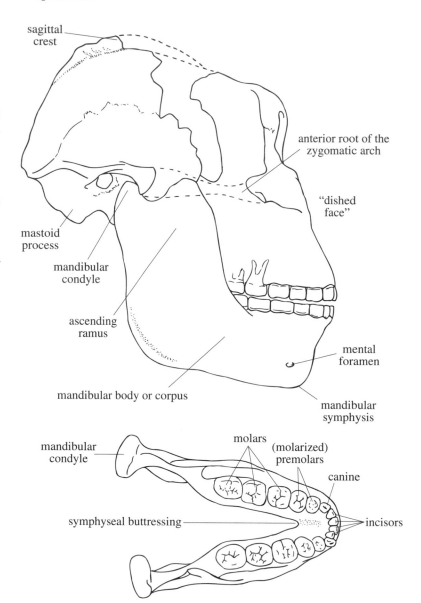

marked hollowing of the buccal wall of the mandibular corpus (fig. 4.28), with a mental foramen that opens anteroposteriorly (rather than directly outward); frequent presence of a diastema between the lower canine and P_3; lower canine projecting, often with a wear facet from contact with the upper canine on the rear (distal) edge; and P_3 relatively long and narrow, oval in outline, with a very small lingual cusp or none at all (fig. 4.29).

The partial mandible from Chad assigned to *Australopithecus bahrelghazali* is broadly similar to mandibles of *A. afarensis* but differs from typical *A. afarensis* mandibles in

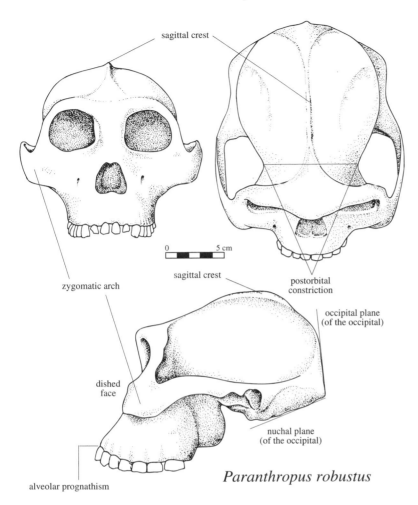

sagittal crest

zygomatic arch

sagittal crest

postorbital constriction

occipital plane (of the occipital)

dished face

nuchal plane (of the occipital)

alveolar prognathism

Paranthropus robustus

0   5 cm

Figure 4.23. Front, top, and side views of a reconstructed skull of *Paranthropus robustus* (redrawn after 1073, fig. 10.7). Note the anteriorly placed sagittal crest, the powerfully built, widely flaring zygomatic arches, pronounced postorbital constriction, dish-shaped face, and sharply angled occipital bone, which singly and together characterize the robust australopithecine skull.

the presence of fully bicuspid P_3s, three-rooted (versus two-rooted) premolars, divergent (versus parallel) cheek tooth rows, and a somewhat flatter, more vertical mandibular symphysis (chin region) (379, 380). Additional fossils are required to determine if *A. bahrelghazali* should be lumped with *A. afarensis* or retained as a separate species.

In general, the morphology of the *A. afarensis* braincase reflects small brain size, combined with a powerful masticatory apparatus and extensive pneumatization. The morphology of the face reflects the large size of the anterior teeth, with their very robust, curved roots. In both braincase and face, *A. afarensis* broadly recalls Miocene to Recent apes and demonstrates beyond all doubt the essentially ape ancestry of the hominids. In the skull, its most significant departure from the ape condition may have been in the organization of the brain. The endocast morphology of one Hadar specimen suggests enlargement

Figure 4.24. Facial and occipital views of *Pan troglodytes* (chimpanzee), *Australopithecus afarensis*, *A. africanus*, *Paranthropus robustus*, *P. boisei*, and *Homo habilis* (redrawn after 2459, figs. 9, 10). Note how *A. afarensis* and the chimpanzee are alike in their pronounced subnasal prognathism, relatively large anterior teeth, diastema between the lateral incisor and the canine, confluence of the temporal and nuchal lines, great breadth of the cranial base, and other features. Note also how *A. afarensis* differs from other hominids in all these respects. (AL = Hadar; STS = Sterkfontein; SK = Swartkrans; KNM-ER = Kenya National Museum–East Rudolf.)

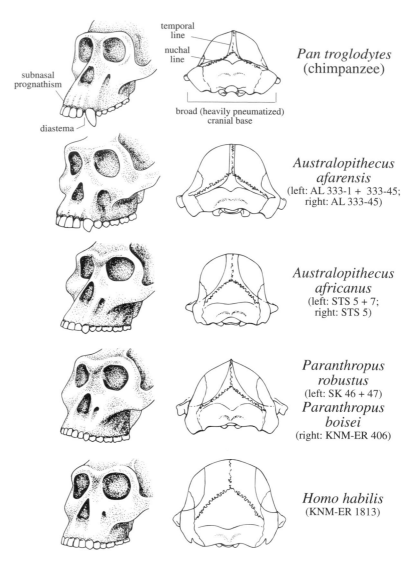

of the association areas in the parietal cerebral cortex (1050, 1057). This is a distinctly human feature, though its expression on the Hadar endocast has been challenged (701–703).

Postcranially, *Australopithecus afarensis* also possessed some remarkably apelike features, including very short thighs, extraordinarily powerful arms, forearms that were probably very long relative to the upper arms (recalling the condition in chimpanzees) (1220), possibly a more cranial orientation of the glenoid cavity on the scapula (allowing the forearm to be directed upward with relatively little rotation of the scapula) (2038), curved (vs. relatively straight) foot and hand phalanges, relatively long toes, and probably a cone-shaped rib cage (vs. the barrel-shaped cage of *Homo ergaster* and later humans). The

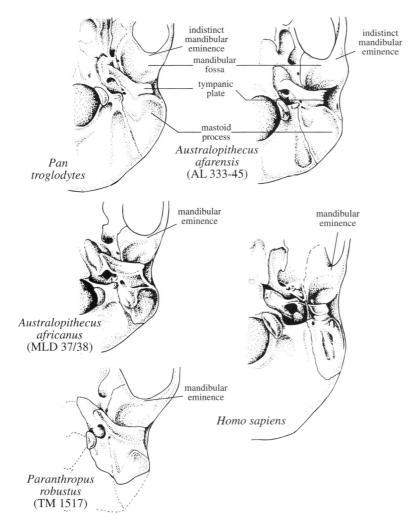

Figure 4.25. Left basal views of the skulls of a chimpanzee (*Pan troglodytes*), *Australopithecus afarensis*, *A. africanus*, *Paranthropus robustus*, and modern *Homo sapiens* (redrawn after 2459, fig. 11). Note that in tympanic and mastoid morphology *A. afarensis* resembled the chimpanzee more closely than it did the other hominids. (AL = Hadar; MLD = Makapansgat; TM = Transvaal Museum.)

implication is that *A. afarensis* was adept at climbing trees, perhaps to obtain food or to avoid predators, and that it was not fully committed to terrestrial bipedalism (2044, 2107, 2108). On the ground, however, it was unquestionably bipedal (1310), as indicated by such characters as the forwardly placed, downwardly directed foramen magnum; the structure of the foot with its basically human arch and nonopposable big toe; the distinct angle between the distal femur and the proximal tibia (fig. 4.31) (the valgus knee, which helps to center the body over one leg while the other is in motion); the orientation of the distal tibia articular surface nearly perpendicular to the long axis of the tibia shaft (a correlative of the valgus knee with broadly the same significance); and above all the pelvis with its short, broad, backwardly extended iliac blade (which centers the trunk over the hip joints, reducing fatigue during upright, bipedal

Figure 4.26. Palates of a chimpanzee, various australopithecines, and a modern human (top row redrawn after 1148, p. 367; bottom row redrawn after 2459, fig. 9). Note the presence of a diastema between the lateral incisor and the canine in both the chimpanzee and *Australopithecus afarensis,* and note also the enlargement of the premolars versus the other teeth in *A. africanus* and especially in *Paranthropus robustus.* (AL = Hadar; STS = Sterkfontein; SK = Swartkrans.)

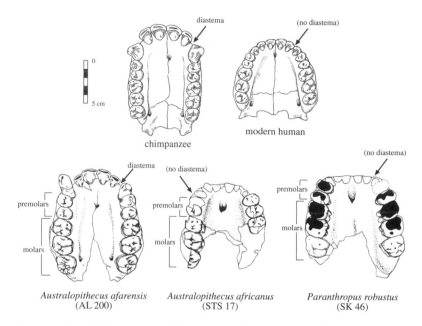

locomotion) (fig. 4.32 illustrates the very similar pelvis of *A. africanus*). In these latter respects, *A. afarensis* was very similar to modern people, and the interpretation is relatively unambiguous.

More problematic are features such as the more posterior orientation of the anterior portion of the iliac blade (vs. the more lateral orientation in modern humans); the relatively limited difference in size between the femoral condyles (in modern people, the lateral condyle is distinctly larger than the medial one); the shape and depth of the notch for the patella (kneecap) between the femoral condyles (intermediate in shape between the shallow notch of apes and the deeper one of people); and·the shape of the lateral condyles (less oval than in *Homo*) (2044). Depending on how these features are interpreted, *Australopithecus afarensis* may have had a slightly bent-hip, bent-kneed gait, something like that of a chimpanzee on two legs (2044), or a striding gait, essentially indistinguishable from that of modern people (1395). Arguably the heel strike, nondivergent big toe, distinct arch, and other unmistakably human features of the Laetoli footprints imply the modern striding gait (592, 2266, 2447), while obstetrical considerations—rather than locomotion—may explain some of the pelvic differences between *A. afarensis* and modern humans (2132). This is particularly likely since significant brain expansion requiring an enlarged birth canal occurred only beginning about 2.5–2.4 my ago, in the evolution of the genus *Homo.*

The muscle and tendon insertions on postcranial bones indicate that *Australopithecus afarensis* individuals were heavily

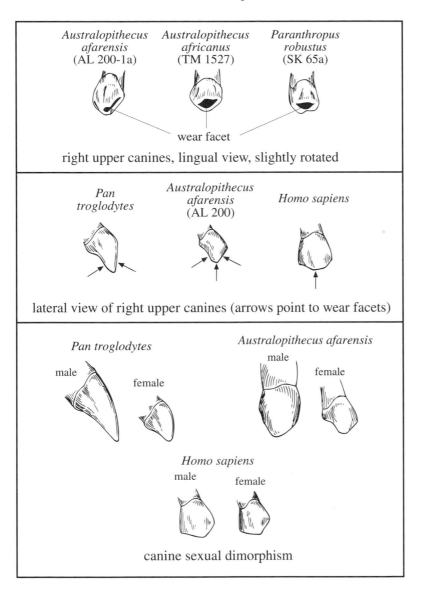

Figure 4.27. Upper canine morphology of various australopithecines, the chimpanzee, and modern humans (top redrawn after 2459, fig. 14; middle and bottom redrawn after 1148, p. 268). In *Australopithecus afarensis* the canine wore not only at the tip, as in later hominids, but also on the fore and rear (mesial and distal) surfaces, as it does in chimpanzees. *A. afarensis* was also more apelike in the degree of canine size difference between the sexes. (AL = Hadar; TM = Transvaal Museum; SK = Swartkrans.)

muscled. However, they varied considerably in body size (table 4.2) and stature, from perhaps 1 m to 1.7 m (3′3″ to 5′7″). Size variation was especially marked in the forelimb (1508), and it has been attributed to exceptional sexual dimorphism. If this is accepted, however, *A. afarensis* was much more dimorphic

Figure 4.28. Cross sections of the mandibular body below P_4 in various australopithecines and in *Homo habilis* (redrawn after 2459, fig. 13). Note the buccal inflation (thickening) of the mandibular body in *Australopithecus africanus* and *Paranthropus boisei*. (LH = Laetoli; MLD = Makapansgat; KNM-ER = Kenya National Museum–East Rudolf.)

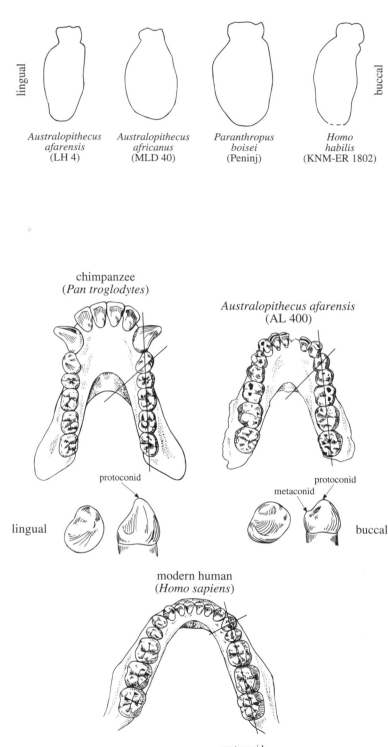

Figure 4.29. Lower third premolar (P_3) morphology in the chimpanzee, in *Australopithecus afarensis*, and in modern humans (redrawn after 1148, p. 269). Note that in its P_3 *A. afarensis* was intermediate between the chimpanzee and modern people. Thus it maintained roughly the same angle between the P_3 axis and the rest of the tooth row as is seen in the chimpanzee, but the P_3 itself was somewhat rounder and sometimes had a small inner or lingual cusp (metaconid), anticipating that of later people.

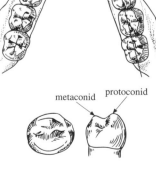

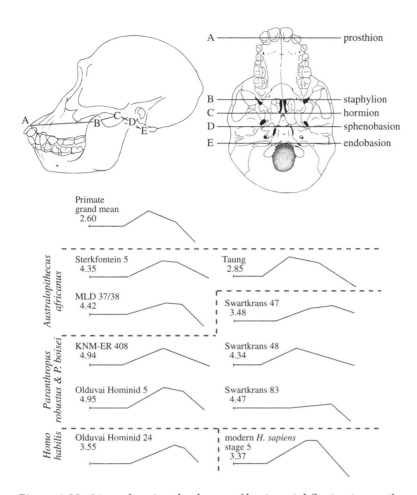

Figure 4.30. Lines showing the degree of basicranial flexion in gracile australopithecines, robust australopithecines, *Homo habilis,* and modern *H. sapiens* full adults (stage 5) (redrawn after 1302, figs. 1, 2, 4). The anatomical points that define each line are shown on a chimpanzee skull. The greater the angularity between the segments of a line, the greater the basicranial flexion. The lines show that the australopithecine cranial base was generally much less flexed than that of *H. sapiens.* Since the degree of flexion shapes the roof of the voice box, the less flexed cranial base of the australopithecines may imply that they could not make the full range of sounds that modern people can (1302). This in turn may mean they were more limited in their ability to communicate orally. (MLD = Makapansgat; KNM-ER = Kenya National Museum–East Rudolf.)

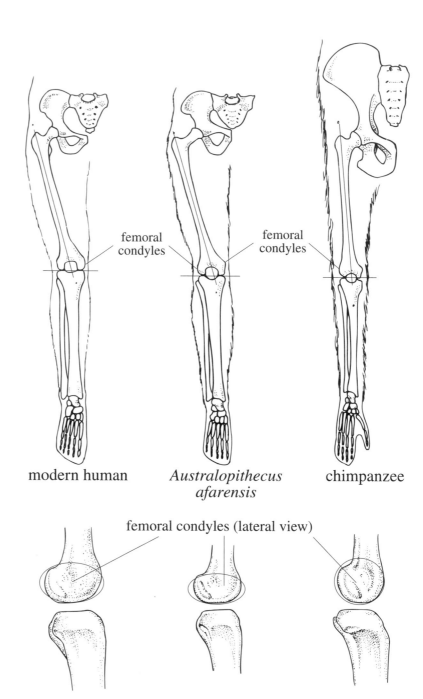

femoral
condyles

femoral
condyles

modern human   *Australopithecus*   chimpanzee
                        afarensis

femoral condyles (lateral view)

Figure 4.31. *Above:* Lower limbs of a modern human, *Australo-pithecus afarensis,* and a chimpanzee, with emphasis on the knee joint (redrawn after 1148, p. 157). *Below:* Lateral views of the knee joint (distal femur and proximal tibia) in the same three species. The distinct angle between the distal femur and proximal tibia, as well as the oval femoral condyles shared by modern people and *A. afarensis,* is an adaptation to habitual bipedal locomotion.

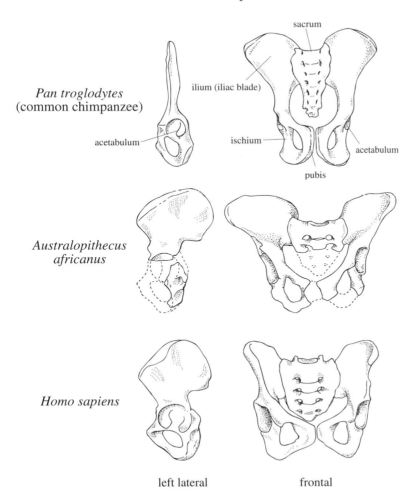

Pan troglodytes
(common chimpanzee)

*Australopithecus
africanus*

Homo sapiens

left lateral   frontal

Figure 4.32. Left lateral and full frontal views of pelvises of a chimpanzee, *Australopithecus africanus*, and a small modern human (redrawn after 1317, pp. 160, 161). Note the basic similarity between the australopithecine and modern human pelvises, reflecting a shared adaptation to bipedalism.

than living humans and even more than chimpanzees. We must thus consider the possibility that two (or more) species have been mistakenly amalgamated (745, 2535). The morphology of the basicranium, nose, jaws, and teeth has also been used to suggest two species—an early robust australopithecine and an early species of *Homo* (including *A. africanus* of other authors) (1638, 1639). The robust australopithecine would be the larger species, which is more abundant in the fossil samples. Two species may also be indicated by morphological variability within postcranial categories, especially among femurs (1915), and it is questionable whether feet with long, curved toes such as those of Hadar *A. afarensis* could have made the totally human footprints found at Laetoli (site G) (2263, 2264). However, the cranial arguments for two species have been systematically refuted (1223, 1225), and the postcranial variability could reflect sexual dimorphism if *A. afarensis* males, being much larger than females, were also significantly more terrestrial (2044).

A larger sample of more complete fossils may be necessary to reveal the full meaning of variability within *A. afarensis*, but on present evidence it need not mean that two species have been compounded.

Australopithecus africanus

Unlike the earlier australopithecines, *Australopithecus africanus* shared many unique cranial traits with *Homo*, but it was still very primitive by modern standards. Endocranial capacity varied from 430 to 520 cc with an average near 440 cc (1045, 1051), and it was perhaps only 10% greater than in *A. afarensis.* The cranial base was not significantly more flexed (fig. 4.30). However, the lateral portions of the base were much less heavily pneumatized, the frontal (forehead) was more arched, and the temporal and nuchal lines did not meet to form a compound crest at the rear of the skull (fig. 4.24). The configuration of the temporal lines indicates a greater emphasis on the anterior (vs. posterior) fibers of the temporalis muscle. In some individuals the anterior parts of the left and right temporalis muscles may have met at the top of the skull, promoting the development of a sagittal crest (bony ridge) along the midline. (This feature is illustrated in figs. 4.22, 4.23, and 4.33 on skulls of robust australopithecines, most of which possessed it. The possible exceptions were probably all small females.) The nuchal plane was more horizontally oriented, and there was a distinct articular eminence in front of a relatively deep mandibular fossa (fig. 4.25).

The face was notably shorter than in *Australopithecus afarensis*, with less subnasal prognathism linked to a reduction in the size of the anterior dentition. Shortening of the premaxilla was accompanied by upward extension of the canine juga to flank the sides of the nasal aperture. The zygomatic arch was long and strongly built, and it arose vertically on the maxilla to provide an attachment for the masseter muscles that was more anterior than that in *A. afarensis.* The palate was relatively deep, with clear premaxillary shelving.

The upper medial and lateral incisors were subequal in size, and the canines were relatively short, wearing mainly at the tips and not on the mesial and distal edges (fig. 4.27). There was apparently little or no sexual dimorphism in canine size. Upper second incisor/canine and lower canine/third premolar diastemata were rare. The cheek teeth were enlarged, with some tendency for molarization of the premolars (fig. 4.27). The mandibular body was significantly more robust than in *Australopithecus afarensis*, with a more heavily reinforced symphysis.

The ascending ramus was broader and taller than in *A. afarensis* and arose lower and more to the side on the mandibular body.

Most of the cranial and dental differences between *A. africanus* and *A. afarensis* can be explained by a reduced emphasis on the anterior dentition and a greater emphasis on the cheek teeth in mastication, coincident with a reorganization of the temporalis and masseter muscles that control jaw movement (2459). In this respect *A. africanus* appears to have diverged from *A. afarensis* in the direction of *Paranthropus robustus* and *P. boisei*.

The cranial and dental differences between *Australopithecus africanus* and *A. afarensis* are matched by some surprising postcranial differences. In fundamental form, the postcranial bones of the two species are very similar (1504), but specimens from Sterkfontein Member 4, including a partial skeleton (Stw 431), show that *A. africanus* had a significantly larger forelimb relative to its rear limb (1513). In this respect it was actually more apelike than *A. afarensis*, despite its more humanlike skull and dentition. Arguably, four associated foot bones from Sterkfontein Member 2 also imply that *A. africanus* also had an apelike divergent big toe (hallux) (509, 678). The Laetoli footprints strongly suggest nondivergent, humanlike big toes for *A. afarensis*. The pelvis and other lower limb bones show that *A. afarensis* and *A. africanus* were both capable bipeds, but the remarkably large size of the forelimb suggests that *A. africanus* may have been an even more proficient tree climber. A heavy reliance on tree climbing may also be indicated by the dimensions of the semicircular canals in the bony labyrinth of its inner ear (2031, 2032). The relative dimensions of different labyrinthine canals in living people are tied to the modern ability to maintain balance during bipedal locomotion, but scaled for body size, canal proportions in *A. africanus* were more apelike than humanlike. In fact *Homo ergaster*, after 1.7 my ago, is the oldest known hominid species to exhibit the fully modern labyrinthine configuration (2031), and it was also the first to exhibit fully human (as opposed to apelike) limb and body proportions.

The remarkably apelike limb proportions of *A. africanus* are puzzling, and they complicate an understanding of how *A. africanus* was related to the other australopithecines and to early *Homo*. In this book and elsewhere, cranial features dominate assessments of ancestor-descendant relationships. *A. africanus* was clearly advanced over *A. afarensis* and approached *Homo* in its somewhat enlarged endocranial capacity, reduced canine size, fully bicuspid P_3, greater basicranial flexion, deeper mandibular (glenoid) fossa, and other craniodental features. On this basis, it is often placed near the ancestry of *Homo.* However, if *A. africanus* was more primitive postcranially than *A. afarensis*, we must consider the possibility that it evolved its craniodental similarities to *Homo* independently (convergently). Alternatively, *A. afarensis* may have evolved its more advanced postcranium independently. This possibility perhaps becomes particularly likely if we accept that *Homo habilis*, as described below, actually had a postcranium that was more like that of *A. africanus* in its apelike limb proportions. Much larger postcranial samples, ideally with fore and rear limb bones from the same individuals, will be necessary to narrow the alternatives.

Australopithecus aethiopicus

The type specimen of *A. aethiopicus* is a robust, fragmentary, toothless mandible from Shungura Formation Member C (68, 69). Members C through lower G have provided other partial mandibles and isolated teeth that have also been assigned to this species (2110, 2112, 2504), but the most diagnostic specimen is the now famous Black Skull (KNM-WT 17000) (fig. 4.33) from the Lokalalei Member of the Nachukui Formation, West Turkana (1226, 2363, 2373). It is primarily on this specimen that the species description is based.

In its cranium, *A. aethiopicus* shared several key primitive features with *A. afarensis,* including a flat (unflexed) cranial base, extensive pneumatization of the lateral structures of the base, great forward projection of the face below the nose (alveolar prognathism), a shallow glenoid (mandibular) fossa without a distinct articular eminence, and very small cranial capacity (roughly 410 cc). However, it departed from *A. afarensis* in its "dish-shaped" midface with forwardly positioned cheek (zygomatic) bones, the smooth transition between the nasal floor and the premaxillary bone just below, the great thickness of the palate, the heart shape of the foramen magnum, the very large size of the molars and especially the premolars, and the nearly linear arrangement of the incisors and canines. In all these respects it approached *Paranthropus robustus, P. boisei,* or both, and the Black Skull was initially regarded as an early stage of *P. boisei* (2363). The features it shares with *P. boisei* and *P. robustus* relate mostly to a shared emphasis on heavy chewing between the cheek teeth, however, and they may reflect a shared dietary adaptation rather than closely shared descent. This is the position tentatively adopted here, as discussed in the section on australopithecine phylogeny below.

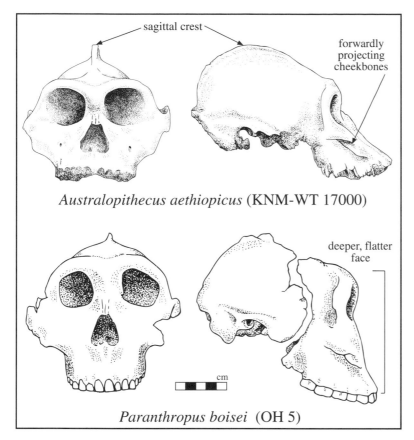

Australopithecus aethiopicus (KNM-WT 17000)

Paranthropus boisei (OH 5)

Figure 4.33. *Top:* a skull of *Australopithecus aethiopicus* dated to about 2.5 my ago at West Turkana, Kenya. *Bottom:* a skull of the "robust" australopithecine, *Paranthropus boisei* (= "*Zinjanthropus*"), dated to about 1.8 my ago at Olduvai Gorge (drawn by Kathryn Cruz-Uribe from casts and photos; © 1999 by Kathryn Cruz-Uribe). *A. aethiopicus* combined very primitive features like small endocranial capacity and pronounced alveolar prognathism with specialized "robust" features like a prominent sagittal crest and forward projection of the cheekbones. This suggests that *A. aethiopicus* could have been ancestral to *P. boisei*. Alternatively, the shared specializations may imply only a shared, independently evolved dietary adaptation, emphasizing heavy chewing between the upper and lower cheek teeth. (KNM-WT = Kenya National Museum–West Turkana; OH = Olduvai Hominid.)

The type mandible of *A. aethiopicus* is much too small for the Black Skull, and this suggests a high degree of sexual dimorphism. There are no postcranial bones that can be unambiguously assigned to *A. aethiopicus*, but a calcaneum and a large ulna from Shungura Formation Member E are likely candidates. The calcaneum implies bipedalism (635), while the ulna is notable for its apelike length and degree of dorsoventral (back-to-front) curvature (1515). Like the other australopithecines,

A. aethiopicus may thus have combined humanlike bipedal locomotion with an apelike ability to climb.

Paranthropus robustus and P. boisei

P. robustus and *P. boisei* are commonly lumped as the "robust"
australopithecines. A reasonable case can be made for retaining them in *Australopithecus,* but they are assigned here to
Paranthropus to reflect their unequivocally close kinship and a
growing consensus that they represent a highly specialized
branch of the hominid family. They differed from one another
in degree rather than in kind, and to the extent that they were
"robust" (in the development of cranial structures that allowed
great force to be applied between huge upper and lower cheek
teeth), *P. boisei* was generally more so. The number of traits
that distinguish *P. boisei* and *P. robustus* has shrunk as the
P. boisei sample has grown (368, 643, 2111), and arguably the
two forms were simply geographic variants of a single widespread species to which the name *P. robustus* would apply.
They existed contemporaneously between 1.8 and perhaps
1.2 my ago, when *P. boisei* was exclusively east African and
P. robustus was strictly southern African. However, they could
both derive from a *P. boisei* morphotype that emerged in eastern Africa about 2.3 my ago (2112, 2504) and that could have
spread to southern Africa between 2.3 and 1.8 my ago. No
southern African site unequivocally records this period. Gene
drift, superimposed on isolation by distance, could then have
produced the relatively minor distinctions that marked *P. robustus* by 1.8 my ago.

Both *P. boisei* and *P. robustus* overlapped with early *Homo,*
and in the period between 1.8 and 1.5 my ago for which the
overlap is best documented, the *Paranthropus* species are much
better known as fossils. This may partly reflect the greater
likelihood of fossilization in the places where they died, but it
may also mean that they were more numerous in life, perhaps
because they occupied an ecological niche that was more her-
bivore- than carnivorelike. Both species seem to have been
remarkably stable once they emerged (901, 2504), and this, together with their fossil abundance, makes them relatively easy
to describe.

P. boisei and *P. robustus* shared the largest cranial capacity of all the australopithecines (range 500–545 cc, with a mean
of perhaps 520 cc [368, 1045, 1051, 2111]), and the difference
implies greater encephalization, since on average individuals of
P. boisei and *P. robustus* were no larger than other australopithecines (table 4.2). By modern standards their limb bones

were stout, with thick shafts, but they were remarkably small (1507). This means that *P. robustus* and *P. boisei* were powerfully built, but they were very small-bodied, even petite, and their small size contrasts starkly with the enormous size of their cheek teeth. In tooth and jaw size, they exceeded the largest male gorillas, and this underscores the need for caution in estimating body size from different bony parameters. From tooth or jaw dimensions one might conclude that *P. robustus* and *P. boisei* were giants, whereas in reality limb bones show that they were more like pygmies. Only in their masticatory apparatus are they properly called "robust."

Both species differ from *Australopithecus afarensis* primarily in the same craniodental features as does *A. africanus*, but to a much greater extent. Their incisors and canines were relatively small, while their cheek teeth were greatly expanded, and their premolars (both deciduous and permanent) were almost fully molarized (figs. 4.22, 4.27). Their long, powerfully built zygomatic arches arose far forward on the maxilla and flared outward, causing the face to appear concave or dish shaped. Their mandibles were very thick and deep (figs. 4.22, 4.28), with heavy buttressing in the symphyseal region and with tall, broad ascending rami arising much more laterally and inferiorly than those in *A. afarensis*. Their anteriorly placed sagittal crests and marked postorbital constriction reflect even greater emphasis on the anterior fibers of the temporalis muscles than is seen in *A. africanus*. Overall, their distinctive craniofacial architecture suggests they were specialized for applying substantial vertical force between the upper and lower cheek tooth rows during mastication (1726).

Robinson (1793, 1794) proposed that the powerful masticatory apparatus of *Paranthropus robustus* was adapted for an exclusively vegetarian diet and that the less specialized dentition of *A. africanus* implied a more omnivorous diet. In support of this hypothesis, he suggested that the cheek teeth of *A. robustus* showed more enamel chipping from soil grit that clung to bulbs or other subterranean plant foods. Subsequent research failed to confirm a difference in the degree of enamel chipping (2192, 2378), but a dietary difference is still implied by the lower, less pointed molar cusps of *P. robustus*; by the tendency for *P. robustus* molars to wear flat (vs. the tendency for *A. africanus* molars to be buccolingually beveled in advanced wear); and by microscopic differences in occlusal wear between the species. These dental contrasts need say nothing about the extent to which either *P. robustus* or *A. africanus* ate meat, but they suggest that *P. robustus* often consumed harder, more fibrous vegetal foods (894, 897, 898, 1189). The microwear on

P. boisei molars indicates that fruits, masticated whole, were probably the preferred vegetal foods and that leaves and grass seeds were avoided (2367). The molar microwear specifically rules out bone crunching.

Diet may also be revealed by the physical elements and isotopes (varieties of elements) from which a creature builds its tissues. In tropical savanna environments like those where *Paranthropus* lived, trees, shrubs, forbs, and tubers (which are C_3 photosynthesizers) have a significantly higher $^{12}C/^{13}C$ ratio than grasses (which synthesize C_4). Nongrassy plants also tend to be significantly poorer in strontium relative to calcium. The carbon isotope and strontium/calcium ratios in herbivores tend to broadly reflect the plants they eat, and the ratios in carnivores reflect those of their prey. Thus a high $^{12}C/^{13}C$ ratio in an herbivore suggests a focus on nongrassy plants, while a high strontium/calcium ratio in a carnivore implies a focus on grazing herbivores. If we can assume that the original elemental and isotopic signals in ancient Swartkrans teeth were not irretrievably obscured after burial, then the carbon isotope and strontium/calcium ratios mean that *P. robustus* enjoyed a catholic diet, including nongrassy plants, grasses, or meat, and perhaps insects (1354, 1355, 1949, 1950, 1952). The possibility of meat eating is most striking, but the amount need not have been great.

In the form of their postcranial bones, *Paranthropus robustus* and *P. boisei* were generally similar to *A. africanus*. They were probably also about the same size, and they were equally dimorphic (table 4.2). All known foot and leg bones unequivocally demonstrate bipedalism (880, 1795). Relatively few upper limb bones are known, but proximal radii of both *P. robustus* and *P. boisei* resemble those of the great apes and of earlier australopithecines in ways that suggest the same apelike ability to climb trees (880, 908). The apelike dimensions of the semicircular canals within the inner ear also imply that *P. robustus* relied less on habitual bipedalism than later humans (no observations are available for *P. boisei*) (2031). In contrast, a thumb metacarpal from Swartkrans Member 1 closely recalls those of *Homo* in precisely the way that those of *Homo* differ from thumb metacarpals of apes (and of *Australopithecus afarensis*), and this may imply that *P. robustus* made the stone tools found in the same deposits (2103, 2104). However, the metacarpal has been assigned to *P. robustus* because 97% of the associated hominid cranial elements come from this species. About 3% come from *Homo*, and it is possible that the metacarpal belongs with these. The implications of thumb metacarpal shape for tool use have also been questioned (1473).

I argue below that *Homo* produced most if not all of the stone tools found at Swartkrans, Olduvai Gorge, and other sites where *Paranthropus* and *Homo* overlapped.

As discussed below, the phylogenetic relation between the robust australopithecines and earlier hominids is still being debated, but their highly derived craniodental features clearly eliminate them from the ancestry of any subsequent humans. Their distinctiveness is further underlined by a unique dental development and eruption sequence that differentiates them not only from other hominids but also from apes (1995).

Taxonomy and Morphology of Homo habilis

Homo habilis is best known from deposits dated to roughly 1.9–1.8 my ago at Olduvai Gorge and Koobi Fora (East Turkana), but individual specimens from the Kada Hadar Member in Hadar, Ethiopia, the Chemeron Formation at Lake Baringo, Kenya, and the Chiwondo Beds at Uraha, Malawi, indicate that it emerged much earlier, perhaps about 2.5–2.4 my ago. It is probably not coincidental that the oldest stone artifacts date from roughly the same time.

In the original definition (1325), *Homo habilis* incorporated specimens from Olduvai Bed I and Lower Bed II that appeared to have larger cranial capacities and smaller (especially narrower) cheek teeth than did known *Australopithecus* (including *Paranthropus* as used here). However, later discoveries, especially at Koobi Fora, showed that during the period including Bed I and Lower Bed II (roughly 2–1.6 my ago), some individuals (most notably KNM-ER 1470) had relatively large skulls combined with large *Australopithecus*-sized teeth (directly observed or inferred from socket size), whereas others (most notably KNM-ER 1813) had small *Australopithecus*-sized skulls combined with relatively small, *H. erectus*-sized teeth (fig. 4.34). Added to this, postcranial bones imply very large differences in body size (table 4.2). To many authorities (214, 444, 911, 1375, 1776, 2080, 2368, 2369, 2372) the overall variability implies at least two contemporaneous species, now commonly divided between *Homo habilis* narrowly understood for the smaller-brained, smaller-toothed form, most clearly exemplified by four specimens from East Turkana (KNM-ER 1813, 1805, 1501, and 1502) and five from Olduvai Gorge (OH [Olduvai Hominid] 7, 13, 16, 24, and 62); and *H. rudolfensis* for the larger-brained, larger-toothed variety, most clearly exemplified by five specimens from East Turkana (KNM-ER 1470, 1590, 3732, 1801, and 1802) (1377, 2495, 2499, 2501, 2502).

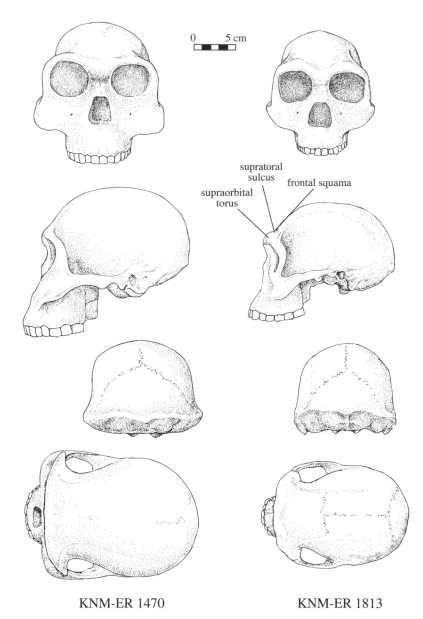

supratoral
sulcus

frontal squama

supraorbital
torus

KNM-ER 1470   KNM-ER 1813

Figure 4.34. Reconstructed skulls of *Homo habilis* (broadly understood) from Koobi Fora (redrawn after 1074, fig 10.9). Both skulls date from roughly 1.9–1.8 my ago, but they differ sharply in size and more subtly in shape. KNM-ER 1470 has an endocranial capacity of about 750 cc and large cheek teeth (inferred from the alveoli). Its face is relatively long (from the bridge of the nose to a point between the upper central incisors), broad across the orbits, and flattened below the nose. KNM-ER 1813 has an endocranial capacity of about 510 cc and much smaller cheek teeth. Its face is shorter, narrower across the orbits, and more projecting below the nose. It also has a more developed supraorbital torus (browridge). Many authorities believe the difference could reflect sexual dimorphism within a single species (male on the left, female on the right) but a growing number think it reflects the existence of two separate, contemporaneous species (*Homo rudolfensis* on the left, *Homo habilis* narrowly defined on the right). (KNM-ER = Kenya National Museum–East Rudolf.)

In the two-species scheme, the principal South African skull of very early *Homo*—Stw 53 from Sterkfontein Member 5—fits most comfortably within *H. habilis* narrowly understood, though it could represent yet a third, geographically distinctive, relatively small-brained, small-toothed species (902, 903). It probably dates from the period 1.9–1.8 my ago when *H. habilis* narrowly understood existed at Koobi Fora and Olduvai. *H. rudolfensis* is not well documented outside Koobi Fora, but it could be represented by isolated teeth from Members E–G of the Shungura Formation, southern Ethiopia (2112), by a temporal bone (KNM-BC 1) from the Chemeron Formation, central Kenya (2494, 2500), and most persuasively by a mandible (UR 501) from the Chiwondo Beds at Uraha, Malawi (360, 1883). The mandible extends the geographic range of *H. rudolfensis* far to the south, while the teeth, temporal bone, and mandible together extend its time range from 1.9–1.8 my ago (at Koobi Fora) to 2.4 my ago (at all three other sites). If the teeth, temporal, and mandible have been correctly assessed, *H. rudolfensis* would then be the oldest known species of *Homo*, perhaps joined by *H. habilis* only after 2 my ago.

H. habilis and *H. rudolfensis* unquestionably coexisted with *P. boisei/P. robustus* about 2–1.8 my ago, but if they were truly distinct species, only one could be ancestral to *H. ergaster* (or African *H. erectus*), which appeared in eastern Africa approximately 1.8–1.7 my ago. If brain expansion is emphasized, then *H. rudolfensis* is the more plausible ancestor for *H. ergaster*, but if facial and dental reduction is accentuated, then *H. habilis* is the better candidate. This kind of contradiction commonly emerges when specialists attempt to reconstruct evolutionary relationships from fossils, and it is partly to circumvent it that authorities turn to cladistic analysis. The logic of cladistic analysis is explored in the first chapter and again in the next section, but it considers multiple characters simultaneously in an effort to separate those that reflect parallel or convergent evolution (homoplasy) from those that indicate true evolutionary relations. A cladistic analysis of forty-eight craniodental characters that include indicators of endocranial volume and of facial and dental size suggests that *H. habilis* and *H. ergaster* are more closely related to each other than either is to *H. rudolfensis* (1377). The implication is that *H. rudolfensis* and *H. ergaster* developed their large brains independently, and the result could be used to argue that *H. rudolfensis* descended from a separate australopithecine ancestor and should be removed from *Homo* altogether.

The variability within *H. habilis* broadly understood is extraordinary, and the idea that it actually confounds two species is growing in popularity. Accepting two species raises new problems, however, including the difficulty of understanding how they could have shared the environment, or more generally what selective factors could have produced their divergence. A more fundamental problem is that authorities disagree on how to sort the relevant fossils between them (903); and even before sorting, the total sample is very small. There are, for example, only eight skulls or partial skulls on which endocranial volume can be estimated, and most are too fragmentary to provide precise results (1559, 1776). In these circumstances it is impossible to confirm the sexual dimorphism that would be expected within *H. habilis* narrowly understood or within *H. rudolfensis* if they were indeed separate species. Thus, for the moment, at least provisionally, the presentation here follows those (including 1073, 1147, 1149, 1559, 2199, 2459) who implicitly or explicitly attribute the variability in skull and dental size to extreme sexual dimorphism within a single species, designated *Homo habilis.*

If a single species is accepted, then in general *Homo habilis* can be distinguished from the australopithecines by larger endocranial volume (range 510 to 750 cc, with a mean about 630 cc) (1051, 1559, 2080), expanded frontal and especially parietal regions, a gutter or sulcus (distinct inflection point) between the supraorbital torus and the frontal squama, a more rounded occipital contour with enlargement of the occipital plane (vs. the nuchal one), a reduction in basicranial pneumatization and ectocranial rugosity (that is, smaller air pockets in the cranial base and less prominent cranial crests), a more parabolic dental arcade, reduced alveolar prognathism (less forwardly projecting jaws), a less robust premaxilla lacking canine juga, smaller jaws, and generally smaller cheek teeth with thinner enamel (214, 2459). The increase in average endocranial volume is particularly striking, since it was accompanied by little if any increase in mean body size compared with the australopithecines (1502, 1503, 1505). In addition, it may correlate with a restructuring of the brain, producing a distinctly human cortical sulcal pattern as opposed to the apelike pattern that apparently characterized the australopithecines (699, 703, 705, 706). The implications of this restructuring are debatable, but it could be related to a developing capacity for articulate speech.

The postcranial skeleton of *Homo habilis* is not well known, and some postcranial bones assigned to the species could even derive from a contemporaneous robust australopithecine, since they were not directly associated with diagnostic skull or jaw bones and their species assignment was based mainly on small size or on a priori assumptions about the probable postcranial distinctions between *H. habilis* and the robust australopithecines. Thus, at Olduvai Gorge, a partial foot skeleton (OH 8) from Bed I has been attributed to *H. habilis* at least partly because its morphology implies the completely modern kind of bipedalism (592). The attribution would be more secure if the foot could be shown to represent the same individual as does a partial juvenile skull of *H. habilis* (OH 7) found at the same site (FLKNN I).

The least equivocal postcranial bones of *Homo habilis* are hand bones that have been assigned to the same individual as has the partial juvenile skull (OH 7) just mentioned (1323) and, more important, portions of the right arm and both legs associated with fragmentary adult cranial bones and teeth (OH 62) at another Bed I locality (DDH) (1147, 1149). The leg bones confirm bipedalism but, contrary to expectations, suggest very short stature. The individual was probably only about 1 m (3′3″) tall, perhaps even shorter than the shortest known australopithecine, the famous "Lucy" (*Australopithecus afarensis*). Moreover, the arms were remarkably long relative to the legs, perhaps even longer than those of Lucy (974). Similar apelike body proportions are implied by a very fragmentary partial skeleton from Koobi Fora (KNM-ER 3735) that may also derive from *H. habilis* (1345). Together with the (OH 7) hand bones, which imply an apelike ability for underbranch suspension (2099, 2106), the long arms suggest that *H. habilis* retained the tree-climbing agility of the australopithecines. This might thus have been lost only with the evolution of *H. ergaster* roughly 1.8–1.7 my ago.

Future discoveries may show that some individuals of *H. habilis* possessed distinctly less australopithecine-like postcraniums, more closely resembling those of *H. ergaster*. This would significantly strengthen the case for dividing the present sample of *H. habilis* into two species, of which only the more derived would be relevant to later human evolution. Arguably, a more *H. ergaster*-like postcranium is already implied by some large isolated limb bones, like the femurs KNM-ER 1472 and KNM-ER 1481A that were found near the KNM-ER 1470 *H. rudolfensis* skull (2501). If the femurs do represent *H. rudolfensis*, then the postcranium may contradict the cranium and suggest that *H. rudolfensis*, rather than *H. habilis* narrowly understood, enjoyed the closer relationship to *H. ergaster*. This may become yet another example of the kind of contradiction that often plagues evolutionary reconstructions, as discussed again in the next section.

The Phylogeny of the Australopithecines and Early *Homo*

The phylogeny of the australopithecines and early *Homo* is controversial, partly because specialists cannot agree on which species and how many species existed at any one time. And even when specialists acknowledge the same species, there is still the problem of distinguishing between species similarities that were inherited from a common ancestor and those that may reflect only adaptation to similar circumstances. Recall that inherited similarities are said to be homologous, while those due to parallel adaptation are referred to as analogous or homoplastic. Only homologies are phylogenetically relevant.

To illustrate the problem of species definition, most specialists recognize only one species, *Australopithecus afarensis*, between 3.8 and 2.9 my ago, but some authorities perceive two (911, 1638, 1639, 1915, 2535), generally distinguished as an early robust australopithecine and an early representative of *A. africanus* or of *Homo* (including *A. africanus* of other authors). Similarly, many specialists regard *Paranthropus robustus* and *P. boisei* as closely related geographic variants (perhaps only subspecies) derived from a common robust ancestor between 2.5 and 2 my ago, while others believe they were more distinct, with separate evolutionary histories extending from a nonrobust ancestor that existed at or before 3 my ago. Perhaps most crucial of all, authorities differ on the relation between the different australopithecines and earliest *Homo*. None would place a robust australopithecine (*P. boisei* or *P. robustus*) on the line to *Homo*, but there is considerable disagreement about whether *A. africanus* should also be ruled out. This view is contested in part because some believe that *A. africanus* may itself comprise two species (1223, 2055), *A. africanus* narrowly understood and a second species that may be more *Homo*-like (1221). An added complication is that authorities also disagree strongly on the taxonomy of very early *Homo*. As already indicated, a growing number believe that the lone stem species that is commonly recognized—that is, *H. habilis*—actually comprises at least two species.

The problem of distinguishing inherited similarity from homoplastic (independently evolved) similarity is perhaps best illustrated by the "robust" australopithecines, *Paranthropus robustus* and *P. boisei* and their erstwhile ancestor, *Australopithecus aethiopicus* (1512). *A. aethiopicus* resembles *A. afarensis* in key cranial features, including very small average endocranial capacity, a shallow mandibular fossa fronted by a weakly defined articular eminence, a flat, shallow palate, the presence of a compound temporal-nuchal crest, and the relative lack of basicranial flexion. In these respects and others, *A. aethiopicus* differs from *P. robustus* + *P. boisei*, which it nonetheless resembles in several other distinctive characters, including the massiveness of the cheek teeth and jaws, the heart-shaped foramen magnum (shared with *P. boisei*), and the dish-shaped face with forwardly positioned zygomatic arches. The sum has convinced many authorities that *A. aethiopicus* was on or near a line between *A. afarensis* and *P. robustus* or *P. boisei* (or both). The single most significant *A. aethiopicus* fossil—the Black Skull (KNM-WT 17000) from northern Kenya—was even assigned by its discoverers to an early stage of *P.* (or *A.*) *boisei* (2363). Unlike *A. aethiopicus*, however, *P. boisei* + *P. robustus* share many notable features with early *Homo*, including brain expansion, a deep mandibular fossa associated with a prominent articular eminence, increased basicranial flexion, and reduced subnasal prognathism. Since *A. aethiopicus* was surely not ancestral to *Homo*, it could be ancestral to *P. boisei* + *P. robustus* only if the numerous features that *P. boisei* + *P. robustus* share with *Homo* are homoplastic, that is, if they developed independently in the two lines. The alternative is that *P. boisei* + *P. robustus* did not descend from *A. aethiopicus* but derived instead from another species that was also ancestral to *Homo*. Temporally and morphologically, *A. africanus* could be this species, but if *P. boisei* + *P. robustus* descended from *A. africanus* or a similar species, the features that *P. boisei* + *P. robustus* share with *A. aethiopicus* must be homoplastic. Except for the heart-shaped foramen magnum, these features mainly reflect heavy reliance on the cheek teeth in mastication, and they may indicate that *A. aethiopicus* and *P. boisei* + *P. robustus* are alike only because their diets were alike. A similar morphological response to similar diets is perhaps particularly likely in species like *A. aethiopicus* and *P. boisei* + *P. robustus*, whose potential for parallel evolution was surely enhanced by many shared genes.

There is no sure way to separate homoplasy from inherited similarity, but cladistic methodology provides a partial solution. The basic procedure is to arrange species in a treelike cladogram according to the number of derived features they share. Species sharing the greatest number of derived features reside on branches that connect to one another before they connect to other branches farther along the tree. The principle can be illustrated with single features like endocranial capacity ("brain size") and cheek tooth occlusal area. If the African great apes are assumed to exhibit the primitive condition for each of these traits, greater endocranial capacity and increased cheek

Figure 4.35. Clado-grams showing the evolutionary rela-tionships of the australopithecines and *Homo* sug-gested by endocra-nial capacity (*left*) and cheek tooth area (*right*). The contrast implies that similarities in endocranial capac-ity, cheek tooth area, or both must partly or wholly reflect parallel evo-lution (homoplasy) rather than true evolutionary rela-tionships (redrawn after 1512, p. 83).

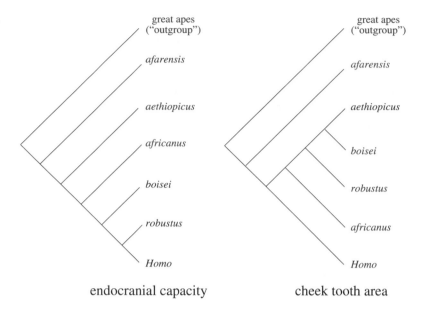

endocranial capacity   cheek tooth area

tooth size are derived conditions in hominids. Figure 4.35 (from 1512) illustrates the cladograms that result from a separate consideration of endocranial capacity (left) and of cheek tooth area (right) in australopithecine species and *Homo*. The clado-gram based on endocranial capacity links *Homo* most closely to *P.* (or *A.*) *robustus* + *P. boisei*, but the one based on cheek tooth area separates *Homo* from a clade in which *P. boisei* + *P. robustus* are more closely linked to *A. africanus*.

The failure of the cladograms to coincide stems partly from an evolutionary reversal in cheek tooth size, which in-creases from the apes through the australopithecines (culmi-nating in the huge cheek teeth of *P. boisei* + *P. robustus*) but declines in *Homo*. Cladograms based on other traits and espe-cially on combinations of traits suggest yet other possible evo-lutionary relationships, and a decision on which cladogram to accept is not straightforward. There is general agreement, how-ever, that when cladograms conflict, the one that requires the least amount of homoplasy (parallelism) is most likely to re-flect real evolutionary relationships. Such a cladogram is said to exhibit maximum parsimony, and figure 4.36 (left) illustrates the most parsimonious cladogram that Skelton and McHenry (1512, 1984) calculated from seventy-seven craniodental traits. Figure 4.36 (right) shows the phylogeny this cladogram implies. It explicitly eliminates *A. aethiopicus* from the ancestry of *P. boisei* + *P. robustus* and proposes that *P. boisei* + *P. robus-tus* instead share a closer (*A. africanus*-like) common ancestor with *Homo*.

Unfortunately, the principle of parsimony cannot overcome differences in the number of species that various authorities recognize, in what fossils they assign to these species, in what traits they choose to analyze cladistically, or in how they score these traits. Even more fundamentally, it cannot overcome the fragmentary nature of the human fossil record, which underlies much of the uncertainty over species numbers and content. The result is that different specialists can produce discordant cladograms that have equal claims to parsimony. Figure 4.37 (left) illustrates a highly parsimonious cladogram that Strait, Grine, and Moniz (2055) obtained from sixty craniodental characters, while figure 4.37 (right) presents a phylogeny that the cladogram implies. The contrast with the cladogram and phylogeny in figure 4.36 is clear, and the difference stems largely from a difference in the features that were analyzed and the way they were scored (1985, 2054). The most important difference is probably that figure 4.37 places *A.* (now *P.*) *aethiopicus* in an enlarged "robust" clade that is separated generically (as *Paranthropus*). It also removes *A. africanus* from the direct ancestry of any later hominids, although strictly speaking the associated cladogram does not demand this. The problem is that a phylogeny in which *P. aethiopicus* derives from *A. africanus* would require a remarkable number of implausible evolutionary reversals, including a reduction in cranial capacity.

Finally, figure 4.37 differs from figure 4.36 in replacing *Australopithecus afarensis* with *Praeanthropus africanus*. The fossils involved are the same, but Strait, Grine, and Moniz argue that if the cladograms in either figure are correct, *A. afarensis* must be removed from the genus *Australopithecus*, since it is less closely related to *Australopithecus africanus* than *A. africanus* is to all other hominids. (In cladistic terminology, *A. afarensis* is the sister taxon of *A. africanus* and all other hominids; that is, *A. afarensis* forms one clade, while *A. africanus* and all other hominids form a second.) To circumvent this problem, they employ the rules of zoological nomenclature to suggest that the fossils previously assigned to *A. afarensis* now be assigned to *Praeanthropus africanus*. If their argument is accepted, even the term australopithecine is problematic, since it is applied mainly to distinguish various early hominid species from *Homo*, when some of these species may be more closely related to *Homo* than they are to each other.

The schemes in figures 4.36 and 4.37 and others that have been proposed are working hypotheses that may be significantly altered by new fossil discoveries. As indicated in the description of *Australopithecus africanus* above, some changes are likely

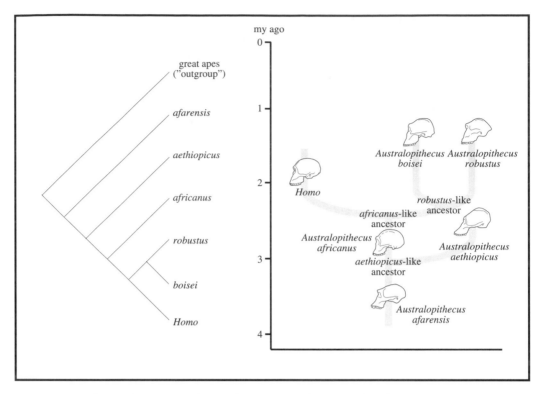

Figure 4.36. The evolutionary relationships among the australopithecines and *Homo* that Skelton and McHenry have inferred from seventy-seven craniodental characters. *Left:* the cladogram requiring the least amount of parallel evolution (homoplasy) in these characters. *Right:* the phylogeny this cladogram implies (redrawn after 1512, pp. 84, 85).

when the cladistic analysis can be extended to postcranial as well as cranial features and when it can be expanded to include *Ardipithecus ramidus* and *Australopithecus anamensis.* At the moment, postcranial characters are too poorly known for most species, *A. anamensis* is too sparsely known overall, and *A. ramidus* awaits sufficient description. At least provisionally, most authorities would probably place *A. ramidus* at the very base of the known human family tree, followed first by *A. anamensis* and then by *A. afarensis* (or *Praeanthropus africanus*). However, some specialists have suggested that the tree was already "bushy" by 4 my ago (1180), in which case *A. ramidus, A. anamensis,* or both may ultimately be removed to side branches.

It is important to emphasize that, despite continuing controversy over australopithecine phylogeny, the known fossil record nonetheless permits reconstruction of the major patterns in early hominid evolution. Thus the earliest well-documented hominids, dating from between 4.5 and 2.9 my ago in Ethiopia and Kenya, still bore the clear stamp of ape ancestry in their skulls and teeth and—to a lesser but still significant extent—

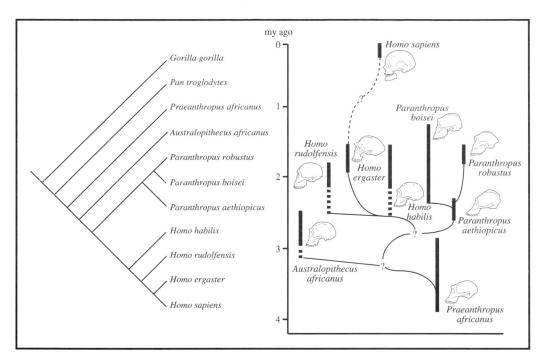

Figure 4.37. The evolutionary relationships among the australopithecines and *Homo* in-ferred that Strait, Grine, and Moniz have inferred from sixty craniodental characters. *Left:* the cladogram requiring the least amount of parallel evolution (homoplasy) in these characters. *Right:* the most plausible phylogeny that this cladogram implies (redrawn after 2055, p. 55).

in their postcranial bones. They were both proficient bipeds and agile tree climbers. By 2.5 my ago, at least two separate hominid lineages had emerged. In both the teeth were totally human and not apelike, particularly in the reduced size of the incisors and canines, but in one lineage—known popularly as the "robust australopithecines," the premolars and molars en-larged dramatically, and craniofacial form allowed powerful grinding between the upper and lower cheek tooth rows. This lineage experienced only minimal brain expansion, and it be-came extinct about 1 my ago. In contrast, in the other lineage, which includes the ancestors of living people, the cheek teeth and associated masticatory apparatus were progressively re-duced, and brain size was significantly expanded. This lineage probably produced the flaked stone tools that also appear about 2.5 my ago, and its increased reliance on technology was prob-ably tied to both its smaller cheek teeth and its enlarged brain. Both lineages may have retained apelike limb proportions and an apelike ability to climb trees, but by 1.8–1.7 my ago the lin-eage that was marked by brain expansion, smaller cheek teeth, and stone tool use developed modern body proportions and other morphological and behavioral features that indicate the modern commitment to a fully terrestrial lifestyle.

Geologic Antiquity of the Earliest Artifacts

Figure 4.5 summarizes the known or inferred dates for sites with very early artifacts. The oldest known stone artifacts come from Kada Gona, Hadar, where they are fixed by ^{40}Ar/^{39}Ar and paleomagnetism at 2.6–2.5 my ago (959, 963, 1912). Broadly similar artifacts dated to 2.4–2.3 my ago have been found at Afar Locality 666, also Hadar (1222); in Member F of the Shungura Formation, lower Omo River Valley (1074, 1079); and at Lokalalei (Kalochoro Member of the Nachukui Formation), West Turkana (1216). Based on associated faunal elements, artifacts from Senga 5A in the Upper Semliki Valley of eastern Zaire may also date to about 2.3 my ago (965), but the faunal elements and the artifacts have been reworked from the Lusso (formerly Kaiso) Beds into a much more recent deposit (247, 964, 1006). Somewhat later occurrences, dating between 2 and 1.6 my ago, are well established at Olduvai and Koobi Fora (1121, 1327), in the Kaitio Member of the Nachukui Formation, West Turkana (957, 1217), and at Chemoigut (Chesowanja), Kenya (876), Melka Konturé, Ethiopia (458), Fejej, Ethiopia (87), Mwimbi (Chiwondo Beds), Malawi (1185), Swartkrans (Member 1) (323, 490), and Sterkfontein (Member 5) (1284, 2050). In sum, there is good evidence that stone artifacts were being made in eastern Africa by 2.6–2.5 my ago and that the practice had spread to southern Africa by 2 my ago. If stone artifacts existed before 2.6 my ago, they may prove difficult to find since, unlike later artifact makers, the earliest ones may have been too mobile to accumulate archeologically visible clusters of debris. The archeological record would be largely invisible if people had not developed the uniquely human habit of returning to the same site for at least a few days (or nights).

Some controversy surrounds the oldest known bone artifacts. Dart (575) argued that numerous animal bones from Makapansgat Member 3 (the "gray breccia") were in fact tools of an "osteodontokeratic" (bone-tooth-horn) culture practiced by the australopithecines. Member 3 is now estimated to be between 3 and 2.5 my old, and Dart's bone tools would thus be the oldest known artifacts of any kind. In reaching his conclusion, Dart was especially impressed by the disproportionate abundance of certain antelope skeletal parts, such as the distal humerus and the mandible, which he felt were chosen by the australopithecines as clubs, saws, or other useful artifacts. However, the unequal pattern of skeletal part representation that Dart observed tends to characterize almost all fossil assemblages, mainly because some skeletal parts are more durable than others and are thus more likely to escape pre- and postdepositional destruction (320). The Makapansgat bones reveal no damage that can be specifically attributed to artifact manufacture or use, and it seems increasingly unlikely that they were tools. Instead, they were probably damaged and accumulated almost entirely by hyenas (1421).

The next oldest bone artifacts include 125 flaked, battered, or polished pieces from Olduvai Beds I and II (1327), 68 bone fragments with localized polish from Swartkrans Members 1–3, and a single similar fragment from Sterkfontein Member 5 (323, 325, 331). At all three sites, the bone implements certainly or probably date from between 2 and 1 my ago. Microscopic examination supports the artifactual nature of 41 Olduvai pieces (1933, 1938). Of these, 4 were not tools in the narrow sense but apparently served as anvils or platforms on which soft substances such as skin were repeatedly punctured by sharp-ended stone artifacts. The remaining 37 are large, flaked pieces of bone, including 26 with polish of the kind that forms on experimental pieces used to cut or smooth soft materials like hide and 11 with wear that probably formed from contact with a more abrasive substance such as soil. Experiments show that digging for subterranean plant foods in rocky soil could have polished most of the Swartkrans and Sterkfontein pieces (331). A few might also have been smoothed from repeated contact with a soft substance like hide.

The microscopic and experimental results indicate that the flaking, battering, or polishing on the Olduvai and the South African bones is almost certainly not natural and thus that artifactual use of bone began by roughly 2 my ago. But it is important to stress that at each site the bones identified as artifacts represent only a tiny fraction of the total number of bones recovered. Even more important, for the most part the bone artifacts were minimally shaped before use, and truly formal bone implements, made in advance to a repetitive pattern, appear only much later, in the Eurasian Upper Paleolithic–African Later Stone Age, beginning between 60 and 50–40 ky ago. Until that time, people certainly handled bones regularly and used them occasionally, but they did not recognize bone as a material that could be broken, carved, and ground into a variety of useful artifact types.

Form and Function of the Earliest Stone Artifacts

Arguably, stone artifacts produced before 2 my ago (at Kada Gona and Afar Locality 666 [Hadar], from sites in Member F of

the Shungura Formation [Omo], and at Lokalalei [West Tur-
kana]) reflect a less refined knowledge of flaking technology
than those made shortly afterward (especially at Olduvai Gorge
and Koobi Fora) (1216). However, most authorities assign all
known artifact assemblages antedating 1.7–1.6 my ago to the
Oldowan Industrial Complex, named for Olduvai Gorge, where
some of the largest and best-described assemblages have been
found (1118, 1326, 1327). In this view the Oldowan persisted
for roughly a million years (1912), from about 2.6 until 1.7–
1.6 my ago, when the appearance of hand axes, cleavers, and
other more sophisticated tools marks the beginning of the
Acheulean Industrial Complex (89, 1874).

Oldowan artifacts tend to be remarkably crude and informal, but in general they can be divided among four fundamental categories: *manuports* (pieces of rock that must have been
carried to a locality but that were not artificially modified);
hammerstones (rocks that were battered or pitted by striking
against other hard objects, probably mainly other rocks from
which flakes were struck); *core forms* (pebbles and rock fragments from which flakes were removed); and *flakes*. Core forms
may be further divided into cores that served only as sources of
flakes and core tools that were flaked to produce useful edges
or a desirable shape. In general, Oldowan knappers struck relatively few flakes from core forms, and the flakes tended to be
relatively small (1712). The core forms thus did not differ greatly
in size or shape from the raw material blanks on which they
were made. It was only with the advent of the Acheulean about
1.7–1.6 my ago that knappers often produced more numerous
and larger flakes that masked the size and shape of the underlying raw material.

In addition to dividing Oldowan artifacts among four very
basic categories, most authorities use a scheme such as that of
M. D. Leakey (1327) to assign core tools and modified (retouched) flakes to specific tool types (figs. 4.38, 4.39). The types
are differentiated mainly by how they are flaked or retouched,
their size, and their shape. Thus a flake with intentional retouch along an edge is called a *scraper*; a small scraper is *light
duty*, a large one is *heavy duty*. A core tool on which flaking is
restricted to one edge is called a *chopper*. If the flaking is restricted to one surface, the chopper is *unifacial*; if the flaking
occurs on two intersecting surfaces, the chopper is *bifacial*. A
core tool that has been extensively flaked to resemble a disk is
known as a *discoid*. If it is relatively spherical, it is a *spheroid*,
and if it resembles a crude, multifaceted cube, it is a *polyhedron*. A *bifacial chopper* on which the flaking tends to extend

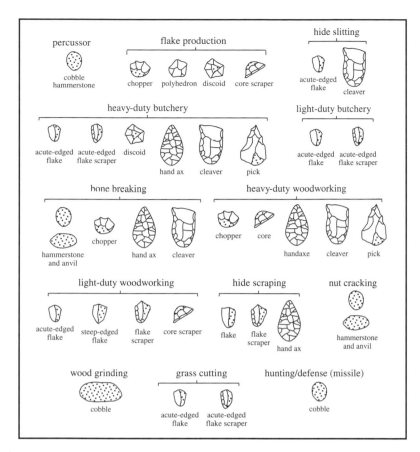

Figure 4.38. The basic types of stone artifacts found at Oldowan and Acheulean sites in Africa (adapted from 2208, fig. 7). The Acheulean is usually distinguished from the preceding Oldowan by the presence of hand axes, cleavers, and other large bifacial tools. Large bifacial tools do, however, occur in some "Developed Oldowan" assemblages. The suggested uses are based on feasibility experiments with replicas.

around the entire periphery is a *protobiface,* which grades into a true biface (hand ax) on which the flaking generally covers both surfaces and the entire periphery. Bifaces are unknown in the Oldowan proper; they appear about 1.7 my ago in assemblages variously labeled Developed Oldowan or Early Acheulean.

Further types and subtypes have been recognized in addition to these basic ones, but even the basic types may overemphasize the formality of Oldowan assemblages, where variation among artifacts tends to be continuous rather than discrete. Experimental replication, in fact, suggests that much of the variation among Oldowan core tools was controlled by the shape of the initial blank, not by a template in the maker's head (2208). This helps explain why core forms tend to intergrade and why archeologists find it difficult to sort them into discrete types.

By later human standards, Oldowan stoneworking technology was very primitive, and it is reasonable to ask if it exceeded the capability of an ape narrowly understood. The answer is probably yes, based on observations of Kanzi, a "pygmy" chimpanzee or bonobo at the Yerkes Regional Primate Center

Figure 4.39. A range of typical Oldowan stone tools and their conventional typological designations (redrawn after originals by B. Isaac and J. Ogden in 2208, fig. 1).

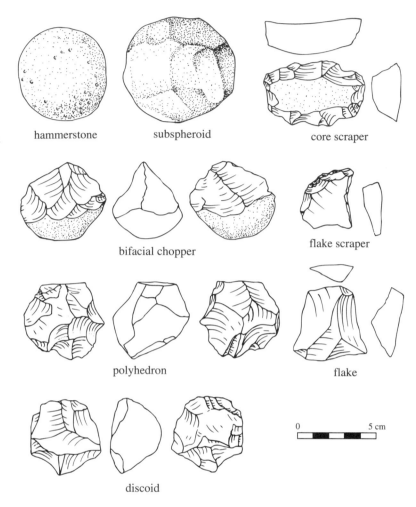

hammerstone   subspheroid   core scraper

bifacial chopper   flake scraper

polyhedron   flake

discoid

0   5 cm

in Atlanta, Georgia (1876, pp. 135–140; 2210). Kanzi is well known among psychologists for his ability to communicate with humanly designed symbols, and he is especially responsive to his human handlers. Experimenters showed him how to use a sharp stone flake to cut a cord encircling a box with food treats and also how to strike a suitable flake from a core. He quickly mastered the purpose and the general idea, but many months later his attempts to strike useful flakes were largely futile, and he had not produced a single piece that was as conspicuously artifactual as many Oldowan pieces. To some extent the difficulty may have been the typical ape structure of his hands, with the thumbs very short relative to the other fingers. This limits the extent to which Kanzi (and other apes) can grasp objects precisely (2105). But a bigger problem was his apparent inability to understand that useful flakes will result only if a hammer strikes a core at the right point and the right angle with the right force. In contrast, even the earliest Oldowan people fully understood this principle, and their artifact assemblages reflect an ability to produce useful flakes more or less at will.

The crudeness and typological poverty of the Oldowan are related to a third important feature—remarkable uniformity through time and space. Most differences between Oldowan assemblages from different times and places can probably be attributed to the stone available. Thus, in the lower Omo River Valley, where small quartz pebbles were the most readily accessible raw material, the range of artifact types is restricted mainly to small sharp-edged quartz splinters produced by smashing (1555). At Koobi Fora and Olduvai, where large pieces of volcanic rock were more abundant, various core forms are much more common (1118, 2208).

Koobi Fora also shows that the size of available stone blanks was probably an important determinant of interassemblage variability. The Oldowan toolmakers found their blanks in stream gravels that came originally from volcanic uplands to the east of Lake Turkana. Understandably, the average gravel clast (particle) size declines away from the uplands, and sites in areas with smaller clasts tend to contain a smaller range of core forms, since smaller pieces cannot be so extensively flaked (2208). The basic pattern is confirmed at Fejej, northeast of Koobi Fora, where the large size of Oldowan artifacts is clearly linked to the large size of the available clasts (87).

Until recently, archeologists describing and interpreting Oldowan assemblages focused on the modified flakes and on the core-tool category or subcategories. It was assumed that the accompanying unmodified flakes were mainly waste (*débitage*) from core-tool manufacture. This assumption can be tested under the microscope, where unmodified flakes may show wear polishes that formed during use (1196). Unfortunately, only flakes made in relatively homogeneous, nongrainy rocks are likely to preserve such polishes clearly, and most Oldowan flakes were made in inappropriate, coarse volcanic rock types. However, there are a small number of Oldowan flakes and flake fragments in suitable stone, and microscopic examination of fifty-six unmodified specimens from Koobi Fora isolated nine with unmistakable use wear or polish (1197). Comparison with wear created on experimental tools shows that three of the Koobi Fora pieces were used for scraping or sawing wood, four for cutting meat, and two for cutting grass stems or reeds. Sharp, unretouched flakes probably produced some of the cut marks found on animal bones at Oldowan sites (383, 1713), and in modern experiments such flakes have proved especially useful for penetrating hides (2208). Experimental knapping further suggests that many Oldowan core tools were only incidental by-products of flake production (2208, 2209).

All this is not to say that core tools were unimportant. Modern experiments show that sharp-edged core tools are generally more efficient than average-sized flakes for prolonged or heavy-duty butchering, because they are heavier, have longer cutting edges, and are easier to hold (1157, 1158). Blunter core tools were probably ideal for fracturing animal bones to obtain marrow, while spheroids or polyhedrons could have been effective, even lethal projectiles (1118). That core tools were not simply sources of flakes is shown in Olduvai Beds I and II, where core tools tend to be made from lava and flakes tend to be made from quartz (1327). Nonetheless, the totality of evidence indicates that Oldowan stone knappers were at least as interested in fresh flakes as in core tools.

Unexpectedly, perhaps, the same replication experiments that emphasize the probable importance of flakes to Oldowan people also suggest they preferred one hand over the other (1876, pp. 140–142; 2207). To produce flakes, a knapper usually holds the hammerstone in the dominant or active hand and the core in the less favored or passive hand. The first flakes that are struck from the core generally bear cortex (weathering rind) whose position tends to reflect which hand held the hammerstone and which hand held the core. Position is judged when the flake is viewed from dorsal (cortical) surface with the butt or striking platform up (fig. 4.40). A modern right-handed knapper produces roughly fifty-six flakes with cortex on the right side for every forty-four with cortex on the left; a left-handed knapper produces more or less the reverse ratio. In Oldowan assemblages from Koobi Fora the ratio is fifty-seven right to forty-three left, which implies that the Koobi Fora Oldowans were predominantly right-handed. In this regard they anticipated modern people, among whom roughly 90% are right-handed, and they were unlike other animals, including chimpanzees, who are divided about equally between left- and right-handers. In modern people the tendency for right-handedness correlates with pronounced lateralization of the brain and marked functional separation of the cerebral hemispheres, and the Oldowan preference for the right hand may indicate that these distinctively human features had already developed by 2–1.5 my ago (2207). The same inference may be drawn from the typically human, cerebral cortical asymmetries that characterize endocasts of the australopithecines and, more certainly, those of *Homo erectus* (1047, 1056). At least incipient human cognitive and communicational abilities are almost certainly implied.

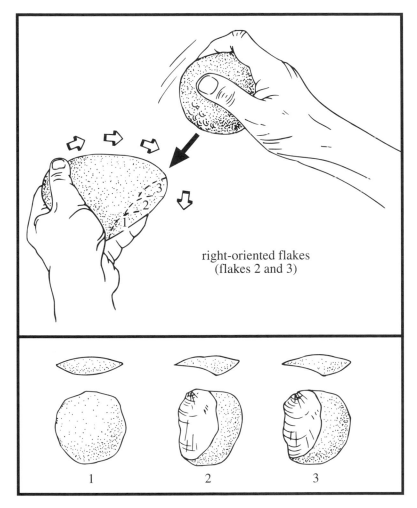

right-oriented flakes
(flakes 2 and 3)

1   2   3

Figure 4.40. *Top:* A right-handed knapper using the preferred right
hand to strike flakes from a pebble held in the left hand. *Bottom:*
Some struck flakes, each pictured from the dorsal (cortical) face with
the striking platform up. Flakes 2 and 3 illustrate the tendency for
flakes struck by a right-handed person to show a previous flake scar
on the left and cortex on the right. Oldowan assemblages from Koobi
Fora contain a preponderance of such "right-sided" flakes, implying
that the Oldowan knappers were predominantly right-handed (redrawn after 1876, p. 141).

Who Made Oldowan Artifacts?

No one doubts that *Homo habilis* made Oldowan tools, since it
(or one of the constituents into which it may ultimately be
split) was ancestral to later toolmaking hominids. However,
the robust australopithecines were also broadly contemporary
with the Oldowan, and their remains significantly outnumber
those of early *Homo* in the 2.5–1.7 my time range of the
Oldowan. Bones of both *P. robustus* and early *Homo* accompany

Oldowan artifacts in Members 1 and 2 at Swartkrans Cave, South Africa (330), and bones of *Paranthropus boisei* and of *H. habilis* have even been found on the same Oldowan "living floor" at the FLK Zinj site in Bed I at Olduvai Gorge (1327). Conceivably, then, robust australopithecines also made some of the tools.

The rudimentary use and manufacture of tools by chimpanzees suggests that robust australopithecines probably inherited a similar proclivity from the common ancestor of chimpanzees and hominids. In addition, like hand bones of *Homo habilis* and later members of the genus *Homo*, hand bones assigned to *Paranthropus robustus* from Swartkrans Cave include distal phalanges with broad tips (apical tufts), proximal phalanges that are straight, and a first (thumb) metacarpal with a relatively broad head (distal epiphysis) (2101, 2102, 2104, 2105). In these features, *Homo* and *P. robustus* differ from chimpanzees and *Australopithecus afarensis*, in which the distal phalanges have narrow tips, the proximal phalanges are curved, and the first metacarpal has a relatively narrow head. The morphology of the first metacarpal head has been especially emphasized, since relatively great breadth is associated with the precision grip in living humans (as opposed to the power grip in apes). The precision grip involves an enhanced ability to grasp objects between the tip of the thumb and the tips of the other fingers, and it is arguably crucial for stone flaking. Its absence may partly explain why Kanzi the bonobo cannot readily manufacture sharp-edged stone flakes like those that typify the Oldowan.

There is the problem, however, that the assignment of Swartkrans hand bones to *P. robustus* is purely circumstantial. It follows from the observation that 97% of the associated hominid craniodental fragments represent *P. robustus* while only 3% come from *Homo*. Metacarpal similarity between *P. robustus* and *Homo* depends on a single specimen that might actually represent *Homo*. Arguably, too, the metacarpal and other available hand bones are insufficient to document precision grasping, and a much larger portion of the hand skeleton is required (1473).

Finally, if both early *Homo* and the robust australopithecines made tools, we might expect to find two distinct, contemporaneous toolmaking traditions. The amorphous nature of Oldowan assemblages might obscure two separate traditions, but this is not true of the more patterned, less amorphous Acheulean assemblages that succeeded Oldowan ones about 1.7 my ago. There is broad agreement that the early Acheulean was produced by the *Homo* lineage, since it plainly anticipates the later Acheulean when only this lineage persisted. The robust australopithecines certainly survived after the emergence of the Acheulean, perhaps for more than 500 ky, yet there is no compelling evidence for a second contemporaneous, divergent artifact tradition. The implication is that the robust australopithecines made relatively few if any chipped stone artifacts. Their limited reliance on stone tools (if it existed) may have been correlated with their emphasis on vegetal food and on powerful jaws to process it, as well as with their failure to develop larger brains. Conversely, an increasing dependence on stone tools may be strongly related to the reduced jaws, enlarged brain, and probable greater carnivory of early *Homo*. Very likely an enlarged brain, reduced jaws, and increased carnivory also promoted greater stone tool use, which may thus have been both cause and effect in the emergence and ongoing evolution of *Homo*.

Structures and Fire

At site DK in Olduvai Gorge Bed I, Oldowan artifacts and fragmentary animal bones were associated with natural lava blocks, 10–25 cm across, clustered in a circle 4–5 m in diameter (fig. 4.41) (1327). A partial skull of *Homo habilis* was found at the same level nearby. Conceivably the circle marks the position of a stone windbreak or the base of a brush superstructure, but it could also have been created naturally when the radiating roots of a tree fractured a lava layer that lies directly underneath (1709). Modern trees are known to do this, and no truly comparable structural remnants have been reported from other Oldowan sites. Concentrations of large natural stones, Oldowan artifacts, and animal bones have been recorded at Melka Kunturé (458), but the sedimentary context suggests that stream flow could be responsible.

The evidence for Oldowan use of fire is equally ambiguous. Most compelling would be concentrations of charcoal surrounded by stone artifacts and other cultural debris, but charcoal does not survive long in dry, open-air African sites (1118). Attention has therefore been focused on burned or heated sediments, particularly the reddened patches found at Koobi Fora site FxJj 20 (184) and the fragments of burned clay associated with Oldowan artifacts at Chesowanja (876). Arguably, the depth and degree of reddening (oxidation) of one patch at FxJj 10 is best explained by a campfire that was repeatedly lit or that burned for several days. However, at both FxJj 20 and

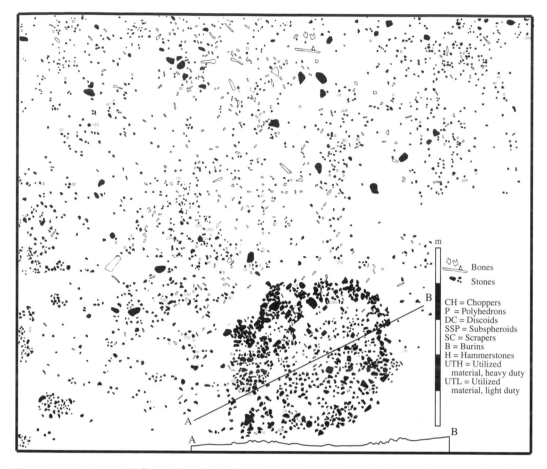

Figure 4.41. A partial floor plan of site DK 1 at Olduvai Gorge (redrawn after 1327, fig. 7). The most striking feature is the large circular concentration of basalt fragments in the right foreground. This may mark the base of the oldest known structure in the world, or more prosaically it may indicate only the location of a tree whose radiating roots fractured pieces of basalt and forced them up from immediately below the occupation surface.

Chesowanja, the burned sediments resemble those that result when vegetation smolders after a brushfire (1117, 1118), and the first appearance of burned bones at Swartkrans only in Member 3 may indicate that control over fire was an Acheulean (post-Oldowan) achievement (326), between 1.5 and 1 my ago. For the moment, the oldest reasonably secure evidence for human use of fire comes from relatively young sites in midlatitude Eurasia, where fire was probably essential for human survival. Unequivocal hearths are commonplace in Middle Paleolithic (Mousterian) sites postdating 200 ky ago, and ash, charcoal, and burned bones may indicate fire use by the Lower Paleolithic people at Zhoukoudian Locality 1 in northern China, putatively dated to about 500 ky ago (1131).

Subsistence and Behavior

Most modern hunter-gatherers eat much more meat than chimpanzees or other nonhuman primates, and paleoanthropologists have long assumed that increased meat eating characterized even the early phases of human evolution. As proof they have cited the fragmentary animal bones found with Oldowan artifacts at many sites, especially at Olduvai Gorge and Koobi Fora. However, archeologists have recently realized that a stratigraphic association between artifacts and bones does not demonstrate a functional link. The problem is that the associations occur almost entirely near ancient lakeside or streamside sites where artifacts and bones could have been brought together by various events or even by chance. For any site, the a priori possibilities include hunting by people, possibly followed by carnivore scavenging; "natural" deaths from carnivore predation, accidents, disease, and so on, followed by human scavenging; human accumulation of bones at a central place, variously known as a living site, occupation site, or home base, perhaps followed by carnivore scavenging; animal accumulation of bones at a den, followed by human scavenging; the coincidental association of artifacts and bones at watering points used by people and animals at separate times; and stream flow concentrating artifacts and bones from different original sites on a sandbar or gravel bar. Stream action or sheet wash could also disturb or blur a bone-artifact association created another way, obstructing interpretation.

The hypothesis that Oldowan artifact-bone concentrations represent ancient home bases was particularly popular in the 1970s, owing to a thoughtful exposition by Isaac (1115, 1116, 1118). He pointed out that only humans (and not apes) repeatedly use a "home base" or "base camp" where garbage accumulates and that the use of a home base is linked to other distinctively human traits, such as food sharing among group members and a division of labor between the sexes. If Oldowan artifact-bone concentrations represent home bases, then an important constellation of typically human behavior patterns had emerged by at least 2 my ago. The home base hypothesis remains intriguing, but it has been placed in limbo, at least temporarily, as archeologists struggle to isolate the actual human contribution to Oldowan bone assemblages.

Separating the human from potential "natural" contributions is difficult, but stream action can generally be ruled out when bones and artifacts appear unabraded (fresh); pieces of different sizes and weights are present and not obviously sorted by size across the surface of a site; and numerous bone or stone fragments in close proximity can be conjoined to reconstruct the larger pieces they came from. Conjoinable bone fragments and artifacts have proved especially useful in demonstrating that some early Koobi Fora sites were probably not seriously disturbed by flowing water (388, 1276, 1277). In addition, of course, the possible or probable effects of flowing water can usually be established from the sedimentologic-geologic context.

Where stream disturbance or transport was unimportant, and where bones were probably concentrated by a biological agent, the problem is to separate the potential role played by people from that of other known bone collectors—carnivores and porcupines. The most useful criteria are the nature of bone damage (fig. 4.42) and the pattern of skeletal part representation. Studies of recent bone assemblages accumulated by porcupines show that they usually leave their distinctive gnaw marks on more than 60% of the bones (320). If this criterion is used, porcupines probably did not accumulate any of the bone assemblages associated with Oldowan tools or early hominid remains in either eastern or southern Africa. In addition, at early east African sites, a porcupine role is ruled out by the sites' location on ancient land surfaces, near streams or lakes. Porcupines accumulate bones in abandoned antbear holes, caves, and other recesses, not in the open air.

In general, a human role in bone collection is clearly implied by the presence of artifacts and, more directly, by bones with damage marks from stone tools. Similarly, a carnivore role is indicated by the presence of fossilized carnivore feces (coprolites) and by bones with damage from carnivore teeth (fig. 4.42). Damage from carnivore chewing is common at southern African australopithecine sites, and at those sites where artifacts are totally absent (Makapansgat and Sterkfontein Member 4) carnivores were probably the sole collectors (320, 1421). This contradicts the claim that these sites contain australopithecine and baboon skulls fractured by clubs before death (573); the skulls were probably damaged after burial, by pressure from hard objects in the ground (318, 320).

Some human role is obviously plausible at those southern African sites with possible Oldowan artifacts (Sterkfontein Member 5 and Swartkrans Member 1), but in each case the tools are relatively rare, and the concreted sediments have precluded excavation to demonstrate the precise nature of bone and artifact associations and to obtain large, unbiased bone and artifact samples. In general, bone damage and the absence or rarity of artifacts imply that carnivores introduced most of the australopithecine bones to South African sites, perhaps when

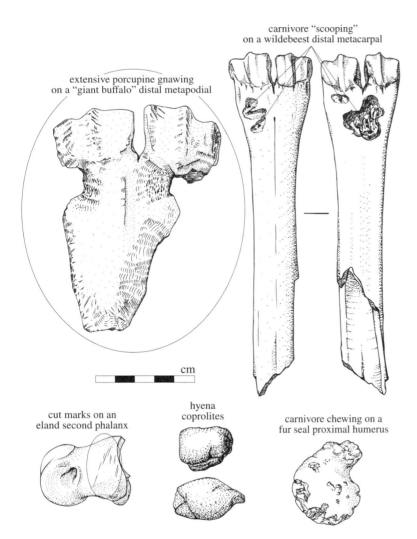

extensive porcupine gnawing
on a "giant buffalo" distal metapodial

carnivore "scooping"
on a wildebeest distal metacarpal

cm

cut marks on an
eland second phalanx

hyena
coprolites

carnivore chewing on a
fur seal proximal humerus

Figure 4.42. *Upper left:* A porcupine-gnawed "giant buffalo" distal metapodial. *Upper right:* A black wildebeest distal metacarpal on which the outer table of the bone has been partially removed ("scooped out") by carnivore gnawing. *Lower left:* Multiple, sub-parallel cut marks on an eland second phalanx. *Lower middle:* Hyena coprolites (fossilized feces). *Lower right:* A carnivore-chewed fur seal proximal humerus. The buffalo and wildebeest metapodials were originally similar in form, and the current difference shows the extensive remodeling that porcupines can produce. The carnivore-damaged bones closely match specimens in collections produced by recently observed carnivores. Cut marks frequently cluster in a pattern like that on the eland second phalanx. The buffalo metapodial comes from Middle Stone Age layer 16 at Klasies River Mouth Cave 1, Western Cape Province, South Africa; the wildebeest metacarpal and the hyena coprolites come from a fossil hyena lair at Deelpan, Free State Province, South Africa; and the eland and fur seal bones come from the Later Stone Age site of Kasteelberg, Western Cape Province, South Africa. (Buffalo metapodial drawn by Katharine Scott. Remaining items drawn by Kathryn Cruz-Uribe; © 1999 by Kathryn Cruz-Uribe.)

carnivores denned in the site (Makapansgat) or when they preyed on australopithecines sheltering on rock ledges at the cave entrance (Swartkrans, Sterkfontein, and perhaps other sites). Carnivore feeding could explain why the sites contain so few australopithecine postcranial bones versus more durable skulls and jaws (320) and also why juvenile skulls of *Australopithecus africanus* are significantly rarer than those of *Paranthropus robustus* (1435, 2195). Juvenile skulls of *A. africanus* were significantly less robust than their *P. robustus* counterparts and were thus more likely to be destroyed. Both kinds of australopithecines are remarkably abundant in the South African sites, and this could mean either that the live populations were relatively large or, more likely, that the sites were places where australopithecines were especially likely to die. In either case the sites still reflect carnivore behavior more than human behavior, and inferences about early human behavioral evolution must thus depend more on east African evidence.

Interpretation of the east African sites is far from straightforward, however, partly because the evidence from bone damage is ambiguous. Large carnivore teeth and stone tools tend to fracture and flake bones in very similar ways, and marks left by stone tools can closely resemble ones made by other agencies, especially carnivore teeth. Some specialists believe that stone tool marks can be identified with the naked eye or a light microscope (384), whereas others think reliable identification requires a scanning electron microscope, with its superb depth of field and high resolution (1935, 1941). A further complication is that contact with sharp particles during trampling or other postdepositional disturbance can simulate cut marks on bones (51, 176). Moreover, at most early archeological sites, even when the sources of bone damage can be reasonably identified, damage or associated objects (artifacts and coprolites) usually implicate both people and carnivores, and the problem is how to estimate their relative contributions. The bone assemblage associated with Oldowan artifacts at the FLK Zinj site in Olduvai Bed I presents a prime case in point. The question of carnivore versus human influence on this assemblage has been debated for nearly twenty years (for example, by 216, 223, 244, 383, 421, 422, 1634, 1711, 1909, 1937), but no consensus is in sight. A comprehensive, seminal analysis by Bunn and Kroll (389) and the responses illuminate the reasons.

Excavations by M. D. Leakey (1327) at the FLK Zinj site exposed an area of about 315 m² on an ancient land surface bordering the Bed I Olduvai Lake. An overlying volcanic ash has been dated to approximately 1.75 my ago. The horizontal and vertical distribution of artifacts and bones shows that flowing water did not materially disturb the site (1276). The bones are generally not very weathered, and weathering differences among them could result either from different times of final burial (389), from varying degrees of shade before burial (320), or from different times of first arrival. Different times of arrival could imply that the bones accumulated over five to ten years (1709–1711), and if people were the main accumulators, they were behaving very strangely by comparison with modern tropical hunter-gatherers, who rarely occupy the same spot for more than a few months and then abandon it for even longer periods. The implication might be that people used FLK Zinj in a unique, nonmodern way, perhaps as a place to cache stones for processing animal bones (1709).

In total, the FLK Zinj excavations produced about 2,500 classic Oldowan artifacts and 60,000 bones. The bones include approximately 16,000 from very small mammals (microfauna) and about 3,500 that can be securely assigned to large mammals (mainly various antelopes; fig. 4.43). The remaining specimens are small fragments (mostly less than 20 mm long) that probably also came from large mammals but that cannot be assigned to skeletal part or taxon. In their analysis, Bunn and Kroll focused on skeletal part representation and on the abundance and placement of stone tool cut marks on the 3,500 indisputable large mammal bones. The sample need represent no more than forty to forty-five individual animals and is admittedly very small compared with many later archeological samples, but it remains the largest and one of the best-preserved bone samples from any Oldowan site (385, 389).

Among the large mammal bones, especially the largest ones, Bunn and Kroll found that craniums, vertebrae, ribs, and metapodials were rare compared with mandibles, humeri, radio-ulnae, femurs, and tibiae. The rare parts are the bulkiest, the least nutritious, or both, and they are therefore commonly left at carcasses by both large carnivores and human hunters. The most abundant parts at FLK Zinj are meaty or marrow-rich bones that predators and people frequently take away. From this Bunn and Kroll conclude that FLK Zinj could not have been a kill or butchery site but must have been a place to which bones were carried. Equally important, they believe that the bone collector must have had early access to carcasses, either as a primary predator or as an efficient and perhaps high-ranking (dominant) scavenger.

Since investigators disagree on the identification of cut marks, Bunn and Kroll dealt only with marks they felt were totally convincing both macroscopically and microscopically. These occurred on 172 specimens, mainly limb bones, where

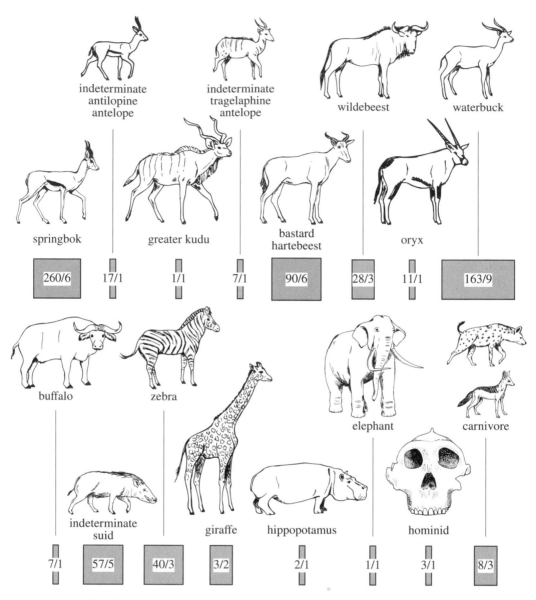

Figure 4.43. The abundance of various large mammals at the FLK Zinj site in Olduvai Gorge Bed I (data from 389). The bars are proportional to the minimum number of individuals by which each species is represented. The numbers superimposed on the bars are the number of bones assigned to each species over the minimum number of individuals. The sample is small, but it still shows the preponderance of antelopes and other medium-sized species that tends to characterize all African Stone Age sites, regardless of age. As discussed in the text, specialists disagree on how the animal bones accumulated at FLK Zinj—whether it was mainly through human activity and, if so, whether the people were hunting or scavenging.

about two-thirds of the marks were on midshafts and only one-third were near joints (epiphyses). This is the pattern Bunn and Kroll predicted for a situation in which the people were not simply disarticulating carcasses but were stripping meat from them afterward; that is, it is the pattern expected where there was still substantial meat to remove. From the abundance of cut marks and artifacts, Bunn and Kroll concluded that Oldowan people must have brought most of the bones to the FLK Zinj site, and from skeletal part representation and the positioning of cut marks, they argued that the people were probably primary predators or efficient scavengers with early access to carcasses.

Bunn and Kroll maintained that the people were probably hunters more than scavengers, since Stone Age people probably could not find fresh carcasses before other scavengers did. In Africa today, early arriving scavengers, particularly hyenas, often quickly consume or remove meaty parts, leaving little for latecomers. Initially, Bunn and Kroll also suggested that hunting was more likely than scavenging because the FLK Zinj sample was dominated by prime-age adult animals and not by the very young and aged individuals whose carcasses are most available to scavengers. On reconsideration, however, they concluded that many individuals they originally called prime were actually past prime, which makes the age (mortality) profile consistent with either scavenging or hunting.

The abundance and placement of cut marks clearly provide the strongest support for Bunn and Kroll's hunting or efficient scavenging hypothesis. However, postdepositional trampling of juxtaposed artifacts and bones could have produced many of Bunn and Kroll's midshaft cut marks (177). Perhaps even more crucial, many of their cut marks do not meet the scanning electron microscope criteria that Shipman (1935) considers crucial for cut mark identification, and a study employing the tighter criteria that Shipman prefers found far fewer cut marks on the same bones (1711).

In addition, a high proportion of cut marks on midshafts may not indicate early carcass access (1934–1936). Arguably, hunting followed by disarticulation would produce relatively more marks near joints, while protective muscle masses would minimize marks on limb bone midshafts. An abundance of midshaft marks might in fact mean that some other creature had already eaten the muscles (meat). This argument runs counter to the relative abundance of midshaft marks on bones of domestic animals from a very Late Stone Age (Pastoral Neolithic) site in Kenya where people clearly had first access to animals (1460), but many more comparative observations will be necessary for firm resolution of the issue.

Finally, a careful examination of bones from various Bed I Oldowan sites found thirteen specimens on which cut marks and carnivore tooth marks overlapped (1934, 1935). In eight cases the tooth marks overlay the cut marks; in five cases the cut marks were on top. The full implications of the overlapping marks are debatable, but at a minimum they indicate that people sometimes did scavenge carcasses that other animals had already fed on, and they suggest that carnivores contributed significantly to some Bed I assemblages, perhaps including FLK Zinj.

Another possible objection to Bunn and Kroll's interpretation is the implication that the FLK Zinj people were remarkably successful at obtaining meat. This seems at odds not only with expectations for this remote period of 1.75 my ago but also with inferences drawn from much later archeological bone assemblages, which suggest humans had relatively limited ability to hunt large mammals (1243). Additional objections stem from other FLK Zinj data. Bunn and Kroll found that approximately four hundred of the bones were damaged by carnivore teeth, which could mean that carnivores scavenged bones people had abandoned but could also mean that carnivores had access to the bones before people did. Even if it means only that carnivores scavenged bones left by people, it still implies that carnivore feeding or removal helped shape the final pattern of skeletal part survival. This is a prospect that Bunn and Kroll do not explicitly consider. They also do not address the possibility that the pattern of skeletal part composition was modified significantly by leaching, profile compaction, and other postdepositional processes that selectively remove softer or more fragile bones or render them unidentifiable. Selective postdepositional destruction has probably affected skeletal part abundance in most fossil assemblages (1244), but its implications for behavioral reconstruction are rarely considered.

Finally, the abundant microfaunal bones at the FLK Zinj site require special attention. Microfaunal bones are also common at other Bed I Oldowan sites, and they were almost certainly introduced not by people, but by raptors or small carnivores such as genets (43). The implication may be that FLK Zinj and other sites were originally clusters of shade trees or bushes near water that attracted a variety of creatures, some of which brought in bones or found the presence of bones an additional attraction. The trees may have provided people with refuge from predators or other scavengers (1935) or with protected, shady places to cache artifacts for processing scavenged bones (1709). The artifacts and cut marks show that people were important at FLK Zinj, but they need not have been the principal bone accumulators, since clusters of bones occur under shade trees or bushes in Africa today, in the absence of human hunter-gatherers (2137). With regard to FLK Zinj and other Bed I sites, the shade tree–artifact cache hypothesis becomes especially plausible if bone accumulation spanned many years, as one observer has suggested (1709, 1710).

For other Oldowan sites, the basic implications of the animal bones are at least as equivocal as for FLK Zinj, and truly compelling interpretations will probably require many fresh, large, well-preserved samples from new sites. These may permit the controlled interassemblage comparisons that are necessary to separate the roles of people, carnivores, and probable postdepositional destructive factors in shaping the composition of individual assemblages. However, it is possible to overemphasize the problems in interpreting Oldowan bone assemblages, and these difficulties should not be allowed to obscure the genuine knowledge that Oldowan people were processing animal parts in a way and to a degree unknown among chimpanzees or any other nonhuman primates (1118). All commentators agree that stone tool percussion to obtain marrow explains the splintering of many limb bone shafts at FLK Zinj and other Oldowan sites. No ape is known to break bones this way, nor do apes employ sharp flakes to sever connecting tissue or strip away muscle masses. Cut-marked bones have even been found at sites where stone artifacts are absent, and some of the bones come from elephants, hippopotamuses, and other animals that are far larger than those other primates are known to feed on (2035). Whatever the precise meaning of FLK Zinj and other sites, the splintered and cut-marked bones and the artifacts used to process them argue that an increased emphasis on animal food was indeed important in early human evolution.

What remains contentious is how the food was gotten—whether mainly by hunting, by scavenging, or by some mix of the two. It seems likely there was some scavenging, particularly of carcasses of large animals that would be difficult or dangerous to kill. Chimpanzees show little interest in carrion, despite their strong interest in fresh meat (2034), and an inclination to scavenge could have been the key ecological trait that distinguished protopeople from protochimpanzees. Moreover, sedimentary and fossil evidence from Olduvai, Koobi Fora, and other east African sites indicates that ancient human bones, artifacts, or both often accumulated in relatively well wooded water-edge environments where scavenging opportunities may have been particularly abundant (242, 243, 245). In comparable environments today, substantial amounts of scavengeable meat and marrow bone escape hyenas, lions, leopards, and other scavenger-predators and are especially available in the drier parts of the year. The amounts may have been even greater 1.8–2 my ago, when the carnivore guild included saber-toothed cats and perhaps other species whose capacity to clean or crack bones was relatively limited (1438). Scavenging would have been especially productive if early people were even moderately successful at driving off large cats or hyenas, as the Hadza people of northern Tanzania often do today (1617).

Proof to check or supplement the logic of the hypothesis of early hominid scavenging remains elusive, however. It is not the specialist studies that are at fault but the sheer complexity of the problem and the ambiguity of the surviving evidence. An equally compelling question that is even harder to answer concerns how much Oldowan people relied on animal food. Plant tissues that could represent food debris do not survive in such ancient sites, and even if they did it might be difficult to gauge how much food they represent versus the meat represented by bones. Again we must turn to logic, supplemented in this instance by studies of recent hunter-gatherers. These studies suggest that Oldowan people relied mostly on plants and perhaps on other gathered foods such as insects. In this light, their day-to-day food quest was probably far less bloodthirsty than some popular accounts have proposed (70, 71).

Summary and Conclusion:
Natural Selection and Hominid Origins

Numerous gross anatomical similarities long ago revealed the close kinship between people, chimpanzees, and gorillas, but rapid advances in DNA analysis now demonstrate how remarkably close the relationship is. Most genetic studies indicate that chimpanzees and people share an especially recent common ancestor, and if a constant rate of genetic divergence is assumed, protopeople probably split from protochimpanzees only 5 to 7 my ago. The split almost certainly occurred in Africa, though so far the African fossil record between 10 and 5 my ago is too sparse to illuminate it.

Between 10 and 5 my ago, a global trend toward cooler, drier climates promoted the spread of grasses at the expense of trees or bush at lower latitudes throughout the world (440). One indication of this change is that high-crowned, grassland ungulates proliferated while forest creatures declined in numbers and diversity (438). Among the adversely affected animals were various ape species that had prospered between 17 and 10 my ago. As the forests shrank or thinned after 10 my ago, many ape species became extinct, especially in Eurasia; but in Africa at least one ape adapted to the changed conditions by spending more time on the ground, perhaps to move between more widely scattered trees or tree stands. Living on the ground presented new challenges and opportunities that favored those individuals whose morphology and behavior gave them a reproductive edge, however slight, over their peers. The known fossil record implies that ground living especially encouraged a new mode of habitual locomotion, bipedal rather than quadrupedal.

Arguably, the emergence of ground-oriented, fully bipedal apes was sparked by a tendency for particularly harsh cold between 6.5 and 5 my ago. During this time periodic growth in the Antarctic ice cap produced dramatic drops in sea level that repeatedly drained the Mediterranean Sea, robbing the adjacent continents of an important moisture source and thereby accelerating the contraction of moisture-dependent forests. Many species could not cope and became extinct, but the most abundant large African mammals—the bovids (antelope and buffalo)—burgeoned to produce eight of the nine zoological tribes that dominate nearly all subsequent African faunas (2350, 2351). Today the most conspicuous of these tribes is perhaps the Alcelaphini, which includes the wildebeests, hartebeests, and bastard hartebeests, but others such as the Reduncini (reedbucks and waterbucks), the Tragelaphini (eland, kudus, and their allies), and the Cephalophini (duikers) remain locally prominent. Species of the various tribes commonly comprise 60% to 80% of the fossils at early hominid sites, and their habitat preferences are well established. Their fossil representation thus illuminates the environments in which early hominids lived.

The initial advantages of bipedalism may have included the ability to carry meat or other food items to trees or other refuges or to other group members (1019, 1020, 1396); a reduction in the amount of skin surface exposed to direct sunlight at midday, thus reducing the danger of heat stress, particularly for the brain (2431, 2432); the freeing of the hands for tool use (577, 2393) or for carrying the young long distances (1977); a decrease in the energy required to walk at normal (low) speeds (1801, 2046); the ability to see more clearly or farther during passage from place to place (576, 593); or the enhancement of threat displays that reduced violent conflicts over scarce resources (1124).

The various alternatives are difficult to disprove, and one or more could have operated in concert. The least likely is perhaps selection to reduce midday heat stress, since fossil pollens, sediments, geochemistry, and especially animal remains indicate that the earliest hominids inhabited relatively wooded, well-shaded environments (1745). It was probably only about 1.8–1.7 my ago, with the advent of *Homo ergaster*, that hominids first invaded the kind of arid, more open settings where daytime foraging could have produced dangerous heat loads. As discussed in the next chapter, heat adaptation plausibly explains why *H. ergaster* was the first hominid to possess an external nose (753), why it was also the first to achieve essentially modern body size, and why it was distinguished by the same long, linear body form that characterizes historic humans in equatorial grasslands and savannas (1846, 2434). If *H. ergaster* was specially adapted to heat, it may also have been the first hominid to possess a largely hairless, naked skin. A hairless skin is essential to promote heat loss through sweating (2433), and for creatures that must remain active at midday under dry, sunny conditions, sweating is the most effective mechanism for maintaining tolerable body and brain temperatures (446). Effective sweating was probably crucial for the significant brain expansion that followed the appearance of *H. ergaster*.

A link between bipedalism and tool use may seem unlikely, since the fossil record suggests that bipedalism antedates stone tool manufacture by more than 2 my; but like the tools that living chimpanzees use to probe for termites, to crack nuts, and so forth, the earliest human tools may have been made of wood or other highly perishable materials. It is also important to recall that the earliest people (broadly defined) probably lacked the typically human habit of occupying a single locale for several days (or nights), which means they may not have left stone tools in archeologically visible clusters.

To begin with, it was only bipedalism that clearly differentiated people from apes, and fossils dating between 4.4 and 2.5 my ago or even later reveal people who remained remarkably apelike in the small size of their brains, in the tendency for males to be much larger than females, in their relatively long arms and short legs, and in features of the shoulder, the arm, the hand, and even the foot that aided in tree climbing. No early hominid site hints that protochimpanzees or gorillas lived nearby, and early hominid environments must have been less forested than the ones chimpanzees or gorillas prefer. However, fossil pollens, other occasionally preserved plant parts, and above all animal remains show that trees remained important wherever very early people lived (267, 1745), while the morphology and ecology of the earliest people indicate they were effectively bipedal apes who still spent considerable time in trees, feeding, sleeping, or avoiding predation. Their degree of sexual dimorphism may even imply a chimpanzeelike social organization in which groups of closely related males competed intensely with other groups for territories and among themselves for access to estrous females (745, 1508). Broadly the same kind of organization may have persisted to 1.8–1.7 my ago, when a sharp reduction in sexual dimorphism marks the advent of *Homo ergaster* (1510).

In both eastern and southern Africa, plant and animal fossils show that tree cover shrank dramatically between 3 and 2 my ago, in accord with accelerated cooling and drying globally and in Africa (648, 1921). By 2 my ago, many forest- or bush-adapted bovid species that existed at 3 my ago had disappeared, while grassland or savanna species had multiplied to dominate the fauna near all known hominid sites. The turnover in bovid species may have occurred in a relatively sharp pulse between 2.8 and 2.5 my ago (2347, 2348, 2351), or it may have happened more gradually over the entire 3 my to 2 my span (179, 1214). Evolutionary change in other important groups, such as pigs and cercopithecoid monkeys, clearly occurred more gradually, but pulsed turnover may also have affected micromammals (mostly small rodents and insectivores) (2428) and hominids. Evidence from other continents is patchy, but mitochondrial DNA suggests that North American songbirds also experienced a speciation pulse about 2.5 my ago (1250, 1839).

The cooling and drying trend between 3 and 2 my ago was not continuous but occurred in steps, and there were brief episodes when it was even temporarily reversed. More generally, the past 3 million years have witnessed many conspicuous climatic fluctuations that were superimposed on directional trends. The best-documented fluctuations were between glacial and interglacial periods, which differed sharply in average temperatures, particularly in middle and upper latitudes. In lower latitudes, including Africa, contrasts in precipitation were probably more important than differences in temperature, and major shifts in rainfall forced savannas, forests, and other vegetation types to fluctuate greatly in extent through time and space. One could argue that human evolution responded more to this climatic and environmental oscillation than to long-term directional trends (1705, 1706). But glacial/interglacial alternation occurred on a geologic time scale rather than a human one, and in general climate remained relatively stable for thousands or tens of thousands of years before it shifted to another mode. Substantial stretches of relative stasis particularly characterized the period between 3 and 2 my ago, even if the overall trend was toward significantly greater aridity and cooler temperatures. In these circumstances, large-scale climatic variability could have been a selective pressure only if it could operate on populations that had not experienced it and could not anticipate it. Human evolution after 2 my ago does reveal

a remarkable ability to cope with varied environments, but selection for life on the savanna is a sufficient explanation for the
implicit behavioral flexibility.

Whatever selective forces are preferred, by 2.5–2.4 my ago
at least two distinct hominid adaptations had emerged: one in
which subsequent change mainly involved an increased emphasis on the cheek teeth for masticating relatively coarse vegetal
foods, and one in which later change involved dramatic brain
expansion, a reliance on tools far exceeding that of living chimpanzees, and an increase in meat eating. Adaptations stressing
the grinding of vegetal foods with the cheek teeth may have
arisen at least twice, first before 2.5 my ago in the line represented by *Australopithecus aethiopicus* and a second time after
2.5 my ago in a line that culminated in the well-known "robust" australopithecines, *Paranthropus boisei* of eastern Africa
and *P. robustus* of southern Africa. Alternatively, *A. aethiopicus* may have been simply the ancestor of the later "robust"
forms. The adaptation involving radical brain expansion, increased tool use, and increased meat eating produced the genus
Homo. Very possibly, early *Homo* comprised two contemporaneous species, in which case there must have been yet a third
adaptation whose basis remains obscure. The only certainty is
that if this adaptation existed, it was eventually unsuccessful,
like the adaptations of the robust australopithecines.

The temporal coincidence between brain expansion in
Homo and the appearance of the first archeological sites between 2.5 and 2 my ago suggests a functional link between
neural and behavioral change, and the additional brain expansion that occurred about 1.7 my ago in the emergence of *Homo
ergaster* may have been likewise tied to the nearly simultaneous appearance of more sophisticated artifacts, including the
first hand axes. At both 2.5 my ago and 1.7 my ago, not only
may neural change have allowed new behaviors, but the new
behaviors in turn may have selected for more neural change.
Chapters 6 and 7 argue that a linkage between neural and behavioral evolution persisted throughout the entire span of human evolution, to the very time when fully modern humans
emerged 50 ky ago.

The next chapter maintains that *Homo ergaster* was probably the first hominid species whose anatomy and behavior justify the label human. An increase in terrigenous dust in sea bottom deposits off subtropical Africa (647, 648), changes in the
stable-carbon isotope composition of soil carbonates (437), and
shifts in plant and mammal species (266, 267) indicate that
H. ergaster emerged at a time when aridity, rainfall seasonality,
or both increased sharply in eastern Africa. The stable-isotope

analyses are particularly informative, since they show that before the appearance of *H. ergaster* about 1.7 my ago the principal grasses in eastern Africa must have been C_3 species adapted
to relatively cool or shady conditions, but that about 1.7 my
ago C_4 grasses, adapted to great heat and moisture stress, expanded dramatically. It is these drought-tolerant grasses that
dominate east African environments today, and the implication is that arid grasslands and savannas of fully modern aspect
appeared locally only about 1.7 my ago. The physical distinctions of *H. ergaster* suggest that it evolved in response, and it
was probably the first hominid species that could prosper in
truly arid settings with long dry seasons and highly unpre-
dictable rainfall. Its more fully terrestrial (as opposed to mixed
terrestrial-arboreal) adaptation was surely promoted by its enlarged brain, but brain enlargement may itself have depended
on a more fully terrestrial existence (2038), since the additional
brain growth occurred mainly postnatally and thus required a
longer period of infant dependency. During this period infants
could not cling to their mothers, as they do in chimpanzees and
as they probably did in the small-brained australopithecines.
Instead they had to be carried, and carrying became feasible
only when mothers no longer depended on tree climbing.

As larger-brained, tool-using people became more carniv-
orous, they surely reduced the amount of animal protein and
fat available to large carnivores, and the emergence of *Homo*
may partly or wholly account for a decrease in the number of
African carnivore species after 2 my ago (2369). Alternatively,
at least some carnivores may have succumbed to the same increase in aridity that apparently favored *H. ergaster*. Greater
aridity may have particularly jarred the saber-toothed cats,
since they seem to have been highly specialized for ambush
hunting in dense vegetation (1438). Suitable habitat probably
became much rarer and more patchy after 1.7 my ago. Similarly,
either niche expansion in *Homo* or climatic change may have
extinguished the robust australopithecines between 1.2 and
0.7 my ago. A priori, climate change may seem more likely,
since both the amplitude and duration of glacial periods grew
significantly about 1 my ago, and African climates became
even drier on average.

The biological and behavioral advances of *H. ergaster* help
explain how it became the first hominid species to colonize the
southern and northern extremes of Africa and ultimately how
it was able to exit Africa altogether. The dating of the exodus
is uncertain, but people were certainly present in eastern Asia
by 1 my ago, and they probably reached Europe between 1 and
0.5 my ago. The archeological record postdating 1 my ago is

actually better known in Eurasia than in Africa, mainly because there have always been more archeologists in Eurasia than in Africa. The unique selection pressures that Eurasian environments presented may help explain some of the major morphological and behavioral advances that occurred after 1 my ago (discussed in chapters 5–7), but Africa did not become a backwater. On the contrary, evidence is accumulating that it may have remained the main center of human evolutionary change up to and including the emergence of anatomically modern people.

EVOLUTION OF THE GENUS *HOMO* 5

For almost three decades, paleoanthropologists have often divided the genus *Homo* among three successive species: *Homo habilis*, now dated between roughly 2.5 million years (my) and 1.7 my ago; *Homo erectus*, now placed between roughly 1.7 my and 500 thousand years (ky) ago; and *Homo sapiens*, after 500 ky ago. In this view each species was distinguished from its predecessor primarily by larger brain size and by details of craniofacial morphology, including, for example, a change in braincase shape from more rounded in *H. habilis* to more angular in *H. erectus* to more rounded again in *H. sapiens*. An early or "archaic" form of *H. sapiens* before about 250 ky ago was set off by a tendency to retain many *H. erectus* features, including thick cranial walls, a low, flattened frontal (forehead), a broad cranial base from which the sidewalls sloped inward, and a relatively massive, chinless jaw with large teeth. Different variants of *H. sapiens* were sometimes distinguished as subspecies, such as *H. sapiens neanderthalensis* (the Neanderthals) between roughly 127 and 40–30 ky ago, and *H. sapiens sapiens* (fully modern humans) after 40–30 ky ago. Authorities debated whether *H. erectus* evolved into *H. sapiens* in the gradual mode that Darwin proposed (2482) or whether *H. erectus* remained largely unchanged for perhaps a million years (1766, 1770, 1772) and then evolved into *H. sapiens* in a spurt like those that punctuate the evolution of many other species (872, 2037).

The conventional view still has adherents, and one prominent authority has even sunk *H. erectus* into *H. sapiens*, primarily because the boundary between the two is very fuzzy (2486, 2487, 2489, 2491). In his view, *H. sapiens* thus originated about 1.7 my ago and subsequently diversified into regional forms that maintained genetic contact and that gradually evolved into living populations in the same regions. The first edition of this book embraced the conventional *H. habilis* to *H. erectus* to *H. sapiens* sequence, but it argued that only African "archaic" *H. sapiens* evolved into living humans, whereas archaic *H. sapiens* in Europe and Asia became extinct. It also presented evidence, reviewed again below, that *H. erectus* persisted in eastern Asia long after archaic *H. sapiens* had appeared in Africa and Europe. New dates underscore the very late survival of *H. erectus* in Indonesia (2122), and genetic analyses summarized in chapter 7 imply ever more strongly that Eurasian archaic *H. sapiens* had little or nothing to do with living people.

The accumulating evidence has increasingly undermined a scenario based on three successive species or evolutionary stages, and it strongly favors a scheme that more explicitly recognizes the importance of branching in the evolution of *Homo* (fig. 5.1). The revision offered here continues to accept *H. habilis* (or one of the species into which it may eventually be split) as the ancestor for all later *Homo*. Its descendant at 1.8–1.7 my ago may still be called *H. erectus*, but for reasons specified below, another name is preferable. Depending on precisely what fossils are included, *H. leakeyi* is available (507), but *H. ergaster* is more widely accepted and is used here. *H. ergaster* was the first human species to colonize Eurasia, and by 1 my ago it had given rise to *H. erectus* in the Far East. By 600–500 ky ago, it had produced additional lines leading to *H. neanderthalensis* in Europe and to *H. sapiens* in Africa. Arguably, about 600 ky ago *H. neanderthalensis* and *H. sapiens* shared a common ancestor to which the name *H. heidelbergensis* could be applied (914, 1774, 1777, 2085, 2087). *H. heidelbergensis* contains many fossils that were previously central to "archaic" *H. sapiens*.

The reality may be far more complex, and the sparse fossil record may obscure an even greater number of fossil species (2139–2143). For example, cranial variation suggests that Far Eastern *H. erectus* could be divided between two lineages: a more conservative one in Southeast Asia (Indonesia) that retained the classic angular braincase of *H. erectus* from perhaps 1 my to 50 ky ago, and a less conservative line in China that evolved a more rounded braincase broadly like that of African and European *H. heidelbergensis*.

Together the human fossil record and archeology imply that the Far East had diverged sharply from Europe and Africa by 1 my ago, whereas Africa and Europe retained much in common at 500 ky ago and afterward. Everywhere, artifacts are much more abundant than human fossils, and they show that *H. ergaster* and its descendants invaded new environments, including arid, highly seasonal ones in Africa and cool temperate ones in Eurasia. By 600–500 ky ago, the range of primitive *Homo* extended from Spain on the west to northern China on the east and from southern Britain on the north to the Cape of Good

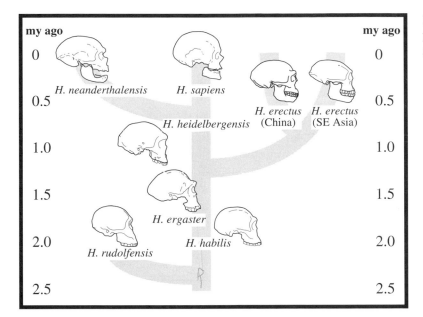

my ago

0

H. neanderthalensis   *H. sapiens*

0.5

H. heidelbergensis   *H. erectus* (China)   *H. erectus* (SE Asia)

1.0

1.5

H. ergaster

2.0   *H. habilis*

H. rudolfensis

2.5

my ago

0

0.5

1.0

1.5

2.0

2.5

Figure 5.1. The phylogeny of the genus *Homo.*

Hope on the south (fig. 5.2). The vast distances that separated far-flung populations and their contrasting environments help explain the morphological differentiation that accelerated after 500 ky ago. This chapter outlines the pattern of human biological and behavioral development between the appearance of *H. ergaster* 1.8–1.7 my ago and the emergence of the classic Neanderthals (*H. neanderthalensis*) and their contemporaries by 127 ky ago. For heuristic purposes, all populations that existed between 1.8 my and 127 ky ago will be referred to as primitive *Homo.*

The Discovery of Primitive *Homo*

The history of primitive *Homo* is usually traced to the discovery of the type Neanderthal fossil near Düsseldorf, Germany, in 1856. The historic significance of this event, described in the next chapter, cannot be overemphasized, but in brain size and other aspects of morphology the Neanderthals were very similar to living people. The similarity is so great that early on they were often seen not as fossil humans, but as pathological recent ones. The first human fossils whose profound evolutionary significance could not be so easily dismissed were those of *Homo erectus.* Their story begins with the Dutch physician and visionary Eugène Dubois (677, 1087, 1626, 2167, 2168).

Dubois was born in 1858, a year before the publication of Darwin's great classic, *The Origin of Species.* He developed a passion for human evolution, and he effectively founded the

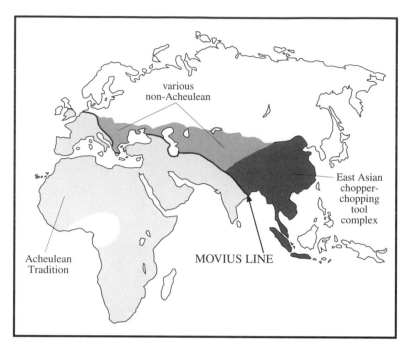

Figure 5.2. The occupied world roughly 500 ky ago (redrawn after 1873, p. 571). Only the shaded portions of the map were inhabited. Occupation is indicated mainly by stone artifacts, which divide between two distinct geographic traditions: the hand ax or Acheulean Industrial Tradition in Africa, western Asia, and Europe and the chopper–chopping tool (or flake-and-chopper) complex in eastern Asia. The chopper–chopping tool complex may have extended westward across south-central Asia to eastern Europe. The boundary separating the Acheulean and east Asian chopper–chopping tool complexes in northern India is commonly known as the Movius Line after the person who first defined it (1585–1587, 1591).

field of paleoanthropology when he decided to search full-time for human fossils. He focused on Indonesia, which was then a Dutch colony and which he and others reasoned was a logical place to start, since it still contained apes that might broadly resemble protohumans. He obtained a medical appointment in the Dutch East India Army and arrived in Indonesia in December 1887. Initially he explored limestone caves on Sumatra, but these supplied only relatively recent animal fossils. In 1890 he turned his attention to Java, where a rugged human skull had been found in a quarry at Wajak (Wadjak) in 1888. His visit produced a second, similar skull, but even though both skulls were very robust, they were unequivocally modern (676). Today they are widely regarded as fossil representatives of anatomically modern people.

In Java, Dubois again focused initially on caves, but he soon switched to fossiliferous alluvial deposits near the center

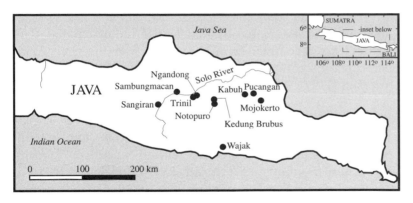

Figure 5.3. Major Quaternary fossil localities of Java (redrawn after 2168, fig. 1).

of the island (fig. 5.3). In November 1890, in deposits at Kedung Brubus, he found a very fragmentary human mandible, and in October 1891, near the village of Trinil on the Solo River, he recovered a low-vaulted, angular, thick-walled skullcap with large shelflike browridges. The skullcap was clearly far more primitive than the Wajak skulls, and the associated animal fossils indicated that it was also much more ancient. In August 1892, in what appeared to be the same bed at Trinil, Dubois unearthed a nearly complete human left femur that was morphologically modern in every respect. The femur and skullcap convinced him that he had discovered an erect, apelike transitional form between apes and people, in keeping with Darwin's hypothesis that upright posture preceded brain expansion in human evolution (577). In 1866 the German biologist Ernst Haeckel had proposed the generic name *Pithecanthropus* for a hypothetical missing link between apes and people, and in 1894 Dubois decided to call his fossil species *Pithecanthropus erectus* ("erect ape-man").

Ironically, it is now clear that Dubois did not appreciate the stratigraphic complexity of the Trinil locality, and the femur may have come from a much younger horizon than did the skullcap (148). In some important respects it looks much more modern than other, undoubted *Homo erectus* femurs (1206), and trace element composition may ultimately show that it is much younger (591). However, early debate on Dubois's finds centered not on their stratigraphic association but on whether they really represented a single species and, beyond this, on whether the skullcap was from an early human as opposed to an ape.

Dubois found no additional human fossils, and he returned to the Netherlands in 1895. Others subsequently explored the same deposits, but new human remains appeared only between 1931 and 1933. These included twelve partial or nearly complete skullcaps and two tibia fragments from ancient alluvium of the Solo River near Ngandong. The craniums were larger than Dubois's *Pithecanthropus* specimen (1863, 2404), and the associated animal fossils indicated that they were also much younger. From the time of their discovery, their evolutionary significance and taxonomic status has been controversial (1127). They have often been assigned to their own species, *Homo soloensis*, but they are regarded here as a very late form of *H. erectus.*

G. H. R. von Koenigswald (1255, 1256) was the first to report fresh Javan fossils that more closely matched Dubois's. The specimens comprised a partial skull found near Mojokerto (= Modjokerto = Perning), eastern Java, in 1936 and three partial craniums ("*Pithecanthropus* II–IV"), some partial jaws, and several dozen isolated teeth found near Sangiran, central Java, between 1937 and 1941. The Mojokerto skull came from cemented alluvial sands, but its precise find spot is now uncertain. It represented a very young individual, probably between four and six years old (62), but it still exhibited, in incipient form, the large browridges, flat receding forehead, narrowing (= constriction) behind the orbits, and angulated occipital of Dubois's *Pithecanthropus*. The Sangiran specimens probably also came from alluvial deposits, but they were recovered mostly by local residents, who sold them to von Koenigswald. All three skullcaps were essentially identical to Dubois's 1891 specimen. Two of them were accompanied by the same animal fossils (the "Trinil Fauna") as Dubois's find. The third occurred with different fossils (the "Djetis Fauna") that suggested a somewhat greater age.

The Sangiran discoveries validated Dubois's claim for a primitive human species, and the case was further reinforced by nearly simultaneous discoveries from near Beijing (formerly Peking), China (1088, 1146, 1925, 1926). The detection of primitive *Homo* in China can be traced to an age-old Chinese custom of pulverizing fossils for medicinal use. In 1899 a European doctor found a probable human tooth among fossils in a Beijing drugstore, and the quest for its origin led paleontologists to a rich complex of fossiliferous limestone caves and fissures on the slopes of Longgu-shan ("Dragon Bone Hill"), about 40 km southwest of Beijing, near the village of Zhoukoudian (formerly transliterated as Choukoutien) (fig. 5.4). In 1921 the Swedish geologist J. G. Andersson began excavating in a collapsed cave at Zhoukoudian that was particularly promising not only for its fossils, but also for quartz fragments that prehistoric people must have introduced. The cave became known as Locality 1 to distinguish it from other fossil-bearing caves nearby. In 1923 Otto Zdansky, an Austrian paleontologist who was working

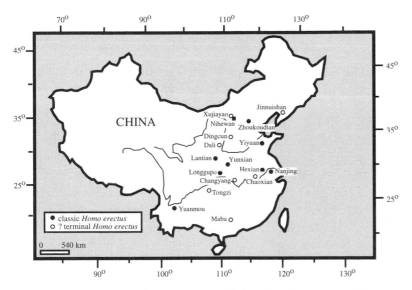

Figure 5.4. The main Chinese sites with fossils of primitive *Homo* (modified after 2533, fig. 1, and 695, fig. 2).

with Andersson, discovered a human molar among the Locality 1 fossils, and in 1926, after the excavations had ceased, he found a human premolar among specimens that had been shipped to Sweden.

The teeth inspired Davidson Black, a Canadian anatomist at the Peking Union Medical School, to seek support from the Rockefeller Foundation for expanded excavations. These began at Locality 1 in 1927, under the field direction of Birgir Bohlin, and they quickly produced a large human molar with a wrinkled crown, for which Black erected a new species and genus of fossil people, *Sinanthropus pekinensis* ("Peking Chinese man"). Many specialists felt Black had been too hasty, but in 1928 continuing excavations revealed two primitive partial human mandibles and some skull fragments. In 1929 W. C. Pei, the new field director, found a skullcap that eliminated all remaining doubt. Black died in 1933 and was succeeded in 1935 by Franz Weidenreich, a distinguished German anatomist from the University of Chicago. Excavations continued at Zhoukoudian until 1937 and eventually produced 5 more or less complete skullcaps, 9 large skull fragments, 6 facial fragments, 14 partial mandibles, 147 isolated teeth, and 11 postcranial bones. These represented more than forty *Sinanthropus* individuals, including men, women, and children.

In 1939 von Koenigswald and Weidenreich compared their fossils in China and concluded that they were extremely similar (1257). Weidenreich (2401) even proposed that *Pithecanthropus* and *Sinanthropus* were only subspecies of a single species, *Homo erectus,* though he continued to use the original generic names as labels.

The Japanese Imperial Army occupied northern China in 1937, and in 1941, anticipating intensified hostilities, Weidenreich moved to New York. For safety's sake, the Chinese authorities also decided to ship the *Sinanthropus* fossils to the United States, and they were entrusted to marine guards who were helping to evacuate the American embassy from Beijing. With their marine caretakers, the fossils arrived at the port serving Beijing on 7 December 1941, the day that Imperial Japan attacked Pearl Harbor. The marines were interned by the Japanese, and the fossils vanished (1134). There has been much speculation on their whereabouts, but they were probably simply discarded by persons who did not realize their value. Fortunately Weidenreich had prepared an excellent set of plaster replicas, which are now housed at the American Museum of Natural History in New York City. Along with his detailed monographic descriptions (2399, 2400, 2402, 2403), the replicas permit continued analysis of the fossils.

Von Koenigswald successfully hid his fossils with friends in Java during the war, and afterward he took them to Frankfurt. Some remain there, but many important specimens have been returned to Indonesia. Dubois had taken his specimens to the Netherlands, and they are still housed in Leiden.

In 1950 the evolutionary biologist Ernst Mayr (1481) argued that *Pithecanthropus erectus* ("Java man") and *Sinanthropus pekinensis* ("Peking man") were too similar to be placed in different genera, and like Weidenreich before him, he suggested that they should join living people in the genus *Homo.* In 1955 the eminent British anatomist W. E. Le Gros Clark sunk *Sinanthropus* into *Pithecanthropus* (1315), and in 1964 he sunk *Pithecanthropus* into *Homo* (1317). Other influential authorities agreed (1071, 1080), and by the late 1960s the concept of *Homo erectus* was widely accepted. It readily accommodates additional specimens from both Java and China.

In Java the new specimens were unearthed mainly between 1952 and 1977, and they include three craniums, some skull fragments, six partial jaws, and five isolated teeth from near Sangiran and a cranium and tibial fragment from Sambungmacan (Sambungmachan), between Sangiran and Trinil (98, 594, 1126, 1128). One of the Sangiran fossils (Sangiran 17 = "*Pithecanthropus* VIII") is the most complete *H. erectus* skull known (1870). The Sambungmacan skull is often likened to the late *H. erectus* (or *H. soloensis*) fossils from Ngandong, and it may be similar in age. Two additional partial skulls and some pelvis fragments were found near Ngandong between 1976 and 1980 (1127, 2122).

The additional Chinese fossils include two skull fragments (that attach to the replica of a skull found in 1934), a fragmentary mandible, six isolated teeth, and two fragmentary limb bones from renewed excavations at Zhoukoudian between 1949 and 1966; a mandible found at Chenjiawo (formerly Chenchiawo), Lantian County, in 1963; a skull found at Gongwangling (formerly Kungwangling), Lantian County, in 1964; a partial skull, two skull fragments (from a second individual), a fragmentary mandible, and nine isolated teeth found in Lontandong Cave, Hexian County, in 1980–81; a highly fragmented skullcap and seven isolated teeth found in a fissure deposit on Qizianshan Hill, Yiyuan County, in 1981; two badly crushed, partial skulls found in alluvial deposits at Quyuankekou, Yunxian County, in 1989–90; two skulls unearthed in a cave near Tangshan, Nanjing County, in 1993; two isolated incisors found at Yuanmou in 1965; three molars found at Jianshi Cave (Longgudong) in 1970; and isolated teeth found at perhaps five other localities after 1975 (2511, 2513, 2516). Chinese sites are often known alternatively by the name of the county in which they occur, and Chenjiawo and Gongwangling are thus frequently lumped together as the Lantian sites.

Excepting Zhoukoudian, the Chinese sites appear to be mainly places where flowing water or carnivores accumulated the human remains. In general nonspecialists made the discoveries, and stratigraphic control is weak. As discussed below, dating is mainly tentative. In south Chinese caves like Longgupo (Wushan) (2381), Jianshi (2532), and the *Gigantopithecus* Cave at Liucheng (2532), there is the added difficulty that teeth of *H. erectus* (or *Homo* sp.) must be separated from closely similar teeth representing other large-bodied hominoids, including one or more species of the orangutan, *Pongo*, and the extinct Asian great ape, *Gigantopithecus blacki*. The same problem of separation affects proposed *H. erectus* teeth found in 1965 at Tham Khuyen Cave, 30 km west of the Chinese border in northeastern Vietnam (469, 472, 1898). Assuming that the human teeth have been properly identified at Tham Khuyen and the Chinese caves, they indicate that people and *Gigantopithecus* overlapped in Southeast Asia from 1 my ago or before until 500 ky ago or later.

To the east Asian inventory of *H. erectus*, many authorities would add European and especially African specimens that are likewise characterized by large browridges, low, flattened frontal bones, thick cranial walls, and massive chinless jaws with large teeth. The European and African specimens were almost all found after Weidenreich and von Koenigswald essentially defined *H. erectus* in 1939 and mostly after *H. erectus* became widely accepted in the 1960s. If the European and African fossils are included in *H. erectus*, they substantially extend its geographic and temporal range. For reasons presented below, however, the relevant African specimens are divided here mainly between *Homo ergaster* (earlier) and ancestral *H. sapiens* (later), while the European specimens are assigned mostly to ancestral *H. neanderthalensis*. A small group of similar and possibly like-aged European and African specimens that may represent the last shared ancestor of *H. sapiens* and *H. neanderthalensis* can be alternatively assigned to *H. heidelbergensis*. Figure 5.1 outlines the proposed relationships of *H. ergaster*, *H. heidelbergensis*, *H. sapiens*, and *H. neanderthalensis* to each other and to *H. erectus*.

The use of *H. sapiens* and *H. neanderthalensis* for evolving lineages that contain more primitive (ancestral) and more advanced (descendant) populations is problematic. In addition, chapters 6 and 7 argue that there may have been a speciation event within the *H. sapiens* lineage that produced fully modern humans about 50 ky ago. If this occurred, only fully modern people after 50 ky ago should be called *H. sapiens*, and older African populations would require another name. The most appropriate is probably *Homo helmei* (675), originally proposed for a skull of "primitive *H. sapiens*" from Florisbad, South Africa, discussed below. In short, the names attached to the three lineages in figure 5.1 may violate proper taxonomic practice, but for the moment there is no obvious, less contentious alternative nomenclature. Tables 5.1, 5.2, and 5.3 list the dates of discovery and other summary information for fossils of *Homo ergaster*, later African fossils assigned to *H. sapiens* or *H. heidelbergensis* or both, and European fossils assigned to *H. neanderthalensis* or *H. heidelbergensis*. Boldface in tables 5.2 and 5.3 distinguishes putative specimens of *H. heidelbergensis*.

Geologic Antiquity

Most fossils of primitive *Homo* are very imprecisely dated, because their stratigraphic position and associations were not carefully recorded, they were not accompanied by materials that are suitable for radiometric methods, or both. The difficulty of placing many key fossils in time, combined with the relatively small size of the fossil samples, explains why reasonable

Table 5.1. African Sites with Fossils of *Homo ergaster* ("African *H. erectus*")

Site	Fossils	Date of Original Discovery	References
Ternifine (Tighennif = Palikao), Algeria	Three mandibles and a parietal fragment of "*Atlanthropus mauritanicus*"	1954–1955	65, 66, 114, 805, 1071
Aïn Maarouf (El Hajeb), Morocco	A left femoral shaft	1955	803
Sidi Abderrahman (Littorina Cave), Morocco	Mandibular fragments	1954	67, 1071
Thomas Quarries, Morocco	A mandible and cranial fragments	1969, 1972	333, 1073, 1090, 1765
Buia, Danakil Desert, Eritrea	A nearly complete skull, two lower incisors, and two pelvic fragments	1995–1997	1, 1575
Gomboré II (Melka Kunturé), Ethiopia	Cranial fragment	1973	457, 1073
Konso, Ethipia	A nearly complete left mandible and an upper third molar	1991	89, 2110
Omo, Shungura Formation, Member K, Ethiopia	Parietal and temporal fragments (L 996-17)	??	1073
Nyabusosi, Uganda	Cranial fragments	1986	1914, 2163
Koobi Fora (East Turkana), Kenya	Partial skeleton (KNM-ER 1808); two nearly complete skulls (KNM-ER 3733 and 3883); parts of three other skulls, nine partial mandibles, and some isolated limb bones	1973–1975	1349, 1352, 2372, 2376
Nariokotome (West Turkana), Kenya	Nearly complete skeleton (KNM-WT 15000)	1984	373, 1351, 2360
Olduvai Gorge, LLK II, Tanzania	Cranium (Olduvai Hominid [OH] 9)	1960	1323
Olduvai Gorge, VEK IV, Tanzania	Fragmentary cranium (OH 12)	1962	1763
Olduvai Gorge, WK IV, Tanzania	Partial pelvis and a femur shaft (OH 28)	1970	588
Olduvai Gorge, VEK/MNK III/IV	Partial mandible (OH 22)	1968	1774
Olduvai Gorge, GTC, Bed III	Partial mandible (OH 51)	1974	1774
Olduvai Gorge, Lower Masek Beds	Partial mandible (OH 23)	1968	1774
Swartkrans, South Africa	Various bones from Members 1–3, including especially two fragmentary mandibles, a fragmentary maxilla, a crushed child's skull (SK 27), and the partial adult skull (SK 847) of "*Telanthropus capensis*" from Member 1	1949 and later	502, 507, 508, 899

Table 5.2. African Sites with Fossils of Early *Homo sapiens*

Site	Fossils	Date(s) of Discovery	References
Jebel Irhoud (Ighoud), Morocco	A nearly complete skull (Irhoud 1), a skullcap (Irhoud 2), a juvenile mandible (Irhoud 3), a juvenile humerus shaft (Irhoud 4), and a partial adult mandible	1961, 1963, 1968, 1969	30, 689–692, 1095, 1100
Kébibat (Mifsud-Giudice Quarry, Rabat), Morocco	Mandible, left maxilla, and occipital fragments	1933	1090, 1859, 2040
Salé, Morocco	Partial skull	1971	1090, 1130
Aïn Maarouf (El Hajeb), Morocco	Femur shaft	1950–1951	803, 1094
Wadi Dagadlé, Djibouti	Left maxilla	1983	259, 260
Singa, Sudan	A skull	1924	151, 489, 924, 1495, 2074, 2090
Bodo, Ethiopia	Partial skull, parietal of a second skull, and a distal humerus shaft	1976, 1981, 1990	85, 492, 494, 538, 539, 1170, 1777
Garba 3 (Melka Kunturé), Ethiopia	Cranium fragments	1976	457
Kapthurin (Baringo), Kenya	Two mandibles, two phalanges, a right metatarsal, and a right ulna	1966, 1983	557, 594, 1320, 1485, 2013, 2297
Lainyamok, Kenya	Three associated maxillary teeth and a femur shaft fragment	1976, 1984	1707, 1708, 1940
Lake Ndutu, Tanzania	Skull	1973	501, 505, 1336, 1598, 1767
Eyasi, Tanzania	Cranium fragments from two or three individuals	1935, 1938	1524, 1525
Broken Hill (Kabwe), Zambia	A nearly complete skull, a cranium fragment, a right maxilla, a fragmentary humerus, two pelvises, six femur fragments, and a tibia fragment	1921–1925	481, 493, 498, 1231, 1718, 1978, 2079, 2505
Berg Aukas (Grootfontein, Otavi Mountains), Namibia	Proximal femur	1965	904
Cave of Hearths, Northern Transvaal, South Africa	Right mandible fragment with three teeth	1947	1475, 1476, 2194
Florisbad, Free State Province, South Africa	Partial skull and an isolated upper third molar	1932	503, 675, 921
Elandsfontein (Hopefield = Saldanha), western Cape Province, South Africa	Skullcap and mandible fragment	1953, 1954	674, 1246, 1978, 1983

Note: Boldface designates sites where the fossils might be alternatively assigned to *H. heidelbergensis.*

authorities can disagree on the course of human evolution between 1.8 my ago and 127 ky ago. The discussion here addresses
geologic age by species, which means in effect by region, starting with *Homo erectus* in the Far East. Figure 5.5 summarizes
the estimated ages of all the main fossils, regardless of species.

Homo erectus

The Chinese *Homo erectus* fossils are ostensibly the best dated,
thanks primarily to more carefully documented stratigraphic
provenance and to paleomagnetic determinations. Within the
framework provided by the paleomagnetism, Chinese scientists
have attempted more precise dates by correlating the climatic
fluctuations recorded at important sites (especially Zhoukoudian) with the dated oxygen-isotope stages of the global marine
stratigraphy (1388, 2510). The correlation procedure is open to
many possible errors, so the results can be only provisional. For
the moment, they suggest that the *H. erectus* fossils at Zhoukoudian Locality 1 accumulated between 500 and 240 ky ago;
that the Hexian fossils are roughly 280–240 ky old; that the
Gongwangling (Lantian) skull is perhaps 800–750 ky old; that
the Chenjiawo (Lantian) mandible is approximately 590–500 ky
old; and that the isolated teeth from Yuanmou and other sites
are all younger than 780 ky. The Yuanmou teeth could in fact
be much younger, since they were surface finds (1701).

The dates for Zhoukoudian have been questioned, because associated fauna, pollen, and sediments suggest that the
human fossils may have accumulated within a single interglacial, perhaps closer to 500 ky ago than to 240 ky ago (15).
However, a single interglacial is difficult to reconcile with the
thickness of the deposits, which exceeds 40 m. In addition,
uranium-series, thermoluminescence, and ESR determinations
all support a 200–300 ky depositional span, beginning 550–
500 ky ago (461, 462, 922, 2533). ESR also broadly confirms an
age near 300 ky ago for Hexian (695). In the absence of correlations to the global climatic stratigraphy, paleomagnetism and
ESR place the Yunxian skulls between 800 and 500 ky ago
(463), while the fauna from Nanjing implies an age between 400
and 200 ky ago (695). ESR brackets the probable *H. erectus*
teeth from Tham Khuyen Cave, Vietnam, sometime between
600 and 400 ky ago (469).

Future research may push the lower age limit for *H. erectus* in China to 1 my ago or beyond. Arguably, paleomagnetism
already implies that the Gongwangling skull dates to 1.15 my

Table 5.3. European Sites with Fossils of Early *Homo neanderthalensis*

Site	Fossils	Date(s) of Discovery	References
Swanscombe (Barnfield Pit), England	Occipital and right and left parietals of a single skull	1935, 1936, 1956	143, 545, 1461, 1623, 1624, 2095, 2520
Boxgrove, West Sussex, England	Tibia shaft, two isolated incisors	1993, 1995	1785, 1787, 2086
Pontnewydd, Wales	Seven isolated teeth and a right maxillary fragment with two teeth	1978–1985	884–886
Montmaurin, Haute Garonne, France	A nearly complete mandible, and some isolated teeth	1949	212, 213, 2095
La Caune de l'Arago (Tautavel), France	More than fifty specimens, including a facial skeleton (Arago 21) and a parietal (Arago 47) possibly from the same individual, a mandible (Arago 2), a half mandible (Arago 13), and a left pelvis (Arago 44)	1964–1974	1403, 1406, 2095
Orgnac 3, Ardeche, France	Seven isolated teeth	1962, 1968–1971	527, 548
Le Lazaret, Nice, France	A juvenile right parietal, a deciduous upper incisor, and an adult lower canine	1953–1964	698, 1400, 1401, 2095
Le Chaise de Vouthon (Bourgeois-Delaunay and Suard Caves), Charente, France	Approximately eighty specimens, mainly isolated teeth and cranium fragments, and a few fragmentary postcranial bones	1960s and 1970s	240, 620, 621, 624, 1692, 2095
Fontéchevade Cave, Charente, France	A frontal fragment (Fontéchevade I), a skullcap (II), and a parietal	1947, 1957	1015, 2095, 2308
Biache-Saint-Vaast, Pas-de-Calais, France	The rear half of a cranium, a fragmentary maxilla with all six molars, and five isolated teeth	1976	2095, 2245, 2247
Sima de los Huesos (Cueva Mayor/Ibeas), Atapuerca, Spain	Three nearly complete craniums, six partial craniums, and more than 1,600 other cranial and postcranial bones representing at least thirty-three individuals	1976 to present	8, 9, 77, 80, 81, 227
Notarchirico, Venosa, Basilicata, Italy	A femur shaft	1986	183, 532, 1691
Fontana Ranuccio, Anagni, Latium, Italy	Four isolated teeth	1970s and 1980s	532, 1907

Site	Fossils	Date(s) of Discovery	References
Castel di Guido, Latium, Italy	Three cranium fragments (from a parietal, an occipital, and a temporal), a right maxillary fragment, and two femur shaft fragments	1980–1986	286, 532, 1425, 1426
Rebibbia-Casal de' Pazzi, Latium, Italy	Parietal fragment	1983	210, 1436
Cava Pompi, Fofi, Latium, Italy	Cranium fragments, a proximal ulna fragment, and a tibia shaft	1960s	532, 548
Visogliano Shelter, Trieste, Italy	A mandible fragment, an isolated upper premolar, and an isolated upper molar	1983, 1985	285, 434, 532
Grafenrain Sand Quarry, Mauer, Heidelberg, Germany	A mandible	1907	381, 1071, 1258, 2095
Reilingen, Baden-Württemberg, Germany	Partial occipital, right temporal, and both parietals of a single skull	1978	568, 618
Steinheim an der Murr, Württemberg, Germany	A nearly complete skull	1933	7, 1071, 2095
Bilzingsleben (Steinrinne Quarry), Thuringia, Germany	Approximately fifteen skull fragments from two or three adult individuals; one deciduous tooth and six isolated permanent teeth	1972–1993	952, 1428, 1431, 1433, 1892, 2095
Weimar-Ehringsdorf, Germany	An adult skullcap, an adult mandible, a juvenile mandible, a partial child's skeleton, and other fragmentary cranial and postcranial bones from perhaps nine individuals	1908–1925	239, 548
Vértesszöllös, Hungary	An adult occipital bone, two isolated deciduous teeth, and two deciduous dental fragments	1964–1965	669, 1895, 2174, 2175, 2478
Petralona Cave, Halkidiki, Greece	Skull	1960	264, 548, 919, 1291, 1309, 1649, 1714, 2076, 2094, 2473, 2480

Note: Boldface designates sites where the fossils might be alternatively assigned to *H. heidelbergensis.*

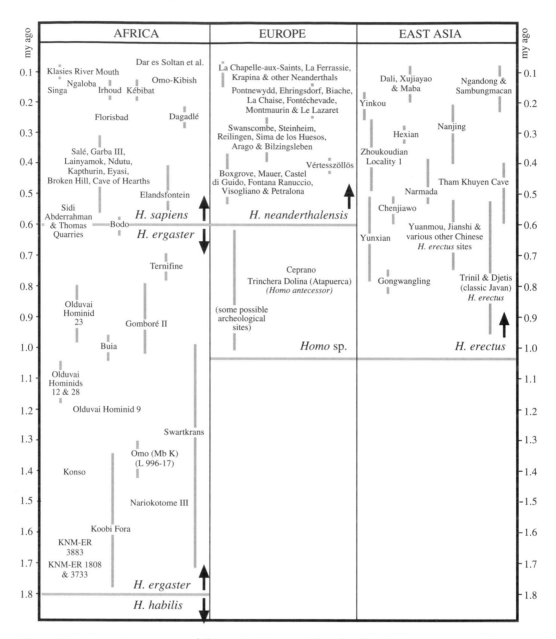

Figure 5.5. Approximate ages of the main sites providing fossils of *Homo ergaster, H. erectus,* early *H. neanderthalensis,* and early *H. sapiens.* The shaded vertical lines indicate time ranges in which fossils may lie. The datings are based variously on faunal correlations; U-series, ESR, or TL determinations; and presumed correspondences between the sequences of glacial-interglacial events recorded at the sites and the global oxygen-isotope stratigraphy.

ago rather than 800–750 ky ago (695, 2516). Human presence by 1.9 my ago may even be demonstrated by an isolated incisor, a mandible fragment containing a premolar and molar, and two possible stone artifacts found at Longgupo Cave, south-central China, in 1987–1988 (1307, 2381, 2497). The Longgupo date stems from the application of paleomagnetism to a very complex stratigraphy, however, and the actual age of the human fragments may approximate that of a deer tooth from a higher-lying unit. This tooth has been bracketed between 1.02 and 0.75 my ago by ESR analysis. There is also the problem that the mandible fragment is not unequivocally human (as opposed to ape) (1900). At present, archeology provides the most secure estimate for the time of human arrival in China. The oldest well-documented archeological sites are in the Nihewan (Nihowan) Basin, approximately 150 km west of Beijing (1875, 1877). At the Donggutuo site, more than 10,000 indisputable artifacts have been recovered from fluvial deposits that accumulated just before the Jaramillo Normal Subchron within the Matuyama Reversed Paleomagnetic Chron. This places Donggutuo at or shortly before 1 my ago, when the Jaramillo Subchron began (2135, 2145).

The upper age limit for *Homo erectus* in China is a matter of definition. Virtually everyone agrees that Hexian and especially Zhoukoudian extend the span of *H. erectus* to roughly 300 ky ago. Fossils putatively dated between 300 and 100 ky ago, including especially those from Jinnuishan (Yinkou), Dali, and Maba (Mapa), are commonly regarded as more advanced (460, 695, 2516). As discussed below, if they had been found in Europe or Africa, they might have been assigned to "archaic" *H. sapiens* or to *H. heidelbergensis*. Arguably, however, they differ from the Zhoukoudian representatives of *H. erectus* little more than the Zhoukoudian representatives differ from those of Gongwangling, and they may represent a further development along the same lineage. If this is accepted, then Chinese *H. erectus*, broadly understood, survived to 100 ky ago or later, until it was eventually replaced by or perhaps mixed with immigrant *H. sapiens* from Africa. A late persistence for Chinese *H. erectus* is tentatively accepted here, pending a denser, more securely dated Chinese fossil record that could demonstrate otherwise.

The Javan specimens of *Homo erectus* remain the most poorly dated. In the Sangiran-Trinil area of central Java, they are conventionally said to come from two rock-stratigraphic units, each with its own distinctive fauna (144, 1255, 1357, 2168). The older unit, which provided relatively few *H. erectus* remains, comprises the Pucangan (or Putjangan) Beds with the Djetis (Jetis) fauna; the younger unit, which has provided most of the *H. erectus* fossils, comprises the Kabuh Beds with the Trinil fauna. In fact, both the lithostratigraphy and the biostratigraphy are probably far more complex (2020), and the precise stratigraphic provenance of most of the *H. erectus* fossils is uncertain in any case. Fluorine analysis has been used to narrow the provenance possibilities (1477), but it is still difficult to relate the fossils to paleomagnetic determinations or to radio-potassium and fission-track dates that have been obtained on volcanic tuffs in the Pucangan Beds and on tuffs, pumices, and tektites in the Kabuh Beds (1075, 1910). In the Pucangan Beds, the dates range from 2 my ago to 570 ky ago, and in the Kabuh Beds they range from 1.6 my ago to 470 ky ago. The obvious discrepancy is underlined by contradictions between some of the dates and paleomagnetic readings on the same sediments. There are many possible reasons for the inconsistencies, including the dating of samples that were redeposited from older sediments. For the moment, the radiometric dates are obviously not very helpful for defining the antiquity of Javan *H. erectus*, though it may be meaningful that most specimens are less than 800 ky old.

The problem of association pertains equally to recently obtained ^{40}Ar/^{39}Ar determinations that place the Mojokerto juvenile skull at roughly 1.8 my ago and that fix two partial adult skulls from Sangiran at about 1.65 my ago (2123). The relevant Mojokerto and Sangiran fossils are usually said to have originated from geographically separate outcrops of the Pucangan Beds. The exact find spots of the fossils are unknown, but the dated volcanic fragments were taken from what are believed to be like-aged exposures nearby. Mineral particles at the Mojokerto collection site closely resemble particles still clinging to the juvenile skull. However, the dates contradict biostratigraphic and paleomagnetic estimates for both sites that are partly controlled by fission-track dates (603). If the fission-track estimates are accepted, the fossils can be no more than 1 my old. The inconsistency could result if the ^{40}Ar/^{39}Ar samples were not actually taken from the layers that contained the human and animal fossils, or if the samples were reworked and redeposited into the deposits from much older layers.

An additional problem is that the earliest Javanese must have come from mainland Southeast Asia, and there is no firm evidence for people on the east Asian mainland before 1 my ago. The reported date of 1.9 my ago for human fossils at Longgupo Cave, China, may appear to support the early Javan dates, but even if the Longgupo dating is accepted, the principal human fossil is a mandible fragment that is not like that of *H. erectus*. It contains a fourth premolar and first molar that have been compared with those of *H. habilis* (1307, 2381), which means that support from Longgupo is contingent on a local east Asian evolution from *H. habilis* to *H. erectus* about 1.8–1.6 my ago. The controversy surrounding the Longgupo and early Javan dates underscores an epistemological maxim in paleoanthropology: Consensus depends on the repeated discovery of substantial fossils in well-documented geologic contexts. This is illustrated by widespread agreement that *H. ergaster* appeared in eastern Africa 1.8–1.6 my ago.

The first inhabitants of Java surely arrived during a Quaternary glacial period, when lowered sea level had exposed the shallow (Sunda) continental shelf that separates Java from mainland Southeast Asia. The amplification of glacial pulses near the beginning of the middle Quaternary, roughly 900–800 ky ago, increased the amount of shelf exposed, and it also provided the most favorable environmental conditions for crossings by a wide range of species, including open-country forms such as people (1615, 2021). In sum, when all factors are considered it is reasonable to conclude that the known Javan *H. erectus* fossils are probably all younger than 1 my (603, 1701).

The upper age limit of Javan *Homo erectus* is also debatable. If the Ngandong and Sambungmacan specimens are included, as their morphology argues they should be, then *H. erectus* persisted in Java until at least 300–250 ky ago and perhaps until sometime between 53 and 27 ky ago. The older age is based on faunal associations, a uranium-series date on one of Ngandong skulls, and a fission-track date from the Notopuro Beds that have sometimes been equated with the fossiliferous sediments at Ngandong (810, 1127, 1910). The younger date is based on ESR and uranium-series analysis of bovid teeth that accompanied the human specimens at both Ngandong and Sambungmacan (926, 2122). If the younger date is correct, it would indirectly support the local replacement of *H. erectus* by *H. sapiens* spreading from Africa after 50 ky ago. However, even the older date supports Chinese evidence (mainly from Zhoukoudian and Hexian) that *H. erectus* persisted in eastern Asia after other kinds of people had appeared in Africa and Europe. The Africans are referred here to the lineage that culminated in living *H. sapiens*, while the Europeans are referred to the antecedents of classic *H. neanderthalensis*.

Homo ergaster

The separation of *Homo ergaster* from *H. erectus* is controversial, and even advocates commonly apply *H. ergaster* only to a small number of relatively complete African fossils dated to roughly 1.8–1.5 my ago. They assign later African specimens, dated between perhaps 1.2 and 0.6 my ago, to *H. erectus*, implying either that *H. erectus* evolved from *H. ergaster* in Africa and then spread to eastern Asia or that *H. erectus* evolved in

Asia and then spread back to Africa. The position taken here is that *H. erectus* developed exclusively in eastern Asia, beginning 1 my ago or before, and that *H. ergaster* persisted in Africa until the emergence of *H. sapiens* or *H. heidelbergensis* 600–500 ky ago. The issue can be decided only by the recovery of a much larger number of diagnostic, well-dated fossils, however, and readers should be aware that the present use of *H. ergaster* is not commonplace, particularly for fossils that postdate 1.2 my ago. In recognition of this, in what follows *H. ergaster* is often designated alternatively as "African *H. erectus.*"

The least equivocal fossils of *H. ergaster* are also the best dated, based on their stratigraphic relation to volcanic tuffs whose radiopotassium or ^{40}Ar/^{39}Ar age is firmly established (370, 373, 717). They include a nearly complete skull (KNM-ER 3733) and a partial skeleton (KNM-ER 1808) from Koobi Fora, East Turkana, that are roughly 1.8–1.7 my old; a second skull (KNM-ER 3883) from Koobi Fora that is only slightly younger; and a skull and associated skeleton (KNM-WT 15000) from Nariokotome III, West Turkana, that is about 1.5 my old. Other less complete specimens from Koobi Fora are fixed between roughly 1.6 and 1.4 my ago. Possible *H. ergaster* cranial fragments from the lower Omo River Valley (Shungura Formation Member K) are dated to 1.4–1.3 my ago.

There are no firm radiometric dates for *H. ergaster* (or African *H. erectus*) fossils from Olduvai Gorge, but previous estimates based on paleomagnetism and presumed sedimentation rates placed a skullcap (OH [Olduvai Hominid] 9) from Upper Bed II near 1.2 my ago and a partial skull (OH 12) and other fragmentary fossils from Beds III/IV between 780 and 620 ky ago (1336). A revision of the Olduvai paleomagnetism need not alter the estimated age of Upper Bed II, but it places the Bed III/IV fossils also before 1.1 my ago, and it brackets a partial mandible (OH 23) from the Lower Masek Beds between 970 and 780 ky ago (2135) (fig. 4.9 in the previous chapter). Paleomagnetism and radiopotassium together bracket a skull fragment from Gomboré II (Melka Kunturé), Ethiopia, between 1.3 my and 780 ky ago (455, 456). ^{40}Ar/^{39}Ar determinations place a mandible from Konso, Ethiopia, at about 1.4 my ago (89, 2111). Finally, paleomagnetism and associated mammalian species suggest that a skull from Buia, Eritrea, is about 1 my old (1).

At the remaining sites with *H. ergaster* (or African *H. erectus*) (table 5.1), associated animal species provide the main basis for dating. Associated fauna at Swartkrans indicates that craniodental fragments of primitive *Homo* approach the 1.8–1.5 my age of Koobi Fora and Nariokotome *H. ergaster* (2347).

A partial skull (SK 847) from Swartkrans Member 1 closely re-sembles the Koobi Fora skulls and could join them in the core sample of *H. ergaster* (507, 508). The fauna found with human mandibles and a skull fragment at Ternifine suggests an age near that of Olduvai Upper Bed IV (805), but the actual age must be somewhat younger, since the Ternifine deposits are normally magnetized. A time just after 780 ky ago is reasonable. Finally, fauna implies that the Sidi Abderrahman and Thomas Quarries sites postdate Ternifine (798, 804). The Sidi Abderrahman and Thomas Quarries human fossils are actually too incomplete for compelling species assignment, and if they dated to after 600 ky ago, they could be arbitrarily transferred to the sample of early *Homo sapiens* discussed below.

Homo sapiens

In the view that is tentatively accepted here, *Homo sapiens* is a strictly African offshoot of *H. ergaster*. The earliest speci-mens are distinguished from *H. ergaster* (and *H. erectus*) by a number of advanced features, including larger endocranial vol-ume (table 5.4), a more rounded occipital, expanded parietals, and a broader frontal with more arched (vs. more shelflike) browridges. These features are not expressed equally in every specimen, perhaps in part because the specimens vary in age. Among the fossils listed in table 5.2, the principal specimens on which one or more advanced features can be identified are the skulls or partial skulls from Salé, Kébibat, and Jebel Irhoud in Morocco, Singa in Sudan, Bodo in Ethiopia, Lake Ndutu and Eyasi in Tanzania, Broken Hill in Zambia, and Elandsfontein and Florisbad in South Africa. The remaining fossils are listed in table 5.2 on the admittedly circular grounds that they fall in the same time range as the diagnostic specimens.

Among the diagnostic specimens, those from Bodo, Kébibat, Irhoud, Singa, and Florisbad set the time frame. The principal Bodo skull is included here because its relatively broad frontal resembles that of Broken Hill or Elandsfontein, but its face is remarkably massive, particularly in the midre-gion. On these grounds it might be expected to be the most an-cient fossil in the group, and ^{40}Ar/^{39}Ar dates fix it near 600 ky ago (494). Faunal associations and a thorium-uranium date on marine shell from an overlying bed place the Kébibat occipital and associated dentitions between 200 and 127 ky ago (1859). ESR brackets the Irhoud fossils between 190 and 90 ky ago (30, 924); in combination with thorium-uranium, it places the Singa skull between roughly 170 and 150 ky ago (1495); and it

Chapter Five

Table 5.4. Estimated Endocranial Capacities of Skulls Assigned to the Early Representatives of *Homo sapiens* and *H. neanderthalensis*

Skull	Estimated Endocranial Capacity (cc)	Source
H. sapiens		
Jebel Irhoud 1	1,305	1048
Jebel Irhoud 2	1,450	1052
Salé	860	1048
Singa	1,340	2090
Bodo	1,300	1777
Lake Ndutu	1,100	1767
Broken Hill	1,285	1049
Elandsfontein	1,225	674
Average and standard deviation	1,233 ± 180	
H. neanderthalensis		
Swanscombe	1,325	1314
Arago 21/47	1,166	1049
Biache-Saint-Vaast	1,200	2095
Atapuerca SH 4	1,390	82
Atapuerca SH 5	1,125	82
Reilingen	1,430	618
Steinheim	1,150	1071
Petralona	1,230	2094
Average and standard deviation	1,252 ± 115	

Note: The averages should be regarded only as rough guides, since the fossils are members of lineages in which average endocranial capacity may have been increasing with time. Useful averages and standard deviations for comparison are *Homo ergaster* (or early African *H. erectus*), 907 ± 115 cc (N = 4) (2361, 2362); classic Indonesian *H. erectus* ("*Pithecanthropus*"), 934 ± 101 cc (N = 5) (1049); Zhoukoudian *H. erectus*, 1,043 ± 113 cc (N = 5) (2403); Ngandong late *H. erectus*, 1,151 + 99 cc (N = 5) (1046); European classic Neanderthals, 1,507 ± 116 cc (N = 10) (1052); west Asian (Skhul-Qafzeh) "near-moderns," 1,545 ± 27 (N = 5) (2216); and early Upper Paleolithic (fully modern) humans, 1,577 ± 135 cc (N = 11) (2216).

dates the Florisbad skull to perhaps 260 ky ago (921). The especially diagnostic Irhoud, Singa, and Florisbad fossils are reconsidered in the next chapter (on the Neanderthals and their contemporaries) because they surely overlap in time with European fossils that exhibit derived traits of the Neanderthals. It is in fact this overlap that makes them especially significant. They demonstrate a mid-Quaternary African trend away from the Neanderthals and toward living humans. Earlier African specimens like those from Broken Hill and Elandsfontein are more difficult to distinguish from their European contemporaries.

The remaining African fossils are dated primarily by their faunal associations. In this regard the skull from Elandsfontein is probably the best dated. Eighteen of the forty-five associated mammalian species are extinct, and nine occur in extinct genera (1246). Comparisons with dated faunas from Olduvai, Olorgcsailie, Lainyamok, and Bodo in East Africa (494, 1707) suggest that the Elandsfontein mammal bones probably accumulated sometime between 700 and 400 ky ago. Absolute age aside, the Elandsfontein fauna is clearly much older than the one from Florisbad, where only five of the twenty-six mammalian species are extinct and only two are in extinct genera (354). The difference is consistent with the much more primitive morphology of the Elandsfontein skull.

In sum, early *H. sapiens* as understood here was present in Africa by at least 600 ky ago, and its separation from contemporaneous *H. neanderthalensis* was well advanced by 260 ky ago.

Homo neanderthalensis

As conceived here, *Homo neanderthalensis* includes the classic Neanderthals, who occupied Europe and southwestern Asia between roughly 127 and 40 ky ago, and their antecedents, who appear to have been strictly European. Cranial or dental remains that are associated with ESR, TL, or U-series determinations at Ehringsdorf, Germany (239), La Chaise de Vouthon (240) and Biache-Saint-Vaast (17, 2017), France, and Pontnewydd, Wales (17, 885), argue that the Neanderthal lineage was in place between 250 and 127 ky ago. Faunal, palynological, and archeological associations place a Neanderthal-like mandible from Montmaurin, France, within the same period (548). There are no like-aged European fossils that are clearly non-Neanderthal, though some, like those from Fontéchevade and Le Lazaret Caves, France, are too incomplete for secure diagnosis (2095).

Yet older European fossils are more variable. Among those that are sufficiently complete for diagnosis, some, like those from the Sima de los Huesos ("Pit of the Bones"), Atapuerca, Spain (8, 9, 79, 81), Swanscombe, England (2086, 2095), Reilingen, Germany (568, 618), Petralona Cave, Greece (2076, 2094), and Arago, France (1403), anticipate the Neanderthals in one or more derived traits, variably including a tendency to midfacial projection, an elliptical area of roughened bone (the suprainiac fossa) overlying the occipital torus, or a dentition in which the rear teeth are significantly diminished in size relative to the front teeth. Others fossils, including especially those from Bilzingsleben, Germany (1431, 2086, 2095), and Vértesszöllös, Hungary (2095, 2174), show no conspicuous Neanderthal specializations and are variously characterized instead by a shelf-like browridge, an angulated occipital with a strong, centrally developed torus, very thick cranial walls, or yet other features that closely recall *Homo erectus*.

The division into two groups might make sense if it could be shown that the more Neanderthal-like fossils postdated the others, but on present evidence we can say only that both groups probably date between 500 and 250 ky ago. An age of this order is implied by ESR and U-series analyses for Bilzingsleben (952, 1431, 1892) and by faunal associations or geomorphic context for most of the remaining sites (references in table 5.2). Widely discrepant U-series and ESR determinations on sediments suggest that the Petralona skull could be younger than 250 ky (919, 1013, 1309, 2473), but the stratigraphic origin of the skull is uncertain (548), and its relation to the dated samples is debatable. Faunal remains that may have been deposited contemporaneously surely antedate 250 ky ago (1291). A firm U-series determination of more than 300 ky ago for the Sima de los Huesos fossils (80) has now been set aside, because it was based on a fragment of travertine (cave limestone) that was reworked from an older horizon. Less secure U-series dates on human bones and U-series and ESR dates on overlying bear bones indicate that the human bones must be at least 200 ky old and could antedate 320 ky ago (227). An antiquity of this order is compatible with the morphology of the bear species (*Ursus deningeri*, a lineal ancestor of the later Quaternary cave bear, *U. spelaeus*) (781) and with associated rodents (567).

Only two European sites have provided human fossils that are likely to be much older than 500 ky. These are the Trinchera (Gran) Dolina, Atapuerca, Spain (200, 424), and Ceprano, Italy (83, 84). The dating at the Trinchera Dolina relies primarily on paleomagnetic readings, which place the fossiliferous unit (TD 6) at perhaps 800 ky ago, shortly before the beginning of the Brunhes Normal Reversed Chron. The dating at Ceprano depends on radiopotassium analysis of volcanic particles from an overlying sandy gravel. The oldest of these particles have been dated to about 700 ky ago, which implies that the skull could be as old as TD 6. If the age estimates at both sites are correct, they could imply a long, pre-Neanderthal occupation of Europe, perhaps extending back to 1 my ago or before, or they might indicate the occurrence of sporadic colonization failures before 500 ky ago. The second alternative is arguably more compatible with the European archeological record, which suggests continuous human occupation only after about 500 ky ago (649, 651, 1806, 1811).

In sum, dating and morphology together indicate that the Neanderthals developed in Europe, beginning perhaps 500 ky ago, but the available evidence could also be used to argue that more primitive populations with few or no Neanderthal features persisted locally until 250 ky ago. Alternatively, and perhaps more likely, Neanderthal features appeared piecemeal, mainly as a consequence of genetic drift in small, isolated populations, and they varied conspicuously in frequency from place to place until after 250 ky ago. A denser, far better dated fossil record will be necessary to choose between these and other possibilities.

Homo heidelbergensis

Homo heidelbergensis is difficult to date, in large part because it is difficult to define. The nomenclatural type is a massive, chinless human mandible recovered in 1907 from quarry deposits at Mauer, about 10 km southeast of Heidelberg, Germany (1882). Associated microfauna implies that the deposits may have formed about 500 ky ago (290, 1258, 1811). The mandible exhibits moderate expansion of the molar pulp cavities (taurodontism) and a very long cheek tooth row, which implies a long, forwardly projecting face, and it could represent a population near the common ancestry of Neanderthals and modern humans. If we assume that this ancestor existed sometime between 600 and 400 ky ago, it could also be represented by the skulls from Petralona, Greece, Arago, France, Bodo, Ethiopia, Lake Ndutu, Tanzania, Broken Hill (Kabwe), Zambia, and Elandsfontein, South Africa. These are sufficiently similar and morphologically generalized to represent a common ancestor for Neanderthals and modern humans (1777, 2085, 2087). However, except for Bodo and Elandsfontein, the dating depends at least as much on skull form as it does on nonmorphological criteria. If it should turn out that some of the skulls fall significantly outside the period 600–400 ky ago, the concept of *H. heidelbergensis* might have to be abandoned.

If *H. heidelbergensis* is retained, its easternmost representative could be a skullcap found at Hathnora in the Narmada Valley, north-central India, in 1982 (1212, 1407, 1409, 2018, 2019). The associated fauna and Acheulean artifacts indicate a possible age between 600 and 400 ky ago, and like European and African skulls that some assign to *H. heidelbergensis*, the Narmada specimen combines primitive features (including a massive, forwardly projecting supraorbital torus, thick cranial walls, and great basal breadth) with derived traits (including an endocranial capacity of more than 1,200 cc, a relatively steep [nonreceding] frontal, and a relatively rounded occipital). Based on a broadly similar morphological mix, Chinese fossils from Jinnuishan (Yinkou), Xujiayao, Dali, and Maba (Mapa) might also be added to *H. heidelbergensis* (914, 1777, 1778), but they are almost certainly much younger than 400 ky ago. Their inclusion would thus imply that west-to-east gene flow intensified after 300 ky ago or that the same evolutionary trends (particularly brain expansion) occurred later on the east than on the west. If *H. heidelbergensis* were to include Chinese fossils that evolved their diagnostic features independently, it would become a grade concept similar to "archaic *Homo sapiens*," with no phylogenetic utility.

Morphology

Morphologically, *Homo ergaster*, *H. erectus*, and the early representatives of *H. sapiens* and *H. neanderthalensis* differed in relatively subtle cranial characteristics, and their recognition arguably turns as much on geography and dating as it does on morphology. They shared dimensions of the semicircular canals within the inner ear like those of modern humans and a functionally identical postcranium, which combine to imply an exclusive reliance on bipedalism. In this they differed from *H. habilis* (or one of its variants) and the australopithecines, all of which probably depended more on a mix of bipedalism and tree climbing (2031), and they were fundamentally similar to living people, from whom they differed mainly in the great structural strength of their limb bones (1845). This strength is shown most notably and consistently in the constriction (stenosis) of the medullary (marrow) cavities and in the concomitant thickness of the surrounding (cortical) walls. The ruggedness of the lower limbs (known mainly from the pelvis and the femur) implies a high level of recurrent stress during extended bouts of heightened muscular exertion, while the powerful muscle markings on their arm bones indicate phenomenal upper body strength (muscular hypertrophy) (2222). The bodily robusticity of early *Homo* combines with archeology to argue that they relied much less on technology for survival than modern humans do.

Postcranial bones show also that, unlike *H. habilis* and the australopithecines, which tended to be small by modern standards, *H. ergaster*, *H. erectus*, and the early representatives

of *H. sapiens* and *H. neanderthalensis* all approximated living
people in size or, perhaps more precisely, that they varied in
size in about the same way as living people. Thus, like living
humans in sunny, tropical climates, individuals of *H. ergaster*
tended to be tall and long limbed. A partial skeleton from
Nariokotome, Kenya, suggests that adult males sometimes
reached 1.85 m (6'1"), and femurs from a mixed-sex sample of
six individuals imply an average adult height of 1.7 m (5'7")
(1846). In contrast, like living people in much cooler climates,
the north Chinese representatives of *H. erectus* were stockier,
and their femurs suggest that on average they were shorter than
individuals of *H. ergaster*. The difference probably reflects the
adaptive advantage of long, linear bodies in sunny, tropical
climes where dissipating heat was critical as opposed to the advantage of shorter, squatter bodies in cooler settings where retaining heat was more important. Since such ecogeographic
variation in body size is observable within living humans (and
other widespread mammalian species), it does not by itself indicate species distinctions, and it would be surprising if it had
not developed once populations of primitive *Homo* succeeded
in colonizing temperate latitudes.

Postcraniums thus not only fail to differentiate *H. ergaster*,
H. erectus, and the early representatives of *H. sapiens* and
H. neanderthalensis but also imply a shared fundamental human adaptation to terrestrial living. Archeology shows that
this adaptation also included a shared, if rudimentary, reliance
on technology and carnivory. It follows that species divergence,
if it is accepted, may have resulted less from natural selection
than from chance gene drift in small, widely separated populations. In these circumstances the species would tend to differ
more quantitatively than qualitatively, and they would overlap
extensively in morphology, particularly early on. In short, the
differences among them would be relatively subtle, and compelling species recognition would depend on large samples from
which both means and ranges of variation could be established.

The unusually large sample from the Sima de los Huesos,
Atapuerca, Spain, illustrates the point. The Atapuerca sample
comprises more than two thousand specimens, carefully recovered from a single deposit that probably accumulated about
300 ky ago. The specimens include more postcranial bones than
the combined samples from all other sites of like age (table 5.5),
together with three well-preserved skulls and fragments of more
than six others (79, 80, 82). Both the postcranial bones and the
skulls are remarkable, because they combine primitive features, including ones that are retained in *Homo sapiens*, with

Table 5.5. The Minimum Numbers of Postcranial Elements Recovered from the Sima de los Huesos, Atapuerca, Spain, through 1995

Anatomical Element	Sima de los Huesos	Rest of the World
Sternums	1	0
Vertebrae	54	9
Clavicles	16	2
Ribs	28	13
Scapulae	14	0
Humeri	18	7
Radii	14	1
Ulnae	15	3
Carpals	61	10
Metacarpals	23	2
Manual phalanges	162	10
Pelvises	18	8
Sacra	7	1
Femurs	29	27
Patellae	14	1
Tibiae	20	6
Fibulae	15	1
Tarsals	65	11
Metatarsals	26	3
Pedal phalanges	113	13
Total	713	126

Source: After 81, p. 122.
Note: Numbers are compared with the total numbers from like-aged (mid-Quaternary) sites elsewhere in the world. The SH sample has increased significantly in the meantime, creating an even greater disparity.

derived features that uniquely distinguish *H. neanderthalensis* and because they vary in the extent to which the primitive-derived mix is expressed (79, 2088). Neither the combination nor its variability could have been anticipated in advance, and it suggests that Neanderthal specializations evolved not as an integrated complex, but perhaps in an accretional fashion, as might be expected if gene drift played a larger role than natural selection.

Unfortunately, the Sima de los Huesos sample is as unique as it is informative, and even if multiple species within primitive *Homo* are accepted, small sample size impedes their description. The logical species to start with is *Homo erectus*,

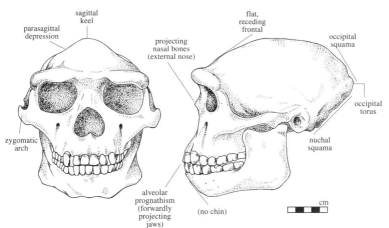

Indonesian *Homo erectus* (Weidenreich reconstruction)

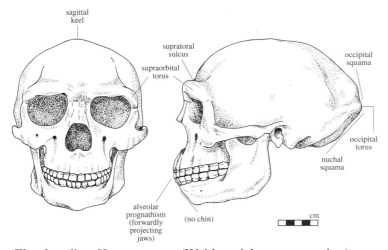

Zhoukoudian *Homo erectus* (Weidenreich reconstruction)

Figure 5.6. Franz Weidenreich's restorations of *Homo erectus* skulls from Sangiran in Java and Zhoukoudian Locality 1 in China (redrawn by Kathryn Cruz-Uribe partly after originals by Janis Cirulis in 1081, pp. 156, 169; © 1999 by Kathryn Cruz-Uribe). Typical *H. erectus* features that are visible in both restorations include a large forwardly projecting, shelflike supraorbital torus, a distinct supratoral sulcus, a low, receding frontal bone, sagittal keeling, a highly angulated occipital with a very prominent occipital torus, and pronounced alveolar prognathism.

since it was the first to be described, is the most widely accepted, and remains the species into which many specialists would sink one or more of the others. It thus provides the standard against which the validity of the others must be judged.

Homo erectus

The description here relies on the type Javan and Chinese fossils, which are the only ones that all specialists assign to *H. erectus*. It depends heavily on characterizations in references 1073, 1084, 1091, 1317, 1768, 1770, 1772, and 1774. It defines an average, to which not all specimens conform exactly. The anatomical features it refers to are illustrated in figures 5.6–5.9.

Braincase. Long, low vaulted, and thick walled; average cranial capacity slightly more than 1,000 cc, probably increasing

Figure 5.7. Skulls of Javan *Homo erectus* and a robust modern person, illustrating key differences, including the more prominent supraorbital torus, lower, more receding frontal, more highly angulated occipital, and greater alveolar prognathism of *H. erectus* (redrawn after 1317, p. 98).

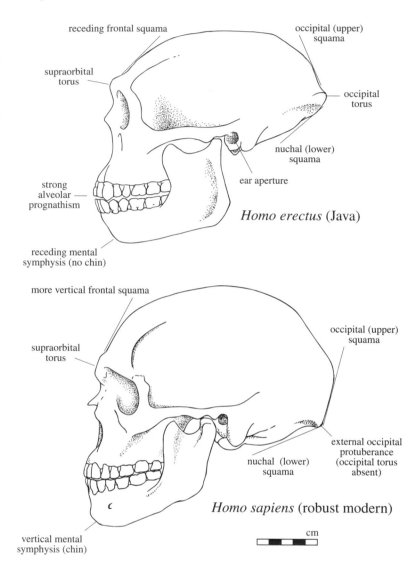

through time (1766, 1770, 1774), particularly if the sample is expanded to include the later Quaternary specimens from Ngandong (fig. 5.9) and Sambungmacan, Java; greatest breadth near the base, often coincident with biauricular breadth (breadth between the ear apertures); frontal bone low and receding, with a large supraorbital torus (or bar) (browridge) that tends to be straight when viewed from in front or above and that is equally well developed in the middle and at the sides; variable development of a supratoral sulcus (an inflection in the frontal profile just behind the torus) and also of sinuses within the torus; pronounced postorbital constriction.

Variable development of a sagittal keel along the midline at the top of the braincase, associated with (parasagittal) de-

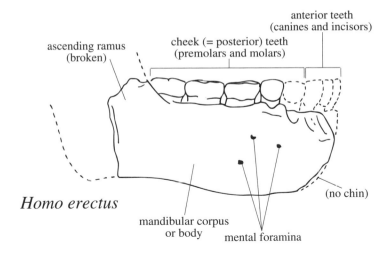

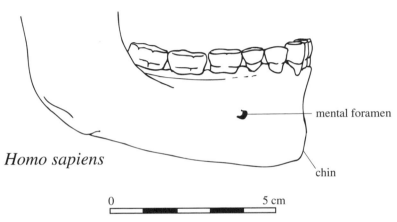

Figure 5.8.
Mandibles of Javan
Homo erectus and
a robust modern
person (redrawn
after 1317, p. 94).
Unlike modern
mandibles, those
of *H. erectus* lacked
a chin, and they
often had multiple
mental foramina.

pressions on either side of the midline and a thickening of the bone at and near bregma along the coronal suture; occipital bone always sharply angled in profile, with the upper (occipital) plate (scale or squama) usually smaller than the lower (nuchal) one; at the juncture of the occipital and nuchal plates, a conspicuous mound or bar of bone (the transverse occipital torus), usually projecting farthest near the midline, where it can form a blunt triangular eminence; basicranium moderately flexed or arched between the hard palate and foramen magnum (1301).

Face. Short (from top to bottom), but massive and relatively wide, with the nasal aperture projecting forward relative to the adjacent maxillary and zygomatic regions; pronounced alveolar prognathism (forward projection of the jaws); mandible robust and backward sloping below the incisors, resulting in the absence of a chin ("mental eminence").

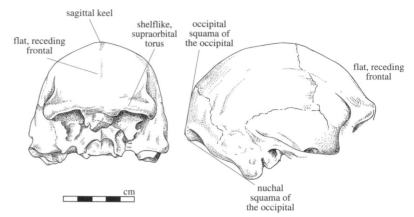

Ngandong XI (late Indonesian *Homo erectus*)

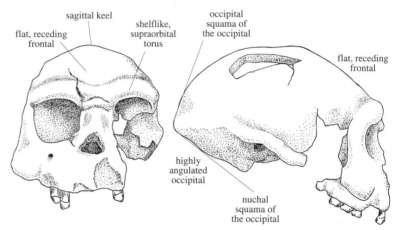

Sangiran 17 (classic Indonesian *Homo erectus*)

Figure 5.9. *Top:* Front and side views of skull XI from Ngandong (Solo), Java (redrawn by Kathryn Cruz-Uribe mainly after originals by Janis Cirulis in 1081, p. 160; © 1999 by Kathryn Cruz-Uribe). *Bottom:* Front and side views of skull 17 (*Pithecanthropus* VIII) from Sangiran, Java (drawn by Kathryn Cruz-Uribe from photos; © 1999 by Kathryn Cruz-Uribe). The Sangiran skull typifies classic Indonesian *Homo erectus*, and it probably antedates the Ngandong skull by several hundred thousand years, yet the two skulls share many conspicuous morphological features, including thick cranial walls, a low, flat frontal, a sagittal keel, a prominent occipital torus, an angulated (flexed) occipital, and the tendency for the skull walls to slope inward from a broad base. The Ngandong skulls have a larger endocranial capacity (1,150–1,300 cc) than those of classic Javan *H. erectus* (average of 934 cc), their browridges are thinner in the central part, and they exhibit less postorbital constriction, but the differences are small compared with the similarities. The Ngandong people are thus often regarded as a late variety of *H. erectus* that occupied Indonesia long after *H. sapiens* and *H. neanderthalensis* had appeared in the West.

Dentition. Cheek teeth large relative to those of modern people, but generally reduced compared with those of *Homo habilis* or the australopithecines; sporadic occurrence (in Javan specimens) of a diastema (gap) between the upper canine and lateral incisor, associated with a projecting, though spatulate, upper canine; third molar generally smaller than the second, as in like-aged or later species of *Homo* but not in the australopithecines; incisors large and often spatulate or shoveled (curled inward at the lateral edges).

Postcranium. Known mainly from the femur and the pelvis, which are generally similar to those of modern people (591, 595). However, the pelvis was remarkably robust and distinguished especially by a thick buttress (or pillar) of cortical (outer) bone arising vertically above the acetabulum (the socket for the femur). The femur was equally robust, with extraordinarily thick external (cortical) bone, a relatively narrow internal (medullary or marrow) canal, and pronounced muscle markings. The femoral shaft differed from that of later people in its greater fore-to-aft compression (greater platymeria, resulting in a more oval, less round shaft circumference), in the absence of a longitudinal bony ridge or pilaster on the rear surface, and in the more distal position of minimal shaft breadth (1206, 1208, 2402).

Homo ergaster

In the scheme that is proffered here, *H. ergaster* is a strictly African species that existed between 1.8–1.7 my ago and perhaps 600 ky ago. As indicated in the section above on geologic age, it is founded mainly on fossils from 1.8–1.5 my old deposits in the Lake Turkana Basin of northern Kenya (2361, p. 421). The principal specimens are two skulls (KNM-ER 3733 and 3883) (fig. 5.10) and a partial skeleton (KNM-ER 1808) from Koobi Fora, East Turkana, and a skull and associated skeleton (KNM-WT 15000) (figs. 5.11, 5.13) from Nariokotome III, West Turkana. To these can be added a skull (SK 847) (fig. 5.12) from Swartkrans Cave, South Africa, that is morphologically similar to the Turkana Basin specimens and that probably also dates between 1.8 and 1.5 my ago.

    H. ergaster was ancestral to all later species of *Homo*, including *H. erectus*, so it is not surprising that it resembled *H. erectus* in many key features. Particularly notable is a suite of shared, derived similarities that separate *H. ergaster* and *H. erectus* jointly from preceding *H. habilis* (broadly understood). These shared, derived distinctions include an increase

Figure 5.10. Skull KNM-ER 3733 from Koobi Fora, East Turkana (formerly East Rudolf), northern Kenya (redrawn by Kathryn Cruz-Uribe after 1073, fig. 10.10; © 1999 by Kathryn Cruz-Uribe). The skull exhibits several features that distinguish it from preceding *Homo habilis*, including a large, forwardly projecting supraorbital torus, a supratoral sulcus separating the torus from a low, receding frontal, a highly angulated occipital bone with a pronounced occipital torus, and an expanded braincase (with an estimated endocranial capacity of 848 cc). It is one of three north Kenyan fossils documenting the emergence of *Homo ergaster* about 1.7–1.6 my ago. (KNM-ER = Kenya National Museum–East Rudolf.)

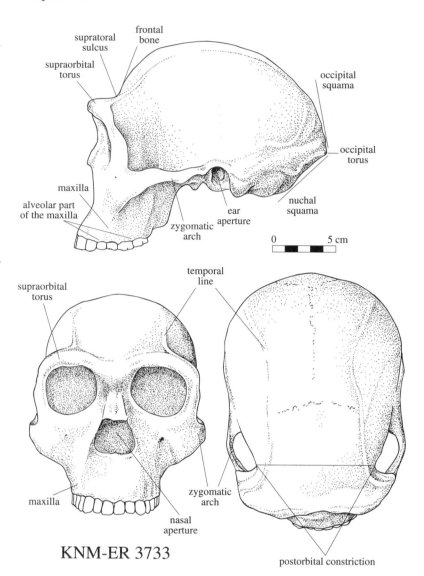

KNM-ER 3733

in the size of the brain (from a mean of roughly 630 cc in *H. habilis* to about 900 cc in *H. ergaster* to more than 1,000 cc in *H. erectus*); a reduction in the postcanine dentition (and in the robusticity of the associated jawbones); a vertical shortening of the face, the formation of a pronounced supraorbital torus with supratoral sulcus; the development of an occipital torus; the forward projection of the nasal aperture; a reduction in relative arm length leading to arm/leg proportions like those in living humans; an increase in overall body size to approximately that of living humans; and a reduction of sexual dimorphism to essentially the modern level. In addition, the nearly complete skeleton of specimen KNM-WT 15000 (fig. 5.13) shows that

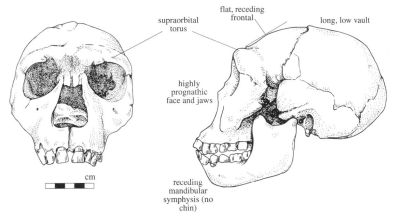

KNM-WT 15000 (Nariokotome III)

Figure 5.11. Skull of KNM-WT 15000 from Nariokotome III, West Turkana, northern Kenya (drawn by Kathryn Cruz-Uribe from photos in 2362; © 1999 by Kathryn Cruz-Uribe). The skull is roughly 1.5 my old, and it exhibits the same features that distinguish other skulls of *Homo ergaster* (or early African *H. erectus*) from those of *H. habilis*, including a distinct supraorbital torus, an enlarged braincase (with an endocranial capacity of 880 cc), and a reduction in the postcanine dentition. (KNM-WT = Kenya National Museum–West Turkana.)

H. ergaster was further distinguished from the australopithecines and possibly *H. habilis* by a very narrow hip region (the pelvis was remarkably narrow between the crests of the iliac blades) (2364) and by a barrel-shaped chest (rib cage) (1142). In contrast, the australopithecines (as known especially from AL 288-1, the famous Hadar skeleton of "Lucy") had relatively much broader hips (pelvises), and they had cone- or funnel-shaped rib cages like those of chimpanzees.

Homo ergaster* may have evolved from *H. habilis* in a burst of rapid evolution (in the mode predicted by punctuated equilibrium), or it may have emerged more gradually. A final decision probably depends on whether *H. habilis* is retained as one species or split in two, and this in turn depends on the accumulation of many new, well-dated fossils (2361). Arguably, the nearly simultaneous emergence of *H. ergaster* and of more sophisticated artifacts (the Acheulean Industrial Tradition discussed below) supports a relatively abrupt origin.

The increased body size of *H. ergaster* could account for most of its brain enlargement, and when its larger mass is considered, *H. ergaster* was only slightly bigger brained (more encephalized) than *H. habilis* (1510, 1511). Substantially greater encephalization occurred only after 500 ky ago in people who resembled *H. ergaster* in size but had much larger endocranial

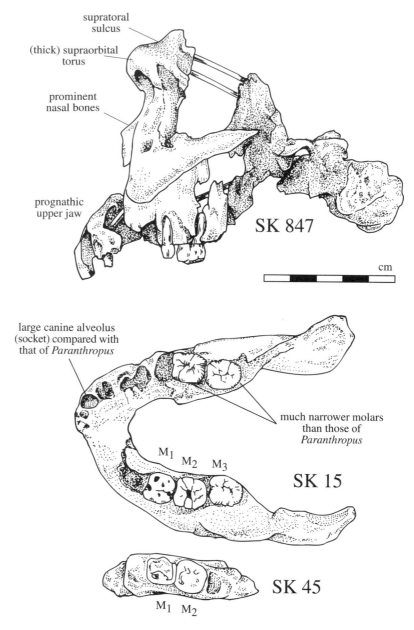

supratoral
sulcus

(thick) supraorbital
torus

prominent
nasal bones

prognathic
upper jaw

SK 847

cm

large canine alveolus
(socket) compared with
that of *Paranthropus*

much narrower molars
than those of
Paranthropus

M$_1$ M$_2$ M$_3$

SK 15

SK 45

M$_1$ M$_2$

Figure 5.12. Fossils or *Homo ergaster* (or early African *H. erectus*) from Swartkrans Cave (SK), South Africa (drawn by Kathryn Cruz-Uribe from photos in 507, p. 187; © 1999 by Kathryn Cruz-Uribe). Faunal associations indicate that the fossils are between 1.8 and 1.6 my old, and they may join very similar specimens from the Lake Turkana Basin, Kenya, in the core sample of *H. ergaster*. The partial skull labeled SK 847 actually comprises three separately numbered pieces that were reassembled by R. J. Clarke. The fragmentary mandible SK 45 may represent the same individual. All the fossils are readily distinguishable from those of the robust australopithecine, *Paranthropus robustus*, which occur in the same deposit.

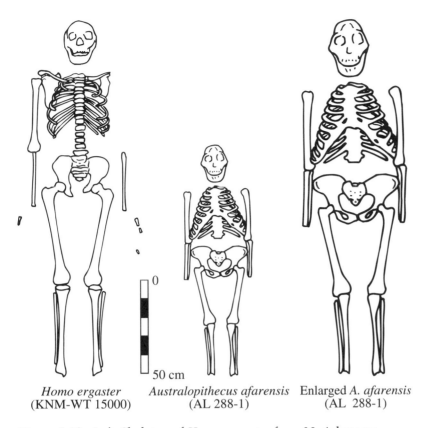

Homo ergaster	*Australopithecus afarensis*	Enlarged *A. afarensis*
(KNM-WT 15000)	(AL 288-1)	(AL 288-1)

Figure 5.13. *Left:* Skeleton of *Homo ergaster* from Nariokotome, West Turkana, northern Kenya. *Middle:* Skeleton of *Australopithe-cus afarensis* (AL 288-1, "Lucy") from Hadar, Ethiopia. *Right:* The *A. afarensis* skeleton scaled to the height of the *H. ergaster* skeleton (redrawn after 1843, p. 55). *H. ergaster* was the first hominid species to achieve the stature and bodily proportions of living humans. The *H. ergaster* individual whose skeleton is pictured was probably an eleven- to twelve-year-old boy who stood about 1.60 m (5' 3") tall and who might have reached 1.85 m (6' 1") at adulthood (1846). His body was long and linear like the bodies of living humans who inhabit similar hot, dry savanna environments. (KNM-WT = Kenya National Museum–West Turkana; AL = Hadar.)

volumes (generally exceeding 1,200 cc) (1511, 2361). The difference from later people implies that *H. ergaster* was behaviorally more primitive, but its increased brain size may still explain the appearance 1.7–1.6 ky ago of hand axes and other Acheulean artifacts. Bigger brains and more advanced technology in turn could help explain how *H. ergaster* became the first human species to colonize truly arid, highly seasonal settings in Africa. Adaptation to aridity may also underlie the change in the position of the nasal aperture, which marks the appearance of the typically human external nose with downwardly facing nostrils (vs. the relatively flat nose, with more forwardly facing nostrils, of the apes and of hominids before *Homo ergaster*) (753). The external nose reduces moisture loss during periods of heightened activity, since it is usually cooler than the central body and moisture thus condenses inside during exhalation (2222). Adaptation to hot, arid environments probably also required new mechanisms for cooling the body and especially for preventing the brain from overheating (704). The most important mechanism may have been an enhanced capacity to sweat, which would have been effected most efficiently by a substantial reduction in body hair. *H. ergaster* may thus have been the first hominid species to possess a largely hairless, naked skin.

Reduced arm length in *H. ergaster* probably signals the final abandonment of any apelike dependence on trees for feeding or refuge and, conversely, the development of the more thoroughly terrestrial adaptation to savanna life witnessed among living humans. A more exclusively terrestrial lifestyle and a greater reliance on bipedalism probably also drove the narrowing of the pelvis and the change from a cone-shaped to barrel-shaped chest (1142). The narrower pelvis increased the efficiency of muscles that operate the legs during bipedal locomotion, and it would have forced the lower part of the rib cage to narrow correspondingly. To maintain chest volume (pulmonary or lung function), the upper part of the rib cage would have had to expand, and the modern barrel shape would follow. The narrowing of the pelvis also constricted the birth canal, and this must have forced a reduction in the proportion of brain growth that occurred before birth (2361). The implication is a longer period of infant dependency, foreshadowing the remarkably long period that distinguishes living humans from other species.

Pelvic narrowing probably also reduced the volume of the intestines (gut), but this could occur only if food quality improved simultaneously (12). Circumstantially, then, pelvic narrowing might imply increased consumption of meat and marrow, more effective processing of tubers and other vegetal foods, or both. It might especially mean the invention of cooking, but as indicated below, the oldest reasonably secure evidence for fire use dates from only about 500 ky ago.

A more complete adaptation to terrestrial life probably also selected for increased body size, and it could even explain the decrease in sexual dimorphism. In extant higher primates, strong sexual dimorphism tends to reflect polygynous mating systems in which males compete vigorously for females, while reduced sexual dimorphism is associated with monogamous mating systems in which males and females pair for long periods. Decreased dimorphism in *H. ergaster* may thus signal the beginnings of the distinctively human pattern of sharing and cooperation between the sexes that promoted historic hunter-gatherer survival in savanna environments.

Reduced cheek tooth size in *H. ergaster/erectus* probably reflects a greater reliance on technology (artifacts) to process food. There may have been a simultaneous increase in the use of the incisor teeth for biting, gripping, and tearing (2481, 2484), culminating in the heavily shoveled upper incisors of *H. erectus*. Shoveling greatly increases the area of the occlusal surface, and it could thus prolong incisor life under heavy use. If increased emphasis on the front teeth is accepted, it could further explain the development of the forwardly projecting supra-orbital torus "to resist bending stresses which concentrate on the frontal bone above the orbit during habitual anterior tooth loading" (1852, p. 350). However, gauges attached above the orbits of living macaque monkeys and baboons show that incision (biting between the front teeth) does not significantly strain the frontal bone (1105), and the supraorbital torus may have developed simply to connect the braincase and face in fossil people whose faces grew entirely in front of the brain (rather than partly below it as in living humans) (1373).

The derived features that *H. ergaster* and *H. erectus* shared are striking, and they explain why many authorities prefer to regard *H. ergaster* simply as early *H. erectus* (337, 342, 1774, 2361). Separation depends on the argument that *H. erectus* was also derived relative to *H. ergaster* in some key features (44, 507, 911, 914, 2077, 2498), including the tendency for *H. erectus* to have a lower, less domed cranium; thicker cranial walls; an interparietal sagittal keel (an elongated bony mound between the parietals at the top of the skull); a more projecting supraorbital torus (browridge) that is as thick laterally as in the middle; a thickened tympanic plate (the bony enclosure of the middle ear that supports the eardrum and is fused to the temporal bone at the base of the skull); a significantly reduced mound of bone (postglenoid process or eminence) behind the articular depression (fossa) for the mandibular condyle on the base of the temporal bone; and a more massive face.

The differences between *H. ergaster* and *H. erectus* are admittedly more subtle than the similarities, particularly since they are mainly quantitative, and they must be assessed on a relatively small number of specimens. If the differences are accepted, however, *H. erectus* is a plausible descendant of *H. ergaster*, and it a less likely ancestor for *H. sapiens* and *H. neanderthalensis*, both of which tend to be derived away from *H. ergaster* in features that also differentiate them from *H. erectus*. These include a supraorbital torus that tends to be thicker

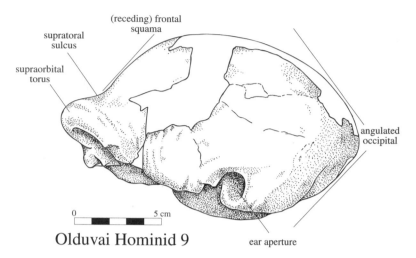

Olduvai Hominid 9

Figure 5.14. Skullcap of Olduvai Hominid 9 from Upper Bed II, Olduvai Gorge (drawn by Kathryn Cruz-Uribe from a cast and slides; © 1999 by Kathryn Cruz-Uribe). The skullcap is roughly 1.2–1.1 my old. It is assigned here to *Homo ergaster*, but among all African fossil skulls, it is the most difficult to separate from skulls of classic Far Eastern *H. erectus*. Conspicuous features it shares with *H. erectus* include a massive, forwardly projecting supraorbital torus, a low, receding frontal bone, a highly angulated occipital, and thick cranial walls.

in the middle than at the sides, parietals that usually lack a midline keel and tend to bulge (or boss) outward, and occipitals that are relatively rounded. Primitive morphology relative to the other three species and the geologic antiquity of *H. ergaster* together explain its basal position in figure 5.1.

On strictly morphological grounds, the strongest objection to *Homo ergaster* as advocated here is probably OH 9 (1323), a previously mentioned skullcap from Upper Bed II at Olduvai Gorge (fig. 5.14). OH 9 is believed to be roughly 1.2 my old, and it is distinguished from the 1.8–1.5 my old Turkana skulls by some of the same features that have been said to distinguish *H. erectus* from *H. ergaster*, namely, thicker cranial walls, a massive supraorbital torus, and a more angulated occipital. Even strong proponents of *H. ergaster* at Lake Turkana have placed OH 9 in *H. erectus* (911, 2499). But like the Turkana specimens of *H. ergaster* and also *H. sapiens*, OH 9 lacks the interparietal keel that marks classic (Far Eastern) *H. erectus*, and it resembles both Turkana *H. ergaster* and *H. sapiens* in details of the tympanic plate (505). At least tentatively, then, OH 9 can be assigned to *H. ergaster*, and it illustrates the difficulty of separating closely related species on quantitative characters alone when the mean conditions and variation around the means are poorly established. A special irony is that if OH 9 is included in *H. ergaster*, *H. ergaster* should perhaps be renamed *H. leakeyi*, since *H. leakeyi* was applied to OH 9 (994) before *H. ergaster* was proposed for Turkana Basin fossils (915).

In sum, *H. ergaster* and *H. erectus* were clearly very similar, and reasonable specialists can disagree on whether they should be separated. The issue can be resolved only with a much larger number of well-dated fossils, but in their absence, specialist disagreement should not be allowed to obscure the fundamental evolutionary implications of the African fossils. Recall that *Homo habilis*, or the constituents into which it may ultimately be divided, was variable in the extent of its departure from the australopithecines. Some specimens retained large cheek teeth and an australopithecine-like chewing apparatus, while others were not especially large brained and some also retained postcranial features suggesting they combined bipedalism and tree climbing in an australopithecine fashion. In contrast, *H. ergaster* or African *H. erectus* was uniformly advanced in brain enlargement, reduced cheek tooth size, and a postcranial structure that reflects an exclusive reliance on bipedalism. Archeology also shows that *H. ergaster* or African *H. erectus* was the first hominid species to colonize arid, seasonal environments like those in which historic hunter-gatherers are well known. If *H. habilis* may be seen as transitional between the australopithecines and later humans, then *H. ergaster* may be regarded as the first true human species.

Homo sapiens, H. neanderthalensis, and H. heidelbergensis

Like *Homo ergaster*, the early representatives of *H. sapiens* (Africa) and *H. neanderthalensis* (Europe) shared many features with *H. erectus*, including large browridges; a low, flattened frontal bone; a relatively broad cranial base; thick cranial walls; relatively massive, chinless mandibles with large teeth; and powerfully constructed postcraniums. However, they diverged from both *H. ergaster* and *H. erectus* in endocranial volumes that generally far exceeded the average (of just over 1,000 cc) for the *H. erectus* core (table 5.4) and in a shared tendency to more rounded occipitals, expanded parietals, and broader frontals with more arched (vs. more shelflike) browridges. The similarities among early *H. sapiens*, early *H. neanderthalensis*, and *H. erectus* are probably primitive features retained from a shared *H. ergaster* ancestor, whereas the characters that separate *H. sapiens* and *H. neanderthalensis* from *H. erectus* are derived traits. They indicate that the line or lines leading to *H. sapiens* and *H. neanderthalensis* had diverged from the line that produced *H. erectus* by 1 my ago or before.

Arguably, *H. sapiens* in Africa and *H. neanderthalensis* in Europe inherited their shared, derived traits from a common ancestor that lived between 600 and 400 ky ago and that is represented by the skulls from Bodo, Ndutu, Broken Hill (Kabwe), and Elandsfontein in Africa, by those from Petralona and Arago in Europe, and perhaps by the like-aged skull from Narmada in north-central India (fig. 5.15). Each of these skulls tends to exhibit roughly the same mix of primitive features that are shared with *H. erectus* and of more advanced features that jointly distinguish both later *H. sapiens* and *H. neanderthalensis*. To the extent that the African and European skulls cannot be separated morphologically, they are reasonably joined in a common species for which the name *Homo heidelbergensis* has been most commonly proposed. Artifactual contrasts discussed below underscore the divergence of eastern Asia from Africa and Europe by 1 my ago, while artifactual similarities indicate the possibility of a population dispersal from Africa to Europe perhaps 500 ky ago. The concept of *H. heidelbergensis* remains problematic, however, partly because it includes fossils that may vary greatly in age and partly because some putative European specimens (especially Petralona) may anticipate the classic Neanderthals (after 200–130 ky ago) in some features (like midfacial projection). In this event the fossils should probably be reassigned to *H. neanderthalensis*. Even if *H. heidelbergensis* is retained, its morphology can be adequately understood from a consideration of *H. sapiens* and *H. neanderthalensis*, and it will not be considered further here.

The case for separate *H. sapiens* and *H. neanderthalensis* lineages rests primarily on European fossils that date between 500 and 200 ky ago and that combine primitive features of the genus *Homo* with evolutionary novelties that uniquely mark the classic Neanderthals who occupied Europe after 127 ky ago (79, 82, 1092, 1096, 2083, 2086, 2306). In contrast, African fossils that are between 500 and 200 ky old exhibit no Neanderthal specializations, but they plausibly anticipate the modern or near-modern people who inhabited Africa and its immediate southwest Asian hinterland after 127 ky ago (333). The relevant European fossils include the skulls from Swanscombe, England (1096), and Reilingen, Germany (618), and above all the extraordinary sample of human remains from the Sima de los Huesos, Atapuerca, Spain (79). The African fossils include the skulls from Salé and Jebel Irhoud, Morocco (1095), Singa, Sudan (1495),

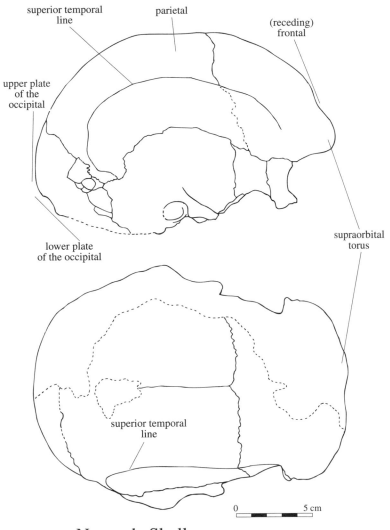

superior temporal
line

parietal

(receding)
frontal

upper plate
of the
occipital

lower plate
of the occipital

supraorbital
torus

superior temporal
line

0   5 cm

Narmada Skull

Figure 5.15. Skullcap of primitive *Homo* from the Narmada Valley,
north-central India (redrawn by Kathryn Cruz-Uribe after 1409,
pp. 26, 30; © 1999 by Kathryn Cruz-Uribe). Associated animal bones
and "Upper" Acheulean artifacts imply the skull is between 600 and
400 ky old. It combines primitive features like a thick, forwardly projecting supraorbital torus, thick cranial walls, and great basal breadth
with advanced features like expanded parietals, a less angulated
(flexed) occipital, and a relatively large endocranial capacity (> 1,200
cc). In its mix of primitive and derived characters, it recalls like-aged
African and European specimens that are variably assigned to early
Homo sapiens, early *H. neanderthalensis,* or *H. heidelbergensis.*

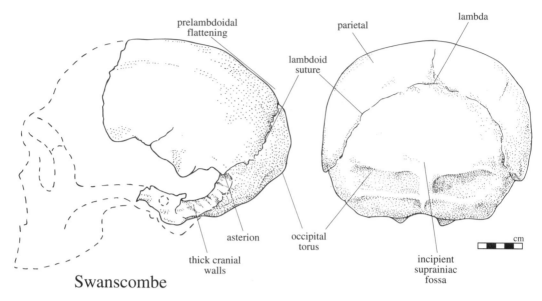

Swanscombe

Figure 5.16. Fossil skull from Swanscombe, England (drawn by Kathryn Cruz-Uribe partly after originals by Janis Cirulis in 1081, p. 218, and photos in 2095; © 1999 by Kathryn Cruz-Uribe). The skull is probably between 400 and 250 ky old. In its thick cranial walls and maximum breadth near the base, it recalls skulls of *Homo erectus*. However, it anticipates skulls of the classic Neanderthals in the flattening of the parietal region just before lambda (a point on the rear of the skull where the lambdoid and sagittal sutures join) and especially in the presence of an incipient suprainiac fossa (an elliptical, depressed area of roughened bone) over the occipital torus. It is assigned here to early *H. neanderthalensis*.

Bodo, Ethiopia (1777), Lake Ndutu, Tanzania (505), Broken Hill (Kabwe), Zambia (1718), Elandsfontein, South Africa (674), and Florisbad, South Africa (921).

The Swanscombe and Reilingen skulls both lack faces (fig. 5.16), but they are very similar in the position of maximum breadth near the base of the skull and in occipital morphology, including a torus that is better developed at the sides than at the center and is surmounted by an elliptical depression of roughened bone (the suprainiac fossa, illustrated in figs. 6.5 and 6.6 in the next chapter). Maximum breadth near the cranial base is a primitive feature that is not retained in Neanderthal skulls, whose maximum breadth is at midparietal level (fig. 5.17). The Reilingen skull is also primitive in the large size of its mastoid process, which tends to be significantly smaller in Neanderthals (the mastoid region is not preserved on the Swanscombe skull). In contrast, the occipital morphology that Reilingen and Swanscombe share is a derived feature that is expressed more fully in the classic Neanderthals.

The Sima de los Huesos (SH) sample includes three nearly complete skulls, large fragments from at least six other skulls,

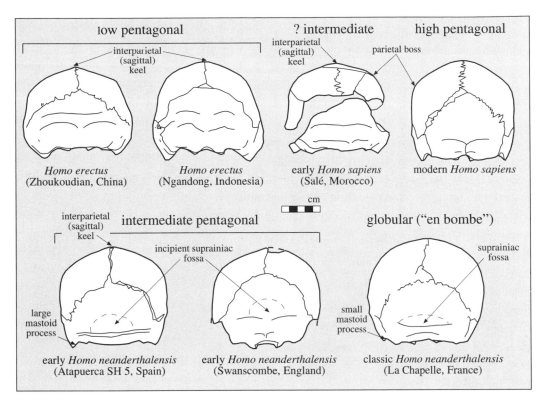

Figure 5.17. Skulls of *Homo erectus*, early *H. sapiens*, modern *H. sapiens*, early *H. nean-derthalensis*, and classic *H. neanderthalensis* viewed in occipital (rear) view (redrawn after 1096, p. 41, and 79, photo on p. 224). From this perspective, skulls of *H. erectus* are pentagonal with sidewalls that slope sharply inward from near the base ("low pentagonal"); skulls of early *H. sapiens* and early *H. neanderthalensis* are pentagonal with walls that tend to rise more vertically; skulls of classic *H. neanderthalensis* are globular with walls that bulge outward at midparietal level; and skulls of modern *H. sapiens* are pentagonal with walls that rise vertically to a point high on the parietals before sloping inward ("high pentagonal"). Skulls of modern *H. sapiens* also tend to show a boss or bulge at the point where the parietals turn inward. Arguably, the early *H. sapiens* skull from Salé, Morocco, shows an incipient boss.

numerous small craniofacial fragments, more than forty-one complete or partial mandibles and numerous isolated teeth from at least twenty-four individuals, and hundreds of postcranial bones. To an even greater extent than the Reilingen and Swanscombe skulls, the SH sample is remarkable for its combination of primitive features that are also seen in *H. erectus* or *H. sapiens*, derived features that are shared between *H. sapiens* and *H. neanderthalensis*, and derived features that are unique to *H. neanderthalensis* (table 5.6).

   The most conspicuous Neanderthal specializations in the SH skulls (79, 80, 82, 1470) are a face that projects far forward along the midline and an oval area of roughened or porous bone overlying the occipital torus (fig. 5.18). Midfacial projection and

Table 5.6. Characters that Fossils from the Sima de los Huesos, Atapuerca, Spain, Share with *H. erectus*, the Classic Neanderthals (after 127 ky ago), and *H. sapiens*

Character	H. erectus	Classic Neanderthals	H. sapiens
Vault broadest near the base	X		
Substantial, overall facial projection (prognathism)	X		
Laterally thick supraorbital torus	X		
Relatively flat occipital squama	X		
Interparietal (sagittal) torus	X		
Mandibular robusticity	X	X	
Lower limb robusticity	X	X	
Cranial capacity range	X	X	X
Double-arched supraorbital torus		X	X
Incipient suprainiac fossa		X	
Substantial midfacial projection		X	
Anterior teeth large relative to posterior ones		X	
Large retromolar space, an asymmetric sigmoid notch, and a large, posteriorly projecting coronoid process on the mandible		X	
Lateral occipital profile relatively rounded		X	X
Relatively thin tympanic bone		X	X
High cranial vault		X	X
Temporal squama high and rounded		X	X
Rear parietal profile	X		
Adult mastoid process large			X

Source: Modified after 2083, p. 502.
Note: The features shared with *H. erectus* are primitive features for the genus *Homo*, whereas those shared exclusively with *H. sapiens* probably characterized a common ancestor of *H. sapiens* and the Neanderthals. The features shared only with the Neanderthals are uniquely derived, which suggests that the Sima de los Huesos population should be placed on or near the line leading to the Neanderthals.

other Neanderthal facial features are particularly impressive in the most complete skull (SH 5) (fig. 5.19). Some other facial fragments are less obviously Neanderthal-like, and one large piece, AT (Atapuerca) 404, exhibits a distinctly non-Neanderthal canine fossa (deflation or hollowing of the bone above the canine tooth between the nasal aperture and the zygomatic arch). Such a fossa also characterizes the well-known mid-Quaternary skull from Steinheim, Germany, which is arguably Neanderthal-like in its relatively curved occipital. The roughened, oval area on the SH occipitals varies from slightly less depressed than on the Swanscombe or Reilingen skulls to flat or even convex, but it still clearly foreshadows the classic Neanderthal suprainiac fossa. The numerous primitive features on the SH skulls include a maximum breadth near the base (versus a maximum breadth at midparietal level in Neanderthals); a horizontal occipital torus that is well developed centrally (versus the double-arched torus of Neanderthals which is better developed laterally); a tendency for the rearmost point on the cranium to lie on the occipital torus (rather than on the occipital plate above the torus in Neanderthals); a supraorbital torus that is double arched (as in Neanderthals and early *H. sapiens*) but that is very thick laterally (as in *H. erectus*); a large mastoid process; and an endocranial capacity that not only was below the Neanderthal average of 1,450 cc (judged from estimates for the two most complete adult SH specimens) but encompassed the entire range observed in other European mid-Quaternary specimens (table 5.4). Two skulls also exhibit an upraised mound or keel of bone between the parietals that is otherwise common only in *Homo erectus*.

The SH mandibles (1822–1824) and dentitions (197, 198, 201) display a similar amalgam of primitive and derived features, but the mandibles are highly variable in size and shape. Like adult Neanderthal mandibles, the SH specimens all exhibit a marked gap (the retromolar space) between the rear edge of the third molar and the leading edge of the ascending ramus, but they vary in how far they show the highly asymmetric sigmoid notch and large, posteriorly projecting coronoid process that mark most Neanderthal mandibles (fig. 5.20). The SH dentitions resemble those of the Neanderthals (and of other mid-Quaternary Europeans, so far as these are known, but not those of *H. erectus* or *H. sapiens*) in the large size of the first incisor through the third premolar (I_1 through P_3) relative to the small size of the fourth premolar through the third molar (P_4 through M_3). In addition, the SH molars tend toward the taurodontism that is common in the Neanderthals. However, the SH P_4

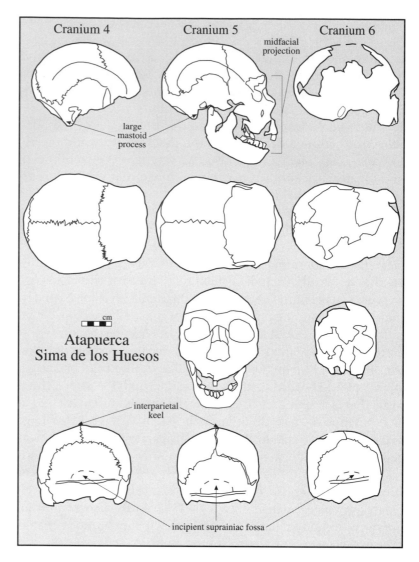

Figure 5.18. Skulls 4, 5, and 6 from the Sima de los Huesos (SH), Atapuerca, Spain (drawn from photos in 82, figs. 2, 4, 6). The skulls are probably about 300 ky old, and they combine characters that are primitive for the genus *Homo* with derived features that indicate a special relation to the classic Neanderthals who inhabited Europe after 127 ky ago. The primitive features include a large mastoid process and a pentagonal shape in rear view (versus the typical Neanderthal condition, in which the mastoid is very small and the skull is globular in rear view). The derived Neanderthal features include the pronounced forward projection of the face along the midline and the presence of an oval area of roughened or porous bone overlying the occipital torus. The roughened area anticipates the suprainiac fossa of the classic Neanderthals. Skull 6 contrasts somewhat with skulls 4 and 5 in its greater occipital rounding and with skull 5 in its more flattened face, but this is probably mainly because skull 6 came from a juvenile, whereas skulls 4 and 5 came from adults. However, the larger SH sample exhibits substantial variability in the expression of Neanderthal facial and occipital morphology. The sum suggests that Neanderthal features evolved piecemeal rather than as an integrated complex. This may mean that Neanderthal evolution owes more to gene drift than to natural selection.

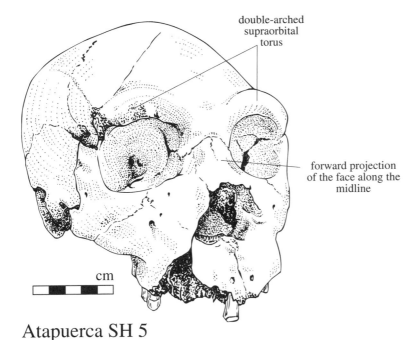

double-arched
supraorbital
torus

forward projection
of the face along the
midline

Figure 5.19. Facial view of skull 5 from the Sima de los Huesos (SH), Atapuerca, Spain (drawn by Kathryn Cruz-Uribe from a photograph; © 1999 by Kathryn Cruz-Uribe). SH 5 anticipates classic Neanderthal skulls in its double-arched browridge and strong midfacial prognathism.

cm

Atapuerca SH 5

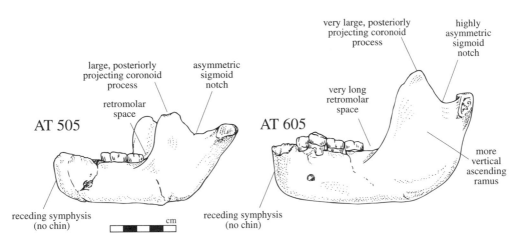

large, posteriorly
projecting coronoid
process

asymmetric
sigmoid
notch

retromolar
space

AT 505

receding symphysis
(no chin)

cm

very large, posteriorly
projecting coronoid
process

highly
asymmetric
sigmoid
notch

very long
retromolar
space

AT 605

more
vertical
ascending
ramus

receding symphysis
(no chin)

Figure 5.20. Mandibles AT 505 and AT 605 from the Sima de los Huesos, Atapuerca, Spain (drawn by Kathryn Cruz-Uribe from photos in 1823, p. 329; © 1999 by Kathryn Cruz-Uribe). The mandibles are both adult, but they differ in size. They both show typical Neanderthal features, including a prominent retromolar space, a large posteriorly projecting coronoid process, and an asymmetric sigmoid notch, but the larger mandible expresses these features more strongly. It also has a somewhat steeper mental symphysis and a more vertical ascending ramus. The difference in size presumably reflects sexual dimorphism, and it shows that sex (size) may also influence the expression of phylogenetically diagnostic morphological features. (AT = Atapuerca.)

through M_3 are unique for their small average size, in which they resemble only recent humans.

Finally, the SH limb bones are also notable for their combination of primitive and Neanderthal features, including the extreme robusticity that marks primitive *Homo* everywhere and a specifically Neanderthal tendency for a dorsal (as opposed to a ventral) sulcus or groove on the axillary margin of the scapula (illustrated in fig. 6.13 in the next chapter) (78, 428). The SH pelvises and femurs have not been described in detail, but like-aged specimens from other sites closely resemble those of more ancient *Homo* and exhibit no conspicuous derived Neanderthal features. Thus, unlike Neanderthal pelvises, a pelvis from Arago, France, resembles much older pelvises from Olduvai Gorge and Koobi Fora (East Turkana) in its great robusticity, including the presence of a thick buttress or pillar of cortical (outer compact) bone above the acetabulum (590, 591). Other lower limb bones, such as three femoral fragments from Arago (1947), a femoral shaft from Notarchirico, Venosa, Italy (532), and a tibia shaft from Boxgrove, England (1787, 2086), are also striking for their great robusticity and for the lack of any specific Neanderthal features. Neanderthal femurs are particularly distinguished by rounded shafts that are strongly bowed from front to back, while the mid-Quaternary specimens all have relatively straight, flat shafts.

In their mix of primitive features and Neanderthal novelties and in the variable expression of the mix, the SH specimens recall not only Reilingen and Swanscombe, but other well-known European fossils that probably date between 500 and 200 ky ago. The others encompass some skulls or skull fragments that are remarkably primitive in occipital or frontal form (especially at Bilzingsleben and Vértesszöllös) (fig. 5.21), some that are arguably Neanderthal-like in facial or occipital features (but not both) (at Arago, Petralona, and Steinheim) (fig. 5.22), and some that clearly anticipate the Neanderthals in key respects (especially at Swanscombe and Reilingen). Like-aged mandibles and teeth are similarly variable, and a Neanderthal-like retromolar space, for example, is apparent on one mandible from Arago (no. 2), absent in a second (no. 13), and only tenuously (incipiently) expressed in yet a third from the site of Montmaurin, France (fig. 5.23). Since the SH sample surely represents a single population, it implies that the entire European sample could represent a single Neanderthal lineage, along which Neanderthal features developed with time, not as an integrated functional complex but in an accretional fashion, perhaps more as a result of gene drift than of natural selection.

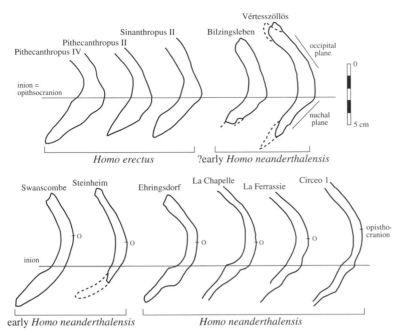

Figure 5.21. Midline (midsagittal) sections through the occipital bones of some key fossils assigned here to *Homo erectus* ("*Pithecanthropus* IV," "*Pithecanthropus* II," and "*Sinanthropus* II"), early *H. neanderthalensis* (Bilzingsleben, Vértesszöllös, Swanscombe, and Steinheim), and later *H. neanderthalensis* (Ehringsdorf, La Chapelle, La Ferrassie, and Circeo 1) (redrawn after 2331, p. 244). Note that with respect to occipital flexion (angulation between the occipital and nuchal planes), the Vértesszöllös and especially the Bilzingsleben fossils are more like *H. erectus* than they are like the Swanscombe, Steinheim, and later Neanderthal fossils.

Determining the actual evolutionary pattern will require a much larger sample of well-dated fossils scattered through time. However, if it is accepted that the fossils from Arago, Vértesszöllös, and Bilzingsleben are older than those from Atapuerca SH and that these in turn are older than those from Swanscombe, Steinheim, and Reilingen, one could argue that Neanderthal facial morphology evolved before the typically Neanderthal occipital region and that the occipital region evolved before the characteristically large and globular Neanderthal vault, which is conspicuous only after 127 ky ago (1096).

The more ancient African skulls—from Bodo (fig. 5.24), Lake Ndutu (fig. 5.25), Broken Hill (fig. 5.26), and Elandsfontein (fig. 5.27)—exhibit no uniquely derived features of modern humans, but they equally lack any Neanderthal specializations. The younger skulls—from Irhoud (fig. 5.28), Singa (fig. 5.29), and Florisbad (fig. 5.30)—also lack Neanderthal novelties, and

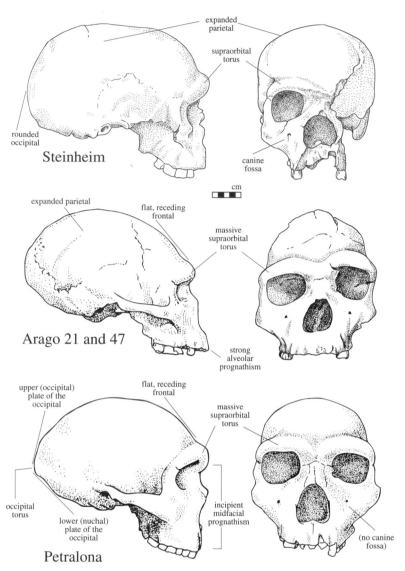

Figure 5.22. Fossil skulls from Steinheim, Germany, Arago, France, and Petralona, Greece (Steinheim drawn by Kathryn Cruz-Uribe from originals by Janis Cirulis in 1081, p. 218; © 1999 by Kathryn Cruz-Uribe; Arago from casts and from a photo in 594, p. 48; and Petralona from photos in 2076, p. 73). None of the skulls is reliably dated, but each is probably between 500 and 250 ky old. The Arago skull is a composite based on the face (*right*) (Arago 21) and a large parietal fragment (Arago 47) that may not come from the same individual. If the combination is accepted, however, the Arago braincase exhibited parietal expansion that might anticipate such expansion in the classic Neanderthals, after 127 ky ago. In other respects, including its flat, receding frontal, massive supraorbital torus, and strong alveolar prognathism, the Arago skull more closely recalls those of *Homo erectus*. The Steinheim and Petralona skulls also foreshadow Neanderthal skulls, but in different, contrasting ways. Thus, in Steinheim the resemblance is primarily in the expansion and rounding of the braincase. The face is flat with a canine fossa more in the manner of *H. sapiens* than of *H. neanderthalensis*. In Petralona the resemblance is primarily in the face, which anticipates those of Neanderthals in its incipient forward projection along the midline. The braincase is more like that of *H. erectus*, particularly in the angulation of the occipital. All three skulls are tentatively assigned here to the Neanderthal lineage.

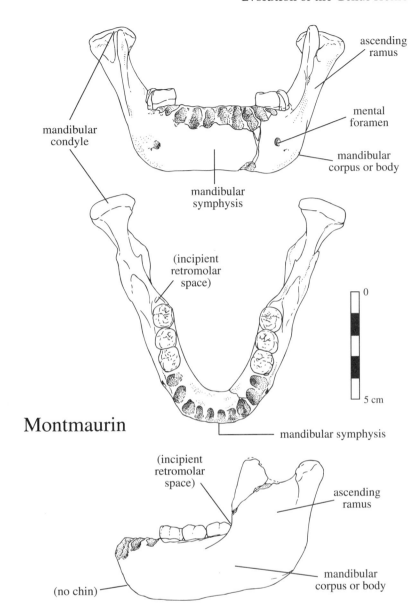

Figure 5.23. Fossil human mandible from Montmaurin, France (redrawn by Kathryn Cruz-Uribe after 213, p. 280; © 1999 by Kathryn Cruz-Uribe). Associated animal remains and pollen suggest that it is between 186 and 127 ky old. It combines large size, the absence of a chin, and other features that are primitive for the genus *Homo* with molar taurodontism and an incipient retromolar space that probably imply a special relation to the classic Neanderthals, after 127 ky ago.

they anticipate the derived modern human tendency for the face to be tucked below the forepart of the brain (rather than mounted in front of it) (1372). Arguably the skull from Salé, which could be chronologically intermediate, is also morphologically transitional in its outwardly bulging (bossed) parietals (fig. 5.17) and in its weakly expressed occipital torus, but the occipital shows signs of pathology and its phylogenetic significance is therefore dubious (1090). The Salé skull also exhibits an interparietal (sagittal) keel. An isolated right maxilla from Broken Hill is much less robust than the maxilla on the famous

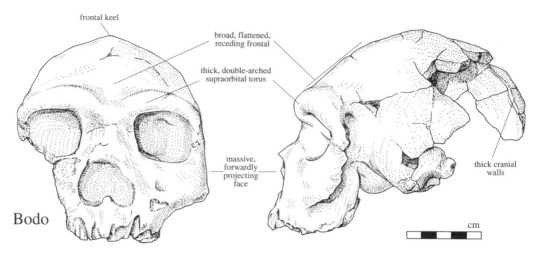

Figure 5.24. The principal fossil skull from Bodo, Middle Awash, Ethiopia (drawn by Kathryn Cruz-Uribe from photographs in 1777, pp. 24–25; © 1999 by Kathryn Cruz-Uribe). Radiometric determinations and associated fauna indicate it is probably about 600 ky old. In its massive, forwardly projecting face, thick supraorbital torus, flat, receding frontal, frontal keel, and thick cranial walls, the skull recalls those of *Homo erectus*. However, it is distinguished from *H. erectus* by its great frontal breadth, the double arching of its supraorbital torus, and its relatively large braincase. It is assigned here to early *H. sapiens*.

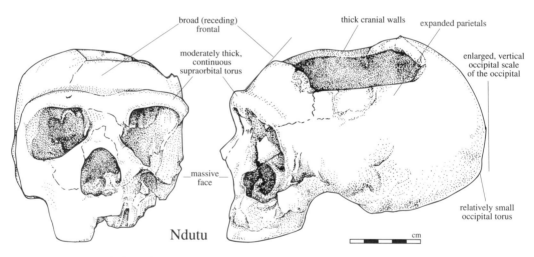

Figure 5.25. Fossil skull from deposits of seasonal Lake Ndutu near the western end of the Main Olduvai Gorge, northern Tanzania (drawn by Kathryn Cruz-Uribe from photos of a reconstruction in 505, pp. 705, 713; © 1999 by Kathryn Cruz-Uribe). The skull is thought to be about 400 ky old. Like the Broken Hill and Elandsfontein skulls (in figs. 5.26 and 5.27), it combines primitive features, including a large, projecting supraorbital torus, a low receding frontal, and thick cranial walls, with more advanced features, including expanded parietals and a nonangulated (rounded) occipital. The occipital is also notable for the large size of the upper (occipital scale), which arises vertically (instead of sloping forward as in *Homo erectus*). Ndutu differs from the Broken Hill and Elandsfontein skulls primarily in its smaller size (with an endocranial capacity of about 1,100 cc) and its weaker occipital torus. If Ndutu represents a female and the others represent males, the difference could reflect sexual dimorphism in the same fundamental population. All three skulls are tentatively assigned here to early *Homo sapiens*.

Broken Hill (Kabwe)

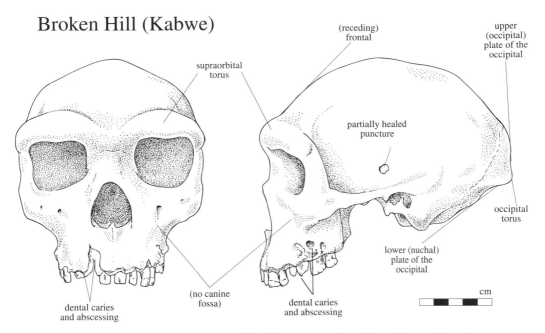

Figure 5.26. Fossil skull from Broken Hill (Kabwe) in Zambia (drawn by Kathryn Cruz-Uribe from a cast and photographs; © 1999 by Kathryn Cruz-Uribe). Possibly associated animal remains suggest the skull may be about 400 ky old. In its massive face, thick brow-ridges, flat, receding frontal, and relatively great basal breadth, the skull recalls those of *Homo erectus*. However, it is readily distinguished from *H. erectus* skulls by its large en-docranial capacity (about 1,280 cc) and by the vertical orientation of the upper or occipital plate of the occipital bone above the torus. It is assigned here to early *H. sapiens*.

complete skull, and unlike the maxilla on the complete skull, it exhibits a canine fossa. This could imply that the Broken Hill people were as variable as those from the Sima de los Huesos, or it could record change through time toward the more modern condition. The stratigraphic relationships of the Broken Hill bones are uncertain, but the specimens are known to have come from different parts of the cave (493).

The small number of African limb bones generally resemble their European counterparts in their great robusticity. Only an ulna from Kapthurin, Kenya, is described as gracile in the manner of modern specimens (2013). In contrast, a pelvis from Broken Hill (Kabwe), Zambia, preserves the same cortical buttress over the acetabulum found at Arago and in specimens from much earlier *Homo* (2079). Femoral fragments from Broken Hill (1208), a proximal femur from Berg Aukas, northern Namibia (904), and a femoral shaft from Lainyamok, Kenya (1708), are variable in morphological detail, but they tend to have very thick cortical bone by modern standards. A tibia from Broken Hill also has thick cortical bone, but it is very long

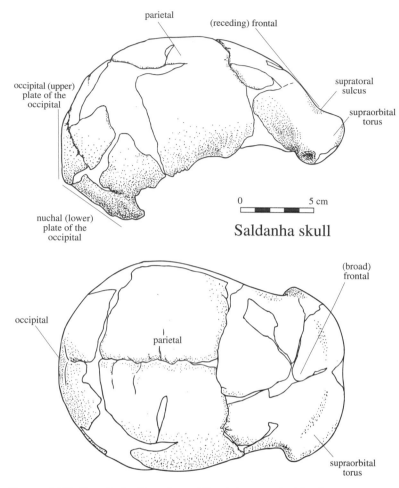

Figure 5.27. Fossil skullcap from Elandsfontein (Saldanha = Hope-field), South Africa (drawn by Kathryn Cruz-Uribe from a cast and photos; © 1999 by Kathryn Cruz-Uribe). Faunal associations suggest an antiquity between 700 and 400 ky ago. The skull has a massive supraorbital torus, a low, receding frontal, thick cranial walls, and other features that tend to characterize primitive *Homo* everywhere, together with expanded parietals, a relatively rounded occipital, and a broad frontal that suggest it is derived in the direction of *H. sapiens*. It is placed here in early *Homo sapiens*.

(1208), implying that the distal leg (between the knee and ankle) was long, as in most living tropical or subtropical Africans. In contrast, the Neanderthals had very short distal legs, like those of modern arctic peoples, whose short limbs help to reduce heat loss. The Cro-Magnons who succeeded the Neanderthals in Europe had tropical limb proportions, and this supports cranial evidence that they evolved in Africa. In general, body form says more about ecology than about phylogeny, but a larger limb

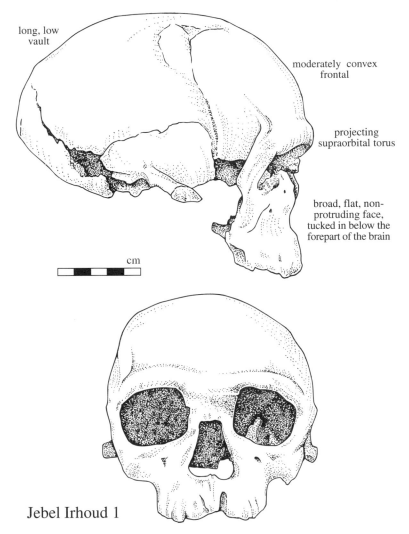

long, low vault

moderately convex frontal

projecting supraorbital torus

broad, flat, non-protruding face, tucked in below the forepart of the brain

cm

Jebel Irhoud 1

Figure 5.28. Right lateral and facial views of Jebel Irhoud skull 1, Morocco (drawn by Kathryn Cruz-Uribe from photographs; © 1999 by Kathryn Cruz-Uribe). The skull is estimated to be between 190 and 90 ky old, yet it anticipates skulls of living people in its relatively convex frontal (forehead) and in its relatively short, flat face tucked beneath the forepart of the brain. It is less conspicuously modern in the thickness and projection of the supraorbital region and in the low height of the braincase relative to length and breadth. Conceivably it represents a population that was near the ancestry of the near-modern or early modern people from Skhul and Qafzeh Caves, Israel.

bone sample from both Europe and Africa may nonetheless help to illuminate the divergence of early *H. neanderthalensis* from contemporaneous *H. sapiens*.

In sum, the African fossil record is sparse between 500 ky ago and 200 ky ago, but the contrast with the European record

Figure 5.29. Left lateral view of the skull from Singa, Sudan (drawn by Kathryn Cruz-Uribe from a photograph in 2074, p. 77; © 1999 by Kathryn Cruz-Uribe). The skull is estimated to date from 170–150 ky ago, yet it resembles skulls of living people in its highly convex frontal, in its high, short vault, and in the placement of the face beneath the forepart of the brain. Arguably, only the thickness of the browridge, particularly at the sides, is a distinctly archaic feature.

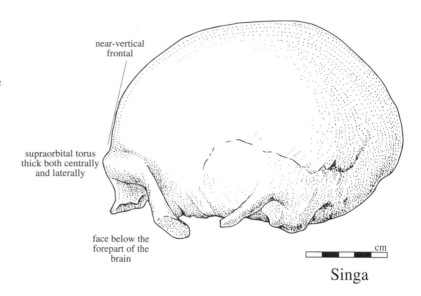

is still clear. Together, the African and European fossils document an evolutionary separation that by 127 ky ago produced the classic Neanderthals in Europe and near-modern or modern people in Africa.

Late *Homo erectus* or *H. heidelbergensis* in China?

The next two chapters argue that fully modern *Homo sapiens* spread from Africa 50–40 ky ago to replace *H. neanderthalensis* in western Eurasia and late surviving *H. erectus* in the Far East. However, the replacement hypothesis is truly compelling only for western Eurasia with its relatively dense and well-dated fossil and archeological records. It is much harder to sustain for the Far East, where the corresponding fossil record is much poorer and where the associated archeological record is almost nonexistent. Within the Far East, replacement is most plausible for Southeast Asia, or more precisely Java, where classic *H. erectus* fossils (from the Pucangan and Kabuh Beds) and later ones (from Ngandong and Sambungmacan) document a single evolving lineage that may have survived to 50 ky ago or later (2122). It is much less persuasive for China, where the evolutionary connection between classic *H. erectus* fossils (especially from Lantian, Zhoukoudian, and Hexian) and later fossils (from Dali, Yinkou [Jinniushan], Xujiayao, Chaoxian [Chaohu], Maba [Mapa], Changyang, Dingcun, Miaohoushan, and Tongzi) is more equivocal.

Among the later Chinese fossils, the most useful are nearly complete skulls from Dali (fig. 5.31) and Yinkou (fig. 5.32), a

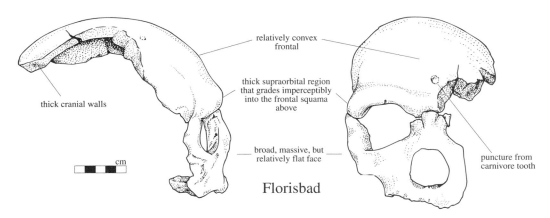

relatively convex frontal

thick supraorbital region that grades imperceptibly into the frontal squama above

thick cranial walls

broad, massive, but relatively flat face

cm

Florisbad

puncture from carnivore tooth

Figure 5.30. Right lateral and facial views of the skull from Florisbad, South Africa (drawn by Kathryn Cruz-Uribe from photos in 503, p. 303; © 1999 by Kathryn Cruz-Uribe). ESR analysis of an associated molar suggests the skull is about 260 ky old. Like most modern human skulls, Florisbad lacks a true supraorbital torus (there is no supratoral sulcus or inflection point separating the supraorbital region from the frontal squama immediately above), and the frontal is relatively convex. However, the supraorbital region is heavily thickened, particularly at the sides, the face is very broad, and the cranial walls are very thick. Arguably, when all features are considered, Florisbad is morphologically intermediate between older, more archaic African skulls like those from Bodo, Ethiopia (fig. 5.24), and Kabwe, Zambia (fig. 5.26), on the one hand, and later, more modern-looking skulls like the one from Singa (fig. 5.29) and other sites considered in the next chapter.

partial skullcap and upper facial skeleton from Maba (fig. 5.33), parietal and occipital bones from Xujiayao, and an occipital from Chaoxian (2516, 2517). The remaining fossils are mainly isolated teeth or jaw fragments of limited comparative value. Dating in each case depends mainly on U-series analysis of associated animal bones, but the reliability of the results is questionable, because the bones may have exchanged uranium with their burial environment (461). The most secure dates are probably those for Yinkou, where a combination of U-series and ESR determinations on animal teeth suggests an age near 200 ky ago (460). If the dates are all taken at face value, they bracket the various fossils between roughly 230 and 100 ky ago. Recall that African fossils from this period foreshadow living humans, while European ones clearly anticipate the Neanderthals.

The Chinese fossils variably combine massive, uninterrupted browridges, keeled, flat, receding frontal bones, low vault heights or other primitive features that characterize *H. erectus* with larger, more rounded braincases, less massive faces, and other advanced features that mark *H. sapiens*. They differ from both their African and their European contemporaries, but in their mix of archaic and derived features, they recall the older African and European fossils that are sometimes grouped in

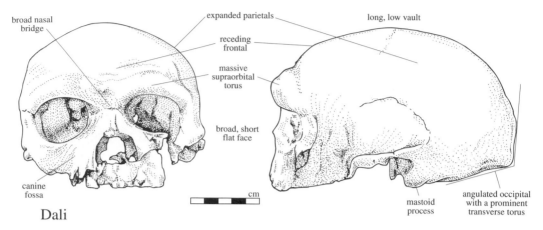

Dali

Figure 5.31. The skull from Dali County, Shaanxi Province, northern China (drawn by Kathryn Cruz-Uribe from photos in 2516, p. 116; © 1999 by Kathryn Cruz-Uribe). U-series analyses of associated animal bones indicate that the skull may be between 230 and 180 ky old. It shares many primitive features with *Homo erectus*, including a massive supraorbital torus, a flat, receding frontal, an angulated occipital with a prominent transverse torus, thick cranial walls, and a relatively small endocranial capacity (1,120 cc). However, it departs from classic *H. erectus* in its more expanded parietals and much more limited postorbital constriction. The face is flat with a canine fossa and without the midfacial projection that marks the Neanderthals. The skull is tentatively assigned here to late Chinese *H. erectus*.

H. heidelbergensis (1778). This could imply that *H. heidelbergensis* extended eastward to China (but not to Java), or it could imply that Chinese *H. erectus* and *H. heidelbergensis* evolved similar characters in parallel. Parallel evolution is tentatively favored here, partly because it is more parsimonious and partly because the Chinese archeological record so far reveals no evidence for a population intrusion. In addition, the later Chinese fossils tend to resemble Chinese *H. erectus* in a handful of features, including a short maxilla, flat, horizontally oriented cheekbones, a very broad nasal bridge, strong shoveling of the upper incisors, and small third molars (M3s). If parallel evolution is accepted, then Chinese and Javan *H. erectus* followed very different evolutionary pathways, and Chinese *H. erectus* might have to be removed to a separate species. Many new fossils, dates, and archeological finds would be necessary to establish this, and fresh fossils and dates are also required to determine whether the "Out of Africa" scenario applies to China in the same way it does to Europe.

Geographic Distribution

Archeology shows that, unlike the australopithecines or even *Homo habilis*, *H. ergaster* (or early *H. erectus*) was distributed

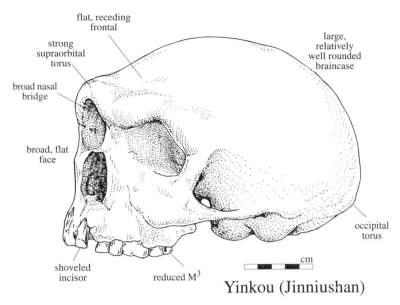

flat, receding frontal

large, relatively well rounded braincase

strong supraorbital torus

broad nasal bridge

broad, flat face

occipital torus

shoveled incisor

reduced M³

cm

Yinkou (Jinniushan)

Figure 5.32. The skull from Jinniushan, near the city of Yinkou, Liaoning Province, northeastern China (drawn by Kathryn Cruz-Uribe from a photo in 2516, p. 121; © 1999 by Kathryn Cruz-Uribe). Much of the postcranial skeleton was associated, but it has not been described in detail. ESR and U-series dates on associated animal teeth suggest the skull may be about 200 ky old. It combines primitive features, including a large supraorbital torus, a flat, receding frontal, and a prominent occipital torus, with advanced features, including a relatively rounded braincase that is broadest well above the base, thin cranial walls, and a large endocranial capacity (1,390 cc). In its primitive-advanced mix, it broadly recalls 500–400 ky old African and European skulls that are sometimes assigned to *Homo heidelbergensis*. However, it is tentatively assigned here to late *H. erectus*.

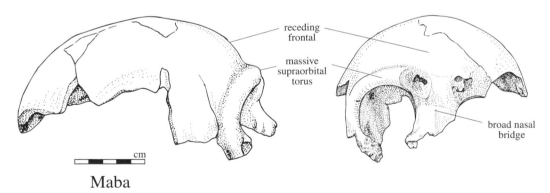

receding frontal

massive supraorbital torus

broad nasal bridge

cm

Maba

Figure 5.33. A partial skull found in a limestone cave near the village of Maba, Guandong Province, southern China (drawn by Kathryn Cruz-Uribe from photos in 2516, p. 139; © 1999 by Kathryn Cruz-Uribe). U-series analyses of associated animal bones indicate the skull may be between 140 and 119 ky old. It exhibits more postorbital constriction than either the Dali or the Yinkou skull (figs. 5.30 and 5.32), but it broadly resembles them in its large supraorbital torus and receding frontal. With them, it is tentatively assigned here to late Chinese *Homo erectus*.

throughout Africa, excepting only the extreme deserts of the north and south and the lowland rain forest of the Congo-Zaire Basin. Archeological discoveries indicate that about 1.5 my ago, shortly after the emergence of *H. ergaster*, people more intensively occupied the drier peripheries of sedimentary basins on the floor of the eastern Rift Valley (959), and they colonized the Ethiopian high plateau (at 2,300–2,400 m) for the first time (497). By 1 my ago, they had extended their range to the far northern and southern margins of Africa (1243, 1740). Range extension north inevitably led to dispersal out of Africa, probably mainly around the eastern end of the Mediterranean and thence eastward to China and Indonesia and westward to Europe.

The earliest emergence from Africa may be signaled at 'Ubeidiya in the Jordan Valley of Israel (115, 116, 123, 127, 862, 927) or at Dmanisi in southeastern part of the Republic of Georgia (123, 293, 680, 775) (fig. 5.34). At 'Ubeidiya, lake and river deposits contain bifaces and other Acheulean or Developed Oldowan artifacts that probably accumulated between 1.4 and 1 my ago. The age estimate is founded on the associated mammalian species, and it is consistent with the reversed magnetism of the deposits and with radiopotassium determinations of about 0.84 my ago on a superimposed basalt (928, 2151, 2153, 2154, 2156, 2157, 2317). The mammals are mainly Eurasian (Palearctic), but some are African (Ethiopian), and the artifacts closely resemble broadly contemporaneous pieces from Upper Bed II at Olduvai Gorge. 'Ubeidiya might thus reflect a slight ecological enlargement of Africa more than a true human dispersal. Israel is only technically outside Africa today, and it was repeatedly invaded by African species during the Quaternary (2154). The invasions occurred mainly during interglacials, including the last one (between roughly 127 and 80 ky ago), when the African invaders included early modern or near-modern humans.

At Dmanisi, alluvial deposits have produced a human mandible, Oldowan-like pebble tools and flakes, and animal bones that may antedate those at 'Ubeidiya. A migration from Africa is indisputable, since Dmanisi is at 44° N, on the southern slope of the Caucasus Mountains, roughly 1,500 km northeast of 'Ubeidiya, and its fossil fauna is exclusively Eurasian. K/Ar and ^{40}Ar/^{39}Ar determinations fix an underlying basalt at 2–1.8 my ago, within the Matuyama Reversed Paleomagnetic Chron, but both the basalt and the overlying alluvium exhibit normal magnetization. The mammalian species constrain the age to between 2 and 1 my ago, and the normal magnetization thus implies a date within the Olduvai Normal Subchron

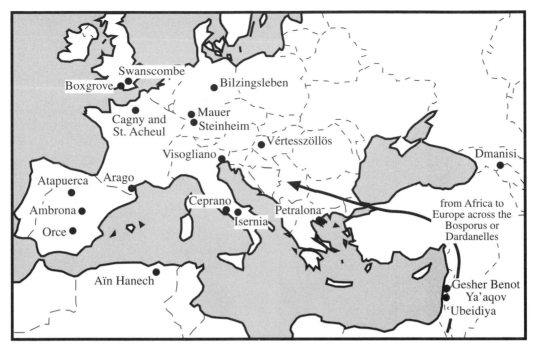

Figure 5.34. The approximate locations of key early archeological or human fossil sites in western Asia and Europe (modified after 1096, p. 39). The artifacts from 'Ubeidiya, Israel, show that people were living outside Africa by 1 my ago or before. A human mandible and artifacts from Dmanisi, Republic of Georgia, may be as old or older. Arguably, the oldest European sites date to only 600–500 ky ago, but discoveries at Ceprano, Italy, Orce, southern Spain, and especially Atapuerca (Trinchera Dolina [TD] 6), northern Spain, may document sporadic human presence between 900 and 600 ky ago. Humans probably first penetrated Europe across the Dardanelles or Bosporus, the narrow straits that now separate Asian and European Turkey.

between roughly 1.95 and 1.77 my ago or within the Cobb Mountain + Jaramillo Normal Subchrons between roughly 1.19 and 0.99 my ago. The human mandible itself may argue for the more recent age, since it differs significantly from mandibles of *H. ergaster* (or early African *H. erectus*) and resembles those of later *Homo* (after 1 my ago) in morphology and in dental proportions (338, 345, 617). Particularly striking is the rounded contour of the symphysis (the bony region below the incisors) and the notable reduction in size from the first molar to the second to the third (fig. 5.35). In *H. ergaster* (or early *H. erectus*), the symphysis slopes backward more steeply, and the first molars tend to be smaller than the second or the third. In its strong progressive reduction in molar size, the Dmanisi mandible in fact recalls mandibles of early *H. neanderthalensis* from the Sima de los Huesos, Atapuerca (1824). In sum, pending additional radiometric determinations and analyses of the fauna,

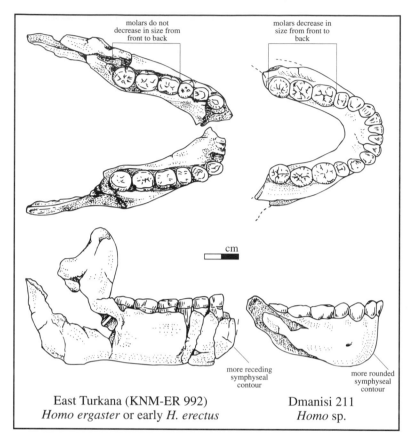

Figure 5.35. A mandible of *Homo ergaster* (or early *H. erectus*) from East Turkana, northern Kenya (drawn by Kathryn Cruz-Uribe from photos in 2499, p. 148; © 1999 by Kathryn Cruz-Uribe) and the mandible of *Homo* sp. from Dmanisi, Republic of Georgia (redrawn after 775, p. 511). The East Turkana mandible is reliably dated to about 1.6 my ago and is the nomenclatural type for *H. ergaster* (915). The Dmanisi mandible may be equally old, but a time closer to 1 my ago is suggested by advanced features of the mandible itself. These include the rounded (as opposed to more strongly receding) contour of the bone below the incisors and the decrease in occlusal area along the molar row, from M_1 to M_3. The typical condition *H. ergaster* is $M_2 > M_1 > M_3$.

Dmanisi does not show that people had penetrated the farther reaches of Eurasia long before 1 my ago.

The timing of human arrival in both eastern Asia and Europe is controversial and may always remain so, since a lack of evidence can never prove human absence. In addition, specialists disagree on what is acceptable evidence, and no one can deny that a very ancient site with universally acceptable evidence might await discovery. As noted in the section above on dating, some authorities have argued that people reached both

China and Indonesia by 1.8 my ago. Such an ancient arrival is appealing in part because it could explain why Far Eastern artifact assemblages always lacked hand axes and other markers of the Acheulean Industrial Tradition (504, 2123). The Acheulean developed in Africa from the preceding Oldowan Tradition only after 1.8 my ago (1797), and if people penetrated eastern Asia at 1.8 my ago or before, they would have arrived without Acheulean tools. Nonetheless, at present the oldest unambiguous evidence for human presence in the Far East is no older than 1 my (1701).

After *H. erectus* became established in eastern Asia, it was apparently limited to middle and lower latitudes, even during interglacials, and archeological sites at Donggutuo, Xiaochangliang, Maliang, and Cenjiawan in the Nihewan Basin (approximately 41° N) are the northernmost well-established occurrences (1875, 1877). They are only slightly north of Zhoukoudian (approximately 39.50° N), where fossil pollen indicates that *H. erectus* was present only under interglacial conditions (15). Siberian artifacts that might be attributed to *H. erectus* either are not clearly artifactual or are almost certainly too young (2527 and the summary of early Siberian prehistory in chapter 7).

It has commonly been assumed that people reached Europe as early as they reached the Far East, and Dmanisi, at the "gates of Europe" could be regarded as confirmation, even if it is no more 1 my old. The first edition of this book accepted an initial European penetration between 1 my and 780 ky ago, based on sites like Le Vallonet Cave near Nice, France (693, 1401, 1402, 1405), the Soleihac open-air (lake) site near Le Puy in the Massif Central of southeastern France (255–257, 747, 2185), the Kärlich A alluvial site in the Neuwied Basin of west-central Germany (290, 2518), the alluvial site of Isernia La Pineta in central Italy (525, 526, 1671, 1672, 2328), the Prezletice lakeside site near Prague, Czech Republic (720, 768–770, 2282, 2284, 2286), the Stránská Skála cave sites near Brno, Czech Republic (768, 1601, 2282, 2286–2288), and the streamside site of Korolevo in southwestern Ukraine (844) (fig. 5.36).

The dating at each site is based on paleomagnetism, radiopotassium, ESR or luminescence determinations, associated mammal fossils, or some combination of these. However, a critical review of these sites and most others that are supposed to be equally old suggests that none are valid and that the oldest persuasive evidence for the occupation of Europe postdates 600–500 ky ago (1811). The oldest unambiguous sites include both Acheulean examples like Boxgrove, England (1785), Ambrona

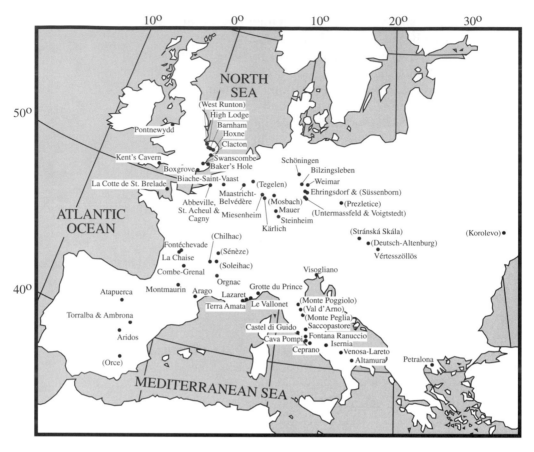

Figure 5.36. The approximate locations of early and mid-Quaternary archeological and human fossil sites in Europe. Some specialists believe that Korolevo in Ukraine, Stránská Skála and Prezletice in the Czech Republic, Kärlich in Germany, Le Vallonet, Chilhac, and Soleihac in France, Isernia and Ceprano in Italy, and Atapuerca and Orce in Spain all document human presence before 500 ky ago, but the case is truly compelling only for Atapuerca (Trinchera Dolina). Sites shown in parentheses are older than 500 ky ago but lack firm evidence for human presence.

and Torralba, Spain, (764, 1072, 1076), Cagny-la-Garenne, France (300, 2240, 2246, 2326), and Fontana Ranuccio, Italy (1907), and non-Acheulean ones like Kärlich G and Miesenheim 1, Germany (290, 2259). The earliest artifact makers, at roughly 500 ky ago, would be represented by the tibia shaft from Boxgrove, by a mandible fragment and isolated teeth from Visogliano, Italy (434), and by the mandible from Mauer, Germany, that is the nomenclatural type for *Homo heidelbergensis*.

At Soleihac, Le Vallonet, Kärlich A, Prezletice, Stránská Skála, and most other sites that supposedly antedate 600–500 ky ago, the key question is the origin of the artifacts. In general, their number is small and they were selected from a much larger number of nonartifactual stones. This suggests that they could be geofacts or artifact mimics that natural impacts of rock upon rock will occasionally produce (291). Such natural flaking is now generally accepted as the explanation for artifacts of the "Kafuan" tradition that was once thought to antedate the Oldowan in Africa (480). It must also account for the occasional flaked pieces known as "eoliths" from early Tertiary deposits in Britain (1625, 2392).

At Isernia La Pineta, the human origin of the artifacts is not disputed, but the dates are problematic. Isernia has provided more than five thousand flaked pieces of flint and limestone from three or four superimposed layers within finegrained alluvial sediments, where natural flaking is highly unlikely. In places the artifacts are associated with abundant animal (mainly bison) bones, some of which appear to have been artifactually cracked for their marrow (34). K/Ar determinations on associated volcanic particles place the archeological layers at roughly 730 ky ago, but the sediments exhibit reversed magnetism, which implies an age beyond 780 ky ago, while the fauna indicates a much younger age, near 500 ky ago. The absence of the primitive (rooted) water vole, *Mimomys*, and the presence of its advanced (rootless) descendant, *Arvicola*, is particularly significant, since the *Mimomys-Arvicola* transition is well fixed elsewhere in Europe at roughly 500 ky ago (fig. 5.37). Given the obvious contradictions and the likelihood that the *Mimomys-Arvicola* transition occurred more or less simultaneously everywhere in Europe, a date for Isernia near 500 ky ago seems most probable (1811).

Initial human colonization of Europe only about 500 ky ago could explain why compelling evidence for human presence is absent at rich, well-known paleontological sites that date from between 1.5 and 0.5 my ago. The sites include the Tegelen pits in the Netherlands, Untermassfeld, Voigtstedt, and Süssenborn in Germany, West Runton in England, Sénèze in France, Deutsch-Altenburg in Austria, and the Val d'Arno in Italy (1811). In contrast, paleontological sites that postdate 0.5 my ago also commonly contain unequivocal stone artifacts and humanly broken or cut animal bones. As discussed in the section on subsistence below, the issue at most sites that postdate 500 ky ago is not whether people were present but whether they killed or butchered the animals. The same kind of contrast before and after 12 ky ago can be used to argue that people reached the Americas only after 12 ky ago (discussion in chapter 7).

There is the difference, however, that there are highly plausible geographic or ecological factors that could have prevented people from reaching the Americas before 12 ky ago, whereas

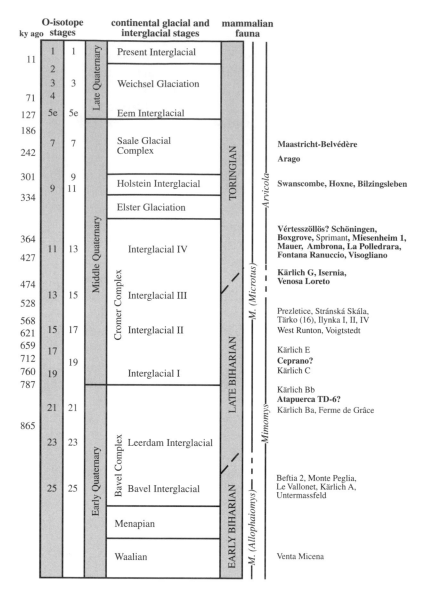

Figure 5.37. Tentative correlation of global marine oxygen-isotope stages, north European glaciations and interglacials, European mammal stages, and key European fossil and archeological sites (modified after 1811, p. 492). Sites shown in boldface have provided human fossils, indisputable artifacts, or both. Odd-numbered isotope stages ordinarily correlate with continental interglacials, but the continental (terrestrial) record for glaciations and interglacials is very fragmentary and discontinuous, and the chart surely underestimates the number of continental glacial/interglacial cycles. Correlations between the continental and marine records are therefore tentative, and the diagram shows two possible schemes. The left margin (based on 150) shows estimated initial and terminal ages (in thousands of years before present) for the odd-numbered isotope stages in the adjacent column. The mammalian faunal record is relatively coarse, but the portion of the Quaternary period represented in the chart comprises at least three recognizable mammalian stages or substages. Stage definition depends especially on shifts from *Microtus (Allophaiomys)* to *Microtus (Microtus)* roughly 1 my ago and from *Mimomys* to *Arvicola* roughly 500 ky ago. The diagram shows that human fossil and archeological sites are rare before 500 ky ago, which may mean that Europe was only sporadically occupied before that time.

there are no obvious factors that could have excluded people from Europe for 500 ky after they had colonized northern China. One possibility is that it was only about 500 ky ago that people developed the means to adapt to the dense deciduous forests that covered temperate Europe during interglacial peaks and to the grassy steppes or steppe-tundras that replaced the forests during glaciations (1807). Another is that they could not successfully compete for carcasses with species of hyenas that finally disappeared from Europe only about 500 ky ago (2251). In either case, it is possible that people were present before 500 ky ago, but only sporadically, and that to begin with they were restricted to warmer climes south of the Pyrenees and the Alps (649). An earlier presence that matches this criterion may be indicated at Ceprano, Italy (83, 84), at Orce, near Grenada, southern Spain (815, 1472, 1803, 2191, 2262), and most persuasively, in the "Aurora Stratum" of layer 6 in the Trinchera (Gran) Dolina site, less than one kilometer from the Sima de los Huesos, Atapuerca, northern Spain (424).

At Ceprano, deposits below a sandy volcanic gravel that may be 700 ky old have produced a human skull with many *Homo erectus*–like features, including a massive, shelflike browridge, an occipital torus that is equally developed at the sides and the center, a sharply angled occiput, thick cranial walls, and a relatively small endocranial capacity of 1,185 cc. The frontal and occipital recall fragments of the same bones from Bilzingsleben, Germany. Among all the European fossils that postdate 500 ky ago, the Bilzingsleben specimens are perhaps the hardest to relate to the Neanderthal lineage.

At Orce, the localities of Fuente Nueva 3 and Barranco Léon 5 have produced crudely flaked pebbles and flakes in lake-edge deposits that may date from 1 my ago. The dating is based on the assumption that the deposits formed shortly before or during the Jaramillo Normal Polarity Subchron, but additional stratigraphic and paleomagnetic research is needed for confirmation (1803). The nearby Venta Micena locality has provided a small cranial fragment and two fragmentary humeri that some have identified as human (816–819). The paleomagnetism of the deposits and the rich associated fauna imply an age near 1.6 my ago, and the fauna contains African carnivores and ungulates that probably first reached Europe about this time (1471, 1647). However, the cranial fragment almost certainly comes from a young horse (1597); the larger of the two humeral fragments is clearly not human (personal observation); and the smaller is perhaps identifiable only as mammal. None of the Venta Micena animal bones was unequivocally damaged by stone tools, and the site should probably join the list of rich

European paleontological localities that antedate 600–500 ky ago and suggest that people were either absent or uncommon in Europe before that time.

At the Trinchera Dolina (TD), a railway trench through a sinkhole infilling has exposed 18 m of sediment, divided among eleven layers, numbered 1 to 11 from bottom to top. The oldest artifacts occur in layer TD 4 (426), but TD 6 is at present the most important. It has produced more than one hundred flaked stones and roughly eighty fragments of human bone in sediments that lie roughly 1 m below the boundary between the Matuyama Reversed Paleomagnetic Chron and the Brunhes Normal Chron (1650). TD 6 must thus antedate 780 ky ago, and an age of about 800 ky ago is the current best estimate. The associated fauna includes bones of the primitive water vole, *Mimomys*, which is sufficient in itself to place TD 6 before any other widely accepted European archeological or human fossil site. The other sites are all marked by the more advanced (rootless) *Arvicola*. The overall composition of the TD 6 fauna, however, suggests an age closer to 500 ky ago than to 800 ky ago, and the inconsistency remains unexplained (651). The problem may lie in the identification of the Brunhes/Matuyama Boundary, which was originally isolated much lower in the section, at the level of TD 2/3 (426).

The TD 6 (and older) artifacts comprise crudely flaked pebbles and simple, nondescript flakes that by themselves might be dismissed as geofacts, but the human remains are incontrovertible. They represent at least six individuals, and the craniofacial fragments are striking for derived features that differentiate them from *H. ergaster–H. erectus* but do not ally them specially with either *H. neanderthalensis* or *H. sapiens* (200). Particularly significant is a fragmentary juvenile maxilla (ATD 6–69) (fig. 5.38) that suggests the midface was fundamentally modern in its lack of strong forward projection (particularly along the midline as in Neanderthals) and in the deep hollowing (canine fossa) between the nasal aperture and the cheek bone (absent in Neanderthals). The same hollowing is conspicuous, if less marked, on an adult maxillary fragment (ATD 6–58). The TD 6 fossils have been assigned to a new species, *Homo antecessor,* on the grounds that they exhibit an otherwise unknown combination of traits and that they are less derived in the Neanderthal direction than later mid-Quaternary European specimens assigned to *H. heidelbergensis.* The TD 6 fossils may thus represent a more plausible shared ancestor for *H. neanderthalensis* and *H. sapiens.* In fact the TD 6 fossils are arguably too fragmentary for species diagnosis or for such a far-reaching interpretation, but they clearly do not exemplify the

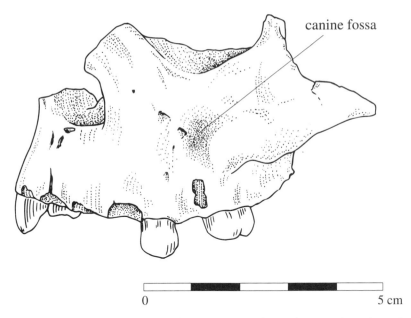

canine fossa

0　　　　　　　　　　　5 cm

Figure 5.38. A juvenile maxilla (ATD 6–69) from the Trinchera (Gran) Dolina site, Atapuerca, Spain (drawn by Kathryn Cruz-Uribe from a photograph in 200; © 1999 by Kathryn Cruz-Uribe). The maxilla is readily distinguishable from those of juvenile Neanderthals by the presence of a canine fossa and by the absence of strong forward projection along the midline. It is tentatively dated to about 800 ky ago, and it could represent the last shared ancestor of Neanderthals and living humans. More securely, together with other human fossils from the same layer, it shows that people were present in Europe before 500 ky ago.

same kind of human found at Ceprano. With Ceprano, then, TD 6 may provide evidence for two separate, ultimately unsuccessful early movements from Africa to Europe.

TD 6, the Orce sites, and other European localities that arguably indicate sporadic European occupation before 500 ky ago all lack hand axes or other markers of the Acheulean Tradition. Conceivably, then, the first permanent inhabitants of Europe were Acheuleans, and similarities between Acheulean assemblages in northern Africa and southern Europe may imply that Acheulean people dispersed across the Straits of Gibraltar or even the central Mediterranean from Tunisia to Italy (20). However, these routes would always have required people to cross open water, even during glacial periods when sea level was lower. It is more likely that movements from Africa to Europe occurred around the eastern shore of the Mediterranean to Anatolia and from there to southeastern Europe across the Dardanelles or Bosporus, which were often dry during glacial periods. A movement 600–500 ky ago that could have produced the earliest European Acheuleans may be signaled at the site of Gesher Benot Ya'aqov (Jisr Banat Yacub), Israel (866, 868, 869). In both form and mode of manufacture, the bifaces and flakes from Gesher Benot Ya'aqov are strikingly similar to contemporaneous African ones (871). The most telling pieces are perhaps large basalt flakes that were produced in a way that left a bulb of percussion on both surfaces. These are otherwise common only in eastern and northern Africa, where they are often called Kombewa flakes. At the African sites and at Gesher Benot Ya'aqov, Kombewa flakes were often used to produce the guillotine-edged bifaces known as cleavers, and they place Gesher Benot Ya'aqov firmly within the African Acheulean tradition. It remains possible, however, that Gesher Benot Ya'aqov represents not so much an African exodus as another in a long series of episodes when Africa expanded ecologically to incorporate its southwest Asian periphery.

Even after people successfully colonized Europe, truly continuous occupation was probably restricted to the warmer Mediterranean borderlands. At sites farther north, plant fossils, animal remains, or both indicate that occupation occurred mainly during interglacials before the beginning of (glacial) isotope stage 6, roughly 186 ky ago (1807). Prominent sites that probably date from the immediately preceding interglacial (isotope stage 7), between roughly 245 and 186 ky ago, include Pontnewydd Cave (885) in Wales, Maastricht-Belvédère (1804, 1808, 1809) in the Netherlands, and Weimar-Ehringsdorf (240, 548) and Stuttgart–Bad Cannstatt (2357, 2358) in Germany. Interglacial conditions are indicated at all north European sites that precede the beginning of (glacial) isotope stage 8 at roughly 301 ky ago, including, for example, Boxgrove (1785, 1787), Clacton-on-Sea (1982, 2522), Hoxne (1979, 1980), High Lodge (547), and Swanscombe (2086, 2095, 2356) in England; Cagny-l'Épinette in France (2246); Kärlich-Seeufer (290, 787), Kartstein (290), Miesenheim 1 (290, 2259), Mauer (548), Bilzingsleben (1428, 1431), and Schöningen (2171) in Germany; and Vértesszöllös (669) in Hungary. The oldest known full glacial occupations—registered, for example, at Ariendorf 1 (2258) in Germany and Mesvin IV (410) in Belgium—probably date from stage 8, between roughly 301 and 242 ky ago, but glacial sites become relatively common only in deposits correlated with stage 6, between about 186 and 127 ky ago. Stage 6 occupations are particularly well established at La Cotte de St. Brelade on the Channel Island of Jersey (413), Biache-Saint-Vaast (2017, 2243) in France, and Ariendorf 2 (2258) and Tönchesberg 1 and 2 (531) in Germany. At La Cotte, a fauna with mammoth, woolly rhinoceros, horse, and reindeer not only implies cold conditions by itself (1905), it indicates lower (glacial) sea level, since the animals could not have existed on Jersey unless Jersey were connected to the French mainland.

To some extent the absence of early, glacial-age sites may reflect erosional destruction during subsequent glaciations. The ice sheets themselves were particularly destructive, and glacial scouring would have removed virtually all traces of very early sites from Scandinavia and other often-glaciated parts of northern Europe. However, in areas like northern France that the ice sheets never reached, erosion is unlikely to have singled out glacial (vs. interglacial) sites, and ancient, glacial-age deposits (especially, wind-borne dust or loess) often persist, but without traces of human activity. Thus the absence of very early, glacial-age sites in northern Europe probably means that the earliest Europeans had to abandon the region during glacial peaks. In this connection it is also noteworthy that in the more oceanic parts of Europe, both glacial and interglacial sites older than 127 ky ago occur only south and west of a line roughly through Bilzingsleben and Vértesszöllös (fig. 5.34). The overall pattern suggests that, like *H. erectus* in the Far East, the earliest Europeans could not cope with truly continental environments, even during interglacials. The Neanderthals and their contemporaries after 127 ky ago did scarcely better, and it was only modern *H. sapiens*, after 40 ky ago, that successfully colonized the harshest environments Eurasia had to offer.

Artifacts

Stone artifacts far outnumber the physical remains of primitive *Homo*, but assemblage variation is even harder to interpret, and artifacts are often beset by the same problems of insecure dating. It follows that interpretation is most compelling when it focuses on mean differences and similarities among assemblages over very long time periods and across broad regions. The main purpose of this section is to describe fundamental geographic and temporal patterning, with special emphasis on the relation between artifact assemblage composition and human type. The principal conclusions are that the singularity of east Asian artifact assemblages underscores the probable divergence of east Asian *Homo erectus* from African and European populations by 1 my ago or before and that the similarity of African and European assemblages supports fossil evidence that Africans and Europeans shared a common ancestor near 500 ky ago.

African artifacts also imply that the putative transition from *Homo ergaster* (or African *H. erectus*) to *H. sapiens* about 600–500 ky ago was not particularly important behaviorally, since the same kinds of (Acheulean) artifacts were made before and after. In fact, artifacts suggest that a much more significant behavioral change occurred about 250 ky ago, when people in Africa, western Asia, and Europe more or less simultaneously stopped making large (Acheulean) bifaces. Population interchange is an unlikely explanation, since human fossils reveal a growing morphological gulf between African and European populations at and after 250 ky ago. A more promising explanation is the appearance of a new technology that made large bifaces obsolete. One obvious possibility is the invention of methods for mounting stone flakes on wooden or bone handles to produce composite tools that performed the same tasks as large bifaces, but more effectively or with less effort (1876). Efficient hafting methods could have spread across continents simply by diffusion, but so far they are well documented only after 50 ky ago.

Since the focus here is on gross geographic and chronological patterning, the discussion is organized along continental lines.

Eastern Asia

No artifacts have been reported in direct association with *Homo erectus* fossils in Java, nor are they unquestionably known from the same stratigraphic units. It has long been assumed that classic *H. erectus* or the succeeding Solo (Ngandong) people made the choppers, chopping tools, flakes, and occasional hand ax–like tools assigned to the Pacitanian (Patjitanian) Industry in south-central Java, but it is now known that Pacitanian tools in the type area come from alluvial terrace fills probably dating to the late Quaternary (145). The Pacitanian Industry therefore probably postdates *H. erectus*, and it may be coeval with the typologically similar end-Pleistocene/early Holocene Hoabinhian Industry of mainland Southeast Asia. A flake and a chopper from alluvial deposits near Sambungmacan probably also date from the late Quaternary (146), rather than from the mid-Quaternary as first suggested (1129). At present, the oldest known tools in Java may be some small, relatively amorphous cores and flakes found in alluvial sediments at Ngebung near Sangiran in the 1930s (146, 147) and again more recently (1911). In both cases the actual age of the artifacts is uncertain, and they could have been made either by classic *H. erectus* or by the succeeding Ngandong (Solo)/Sambungmacan people.

The absence of archeological sites that are unambiguously associated with *H. erectus* on Java may reflect a shortage of highly suitable, low energy sediment traps, together with limited prospection. Twenty basalt flakes and fractured pebbles from fine-grained alluvial deposits at Mata Menge on the island of Flores, east of Java (2022, 2292), may show what Java will eventually provide. The artifacts were associated with bones of the extinct proboscidean *Stegodon trigonocephalus*, which apparently arrived in Java about the same time as *H. erectus*. Paleomagnetic analysis suggests that the find horizon at Mata Menge shortly postdates the shift from the Matuyama Reversed Chron to the Brunhes Normal Chron at roughly 780 ky ago. The magnetism is inconsistent with fission-track dates that suggest an age of 800–880 ky ago (1580), but if either dating is correct and if the stones are genuinely artifactual, the likely maker was classic *H. erectus*. One intriguing aspect is that, unlike Java, Flores was never linked to the Southeast Asian mainland by dry land, and to reach it people may always have had to make at least three voyages of 19 km or more. This implies that *H. erectus* had seaworthy watercraft, which are otherwise inferred only for *H. sapiens* after 60–50 ky ago (chapter 7). However, *Stegodon* and some rodents also had to make the crossing, which probably means narrower straits between islands or even an undetected transitory land bridge (913).

Artifacts that are demonstrably as old as the oldest *Homo erectus* are unknown on the Southeast Asian or south Asian mainland, with two possible exceptions. The oldest and potentially the most important case concerns three flaked quartzite cobbles and two quartzite flakes from Riwat, near Rawalpindi in northeastern Pakistan (653). The containing deposit has been dated to roughly 2 my ago by a combination of paleomagnetism and geologic correlation with a fission-track-dated layer in the same region. Unfortunately, however, the deposit is a consolidated gravel (conglomerate) in which it is difficult to separate flaking produced when cobbles collided naturally from flaking produced by humans, and the presumed artifacts have been distinguished from other, apparently natural flakes or flaked cobbles mainly by quantitative criteria, including more (but still scant) flake scars, a smaller amount of remaining cobble cortex, and clearer bulbs of percussion. If the supposed artifacts are genuine, their 2 my age has profound implications for the history and nature of human dispersal to Eurasia, but it remains possible that they represent simply one extreme along a continuum of naturally flaked pieces.

The remaining exception is less momentous. It involves three flaked pebbles from alluvial deposits near Ban Mae Tha, northern Thailand (1702). The alluvium is overlain by a paleo-magnetically reversed basalt radiopotassium-dated to 800 ± 300 and 600 ± 200 ky ago. Together the paleomagnetic and radiopotassium readings imply an age near the end of the Matuyama Reversed Polarity Chron, approximately 780 ky ago. If the stratigraphic position of the pebbles has been properly assessed, and if they are genuinely artifactual, they would be among the oldest well-dated artifacts in eastern Asia. In southern Asia, the next oldest well-dated pieces are probably Acheulean hand axes and other artifacts from alluvial gravels near Dina, northeastern Pakistan, and at Bori, Maharashtra State, west-central India. At Dina, paleomagnetism and stratigraphic correlation with radiometrically dated sediments bracket the artifacts between 780 and 400 ky ago (1750). At Bori, ^{40}Ar/^{39}Ar determinations place the artifacts at roughly 670 ky ago (1568).

The Narmada skullcap, from Hathnora, north-central India, may represent the people who made the Dina and Bori hand axes. Acheulean hand axes occurred nearby, and the associated mammal fauna suggests broad contemporaneity with Dina and Bori (1212). Morphologically, the Narmada skull recalls like-aged European and African specimens that are sometimes lumped in *Homo heidelbergensis*, and it may strengthen the case for an African exodus roughly 600–500 ky ago that brought Acheulean hand ax makers to both Europe and southern Asia. Alternatively, Bori may imply that Acheulean people penetrated southern Asia first, perhaps because it resembled Africa environmentally.

In China, artifacts directly associated with *Homo erectus* have been found at Yuanmou, Lantian (the Gongwangling and Chenjiawo localities), at Zhoukoudian, and also at several sites where geologic or paleontologic context implies that *H. erectus* was the artifact maker, although *H. erectus* fossils are absent (13, 14, 1075, 1144, 2527, 2531). The oldest unquestionable artifacts in firm stratigraphic context come from Donggutuo and Xiaochangliang in the Nihewan Basin (1873, 1877), 150 km west of Beijing. Paleomagnetism suggests the artifacts date from about 1 my ago. The next oldest artifacts outside the Nihewan Basin are probably those from the Lantian sites (2531), which are tentatively bracketed between 800 and 500 ky ago.

The artifact assemblages produced by Chinese *Homo erectus* share many features that are particularly well exemplified by the very large and relatively well described artifacts from Zhoukoudian Locality 1 (1668, 2512, 2531). The Locality 1 collection comprises perhaps 100,000 pieces, but it lacks hand axes and other well-made bifacial tools that characterize many

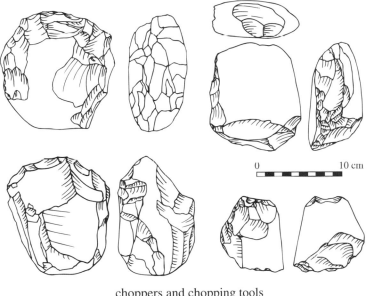

0   10 cm

choppers and chopping tools

Figure 5.39. Sandstone and quartz artifacts associated with *Homo erectus* at Zhoukoudian Locality 1 (redrawn after 1587, figs. 22, 23). The Zhoukoudian assemblage lacks hand axes and other typical Acheulean tools found at many sites in Africa, western Asia, and Europe and probably represents a totally distinct mid-Quaternary artifact tradition that was widespread in eastern Asia.

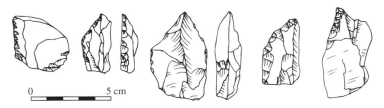

0   5 cm

retouched or utilized flakes

Zhoukoudian Locality 1

contemporaneous assemblages in Africa and Europe. It has long been taken as the type assemblage for the so-called Choukoutienian (Zhoukoudianian) chopper–chopping tool industry, but it actually contains relatively few choppers and chopping tools (bifacial choppers). Instead, it is heavily dominated by flakes, a small proportion of which are modified by retouch (fig. 5.39). In systematic overviews of the east Asian Paleolithic, H. L. Movius (1585–1587, 1591) emphasized the absence of hand axes in the Choukoutienian and likened it to the Pacitanian of Java, the Anyathian of Burma, the Tampanian of Malaysia, and the Soan (or Soanian) of Pakistan, in all of which hand axes are rare or absent. Movius assumed that these industries were all broadly contemporaneous with each other and with industries in Africa and Europe where hand axes abound, and he concluded that there were two great early-Paleolithic culture areas—the hand ax (Acheulean) area from peninsular India westward across southern Asia into Europe and Africa (fig. 5.2) and the chopper–chopping tool area eastward from northern India through eastern Asia and Southeast Asia.

Movius's formulation is questionable today on three counts. First, the east Asian industries assigned to the chopper–chopping tool complex are not necessarily contemporaneous with each other or with hand ax industries to the west (1103). Some, like the Pacitanian, are clearly later. Second, many important early Paleolithic assemblages in the west also lack hand axes. Examples include the assemblages from Atapuerca TD 6 (200, 424) in Spain; Isernia La Pineta (526, 1671, 1672) and La Polledrara (64) in Italy; Bilzingsleben (1428, 1431) and Schöningen (2171) in Germany; Vértesszöllös (669, 2319, 2320) in Hungary; and Clacton (524, 1631, 1632, 1982) and the lower horizons at Swanscombe (545, 2522) in England. Third, hand ax–like bifacial artifacts have now been recorded at several east Asian early Paleolithic sites, including especially Lantian, Dingcun, and Kehe (K'oho) in China and Chon-Gok-Ni in South Korea (2527).

The sum suggests that the rarity or absence of hand axes and similar bifacial tools at Zhoukoudian and other east Asian sites could reflect the general rarity of suitable raw material and that Movius's distinction between east and west was mistaken. At Zhoukoudian, for example, the artifacts are made primarily of quartz, which rarely occurs in large enough fragments for hand ax manufacture. However, even where cobbles or other large fragments of quartzite, chert, and siliceous limestone occurred in eastern Asia, local people rarely if ever made true hand axes and other characteristic Acheulean bifaces. Pieces that broadly fit the definition of a hand ax are generally scarce, and where they do occur, as at Dingcun, Kehe, or Chon-Gok-Ni, they tend to be thicker and more crudely made than typical African or European hand axes (1873). Some more closely recall the less formal, three-sided (trihedral) core tools or "picks" that typify some post-Acheulean (Sangoan) assemblages in central Africa. In addition, the large core tools at Dingcun and Chon-Gok-Ni have now been dated to 200 ky ago or later (1873), and they thus postdate the Acheulean in the west. In fact, true Acheulean hand axes and related tools remain unknown north and east of northern India (491), and fifty years after the "Movius line" was first formulated, it still implies a fundamental geographic division in the early and middle Quaternary. Possible early or mid-Quaternary flake-and-chopper assemblages at Kuldara and other sites in Tadzhikistan and Kirgizstan (584, 1731–1733) suggest that the boundary between the Acheulean and non-Acheulean assemblages extended westward across south-central Asia (fig. 5.2). As discussed below, it may even have reached Europe, where it separated a non-Acheulean region in the east and center from Acheulean areas in the south and west (291).

In China, people like those represented at Dali, Mapa, and Yinkou after 300–200 ky ago produced artifact assemblages that closely resembled those of classic *Homo erectus* before them (1873). The principal tools remained flakes, accompanied by choppers and chopping tools (1719, 1720). Arguably, stone artifact assemblages changed little throughout most of China and adjacent Southeast Asia even after the appearance of fully modern people like those represented at Zhoukoudian Upper Cave. Radiocarbon tentatively brackets the Upper Cave deposits between perhaps 30 and 11 ky ago (1174, 1175). However, the Upper Cave people produced formal bone artifacts and personal ornaments that broadly resemble like-aged items in the Far West, and future research may show that a major artifactual transformation occurred in the Far East 50–40 ky ago, just as it did in Africa and Europe.

Africa

In Africa, stone artifacts are stratigraphically associated with possible or probable fossils of *Homo ergaster* (or African *H. erectus*) primarily at Ternifine (114), Sidi Abderrahman (208, 1130, 1740), the Thomas Quarries (804, 1757), Gomboré II (Melka Kunturé) (455), Konso (formerly Konso-Gardula) (2111), Olduvai Gorge Beds II and IV (1328, 1329, 1338), and Swartkrans (490, 2339) (fig. 5.40). In each case the artifacts include hand axes and other bifacial tools (fig. 5.41) that are the hallmark of the Acheulean Industrial Tradition, named after the site of St. Acheul in northern France, where numerous hand axes were recovered in the nineteenth century. Radiopotassium dates from West Turkana in northern Kenya (1797) show that the Acheulean appeared about 1.65 my ago, at roughly the same time as *H. ergaster*. It is also well documented by 1.5–1.4 my ago at Konso in southern Ethiopia (89), on the Karari Escarpment at East Turkana (1121), and at Peninj in northern Tanzania (1113, 1120).

From a typological and technical perspective, Acheulean hand axes and other bifaces are a logical development from preceding Oldowan bifacial choppers, and the Oldowan and Acheulean are commonly lumped together in the Early Stone Age. Oldowan tool types continued on in the Acheulean, where they sometimes dominate assemblages at the expense of hand

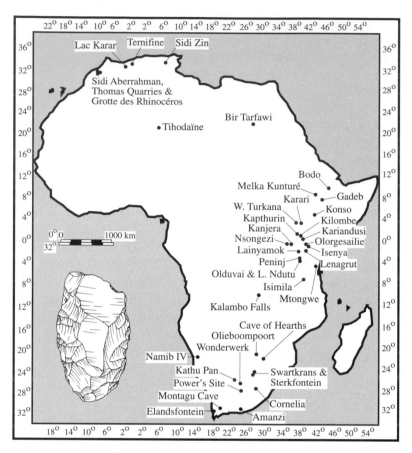

Figure 5.40. Map showing the approximate locations of major African Acheulean sites.

axes and other bifaces. At Olduvai Gorge and elsewhere in eastern Africa, such biface-poor assemblages have been assigned to the Developed Oldowan B, which is presumed to be an outgrowth of the Developed Oldowan A, a late facies (variant) of the Oldowan proper (1327, 1328). However, the differences between Developed Oldowan B assemblages and contemporaneous Acheulean ones may simply reflect differences in the activities carried on by the same people at different sites or, in some cases, differences in the availability of raw material. Most authorities thus subsume the Developed Oldowan B within the Acheulean Tradition.

Acheulean artifacts are also well known from African sites that lack human remains but that have been dated by radiopotassium, paleomagnetism, or faunal associations to before 600 ky ago, when *Homo ergaster* was the probable artifact maker. Prominent examples include Gadeb 2 and 8 (488, 497) and Melka Kunturé (the Garba II, Garba XII, and Simbirro III localities) (455) in Ethiopia and Kariandusi (696, 875, 1249, 1321),

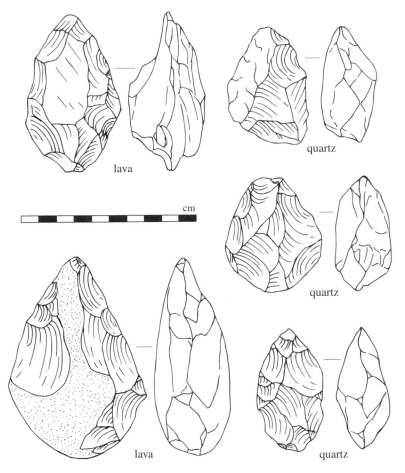

Figure 5.41. Bifaces from site TK in Upper Bed II, Olduvai Gorge (redrawn after 1327, p. 190). The artifacts are probably about 1.2 my old, and they are among the oldest known markers of the Acheulean Industrial Tradition, which began in Africa 1.7–1.6 my ago.

Olduvai Gorge Upper Bed II

Kilombe (873, 874), and Olorgesailie (407, 631, 1704) in Kenya (fig. 5.40). Acheulean artifacts are further prominent at sites that probably or certainly postdate 600 ky ago, including the Grotte des Rhinocéros (Oulad-Hamida 1 quarry) (1739, 1757) in Morocco, Lac Karar (113) and Erg Tihodaïne (2176) in Algeria, Sidi Zin (113, 848) in Tunisia, Bir Tarfawi (2417) in Egypt, Bodo (Middle Awash) (488, 492, 494, 1170) in Ethiopia, Nsongezi (520) in Uganda, Isenya (378, 1798, 1799) and Kapthurin (557, 1320, 1485) in Kenya, Lake Ndutu (1598) and Isimila (1077, 1078) in Tanzania, Kalambo Falls (485) in Zambia, and Elandsfontein (fig. 5.42) (1235, 1246, 1983), Montagu Cave (1199), Wonderwerk Cave (161, 163, 1422), Kathu Pan (163, 1241), and the Cave of Hearths (1475, 1476) in South Africa. At Bodo, Lake Ndutu, and Elandsfontein, the hand axes are accompanied by diagnostic remains of early *Homo sapiens* (or *H. heidelbergensis*), and it is clear that *H. sapiens* made precisely the same kind of artifacts as preceding *H. ergaster*. This means that the identity of

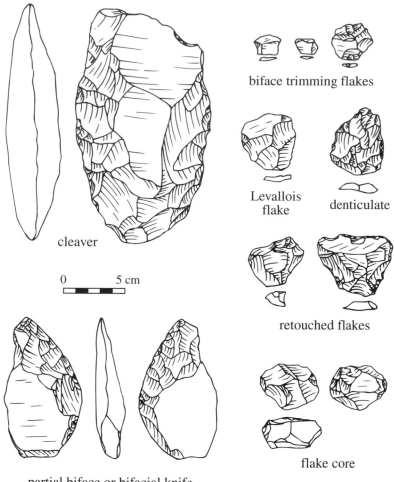

biface trimming flakes

Levallois flake

denticulate

cleaver

0   5 cm

retouched flakes

flake core

partial biface or bifacial knife

Elandsfontein Cutting 10

Figure 5.42. Late Acheulean artifacts from Elandsfontein Cutting 10, South Africa (redrawn from originals provided by T. P. Volman). All are in silcrete, a very fine-grained, locally available siliceous rock. Elandsfontein has furnished numerous typical Acheulean large bifacial tools, exemplified here by a cleaver and partial biface or bifacial knife. Also present are small, thin flakes that were probably produced during biface manufacture or trimming, larger, thicker flakes that were struck for use as tools, and some of the cores the larger flakes came from. A radial pattern of scars on the dorsal surfaces of some flakes indicates they were struck by the so-called Levallois technique, which permitted a knapper to predetermine flake size and shape. The Levallois technique was probably a late Acheulean invention, after 500 ky ago. Most of the flakes are unretouched or only lightly retouched, perhaps partly from use. Some were retouched to produce a serrated or denticulated edge, perhaps for sawing or shredding.

the hand ax maker is indeterminate at sites like Amanzi Springs (605) in South Africa, where the dating is highly uncertain.

The end of the Acheulean in Africa is not firmly dated, but the ensuing Middle Stone Age/Middle Paleolithic probably was in place by 250 ky ago (1485) and certainly was by 200 ky ago (491, 696, 1289, 2165, 2339, 2410). The Acheulean thus persisted for 1.4–1.5 my. Although many crucial sites remain poorly dated within this long period, Acheulean artifacts appear to have changed remarkably little through time. In eastern Africa, later Acheulean assemblages may contain a wider variety of artifact types than do earlier ones, and later hand axcs perhaps tend be thinner and to possess more trimming scars than do earlier ones (1113). Later Acheulean people, after 500 ky ago, also invented the Levallois and other prepared-core techniques for predetermining flake size and shape, but indisputable Levallois flakes are common only in the final Acheulean and in the succeeding Middle Stone Age/Middle Paleolithic (291). In sum, trends within the Acheulean are weak, and it is difficult if not impossible to date most Acheulean assemblages on typological or technical grounds alone. There is also little if any regional differentiation within the African Acheulean that cannot be explained by differences in raw material availability, and the overall impression is one of remarkable behavioral conservatism through time and space. Artifactual continuity further suggests that the emergence of *Homo sapiens* was not a behaviorally significant event and that the transition from *Homo ergaster* was gradual.

The Middle Stone Age/Middle Paleolithic artifact assemblages that appeared between 250 and 200 ky ago differ from Acheulean ones mainly by the absence of large, carefully trimmed hand axes and other bifaces. They are described more fully in the next chapter, but their appearance, or perhaps more precisely the disappearance of hand axes, remains a mystery. It is tempting to attribute this to a major technological innovation that made hand axes obsolete and that might be linked to the appearance of more advanced people, like those represented at Florisbad, South Africa, Singa, Sudan, and Jebel Irhoud, Morocco. It is likely that the human fossils from these sites all date between 260 and 127 ky ago, and they clearly anticipate the near-modern humans who inhabited Africa immediately afterward. Archeological support for a significant innovation remains elusive, however.

The overwhelming majority of African Acheulean sites are in ancient stream or lake deposits, and this suggests that Acheuleans were closely tied to water. In contrast, Middle Stone

Age and later sites are more commonly in caves, and the connection to water is not so clear. The difference is probably a result of preservation, however, since most caves of Acheulean age collapsed long ago and their deposits were eroded away or are at present inconspicuous. The same tendency for open-air sites to be more common early on is also obvious in Eurasia, and the reason is the same.

Western Asia and Europe

The archeological record of primitive *Homo* in western Asia closely parallels that of Africa and is similarly dominated by Acheulean sites before about 250 ky ago and by Middle Paleolithic ones afterward (123). The Acheulean people who inhabited 'Ubeidiya, Israel, by 1 my ago or before may have been among the first people to exit Africa. They probably belonged to the species *Homo ergaster* as understood here, but their physical remains are unknown. *Homo ergaster* may also have produced artifacts at somewhat younger Acheulean sites like Evron Quarry, Israel (2157), and Latamne, Syria (482, 484, 1862), and either *H. ergaster* or early *H. sapiens* could have made the hand axes at Gesher Benot Ya'aqov, Israel (871). Two human femoral shafts show only that the Gesher Benot Ya'aqov people were extremely robust (806). Early *H. sapiens*, *H. heidelbergensis*, or less likely early *H. neanderthalensis* was probably responsible for the hand axes at later Acheulean sites like Berekhat Ram (863), near the Israeli/Syrian border, or Azykh Cave (layers VI and V) (1410), Azerbaijan, that probably postdate 500 ky ago. Truly diagnostic human remains are known in western Asia only from sites that postdate 127 ky ago. They variously represent near-modern humans (*H. sapiens*) or Neanderthals (*H. neanderthalensis*), as discussed in the next chapter.

   In Europe, the oldest proposed archeological assemblages lack hand axes and other Acheulean markers, and they instead comprise choppers, chopping tools, and flakes like those made by Oldowan people in Africa before 1.7–1.6 my ago. Often-noted early nonbiface sites include Le Vallonet Cave and Soleihac in France that date to about 1 my ago and the lower layers at Atapuerca Trinchera (Gran) Dolina, Spain, the lower horizons at Kärlich, Germany, and the sites of Stránská Skála and Prezletice, Czech Republic, which date variously between 1 my ago and 500 ky ago. In each case, however, there is the problem that the flaked stones may not be artifactual (1811), and human remains to confirm an artifactual origin occur only at Atapuerca Trinchera Dolina. If the validity of all the sites is accepted, however, the earliest Europeans either did not bring hand axes from Africa or, for reasons that remain obscure, developed a separate Oldowan-like tradition that persisted for perhaps 500 ky. Alternatively, it is possible that the earliest Europeans sometimes made hand axes but failed to leave them at the known sites, that the environs of the known sites lacked suitable raw material for hand ax manufacture, or that hand axes are absent solely by chance. Chance could perhaps particularly explain the lack of hand axes at Atapuerca Trinchera Dolina, where the total number of artifacts is less than two hundred.

Achculcan assemblages that broadly resemble African ones occur at numerous European sites that date from roughly 500–400 ky ago, including Torralba (764) and Ambrona (1076) in Spain (fig. 5.43); Fontana Ranuccio (1907), Venosa-Notarchirico (183), Castel di Guido (286), and Torre in Pietra (1423) in Italy; Abbeville (Porte du Bois) (351, 1072), St. Acheul (529, 939, 2240), and Terra Amata (1399, 2323) in France; and Boxgrove, Swanscombe (middle or upper horizons) (545, 1802, 2356, 2521), and Hoxne (1979) in England. A partial human skull from Swanscombe shows that the hand ax makers included early members of the Neanderthal lineage. The European Acheulean undoubtedly came from Africa, and its appearance about 500 ky ago underscores the likelihood that *H. neanderthalensis* and *H. sapiens* shared an ancestor at about the same time.

Although most European assemblages that date between roughly 500 and 250–200 ky ago contain hand axes or other large bifaces (fig. 5.44), there are also assemblages where the tools comprise a variable mix of flakes and flaked pebbles (choppers and chopping tools) without bifaces (fig. 5.45). Prominent nonbiface sites that fit squarely within the Acheulean time span include Clacton (1631, 1632, 1982) in England; most layers at Arago Cave, near Tautavel, France (1401, 1406); Bilzingsleben (1428, 1431, 1433), Bad Cannstatt (290, 2357), and Schöningen (2171) in Germany; Vértesszöllös in Hungary (669, 2319, 2320), and La Polledrara (64) and Isernia La Pineta (1671, 2328) in Italy.

The nonbiface assemblages have often been assigned to separate named traditions like the Clactonian, Tayacian, proto-Mousterian, or Taubachian that are presumed to have paralleled the Acheulean and to have been produced by different people (278, 1401, 2114, 2285). Alternatively, in some cases the absence of large bifaces could simply reflect the local absence of suitable raw material, the conduct of activities that did not require bifaces, or some combination of these factors (2323). Raw material or activity differences are perhaps particularly likely to explain the absence of large bifaces at sites within restricted regions like southern France or central Italy, where Acheulean

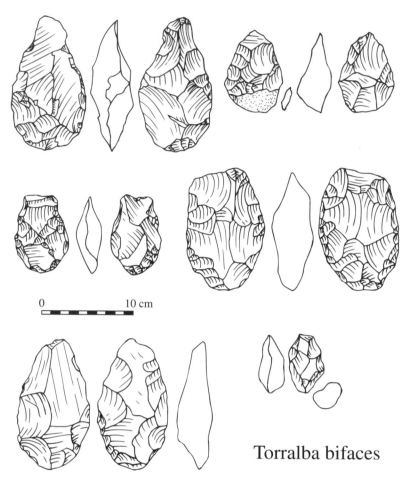

Torralba bifaces

Figure 5.43. Acheulean bifaces from Torralba, Spain (redrawn from originals provided by F. C. Howell and L. G. Freeman). The artifacts probably date from sometime between 600 and 400 ky ago, and they were stratigraphically associated with bones of elephants, horses, and other large animals. As at most sites in the same time range, the extent to which the bifaces and other tools were used to kill or butcher the animals remains unclear.

and non-Acheulean assemblages existed more or less side by side. They are less likely to explain the absence or near absence of hand axes over large regions like east-central Europe, where only flake-and-chopper assemblages occur. Arguably, then, east-central European flake-and-chopper sites like Vértesszöllös in Hungary and Bilzingsleben in Germany reflect a separate non-Acheulean tradition with roots in eastern or central Asia (1818). The implication could be for two population movements to Europe roughly 500 ky ago: one from Africa that introduced hand axes to western and west-central Europe and a second from Asia that introduced the flake-and-chopper tradition farther east. Such a scenario might seem implausible on its face,

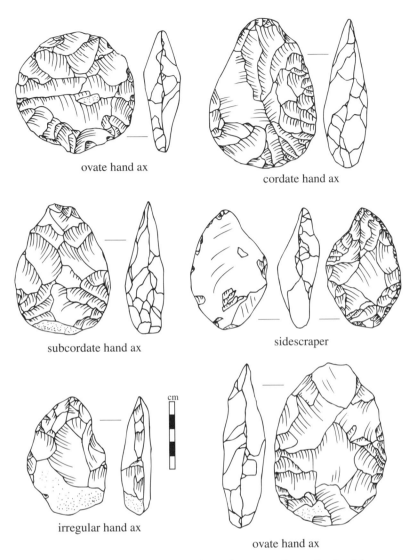

ovate hand ax

cordate hand ax

subcordate hand ax

sidescraper

cm

irregular hand ax

ovate hand ax

Figure 5.44. Late Acheulean artifacts from southern England (redrawn after 2519, p. 147). Finely crafted hand axes of various shapes and sizes are the hallmark of the late Acheulean in Europe, but biased collection probably artificially raised their frequency in most assemblages obtained in the later nineteenth and early twentieth centuries. Fine hand axes tend to be much less abundant in samples that have been carefully excavated in more recent decades.

but it could further explain why the human remains from Vértesszöllös and Bilzingsleben resemble those of *H. erectus* more closely than any others that are tentatively assigned here to the Neanderthal lineage.

As in Africa and western Asia, the Acheulean in Europe was succeeded 250–200 ky ago by (Middle Paleolithic or Mousterian) flake industries that lack large bifaces (2326). Diagnostic human remains are rare at early Middle Paleolithic sites, but

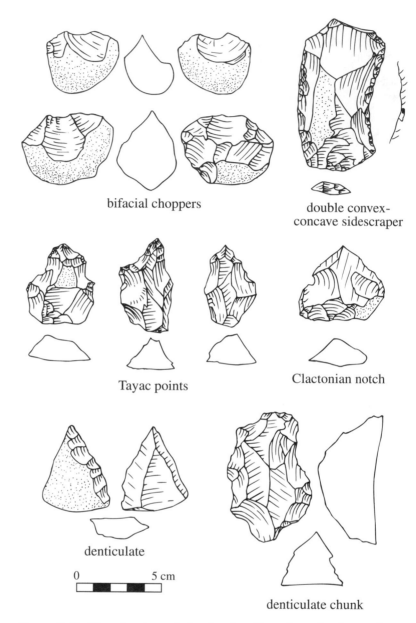

bifacial choppers

double convex-
concave sidescraper

Tayac points

Clactonian notch

denticulate

0   5 cm

denticulate chunk

Figure 5.45. Tayacian (non-Acheulean) artifacts from La Caune de l'Arago, southern France (redrawn after 1401, figs. 7, 8). The term Tayacian has been applied to a wide range of European assemblages that lack hand axes but that were broadly coeval with Acheulean assemblages between roughly 500 ky and 250–200 ky ago. Among competing explanations for the difference, the most popular are that Tayacian and Acheulean assemblages were made by different people and that both kinds of assemblages were made by Acheulean people who sometimes did not need hand axes or who sometimes could not find suitable raw material nearby. Both explanations may be valid, depending on the site or assemblage.

where they occur—mainly at Biache-Saint-Vaast, France, and Ehringsdorf, Germany—they contrast sharply with contemporaneous (early Middle Paleolithic/Middle Stone Age) human remains in Africa. Whereas the African fossils anticipate modern people, the European ones anticipate the classic Neanderthals who occupied Europe after 127 ky ago. The contrast in human types could not have been anticipated from the associated artifacts, which are remarkably similar between Europe and Africa. A conspicuous artifactual difference between Africa and Europe appears only after 50 ky ago with the emergence and spread of modern humans from Africa. The next chapter argues that the similarity before 50 ky ago implies a shared primitive behavioral repertoire in which innovation was unusual, whereas the difference after 50 ky ago implies a shared derived behavioral repertoire in which innovation was commonplace.

Other Aspects of Behavior

In total, the fossil sample of *Homo ergaster, H. erectus,* early *H. sapiens,* and early *H. neanderthalensis* comes from perhaps 130 individuals, represented mainly by bits and pieces, and the behavior of the species must be inferred from fewer than fifty reasonably well excavated archeological sites. The skeletal remains and sites are widely scattered in time and space, and the time dimension is poorly controlled, particularly outside Africa. Compounding these problems of small sample size and poor temporal control, the archeological evidence for behavior is often highly ambiguous, and nontrivial behavioral inferences are mostly tentative.

Artifact Function

Primitive *Homo* undoubtedly used stone artifacts for many purposes, including cutting, whittling, scraping, shredding, and butchering. Damage and wear polish on tool edges can sometimes reveal the mode of use and, more often, the material to which an artifact was applied (1195, 1196). Wear polish analyses show that *H. ergaster,* early *H. sapiens,* and early *H. neanderthalensis* applied tools to essentially the entire range of materials (bone, antler, meat, hide, wood, and nonwoody plant tissue) that the methodology can differentiate. However, only a small proportion of tools preserve diagnostic use traces, and trace identification is very time consuming. Since there is also no obvious one-to-one relation between tool form and inferred function, the functional significance of whole assemblages remains obscure. Flakes and other light-duty tools are much more common than hand axes at some sites where animal bones suggest that butchering was important, which may mean that flakes, rather than hand axes, were the primary butchering tools (486, 496). But experiments show that hand axes or other large bifacial cutting tools are more efficient than flakes for butchering large animals (1157), and probably nothing about butchery can be inferred from either the rarity or the abundance of hand axes. In truth, it may be that hand axes were used for butchering and for many other purposes, perhaps sometimes even as cores to obtain small, sharp flakes (1136). An additional puzzle, which may hold clues to the function of hand axes, is why they were sometimes discarded in vast numbers as, for example, at Melka Kunturé, Olorgesailie, Isimila, Kalambo Falls, and other Acheulean sites in eastern and southern Africa. One intriguing speculation is that large accumulations served as caches for future artifact use or manufacture (1136).

Raw Material Use

Early Paleolithic people preferentially selected locally available rock types with the best flaking properties, and they commonly used different rock types to produce different kinds of tools. Thus the occupants of many sites tended to make large, heavy-duty core tools out of softer rocks like limestone and basalt, while they produced smaller, more fully shaped core tools and flakes out of harder rocks like flint (2328).

Stone artifacts dominate the Paleolithic record because of their durability, but early people surely used other raw materials, including bone and more perishable substances like wood, reeds, and skin. Animal bones survive at many sites of primitive *Homo* in Africa, Europe, and Asia, and percussion marks or cut marks sometimes suggest that people used individual bones as hammers, retouchers, anvils, or cutting boards. A few, mainly European, sites have also produced bones that were flaked, like stone, by percussion. The most striking examples are elephant-bone bifaces from Fontana Ranuccio, Malagrotta, Castel di Guido, and other Italian Acheulean sites (1426, 2326) and elephant-bone scraperlike tools from the non-Acheulean site of Bilzingsleben, Germany (1434) (fig. 5.46). However, neither these sites nor any like-aged ones contain bones that were cut, carved, or polished to form points, awls, borers, and so forth. Such artifacts are common only in sites of fully modern humans after 50–40 ky ago, and chapter 7 argues that fully modern people were the first to shape bone by cutting, carving, and polishing.

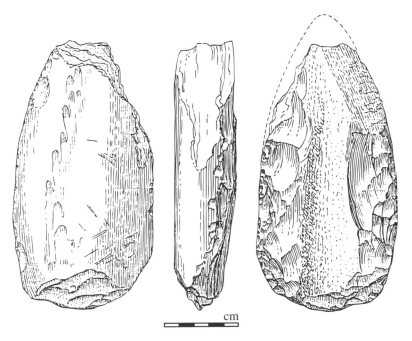

cm

Figure 5.46. A flaked fragment of elephant bone from Bilzingsleben, Germany (after 1434, fig. 4). The Bilzingsleben site is probably about 350 ky old, and it has provided other flaked bone artifacts in addition to numerous stone artifacts and some fragmentary human remains. The Bilzingsleben bone artifacts and similar ones from other sites (particularly in Italy) show that early people sometimes flaked bone in the same way they flaked stone. However, only people after 50 ky ago routinely carved or ground bone to produce formal artifacts.

Artifacts in the most perishable raw materials are scarce almost by definition. Bamboo makes edges that rival or exceed those of stone in sharpness and durability, and it is especially likely to have been used in eastern Asia, where it enjoys roughly the same distribution as the chopper–chopping tool tradition defined by Movius. An emphasis on bamboo could explain why east Asian stone-artifact assemblages often appear typologically impoverished next to contemporaneous Acheulean ones from Africa, western Asia, and Europe (280, 1698, 1699, 1703), but actual bamboo artifacts remain unknown.

Reed and leather artifacts are similarly lacking, and wooden artifacts have been found at only a handful of African, west Asian, and European sites, where they were putatively produced by early *H. sapiens*, early *H. neanderthalensis*, or *H. heidelbergensis*. The sites are distinguished by unusually dense, anaerobic deposits that inhibited bacterial decay. The artifacts all date between roughly 600 and 300 ky ago, and they include a possible club and other shaped wooden pieces from Kalambo Falls, Zambia (485), a wooden fragment apparently polished by use from Gesher Benot Ya'aqov, Israel (181), a wooden spear point from Clacton-on-Sea, England (1629, 2391, 2519), and three complete spears from Schöningen, Germany (650, 2171). Schöningen has also provided a shorter double-pointed wooden stick of uncertain function and three shaped and grooved wooden branches that may have been handles for stone tools. The Clacton spear point might actually have tipped a digging stick, a stake, or even a snow probe (778, 779), but the Schö-ningen spears are unequivocal (fig. 5.47). They are over 2 m long, and they are thickest and heaviest near the pointed end, like modern javelins or throwing spears. They were directly associated with hundreds of horse bones, many of which were cut or fractured by stone artifacts, and they arguably present the oldest, most compelling case for concerted early human hunting.

Artifact Form and Cognition

Modern novices who attempt to produce hand axes commonly fail because they remove too many flakes from the sides of the blank or preform before they begin to thin it, because they trim some parts of the periphery prematurely (producing an undesirable sinuous edge), because they overcorrect as they attempt to maintain a symmetrical outline (resulting in a highly asymmetrical one), because they cannot extend the peripheral edge through the thickest parts of the blank, or because they break a thinned biface in two by too strong a blow (1873). The same problems do not affect the production of simple flakes and choppers, and we may reasonably ask whether Acheulean people possessed cognitive abilities beyond those of their flake-and-chopper contemporaries. The answer is, not necessarily. To begin with, the principles involved in hand ax manufacture are easily transmitted and not particularly complex. They can be learned entirely by imitation and repetition, and they are far simpler, for example, than the intricate and subtle grammatical rules than underlie all known languages (2523, 2524). Second, and more important, stone artifacts surely formed only a small part of ancient material culture, and there is at present no basis for assessing the sophistication of the rest. The well-made wooden spears from Schöningen, discussed immediately above, illustrate the point, since they are associated with a flake-and-chopper stone tool assemblage yet reflect at least as much forethought and control in manufacture as any hand ax.

Language

The complexity of hand ax manufacture and the sophistication of the Schöningen spears show that early *Homo sapiens*

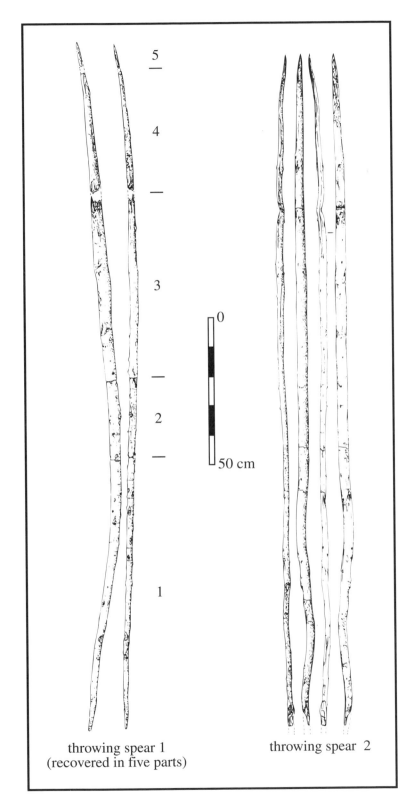

Figure 5.47. Multiple views of two of the three wooden throwing spears recovered from 400 ky old lakeside deposits at Schöningen, Germany (redrawn after 2170, fig. 9). The spears are the oldest known unequivocal human hunting weapons. They were probably used to hunt wild horses, whose bones abound in the same deposits.

5

4

3

2

1

0

50 cm

throwing spear 1
(recovered in five parts)

throwing spear 2

and *H. neanderthalensis* were cognitively far advanced over the australopithecines, and they probably possessed a much more advanced form of communication. But direct evidence for language is lacking, and it is difficult even to imagine what intermediate stages might have been like. Historically observed languages provide no clues, since they are all equally complex and equally far removed from the communication systems of other animals.

The next two chapters employ archeological evidence to argue that only fully modern humans after 50 ky ago possessed fully modern language ability, and that the development of this ability may underlie their modernity. By extrapolation, one could argue that the anatomical and behavioral advances that mark *Homo ergaster* as the first truly human species imply that it was also the first to possess rudimentary language. Three kinds of anatomical evidence can be brought to bear on this. First is the size and configuration of the brain. Size speaks mainly to potential, and on this basis *H. ergaster* is perhaps more likely to have had language than earlier species. Configuration is arguably less ambiguous, because two regions (Broca's and Wernicke's areas) on the external cerebral cortex are linked to motor control over speech, and both are detectable on endocranial casts. Endocast incompleteness or distortion often hinders observation, and reliable detection is particularly difficult for Wernicke's area (2361). Wernicke's area also appears to be well developed in common chimpanzees (780), which of course totally lack the human ability to speak. Broca's area is less problematic, and it appears to have been absent or relatively undeveloped in the australopithecines and much better developed in *H. habilis* and *H. ergaster* (168, 2199). Some degree of humanlike speech may be implied.

The remaining two lines of anatomical evidence are contradictory. On the negative side, there is the small diameter of the neural canals on the vertebrae of an adolescent skeleton (KNM-WT 15000) from Nariokotome, West Turkana, Kenya (1415, 2370). The neural canals enclose the spinal cord with its nerve cells and nerve fibers, and small neural canal size in thoracic (chest) vertebrae implies limited nervous control over movements of the rib cage. Fine control is essential for articulate speech.

On the positive side, there is the moderate degree of basicranial flexion (upward arching between the posterior edge of the palate and the anterior margin of the foramen magnum) that has been observed on *H. ergaster* skulls (1301). In the degree of basicranial flexion, *H. ergaster* was apparently intermediate between the australopithecines, who had very flat cranial bases (like those of extant apes and very young humans), and modern adult humans, who have highly flexed bases. In extant primates, the degree of flexion is related to the anatomy of the upper respiratory tract, including the position of the larynx, which rests much lower in the neck of modern humans (with pronounced flexion) than in the neck of apes (with limited flexion). Since the low position of the larynx permits people to make crucial vowel sounds found in all languages, the moderate basicranial flexion in *H. ergaster* suggests at least an incipient capacity for true speech. Since the basicranium is unknown in unequivocal specimens of *H. habilis*, *H. ergaster* is the earliest hominid for which basicranial flexion suggests a degree of articulate speech.

Site Modification

No archeological site that antedates 127 ky ago has provided persuasive evidence for structures. Seemingly patterned arrangements or concentrations of large rocks at sites like Soleihac (256) and Terra Amata (1399) in France, Latamne (482, 484) in Syria, Melka Kunturé in Ethiopia (454, 455), and Olorgesailie (1114) in Kenya may mark the foundations of huts or windbreaks, but in each case the responsible agent could equally well be stream flow, slope creep, or some other natural process. The likelihood of a natural origin is particularly strong at Soleihac, where a 20 m long line of basalt blocks is associated with flaked stones whose artifactual origin is itself dubious (1811).

Many more sites contain clusters of artifacts, bones, and other debris that could mark hut bases or specialized activity areas. The evidence is more conjectural than compelling, but the most noteworthy examples perhaps include those from Bilzingsleben (1428, 1429, 1431) and Ariendorf 1 (289, 2258) in Germany and from the Le Lazaret Cave (1400, 1401, 1404) in southern France. The sites vary in age from perhaps 350 ky ago at Bilzingsleben to a time within global-isotope stage 8 (301–242 ky ago) at Ariendorf 1 to a time within isotope stage 6 (186–127 ky ago) at Le Lazaret.

At Ariendorf 1, several large quartz and quartzite blocks (up to $60 \times 30 \times 30$ cm) occur in an otherwise sandy-silty (loess) deposit where they must have been introduced by people. Conceivably they mark the base of a structure, and they partly surround a scatter of artifacts and fragmentary bones. Refittable artifacts and conjoinable bone fragments imply that many pieces were deposited at more or less the same time.

At Bilzingsleben, the ancient inhabitants camped beside a stream flowing from a nearby spring to a small lake. They also settled on parts of the streambed that were periodically dry. The evidence for structures consists of one circular and two oval concentrations of artifacts and fragmentary animal bones (probable food debris) that co-occur with large stones and bones that could have been used to build walls. The concentrations are 2–3 m in diameter, and each is immediately adjacent to a spread of charcoal interpreted as a fireplace. Also nearby are clusters of artifactual "waste" that may represent "workshops" (fig. 5.48).

At Le Lazaret, the presence of a structure is suggested by an 11 × 3.5 m concentration of artifacts and fragmented animal bones bounded by a series of large rocks on one side and by the cave wall on the other (fig. 5.49). The area also contains two hearths, as well as numerous small marine shells and carnivore teeth that could derive respectively from seaweed and skins that were introduced as bedding. The rocks could have supported poles over which skins were draped to pitch a tent against the wall of the cave. Like-aged concentrations of artifacts and other cultural debris surrounded or accompanied by natural rocks may represent structure bases in other French caves (especially La Baume Bonne and Orgnac) (2322), but the patterning is less compelling.

In sum, the evidence for housing before 127 ky ago is remarkably sparse and equivocal, and the next two chapters show that it becomes abundant and compelling only after the advent of fully modern humans between 50 and 40 ky ago. People before 50 ky ago must have built shelters, particularly at open-air sites in midlatitude Europe and Asia, but the structures were apparently too flimsy to leave unmistakable archeological traces. The more substantial structures that appeared after 50–40 ky ago help explain how fully modern humans were able to colonize the most continental parts of Eurasia where no one had lived before.

Fire

Homo ergaster (or *H. erectus*) may have had to master fire for warmth and cooking before colonizing Eurasia, but direct archeological support is tenuous. The oldest evidence (noted in chapter 4) may be patches of baked earth in deposits dated to 1.5–1.4 my ago at Koobi Fora and Chesowanja in Kenya, but naturally ignited, smoldering vegetation might have produced the same results (479, 876, 1117). Natural fires or ignition could also account for burning at most other early Paleolithic sites, including burned bones associated with Developed Oldowan or early Acheulean artifacts in Member 3 at Swartkrans Cave in

South Africa (326, 332, 1951), patches of burned earth at the Olorgesailie Acheulean site in Kenya (1114), burned bat guano below the Acheulean sequence at the Cave of Hearths in South Africa (1475), dispersed ash in the Acheulean deposits at the Cave of Hearths and Montagu Cave, also in South Africa (394), burned flints in the Acheulean site of Terra Amata in France (1399, 2323), and dispersed charcoal in the Acheulean layers of Torralba and Ambrona in Spain (1072) and in the non-Acheulean site of Prezletice in the Czech Republic (768).

The Prezletice charcoal may be irrelevant if, as noted above, the associated flaked stones are not artifactual. At the other extreme, the Swartkrans burned bones may be particularly noteworthy, because they are absent in older Swartkrans sediments where they might also be expected if they were produced by natural burning. If the Swartkrans bones reflect human burning, they would imply that *H. ergaster* possessed fire between 1.5 and 1 my ago, and the use of fire by its geographically scattered descendants would not require independent invention.

Swartkrans Member 3 aside, the oldest reasonably secure evidence for fire comes from the classic *H. erectus* site of Zhoukoudian Locality 1 in China. Recall that Locality 1 has been tentatively dated to between 500 and 240 ky ago. The deposits contain burned bones and stones, thick ash beds, and thinner lenses of ash, charcoal, and charred bone. Some of the thicker beds may represent naturally ignited organic detritus, and some charring may actually be mineral staining (222), but some of the thinner ash lenses almost certainly represent true fossil hearths (1668, 2537).

In Europe, equally old hearths may be marked by concentrations of charred bones in shallow depressions at Vértesszöllös, Hungary (669, 2320), clusters of wood charcoal at Bilzingsleben, Germany (1428, 1431), and a reddened patch of sediment at Schöningen, Germany (2171). Uranium-series determinations, ESR dates, mammalian species, fossil pollen, or a combination of these tentatively place all three sites between 400 and 300 ky ago (1811). Otherwise the oldest European evidence for fire dates from the Penultimate Glaciation (isotope stage 6), between 186 and 127 ky ago. It comprises occasional patches of ash and charcoal at Pech de l'Azé and perhaps other French caves (2322) and abundant carbonized bones, less abundant charcoal, and numerous burned flint artifacts at the cave of La Cotte de St. Brelade (Jersey) (414).

Unequivocal fireplaces are conspicuous in many sites occupied by the European Neanderthals and their near-modern African contemporaries after 127 ky ago, and it is reasonable to

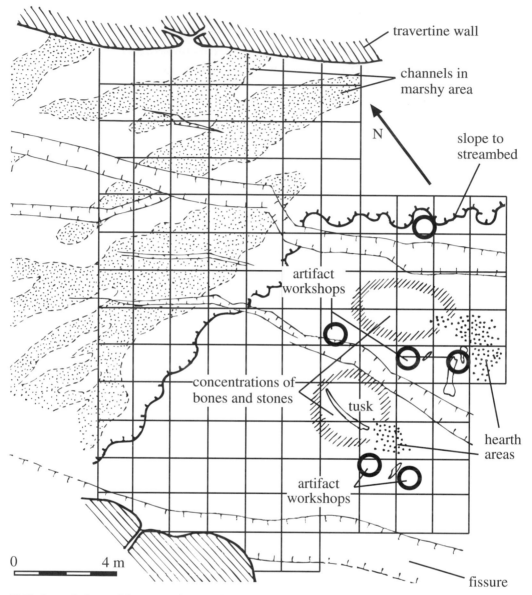

Bilzingsleben Excavation Plan

Figure 5.48. *Left:* Locations of bone and artifact concentrations at the mid-Quaternary site of Bilzingsleben, Germany (adapted from 1428, p. 244). *Right:* Close-up of the oval accumulation of stones and bones whose position is indicated to the left (adapted from 1428, p. 245). The accumulation may mark the base of a structure in which large stones and bones were used to build walls, though the spatial patterning is less convincing than it is in "ruins" at many sites that postdate 40 ky ago. The location of a second, similar accumulation is not shown on the plan (*right*), because it was excavated only after the plan was drawn.

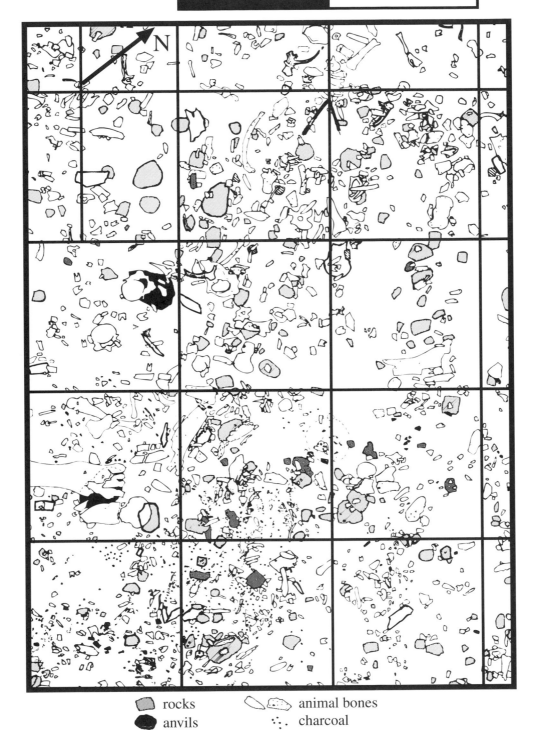

	rocks		animal bones
	anvils		charcoal
	pebble tools		charred wood and charcoal

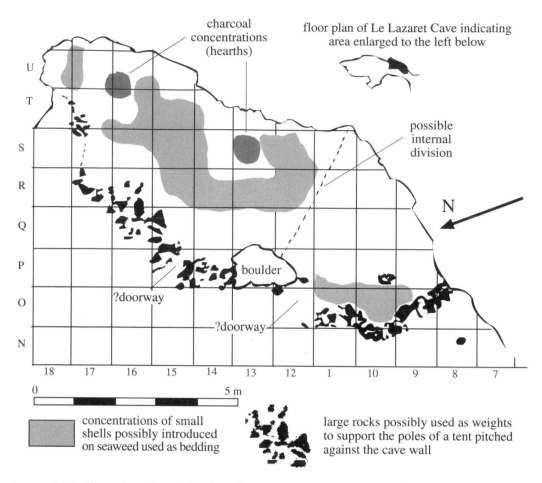

Figure 5.49. Floor plan of an Acheulean layer at Le Lazaret Cave, southern France (adapted from 1404, p. 639). Stone artifacts and fragmentary animal bones were heavily concentrated between the cave wall and a line of large rocks that may have supported the poles of a tent pitched against the wall. Artifacts and bones spilling out from between the rocks at two points might mark former doorways, while concentrations of small seashells could come from seaweed introduced as bedding. Two roughly circular concentrations of charcoal perhaps delineate former fireplaces (hearths).

ask whether the paucity of evidence in older sites indicates a more limited ability to make or control fire. The answer is, not necessarily, since the evidence after 127 ky ago comes mainly from caves, whereas most well-known older sites are open-air occurrences near ancient water sources. Both individual charcoal fragments and lenticular concentrations that unambiguously mark hearths are much less likely to survive at open-air sites, and it is probably no accident that the oldest widely accepted evidence before 127 ky ago comes from the Locality 1 cave. Unfortunately, caves of similar or greater antiquity are exceptionally rare, and reconstructing the history of human fire use may thus always depend more on speculation than on archeological discovery.

Subsistence

Like fire, an advance in subsistence ecology could explain how *Homo ergaster* and its descendants managed to colonize new regions. But the evidence bearing on subsistence is very limited. Both logic and observations on historic hunter-gatherers suggest that, in general, early people everywhere depended more on plants than on animals. Tubers and other underground storage organs may have been especially important, since they were staples among historic low- and midlatitude hunter-gatherers. It may not be coincidental that *H. ergaster* emerged at a time 1.8–1.7 my ago when tubers had probably become more abundant, following a shift to yet drier, more seasonal climate over much of Africa. As indicated previously, key features of *H. ergaster* anatomy and behavior suggest it was especially adapted for foraging in arid, highly seasonal environments. The evolution of *H. ergaster* (or African *H. erectus*) has sometimes been tied to males' enhanced ability to hunt, but it may actually have depended more on an females' enhanced ability to locate, excavate, and process tubers (1621).

Unfortunately, plant remains that could illuminate the importance of tubers or other plants in early human diets have rarely survived. In addition, in the few very early sites where parts of edible plants have been found, they cannot be unequivocally linked to human activity. The sites include Locality 1 at Zhoukoudian in China (465); the Acheulean layers at Kalambo Falls in Zambia (485); Gesher Benot Ya'aqov in Israel (866, 867); and Kärlich-Seeufer in Germany (787); and the non-Acheulean deposits of Bilzingsleben in Germany (1431) and Vértesszöllös in Hungary (669).

In the absence of indisputable plant food residues, plant consumption might be demonstrated by a chemical analysis of early human fossils, assuming that an antemortem dietary signal can be detected in bones whose chemical composition has been altered in the ground (1948). So far, the only fossil of primitive *Homo* that may provide a hint of diet is a partial skeleton of *H. ergaster* (KNM-ER 1808) from Koobi Fora, dated to roughly 1.7–1.6 my ago. The long bone shafts of this individual are covered by a layer of abnormal, coarse-woven bone as much as 7 mm thick (2376). Observations on living people suggest this could mark the oldest known case of yaws (1840), an infectious disease related to syphilis that induces similar bone growth in its final stage. However, it could also reflect a toxic excess of vitamin A (hypervitaminosis A). If vitamin A poisoning was the cause, it could be explained by an overindulgence either on carnivore livers (2376) or on honeybee eggs, pupae, and larvae (1991).

Except perhaps at Koobi Fora, inferences about the subsistence of primitive *Homo* must be based almost entirely on the animal bones associated with artifacts at sites such as Elandsfontein (1235, 1246, 1563, 1983) and Kathu Pan (163, 1241) in South Africa; Olorgesailie (1114, 1939) in Kenya; Ternifine (805) in Algeria; Revadim in Israel (1437); Torralba, Ambrona, and Aridos in Spain (761, 764, 1072, 1076, 1865, 2325); Notarchirico (183), La Polledrara (64), Isernia la Pineta (34), and Rebibbia (210) in Italy; Bilzingsleben (1434), Schöningen (2171), Miesenheim 1 (290, 2259), and Bad Cannstatt (2336, 2358) in Germany; and Boxgrove (1785) in England. Until the 1970s, most archeologists simply assumed that the artifacts at such sites were used to kill and butcher the animals. It followed that the people were successful big-game hunters, since the animals often included elephants, rhinoceroses, buffalo, and other formidable prey.

In the 1970s, however, a growing number of archeologists began to specialize in the analysis of animal bones, and after close scrutiny it was soon realized that a stratigraphic association between bones and artifacts need not imply a functional relationship. Most early archeological sites, including all those listed above, were in the open air near ancient springs, streams, or lakes that naturally attracted both people and animals. The animal bones at such sites could represent human kills, or they could represent carnivore kills or even natural deaths (from starvation, disease, etc.) that totally escaped human notice or that were subsequently scavenged by people (or carnivores). The famous Torralba and Ambrona Acheulean sites in Spain provide good cases in point (1240).

Torralba and Ambrona are 2 km apart, about 150 km northeast of Madrid, on opposite sides of the Masegar River Valley (fig. 5.50). Stratigraphic context and fauna (including bones of the advanced, rootless vole, *Arvicola*, at Ambrona) indicate that the Acheulean artifacts accumulated roughly 500 ky ago. This may have been only shortly after the initial Acheulean penetration of Europe, which could explain why the Torralba and Ambrona bifaces closely resemble like-aged African ones. The sediments and fauna also show that the sites were on the margins of a shallow, marsh-edged lake, which filled the Masegar Valley. Neither site has produced human remains, but if diagnostic remains were present, they would probably be assigned to early *Homo neanderthalensis* or *H. heidelbergensis*,

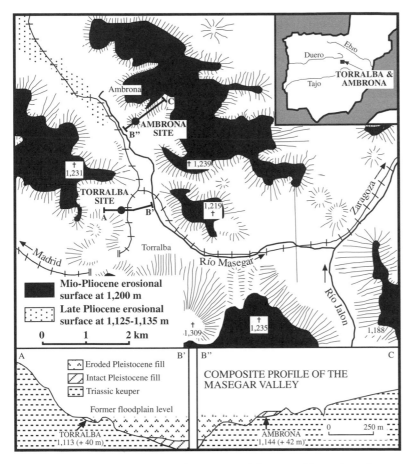

Figure 5.50. The approximate locations of the Torralba and Ambrona Acheulean sites, north-central Spain (modified after 392, fig. 1).

as defined here. The large animal remains are heavily dominated by bones of elephants and horses (fig. 5.51).

At both sites, the stone tools and occasional tool-damaged bones imply some human role in the bone accumulations, but the tools are thinly scattered among the more numerous bones, and there are also carnivore coprolites (fossilized feces), bones damaged by carnivore teeth, and numerous bones abraded by flowing water. Among the bones of most species, the relative abundance of axial elements (skulls, vertebrae, and pelvises) versus limb bones may indicate that predators or scavengers often removed meatier parts from the site, but either people or large carnivores would create the same pattern. Similarly, people, large carnivores, or—especially—natural mortality could account for the apparent overrepresentation of older adults among the elephants (other species are too poorly represented for the construction of age profiles). In sum, the available

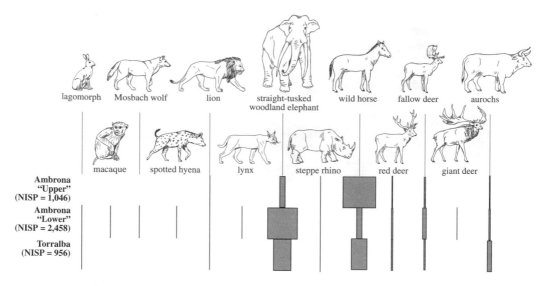

Figure 5.51. Abundance of large mammals at the Torralba and Ambrona Acheulean sites, north-central Spain. The bars are proportional to the number of identifiable bones assigned to each species (NISP). Torralba is about the same age as Ambrona "Lower" and has been placed below Ambrona for display purposes only. Like other Acheulean sites, Torralba and Ambrona contain many bones of large mammals, including elephants, but as at other sites, the bones could represent natural deaths or carnivore kills, and so far it has not been possible to determine the hunting capability of Acheulean people.

data do not allow us to isolate the relative roles of humans, carnivores, and factors such as starvation, accidents, and stream action in creating the bone assemblages. In the current state of our knowledge, Torralba and Ambrona need not have differed significantly from the margins of historic African streams or water holes, where the events that produce carcasses can be complex and need not involve people. Certainly, as understood at present, the sites do not tell us how successful or effective early *H. neanderthalensis* was at obtaining meat.

Greater human involvement may be indicated at other presumed early *H. neanderthalensis* sites like Aridos, Isernia La Pineta, Bilzingsleben, Schöningen, and Bad Cannstatt, where stone artifacts, artifact-damaged bones, or both are more abundant, but even at these sites it is impossible to show that people significantly shaped the bone accumulations (788). The human role is similarly ambiguous at like-aged or earlier African sites, including Ternifine, Olorgesailie, and Elandsfontein, where the hunting or scavenging success of *H. ergaster* and early *H. sapiens* remains equally indeterminate. The uncertainty in each case stems partly from a heightened awareness of the natural factors that may have shaped bone accumulations near ancient lakes and streams and partly from the incompleteness and ambiguity of the surviving evidence.

Assessing human ability to obtain animals, whether by hunting or scavenging, requires, at minimum, a site where people were the only—or at least the principal—bone accumulators. In this regard caves are generally more promising than open-air sites, because the main game animals available to early people do not enter caves voluntarily. In most instances, therefore, it is safe to assume that their bones were introduced by a predator or scavenger. If the cave fill contains numerous artifacts, hearths, and so on and little or no evidence for carnivore activity (coprolites, chewed bones, etc.), then people are clearly implicated as the principal bone accumulators. Caves to which people surely brought most bones are well known for the Neanderthals and for their modern or near-modern African contemporaries after 127 ky ago.

Unfortunately, caves themselves have a limited life span, and few survive from the era before 127 ky ago. The famous Chinese *Homo erectus* site at Zhoukoudian Locality 1 is by far the most salient exception, and it was literally filled with bones from many species, ranging from small rodents and insectivores through rhinoceroses and elephants (13, 1075, 1168). Two extinct species of deer were especially common, and it has sometimes been said that the Zhoukoudian people were proficient deer stalkers (2512). But this interpretation is complicated by the presence of numerous hyena bones and coprolites and of bones plainly damaged by hyena teeth, indicating that hyenas sometimes occupied the site. They may have introduced many of the animal bones, and their activities, combined with profile compaction and other postdepositional destructive pressures, could also explain why the Zhoukoudian human craniums lack faces and basal parts and, more generally, why the human fossils are so fragmentary (222). A less plausible alternative is that the Zhoukoudian people were cannibals. The bottom line is that pending the fresh excavation of a large bone sample, Zhoukoudian cannot unambiguously illuminate the predatory or scavenging prowess of *H. erectus*.

With one notable exception, the few European or African caves with deposits older than 127 ky ago have been similarly uninformative, sometimes because the bone samples are small or were biased during excavation, sometimes because they may have been partly accumulated by hyenas or other carnivores, and sometimes because they remain unanalyzed from a human behavioral perspective. The main exception is the cave of La Cotte de St. Brelade on the Channel Island of Jersey (1903, 1904, 1906). Here layers 6 and 3, which probably formed during the Penultimate Glaciation (isotope stage 6), between 186 and 127 ky ago, have provided numerous Middle Paleolithic

(Mousterian) artifacts, together with bones of at least twenty mammoths and five woolly rhinoceroses. There are no securely associated human remains, but the people were probably early *H. neanderthalensis*. Human butchering is strongly implied by the abundance of the associated artifacts (more than 1,100 in level 3), by damage to the bones (especially to skulls, which may have been opened to obtain the brains), and by peculiar patterns of skeletal part representation: skulls, mandibles, scapulae, and pelvises heavily dominate layer 6; scapulae, pelvises, humeri, and femurs dominate layer 3; and radii, ulnae, tibiae, carpals, tarsals, foot bones, and vertebrae are all but absent in both layers. The missing parts are the least bulky ones, which people (and possibly other large predators) would have found easiest to remove. In addition, the tight, vertically restricted packing of bones within each layer suggests rapid accumulation, perhaps when small groups of mammoths or rhinoceroses were simultaneously driven over a headland just above the site. Driving could explain why the mammoths and the rhinos were mainly subadults and prime-age adults. Unlike very young and old individuals, subadult and prime-adult animals would otherwise be difficult to obtain by either hunting or scavenging.

The evidence for butchering at La Cotte foreshadows abundant evidence for butchery by the Neanderthals and their African contemporaries after 127 ky ago (449, 1440, 1563, 1564, 2051). Specialists agree that these people introduced large numbers of bones to many cave sites, where numerous cut marks or percussion marks indicate that the bones represent food debris. The ages of animals at time of death sometimes implies driving in Africa (1236) and in Europe (1363), but authorities disagree vigorously about whether hunting was more common than scavenging (219, 449). The issue may not be resolvable, but data summarized in chapter 6 imply that whatever means southern African people used to obtain large animals between 127 and 60 ky ago, they did not succeed very often (1236, 1243). If it is fair to project backward, the more ancient people on which this chapter focuses were even less successful, and they probably rarely fed on the elephants, rhinoceroses, buffalo, or other large animals that dominate their sites.

Cannibalism? And Interpersonal Violence

Cannibalism has rarely been observed among historic people (72), but some prehistoric agriculturalists may have practiced it routinely. The most compelling evidence comprises numerous cut, smashed, and burned human bones from late prehistoric

Anasazi Pueblo sites in the American Southwest (2456). By analogy, smaller numbers of cut-marked and smashed human bones may imply cannibalism at some Neanderthal sites (discussed in the next chapter) and perhaps at Atapuerca TD 6, Spain. Recall that the TD 6 human bones are tentatively dated to about 800 ky ago. There are eighty or so specimens, and these are said to exhibit numerous cut marks (811). They could thus present the oldest case of cannibalism on record. Progressively younger instances might be indicated by cut marks on the principal Bodo skull, Ethiopia (approximately 600 ky old) (2451, 2453), by arguably intentional removal of the rear part of the Arago skull, France (?400–250 ky old) (1401), and by possibly intentional removal of the cranial bases on the Ngandong (Solo) skulls, Java (?200–50 ky old) (553).

Cannibalism by itself suggests interpersonal violence, and this may be further indicated by depressed fractures and other antemortem damage on the Ngandong skulls (553). Similar damage is relatively common on Neanderthal bones, as discussed in the next chapter. The main Broken Hill (Kabwe) skull (fig. 5.26), Zambia (?400 ky old), also exhibits a partially healed perforation on the left temporal that could have been made by a pointed weapon (1198). Arguably, however, it is more consistent with a carnivore bite (2138) or with the lesion from a small tumor (1573, 1716).

Mortality

It is very difficult to determine the pattern of mortality in early humans, partly because it is difficult to estimate individual age at death, particularly for adults, and partly because the samples comprise individuals that may have died in very different circumstances. Some may have succumbed to starvation, accidents, endemic disease, or other everyday attritional factors that tend to select out the very young and the old, while others could have been the victims of natural disasters or catastrophic interhuman conflict that might affect all age groups equally. A life table showing the likelihood that an individual will survive to a given age (or set of successive ages) can be constructed only when those who died attritionally can be separated from those who died catastrophically. This condition is unlikely to be met in fossil samples that comprise individuals from many different sites. For primitive *Homo,* and almost uniquely for fossil humans, a single site—the Sima de los Huesos (SH) at Atapuerca, Spain—has provided a large sample representing people who probably died at more or less the same time in broadly similar circumstances.

Recall that the SH human bones are probably about 300 ky old and that they have been assigned to the lineage that produced the classic Neanderthals after 127 ky ago. The site is a small chamber measuring about 17 m² (29 sq. ft.) in area that is connected to the surface by a 13 m (43 ft.) vertical shaft (81). Other entrances probably existed in the past but are now blocked. The human bones are virtually the only ones in the layer where they occur, but the overlying layer contains roughly nine thousand bones from cave bears and occasional bones from other carnivores. Stone artifacts are totally absent, and there is no other evidence that people inhabited the site. The way the human bones accumulated remains conjectural, but the presence of virtually all skeletal parts, including tiny ones like terminal phalanges, suggests that whole bodies came down the vertical shaft and disintegrated at the bottom. Bones in anatomical order are rare, most bones are broken, and the breaks are commonly smoothed, suggesting that sediment flow or perhaps cave bear trampling redistributed specimens across the site. The virtual absence of bones from any species except humans essentially rules out accumulation by carnivores, but probable fox tooth marks on more than half the human bones imply that foxes were able to enter and scavenge the bodies (53).

The origin of the cave bear bones in the overlying layer is equally mysterious, since they represent mainly two-year-olds and not the older animals that would be most likely to die during hibernation (781). Possibly the bears fell in accidentally when they attempted to hibernate elsewhere in the cave system.

The human sample comprises more than two thousand bones from at least thirty-two individuals. Dental eruption and wear indicates that seventeen of the thirty-two were adolescents between eleven and nineteen years of age and ten were young adults between twenty and twenty-seven years old (199, 202). Only three individuals were under ten years old, and none were clearly older than thirty-five. Sex determinations based on dental and mandibular size indicate that the sample is about equally divided between males and females. The rarity of children is difficult to explain, since they should dominate whether the sample accumulated as a result of everyday attritional mortality or as the result of a natural or humanly induced catastrophe. The answer may be that children's bones and teeth are particularly fragile and that they were thus selectively destroyed by carnivores and by postdepositional trampling and profile compaction. Selective destruction obviously operated on the SH sample, since the older individuals are represented primarily by the densest elements (above all teeth, but also distal humeri and proximal femurs), and fragile ones (like vertebrae or ribs) are relatively uncommon (53) (table 5.5).

The rarity of older adults is less puzzling if we accept that the sample accumulated catastrophically, that is, as the result of a natural disaster or of intergroup conflict. By definition, a catastrophe will freeze the structure of the live population in the ground, and until very recently, live human populations tended to contain relatively few old adults. The SH bones may themselves rule out intergroup conflict, since none show cut marks or any other humanly inflicted damage (53). For the most part cause of death is unclear, although the most complete skull (no. 5) came from a young adult who suffered from a massive facial abscess that could have produced systemic blood poisoning (septicemia) and the browridge of one juvenile had been severely fractured (1673). Five of the six adult individuals represented by frontal bones also show wormlike vascular channeling in the orbital roof that might reflect nutritional stress earlier in life, and eight of eleven individuals represented by temporal bones show substantial remodeling of the temporomandibular joint. Such remodeling is also common in Neanderthals, and it may be related to heavy chewing between the upper and lower molar rows.

Whatever the specific causes of death, if the age structure of the Sima sample is assumed to reflect the structure of the live population over ten years of age, the absence of older adults is striking, and it may imply that the maximum life expectancy of the Sima people was less than forty years. If so, they probably resembled the classic Neanderthals (discussed in the next chapter), and they differed significantly from fully modern people after 50–40 ky ago.

Disease

Besides the Sima de los Huesos (Atapuerca) fossils that have just been discussed, two other fossils of primitive *Homo* exhibit particularly conspicuous pathologies. These are the KNM-ER 1808 skeleton (*Homo ergaster*) from Koobi Fora, northern Kenya, and the principal Broken Hill (Kabwe) skull (early *H. sapiens*) from Zambia. It has already been noted that the limb bone shafts of 1808 are covered with up to 7 mm of abnormal coarse-woven bone and that possible causes include yaws (1840) or vitamin A poisoning (2376). The Broken Hill skull is remarkable for the extraordinary degree of caries (dental decay) and related abscessing on the adjacent maxillary bone (fig. 5.26). In this regard the skull is virtually unique among all

ancient human skulls, which rarely show caries, probably because human diets were generally low in sugar until the origins
and spread of agriculture, beginning about 10 ky ago.

Conclusion

The emergence of *Homo ergaster* 1.8–1.7 my ago marked a
watershed, for *H. ergaster* was the first hominid species whose
anatomy and behavior fully justify the label human. Unlike the
australopithecines and *Homo habilis*, in which body form and
proportions retained apelike features suggesting a continued reliance on trees for food or refuge, *H. ergaster* achieved essentially modern form and proportions. Members also differed
from the australopithecines and *H. habilis* in their increased,
essentially modern stature and in their reduced, essentially
modern degree of sexual dimorphism. The sum suggests that
this was the first hominid species to resemble historic hunter-
gatherers not only in a fully terrestrial lifestyle, but also in a social organization that featured economic cooperation between
males and females and perhaps between semipermanent male-
female units.

  H. ergaster was also larger brained than earlier hominids,
and this increased brain size was probably linked to the nearly
simultaneous appearance of hand axes and other relatively sophisticated artifacts of the Acheulean Industrial Tradition. The
emergence of *H. ergaster* and the Acheulean recall the appearance of the genus *Homo* and of the first stone tools roughly
800 ky earlier. Both events reveal a close connection between
biological and behavioral change in the early phases of human
evolution. Arguably too, both occurred abruptly, in the punctuational (as opposed to the gradual) mode of evolution, and
each may have been stimulated by a broadly simultaneous
change in global climate.

  The anatomical and behavioral advances that mark *Homo
ergaster* help explain how it became the first hominid species
to invade arid, highly seasonal environments in Africa, and also
how it became the first to colonize Eurasia. Its broad dispersal
greatly enhanced the potential for natural selection, gene drift,
or both to promote genetic divergence among human populations. The eventual result was the emergence of at least three
geographically distinct human lineages: *Homo erectus* in the
Far East, *Homo sapiens* in Africa, and *Homo neanderthalensis*
in Europe. *H. erectus* was probably the earliest to differentiate,
and fossils and artifacts together imply that east Asian populations followed a singular evolutionary course beginning 1 my
ago or before. In contrast, fossils and artifacts indicate that

H. sapiens (Africa) and *H. neanderthalensis* (Europe) shared a common ancestor as recently as 500 ky ago, and their morphological differentiation is manifest only after 400–300 ky ago. Arguably, eastern Asia was more distinctive because it was colonized (from Africa) only once, whereas Europe was colonized several times and only the event at 500 ky ago produced a permanent resident population.

Glacial climates dominated Europe during the 400–300 ky period when *Homo neanderthalensis* was evolving, and adaptation to cold probably explains some distinctive Neanderthal features, particularly in body form. Most key craniofacial characters, however, seem to have developed independently and sequentially rather than as part of an integrated complex, and this may mean they resulted mainly from gene drift in small populations now isolated from their African contemporaries.

By 250–200 ky ago, the craniofacial differences between *Homo erectus*, *H. neanderthalensis*, and *H. sapiens* were highly conspicuous, but all three lineages seem to have shared a tendency toward brain enlargement. By 200 ky ago, brain size, whether measured on its own or relative to body size, roughly approached the modern standard, and only variation in braincase form might imply that the different lineages differed neurologically from each other or from later people, including living ones. The shared natural selective forces that could have driven brain enlargement everywhere are easy to imagine, but the pace of brain enlargement remains uncertain, partly because the relevant fossils are scarce and partly because they are often very imprecisely dated. It is thus impossible to say whether enlargement occurred in sudden spurts or gradually, but a shared trend in three otherwise distinct lineages may argue that the increase was mostly gradual. In addition, the archeological record associated with each lineage nowhere reveals a striking behavioral advance that might reflect abrupt brain enlargement.

Homo erectus, early *H. sapiens*, and early *H. neanderthalensis* all possessed remarkably robust postcranial skeletons that differed little in overall ruggedness from those of *H. ergaster*. The implication is that members of all three lineages engaged in strenuous physical activity and that compared with anatomically modern humans they relied more on bodily power and less on artifactual skill to accomplish essential tasks. Equally important, striking artifactual uniformity over vast distances and long time periods suggests that *H. erectus*, early *H. sapiens*, and early *H. neanderthalensis* were all extraordinarily conservative in their behavior. The next chapter shows that such conservatism persisted until 50–40 ky ago, and it is only after this that significant artifactual change through time and space signals the modern capacity for innovation. The next two chapters argue that this capacity developed abruptly in Africa among *H. sapiens* and that it then allowed *H. sapiens* to replace *H. neanderthalensis* and *H. erectus* in Eurasia. The sudden development of a greatly enhanced capacity for innovation is not easy to explain, but it may reflect a genetic change like those that ushered in earliest *Homo* and *H. ergaster.* The difference is that the last change, 50–40 ky ago, would have affected only brain structure, not brain size, which makes it far more difficult to detect in fossil skulls.

THE NEANDERTHALS AND THEIR CONTEMPORARIES

6

The Neanderthals are the most famous and best understood of all nonmodern fossil people. In their European and west Asian homelands (fig. 6.1), they were the immediate predecessors of fully modern humans, and they were among the last truly primitive human groups. Historically the term primitive has been applied to people whose social organization, technology, economy, and so forth were relatively simple, even though the reason was their history, not their biology. Whether by force or free will, they have repeatedly demonstrated their ability to participate fully in far more complex societies and cultures.

Many important aspects of Neanderthal behavior remain obscure, but enough is known to show that they were primitive not merely in the sense of simple or unsophisticated. The archeological record indicates that they lacked some fundamental behavioral capabilities of living people, probably because their brains were differently organized. The former existence of such truly primitive people is of course totally consistent with the concept of human evolution, but it must also be emphasized that both morphologically and temporally the Neanderthals were far removed from the earliest people, and it is only because they lived so recently and are so abundantly known that their primitiveness is so well documented. The relatively rich record also shows that they resembled living people in many important morphological and behavioral respects and that they were certainly far advanced over earlier peoples.

Because the Neanderthals were morphologically and behaviorally so similar to modern humans, it has sometimes been suggested that they were directly ancestral to living populations. However, at the same time that the Neanderthals occupied Europe and western Asia, other kinds of people lived in the Far East and Africa, and those in Africa were significantly more modern than the Neanderthals. These Africans are thus more plausible ancestors for living humans, and it appears increasingly likely that the Neanderthals were an evolutionary dead

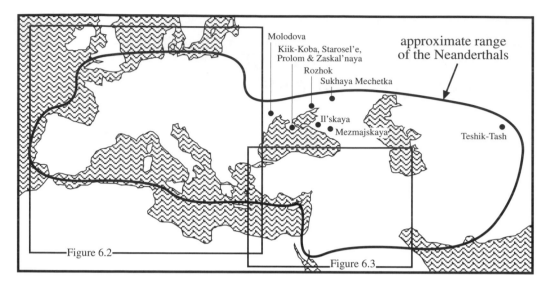

Figure 6.1. The European and west Asian realm of the Neanderthals (circled). The rectangles surround regions shown in greater detail in figures 6.2 and 6.3. The map indicates some key Middle Paleolithic sites that lie outside these regions.

end, contributing few if any genes to historic populations. This chapter summarizes the archeological and fossil evidence for both the Neanderthals and their contemporaries, with particular attention to the fate of the Neanderthals and to the origins of anatomically modern humans.

History of Discovery

The discovery of the Neanderthals was important not only to paleoanthropology but to science in general, for they were the first truly fossil, nonmodern people to gain scientific and popular recognition. In essence they provided the first real evidence for human evolution, and they were found at a time when the concept of evolution itself was still hotly debated.

Strictly considered, the first Neanderthal fossil still on record was a fragmentary child's skull found by P. C. Schmerling at Engis Cave near Liège. Belgium, in 1829–1830 (1209, 1626, 2029, 2095, 2237, 2238). It did not excite great scientific interest, however, partly because its nonmodern morphology was not demonstrated until more than a century later (750). The significance of a partial adult Neanderthal skull found in 1848 in a cave in Forbes' Quarry on the Rock of Gibraltar was also not recognized for decades. As a result, most authorities trace the discovery of the Neanderthals to 1856, when quarry workers uncovered a human skullcap and some postcranial bones at the Feldhofer Cave in the Neander Valley near Düsseldorf,

Figure 6.2. Approximate locations of key European and north African Middle Paleolithic sites mentioned the text.

Germany (1069) (fig. 6.2). In German, valley is *Tal*, then spelled *Thal*, and the kind of people represented at the site therefore came to be known as Neanderthalers—or simply Neanderthals.

The quarry workers may have inadvertently discarded parts of the original Neanderthal skeleton, along with many associated items, possibly including stone tools. Unfortunately there is almost no record of context and associations. The quarry owner believed the bones came from a bear, but he turned them over to a local schoolteacher, J. K. Fuhlrott, who realized they were from an unusual human. The skullcap had large brow-ridges, a low forehead, and bulging sidewalls (vs. the relatively parallel-sided sidewalls of modern skulls). The limb bones were thick and powerfully built. Because Fuhlrott's own expertise was limited, he transferred the bones to the well-known German anatomist, Hermann Schaaffhausen, who decided they represented a race that had inhabited northern Europe before the Germans and the Celts. Although Schaaffhausen was a proponent of organic evolution, he did not explore the evolutionary implications of a Neanderthal race.

Darwin knew of the Neanderthal bones when he published *The Origin of Species* in 1859, but he did not mention them, and he all but avoided the touchy issue of human evolution. In 1863, in *The Geological Evidence of the Antiquity of Man*, Charles Lyell, one of the founders of modern geology, stressed the great antiquity implied by the animal fossils that accompanied Schmerling's Engis fossil and by the degree of fossilization that characterized the Feldhofer Neanderthal. However, at the time Lyell was a nonevolutionist who felt that the Engis and Neanderthal fossils had no evolutionary significance. The first person to discuss this possibility at length was probably Thomas Huxley, Darwin's most eminent early disciple. In his *Evidence as to Man's Place in Nature* (1863), Huxley concluded that the Neanderthal skull represented a morphologically primitive person who was nonetheless far removed from the apes and in no sense a missing link. This is close to the opinion that virtually all paleoanthropologists hold today.

In 1864 the Irish anatomist William King suggested that the Neanderthal fossil represented an extinct human species, for which he coined the name *Homo neanderthalensis*. King's position is close to the one taken here, though King himself eventually came to agree with the prominent Prussian pathologist Rudolf Virchow, who argued that the Neanderthal skull came from a modern person afflicted by disease and antemortem blows to the head. Modern specialists have mostly taken an intermediate position, in which the Neanderthals are regarded as an extinct variety or subspecies of *Homo sapiens*.

To begin with Virchow's opinion was widely accepted, partly because it was cogently presented by an expert, partly because many people found it more comfortable than the idea of human evolution, and partly because there were few facts to contradict it. Perhaps most important, there was little evidence that the Neanderthal skeleton was very ancient. In addition, it was an isolated discovery that could have been idiosyncratic.

The next important discovery was in 1866 at the cave of Trou de La Naulette, Belgium, which provided a robust human mandible, an ulna, and a metacarpal. Associated fossils of mammoth, rhinoceros, and reindeer unambiguously indicated great antiquity. Unlike modern mandibles, the La Naulette specimen lacked a chin, and we know today it belonged to a Neanderthal. This was not obvious at the time, however, partly because there were no other, nonmodern mandibles for comparison and partly because Virchow again dismissed it as pathological. He expressed the same view on a generally similar mandible found in 1880 in Sipka Cave in Moravia (Czech Republic). His opinion was first seriously challenged by the discovery of two adult skeletons in a cave near Spy, Belgium, in 1886.

The Spy skulls closely resembled the original Neanderthal one, and it seemed unlikely that they shared precisely the same pathological history. Equally important, the Spy skeletons were carefully excavated, at least by the standards of the day, and they were unquestionably associated with stone tools and with the bones of mammoth, rhinoceros, reindeer, and other animals that indicated great antiquity.

The Spy skeletons authenticated the Neanderthals as a kind of fossil people, and the general case for human evolution was strengthened in 1891 when Eugène Dubois found the first *Pithecanthropus* specimens in Java. Between 1908 and 1921, any lingering doubts about the Neanderthals were removed by the discovery of a series of relatively complete skeletons in southwestern France. These came from the caves of La Chapelle-aux-Saints (Bouffia Bonneval), Le Moustier, La Ferrassie, and La Quina, all of which are famous today not only for their Neanderthal fossils but also for the light they shed on Neanderthal behavior.

The French sites were complemented by an equally important discovery at the Krapina Rockshelter near Zagreb, Croatia, between 1899 and 1905. Relatively systematic excavations here produced almost nine hundred fragmentary Neanderthal bones variously estimated to represent between fourteen and eighty-two individuals of both sexes and various ages (316, 546, 1724, 2231, 2479). The human fossils were associated with the same kind of stone artifacts found at Spy and in the French caves and also with faunal remains that demonstrated their great age. The Krapina discovery significantly enlarged both the sample of Neanderthal remains and their known geographic distribution.

The pace of discovery slowed after the 1920s, partly because many of the more promising sites had already been excavated and partly because archeologists began to adopt slower, more meticulous excavation methods. Important discoveries continued to accumulate, however, and in the 1930s the demonstrable range of the Neanderthals was extended to western Asia, first (in 1931) at Tabun Cave in what is now Israel and then (in 1938) at the Teshik-Tash Cave in Uzbekistan (west-central Asia) (fig. 6.1). The west Asian sample was significantly augmented at Shanidar Cave, Iraq, between 1953 and 1960, at Amud Cave (in 1961, 1964, and 1991–1992) and Kebara Cave (in 1965 and 1983), Israel, and at Dederiyeh Cave, Syria (in 1993) (fig. 6.3). Among European sites providing Neanderthal fossils after 1920, perhaps the most important are Kiik-Koba (Wild Cave) in the Crimea (1924), Subalyuk Cave in Hungary (1932), the Ehrings-dorf (Weimar-Ehringsdorf) spring site in Germany (1925, also 1908–16), the Saccopastore open-air site (1929 and 1935) and Guattari Cave (Monte Circeo) (1939) in Italy, Mezmajskaya Cave in southeastern Russia (1993–1994), and a series of French caves, including Regourdou (1957), Hortus (1960–1964), and Saint-Césaire (La Roche à Pierrot Rockshelter) (1979) (fig. 6.2). A skull and skeleton found at Altamura Cave, Italy, in 1993 (633, 634) should possibly be added to this list, but the fossils remain en-cased in nodular flowstone, and they have not been studied. Today, with or without Altamura, more than 275 individual Neanderthals are known from more than seventy sites. Most are represented by fragments, ranging from isolated teeth to partial or nearly complete skulls, but in sharp contrast to the situation with regard to earlier people, several are represented by complete or nearly complete skeletons, recovered from the world's oldest known intentional graves.

Sometimes the term Neanderthal (or Neanderthaloid) has been extended to include African and east Asian fossils. One problem with this is that the best-known African specimens for which the term has been used (the Broken Hill and Saldanha skulls) are certainly older than the Neanderthals. More impor-tant, the large browridges, low frontal, low position of maxi-mum cranial breadth, and other features that so-called African and east Asian Neanderthals share with the indisputable Euro-pean and west Asian ones characterize virtually all nonmodern skulls of the genus *Homo.* At the same time, the African and east Asian skulls generally lack a cluster of traits that tend to be unique to the European and west Asian Neanderthals (1863, 2095, 2221). These traits include the subspheroid form of the skull when viewed from behind, the bunlike form of the occip-ital bone when viewed from the side, the extraordinary projec-tion of the midfacial region, distinctly double-arched brow-ridges, a suprainiac fossa, a large juxtamastoid (occipitomastoid) crest, and other characters that are listed below.

Although much remains to be learned about the Afri-can and east Asian contemporaries of the European and west Asian Neanderthals, it is increasingly clear today that they were

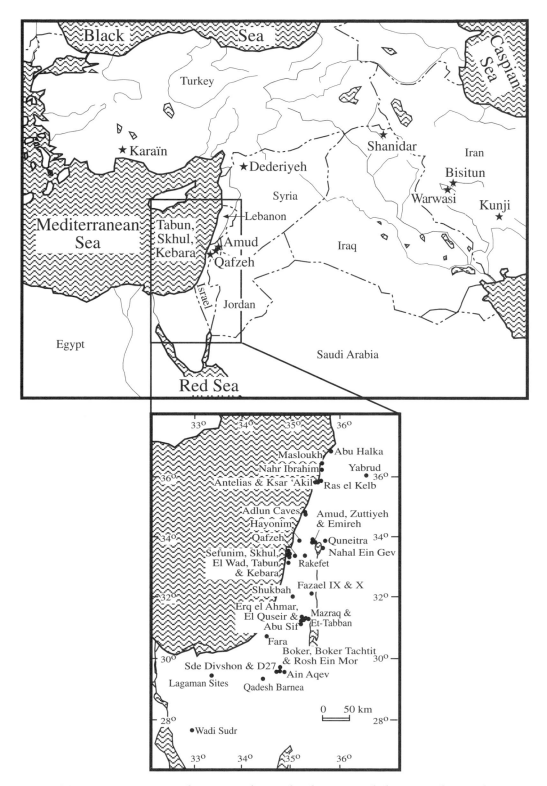

Figure 6.3. *Top:* Approximate locations of sites that have provided near-modern and Neanderthal fossils in western Asia (adapted from 2212). *Bottom:* Approximate locations of principal Middle and Upper Paleolithic sites in the Levant (modified after 115, p. 108).

morphologically different, and calling them Neanderthals only obfuscates their evolutionary significance. They will be discussed here under the heading of Neanderthal contemporaries.

Geologic Antiquity

In every instance where Neanderthal fossils have been dated by radiocarbon (^{14}C) analysis of reliable material, especially charcoal, the dates have been at or clearly beyond the practical limit of the ^{14}C method at 30–40 thousand years (ky) before the present (B.P). The same is true for dates on reliable material accompanying the kinds of (Middle Paleolithic = Mousterian) artifacts that Neanderthals are known to have made. It is therefore clear that, as a group, the Neanderthals antedate 30 ky ago. The youngest known Neanderthal fossils are perhaps those from Zafarraya Cave in southeastern Spain, Arcy-sur-Cure (Grotte du Renne) in north-central France, and Saint-Césaire in southwestern France.

At Zafarraya, uranium-series (U-series) disequilibrium dating of animal teeth and ^{14}C dating of the organic residue in associated bone suggest that a typical Neanderthal mandible and other fragmentary Neanderthal bones are about 33 ky old (1097); at Arcy, ^{14}C dates on the organic fraction in fossil bone indicate that a Neanderthal temporal fragment and three probable Neanderthal teeth are roughly 34 ky old (842, 998, 1098); and at Saint-Césaire, thermoluminescence (TL) dates on burned flint artifacts imply that a partial Neanderthal skeleton is 36 ± 3 ky old (1361, 1550). Together the three sites suggest that Neanderthals survived in parts of Spain and France for up to 7 ky after anatomically modern Cro-Magnon people had appeared nearby. In western Europe the oldest well-documented Cro-Magnon fossils date from only about 30 ky ago, but most specialists assume that the Cro-Magnons produced early Upper Paleolithic Aurignacian artifacts. Radiocarbon (^{14}C) dates on charcoal place the basal Aurignacian Culture near 40 ky ago at Le Trou Magrite Cave in Belgium, Castillo Cave in northern Spain, and at L'Arbreda and Romaní Shelters in northeastern Spain (229, 232, 408, 998, 2061, 2066, 2069). At Romaní, particularly reliable U-series determinations on flowstone strongly support an age near 40 ky ago.

The possibility of a long temporal overlap between Neanderthals and Cro-Magnons is intriguing, but it depends on the assumption that the Zafarraya and Arcy ^{14}C dates are truly finite values. The alternative is that, like many other ^{14}C dates of similar magnitude, they are only minimal estimates. The problem is that the addition of less than 1% of modern carbon could make an object that was actually 40 ky old appear 6–7 ky younger, and the bone organic residues that were dated at both sites could include minute amounts of undetectable younger organic carbon (from soil bacteria or humic acid). Such minute contamination is the most plausible explanation for the inconsistencies or stratigraphic inversions that characterize even the most carefully obtained ^{14}C dates at French early Upper Paleolithic sites (877, 1540). The dates at Arcy are not entirely in stratigraphic order, and they include one of 45.1 ± 2.8 ky on bone from an early Upper Paleolithic (Châtelperronian) level overlying the one that provided the principal Neanderthal specimens (998). At Zafarraya, the U-series determinations are actually somewhat younger than the corresponding ^{14}C dates, although both theory and paired U-series and ^{14}C dates elsewhere suggest they should be 3–5 ky older (136, 137, 229). The reason (discussed in chapter 2) is that uncontaminated U-series readings provide true calendar (solar) ages, whereas ^{14}C ages tend to systematically underestimate them. With regard to Saint-Césaire, the 3 ky error estimate attached to the TL date actually overlaps the basal Aurignacian ^{14}C dates, even when their error estimates are ignored. Thus it remains possible that Neanderthals and Cro-Magnons coexisted in western Europe for only a millennium or two about 40 ky ago. The issue can be resolved only by fresh dating at many additional sites. As indicated below, however, it will still be necessary to explain how or why the Saint-Césaire and especially the Arcy Neanderthals produced Upper Paleolithic artifacts broadly resembling those of the Cro-Magnons.

Zafarraya, Arcy, and Saint-Césaire aside, all other dated Neanderthal fossils are older than 40 ky ago (924, 1782). This is especially true for western Asia (the Near East), where ^{14}C, TL and, ESR (electron spin resonance) dates, sedimentary evidence, associated artifacts, or a combination of these suggest that the last Neanderthals lived between 50 and 45 ky ago (125, 712–714, 1138, 1140, 1548, 1888, 2216, 2225).

The geologic age of the oldest Neanderthals is partly a matter of dating and partly of definition. With regard to dating, calcareous spring deposits (travertines) at Ehringsdorf, Germany, have produced cranial fragments, mandibles, and some postcranial bones from perhaps nine individuals whose Neanderthal status is widely accepted. Associated plant and animal remains have often been taken to suggest a Last Interglacial (isotope stage 5) age, between 127 and 71 ky ago (548), but U-series and ESR dates suggest that the enclosing deposits probably formed within the interglacial before last (isotope stage 7), between roughly 242 and 186 ky ago (239, 923, 924). With regard to definition, a partial skull from Biache-Saint-Vaast (northern France), is relatively small but, in the parts that are preserved, otherwise indistinguishable from typical Neanderthal skulls. Stratigraphic context and associated paleobiological materials imply an age early within the Penultimate Glaciation (isotope stage 6) (2017), which began roughly 186 ky ago, and this is strongly supported by TL dates between 190 and 159 ky ago on associated burned flints (17). If the older Ehringsdorf age is accepted and the Biache skull is included among the Neanderthals, then Neanderthal beginnings would clearly extend back to 190 ky ago or before.

Excluding Ehringsdorf, Biache, and some other probable late middle Quaternary Neanderthals or near-Neanderthals, the oldest specimens commonly assigned to the Neanderthals sensu stricto are probably those from Krapina in Croatia and Saccopastore in Italy. A combination of U-series and ESR determinations suggest that the Krapina deposits accumulated rapidly near the beginning of the Last Interglacial, roughly 130 ky ago (1781). This age is consistent with faunal evidence for a mild climate (1998, 2000). At Saccopastore, two Neanderthal craniums were found in fluvial deposits whose stratigraphic position and floral and faunal contents suggest a Last Interglacial age (2095).

Based on stratigraphic position and faunal and floral associations, all other important European Neanderthals are commonly assigned to the earlier part of the Last Glaciation, between 71 and 50–35 ky ago. ESR, TL, or U-series determinations confirm an antiquity of this order for well-known specimens from the caves of La Chapelle-aux-Saints and Le Moustier, France, and from Guattari Cave, Italy (924). However, some less precisely dated Neanderthal fossils could belong to the colder phases of the Last Interglacial, namely to global marine oxygen-isotope stages 5d and 5b, centering on 109 and 92 ky ago, respectively. These stages probably comprise some of the climatic events commonly attributed to the early part of the Last Glaciation in western Europe (for example, to the Würm I in France) (400, 652, 1539).

The oldest known Neanderthal fossils in western Asia probably date from the earlier part of the Last Glaciation, although the separation of Last Glaciation and Last Interglacial deposits is generally more difficult in western Asia than in Europe (712). The best candidates for pre–Last Glaciation Neanderthal fossils in western Asia are a femoral shaft from Tabun Cave (layer E) and a frontofacial fragment from Zuttiyeh Cave, both in Israel. Both fossils were associated with artifacts of the pre-Mousterian Acheuleo-Yabrudian (or Mugharan)

Industry. At Tabun Cave, the Acheuleo-Yabrudian has now been dated by ESR to more than 150 ky ago and by TL to more than 250 ky ago. U-series determinations on flowstone (travertine) bracket the Acheuleo-Yabrudian at Zuttiyeh between roughly 160 and 97 ky ago, though the results are suspect because the flowstone was contaminated by detrital material (1888). If the ESR, TL, and older U-series dates are accepted, the Tabun and Zuttiyeh fossils would both antedate any generally accepted European Neanderthals. Arguably this reduces the likelihood that they represent Neanderthals, and in any case the femoral fragment is not taxonomically diagnostic. The frontofacial fragment exhibits less midfacial prognathism and perhaps a more vertical frontal (forehead) than are typical for the Neanderthals, and its taxonomic assignment is controversial. It has variably been assigned to an early Neanderthal anticipating those from Shanidar Cave in Iraq (2214, 2216) or to a population anticipating the robust early modern or near-modern people found at Skhul and Qafzeh Caves, also in Israel (2305, 2307). A third alternative is that the Zuttiyeh fossil marks a population of archaic humans with no certain links to either the Neanderthals or modern people (1956).

In sum, all fossils that are indisputably Neanderthal date from the earlier part of the late Quaternary, between roughly 127 and 50–35 ky ago, the exact terminal date probably depending on the place. It is always important to stress that this relatively limited time span is associated with a relatively limited geographic range, including only Europe and western Asia.

Morphology

Morphologically the Neanderthals are a remarkably coherent group, and they are easier to characterize than most earlier human types because they are represented by relatively abundant fossils from virtually all parts of the skeleton. The following description draws heavily on ones in references 553, 1068, 1069, 1083, 1317, 1863, 2000, 2095, 2215, 2216, 2218, 2221, and 2234. It should be read in conjunction with figures 6.4 to 6.16, which illustrate some particularly diagnostic Neanderthal features.

Braincase

Long and relatively low, but not especially thick walled; globular (*en bombe* = subspheroid or oval) when viewed from behind, with maximum breadth at the midparietal level; large endocranial capacity, ranging from 1,245 to 1,740 cc and averaging about 1,520 cc (1048, 1052) (this compares with an average of about 1,560 cc in the earliest anatomically modern people and

Figure 6.4.
Neanderthal skulls
from La Ferrassie
in France and Amud
in Israel, showing
some of the key fea-
tures that charac-
terize Neanderthals
as a group (redrawn
after 1863, pp. 622,
625). Among the
distinctive Nean-
derthal features
illustrated are the
extraordinary for-
ward projection of
the face along the
midline (midfacial
prognathism), the
well-developed
mastoid crest or
bump behind the
auditory aperture,
the large juxtamas-
toid crest, equaling
or exceeding the
mastoid process in
length, the supra-
iniac depression
or fossa above the
occipital torus, and
the globular shape
of the skull when
viewed from
behind.

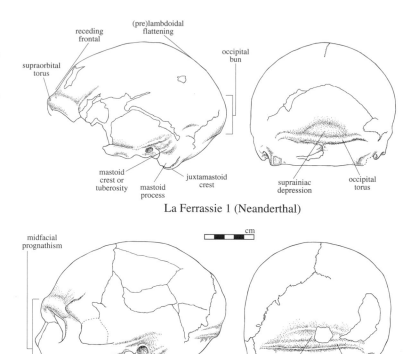

La Ferrassie 1 (Neanderthal)

Amud 1 (Neanderthal)

of about 1,340 cc in living people); frontal low and receding, with a continuous supraorbital torus (browridge) generally forming a double arch above the orbits and usually separated from the frontal scale (or squama) by a sulcus or gutter; frontal sinuses generally large and restricted to the torus; occipital squama usually extended backward to form a distinct bun (or chignon), associated with flattening in the region of lambda (the point on the back of the skull where the lambdoid and sagittal sutures meet); prominent, nearly horizontal transverse occipital torus, surmounted by a conspicuous elliptical depression (the supra-iniac fossa or depression) generally surrounded by a triangular, uplifted area of bone; mastoid process usually small, with a pronounced mound or bump of bone (the mastoid crest or tuberosity) behind the auditory aperture; ventral to the mastoid process, an equally large or larger mound or ridge of bone (the juxtamastoid crest) bounded by distinct depressions; inner ear with remarkably small anterior and posterior semicircular canals and with a uniquely low position for the posterior canal (versus a higher position in all other known humans) (fig. 6.7) (1098); skull base flatter (less flexed) between the hard palate and the foramen magnum than that in modern people (1303).

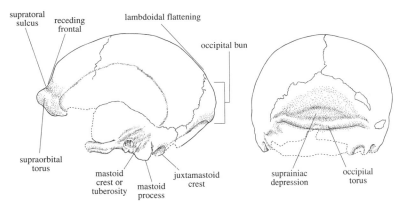

Spy 2 (Neanderthal)

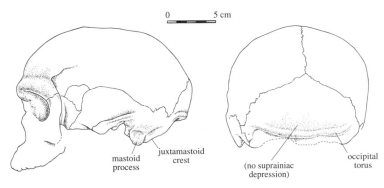

Jebel Irhoud 1 (non-Neanderthal)

Figure 6.5. One of the Neanderthal skulls from Spy Cave, Belgium, compared with the somewhat older non-Neanderthal skull from Jebel Irhoud, Morocco (redrawn after 1863, pp. 624, 627). The Irhoud skull has sometimes been labeled Neanderthal, but it differs from typical Neanderthal skulls in several crucial features, including its shorter and flatter face, its more rectangular orbits, its more parallel-sided (less globular) braincase, and its lack of a suprainiac fossa or depression. In most of these features it anticipates anatomically modern people, near whose ancestry it may lie.

Face

Long and forwardly protruding (prognathous), especially along the midline. Nasal aperture and cavity generally very large; zygomatic arches receding (rather than angled as in modern people); maxillary bone inflated above the canine teeth (no canine fossa); orbits large and rounded; chin (forward protrusion of the mental symphysis) variably developed but usually absent; mandibular sigmoid notch highly asymmetric and associated with a posteriorly projecting coronoid process that is usually much larger than the condyloid process (fig. 6.8).

Dentition

Cheek teeth smaller than in earlier people and overlapping those of early, robust near-modern and modern people in size; incisors as large as or larger than those of earlier people and significantly larger than those of modern people; maxillary incisors usually shovel shaped (fig. 6.9); mandibular and maxillary incisors tending to exhibit distinctive, rounded wear on the labial

Figure 6.6. Neanderthal skull from La Quina in France compared with the non-Neanderthal skull from Broken Hill (Kabwe), Zambia (redrawn after 1863, pp. 623, 626). The Broken Hill skull has sometimes been identified as an African Neanderthal, but it lacks such characteristic Neanderthal features as the suprainiac depression and the globular shape in rear view (it is more angular from behind and has its maximum breadth nearer the cranial base). It also antedates the Neanderthals by a substantial interval and is now widely regarded as a representative of mid-Quaternary African *Homo sapiens,* as discussed in chapter 5.

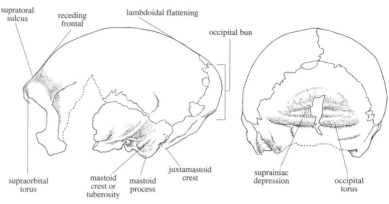

La Quina 5 (Neanderthal)

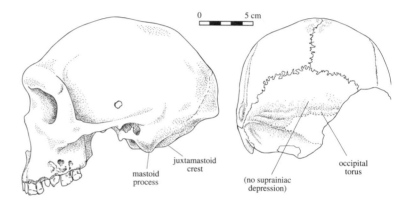

Broken Hill (Kabwe) (non-Neanderthal)

(lip) surface in older individuals (fig. 6.10); cheek teeth usually less worn than incisors and canines (in contrast to most modern humans, whose anterior and posterior teeth tend to be about equally worn); root fusion with enlargement of the pulp cavity (taurodontism) common in the cheek teeth (fig. 6.11); readily discernible gap (retromolar space) between the posterior wall of the lower third molar and the anterior margin of the ascending ramus, resulting from the combination of a long mandible, a short postcanine (premolar and molar) dentition, and a relatively narrow ascending ramus (754); mandibular foramen usually horizontal-oval.

Postcranium

Cervical vertebrae with long horizontal robust spines at or beyond the range of modern human variation; the neural (vertebral) canals that enclosed the spinal column enlarged as in living humans (2232); ribs extraordinarily thick and weakly curved to encircle a broad, thick chest; trunk very broad and

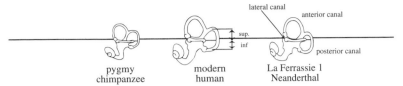

Figure 6.7. Lateral view of the left labyrinthine inner-ear structure in pygmy chimpanzees, modern humans, and the La Ferrassie 1 Neanderthal (redrawn after 1098, p. 224). For the position of the labyrinth within the ear, see figure 3.4. Like all known humans from the time of *Homo ergaster*, relative to apes, Neanderthals had large anterior and posterior semicircular canals. Relative to other humans, however, Neanderthals had smaller anterior and posterior canals, and they were also unique in the low position of the posterior canal. The similarities to other humans imply shared balance requirements during habitual bipedal locomotion. The differences may mean that Neanderthal physiques imposed unique balance requirements, or the differences may have been forced by the pattern of Neanderthal brain growth as it affected the basicranium. The unique morphology of the Neanderthal labyrinth has been used to identify an otherwise equivocal temporal fragment associated with Châtelperronian artifacts at the Grotte du Renne, Arcy-sur-Cure, France.

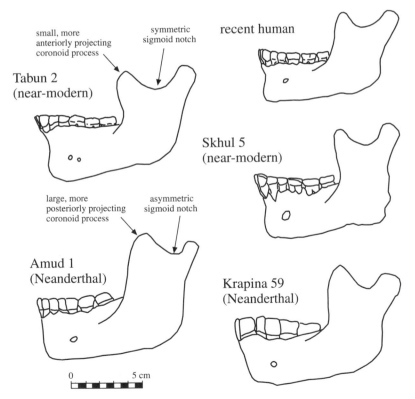

Figure 6.8. Mandibles of a recent human, of near-modern humans from Tabun and Skhul Caves, and of Neanderthals from Amud and Krapina Caves (Tabun 2, Skhul 5, and recent human redrawn after 1491; Amud 1 from a cast; and Krapina 59 after 1724). Like other Neanderthal jaws, those from Amud and Krapina are distinguished by a large, posteriorly placed coronoid process and a highly asymmetric sigmoid notch.

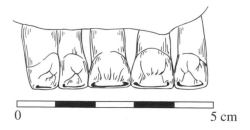

0   5 cm

Figure 6.9. Left upper canine and all four upper incisors of tooth
set K from Krapina, Croatia, viewed from the lingual (inside) surface
(drawn by Kathryn Cruz-Uribe from a photograph in 2481, p. 274;
© 1999 by Kathryn Cruz-Uribe). The lateral edges of the canine and
incisor crowns are curled backward and inward to produce the shovel
shape that characterizes Neanderthal front teeth in general and the
upper front teeth in particular. Shoveling is also common in *Homo
erectus* and in some modern populations, especially in some east
Asians and Native Americans. It increases the occlusal surface of the
front teeth, which may have been particularly significant in the case
of the Neanderthals, who used these teeth as tools.

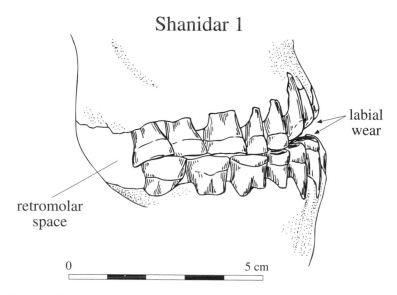

Shanidar 1

labial
wear

retromolar
space

0   5 cm

Figure 6.10. Enlarged view of the dentition of Shanidar skull 1 showing the retromolar space and the rounded wear on the labial surface
of the incisors that are characteristic Neanderthal features (drawn by
Kathryn Cruz-Uribe from a photograph; © 1999 by Kathryn Cruz-
Uribe). The wear resulted from exceptional use of the incisors, perhaps for clamping or gripping skins or other objects.

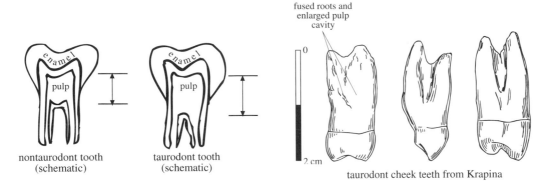

Figure 6.11. *Left:* Schematic cross sections to show the difference between nontaurodont and taurodont teeth in the size of the pulp cavity. *Right:* Upper cheek teeth from Krapina, Croatia, showing the fused roots and enlarged pulp cavity (taurodontism) that is a common feature of the Neanderthals (drawn by Kathryn Cruz-Uribe from a slide; © 1999 by Kathryn Cruz-Uribe).

deep relative to inferred stature; muscle and ligament attachment areas extremely large and well developed on bones of the arms, legs, hands, and feet; shafts of the femur and radius usually somewhat bowed (fig. 6.12); distal limb segments (forearm and lower leg) relatively short; glenoid fossa of the scapula (the socket that articulates with the head of the humerus) long, narrow, and relatively shallow (flat); blade of the scapula very broad, commonly with a deep groove or sulcus on the outer edge of the dorsal surface (vs. the shallower one common on the ventral or rib surface in modern people) (fig. 6.13) (2048); distal phalanx of the pollex (thumb) roughly as long as the proximal one (unlike the situation in modern people, in whom the distal phalanx is usually about one-third shorter than the proximal one); tips (apical tufts) of the hand (manual) phalanges large and round (vs. narrower, hemiamygdaloid-shaped tips in modern people) (fig. 6.15).

Large femoral and tibial epiphyses with robust, cortically thick shafts (diaphyses); femoral shaft round, without the longitudinal bony ridge (pilaster) found on the dorsal surface of modern femoral shafts (fig 6.12); an average angle of about 120° between the shaft of the femur and its neck (the proximal-medial protrusion that articulates with the pelvis), compared with average angles between 124° to 135° in various recent human populations (fig. 6.14) (2229); shafts of the proximal phalanges of the foot expanded mediolaterally; pelvis less massive than in mid-Quaternary hominids and with the same sexual dimorphism seen in the greater sciatic notch of modern humans (2217), but with the pubic bone noticeably lengthened and

Figure 6.12. Right femurs of a Neanderthal and of a modern person, illustrating the much greater anterior-posterior bowing of the shaft that is characteristic of the Neanderthals (redrawn after 299, p. 236). The difference probably reflects the extraordinary muscular strength (muscular hypertrophy) of the Neanderthals. Neanderthal femurs also differ from modern ones in their more rounded cross section and in the absence of a bony ridge (pilaster) on the posterior surface.

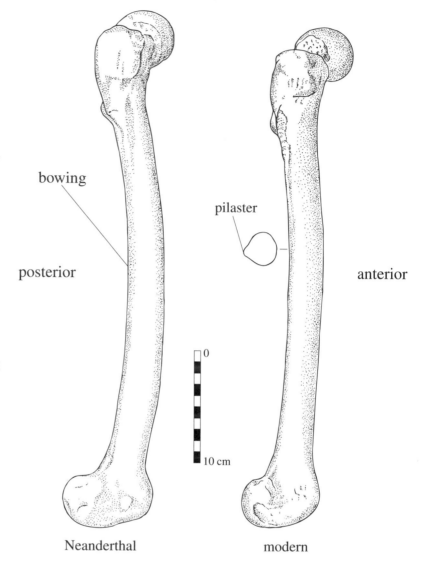

thinned compared with that of modern people (2047); average stature, estimated from long bone lengths, about 166 cm (5′4″) (this compares with an average of approximately 178 cm [5′8″] in the earliest anatomically modern people [2216]).

Overview

The Neanderthals were extremely robust, heavily muscled, barrel-chested people with large, long, relatively low globelike skulls and large, long, prognathic faces. In the first detailed studies, the vertebrate paleontologist Marcellin Boule (296–298) and some other prominent investigators concluded that

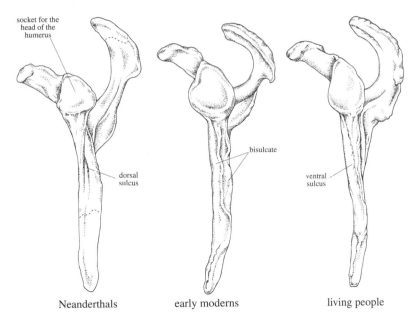

socket for the
head of the
humerus

dorsal
sulcus

bisulcate

ventral
sulcus

Neanderthals   early moderns   living people

Figure 6.13. Variation in the axillary (lateral) border of the scapula in
Homo sapiens (redrawn after 2215, p. 171). In distinction from modern people, in which a ventral sulcus is the norm, Neanderthals tend
to exhibit a dorsal sulcus. The difference may reflect the greater
power of Neanderthal upper arm muscles. The earliest anatomically
modern humans and some living athletes exhibit a composite or
bisulcate pattern.

the Neanderthals were apelike in posture, foot structure, and
other important features. However, it is now clear that they
were indistinguishable from modern people in these characters
and, more generally, in the basic structure and function of their
limbs and muscles. The postcranial differences between the
Neanderthals and modern people relate mainly to the Neanderthals' extraordinary robusticity (muscularity). Overall, they
were in fact only slightly less robust than their mid-Quaternary
predecessors.

Morphologically, the Neanderthals can be distinguished
from all other people, fossil or living, primarily by the skull and
face. The face was particularly striking for the forward projection of the central browridge, the large nose, and the dentition—in short, for its prognathism along the midline, from
which both the orbits and cheeks swept backward. Even when
the craniofacial differences are considered, however, the Neanderthals differed from anatomically modern people only about
as much as do two subspecies in some living mammal species.
The recent recovery of mitochondrial DNA from a Neanderthal
bone (1275), discussed below, underscores the divergence of the

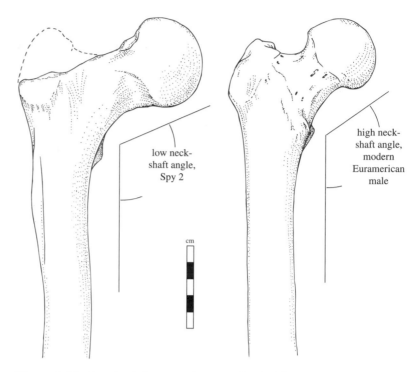

Figure 6.14. Proximal femurs of a Neanderthal from Spy Cave, Belgium, and of a modern Euramerican male, illustrating the low angle between the femoral shaft and the femoral neck in Neanderthals (drawings by Kathryn Cruz-Uribe; © 1999 by Kathryn Cruz-Uribe). The neck angle is lowest in modern human humans who are physically very active, particularly during growth and development, and the remarkably low angle in Neanderthals suggests intense physical activity, essentially beyond the modern range.

Neanderthals from living people, but the difference need not imply two separate species. Consequently, many authorities distinguish the Neanderthals and modern people only at the subspecies level, as *Homo sapiens neanderthalensis* and *H. sapiens sapiens*. Against this, Neanderthals and modern humans overlapped in Europe, at least briefly, and a combination of fossils and archeology indicates that they rarely if ever interbred. The isolating mechanism could have been behavioral (as opposed to genetic), but it would still justify placing the Neanderthals in a separate species, *Homo neanderthalensis*. That is the position adopted here.

The Meaning of Neanderthal Morphology

Many specialists have sought functional (biomechanical or physiological) explanations for distinctive Neanderthal traits. This approach is attractive because it can be grounded in the

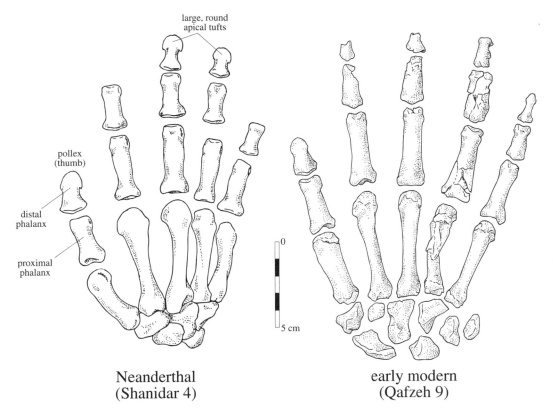

large, round
apical tufts

pollex
(thumb)

distal
phalanx

proximal
phalanx

0

5 cm

Neanderthal
(Shanidar 4)

early modern
(Qafzeh 9)

Figure 6.15. Partially restored hand skeletons of a Neanderthal (Shanidar 4) and of a very
early modern human (Qafzeh 9). Neanderthal hands differed from modern ones in several
respects, including larger, rounder apical tufts on the distal (terminal) phalanges and a relatively long distal phalanx on the thumb. (Shanidar hand drawn from a photograph in 2215,
p. 260; Qafzeh hand redrawn after 2304, p. 226.)

functional interpretation of living human variation. It assumes
only that natural selection guided the development of Neanderthal morphology. As a practical matter, however, this approach
is difficult to pursue, partly because the Neanderthals lay outside the range of living humans and partly because many Neanderthal parts are in very short supply, particularly in complete
specimens. Small sample size makes it difficult to determine
both the mean and dispersion around the mean for many Neanderthal features and thus to determine just how different the
Neanderthals were from other people. There is the further problem of knowing when a particular feature is meaningfully analyzed on its own and when it may be simply a consequence or
correlate of another feature that is not being considered.

Finally, and arguably most important, there is the possibility that many distinctive Neanderthal features might have no
real functional meaning but might have originated instead from
gene drift (random change) in a population that was isolated in

Europe during most of its evolutionary history (1070). This possibility may be implied by the accretional way Neanderthal features seem to have accumulated in ante-Neanderthal populations. It may also be signaled by the striking artifactual similarity between the Neanderthals and their near-modern contemporaries. The artifacts are described below, and the similarity is hard to explain if Neanderthal morphology implies a meaningful behavioral differentiation from their contemporaries. With this and previous caveats in mind, the focus here will be on the best available understanding of Neanderthal morphology in adaptive (biomechanical or physiological) terms. The discussion relies particularly on analyses and syntheses by Trinkaus (2221, 2222, 2225, 2228).

The large, well-developed muscle and ligament markings on the bones of Neanderthal limbs and extremities indicate exceptional muscular strength (muscular hypertrophy). The bowing of the radius and femur has sometimes been attributed to rickets (vitamin or hormone D deficiency) (1123), but it also tends to characterize yet earlier people, and neither they nor the Neanderthals exhibit other diagnostic bony symptoms of rickets. More likely the bowing is related to the extraordinary muscularity of all premodern people, including the Neanderthals.

The proportions and morphology of Neanderthal hand bones imply an extraordinarily powerful grip, especially as exercised by the thumb when opposed to the other fingers. The overall morphology of the scapula reflects the power of the upper arm muscles that attached to it; hypertrophied development of some of these muscles would have destabilized the arm during throwing or striking if another muscle had not shifted its attachment on the scapula, producing the groove on the outer border of the dorsal face (2215, 2224). The cortical thickness of the leg bones reflects an ability to endure long periods of intensive use. The relatively low angle between the neck of the femur and its shaft implies elevated stress (loading) on the hip as individuals matured (2229). In living human populations, the angle is lowest where juveniles are most active and highest where they are most sedentary. The mediolaterally expanded shafts of the pedal phalanges were possibly an adaptation for prolonged movement over irregular terrain.

A full assessment of robusticity requires that metric measures be scaled for body mass, which on average was probably much greater in the Neanderthals than in fully modern people. Long-bone cortical thickness has been scaled for probable body mass in the west Asian Neanderthals (from Kebara, Amud, Tabun, and Shanidar Caves), and the result underscores the remarkable robusticity of Neanderthal arms and legs versus those of fully modern people. When such scaling is applied to the near-modern contemporaries of the Neanderthals from Skhul and Qafzeh Caves, Israel, however, it shows that the near-moderns had less robust arms but equally robust legs (2236). The implication may be that Neanderthals were distinctive primarily in the way they handled tools, weapons, or other objects and less so in their habitual patterns of movement. A unique pattern of Neanderthal object manipulation may also be reflected in the singular structure of the Neanderthal carpal-metacarpal joint (2233) and in the peculiar features of the scapula (468).

Adaptation to cold probably explains the great breadth of the Neanderthal trunk, the shortness of the limbs, and especially the shortness of the forearm and lower leg (1043, 1843, 1844, 2213, 2215). This inference follows from the ecological generalizations known as Bergmann's and Allen's rules, which state that, all other things equal, individuals of a warm-blooded species will tend to have larger body cores (trunks) and shorter limbs in colder climates. A larger body core is advantageous because, as core size increases, volume, which conserves heat, increases more rapidly than skin area, which dissipates it. Shorter limbs, and especially shorter distal limbs, further reduce heat loss. Both Bergmann's and Allen's rules apply broadly to living humans, in whom trunk breadth tends to increase and limb length, especially distal limb length, tends to decrease from the equator to the poles. (In technical terms, the decrease in distal limb length is in the radius/humerus [brachial] and the tibia/femur [crural] ratios or indexes.) European and west Asian Neanderthal limb bones are sufficiently abundant to suggest that European populations had especially short limbs, in keeping with their exposure to greater cold. In their great trunk breadth and short limb length, the Neanderthals as a group equaled or exceeded the Inuit (or Eskimo), although the Neanderthals probably experienced less extreme cold. This suggests that relative to the Inuit, the Neanderthals relied more on physiology and less on culture as a buffer to cold.

The longer, thinner pubis of the Neanderthal pelvis could indicate that the average diameter of the birth canal (pelvic inlet or aperture) was significantly greater in Neanderthals than in modern people. Among living people, birth canal size tends to be largest in populations with the largest head size relative to stature, implying that a large Neanderthal birth canal could be simply a structural reflection of their very large heads and very short stature (1832, 1833, 2485). Alternatively, it could indicate that in utero brain growth was faster in Neanderthals than in modern people (619) or that the Neanderthal gestation period was longer (2215, 2217, 2218).

The hypothesis of a longer gestation period is particularly intriguing for its evolutionary implications. It would have meant greater neuromuscular development at birth and a better-developed immune system, but it would also have kept the neonate from the exposure to environmental stimuli during a period of rapid brain growth. It would also have forced greater spacing between births, reducing the potential for population growth. It might thus have placed the Neanderthals at a selective disadvantage compared with modern humans. The idea of a longer gestation period is controversial, however, not only because there are alternative explanations for a larger Neanderthal birth canal but also because it is not clear that the birth canal really was larger. The inlet diameter of the most complete Neanderthal pelvis, from Kebara Cave in Israel, is not significantly greater than that in modern people (fig. 6.16), though the pubis is typically long and thin as in other Neanderthals (1729). The implication may be that the peculiar shape of the Neanderthal pelvis has nothing to do with reproduction but rather was a correlate of great upper body breadth (related to cold adaptation) or was due to slight differences in habitual posture and locomotion.

The long, low shape of the Neanderthal cranium with its typically large occipital bun may reflect relatively slow postnatal brain growth relative to cranial vault growth (2235). Slower brain growth could have been tied to slower and ultimately more limited intellectual growth and development. The large average size of the Neanderthal brain is the culmination of a tendency toward increasing size throughout the course of human evolution, and it almost certainly means that the Neanderthals were more intelligent than any of their smaller-brained antecedents. Brain enlargement could account for the large diameter of the vertebral canal, which allowed more expansive brain-body connections (enervation), and brain size and shape (flatness) could together explain the singular morphology of the inner ear (1098). Brain expansion beyond the average for modern humans is difficult to explain, but it may simply reflect the greater metabolic efficiency of large brains in colder climates, the large amount of lean body mass (striated muscle) in Neanderthals, or both (1052). Among living people, the largest average brain size, actually about equal to that in Neanderthals, occurs in Inuit, who live in very cold conditions and who also tend to possess a large quantity of lean body mass.

The structure of the Neanderthal brain cannot be examined directly, but Neanderthal endocranial casts fall generally within the modern range of variation and show the same pattern of asymmetries (petalias) that modern brains do (1048, 1052).

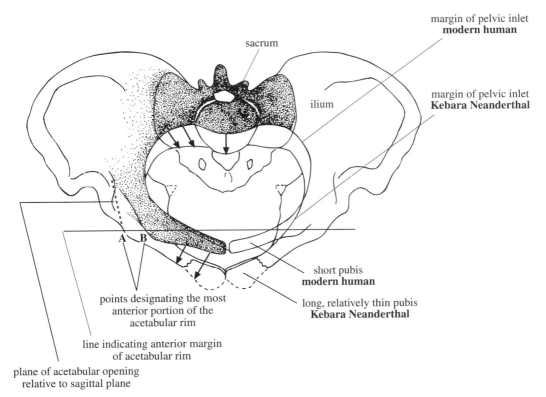

margin of pelvic inlet
modern human

margin of pelvic inlet
Kebara Neanderthal

sacrum

ilium

short pubis
modern human

long, relatively thin pubis
Kebara Neanderthal

A B

points designating the most
anterior portion of the
acetabular rim

line indicating anterior margin
of acetabular rim

plane of acetabular opening
relative to sagittal plane

Figure 6.16. Superior view of the reconstructed Neanderthal pelvis from Kebara Cave, Israel (redrawn after 1729, p. 230). Like other, more fragmentary Neanderthal examples, the Kebara pelvis shows a distinctive lengthening of the pubis compared with the condition in modern people. In spite of this, however, its pelvic inlet (birth canal) was about the same size as in a modern person. Assuming that other Neanderthal pelvises would have similar-sized inlets if they were complete enough to measure, the most parsimonious explanation of their pubic lengthening is probably minor differences in posture or locomotion between Neanderthals and modern people rather than differences in gestation period or other obstetrical factors.

Thus the left occipital lobe tended to be larger than the right, while the right frontal lobe was larger than the left. By analogy with living people, this particular asymmetry implies that the Neanderthals were usually right-handed. Other behavioral inferences are more speculative, and the endocranial casts do not point clearly to behavioral differences, though some average differences in frontal lobe surface morphology and in parietal lobe proportions may be behaviorally significant. One crucial question that is not yet answerable based on the endocranial casts concerns Neanderthal language ability, and this must be inferred from other anatomical regions. On the one hand, the relatively flat (nonflexed) cranial base suggests that Neanderthals could not make the full range of sounds that modern people can (1303). On the other hand, the large diameter of the neural (vertebral) canals implies refined nervous control over phonetically significant movements of the rib cage. Recall from the previous chapter that *H. ergaster* had much smaller neural canals and that this can be used to argue that *H. ergaster* did have fully articulate speech (1415).

The full functional interpretation of Neanderthal facial structure is a matter of ongoing debate (1727, 2223), but there are some tentative points of agreement. The large Neanderthal nose, far in front of the brain, may have functioned to warm cold air before it reached the lungs (553). Alternatively, it is possible that Neanderthal noses served not so much to warm incoming air as to dissipate body heat during frequent periods of heightened activity (2222).

The forward placement of Neanderthal jaws and the large size of the incisors may reflect habitual use of the anterior dentition as a tool, perhaps mostly as a clamp or vise. Such para- or nonmasticatory use is implied by the high frequency of enamel chipping and microfractures on Neanderthal incisors (1999, 2216, 2490), by nondietary microscopic striations on incisor crowns (317, 1999), and by the peculiar, rounded wear seen on the incisors of elderly individuals (1000) (fig. 6.10). Inuit, who also tend to use their front teeth to clamp, also often show similar though less extensive damage (317, 2270).

Biomechanically, the forces exerted by persistent, habitual, nonmasticatory use of the front teeth (anterior dental loading) could account in whole or in part for such well-known Neanderthal features as the long face, the well-developed supraorbital torus, and even the long, low shape of the cranium (1999). Recurrent, rather than particularly powerful, anterior biting would be implied, since Neanderthal jaws and jaw musculature were not efficiently arranged to produce large bite forces between the front teeth (63, 1173). If recurrent anterior dental loading is accepted, it could further explain the large, posteriorly projecting mandibular coronoid process, which anchored the posterior fibers of the temporalis muscle that control anterior bite force (1567), and the unique Neanderthal occipitomastoid region, which perhaps provided the insertions for muscles that stabilized the mandible and head during dental clamping (2221).

The skeletons of Neanderthal children show that many typical Neanderthal features developed very early in life (2092, 2221). These include both primitive features like great skeletal robusticity, low and thick braincases, and large jaws that Neanderthals shared with all premodern humans and also derived features like extraordinary midfacial projection and large endocranial volumes that set the Neanderthals apart. The implication is that Neanderthals differed from their fully modern successors at least partly for genetic reasons. As discussed below, the Neanderthals probably disappeared almost everywhere in a few millennia or less, and the genetic difference thus reduces the likelihood that they simply evolved into modern humans.

The Origin of the Neanderthals

As emphasized in the next chapter, the evolutionary origins and relationships of the Neanderthals have been dramatically clarified by the startling recovery of a mitochondrial DNA fragment from the humerus of the type Neanderthal specimen (from Feldhofer Cave, Germany) (1275). Comparisons show that in mitochondrial DNA, Neanderthals differed from all living humans far more than living humans differ from each other, and the implication is that the lines leading to Neanderthals and to living humans split a very long time ago. Based on the degree of genetic difference, the best available estimate for the split is between 690 and 550 ky ago. This corresponds closely to the time when many specialists think Europe was first permanently inhabited, and it is in keeping with the suggestion, outlined in the last chapter, that the Neanderthals evolved in Europe from preceding local, mid-Quaternary populations of more archaic humans (82, 1096, 2078, 2306).

On morphological grounds, European fossils that date between roughly 500 and 127 ky can be divided between (1) a presumed earlier group comprising the specimens from Petralona, Vertésszöllös, Mauer, Bilzingsleben, and Arago, some of which display some Neanderthal facial characteristics, though as a group they tend to resemble classic *H. erectus* in basic morphology, and (2) a later set from Steinheim, Reilingen, Swanscombe, Montmaurin, La Chaise, Biache, and the Sima de los Huesos at Atapuerca (Atapuerca SH), all of which share one or more unique, derived features with the Neanderthals. Excepting the Atapuerca SH sample, the number of specimens is small, but a reasonable hypothesis is for an evolutionary sequence from the earlier group to the later one to the Neanderthals. The main difficulty with this idea is probably the uncertain dating of several crucial specimens, above all those from Petralona, Arago, and Bilzingsleben, which may be younger than some of their supposed descendants. Small sample size and dating problems aside, however, some European fossils that antedate 127 or even 200 ky ago unequivocally anticipate the Neanderthals, while like-aged African and Asian fossils do not. Clearly, the implication is that the Neanderthals were an indigenous European development.

The Atapuerca SH fossils present the most compelling case for a European origin of the Neanderthals beginning at or before 300 ky ago, but together with other mid-Quaternary specimens, they also suggest that different Neanderthal features appeared at different times in different populations. In particular, the SH skulls exhibit the double-arched supraorbital torus and remarkable midfacial prognathism that mark the Neanderthals, and some possess a rudimentary version of the distinctively Neanderthal elliptical depression (suprainiac fossa) above the occipital torus (82). However, the SH braincases more closely resemble those of earlier people in the tendency for the sidewalls to slope inward or to rise vertically (versus the tendency to bulge outward in the Neanderthals) and in the absence of "occipital bunning" (the backward protrusion of the occipital that characterizes the Neanderthals). Within the SH sample, the precise expression of derived Neanderthal features and their co-occurrence with more primitive features varies from specimen to specimen, and the same variation is found in other European mid-Quaternary fossils, some of which (like Petralona) are more Neanderthal-like in the face whereas others (like Steinheim) are more Neanderthal-like in the braincase. The sum suggests that the distinctive Neanderthal craniofacial complex evolved as a set of disconnected features rather than as a functional complex. The may support the view that the development of Neanderthal morphology owed more to gene drift than to natural selection.

As discussed below, it seems likely that the west Asian Neanderthals originated in Europe, and they may have displaced a previous, more modern population in Israel at the beginning of the Last Glaciation, roughly 71 ky ago (117). The principal fossil that could contradict this is the cranial fragment from Zuttiyeh Cave, Israel, which has been dated to more than 100 ky ago. If it represents an early Neanderthal (2214, 2216), it could imply that western Asia participated in Neanderthal evolution. If, on the other hand, it represents a less specialized form of *Homo* (1956, 2305, 2307), it would be consistent with a Last Glacial Neanderthal incursion.

The Contemporaries of the Neanderthals

During the earlier part of the late Quaternary (between approximately 127 and 50–35 ky ago), when the Neanderthals occupied Europe and western Asia, no one had yet colonized north-easternmost Europe and northern Asia, but people were living more or less throughout Africa and in eastern Asia, from roughly the southern border of Russia (Siberia) southward. Archeology documents the occupation, especially in Africa, but few human fossils that have been positively dated between 127 and 50 ky ago are known from either Africa or eastern Asia. Part of the reason is that archeology began relatively late in both places, and even today many fewer archeologists work in either place than in Europe. In addition, Europe is especially rich in limestone caves that are favorable for bone preservation, whereas many African caves occur in noncalcareous rock types where bones are rapidly destroyed by acid groundwater.

It is also pertinent that the Neanderthals are well known partly because they buried some of their dead in rockshelters and caves. Underground, the bodies were protected from destruction by hyenas and other carnivores. During the Last Glaciation in Europe, frost fracturing (alternate daytime thawing and nighttime freezing of moisture trapped in the cracks of cave walls and ceilings) broke off particles that rapidly accumulated as friable deposit on the floors below, making it relatively easy for the Neanderthals to dig graves. We do not know whether their African and east Asian contemporaries also dug graves, but if they did these were probably often very shallow, because cave floor deposits were thinner in the absence of frost fracturing. It is thus possible that carnivores found it easier to exhume and eat the bodies, leaving nothing or only a few scraps of human bone.

The east Asian contemporaries of the Neanderthals may be represented by the Ngandong (Solo) fossils from Java (1772, 1864) and by the Maba, Changyang, Dali, Dingcun, Xujiayao, and Yinkou (Jinniushan) fossils from China (2514–2517), discussed in the last chapter. ESR and mass-spectrometric U-series dates on associated animal teeth suggest that the Ngandong people survived until sometime between 53 and 27 ky ago (2122), and they may thus have overlapped both the Neanderthals and fully modern humans in time, though not necessarily in space. They had skulls that differed from those of classic Javan *Homo erectus* mainly in larger size, and they can thus be subsumed within *H. erectus* as a late or terminal form (1772).

The Chinese fossils have been less precisely dated, but they are probably mainly between 250 and 100 ky old, which means they were contemporaries of the ante-Neanderthals of Europe, the Neanderthals proper, or both. As discussed in the previous chapter, as a group the Chinese fossils arguably bespeak a population that resembled much older European and African fossils assigned to "archaic" *Homo sapiens* or *Homo heidelbergensis,* but they could conceivably represent an evolved end product of classic Chinese *Homo erectus* (2514–2516). The most secure conclusion is that, together with the Ngandong skulls, they show that eastern Asia diverged significantly from western Asia, Europe, and Africa during the later stages of human evolution.

Table 6.1 lists African fossils that probably date from the Neanderthal time range, between 127 ky ago or somewhat before until 50–40 ky ago. Figures 6.17 and 6.18 show the approximate locations of the fossil sites. The table includes some fossils, particularly from Border Cave and Omo-Kibish (discussed below), whose stratigraphic provenance and dating are problematic. It deliberately excludes fully modern human cranial and postcranial bones from Kanjera, near Lake Victoria, western Kenya. L. S. B. Leakey (1322) believed the Kanjera bones were associated with animal fossils dating from the early or middle Quaternary as defined here. From the time of Leakey's first report, however, other specialists argued that the associations were spurious and that the human bones were much younger (1329, 1695). Trace-element analysis has supported the critics, and the human bones are now thought to come from shallow Holocene graves, postdating 10 ky ago (178, 1694, 1696).

The table also excludes two human footprints found on the shore of Langebaan Lagoon, about 110 km north of Cape Town, South Africa (861, 1784). The prints were made on the slope of a moist dune and then buried by additional sand that was cemented into dune rock (eolianite). Luminescence and U-series dates on the dune rock vary between 228 and 103 ky ago, but none are considered reliable, and the footprints have been tentatively placed at 117 ky ago. This is the earliest time within the Last Interglacial when the enclosing dune would have been above sea level. However, the dune could have formed much later, perhaps during the Last Glaciation, between 71 and 11 ky ago, when significantly lower sea levels encouraged sand mobilization and dune formation on what is now the lagoon edge. The footprints lack diagnostic detail, and they are less compellingly human than nine prints in dune rock at Nahoon Point, about 800 km east of Cape Town (1582). Three of the Nahoon Point prints show distinct toe and heel impressions that are lacking at Langebaan. Shell fragments and secondary carbonates in the same layer at Nahoon Point have been radiocarbon-dated to 29 ky ago (604), but this is probably a minimum age, and the Nahoon prints could be as old as or older than the Langebaan specimens. Not surprisingly, the Nahoon prints reveal feet that are unequivocally human in size and proportions.

Among African fossils with secure provenance, the most questionable dating estimates are probably those for specimens associated with Aterian artifacts at Mugharet el 'Aliya, Dar es Soltan Cave 2, Zouhrah Cave, and Smugglers' Cave in Morocco. Radiocarbon (^{14}C) dates on Aterian layers vary from greater than 40 ky ago to less than 30 ky ago, and some authorities believe the Aterian persisted until 40 ky ago or later (625, 626). However, minute amounts of undetectable younger carbon could make objects that are much older than 40 ky appear younger, and only the older, infinite Aterian ^{14}C dates are probably reliable (416, 417, 2409, 2415). Sedimentological context and faunal or floral remains often imply that the Aterian existed under remarkably moist conditions, especially in the Sahara, and there is growing evidence that northern Africa was significantly wetter mainly during interglacials (435, 746, 2409, 2415). If this is accepted, then the Aterian could largely antedate the start of the Last Glaciation about 71 ky ago. Table 6.1, however, accepts the possibility that the Aterian persisted into the early part of the Last Glaciation. This is perhaps particularly likely in western Morocco, which has provided all the Aterian-associated human fossils.

Among the fossils listed in table 6.1, the more complete skulls and jaws range from clearly archaic (especially Jebel Irhoud [fig. 6.5], Omo-Kibish 2, Eliye Springs, Smugglers' Cave, and Florisbad) to only marginally archaic (Singa, Guomde, and Ngaloba) to essentially modern (Zouhrah Cave, Dar es Soltan Cave 2 [fig. 6.19], Haua Fteah, Omo-Kibish 1, Border Cave, and Klasies River Mouth [fig. 6.20]). A stark morphological contrast between Omo-Kibish 1 and Omo-Kibish 2 (fig. 6.21) may mean that one (or both) were intrusive into the stratigraphic unit they derive from, and Omo 1 (more modern) may be much more recent. Similarly, the remarkably modern Border Cave fossils may derive from relatively recent graves dug into much older deposits. This possibility is especially likely for the adult skull (BC1) and one of the mandibles (BC2), which were found during nonscientific excavations for guano (fertilizer). It also concerns the infant's skeleton (BC3), which was found by scientists but is remarkably well preserved compared with animal bones from the same level. The contrast is striking since the animal bones were initially much more durable than the infant's skeleton. Analysis of bone mineral crystallinity has now resolved the contradiction. Fresh bone is relatively noncrystalline, but crystallinity tends to increase after burial and should be the same for bones from the same layer within a single site (1954). Crystallinity analysis shows that the Border Cave infant's bones must be significantly younger than the associated animal fossils, and

Table 6.1. The African Contemporaries of the Neanderthals

Site	Human Fossils	Age Estimate (basis)	Sources
Morocco			
Mugharet el ʿAliya (Tangiers)	A maxillary fragment and three isolated teeth	Between 127 and 40 ky ago? (Aterian artifact associations)	623, 1067, 1095
Jebel Irhoud Cave	A nearly complete skull, a skullcap, two partial mandibles, and a humerus shaft	Between 190 and 90 ky ago (ESR and associated fauna)	30, 691, 692, 924, 1095, 1099, 1100
Dar es Soltan Cave 2	The partial skull and mandible of an adult, a child's skull, and an adolescent mandible	> 30 ky ago (^{14}C); between 127 and 40 ky ago? (Aterian artifact associations)	344, 623, 625, 626, 725, 1095
Zouhrah Cave (El Harhoura)	A mandible and an isolated canine tooth	> 30 ky ago (^{14}C); between 127 and 140 ky ago? (Aterian artifact associations)	622, 623, 625
Smugglers' Cave (Grotte des Contrebandiers [El Mnasra 2] à Témara)	Occipital, parietal, and frontal fragments and the mandible of a single individual	Between 127 and 40 ky ago? (Aterian artifact associations)	623, 625, 724, 1095, 2276
Libya			
Haua Fteah (Great Cave)	Ascending rami of two mandibles	Between 130 and 50 ky ago? (Mousterian artifact associations)	1486, 1487, 1728, 2193
Sudan			
Singa	A skull	Between 170 and 150 ky ago (Th/U, ESR, and associated fauna)	151, 489, 924, 1495, 2074, 2090
Ethiopia			
Diré-Dawa (Porc-Epic)	A partial mandible	> 60 ky ago (obsidian hydration dating and associated Middle Stone Age artifacts)	487, 489, 499, 2277
Omo-Kibish	A partial skull and associated postcranial bones (Omo 1); a second partial skull (Omo 2) and fragments of a third (Omo 3)	Between 127 and 37 ky ago (^{14}C, U-series, and geologic context)	334, 405, 589, 597, 598
Kenya			
Eliye Springs (KNM-ES 11693) (West Turkana)	A skull	Between 200 and 100 ky ago? (cranial morphology)	340
Guomde (KNM-ER 3884 and ER 999) (Ileret, East Turkana)	A partial skull and a femur	Between 300 and 100 ky ago (geologic context, U-series, and cranial morphology)	341, 348, 717, 1341, 2230

Site	Human Fossils	Age Estimate (basis)	Sources
Tanzania			
Ngaloba (Laetoli 18)	A skull	120 ± 30 ky ago (U-series; geologic context; artifact and faunal associations)	596, 981, 1336, 1419, 1769
Mumba Shelter	Three isolated molars	130–109 ky ago (U-series)	334, 343, 1525
Zambia			
Mumbwa Cave	Fragments of a radius	> 40 ky ago (MSA artifacts associations)	138
South Africa			
Equus Cave	A partial mandible and ten isolated teeth	Between 71 and > 27 ky ago (^{14}C; geologic context and associated fauna)	401, 406, 905, 1248, 2354
Florisbad	Facial, frontal, and parietal fragments of a single skull and a right upper third molar	259 ± 35 ky ago (ESR, geologic context, and fauna)	355, 401, 403, 921, 1287
Border Cave	An infant's skeleton (BC3), an adult skull (BC1), two partial adult mandibles (BC2 and BC5), a right humerus shaft, a right proximal ulna, and right metatarsals IV and V (all unnumbered)	Between 90 and 50 ky ago (^{14}C, ESR, and isoleucine epimerization of ostrich eggshell)	160, 162, 404, 551, 601, 602, 920, 924, 1558, 1578, 1666, 1676, 1764, 1769, 1953
Sea Harvest	An upper premolar and a phalanx	Between 127 and > 40 ky ago (^{14}C and geologic inference)	906, 1007
Hoedjies Punt	Cranium fragments, isolated teeth, and postcranial bones	Between 300 and 71 ky ago (U-series, geologic context, associated fauna)	192, 1238, 2337
Die Kelders Cave 1	Twenty-three isolated teeth and two phalanges	Between 71 and 45 ky ago (ESR, geologic context; faunal associations)	94, 907, 2136
Klasies River Mouth	Five partial mandibles, two partial maxillae, frontonasal, zygomatic, temporal, and other cranium fragments, isolated teeth, an atlas, a lumbar vertebra, a metatarsal, and portions of a clavicle, radius, and ulna	Between 115 and 60 ky ago (ESR, geologic context; faunal associations)	339, 346, 347, 467, 606–609, 1667, 1779, 1981

Note: Age estimates are presented in thousands of years (ky) ago.

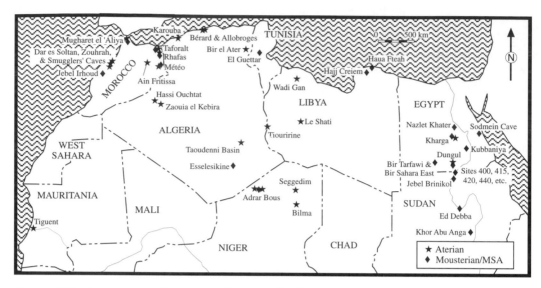

Figure 6.17. Approximate locations of key north African Aterian and Mousterian/MSA archeological sites (modified after 2415, p. 44).

it confirms great (Middle Stone Age) antiquity only for two isolated postcranial elements (the unnumbered ulna and humerus fragments of table 6.1) (1953).

Even when the Omo 1 and the Border Cave specimens are put aside, however, there are no African fossils with indisputably Neanderthal features. Some of the mandibles are large and rugged, but where the appropriate parts are preserved they lack retromolar spaces and usually have distinct chins (fig. 6.20). Together with other facial bones, the mandibles indicate that Africans who lived between 250 and 40 ky ago tended to have significantly shorter and flatter faces than did the Neanderthals. Similarly, some of the skulls (Florisbad, Jebel Irhoud, Omo 2, Singa, and Ngaloba) are ruggedly built, with large browridges and in some cases (Jebel Irhoud, Omo-Kibish 2, and Ngaloba) relatively prominent transverse occipital tori, as well as pronounced bony crests or mounds in the occipitomastoid region. (See fig. 6.5 for an illustration of Jebel Irhoud skull 1 and figs. 5.29 and 5.30 for illustrations of Singa and Florisbad.) In general, however, the African skulls tend to be shorter and higher than classic Neanderthal skulls, and some approach or equal modern skulls in basic vault shape. Where endocranial capacity can be reasonably estimated (for Guomde, Ngaloba, Omo-Kibish 1 and 2, Singa, Eliye Springs, Border Cave 1, and Jebel Irhoud 1 and 2), the African skulls range between roughly 1,370 cc (Ngaloba) and 1,510 cc (Border Cave 1) (334), comfortably within the range of both the Neanderthals and anatomically modern people.

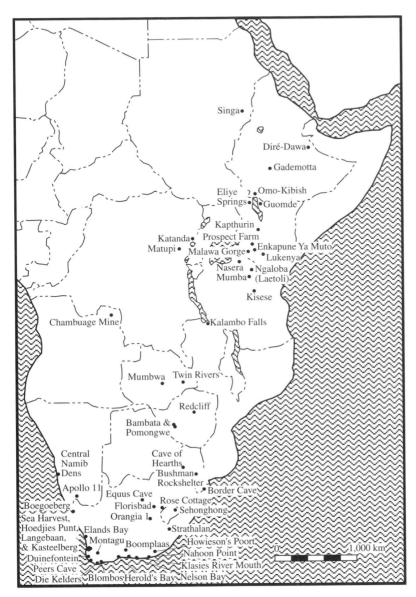

Figure 6.18. Approximate locations of the main southern and east African archeological and fossil sites mentioned in the text.

In sum, the African fossil sample is small and dominated by fragments. It is less homogeneous than the Neanderthal sample, perhaps because it represents a larger time range and geographic area, perhaps because it represents populations undergoing more rapid change, or perhaps because it (erroneously) includes some specimens that actually date from after 40 ky ago. Still, it is clear that, in age, most of the African fossils are similar to the Neanderthals and that none of them exhibit typical Neanderthal morphology. Instead they possess craniofacial features that are near-modern to modern, and as a group they strongly imply that anatomically modern people were

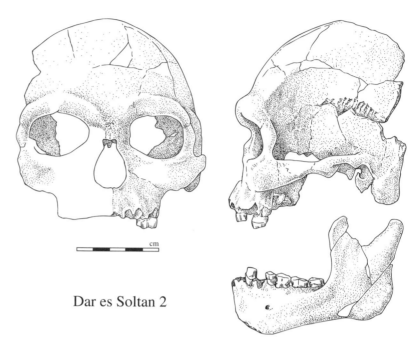

Dar es Soltan 2

Figure 6.19. Modern or near-modern human skull and mandible from the cave of Dar es Soltan 2, Morocco (redrawn after 626, p. 237). The braincase is higher (and probably shorter) than in Neanderthals, and the face is distinctly shorter and less protruding, especially along the midline. Except for the large projecting browridge, the skull exhibits no features that distinguish it significantly from most modern human skulls, yet the population it represents inhabited Morocco at the same time that Neanderthals occupied Europe.

developing in Africa at the same time that Neanderthals were the only people in Europe. The Saint-Césaire face and partial braincase (fig. 6.22) show that even the latest Neanderthals, dated to near 40 ky ago, were as distinctive and nonmodern as the earliest, dated to 71 ky ago or before.

The case for an African origin of modern human morphology becomes particularly strong if the near-modern human fossils from Qafzeh and Skhul Caves, Israel, are added to the African sample. As discussed below, TL and ESR dating bracket the Qafzeh-Skhul fossils between roughly 120 and 90 ky ago, when animal fossils suggest that Israel lay within a slightly expanded Africa (2154, 2155). As a group, the Qafzeh-Skhul skulls are highly variable in their expression of chins, vertical foreheads, rounded occipitals, parietal bossing, and other modern features, and in some important respects, such as strongly developed browridges, large teeth, and a tendency to pronounced alveolar prognathism, they tend to recall more archaic humans (335, 558, 1218, 1491, 1956, 2082, 2087, 2226, 2488). The variability and

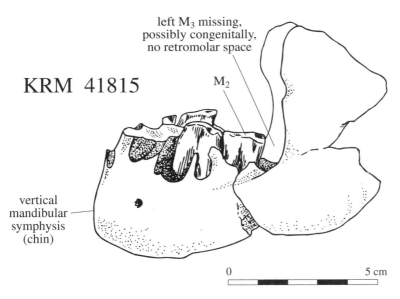

left M$_3$ missing,
possibly congenitally,
no retromolar space

KRM   41815

M$_2$

vertical
mandibular
symphysis
(chin)

0   5 cm

Figure 6.20. Left buccal view of the mandible from Klasies River Mouth Cave 1B, level 10 (specimen 41815) (drawn by Kathryn Cruz-Uribe from a slide and a photograph in 1981, fig. 68; © 1999 by Kathryn Cruz-Uribe). The specimen probably dates from the Last Interglacial sensu stricto (isotope stage 5e), between 127 and 115 ky ago. The right P$_2$ and M$_1$ and the left M$_1$ and M$_2$ are in place. The remaining teeth were not recovered, but the sockets are present for all except left M$_3$, which probably never developed. The teeth are heavily worn, indicating advanced age, and there are signs of periodontal disease, including abscesses at the tips of the M$_1$ roots. Except for the large size of the teeth, the mandible is remarkably modern, with a distinct chin and no retromolar space. The individual probably had a relatively short, broad, flat face, far more like that of modern humans than like that of the Neanderthals. Overall, together with additional human fossils from Klasies River Mouth and from other African Last Interglacial or early Last Glaciation sites, the mandible indicates that the African contemporaries of the Neanderthals had much more modern skulls.

robusticity apparent in the Skhul craniums is so great that when they were first monographically described, they were considered to represent the same population as the Neanderthal from nearby Tabun. Subsequently the Tabun and Skhul people together were sometimes lumped as "progressive Neanderthals," versus "classic Neanderthals" such as those from the caves of southwestern France. The Skhul people were clearly not Neanderthals, however, and their robusticity and variable expression of modern traits is probably better encapsulated in the term near-modern. In overall appearance, they resembled living humans far more closely than the Neanderthals did, but they (and the Qafzeh people) differed significantly not only from living

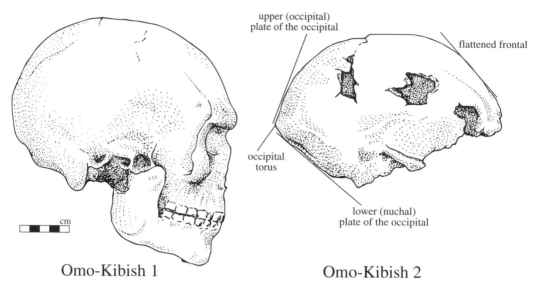

Omo-Kibish 1   Omo-Kibish 2

Figure 6.21. Skulls 1 and 2 from the Omo-Kibish locality, lower Omo River Valley, Ethiopia (drawn from photographs in 598). Skull 1 is partly reconstructed, particularly in the face. Both skulls are thought to date from the same late Quaternary period, between roughly 127 and 37 ky ago, but they contrast morphologically. Omo-Kibish 1 has a steeply rising frontal, inflated parietal regions, a rounded occipital with a relatively small nuchal plane, and other features that are unequivocally modern, whereas Omo-Kibish 2 has a flatter, more receding frontal, inwardly sloping sidewalls, a highly angulated occipital with a very extensive nuchal plane, and a strong occipital torus, all reminiscent of *Homo erectus*. However, it resembles skulls of *Homo sapiens* in its relatively small browridge and in its large internal (endocranial) capacity of about 1,400 cc. The strong contrast between the skulls is puzzling and may imply a difference in age that was not obvious when they were recovered.

humans but also from the earliest universally accepted modern (Upper Paleolithic) inhabitants of both Europe and western Asia (1086).

Both individually and as a group, the African skulls contrast strongly with Neanderthal skulls, and it is pertinent to ask whether postcranial differences were equally marked. Based on Skhul-Qafzeh, the answer is probably yes. The African fossil sample narrowly understood contains few postcranial bones, and only six have been described in detail: a proximal ulna and a proximal radius from Klasies River Mouth (467, 1667, 1779), a proximal ulna and a humerus shaft from Border Cave (1578, 1666, 1676), a femur from Guomde (East Turkana) (2230), and a humerus from Jebel Irhoud, Morocco (1095, 1100). The Klasies River Mouth radius and ulna and the Border Cave ulna broadly resemble their Neanderthal counterparts in overall morphology and robusticity, but they are no further from the modern mean than some terminal Pleistocene and Recent African counterparts. The Border Cave humerus has slightly thicker cortical

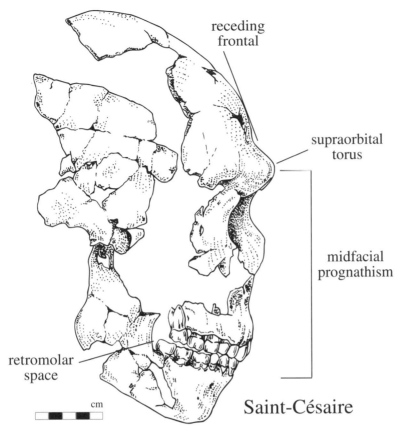

receding
frontal

supraorbital
torus

midfacial
prognathism

retromolar
space

cm

Saint-Césaire

Figure 6.22. The reconstructed face and skull of a Neanderthal associated with Châtelperronian artifacts at La Roche à Pierrot, Saint-
Césaire, France (drawn by Kathryn Cruz-Uribe from a photograph;
© 1999 by Kathryn Cruz-Uribe). The fossil is among the youngest
known Neanderthal specimens, but in all observable features, including the strong browridge, low frontal, and far forward protrusion of
the face along the midline, it is no more modern than much more
ancient Neanderthal skulls. Since it postdates modern or near-modern
human remains elsewhere, it all but precludes the possibility that
Neanderthals evolved into modern humans.

walls and a correspondingly narrower medullary (marrow) canal
than most modern humeri, but it is morphologically unremark-
able. The Guomde femur is robust but exhibits typically modern (and non-Neanderthal) morphology, including a distinct
bony ridge (pilaster) on the posterior surface and a high neck-
to-shaft angle. The Jebel Irhoud humerus has a hyperrobust
(hypertrophied) shaft like those of the Neanderthals and earlier nonmodern humans, but it probably antedates the near-
modern fossils from Klasies River Mouth, Border Cave, and
Skhul-Qafzeh.

The Skhul-Qafzeh postcraniums are unequivocal. They indicate people who were robust, particularly in their legs, but who were fully modern in form. This is particularly clear in the femur and pelvis, which differ from those of the Neanderthals in precisely the way that those of living humans do, but it is also patent in every other anatomical region for which there is evidence. If it is assumed that Neanderthal form largely reflects a reliance on jaws and bodies to accomplish everyday tasks, the Skhul-Qafzeh postcraniums might imply that modern morphology arose when Africans began to rely more on tools (or culture). However, no fundamental behavioral difference is implied by tool kits, which are basically similar between the Neanderthals and their near-modern contemporaries everywhere. Artifactual change that could signal a major behavioral difference from the Neanderthals occurs only after 50 ky ago, with the widespread appearance of fully modern humans. The implication may be that the anatomical differences between Neanderthals and near-moderns have more to do with climatic adaptation and genetic drift than with differences in behavior.

Archeological findings reported below suggest that fully modern humans originated in eastern Africa only after 50 ky ago, and they may have replaced not only the Neanderthals, but also more modern-appearing people like those at Klasies River Mouth, Dar es Soltan, and Skhul-Qafzeh. In this light, the known African and Israeli near-modern fossils are significant not because they represent the lineal ancestors of living people, but because they point to Africa as the birthplace of modern human morphology. Craniofacial similarities between the Florisbad and Omo-Kibish 2 skulls on the one hand and the older (mid-Quaternary) Broken Hill and Elandsfontein skulls on the other (1769, 1771) and between the Qafzeh and Skhul skulls and the more ancient ones from Jebel Irhoud, Morocco (1095), hint that it may one day be possible to trace the origin of modern humans through a succession of progressively less archaic African fossils.

Artifact Industries

Archeologists often lump similar assemblages of like-aged artifacts into an "industry" or "industrial complex," and the artifact assemblages the Neanderthals made are now usually assigned to the Mousterian Industrial Complex. This is named for the rockshelters at Le Moustier (southwestern France) where, in 1863, the pioneer prehistorians Edouard Lartet and Henry Christy initiated a long series of important excavations. In both

Europe and western Asia, the term Middle Paleolithic is often used as a synonym for Mousterian, though it has also been used more loosely for any artifact assemblages that are supposedly coeval with Mousterian ones regardless of their typology. It is in this relatively loose time sense that it has been applied in eastern Asia, where in general the artifact assemblages it refers to are still poorly known and poorly dated (1635, 1719, 1720).

In Africa the contemporaries of the Neanderthals produced artifacts that are very similar to Mousterian ones, and the term Mousterian has been applied directly to some north African assemblages, particularly in the Nile Valley (1442, 2298), the eastern Sahara (511, 2415–2417, 2420), Cyrenaica (northern Libya) (1486), and the Maghreb (northwestern Africa) (113, 416, 2422, 2424). Many other north African assemblages have been assigned to the Aterian Industry, named after the site of Bir el Ater in northeastern Algeria. The Aterian is distinguished from the Mousterian primarily by the presence of stemmed or tanged pieces (271, 625, 726), but this difference is no greater than the difference between European industries assigned to different facies (variants) of the Mousterian, and the separation of the Aterian from the Mousterian owes less to its typology than to the now-abandoned idea that the Aterian postdated the European Mousterian. In fact, like the Mousterian, the Aterian is now known to date from the earlier part of the late Quaternary, at or beyond the 30–40 ky practical limit of ^{14}C dating (416, 417, 2415).

In northern Africa, as in Europe and western Asia, the Mousterian (including the Aterian) was replaced by the Upper Paleolithic industrial complex. The Maghreb and the Sahara were sparsely populated or abandoned when the Upper Paleolithic appeared, but ^{14}C dates from the Haua Fteah and Ed Dabba Cave in Cyrenaica, Libya (1487), and Nazlet Khater in the Egyptian Nile Valley (2314–2316) show that the Upper Paleolithic had appeared by 40–35 ky ago.

In sub-Saharan Africa, the complex of industries that are contemporaneous with the Mousterian is conventionally assigned to the Middle Stone Age, or MSA. Like the Aterian, the MSA was once thought to postdate the Mousterian, but numerous ^{14}C dates, obtained mainly since 1970, show it was broadly coeval with the Mousterian in the early part of the late Quaternary, before 40–30 ky ago (489, 2165, 2332, 2339). Also like the Aterian, on strictly typological grounds the various MSA industries could be regarded simply as facies of the Mousterian. Scholarly tradition and geographic distance are the principal reasons for the distinction.

The time when the Mousterian/MSA began is not firmly established, partly because it lies in the interval between 250 and 127 ky ago that is currently difficult to date radiometrically. There is also the problem that in both Africa and Europe the Mousterian/MSA differs from the preceding Acheulean mainly in the absence of large bifacial tools (hand axes and cleavers). However, people who made large bifaces need not have left them at every site they visited, and large bifaces do not occur at several African and European sites that are clearly contemporaneous with Acheulean sites in the same regions. In addition, at some sites where large bifaces do occur, they are rare, suggesting that chance (sampling error) could explain their absence elsewhere. Still, although the difficulty of distinguishing the MSA from the Acheulean remains, radiopotassium dates on volcanic ash that overlies probable MSA assemblages at Gademotta, Ethiopia (489, 2410), and at Malawa Gorge, Kenya (696); U-series dating of flowstone from the MSA sequence at Twin Rivers Shelter, Zambia (139); optically stimulated luminescence dating of sediments at the Florisbad MSA site, South Africa (921, 1289); and the stratigraphic or paleontologic context of probable MSA artifacts at various sites in South Africa (2165, 2339) suggest the MSA began in the late middle Quaternary, 250–200 ky ago.

The situation in Europe, western Asia, and northern Africa may be more complex, since Mousterian industries may have replaced the Acheulean at different times in different places. However, there is growing evidence that the Mousterian appeared in most places by 250–200 ky ago, during the Penultimate Glaciation or the preceding interglacial (119, 123, 288, 289, 1809, 2240–2242, 2244, 2415–2417).

The various Mousterian/MSA industrial sequences almost certainly ended at different times, depending on the place. In sub-Saharan Africa they may have been replaced between 50 and 45 ky ago (31), in northern Africa and western Asia between 45 and 40 ky ago (125, 1042, 1444, 1486, 1548, 1888, 2014), in eastern Europe shortly before 40 ky ago (1264, 1538, 1539), and in western Europe about 40 ky ago (229, 232, 408, 2061, 2066). In far western Europe (southern Spain) some Mousterians may have persisted until as recently as 30 ky ago (1097).

Although it is heuristically useful to equate the Neanderthals and the Mousterian, it is important to stress that, strictly considered, this is incorrect. The people who made Mousterian/MSA artifacts in Africa were clearly not Neanderthals, and at Saint-Césaire and Arcy-sur-Cure (Grotte du Renne) in France, Neanderthal fossils are associated with early Upper Paleolithic artifacts assigned to the Châtelperronian Industry. Conceivably the association implies that fully modern Châtelperronians sometimes accumulated Neanderthal bones (272), but the Châtelperronian is Mousterian-like in some respects, and most authorities therefore believe that the Saint-Césaire and Arcy Neanderthals represent its makers. Finally, and perhaps most important, the people who made Mousterian artifacts at the Skhul and Qafzeh Caves in Israel were near-modern people. The significance of these departures from the Neanderthal = Mousterian formula will be considered in the section below on the fate of the Neanderthals.

Mousterian/MSA Stone Artifact Technology

Mousterian/MSA people were consummate stone knappers who used a variety of sophisticated techniques to produce their tools. Beginning with the great French prehistorian François Bordes (274, 277), other, mostly French archeologists (249–252, 796, 2261) have illuminated Mousterian technology by analyzing the *chaîne opératoire* (1279, 1539, 1880, 1908). This term can be translated loosely as the stone reduction sequence, or the sequence of actions between the selection of raw material for flaking and the abandonment of exhausted tools. The intermediate steps comprise the primary production of sharp-edged blanks or preforms and their secondary modification, use, or maintenance (refreshing). Broadly speaking, then, the *chaîne opératoire* refers to the entire process of manufacture and use, and the varied *chaînes* employed by Paleolithic people can be reconstructed by experimental replication or by refitting excavated pieces (fig. 6.23). The focus here is on Mousterian/MSA reduction of raw stone nodules or blocks to sharp-edged blanks. Blank modification and use are discussed in the next section on Mousterian/MSA retouched tool typology.

Even the earliest stone tool makers preferentially selected the most desirable rock types at their disposal, and Mousterian/ MSA people were particularly discerning. Generally speaking, the best rock types are relatively hard, fracture easily when struck, and have a smooth, homogeneous internal consistency. Among widespread rock types that best meet these criteria are very fine-grained siliceous varieties such as flint and chert. Flint is common throughout Europe and southwestern Asia (the Near East), where it was used so often that the local Stone Age could almost be called the "flint age." In Africa, where large nodules of flint or related rocks are uncommon in most regions, Stone Age people were often compelled to use other

Figure 6.23. Partial
reconstitution of a
flat, discoidal core
by the refitting of
162 separately
excavated flakes
from Mousterian
Site C, Maastricht-
Belvédère, the
Netherlands
(redrawn after
1809, fig. 7).

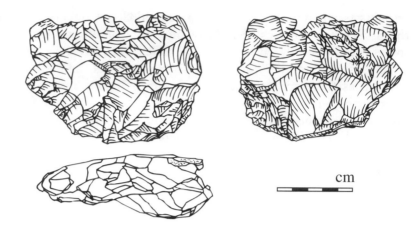

cm

materials, including volcanic rocks and quartzites. Some fine-grained quartzites approach flint in quality, but among volcanic rocks only one—obsidian (volcanic glass)—is truly comparable. It is limited to parts of eastern Africa, where some MSA people used it intensively. Other more widespread volcanic rocks, such as basalt, are generally more difficult to work, and they may therefore yield cruder-looking artifacts.

The reduction process begins when the knapper strikes a *percussor* against a raw stone nodule in order to remove a flake. The percussor can be made of stone or of softer material like wood, antler, or bone. Replication experiments suggest that Mousterian/MSA and earlier peoples mainly used stone, and their percussors are commonly called *hammerstones.* A nodule from which one or more flakes has been struck is called a *core,* and the resultant flakes are commonly said to have three main parts: the striking platform or butt, the ventral surface, and the dorsal surface. The *striking platform* is the part of the flake that was struck by the hammerstone when the flake was detached from the core. Part of the struck area remains on the core, where it forms the striking platform of the core. A given core can have many striking platforms, but a flake will have only one (or very rarely two, formed when two blows hit a core at the same time; this can happen when a struck core bounces against another hard object).

By convention, unless otherwise noted, flakes are illustrated with the striking platform down. The *ventral surface* is the one that was originally inside the core; the *dorsal surface* is the one that was outside. The ventral surface is usually smooth, though it always has a variably pronounced bulge or *bulb of percussion* immediately adjacent to the striking platform. This results from the rapid dissipation of the hammer blow through the interior of the core. Unlike the ventral surface, the dorsal one always exhibits the weathering rind (cortex) of the unflaked core, hollows and ridges (scars) from previous flake removals, or both.

Once a knapper has removed one or more flakes from a core, the negative scars (or facets) left on the core can serve as striking platforms for additional flakes, and the process can be repeated until the core becomes too small for further reduction. A core that has been worked to this extent is said to be exhausted.

Modern replication experiments show that Mousterian/MSA knappers often shaped a core or modified its striking platform to predetermine the size or shape of flakes. Platform modification was commonly done to remove irregularities that would diffuse or misdirect the force of the hammer blow. It results in flakes that have prepared or faceted striking platforms (or butts), as opposed to ones with unprepared or smooth platforms (or butts). A core that was extensively shaped to determine flake size and shape is called a *Levallois core*, after a site in the Levallois suburb of Paris where Paleolithic examples have been known since the nineteenth century. The corresponding flakes are called *Levallois flakes.* In general, Mousterian/MSA knappers selected flattish nodules or blocks for Levallois reduction (fig. 6.24). They first struck downward to remove flakes around the entire periphery of the nodule and then used the peripheral flake scars to strike inward, removing flakes systematically from one surface of the core. When this surface was completely prepared, they again struck inward to remove one or more flakes whose shape and size was determined by the arrangement of previous flake scars on the core surface.

Levallois flakes usually have faceted platforms, but the defining characteristic is the pattern of dorsal scars reflecting deliberate preparation of the core surface. Well-made Levallois flakes and cores appear in some later Acheulean assemblages, dating between perhaps 400 and 200 ky ago, but they become common only in Mousterian and MSA assemblages after 200 ky ago. Not all Mousterian/MSA people produced Levallois flakes, but many did, and Levallois technology is sometimes regarded as a hallmark of the Mousterian/MSA.

By replication and refitting, Boëda (251) has shown that Mousterian knappers actually employed a range of Levallois reduction techniques, variably designed to produce a single large flake (fig. 6.24), multiple smaller flakes, flakes that were naturally pointed at one end, or flakes that were exceptionally long (fig. 6.25). By convention, flake length is measured along a line

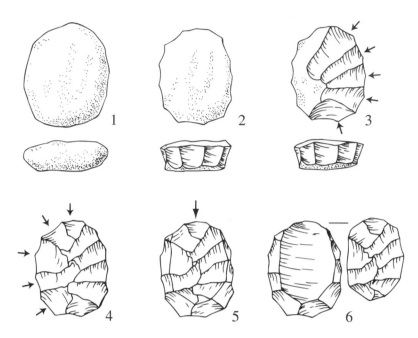

Figure 6.24. Stages in the manufacture of a classic Levallois core (redrawn after 276, fig. 4): (1) raw nodule; (2) nodule with flakes struck off around the periphery; (3) nodule with flakes struck radially inward on one surface by using the peripheral scars as striking platforms; (4) radial preparation completed; (5) final hammer blow (indicated by the arrow); and (6) final Levallois core and flake.

that bisects the butt (or striking platform), and breadth is the maximum dimension perpendicular to this line. Flakes that are longer than they are wide are sometimes called *flake blades*, and ones that are at least twice as long as wide are called *blades*. Any extensive knapping session will produce some blades, but routine blade production requires special expertise in core preparation.

Blades are usually considered a marker of the Upper Paleolithic culture complex that succeeded the Mousterian in Europe, western Asia, and northern Africa, but some Mousterian and MSA people produced numerous blades during the Last Interglacial or even before. As discussed below, blades or flake blades mark the opening phase of the Levalloiso-Mousterian complex in the Levant (the eastern Mediterranean coast and its hinterland) (123). An even earlier Levantine industry has so many blades that it has been dubbed the "Pre-Aurignacian" for its rough resemblance to the very early Upper Paleolithic, Aurignacian Industry. Blades or flake blades also characterize many sub-Saharan MSA industries (2165), and well-made blades are known even in late Acheulean context at Kapthurin near Lake Baringo, Kenya (1485). Blades tend to be rarer in the

UNIPOLAR RECURRENT LEVALLOIS BLADE PRODUCTION

BIPOLAR RECURRENT LEVALLOIS BLADE PRODUCTION

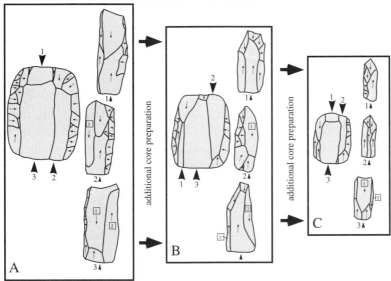

Figure 6.25. Two variants of the Levallois technique for producing elongated flakes or blades (redrawn after 249, fig. 4). Each method involves prior preparation of the core, indicated by small arrows in the flake preparation scars. Each can produce multiple blades, in the first case (the "unipolar" method) from a single striking platform and in the second (the "bipolar" method) from two opposed striking platforms. Continuing blade production requires continuing core preparation, but the process will eventually terminate when the core becomes too small for further working.

European Mousterian, but they are common at some sites (1265), including Tönchesberg and Rheindahlen, Germany, Rocourt, Belgium, Seclin, France, and other northwestern European sites that date to the end of the Last Interglacial or the early part of the Last Glaciation (33, 530, 531, 1640, 2244). In general, Mousterian/MSA people produced blades from one face of specially prepared Levallois cores. Like many Upper Paleolithic people, however, some Mousterian knappers may have struck blades from the top of a core around its entire periphery. Such a core comes to look like a prism, and prismatic blade production is sometimes taken as the pinnacle of knapping refinement, since it maximizes the amount of cutting edge a single core can provide.

As discussed below, many Mousterian/MSA people used Levallois technology almost exclusively, while others used it more rarely or not at all. The reasons for this variability are not well understood, but the Levallois technique generally requires large nodules of high-quality raw material, and it was used most commonly in regions were suitable nodules were extremely abundant, such as parts of northern and southwestern France (658, 1539, 1819), Israel (118, 1527), and the Nile Valley (2298).

A priori, it might seem that Mousterian/MSA people who used non-Levallois technology were less sophisticated than their Levallois contemporaries, but this is not necessarily true. Analysis of the non-Levallois technology employed by the so-called Quina Mousterians of France shows that they reduced elongated flint cores systematically the way a picnicker would slice through a sausage (2261) (fig. 6.26). When they could, they selected appropriately shaped tubular nodules, but when they had to, they flaked nodules or blocks of other shapes into tubular form. Their principal goal was to produce a series of flakes with roughly triangular cross sections on which a sharp edge was opposed by a very thick, dull one. The dull edge could have served as a finger rest during cutting or scraping, or it might have made hafting easier, but whatever its purpose, Quina people showed great skill in producing flakes of the desired form. A distinct advantage of Quina-type reduction is that it generates relatively little waste (*débitage*), that is, flakes that are not intended for use. Experiments show that roughly 60% to 75% of Quina flakes were usable fresh from the core, whereas Levallois reduction generally produces less than 30% immediately useful flakes.

The bottom line is that Mousterian/MSA people not only had deep insight into the mechanics of stone flaking but routinely performed lengthy and complex operations to transform raw stone nodules into desirable flakes and blades. A primary conclusion of this chapter is that they were more primitive than their Upper Paleolithic/Later Stone Age successors in many important behavioral respects, but their primitiveness clearly did not extend to the primary working of stone. In this

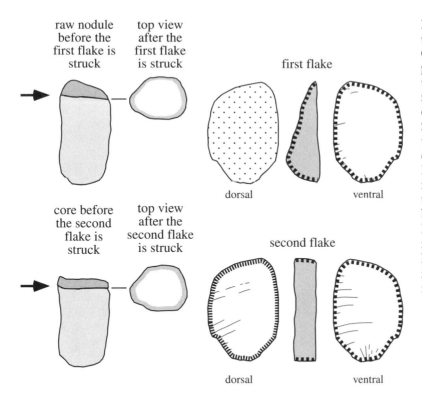

raw nodule before the first flake is struck

top view after the first flake is struck

first flake

dorsal

ventral

core before the second flake is struck

top view after the second flake is struck

second flake

dorsal

ventral

Figure 6.26. The technique that Quina Mousterian people often used to "slice" flakes from an elongated core (redrawn after 2261, fig. 6.5). The technique is less complex than the Levallois method illustrated in figures 6.24 and 6.25, but it is also much less wasteful, since it produces a very high percentage of immediately useful flakes.

respect they were as human as anyone, and they have never been surpassed.

Mousterian/MSA Retouched Tool Typology

Edge damage and use wear show that many Mousterian/MSA flakes and blades were put to immediate use without further working. However, some flakes and blades were further reduced or modified by removing small flakes or chips from one or more edges. Such modification is called *retouch,* and the altered edges are said to be retouched. In most instances retouch chips were struck from the ventral surface, and the chip scars appear on the dorsal surface. Retouching blunts fresh edges, and it was thus done either to modify the shape of an edge, to give it greater stability, or to resharpen it after it had been dulled by use.

Beginning with Gabriel de Mortillet (600) and others in the previous century, European archeologists have emphasized the position and quality of retouch to define different Mousterian tool types, and François Bordes (277) employed these criteria to formalize an influential Mousterian/MSA typology (fig. 6.27). He recognized a total of sixty-three discrete flake tool types, but his scheme centered on twenty-one types of sidescrapers (or

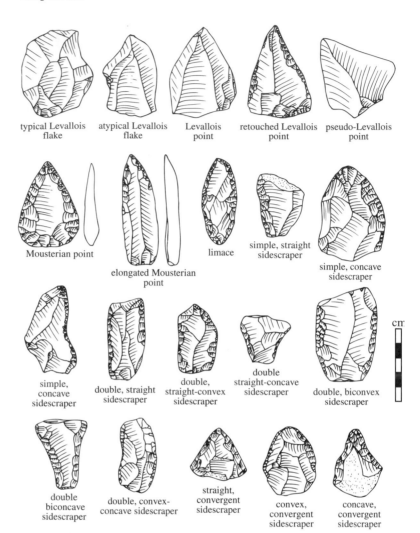

Figure 6.27. Basic Mousterian and MSA stone tool types (as defined by 277). Most assemblages are dominated by scrapers, points, and denticulates. The distinctions among the types are strictly formal rather than functional, and use wear studies show that there is no

racloirs, the French term, sometimes used in English), two or three types of retouched points, and three or four types of notched or denticulated pieces. He stressed these types because the underlying sidescraper, point, and denticulate classes dominate almost all Mousterian assemblages. Hand axes like those that mark the preceding Acheulean Industrial Complex tend to be rare in Mousterian/MSA sites, though small, usually triangular or cordiform (heart-shaped) ones characterize some Mousterian assemblages, mainly in France, and small pointed, bifacially retouched pieces that fit the technical definition of

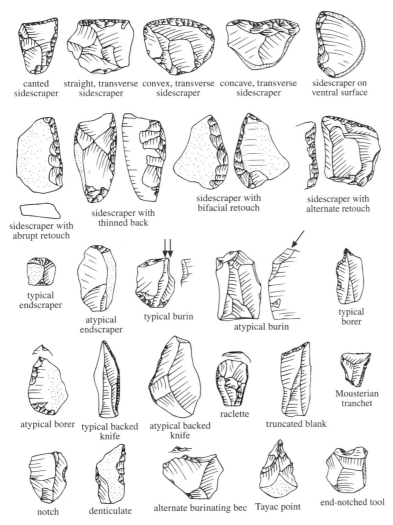

canted sidescraper; straight, transverse sidescraper; convex, transverse sidescraper; concave, transverse sidescraper; sidescraper on ventral surface; sidescraper with abrupt retouch; sidescraper with thinned back; sidescraper with bifacial retouch; sidescraper with alternate retouch; typical endscraper; atypical endscraper; typical burin; atypical burin; typical borer; atypical borer; typical backed knife; atypical backed knife; raclette; truncated blank; Mousterian tranchet; notch; denticulate; alternate burinating bec; Tayac point; end-notched tool

strong relation between type and function. Thus pieces assigned to the same type often seem to have been used for different purposes, while different types often appear to have served broadly the same purpose.

hand axes occur sporadically elsewhere, especially in central Europe. Endscrapers, burins, borers, backed pieces (flakes or blades with one edge intentionally dulled), and other so-called Upper Paleolithic/Later Stone Age (LSA) tool types also tend to be rare in Mousterian/MSA sites, and where they occur they could have been produced unintentionally, since they are usually atypical or poorly made. Numerous endscrapers and burins do mark some west Asian Mousterian assemblages, and well-made backed pieces typify localized Mousterian/MSA variants in both Africa and Europe, but most research on Mousterian/MSA retouched tool form and function has understandably focused on the far more common and widespread sidescrapers, points, and denticulates.

Following long-standing tradition, Bordes defined a *side-scraper* as a flake on which one or more edges bear smooth, continuous retouch. He distinguished different sidescraper types depending on how many edges were retouched, their shape (convex, concave, or straight) or combination of shapes, their position with respect to the line bisecting the bulb of percussion, and so forth. He defined a *point* as a flake on which two continuously retouched edges converged directly opposite the striking platform (or butt). To qualify, the flake also had to be relatively thin; a thick, pointed flake became a "convergent sidescraper." Finally, he defined a *denticulate* as a flake that was retouched to produce a ragged or serrate edge, comprising several adjacent indentations. He designated a denticulate with only a single indentation as a "notch."

Bordes believed that Mousterian people recognized essentially the same types or subtypes of sidescrapers, points, and denticulates as he did. However, this conclusion has been questioned, in part because the types often grade into one another, and sorting retouched Mousterian/MSA pieces among types often requires great tolerance for ambiguity and arbitrariness. The problem may be partly that some types or subtypes represent stages in the resharpening or refreshing of a very small number of basic forms (655, 656, 658, 1817, 1819). Figure 6.28, for example, illustrates how reduction or rejuvenation could transform a relatively simple convex sidescraper (on which a single convex retouched edge is aligned perpendicular to the butt) into either a typical "convergent" form (on which two retouched edges converge to a point) or a classic "transverse" form (on which a single retouched edge lies opposite the butt).

The logic behind figure 6.28 is appealing, because resharpening undoubtedly occurred, and it could have altered sidescraper form in precisely the way the figure suggests. Indisputable rejuvenation flakes that removed the dulled edge of a retouched tool have been identified at Combe-Grenal and La Micoque, southwestern France, at La Cotte de St. Brelade, Jersey, and at other Mousterian sites (1539). Additionally, retouched flakes tend to be especially abundant in regions where high-quality raw material is relatively rare and rejuvenation of used flakes might have been essential. But rejuvenation can explain only a portion of Mousterian retouched tool variability (1539). In the first place, it fails to account for sites where sidescrapers or other reduced forms are very numerous even though high-quality raw material for producing fresh flakes was readily

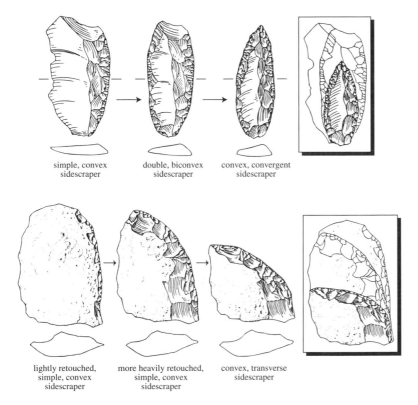

simple, convex
sidescraper

double, biconvex
sidescraper

convex, convergent
sidescraper

lightly retouched,
simple, convex
sidescraper

more heavily retouched,
simple, convex
sidescraper

convex, transverse
sidescraper

Figure 6.28. The way progressive reduction can transform a simple,
convex sidescraper into a double, biconvex form, and finally into
a convex, convergent form (*top*) or into a more heavily retouched
simple, convex form and finally into a convex, transverse form (*bottom*) (redrawn after 656, figs. 1 and 2). The implication is that some
frequency variation in sidescraper types may reflect only the extent
to which Mousterians at different sites refreshed or resharpened used
scrapers.

available. It also neglects the existence of sidescraper-rich assemblages where the sidescrapers are relatively large when they
should be small, assuming that an abundance of sidescrapers
implies heavy reduction. Finally, it contradicts the observation
that where transverse sidescrapers are most common—in the
Quina variant of the French Mousterian (discussed below)—
they rarely resulted from the reduction of other forms but were
made on broad, short flakes that were specially struck to accommodate them (2261). Short, wide flakes were also specially
chosen for transverse sidescrapers at the Bir Tarfawi Mousterian site in the eastern Sahara, Egypt (510), and both in France
and at Bir Tarfawi, transverse sidescrapers commonly lack the
lateral, basal retouch they must have if they derive from simple,
lateral forms.

Whatever the precise reasons for variation in Mousterian/
MSA tool form, Upper Paleolithic/LSA tool types tend to be

much less variable (more standardized) (1279, 1530, 1539, 1640), and the result is that Upper Paleolithic/LSA tools are far easier to sort among a relatively large number of readily defined, discrete types. The contrast may reflect a difference in intentionality; that is, Mousterian/MSA people may have tried less often to impose a preconceived form on a raw flake or blade. Instead, unlike Upper Paleolithic/LSA people, who were clearly concerned about the overall shape of finished tools, Mousterian/ MSA people may have focused mainly on the character of an edge or the sharpness of a point. Another aspect of the difference may be that Mousterian/MSA people were mostly preoccupied with function whereas Upper Paleolithic/LSA people were more concerned with style. In addition, as discussed below, Upper Paleolithic/LSA tools may have served a wider variety of functions, including, for example, the working of bone, antler, ivory, and shell. The bottom line is that Upper Paleolithic tools not only are easier for a modern archeologist to sort, they also tend to vary much more through time and space. In sum, unlike stone flaking technology, which suggests no fundamental distinction between Mousterian/MSA people and their successors, tool typology implies a very significant behavioral difference. Arguably this reflects an underlying biological contrast in basic cognitive capacity.

Mousterian/MSA Tool Function

The names assigned to Mousterian/MSA stone artifact types suggest that their functions are known, but in general this is not the case. As outlined above, the types are defined on morphological grounds, and their functions are speculative, based on resemblances to historic tools of known function or on feasibility experiments with modern replicas. Even the conventional distinction between tools and manufacturing waste is strictly morphological. Thus, by convention, only retouched pieces have been classified as tools, even though feasibility experiments show that many unretouched pieces could have been used effectively as knives, scrapers, and so forth. Additionally, as already noted, many unretouched pieces have macroscopic edge damage or microscopic wear polishes that formed during use.

Polishes, striations, microfractures, edge rounding, and other microscopic wear traces on both retouched and unretouched artifacts can show not only that an artifact was used but also how it was used (to cut, scrape, saw, etc.) and on what substance (wood, bone, hide, etc.) (878, 1166, 1195, 1196, 1913, 1929, 2295). Ancient use is inferred from similarities to wear patterns developed on modern replicas used in known ways.

Microwear analysis has revealed the functions of many individual prehistoric tools, but it has some important limitations: it can be applied only to certain very fine-grained rock types, it is very time consuming, and it is only moderately accurate (1166, 1581). It has not led to a reformulation of artifact typology along functional (vs. morphological) lines, nor is it likely to. This is partly because of its inherent practical and technical constraints and partly because of the loose relation between form and function.

Analyses of French Mousterian tools illustrate both the potential and the limitations of the microwear method. Microwear suggests that, as among the three major Mousterian tool classes, only notches and denticulates were used mainly in just one way on just one material (37, 206, 207). The use was to shave, plane, or whittle, and the material was wood. Most notches and denticulates are ideally suited to woodworking not only by their form, but also by their tendency to have steeply retouched edges. Sidescrapers are more variable in both form and edge angle, and it is therefore perhaps not surprising that they were used more variously, not just to scrape or plane but also to cut or slice. They were applied mostly to wood, followed by flesh, bone, and hide. Points (including convergent sidescrapers) were likewise used mostly on wood and less often on bone or hide, and they differ from common sidescrapers mainly in stronger evidence that some were hafted. The indications are polishes and striations not on the edges or the tip, but on the ventral and dorsal surfaces below the tip, as if they formed by friction against a loosely attached wooden handle (fig. 6.29). So far there is no evidence that European Mousterians hafted stone points on wooden spear shafts.

Mousterian artifacts from Kebara and Qafzeh Caves, Israel, conform broadly to the French pattern (1927, 1928). Most edge-worn Kebara and Qafzeh specimens were apparently used on wood, while smaller numbers were applied variously to flesh, hide, bone, and soft plant material. As in the French case, there was no obvious correspondence between artifact form and use mode. Unlike French points, however, some of the Israeli ones had symmetrically placed wear on their lower lateral edges (fig. 6.30), suggesting they were bound to wooden spear shafts. The use of points as armatures could also explain the scars from hinge flakes or step flakes on the tips of some Israeli and Jordanian examples (1927–1929). Experiments show that such scars often result when a point strikes bone or another hard object, including, for example, a rock near the desired target. Similar "impact fractures" have been observed on Mousterian points from La Cotte de St. Brelade, Jersey (412).

Figure 6.29.
Convergent side-
scrapers from the
Mousterian site of
Biache-Saint-Vaast,
northern France
(after 207, p. 220).
Friction with wood
or hide produced
polish on roughly
the lower two-
thirds of each piece,
suggesting that
each was mounted.

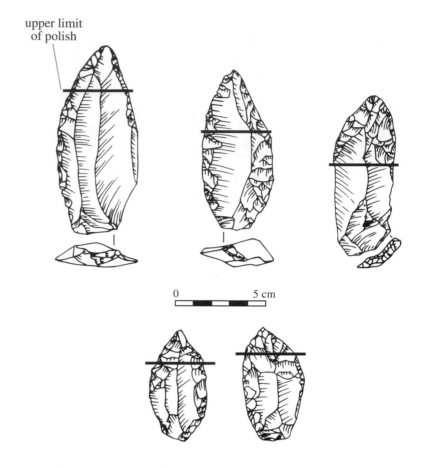

upper limit
of polish

0   5 cm

The use wear studies support three important generalizations. First, to the extent that worked material can be inferred,
Mousterian tools were used primarily on wood. In contrast,
Upper Paleolithic tools were used far more commonly on bone
and flesh and especially on hide (1166). The rarity of evidence
for bone working in the Mousterian is in keeping with the rarity or absence of formal bone, antler, and ivory artifacts, which
become common only in the Upper Paleolithic. Second, compared with Upper Paleolithic people, Mousterians much more
rarely mounted stone artifacts on wooden or bone handles, and
only Upper Paleolithic people regularly and unequivocally used
stone points or other retouched pieces as projectile armatures.
The contrast with the Mousterian is stark (1279), especially
since Upper Paleolithic people were also the first to use bone
armatures. The difference surely implies that Mousterian and
Upper Paleolithic people differed fundamentally in hunting
strategy and, more important, in hunting success. Finally, the
absence of a strong one-to-one association between Mousterian tool form and function underscores the arbitrary nature of
the conventional typology. In contrast, overall tool form and

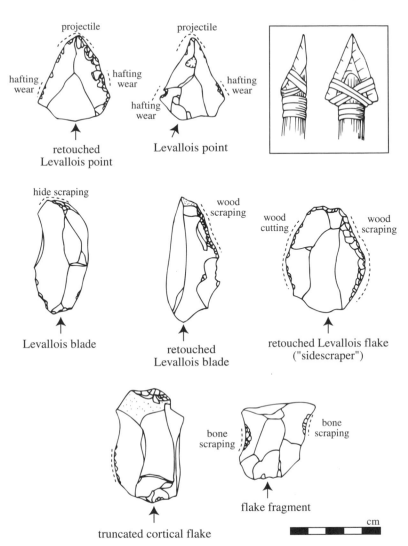

Figure 6.30. Utilized Levalloiso-Mousterian artifacts from Kebara Cave, Israel (redrawn after 1927, p. 447, and 1928, p. 615). Use wear studies indicate that like their European contemporaries, the Kebara Mousterians used their tools mainly on wood, though they also sometimes worked hide, bone, and other materials. Unlike any known European Mousterians, they may have also mounted points on wooden spear shafts, perhaps with leather thongs.

function appear to have been much more closely linked in the Upper Paleolithic (1166). The difference once again suggests that form, as manifested in the eye of the modern archeologist, was much less important to Mousterian toolmakers.

Mousterian/MSA Interassemblage Variability

As a whole, the Mousterian/MSA is remarkably uniform though time and space, and much superficial variability between assemblages from different regions or time periods can probably be ascribed to differences in the size or flaking quality of available stone raw materials (658, 1819). Flint of variable quality was available throughout much of Europe and western Asia but was generally absent in Africa, where quartzite and even volcanic rock types were widely used. Among the few tool types whose localized distribution cannot be explained by differences in raw material are the triangular and cordiform hand axes mentioned above (largely restricted to France, with extensions to Britain, Belgium, and Germany); small cleavers on flakes (cleaver flakes), found mainly in northern Spain; Bockstein knives (small, pointed, hand ax–like objects), found mainly in Germany and adjacent parts of central Europe; bifacially flaked, leaf-shaped points (*Blattspitzen*), restricted mostly to central Europe; backed knives, common in one of the Mousterian variants of southern France and, independently, in an MSA variant in South Africa; and stemmed or tanged pieces (points, sidescrapers, etc.), restricted essentially to the western two-thirds of northern Africa. Except for these types and some others that are less strictly localized, most variability within the Mousterian/ MSA is quantitative, that is, it consists of differences in the percentage representation of the same widespread types of sidescrapers, points, and denticulates.

Interassemblage Variability in Europe

Typological variability among Mousterian assemblages has been studied most thoroughly in Europe, above all in France, thanks to the descriptive artifact typology and analytic methodology developed by François Bordes. Working first with data from such crucial French caves as Le Moustier, La Ferrassie, La Quina, Combe-Capelle, and Pech de l'Azé and later with material from his own meticulous thirteen-year excavations at Combe-Grenal, Bordes recognized four basic Mousterian variants or facies (275, 276, 278, 279):

1. *Mousterian of Acheulean Tradition (MAT).* Characterized by numerous triangular or cordiform hand axes in addition to the sidescrapers, denticulates, and points that are ubiquitous in the Mousterian. In an earlier phase (MAT Type A), backed knives occur but are relatively rare. In a later phase (MAT Type B) they are much more numerous, essentially replacing small hand axes as the type implement of the MAT.
2. *Typical Mousterian.* Characterized mainly by sidescrapers, with some denticulates and points. Hand axes and backed knives are rare or absent.
3. *Denticulate Mousterian.* Characterized by the dominance of denticulates and notches, with much smaller numbers of sidescrapers, points, and such. Hand axes and backed knives are absent.
4. *Quina-Ferrassie Mousterian (or Charentian).* Characterized by a high proportion of sidescrapers; distinguished from the

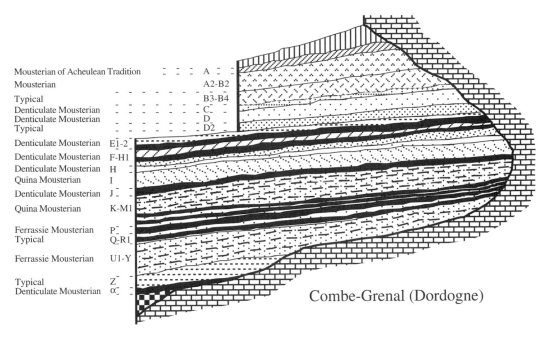

Mousterian of Acheulean Tradition   A
Mousterian   A2-B2
Typical   B3-B4
Denticulate Mousterian   C
Denticulate Mousterian   D
Typical   D2
Denticulate Mousterian   E1-2
Denticulate Mousterian   F-H1
Denticulate Mousterian   H
Quina Mousterian   I
Denticulate Mousterian   J
Quina Mousterian   K-M1
Ferrassie Mousterian   P
Typical   Q-R1
Ferrassie Mousterian   U1-Y
Typical   Z
Denticulate Mousterian   α

Combe-Grenal (Dordogne)

Figure 6.31. Schematic section of Combe-Grenal Cave, southwestern France, showing the partially random interstratification of Mousterian facies (modified after 276, fig. 6). The apparently random pattern has been used to argue that the facies were produced by ethnically distinct Mousterian groups who occupied France side by side for thousands of years.

Typical Mousterian by abundant sidescrapers with distinctive steep, scalar retouch in which successive ranks of flake scars overlap like scales on a fish. It is also differentiated by the relative abundance of limaces—slug-shaped pieces that could be described as double convergent sidescrapers.

Within each variant, Bordes further distinguished between assemblages with many Levallois flakes and ones with few. He assigned Levallois-rich assemblages to the appropriate Levallois facies (for example, Typical Mousterian of Levallois Facies), except within the Quina-Ferrassie variant, where he assigned them to the Ferrassie subvariant, as opposed to the Quina subvariant, in which Levallois flakes are rare.

Apart from the Mousterian of Acheulean Tradition B, which was clearly later than Type A, Bordes argued that there was no chronological order to the facies and that any one of them could overlay any other within a site (fig. 6.31). He thought this was because the facies were produced by separate Mousterian tribes who moved from site to site, randomly replacing each other.

In Bordes's view, the separate tribes coexisted in France for tens of thousands of years, until one or perhaps two evolved into succeeding Upper Paleolithic tribes. He recognized that such prolonged social separation might produce discrete physical types, but he was never able to demonstrate this. Diagnostic human remains are still unknown from the Mousterian of Acheulean Tradition and from the Denticulate Mousterian, and fossils with the Typical Mousterian (at Le Moustier) and the Quina-Ferrassie Mousterian (at La Quina, La Chapelle-aux-Saints, La Ferrassie, Le Regourdou, Spy, and other sites) represent typical Neanderthals.

Perhaps the most serious problem with Bordes's tribal hypothesis is to imagine the social or cultural mechanisms that kept the tribes separate for tens of thousands of years. Another difficulty is that Bordes may have underestimated the amount of typological variability that is linked to time. Thus, when some questionably excavated or analyzed occurrences are ignored, there is a consistent sequence from Ferrassie Mousterian to Quina Mousterian to Mousterian of Acheulean Tradition within key French caves (1533, 1534, 1537, 1539). The principal obstacle to this chronological succession has been the sedimentologic/climatologic correlation of the sequences from Le Moustier (lower shelter) and Combe-Grenal, from which it appeared that the Mousterian of Acheulean Tradition at Le Moustier was as old as or older than the Ferrassie and Quina Mousterian at Combe-Grenal. However, TL dating of burned flints and ESR dating of animal teeth from Le Moustier (924, 2272), combined with climatic correlation of the Combe-Grenal sequence to the global oxygen-isotope stratigraphy (1312), now suggests that the Mousterian sequence at Combe-Grenal largely antedates that at Le Moustier (fig. 6.32), and the Ferrassie to Quina to Mousterian of Acheulean Tradition succession is thus supported (1537, 1539). Moreover, some Mousterian assemblages excavated after Bordes first defined the facies clearly fall between these variants, suggesting that in part the facies variation is continuous rather than discrete (762). Partially in response to this problem, Bordes (281) modified and loosened the facies definitions.

Still, no one doubts there is significant quantitative variability among broadly contemporaneous Mousterian assemblages, not just in France, but elsewhere in Europe where Bordes's typological methods have been applied—for example, in Spain (762) or in Ukraine and European Russia (1229). Unlike Bordes and many French archeologists, however, some prominent Anglo-American investigators believe the variation reflects different tasks carried out by the same people (or their descendants) at different sites or at the same site at different times, perhaps at different seasons (215, 221, 225). Thus denticulate-rich assemblages might indicate an emphasis

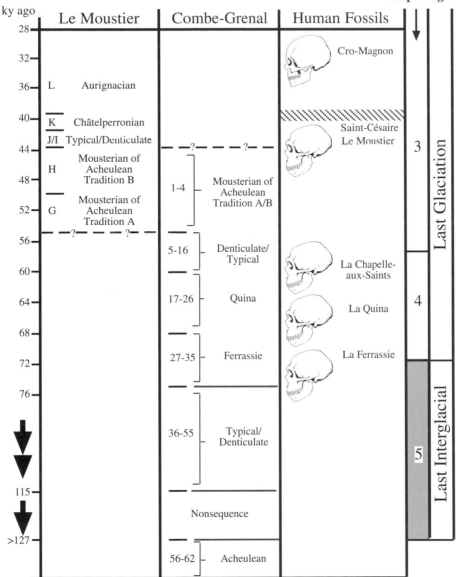

Figure 6.32. Correlation between the archeological sequences at Le Moustier and Combe-Grenal, southwestern France, based on TL dates for Le Moustier and climatic correlation of the Combe-Grenal sequence with the global marine stratigraphy (after 1531, 1537, 2272). The sequence of Ferrassie and Quina Mousterian at Combe-Grenal is interrupted by a single level of Denticulate Mousterian in layer 20 and by three levels of either Typical or "attenuated" Ferrassie Mousterian in layers 28–30, but the correlated sequences nonetheless support the idea that Ferrassie Mousterian, Quina Mousterian, and Mousterian of Acheulean are temporally successive facies. If we assume that this succession is valid and has been properly dated at Combe-Grenal, it permits some important southwestern French human fossils to be roughly fixed in time. It would also imply that much of the variability in the French Mousterian reflects temporal change within a single tradition rather than the co-existence of several separate traditions.

on woodworking (bark stripping, whittling, or shaping), while sidescraper-rich ones might reflect a focus on food preparation or on hide processing (cutting and scraping).

The activity variant hypothesis would clearly be strengthened if it could be shown that the principal tools that characterize each facies had a distinctive function. However, so far, ancient tool use wear suggests just the opposite (37, 205–207). As noted in the previous section, use wear analysis implies that, independent of tool type (denticulate or sidescraper), most tools in each facies were employed for woodworking, whereas relatively few seem to have been used for butchering, for hide working, or for processing nonwoody plant materials. For reasons that may have to do with the geographic locations of the sites the tools come from, this contrasts with the results of use wear studies on (succeeding) Upper Paleolithic tools, most of which appear to have been used for working hides. However, it does not elucidate Mousterian facies variability. It is intriguing that a few Mousterian tools show wear from possible friction with a wooden haft or handle, but there is no indication yet that hafting was more common in one facies than in another. Perhaps most important, there is no evidence for a close relation between Mousterian tool type and tool function. Since there is also no compelling evidence that different tool types or facies tend to associate with different kinds of bones or pollen, in the final analysis the strongest argument for the activity variant model remains its plausibility.

Interassemblage Variability in Southwestern Asia

Southwestern Asia (the Near East) rivals western Europe as a focus of research on the Mousterian, thanks to a multitude of rich sites and to the efforts of many dedicated archeologists for more than half a century. The extraordinary potential of the region was first revealed in the 1930s, when Dorothy Garrod excavated rich Mousterian levels at the caves of Tabun, Skhul, and El Wad in the Wadi el Mughara (Valley of the Caves) on the slopes of the biblical Mount Carmel (783). Today the most informative sites occur in two principal areas. The richer area, with more than fifty excavated sites, including the classic Mount Carmel caves, is the Levant, a relatively narrow strip of land between the Mediterranean coast and the Syro-Arabian deserts. In modern political terms it includes Lebanon, Israel, and the adjacent parts of Syria and Jordan (fig. 6.3). The second, poorer area is the Zagros foothills of northeastern Iraq and northwestern Iran, with perhaps six excavated sites, including, most notably, Bisitun, Warwasi, Kunji, and Shanidar Caves.

The west Asian Mousterian assemblages can be described using Bordes's typology, and they exhibit the same kind of quantitative interassemblage variability seen in Europe, but typological change linked to time is more obvious in western Asia. The sequence that has been established in the deeply stratified deposits of Tabun seems to characterize most of the Levant (118, 119, 121, 123, 129, 555, 1137–1140, 1448, 1526, 1527). Although the Tabun deposits lie entirely beyond the range of ^{14}C dating, they are amenable to the TL and ESR methods (1548, 1552, 1888). Both show that the Mousterian began in the middle Quaternary, before 127 ky ago, but TL implies a particularly long chronology (fig. 6.33). Recall from chapter 2 that TL and ESR differ in the primary material to which they are usually applied—TL to burned flints and ESR to dental enamel. Arguably, the TL dates are more reliable, because unlike enamel, flint neither adsorbs nor loses uranium after burial, and the calculation of TL dates therefore requires no assumptions about the pattern or rate of uranium uptake or loss.

Near the base of the Tabun sequence, there is a late Acheulean industry whose mid-Quaternary age is indisputable. It grades typologically and technically into succeeding Acheuleo-Yabrudian (or Mugharan) assemblages dating to the middle or later part of the middle Quaternary. These assemblages include some that are relatively rich in small hand axes and others that are much richer in sidescrapers, but in the sidescraper component they are all broadly similar to the Quina-Ferrassie (Charentian) Mousterian of France.

In the midst of the Acheuleo-Yabrudian sequence at Tabun, there is an assemblage that is relatively richer in blades (flakes that are at least twice as long as wide) and in burins, endscrapers, and backed knives, alongside more typical Acheuleo-Yabrudian sidescrapers. A similar assemblage occurs in basically the same stratigraphic position at the Zumoffen Shelter (Adlun) in Lebanon, Yabrud Shelter I in Syria, and the Haua Fteah in northern Libya (1487). At Tabun it is usually called the Amudian, but elsewhere it is commonly known as the Pre-Aurignacian, since blades, burins, endscrapers, and backed knives are typical of Upper Paleolithic industries (like the Aurignacian) that succeed the Mousterian in the Near East and Europe. However, the blades and other Upper Paleolithic elements in the Amudian/Pre-Aurignacian tend to be much more casually made than their true Upper Paleolithic counterparts, and careful study of the Amudian at Tabun suggests it was just a facies within a continuum of Acheuleo-Yabrudian variation that also includes a facies very rich in sidescrapers and another relatively rich in small hand axes (1137). Both ESR and TL imply

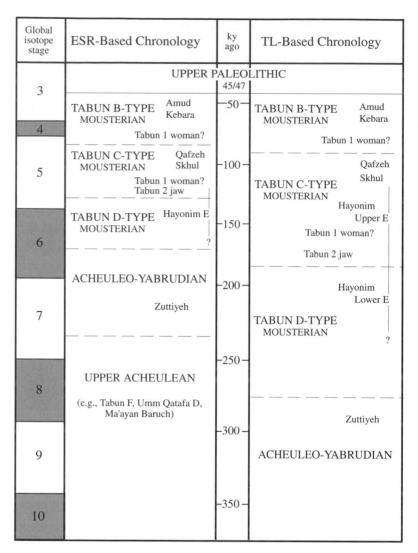

Global isotope stage	ESR-Based Chronology	ky ago	TL-Based Chronology
3	UPPER PALEOLITHIC	45/47	
4	TABUN B-TYPE MOUSTERIAN — Amud, Kebara; Tabun 1 woman?	50	TABUN B-TYPE MOUSTERIAN — Amud, Kebara; Tabun 1 woman?
5	TABUN C-TYPE MOUSTERIAN — Qafzeh, Skhul; Tabun 1 woman? Tabun 2 jaw	100	TABUN C-TYPE MOUSTERIAN — Qafzeh, Skhul; Hayonim Upper E; Tabun 1 woman?; Tabun 2 jaw
6	TABUN D-TYPE MOUSTERIAN — Hayonim E ?	150	
7	ACHEULEO-YABRUDIAN — Zuttiyeh	200	Hayonim Lower E; TABUN D-TYPE MOUSTERIAN ?
8	UPPER ACHEULEAN (e.g., Tabun F, Umm Qatafa D, Ma'ayan Baruch)	250	Zuttiyeh
9		300	ACHEULEO-YABRUDIAN
10		350	

Figure 6.33. Paleolithic cultural stratigraphy and chronology of the Levant (redrawn after 124, p. 515). The figure illustrates a discrepancy in dating between electron spin resonance (ESR) and thermoluminescence (TL) that grows with age. Both methods, however, place the near-modern humans from Qafzeh and Skhul Caves before the Neanderthals from Amud and Kebara Caves, and both show that the Levantine Mousterian began during the middle Quaternary, before 127 ky ago.

that the Amudian antedates the Upper Paleolithic by more than 100 ky, and it clearly also antedates other "precocious" industries, like the Aterian of northern Africa and the Howieson's Poort of southern Africa, discussed below.

At Tabun, at a time that corresponds either to global isotope stage 8 (TL) or to stage 6 (ESR), the Acheuleo-Yabrudian grades typologically and technically into the true Mousterian that succeeds it. Almost everywhere in the Levant, including Tabun, this Mousterian tends to be rich in Levallois flakes, and it is often called the Levalloiso-Mousterian. Broadly speaking, it resembles the French Typical Mousterian of Levallois Facies, but it can be subdivided into three chronologically successive industries that are designated Types D, C, and B, after the successive Tabun Cave layers in which they occur. Type D, the earliest industry, is richest in blades and elongated Levallois points. In contrast, Types C and B emphasize squatter, thinner Levallois flakes, and Type B is distinguished from Type C by greater numbers of triangular points and retouched tools. The difference between Types C and B is in keeping with a difference in Levallois core reduction (flake removal), which tended to be bidirectional or radial in C and unidirectional-convergent in B. The very latest Levalloiso-Mousterian assemblages sometimes contain Levallois points on which the butt has been bifacially thinned, perhaps to make hafting easier. These are known as Emireh points, after Emireh Cave in Israel, and assemblages in which they occur have sometimes been assigned to a separate Emiran phase or industry (1445).

The sequence in the Zagros foothills is much less well documented than the one in the Levant, but infinite ^{14}C dates, together with climatic inferences drawn from the sediments, suggest that the Zagros Mousterian dates primarily from the earlier part of the Last Glaciation (2014). It is therefore broadly contemporaneous with the Type B Levalloiso-Mousterian of the Levant. It varies somewhat from site to site (153, 657, 1986), but compared with the Levalloiso-Mousterian, it is poorer in Levallois flakes and richer in sidescrapers, points, and other characteristic Mousterian retouched pieces. The smaller size of the raw nodules available for flaking may explain both the more limited use of the Levallois technique and the higher frequency of retouch (2014). A similar kind of Mousterian with relatively few Levallois flakes and numerous retouched pieces may have extended from the Zagros in a broad arc westward through the Taurus Mountains of Turkey, where it is prominently represented in Karaîn Cave (1780).

In sharp contrast to Europe, in western Asia the Mousterians included both Neanderthals and near-modern people. Neanderthal-Mousterian associations are particularly well documented in Israel at Tabun (a partial skeleton with skull and some isolated postcranial elements and teeth) (1140, 1491), Kebara Cave (partial skeletons of an infant and an adult, twenty-seven isolated cranial and postcranial pieces from additional individuals) (74, 133, 134, 2005), and Amud Cave (a partial adult skeleton, cranial fragments of two other adults, a partial infant skeleton, and fragmentary remains of perhaps twelve other, mainly very young individuals) (1065, 1066, 1730, 2113); in Syria at Dederiyeh Cave (an infant's skeleton) (19); in Iraq at Shanidar Cave (nine partial skeletons) (2216); and in Uzbekistan at Teshik-Tash Cave (a child's skeleton) (1589). Mousterian artifacts accompany the remains of robust modern or near-modern people at Skhul Cave (seven partial skeletons and the isolated bones of three other individuals) (1491), Qafzeh Cave (five partial skeletons and the cranial or postcranial fragments of as many as ten additional individuals) (2304); and possibly Tabun (an isolated mandible). Conceivably, a cranial fragment from Zuttiyeh Cave was in a Mousterian layer, but more likely it was associated with the hand ax–rich (Acheulean) facies of the Acheuleo-Yabrudian Tradition (843). It is too incomplete for absolutely firm diagnosis, and as indicated previously, it could represent either an early Neanderthal population (2214, 2216), a population broadly ancestral to the robust early near-modern people found at Qafzeh (2305, 2307), or a separate archaic human population linked to neither the Neanderthals nor Qafzeh (1956).

It is certain that the near-modern people at Skhul and Qafzeh are as old as or older than many European Neanderthals, but their chronological relation to the west Asian Neanderthals remains problematic. At Tabun the width/thickness ratio of stone flakes changes progressively upward through the sequence, reflecting a continuing decline in average flake thickness relative to width, and flakes associated with the Qafzeh and Skhul skeletons tend to be thinner yet, suggesting that the Qafzeh and Skhul people postdate the Tabun Neanderthals (1137–1139). However, there are archaic rodent species at Qafzeh that do not occur with Neanderthal fossils at Tabun and a more modern rodent at Tabun that is absent at Qafzeh (128, 130, 2152, 2154, 2155). The Tabun fauna further implies a cooler, moister climate, perhaps because it accumulated early in the Last Glaciation, whereas the Qafzeh fauna accumulated in the preceding interglacial. In sum, supplemented by geologic-sedimentologic analyses (712), the Qafzeh and Tabun faunas suggest that the Israeli Neanderthals at least partly postdate the near-modern Qafzeh/Skhul people. This has now been amply confirmed by multiple broadly concordant TL, ESR, and U-series age determinations (924, 925, 1494, 1548, 1552, 1888, 1889, 1894, 2093, 2273, 2274).

Together, TL and ESR bracket the Qafzeh and Skhul near-modern human fossils between about 120 and 80 ky ago, while they place the Kebara and Amud Neanderthals between roughly

65 and 47 ky ago (fig. 6.33). A reasonable inference is that the Neanderthals actually succeeded near-modern people in Israel, but there is one important complication. This concerns the human remains and dates from Tabun Layer C. The Tabun C fossils include Tabun 1, the partial skeleton of a Neanderthal woman, and Tabun 2, a mandible whose identity is contested.

In its chin development and in some features of the ascending ramus (fig. 6.8), the Tabun 2 mandible appears modern (1721, 1728), but in other features, including especially the large size of its incisors relative to its cheek teeth, it appears more like the Neanderthals (2042). Arguably, it is in fact neither modern nor Neanderthal but "near-modern" in the same sense as the fossils from Qafzeh, Skhul, Klasies River Mouth, and other African sites. ESR and TL dates show that it dates from broadly the same Last Interglacial period as the Skhul/ Qafzeh fossils and before the Neanderthals of nearby Kebara and Amud. The Tabun 1 skeleton is more problematic, not for its physical identity, which is unqualifiedly Neanderthal (vs. near-modern), but for its dating. As already indicated, it is conventionally assigned to layer C, and if this is correct it would represent a Neanderthal population that was broadly contemporaneous with the Skhul-Qafzeh people. However, it in fact came from the interface between layers B and C, and the excavator was uncertain of its precise provenance. If it actually originated from layer B, it could represent the same early Last Glaciation Neanderthal population that is represented at Kebara and Amud. Only direct dating could probably resolve this thorny issue, which is clearly critical for assessing the evolutionary relation between Neanderthals and modern humans, discussed at the end of this chapter.

A further issue that remains unsettled is the relation between the appearance of truly modern people in western Asia and the subsequent appearance of the Upper Paleolithic. At Tabun, an acceleration in the rate of change in flake thickness in the latest Mousterian layers may anticipate the more radical artifactual change that followed (1139). Artifact assemblages that are supposedly transitional between the Mousterian and the Upper Paleolithic have been reported from several sites in the Levant (1445), but in most cases their transitional nature either is not truly obvious or could reflect the admixture, either naturally or during excavation, of Mousterian and Upper Paleolithic layers. Probably the best case for a transition comes from the alluvial site of Boker Tachtit in the Negev Desert, Israel. Here, in a sequence of levels that are tentatively bracketed between 47 and 38 ky ago by ^{14}C and U-series dates, a Levallois point technology progressively gives way to a true blade technology, with a concomitant increase in Upper Paleolithic tool types, especially endscrapers (1444, 1445).

Interassemblage Variability in Northern Africa

In northern Africa, the Mousterian in the narrow sense is best known in Nubia, straddling the border between Egypt and the Sudan along the Nile (1441, 1442, 2413, 2415, 2419). There are no long stratified sequences here, and it is difficult to discern valid time trends. Several kinds of assemblages have been described, including ones that resemble the European Typical Mousterian, others that are more similar to the Denticulate Mousterian, and yet others that are relatively unique, particularly those with numerous burins that were once assigned to a separate "Khormusan" industry. Levallois flakes tend to be common in all Nubian Mousterian facies. Broadly comparable Mousterian assemblages occur in the Egyptian Sahara, where they are often associated with now-defunct springs or shallow lakes (511, 2408, 2416, 2420). These indicate relatively moist conditions, probably mainly in the Last Interglacial, but perhaps also partly in the early part of the Last Glaciation.

In northeastern Africa, the Mousterian narrowly defined is represented at no more than ten to twelve sites, but the Aterian Industry or facies, with its characteristic stemmed pieces (fig. 6.34), is abundant (416, 625, 726, 2298, 2415, 2422). Except for the stems, the tool types involved are primarily the same kinds (sidescrapers, points, denticulates, etc.) that characterize Mousterian assemblages elsewhere, and most Aterian assemblages contain numerous Levallois flakes. Especially in the Sahara, some are also characterized by well-made bifacial leaf-shaped points. The Aterian is best known in the Mediterranean borderlands of northwestern Africa (the Maghreb) from eastern Morocco to Cap Blanc in Tunisia, but it also occurred through much of the Sahara, southward to Mauritania, Mali, and northern Niger, and eastward to Kharga and Dungul Oases in the western desert of Egypt (fig. 6.17). It apparently did not cross the Nile Valley. At Mugharet el ʿAliya, Taforalt (Pigeon Cave), Rhafas Cave, and the Météo alluvial site in Morocco and in alluvial (wadi) sequences at Kharga Oasis and other Saharan localities, the Aterian postdates the local Mousterian, but in Cyrenaica and the Nile Valley, the Mousterian continued long after the Aterian had appeared elsewhere.

Like Mousterian assemblages in northern Africa, Aterian ones often occur in a sedimentologic-paleontologic context indicating relatively moist conditions. This is above all true in

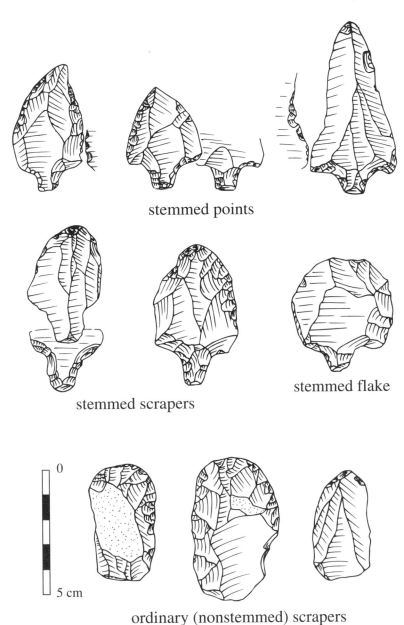

stemmed points

stemmed scrapers

stemmed flake

ordinary (nonstemmed) scrapers

Figure 6.34. Aterian artifacts from Algeria (redrawn after 2311, fig. 44). The tanged or stemmed pieces are the principal artifacts that distinguish the Aterian from the Mousterian.

the Sahara, where Aterian occupations usually occur in lake sediments in areas where standing water is absent today. At stratified caves such as Taforalt and Dar es Soltan 2 in Morocco, Aterian layers are separated from overlying Upper Paleolithic (or Epi-Paleolithic) ones by a long occupation gap, from 40 ky ago or before to 20 ky ago or later. As indicated previously, northern Africa appears to have been significantly wetter only during interglacials (2415), and the occupation hiatus between the Aterian and later Paleolithic industries may thus span the entire early and middle part of the Last Glaciation, from roughly

71 to 20 ky ago. There is little or no indication for significant human populations anywhere in northern Africa during the hiatus, probably because the entire region suffered from extreme aridity in the earlier and middle parts of the Last Glaciation (933).

As indicated above, the relatively sparse human remains associated with Mousterian/Aterian artifacts in northern Africa came from people who were clearly distinct from the Neanderthals and whose appearance approached that of modern people in important craniofacial features.

Interassemblage Variability in Sub-Saharan Africa

MSA artifact assemblages have been recovered in eastern Africa (61, 489, 1474, 1484, 1522, 1523, 1678, 1930, 2412) and western Africa (24), but they are far better known in southern Africa, where the MSA was first defined (859, 860). Here the overall nature and extent of interassemblage variability has been established from a series of stratified sequences, including especially those at Pomongwe, Bambata, and Redcliff Caves in Zimbabwe and at the Cave of Hearths, Bushman Rockshelter, Border Cave, Rose Cottage Cave, the Klasies River Mouth Caves, Nelson Bay Cave, Boomplaas Cave A, Montagu Cave, and Die Kelders Cave in South Africa (32, 607, 1860, 1981, 2164, 2165, 2338, 2339).

In general, by comparison with Mousterian assemblages in Europe, western Asia, and northern Africa, southern African MSA assemblages are poor in retouched pieces. However, the same basic retouched types prevail (scrapers, points, and denticulates), and their relative importance varies among MSA assemblages just as it does among Mousterian ones. Denticulates are the most common retouched type in some assemblages, whereas either scrapers or scrapers and points dominate others. Most specialists in southern Africa have not counted Levallois flakes separately, but MSA people apparently employed the Levallois technique more in the interior than along the southern coast. At least in part, this reflects differences in the rock types available for artifact manufacture. The abundance of blades also varies among southern African MSA assemblages, and both the number and size of blades may increase with time.

There is little other variability that is obviously time linked, except for the dramatic increase in well-made backed pieces, mainly segments and trapezoids, in the later part of MSA sequences at numerous sites south of the Limpopo River. The assemblages with backed pieces are commonly assigned to the Howieson's Poort Industry (fig. 6.35), named for the Howieson's Poort Shelter in the Eastern Cape Province of South Africa (613), and they were once thought to represent an intermediate stage

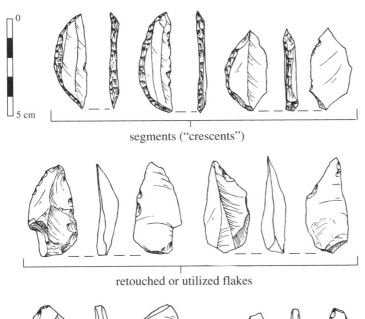

segments ("crescents")

retouched or utilized flakes

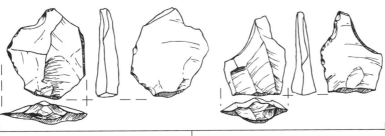

retouched or utilized flakes

Figure 6.35. Howieson's Poort artifacts from Nelson Bay Cave, South Africa (drawings by J. Deacon). The Howieson's Poort Industry is distinguished from other MSA industries mainly by the presence of well-made backed tools, including the segments (or "crescents") pictured here.

between the MSA and LSA. However, at Peers Cave (Skildegat), Border Cave, Klasies River Mouth, Apollo 11 Cave, and other sites, Howieson's Poort assemblages are stratified below typical MSA assemblages, and observations at Boomplaas and Klasies River Mouth suggest that they accumulated mainly during the transition from the Last Interglacial (global isotope stage 5) to the initial phase of the Last Glaciation (isotope stage 4), centering on 71 ky ago (606, 607). Today they are commonly assigned to the Howieson's Poort variant of the MSA.

Unlike artifacts in other MSA variants, which tend to be made on quartzites or other widespread, local rock types, Howieson's Poort backed elements are often made on "exotic" fine-grained types like silcrete and chalcedony. This could mean that Howieson's Poort groups roamed over much larger territories, perhaps in response to early Last Glaciation climatic deterioration that reduced the abundance of stable, predictable plant and animal resources and increased human dependence on highly migratory, gregarious ungulates (32). If this interpretation is correct, the Howieson's Poort variant would provide the oldest demonstrated example of a climatically driven change in human territorial behavior. However, the fine-grained rock types that distinguish the Howieson's Poort were generally available near Howieson's Poort and other MSA sites, though usually only in nodules that were too small for more commonplace MSA flakes and blades. It may therefore be that Howieson's Poort people differed from other MSA groups not so much in their larger territories as in their need or ability to utilize relatively small lumps of high-quality stone.

Howieson's Poort backed pieces are larger than but otherwise similar to backed pieces that local LSA people mounted on shafts or handles with vegetal mastic. If the backing in the Howieson's Poort was to aid in hafting, Howieson's Poort assemblages may present some of the oldest known evidence for composite tools anywhere in the world. The possibility that Howieson's Poort backed pieces were mounted, and even more the degree of standardization they exhibit, is sometimes said to presage the LSA/Upper Paleolithic and perhaps thereby to signal a precocious emergence of fully modern human behavior (1529, 1530). However, the Howieson's Poort Industry lacks other LSA/Upper Paleolithic traits like formal bone artifacts and art, and in every detail of stone artifact technology and typology except backed elements, it closely resembled other MSA variants. There is, moreover, no obvious link between the Howieson's Poort artifacts and the emergence of modern or near-modern human morphology, which appeared in southern Africa several tens of thousands of years before them.

In southern Africa, stratified cave sites such as Montagu Cave, Nelson Bay Cave, Die Kelders Cave 1, and the Klasies River Mouth Caves often exhibit major occupation gaps between the latest MSA, antedating 40–30 ky ago, and succeeding LSA occupations, commonly dating from the very end of the Last Glaciation or even from the Holocene. As in northern Africa, the occupation hiatus probably reflects a reduction in human population density owing to extreme aridity in the middle part of the Last Glaciation. Although few relevant sites have been excavated in eastern Africa, discoveries at Lukenya Hill (1439, 1562) and Enkapune Ya Muto (31) in Kenya, at Nasera (Apis) Shelter (1522) and Mumba Shelter (1523) in Tanzania, and at Matupi Shelter in Zaire (2296) suggest that occupational continuity was greater there, perhaps because of more favorable climatic conditions throughout the Last Glaciation.

Like their north African contemporaries, the people who made MSA artifacts in southern and eastern Africa were near-modern to modern people, as detailed above.

Nonstone Artifacts

As indicated above, use wear analysis shows that Mousterian stone artifacts were often used on wood, and wooden artifacts have survived at a handful of sites. Examples include a remarkable 2.4 m long wooden spear found among the ribs of a straight-tusked elephant at the Mousterian site of Lehringen in Germany (1588, 2172), two possible wooden shovels or trays from anaerobic deposits of Abric Romaní in northeastern Spain (425), and fragments of possible throwing sticks discovered at the MSA sites of Chambuage Mine in Angola (483) and Florisbad in South Africa (479). The tip of the Lehringen spear may have been deliberately hardened by exposure to fire while it was still green. The Romaní deposits also contained flint flakes with typical woodworking micropolish and cavities or hollows suggesting additional, rotted wooden artifacts or firewood (433). Mousterian woodworking could have been anticipated from wooden spears or spear points recovered at the Lower Paleolithic sites of Clacton, England (1629, 2391) and Schöningen, Germany (2171).

Curiously, although bones are often abundant and well preserved at Mousterian/MSA sites, deliberately shaped, formal bone tools are almost unknown (1265, 1539, 2165, 2339). The most conspicuous exceptions are eight whole or partial barbed bone points and four other indisputable bone artifacts from the riverside sites of Katanda 2, 9, and 16, upper Semliki Valley, eastern Zaire (363, 2525, 2526). The accompanying stone artifacts are not classically MSA, but ESR dates on associated hippopotamus teeth and luminescence dates on enclosing sands bracket the sites between 150 and 90 ky ago, within the MSA time range. The dates are particularly striking, because the next oldest barbed points in Africa are no more than 25 ky old and most specimens are younger than 10 ky (363, 1679, 1783). The implication is for a tradition of point manufacture that spanned at least 80 ky.

An alternative possibility is that the Katanda luminescence and ESR dates do not bear on the bone artifacts. The dated sands may not have been fully bleached (or "zeroed") before burial (715), and unlike the bone artifacts, the Katanda mammal bones and teeth tend to be conspicuously abraded and rounded. This suggests they were transported by moving water and that they may rest in a river bar on which much younger bone artifacts were dropped. The contemporaneity of artifacts and bones might be investigated by a comparison of their mineral crystallinity and elemental content or by direct ^{14}C dating of the artifacts. If either analysis should confirm contemporaneity, archeologists will be hard-pressed to explain why other much larger and often younger MSA bone assemblages lack such sophisticated bone artifacts.

The Katanda bone artifacts aside, other Mousterian/MSA bone artifacts comprise fragments that appear to have been casually used as hammers, scrapers, or anvils, although even many of these may have been shaped or damaged by carnivore gnawing or crunching rather than by human use (217). Again, Katanda aside, formal bone artifacts are so rare in Mousterian/MSA contexts that where they do occur they could easily represent undetected—perhaps undetectable—intrusions from higher levels. Certainly it is reasonable to conclude that few if any Mousterian/MSA people recognized bone as a raw material that could be cut, carved, and polished into points, borers, awls, and so on. In this respect they were similar to their precursors but very different from their successors.

Also like earlier peoples and unlike later ones, Mousterian/MSA peoples left little or no evidence for art or ornamentation. The list of proposed art objects from Mousterian/MSA sites depends on the author, but frequently cited examples include a reindeer phalanx and a fox canine, each punctured or perforated by a hole (for hanging?) from La Quina in France; numerous pierced animal phalanges from Prolom II in the Crimea, Ukraine; two grooved or collared cave bear incisors from Scalyn, Belgium; occasional bones with lines that may have been deliberately engraved or incised, from La Quina and La Ferrassie in France, Bacho Kiro Cave in Bulgaria, Molodova I in Ukraine, and a handful of other European sites; a fragment of mammoth molar plate that may have been deliberately carved and polished to an oval shape and an invertebrate fossil (nummulite) on which a supposedly engraved line intersects a natural one to form crosses on both surfaces, both from Tata in Hungary; a juvenile bear femur shaft with four evenly spaced perforations suggesting a flutelike musical instrument from Divje Babe Cave 1 in Slovenia; and a plate of flint cortex with incised semicircles and other lines, from Quneitra on the Golan Heights (106, 166, 167, 452, 464, 571, 1311, 1455, 1456, 1459, 2043, 2249, 2250). In support of Mousterian/MSA art objects, enthusiasts also often cite even older bones with incised lines from the Lower Paleolithic site of Bilzingsleben, Germany (167, 1430), and a lump of volcanic rock broadly recalling a female figure from the Acheulean site of Berekhat Ram, Israel (107, 864, 870). The occupations at both sites probably antedate 300 ky ago.

On many of the proposed art objects, the modification may not be artificial, and on most it is not persuasively artistic. The phalanx "pendant" from La Quina, for example, could have been punctured by a carnivore's bite, while a bone with parallel engraved lines from La Ferrassie could simply represent a cutting board on which skin or some other soft material was repeatedly sliced (452). Similar nonartistic, repetitive cutting could explain the Bilzingsleben incisions (581), and large animal trampling on a sandy or gritty substrate could explain others (569). Elephant trampling of bones near African water holes occasionally produces clusters of subparallel cut mark mimics that broadly resemble those from Bilzingsleben (991, 992). Feeding on coarse plant tissue could have grooved the Scalyn cave bear incisors (789); chewing or partial digestion by carnivores could have perforated phalanges like those from Prolom II and other sites (451, 571, 2446); and natural geologic forces could have shaped the Berekhat Ram figurine (1670). In fact, given the known human and natural actions that can simulate crude human attempts at art and the huge sample of available Mousterian/MSA (and earlier) objects, it would be remarkable if some did not exhibit incising, perforating, or other marking that could be seen as artistic. In sharp contrast, even some of the earliest Upper Paleolithic/LSA sites contain numerous shaped objects whose artistic (as opposed to natural or utilitarian) origin is incontestable (942, 2443, 2444, 2446). The contrast is particularly striking given the vastly longer period spanned by Mousterian/MSA and earlier cultures. The conclusion must be that in art as in so much else, Mousterian/MSA (and earlier) people were qualitatively different from their Upper Paleolithic/LSA successors.

Of course it remains possible that Mousterian/MSA (and earlier) people produced art only in perishable substances. Many Mousterian/MSA sites contain fragments of natural ocherous red or black pigment (iron oxide and manganese dioxide) that do not occur naturally in the deposits and that are sometimes striated or faceted from repeated rubbing (166). Conceivably Mousterian/MSA people used the ocher fragments to decorate their skins or other perishable media (270, 278, 528), but a priori it is at least equally likely that they used the ocher for more mundane purposes like tanning skins (1196). In addition, unlike their successors, Mousterian/MSA people did not concentrate ocher in graves (969), nor did they intentionally burn (oxidize or reduce) ocher fragments to obtain a wider range of colors (1674). More fundamentally, even if the pigments imply some form of decoration, it remains true that Mousterian/MSA peoples differed strikingly from later (Upper Paleolithic/LSA) peoples, who left durable art and ornamentation in almost every region they occupied.

An Overview of Mousterian/MSA Artifacts

The Mousterian/MSA was perhaps characterized by more variability through time and space than were earlier (Lower Paleolithic/Early Stone Age) industries, but it was still remarkably uniform over vast areas and time spans compared with the succeeding Upper Paleolithic/LSA. Part of the reason for this is that, overall, Mousterian/MSA peoples made a much smaller variety of readily definable artifact types than did their successors. Together, the relative homogeneity of the Mousterian/MSA through time and space and the relatively small number of distinguishable MSA/Mousterian artifact types suggest that the behavior of Mousterian/MSA people differed fundamentally from that of their Upper Paleolithic/LSA successors.

As detailed in the section below titled "Settlement Systems," it is further noteworthy that Mousterian people rarely transported raw materials more than a few kilometers, whereas Upper Paleolithic people sometimes transported or imported them over scores of kilometers (778, 1228, 1529, 1535, 1539, 1878, 2007). Some of the items that Upper Paleolithic people moved were nonutilitarian goods, such as marine shells and amber, that may not have interested Mousterian people, but some Upper Paleolithic sites contain large quantities of exotic (nonlocal) stone that was used to manufacture utilitarian artifacts. On the relatively rare occasions when exotic stone occurs in Mousterian/MSA sites, it usually involves isolated, finished artifacts, whereas exotic stone in Upper Paleolithic sites commonly consists of raw nodules and intermediate flaking products. The sum implies that, relative to the Mousterians, at least some Upper Paleolithic people roamed over far larger territories or participated in far more sophisticated social networks.

The rarity or absence of formal bone artifacts and art objects in the Mousterian/MSA only adds to the contrast. The most general implication is that Mousterian/MSA people were behaviorally more conservative than their successors, with a limited ability to innovate even in the face of significant environmental variability through time and space. At least tentatively, it seems reasonable to propose that the behavioral limitations implied by Mousterian/MSA artifacts reflect biological (genetic) differences from later humans.

Site Distribution

The overall distribution of Mousterian/MSA sites closely resembles the distribution of earlier (Lower Paleolithic/Early Stone Age) sites, except that Mousterian sites occur in easternmost Europe (Ukraine and European Russia), where traces of earlier people are at best equivocal (1030, 1232). Since the commercial and archeological activities that promote site discovery have been as intense in eastern Europe as in central and western Europe, where earlier sites are well documented, it seems increasingly likely that the lack of pre-Mousterian sites on the east reflects the absence of people. The most probable reason is that pre-Mousterians could not cope with the especially cold winters of easternmost Europe, whether during glacial or interglacial periods.

At the same time, it is important to stress that even Mousterian sites are restricted to the southern and western margins of easternmost Europe, and no sites have been firmly documented in the most continental—central and eastern parts—in spite of significant commercial and archeological activity there. This activity has revealed numerous rich Upper Paleolithic sites, and it seems increasingly likely that Upper Paleolithic people were the first to occupy these areas, whose climate must have been especially intimidating during the Last Glaciation. Upper Paleolithic people were also apparently the first to inhabit the equally harsh or harsher subarctic and arctic environments of northern Asia (Siberia), where, in spite of dedicated efforts, no convincing Mousterian (or earlier) sites have so far been found.

Site Types

Most known Mousterian/MSA sites are in rockshelters or the mouths of caves, but this does not mean that Mousterian/MSA peoples always preferred caves to open-air sites, even where caves were plentiful. Cave sites are better known today because they constitute obvious places to excavate for artifacts and skeletal remains, and Mousterian/MSA peoples lived so recently that most of their caves are still visible. In addition, caves are more likely to retain artifacts, bones, and other human refuse than are open-air sites, and they are abundant in areas such as southwestern France, where serious prehistoric archeology has been under way for many decades. Most open-air sites were probably long ago destroyed by erosion, and those that remain are harder to locate than caves. Nonetheless, important open-air

Mousterian sites have been found even in southwestern France (711, 929, 1135), and they obviously provide most of the Mousterian/MSA record in those parts of Europe, western Asia, and Africa where caves are rare or absent.

With rare exceptions, open-air sites occur near active or once active springs, lakes, or streams, partly because these were favorable places for camping or obtaining food and partly because sites were more likely to be preserved in places of sediment accumulation as opposed to erosion. Prominent open-air sites in Europe include Mauran and La Borde in southwestern France (710, 711, 1135), Riencourt-les-Bapaume and Seclin in northern France (2244, 2248), Maastricht-Belvédère in the Netherlands (1804, 1805, 1809), Ehringsdorf, Salzgitter-Lebenstedt, Königsaue, Rheindahlen, Wallertheim, and Tönchesberg 2 in Germany (174, 287, 530, 531, 786, 1432, 2169, 2173, 2203), Tata and Érd in Hungary (773, 2318), and Molodova I and V in Ukraine and Rozhok I, Il'skaya, and Sukhaya Mechetka (Volgograd) in European Russia (149, 1035, 1229). Important examples in Africa include El Guettar in Tunisia (916, 917), Hajj el Sidi Creiem (Wadi Derna) in Libya (1489), Bir Tarfawi 14 in Egypt (511, 2416), Seggedim in Niger (2187, 2188), Prospect Farm in Kenya (61, 1557), and Florisbad (1287, 1289), Orangia 1 (1860), and Duinefontein 2 (1234) in South Africa. The most important examples in western Asia are perhaps Quneitra on the Golan Heights (585, 865) and Rosh Ein Mor and Nahal Aqev in the central Negev Desert of Israel (1443, 1444).

Most Mousterian/MSA caves and rockshelters were probably living places or camps rather than places where people killed or butchered animals. This is indicated by their very nature and by the fossil fireplaces or hearths that are a prominent feature in virtually every well-excavated cave where preservation conditions were suitable. At some sites, like Kebara Cave in Israel (1528) and Klasies River Mouth Main Site in South Africa (606, 607), stacked hearths comprise much of the deposit, and they indicate recurrent, intensive fire building over many millennia. The probable uses of fire include the provision of warmth and light, protection from predators, and of course food preparation. At lower latitude sites like Kebara and Klasies River Mouth, plant-food preparation may be especially indicated (606). However, even where hearths are especially abundant in Mousterian/MSA contexts, they tend to be relatively simple lenses of ash and charcoal 50 cm to 1 m across and few centimeters thick. Some are surrounded or underlain by rocks (1674), but none are as elaborate as some Upper Paleolithic hearths, which were fitted with stone liners, air-intake ditches, or other features to control airflow and heat dissipation.

In contrast to caves and rockshelters, most Mousterian/ MSA open-air sites are more difficult to interpret. Some—for example those in the Negev—are near sources of desirable stone raw material, and the abundance of flaking debris implies they were stone tool workshops. Most open-air sites, however, are scatters of artifacts or of artifacts and fragmentary animal bones near water sources where people might have camped, killed or butchered animals, or both. Distinguishing campsites from kill/butchery sites is difficult because few sites preserve distinctive features, such as partial animal skeletons, which might indicate butchering, or traces of possible structures, which would imply a camp. Further, at sites such as Duine-fontein 2 where partial skeletons do occur, there is usually the problem of determining who was responsible for skeletal disarticulation—people or other predators such as hyenas or lions. Virtually the only exception to this problem is Lehringen, where a wooden spear was actually found among the ribs of an elephant.

Aside from hearths, indisputable structural traces are very uncommon in Mousterian/MSA sites. Postdepositional blurring or destruction cannot be the entire reason, since relatively fragile hearths are preserved at many sites. More likely, Mousterian/MSA people rarely built structures that were substantial enough to leave clear traces. It is possible that structure bases are marked by distinct concentrations of artifacts and other cultural debris found at many sites, but more compelling structural traces have rarely been detected. The most notable exceptions include possibly artificial depressions containing artifacts and other cultural materials at the open-air sites of Ariendorf (find level II) (fig. 6.36) and Rheindahlen in Germany (292, 2169, 2173, 2258), and perhaps more convincingly, the apparent structural features at Cueva Morín in Spain (763), at the caves of La Ferrassie, Pech de l'Azé I, and Combe-Grenal in France (279, 1404), at the open-air sites of Molodova I and V in Ukraine (464, 1229, 2007), and at Mumbwa Cave in Zambia (138). Equally unequivocal structural traces from the MSA site of Orangia 1 in South Africa (1860) have been ignored here, because the excavator now believes they postdate the MSA artifacts at the site.

Cueva Morín contained a dense accumulation of cultural debris bounded by a line of stone piles that may represent the remnant of a low stone wall. La Ferrassie contained a 5 m by 3 m spread of limestone fragments possibly marking the base of a tent built within the cave. Pech de l'Azé 1 preserved a small part of a low drystone wall, and Combe-Grenal contained an apparent posthole. An MSA horizon at Mumbwa Cave contained

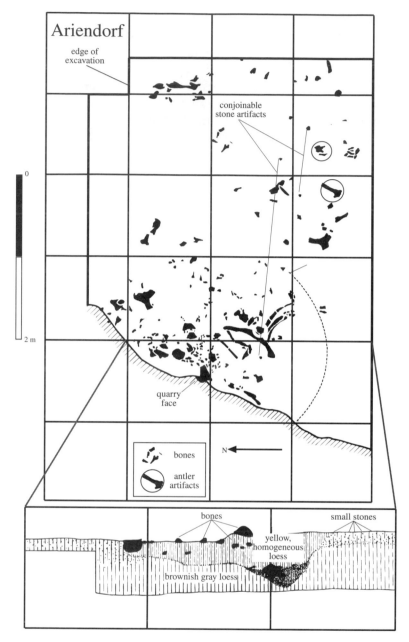

Figure 6.36. Floor plan (*top*) and section (*bottom*) through "find level II" at the Ariendorf open-air site, Germany (redrawn after 292, pp. 163, 164). The occurrence of bones and artifacts was revealed by quarrying through windblown silt (loess). The quarrying unfortunately destroyed part of the site, but careful excavation showed that elephant ribs, other large bones, and stone artifacts were concentrated in and near an ancient depression 2–3 m across. The depression may have been dug by early Middle Paleolithic (Mousterian) people to level the floor of a hut whose frame was made partly of elephant ribs. Alternatively, it could be a natural feature, perhaps created when a large tree was uprooted. Less equivocal structural traces are remarkably rare in Middle Paleolithic and earlier sites, suggesting that the people rarely built substantial houses.

three possible postholes on the margin of a dense semicircular concentration of ash, sediment, quartz artifactual debris, and fragmentary animal bones.

The Molodova sites contain large rings or partial rings of large (mainly mammoth) bones surrounding dense concentrations of artifacts, fragmentary bones, and ash spreads (hearths). The bones are thought to represent weights that held down skins stretched over a wooden framework that disintegrated long ago. Although the bone rings appear deliberate, the structural reconstruction is problematic because of the large areas they enclose. In the case of the most convincing rings (in Molodova I, level 4) (fig. 6.37), these are ovals roughly 8 m long by 5 m across, which would have required considerable architectural skill to cover. The full meaning of the Molodova features remains unclear, and it is possible that slope wash or some other natural process is largely responsible for the patterning.

In sum, Mousterian/MSA caves and open-air sites contain limited evidence for structural modification. This is particularly striking in Europe, where many Mousterian sites have now been carefully excavated and where the contrast with subsequent Upper Paleolithic sites is particularly sharp (1232, 1279, 1535, 1539). Well-excavated Upper Paleolithic sites almost always contain unambiguous and often spectacular evidence of structures, in the form of artificially excavated depressions and pits, patterned arrangements of large bones or stones, postholes, or some combination of these. Even Upper Paleolithic caves commonly preserve structural traces such as stone walls, artificial pavements, and pits, and to the annoyance of archeologists, Upper Paleolithic cave dwellers sometimes dug out preexisting deposits, perhaps to enlarge the living area or to level out a floor. As in many other aspects of culture, in site modification Mousterian/MSA people seem to have been quantitatively, if not qualitatively, different from their successors. At least in part, their limited architectural ability may explain why they were unable to occupy the most continental parts of Eurasia.

Settlement Systems

Mousterian/MSA peoples were certainly nomadic to some extent, occupying different sites for different purposes or at different seasons to take advantage of seasonal changes in resource availability. However, in most regions the distribution of known sites probably does not closely mirror the distribution of utilized sites. More likely it reflects the distribution of caves and

rings of large bones

N

m

mammoth tusks ⬭ hearth

mammoth molars natural depression with bones

other animal bones ᴺᴺ cores

bones with stone tool cut marks ⋯ other flint artifacts

Molodova I, level 4

Figure 6.37. Plan of the excavations in Molodova I, level 4, Ukraine (redrawn after 464, fig. 8). The two large rings of large bones may represent the foundations of structures built by the Mousterian occupants of the site. Alternatively, they may have been formed partly by a natural process such as slope wash.

other places that favor preservation and the distribution of modern settlements and roads used by archeologists. Since it is impossible to determine whether different sites were occupied by essentially the same people, in most instances it is also virtually impossible to reconstruct Mousterian/MSA settlement systems.

However, the stone raw materials that Mousterian/MSA peoples used may provide some insight into their settlement patterns or at least into the extent of their movements, since wherever Mousterian/MSA stone sources have been established, they turn out to be highly local. In southwestern France, where the most detailed studies have been conducted (716, 794–796, 1539, 2260), 70–98% of the stone in various Mousterian sites came from less than 5 km away. Only 2–20% originated between 5 and 20 km away, and less than 5% traveled more than 30 km. A broadly similar pattern has been found in eastern Morocco (2423–2425), and it also appears to characterize eastern Africa (Kenya and northern Tanzania) (1553, 1554) and central and eastern Europe (716, 1810). There is the difference, however, that in eastern Africa and in central Europe (Poland, Czech Republic, Slovak Republic, and Hungary), desirable stone occasionally traveled greater distances, even exceeding 100–200 km. The contrast with France may mean that Mousterian/MSA populations in eastern Africa and central Europe occupied larger territories because environmental conditions were less favorable, or perhaps more likely, it could mean that highly desirable raw material sources were more widely dispersed (less tightly packed) in eastern Africa and central Europe.

The interplay between environment and raw material dispersion is clear in the eastern Sahara, where the Mousterians of Bir Tarfawi and Bir Sahara East quarried raw chunks of suitable stone at outcrops, transported the chunks to lake edges for core manufacture, and then took the cores elsewhere (possibly to nighttime sleeping sites) to produce tools (2415). Like Mousterians elsewhere, however, those in the Sahara rarely moved stone raw material more than 30–40 km. Everywhere, Mousterians showed their appreciation for high-quality exotic stone by conspicuously transforming it into tools, but they still obtained it relatively infrequently. In contrast, Upper Paleolithic people often acquired 20–25% of their stone from more than 30 km away (1265, 1539, 2006, 2007), and they frequently transported not just desirable flakes but also cores of nonlocal stone. The sum suggests that Upper Paleolithic groups routinely roamed over larger territories or that they interacted (traded) more intensively with other groups. Whichever possibility is accepted, it implies a significant shift in settlement patterns, social organization, or both.

So far, stone raw materials provide no evidence for a difference between Mousterian and Upper Paleolithic settlement patterns in western Asia, but analyses of growth increments (or annuli) in animal teeth suggest a possible difference between the settlement patterns of west Asian near-modern people and their Neanderthal successors. The relevant growth increments

occur in acellular cementum that anchors tooth roots to the
surrounding socket and that forms in successive bands, each
comprising a thicker subband formed during the season of maximum growth and a thinner subband formed during the season
of reduced growth. Under transmitted light, the thicker bands
tend to be translucent and the thinner bands opaque. When the
cementum is appropriately sectioned and examined under a
microscope, the total number of bands can reveal the age of an
animal at time of death, and the translucency or thickness of
the last (outermost) subband can reveal the season of death.
Analysis of increments on the roots of Qafzeh and Kebara Cave
gazelle teeth imply that Qafzeh was occupied seasonally (the
gazelles died only in the local dry season), while Kebara was occupied more or less throughout the year (the gazelles died in
both the dry and the wet seasons) (1371, 1374, 1376). This
might mean that the Qafzeh (near-modern) people circulated
from one base camp to another seasonally, whereas the Kebara
Neanderthals tended to occupy the same base camp year-round.
The Neanderthals may have been equally mobile, however, if
they frequently radiated to smaller, more ephemeral camps
that have not yet been identified. The Qafzeh/Kebara contrast
is echoed by a broadly similar one between Skhul and Tabun
Cave level C on the one hand (seasonal) and Tabun level B on
the other (nonseasonal or multiseasonal). Recall that Skhul and
Tabun C broadly parallel Qafzeh in time and in occupation by
near modern people and that Tabun B broadly parallels Kebara
in time and in probable occupation by Neanderthals.

If the seasonality contrast is accepted, it is the only documented behavioral difference between near-modern people and
Neanderthals in western Asia. As discussed above and below,
both groups made broadly similar Levalloiso-Mousterian artifacts, both built simple fireplaces, both buried their dead in un-
complicated graves, and both hunted medium-sized mammals,
at least on occasion. However, a difference in seasonality may
say more about climate than about culture, since near-moderns
occupied Qafzeh, Skhul, and Tabun C during the Last Interglacial and Neanderthals inhabited Kebara and Tabun B during
the early and middle phases of the Last Glaciation. The way
climate could force such a change in seasonality of settlement
is perhaps illustrated by a later, similar change that has been
inferred between Mousterian and Upper Paleolithic patterns
in the Avdat/Aqev area of the central Negev Desert, Israel
(1446, 1450). Here Mousterian sites appear to comprise semipermanent base camps near permanent water and outlying,
more temporary sites near probable sources of food and stone

raw material. Upper Paleolithic sites include no semipermanent base camps, but only "ephemerally occupied multipurpose camps and even more ephemeral loci of unknown function" (1450). The Mousterian pattern suggests a radiating system in which the people moved back and forth between the central base camp and the peripheral resource sites, whereas the Upper Paleolithic pattern suggests a circulating system in which the people simply moved from multipurpose camp to multipurpose camp. The shift from a radiating to a circulating pattern may have been an adaptive response to contemporaneous environmental deterioration (increasing aridity) that has been documented from geomorphic and palynologic evidence in the central Negev (1443).

The adoption of a circulating system could explain why Upper Paleolithic people abandoned the flake technology of the Mousterians (1450), since Upper Paleolithic blade technology reduced the need to deviate from a circulating pattern to renew stone raw material reserves at quarry sites. This is because, by comparison with the Mousterian flake technology, Upper Paleolithic blade technology produced more usable blanks per core, that is, more usable blanks per quarry visit.

Mousterian/MSA Ecology

Mousterian/MSA peoples were restricted to tropical and temperate environments, where they lived entirely by hunting and gathering. In their ecology they probably resembled recently observed tropical or temperate hunter-gatherers such as the Hadza people of northern Tanzania. The Hadza hunt and scavenge medium-sized and large mammals including impala antelope, hartebeest, wildebeest, warthog, and zebra, but for a steady nutrient flow they depend heavily on berries, roots, fruits, honey, tortoises, and other gatherable resources (1617, 1619, 1620). MSA/Mousterian people probably relied at least as much on gathered foods, but unfortunately the gathered items would leave little archeological trace. This is particularly true of plants, and except for nondietary charcoal and pollen, Mousterian/MSA sites have provided few plant fossils. Rare exceptions include some charred seeds or nuts and pieces of corms (tuberlike subterranean stems) found at the South African MSA sites of Bushman Rockshelter, Border Cave, and Strathalan Cave B (2165, 2339). Some MSA/Mousterian sites have also furnished "grindstones" (artificially smoothed or polished pieces of coarse rock) like those that historic people used to crush nuts or seeds. Mousterian/MSA examples could have been used for other purposes, but in southern Africa they tend to concentrate where the most abundant edible plant parts have probably always been seeds or nuts rather than softer items like bulbs and corms (2339). At least circumstantially, this implies that grindstones were used to process hard plant tissues that have generally failed to survive.

In contrast to plant fossils, bones that reflect the meat component of Mousterian/MSA diet occur at many sites. The principal animals in Mousterian/MSA sites are medium-sized and large ungulates that were doubtless most common near the sites. In Eurasia, depending on the time and place, these animal species include red deer, fallow deer, reindeer, bison, wild oxen (aurochs), wild sheep and goats, gazelles, and horses (including asses and onagers). In Africa the ungulates include various antelopes, zebras, and wild pigs. The largest available species—elephants and rhinoceroses, present in both Eurasia and Africa—tend to be rare, and in sites such as those at Molodova (Ukraine) where they do abound, their bones may have been used in construction (1232, 2006, 2007). Thus, they do not necessarily represent animals that people ate.

Carnivores, particularly large ones such as lions, hyenas, bears, and wolves, also tend to be rare in Mousterian/MSA sites, probably because they avoided people and vice versa. In apparent contradiction, a Neanderthal bear cult has been inferred from apparently patterned arrangements of cave bear bones in European caves such as Le Regourdou (France), Drachenloch (Switzerland), and Petershöhle (southern Germany) (196). In these caves and others, complete bear bones, especially skulls, tend to occur near the walls or near large limestone slabs that have been imaginatively interpreted as stone lockers built by the Neanderthals. However, there are no stone artifacts or cut-marked bones to demonstrate human activity, and bear trampling could have destroyed many complete bones and displaced others to cave walls or to large fallen slabs. Almost certainly the bones accumulated naturally from hibernating bears that occasionally died in the caves (1290). Cave bear bones become much less frequent with the advent of the Upper Paleolithic, and sometime before the Upper Paleolithic concluded, the cave bear became extinct. The pattern suggests that, unlike the Mousterians, Upper Paleolithic people excluded cave bears from favored hibernation spots and may thus have precipitated their extinction.

In general it is difficult to estimate how successful the occupants of any particular Mousterian/MSA site were at obtaining animals. At sites where artifacts, humanly damaged bones, or other clear evidence of human presence are sparse, it is even difficult to tell what proportion of the animals were obtained by people and what proportion may have been killed by carnivores or died naturally. The problem tends to be particularly acute at open-air sites, where fossilized feces (coprolites), chewed bones, or other evidence of carnivore activity may be as abundant as cultural debris. Even at a site where context and associations indicate that people were the principal bone accumulators, it is generally impossible to establish whether the relative abundance of various species reflects human practices or preferences, the natural abundance of the species near the site, or some combination of these factors.

Separating the relative roles of human behavior and environment is possible only when species abundance can be compared between two or more sites. In this event it may be possible to control for differences in environment by using palynologic, sedimentologic, or geochemical observations and to control for differences in behavior by using artifacts or other cultural debris. Thus, if pollen and sediment analyses indicate that two sites shared a similar environment, human behavior becomes a likely cause for differences between them in species abundance. Conversely, environment is implicated if artifacts or other cultural remains indicate that the occupants of two sites were behaving very similarly.

In most instances where controlled comparisons are possible, differences in species abundance among Mousterian/MSA sites or between Mousterian/MSA sites and earlier or later ones appear to reflect differences in site environment rather than in occupant behavior. Thus, in France fluctuations in the relative abundance of red deer versus reindeer, traced from the Acheulean of the late middle Quaternary through the Mousterian of the early late Quaternary, correlate with climatic fluctuations indicated by geomorphic and palynologic evidence (273). Red deer peaks are closely associated with warm periods, and reindeer peaks are associated with cold ones. In Spain, Italy, and other parts of Mediterranean Europe, where climatic fluctuations were less extreme, reindeer were always rare or absent and red deer prevailed in both interglacial and glacial periods (28, 29, 2051, 2065). Similarly, in Ukraine and European Russia, where climate was always continental and climatic variation through time was perhaps less important than variation through space, bison are more common in Paleolithic sites on the south and deer and horses in sites on the west, regardless of time or Paleolithic culture (1029, 1035, 1229, 1232, 2007).

In Africa too, most faunal differences or changes among Mousterian/MSA sites relate more clearly to climate than to human behavior. The most dramatic example comes from the

Sahara, where episodes of nonoccupation alternated with periods when Mousterian or Aterian people occurred together with typical African grassland or savanna ungulates (511, 2408, 2411, 2415, 2416, 2420). Associated deposits from now defunct springs or lakes show clearly that climatic change (not human behavior) accounts for the alternation. More subtle climatic change controlled the abundance of grazing versus browsing species within MSA and LSA archeological sequences at the extreme southern tip of Africa (1237, 1238). Here, independent of associated artifacts, deposits that suggest relatively cool, dry (glacial) conditions are dominated by grazers, whereas deposits suggesting relatively warm, wet (interglacial) climate are dominated by browsers. The same pattern characterizes deposits containing bones that were collected by carnivores, confirming a climatic—versus a human behavioral (cultural)—explanation for the species differences.

Among the comparatively rare instances of species differences that are probably due to Mousterian/MSA behavior (rather than to environment), perhaps the best examples are the change from gazelle abundance to fallow deer abundance within the Mousterian sequence at Tabun Cave in Israel and contrasts in fish, bird, and mammal frequencies between coastal MSA and LSA sites in South Africa. Fallow deer prefer moister conditions than gazelle, and the sharp increase in deer late in the Tabun Mousterian sequence was originally thought to reflect a regional increase in moisture. However, local and regional sedimentologic and palynologic data do not indicate a climatic change at the same time, though they do suggest an earlier change toward drier (not wetter) conditions. In fact, analysis of the Tabun sedimentary sequence now suggests that fallow deer increased because of a change in the way people used the cave (1139). In the earlier Mousterian, it was probably a camp to which gazelles were brought from the nearby coastal plain. Later, after a chimney had opened through the roof, it was unsuitable for habitation and became a trap into which deer were driven from the wooded slopes above.

MSA Coastal Ecology in South Africa

MSA and LSA sites on the coast of South Africa have provided one of the richest and most detailed records of Stone Age ecology anywhere in the world. Table 6.2 lists the principal MSA sites and their probable ages. Together with Smugglers' Cave (Témara) (1800), Zouhrah (627), and Mugharet el ʿAliya (1067) in Morocco; Haua Fteah in Libya (1486); the Bérard open air-site in Algeria (1841); Devil's Tower and Gorham's Cave on the Rock

Table 6.2. South African Coastal Middle Stone Age Sites

Site	Geologic Age (basis)	References
Klasies River Mouth Cave complex	Between 127 and 57 ky ago (ESR, geologic context; faunal associations)	606–609, 1233, 1981, 2166, 2334
Herold's Bay Cave	Between 120 and 80 ky ago (U-series dating, geologic context, and faunal associations)	356
Die Kelders Cave 1	Between 71 and 45 ky ago (ESR, geologic context; faunal associations)	94, 907, 1247, 2136
Sea Harvest	Between 127 and > 40 ky ago (^{14}C and geologic inference)	906, 2337
Hoedjies Punt	Between 127 and > 40 ky ago (^{14}C and geologic inference)	J. E. Parkington (personal communication); 2337
Boegoeberg 2	Between 127 and > 40 ky ago (^{14}C and geologic inference)	D. J. Halkett, T. J. Hart, and J. E. Parkington (personal communication); 1247

of Gibraltar (784, 2355); and Moscerini Cave and possibly other sites in coastal Italy (2051), the South African MSA sites provide the oldest known evidence for systematic human use of coastal resources, dating from the Last Interglacial and the early part of the Last Glaciation, between 127 and 57 ky ago.

Table 6.3 lists the LSA sites with which the MSA sites are most fruitfully compared. As a group, the LSA sites date partly from the end of the Last Glaciation, between 24 and 11 ky ago, but mainly from the Present Interglacial (or Holocene), after 11 ky ago. MSA and LSA sites from the middle part of the Last Glaciation, between 57 and 24 ky ago, are rare and usually poor in southern Africa, perhaps because conditions were mainly too dry to support substantial human populations (610). The long occupational hiatus is vexing, because it spans the shift from the MSA to the LSA, which so far is well documented only in eastern Africa (31). However, the richness of Last and Present Interglacial occupations in South Africa offers an ideal opportunity to compare the economies of MSA and LSA people living under very similar conditions.

Table 6.3. South African Coastal Later Stone Age Sites

Site	References
Nelson Bay Cave	612, 1110
Byneskranskop 1	1902
Die Kelders Cave 1	94, 1901
Paternoster Midden	1792
Kasteelberg A and B	1245, 1993, 1994
Elands Bay Cave	1651, 1652
Dune Field Midden	1653

The MSA and LSA sites contain the same species, but often in very different frequencies. The most striking contrast involves fish, which are abundantly represented in all coastal LSA sites but rare or absent in the MSA sites. This is true even though sediments and bones of seals and shore birds indicate that the coastline remained nearby throughout each MSA occupation. Offshore bathymetry further affirms that the coastline was generally within 5 km, even during early Last Glaciation periods of sea level depression (2289). The abundance of fish in LSA sites is paralleled by the presence of likely fishing implements, including grooved stones that were probably net or line sinkers and carefully fashioned toothpick-sized double-pointed bone splinters that could have been tied to a line and baited. Historic people used similar pieces to catch both fish and birds. The MSA sites contain no obvious fishing gear, suggesting that MSA people did not know how to fish.

The MSA layers at Klasies River Mouth and Die Kelders Cave 1 also contrast with coastal LSA layers in the relative abundance of various marine birds. Thus jackass penguins dominate the MSA layers, whereas cormorants and other airborne species heavily dominate the LSA sites (93). The abundance of airborne species in LSA sites is associated with artifactual indicators of the bow and arrow, including bone rods that closely resemble historic linkshafts or foreshafts and tiny backed and nonbacked stone bits (microliths) like those that tipped historic arrows (612). MSA sites lack any evidence for projectile weapons, and in their absence, MSA people probably did little fowling. Recent discoveries at Boegoeberg 1 and 2 may appear to complicate this interpretation (1247), because Boegoeberg 1, a fossil hyena den, contains mostly penguins, whereas Boegoeberg 2, an MSA site, contains mostly cormorants. In the relative abundance of penguins the Boegoeberg 1 hyena site is more extreme than the MSA sites at Klasies River Mouth and Die Kelders, and in the relative abundance of cormorants the

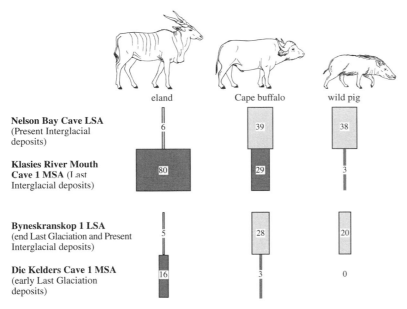

Figure 6.38. *Top:* The minimum numbers of eland, Cape buffalo, and wild pigs in samples from the Last Interglacial deposits of Klasies River Mouth Cave 1 and in the Present Interglacial deposits of nearby Nelson Bay Cave, South Africa. *Bottom:* The minimum number of individuals from the same species in the early Last Glaciation deposits of Die Kelders Cave 1 and in the end Last Glaciation/Present Interglacial deposits of nearby Byneskranskop Cave 1. The relatively greater abundance of buffalo and wild pigs at Nelson Bay and Byneskranskop 1 probably reflects the technological superiority of local LSA hunters over their MSA predecessors.

Boegoeberg 2 MSA site is actually more extreme than any comparable LSA site. The explanation for the contrast between the Boegoeberg sites may be environmental, since other faunal elements imply that Boegoeberg 1 formed under remarkably wet (and cool) conditions in what is now a hyperarid environment. However, even if the Boegoeberg sites suggest that environment may influence the penguin/cormorant ratio, it is difficult to see how environment could explain penguin dominance in the MSA deposits at Klasies River Mouth and Die Kelders 1, which span both interglacial and glacial times.

Finally, the relative numbers of large ungulates differ strongly between MSA and LSA sites where ungulate bones are common (1247). The clearest example concerns the contrast between the Last Interglacial MSA layers at Klasies River Mouth and the Present Interglacial LSA layers at Nelson Bay Cave (fig. 6.38). The environs of both sites are very similar today, and sediments and geochemistry indicate that the environs were similar to the present and to each other during the Last and Present Interglacials. In the historic environment, buffalo and wild pigs significantly outnumbered eland, and this situation is mirrored in the LSA layers at Nelson Bay Cave. However, in the MSA layers at Klasies River Mouth, eland strongly outnumber both buffalo and pigs, and if the MSA environment truly resembled the historic one, the most likely explanation is that MSA people focused selectively on eland. Buffalo and pigs are far more dangerous to hunt than eland, and without the bow and arrow, MSA people may have been forced to concentrate on more approachable species. Assuming that limited MSA technology explains the relatively small numbers of buffalo and pigs at Klasies River Mouth, the MSA disadvantage would have been very great, since buffalo and pigs were probably very common nearby.

Eland similarly dominate buffalo in the MSA layers of Die Kelders Cave 1, whereas buffalo strongly dominate eland in the LSA layers of nearby Byneskranskop 1 (fig. 6.38). In explaining the difference in this instance, environment cannot be fully excluded, since the Die Kelders MSA layers formed in a cooler, wetter climate. However, temperature variation had no obvious impact on historic buffalo and eland numbers in Africa; wherever both species occurred together, buffalo tended to be much more numerous (694, 1988). In addition, the eland/buffalo ratio remained constant when "glacial" conditions replaced "interglacial" ones at Klasies River Mouth. At least circumstantially, then, the much increased abundance of buffalo at Byneskranskop provides additional support for an advance in LSA hunting competence.

The South African evidence for a sharp MSA/LSA contrast cannot be checked elsewhere in Africa, for lack of sufficient MSA and LSA animal remains. In marked distinction to Klasies River Mouth, Die Kelders Cave 1, and other coastal MSA sites in South Africa, however, the Katanda MSA sites on the Semliki River (eastern Zaire) (363, 2526) and Mousterian Site 440 on the Nile River (northern Sudan) (891, 1932) have provided numerous fish bones. As already discussed, the Katanda sites have been dated to at least 90 ky ago, and the fish bones are accompanied by well-made barbed bone points. The points could have tipped fish spears, and the Katanda fish come mainly from two species that could have been speared in seasonally shallow floodplain waters. But as discussed above, the very ancient dating of the barbed points, and by extension the fish bones, requires reconfirmation. At site 440, the bones and stone artifacts were diffused ("mixed") through two sandy layers 30–50 cm thick, and the bone-artifact association might not be the product of human activity. The Katanda sites are clearly more significant, and if an MSA age is confirmed, it will be hard to explain why such a sophisticated fishing technology remained localized for tens of thousands of years. It not only failed to reach southern Africa, but it is also unknown at site 440 and other river or lakeside Mousterian/MSA sites in eastern and northern Africa.

The exploitation of aquatic foods is never as evident in European Mousterian and Upper Paleolithic sites as it is in South African MSA and LSA localities, perhaps because Europe has provided relatively few rich coastal sites. As in South Africa, however, fish bones are totally absent in coastal Italian Mousterian sites, including Moscerini Cave, where shells of intertidal mollusks clearly demonstrate the proximity of the coast (2051). In addition, many European Mousterian and Upper Paleolithic sites are adjacent to rivers, and limited evidence suggests that Mousterian people in such circumstances fished and fowled less regularly than their Upper Paleolithic successors. The best data are perhaps from southwestern France and northern Spain, where advances in fishing and fowling may not have occurred until relatively late in the Upper Paleolithic (1535, 2058, 2059). Sudanese Mousterian site 440 and Katanda aside, it may be difficult to determine when prehistoric Africans first actively engaged in fishing and fowling, because the pertinent sites probably date from the period between 60 and 20 ky ago when much of southern and northern Africa appear to have been depopulated. In addition, this was a time of lower sea levels, and the earliest evidence for coastal fishing may now lie submerged on the continental shelf.

Hunting versus Scavenging

For the sake of argument, it has been assumed so far that Mousterian/MSA people actively hunted large animals, albeit less successfully than their Upper Paleolithic/LSA successors. But we must consider the possibility that Mousterian/MSA groups acquired most large carcasses by scavenging. Most specialists believe that at least some Mousterian/MSA people routinely hunted large animals (448–450, 1034, 1247, 1265, 1640, 2051), but some argue that Mousterian/MSA populations mainly scavenged (218–220). The issue is complicated partly because hunting and scavenging are not as distinct as they might appear. Thus, recently observed Hadza hunter-gatherers in northern Tanzania obtain about 80% of their large mammal carcasses by bow-and-arrow hunting and about 20% by scavenging (1617, 1620). Their scavenging often involves driving leopards, spotted hyenas, or lions off kills, however, and if the predators refuse to leave, they may themselves be killed. In both practice and result, this kind of "confrontational" scavenging may not be usefully separated from active hunting, and a more meaningful distinction is probably between early and late access to carcasses (386, 387, 1709). It is this difference that archeologists must address if they are to evaluate the hunting prowess of MSA/Mousterian people. Among the available clues are the pattern of skeletal part representation, the ages of prey animals at death, and the frequency and positioning of bone damage marks. The MSA animal bones from Klasies River Mouth (KRM) provide a case in point.

At KRM the large animals are mainly buffalo and eland, and these are represented primarily by bones of the feet (metapodials, carpals, tarsals, and phalanges) and of the skull (mainly teeth). Proximal or "upper" limb bones (humeri, radii, femurs, and tibiae) are relatively scarce, although they are very rich in meat, marrow, and other edible tissue and are thus the bones that both people and other large predators tend to carry away first. Their scarcity at KRM might therefore mean that the people were mainly picking over carcasses that other predators had already fed on (218). The implication would be for scavenging in the narrowest sense of the term.

There is the problem, however, that foot and skull (mainly dental) elements from large ungulates tend to outnumber limb bones not only at KRM but at most prehistoric sites throughout the world, including, for example, African Iron Age sites (2335), where the large ungulates were mainly domestic cattle that are unlikely to have been scavenged. The pervasiveness of the pattern suggests it is due to factors that all large ungulates share, and the most conspicuous factors are relative bone durability and identifiability. The skull and foot bones that tend to dominate large ungulate assemblages are among the densest parts of the skeleton, and they are also among the easiest to identify even after they have been fragmented. Their density and relative lack of nutrients probably combined to ensure that, unlike less dense, more nutritious limb bones, they were not extensively butchered and broken during food preparation. They were thus more likely to be buried intact, and this, together with their initial density, helped maintain their identifiability even after relatively intense postdepositional leaching and profile compaction.

At KRM and many other sites, in contrast to large ungulates, small species (like steenbok and bushbuck) are relatively well represented by limb bones (versus skull and foot elements). This could mean that the KRM people hunted small ungulates more often. However, it is also notable that wherever small ungulate limb bones are abundant, they also tend to be much more complete than their large ungulate counterparts. The implication is that people processed smaller bones less intensively, that small size enhanced postdepositional survival, or both. Another possibility is that hyenas selectively removed the epiphyses (ends) of large ungulate limb bones while leaving behind a larger proportion of the smaller ungulate specimens (1440). It is from epiphyses that the small and large ungulate counts were compiled. Hyenas are not known to prefer larger ungulate epiphyses, and there is little or no evidence for hyena activity at KRM and at many other sites that show the same pattern, but the possibility of a hyena role cannot be excluded. The bottom line is that the relative rarity of large ungulate limb bones at KRM and many other sites may mean only that butchering, profile compaction, and so forth selectively destroyed large limb bones or rendered them unidentifiable, not that prehistoric people failed to obtain them.

In contrast to eland and buffalo skeletal part representation at KRM, whose implications are at best equivocal, bone damage marks and death ages of prey point more directly to active hunting. Carnivore tooth marks are extremely rare, and the few that have been observed could reflect occasional carnivore visits when the people temporarily abandoned KRM. Stone tool cut marks are far more abundant, and none overlie tooth marks. Their placement shows that the people were both disarticulating intact buffalo and eland carcasses and defleshing individual bones (1563, 1564). A unique discovery that also implies hunting is the tip of a stone point still embedded in a buffalo vertebra.

The ages of individual buffalo, eland, and other ungulates can be estimated from the heights of their teeth above the roots (1244). When the ages are summed to make an age profile for each species, two patterns emerge. The buffalo exemplify the first pattern, in which most individuals are either very young (less than 10% of maximum life span) or relatively old (more than 50% of life span) (fig. 6.39). Prime-age adults (between 10% and 50% of life span) are very rare, particularly compared with their abundance in live herds. Paleobiologists call such an age profile "attritional," because it could result from normal everyday mortality factors—such as endemic disease, accidents, and carnivore predation—that tend to bypass prime-age individuals and to affect mainly the very young and the old. Since very young and old buffalo are precisely those that die most often of natural causes, it might be argued that their abundance at KRM implies human scavenging. However, it is unlikely that the people could have located so many very young buffalo before

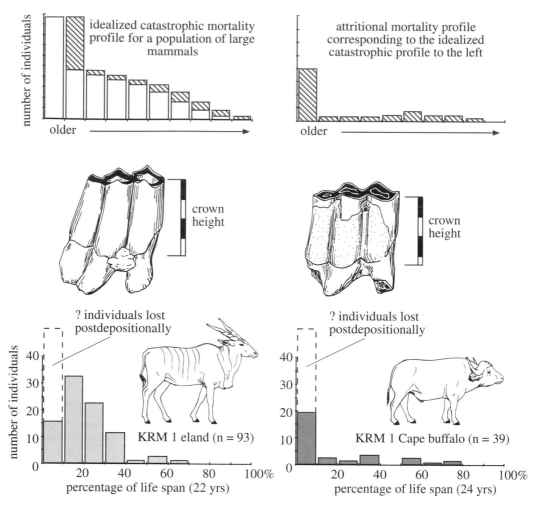

Figure 6.39. *Top left:* Schematic catastrophic age (mortality) profile for a population of large mammals that is basically stable in size and age structure. The open bars represent the number of individuals that survive in each successive age cohort, the hatched bars the number that die between successive cohorts. *Top right:* Separate plot of the hatched bars, showing the corresponding schematic attritional age profile. The basic form of corresponding catastrophic and attritional profiles is the same for all large mammals, but the precise form will differ from population to population depending on species biology and specific mortality factors. *Middle:* Eland and buffalo lower third molars (M_3s) showing the crown height dimension, from which individual age at death may be estimated. *Bottom:* Mortality profiles based on crown heights for eland and Cape buffalo from the MSA layers of Klasies River Mouth Cave 1. It is probable that postdepositional leaching, profile compaction, and other destructive factors have selectively destroyed teeth of the youngest eland and buffalo. When this consideration is taken into account, the eland profile is clearly catastrophic and the buffalo profile is attritional. As discussed in the text, this suggests that the Klasies people obtained eland by driving entire social groups over cliffs or into other traps where differences in individual vulnerability due to age would be meaningless. In contrast, the Klasies people must have obtained buffalo by methods to which prime-age adult buffalo were relatively immune.

hyenas or other scavengers did, and modern observations indicate that these scavengers consume young carcasses quickly and completely. The abundance of very young buffalo is thus more likely to reflect active hunting.

The second age profile is exemplified by the eland (fig. 6.39), in which, relative to the very young and the old, prime-age adults are well represented, roughly in proportion to their abundance in live herds. Paleobiologists call this kind of age profile "catastrophic," because it could be fixed in the ground only by a great flood, volcanic eruption, or other catastrophe that kills individuals in rough proportion to their live abundance, regardless of their age. Buffalo are too poorly represented at other MSA sites to determine the nature of mortality, but eland are abundant enough at Die Kelders Cave 1 to show that they also died "catastrophically." At both KRM and Die Kelders 1, the most likely catastrophe was humans' driving whole groups into traps or over cliffs such as those that are adjacent to both sites. Modern observations show that among all the species represented at KRM and Die Kelders, eland are the most amenable to driving, and it appears that the KRM and Die Kelders people had discovered this special vulnerability.

The KRM and Die Kelders people could not have driven eland very often, however, or the repeated removal of numerous prime adults would have sapped the reproductive vitality of the species, and there is no evidence that eland numbers declined during or after the occupation of either site. Historically eland were extremely widespread in Africa, but they were nowhere very numerous, and eland herds tended to be widely dispersed and difficult to locate. This means that the KRM and Die Kelders people probably rarely found eland in a position suitable for driving—and thus that they probably killed only a small fraction of the available animals. If eland were rare in the environment and if the KRM and Die Kelders people did not obtain them very often, it follows that they must have been even less successful at obtaining other ungulates, such as buffalo, which were probably more common in the environment but are rarer in the site.

In sum, the KRM and Die Kelders prey mortality profiles imply that MSA people actively hunted. Combined with the dominance of eland over buffalo, however, they also suggest that the people obtained very few ungulates overall.

Seasonality of Site Occupation

The analysis of gazelle teeth summarized in the section above titled "Settlement Systems" shows that west Asian near-modern Mousterians moved camp with the seasons. Their motivation was presumably seasonal variability in local resource availability, which often drove movements among historically observed hunter-gatherers (1204, 1205). African fossil ungulate teeth have not yet been analyzed for indications of seasonal bone accumulation, but South African coastal MSA and LSA sites invariably contain bones of the Cape fur seal (*Arctocephalus pusillus*), and these afford an independent opportunity to determine the season(s) of site occupation. The reason is that until 1941, when fur seals were first legally protected from human predation, they bred almost exclusively on offshore rocks (580, 1988). The vast majority of births occurred within a few weeks in late November and early December, and adults forced the young from the rocks about nine months later. In recent times large numbers of nine- to eleven-month-old seals then washed up ashore, exhausted or dead. It is the short fur seal birth season and the consequent seasonal peak in onshore availability that allow estimates of when fur seal bones accumulated at fossil sites. In cases where fossil seal ages cluster tightly around nine to eleven months, bone accumulation probably centered tightly on the August–October period of superabundance of nine- to eleven-month-old seals. Where ages cluster more loosely around nine to eleven months, bone accumulation probably included not only the August–October period but also other times. And where ages fail to cluster near the nine- to eleven-month average, then accumulation probably fell largely outside that period.

Fossil seal bones may be "aged" by comparison with bones from known-age animals. Various skeletal elements are useful for age determination, but the distal humerus is especially apposite, because it is relatively durable and thus tends to dominate fossil samples. The most consistently available dimension is the mediolateral diameter or "breadth" of the distal end. Figure 6.40 uses a boxplot format to summarize breadths of fur seal distal humeri in samples from key MSA and LSA sites, in two samples accumulated by brown hyenas (*Hyaena brunnea*), and in samples of modern newborns and nine-month-olds. In each plot, as described in reference 2312, the vertical line near the center is the median, the open rectangle encloses the middle half of the data (between the twenty-fifth and seventy-fifth percentiles), the shaded rectangle is the 95% confidence interval for the median, and the vertical lines at the ends mark the range of more or less continuous data. Circles or asterisks indicate extreme values (points that are far removed from the main body of data.) The number of specimens in each sample is given in parentheses. Samples for which the 95% confidence limits do not overlap differ significantly in the conventional

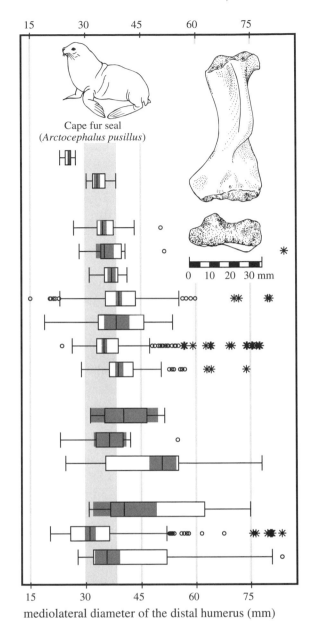

Figure 6.40. Boxplots summarizing distal humerus breadth in known-age fur seals and in fossil or subfossil samples from LSA, MSA, and brown hyena sites on the coast of southern Africa. The key elements of each plot are the vertical line near the middle, which represents the median measurement or fiftieth percentile in each sample, and the open rectangles, which enclose the middle half of the measurements, between the twenty-fifth and seventy-fifth percentiles. The shaded rectangles mark the 95% confidence limits for the medians. When two shaded rectangles fail to overlap, chance is a very unlikely explanation for the difference between the associated medians. The MSA samples from Boegoeberg 2 and Die Kelders 1 are probably too small for meaningful comparisons, but the sample from Klasies River Mouth differs from the various LSA samples in two important ways: it contains significantly larger humeri on average (as indicated by the median and its confidence limits), and it is characterized by greater dispersion around the median (as indicated by the open rectangle). In median size and especially in dispersion, the Klasies River Mouth sample recalls the fossil brown hyena sample from Boegoeberg 1. As discussed in the text, the boxplots suggest that LSA people timed their coastal visits to coincide with the August–October period when nine- to eleven-month-old seals are especially abundant, whereas the MSA inhabitants of Klasies River Mouth followed a less focused seasonal round. It is even possible that MSA people occupied the coast year round, like the Boegoeberg 1 hyenas.

statistical sense. To aid visual assessment, a vertical gray bar extends the range for nine-month-olds through the figure.

Figure 6.40 shows that in most samples, the median breadth of fur seal distal humeri lies within or just outside the range for known nine-month-olds. In addition, breadths in the LSA samples tend to be tightly clustered around the median, and the sum suggests that LSA people focused their coastal visits on the August–October period when they could harvest nine- to eleven-month-old seals on nearby beaches. The most glaring departure from the LSA pattern involves the MSA sample from Klasies River Mouth Cave 1, where distal humerus breadths are much more broadly dispersed and the median is substantially and significantly greater than that for nine-month-olds. This implies that MSA people occupied Klasies River Mouth mainly outside the peak period of availability of nine- to eleven-month-old seals and, more generally, that MSA people followed a very different seasonal round than their LSA successors. From the available data, it could even mean that they did not follow a seasonal round at all. Arguably, the MSA sample from Boegoeberg 2 supports the same inference, while the one from Die Kelders Cave 1 suggests an LSA-like pattern. Unfortunately, however, both samples are too small for any firm conclusion.

As an aid to further interpretation, it is instructive to compare the LSA and MSA boxplots with those for the two brown hyena samples at the bottom of figure 6.40. The sample labeled "Central Namib Dens" was accumulated between 1990 and 1996 in dens adjacent to a fur seal breeding colony about 20 km south of Luderitz, Namibia (1987, 1989, 1990). The other sample was accumulated at the Boegoeberg 1 fossil hyena den mentioned above. In keeping with hyena access to a breeding colony, the Central Namib sample includes fifty-four humeri from newborn individuals. The abundance of newborn specimens explains why Central Namib median distal humerus breadth is so small, and the rarity or absence of newborn specimens in the various fossil sites shows that none were near breeding colonies. For heuristic purposes, the Central Namib sample is plotted twice, with and without the newborn specimens. When the newborn specimens are removed, the median humerus breadth more closely approximates that in the various LSA samples, but the degree of breadth dispersion becomes much greater. This is because the Central Namib brown hyenas remain at the coast throughout the year, including times when the only seals available are older than nine to eleven months. The comparable or greater dispersion in the Boegoe-berg 1 fossil hyena sample also illustrates nonseasonal bone accumulation.

In sum, the boxplots in figure 6.40 imply that LSA people generally timed their coastal visits to include the seasonal peak in availability of young fur seals, whereas the MSA occupants of Klasies River Mouth obtained many, if not most of their seals at other seasons. If larger fur seal samples from other MSA sites replicate the Klasies River Mouth pattern, the implication would be that MSA people occupied the coast nonseasonally, perhaps like brown hyenas. MSA people may have differed from LSA people because they failed to perceive seasonal variability in fur seals and other resources or because they were technologically more limited. In this regard, it may be pertinent that no MSA site has provided evidence for water containers, whereas LSA sites have repeatedly provided fragmentary or even whole ostrich eggshell canteens (612). The contrast is particularly striking, because ostrich eggshell abounds at Boegoeberg 2 and other west coast MSA sites.

Disposal of the Dead

One of the reasons the Neanderthals are so well known is that they buried their dead, at least on occasion. Few excavators have identified burial pits (782), perhaps because these were usually shallow and easily obscured by profile compaction long ago. Some pits may also have been overlooked in the relatively unsystematic excavations that produced important Neanderthal skeletons in the late nineteenth and early twentieth centuries. In addition, there are no instances where Neanderthal skeletons are accompanied by special artifacts or other indisputable grave goods. However, burial pits have been reported in France from La Chapelle-aux-Saints, La Ferrassie, and Le Roc de Marsal (282, 301, 353, 423, 2324), and in Israel at Kebara Cave (131, 133). At other sites, including the original Neanderthal locality (Feldhofer Cave) in Germany, Spy Cave in Belgium, and Le Moustier, La Quina, Le Regourdou, and La Roche à Pierrot (Saint-Césaire) caves in France, deliberate burial is the most plausible explanation for the presence of nearly complete or articulated Neanderthal skeletons (302, 629, 758, 2062, 2303). Intentional burial probably also accounts for the Neanderthal remains found at Kiik-Koba in the Crimea (Ukraine), at Tabun and Amud Caves in Israel, at Dederiyeh Cave in Syria, at Shanidar Cave in Iraq, and at Teshik-Tash Cave in Uzbekistan (19, 224, 629, 969, 2189). La Ferrassie was a veritable cemetery of eight graves and Shanidar may have contained five, although some of the Shanidar skeletons may lie under natural rockfalls (2016).

The near-modern Israeli contemporaries of the Neanderthals at Skhul and Qafzeh Caves also buried their dead, and the graves tend to resemble those of the Neanderthals in their fundamental simplicity (133, 180). A boar mandible found partly under the left radius of Skhul individual 5 (1490) and the antlers of a fallow deer associated with Qafzeh individual 11 (133, 2302) may represent grave goods, but the case depends largely on the completeness of the bones. Conceivably this reflects only their fortuitous inclusion in grave fill, which protected them from the various pre- and postdepositional forces that fractured bones of the same species elsewhere in the deposits. Certainly neither Skhul nor Qafzeh provides the striking evidence for burial ritual or ceremony that marks some Upper Paleolithic/Later Stone Age graves, as discussed in the next chapter. So far there is no evidence that the African contemporaries of the Neanderthals buried their dead, but the site sample is relatively small and, as discussed above, the conditions for grave preservation were generally not as good as in Europe and western Asia.

The Neanderthals and their contemporaries may have been the first people to bury the dead, but earlier burials may be missing because there are so few sites, particularly caves, where graves are most likely to be preserved. Even at most sites occupied by the Neanderthals and their contemporaries, isolated, fragmentary human remains, scattered among artifacts and animal bones, are far more common than intentional burials. At some sites the human fragments are cut marked or burned, and the prospect of cannibalism must be considered. Cut-marked Neanderthal bones have been found at Combe-Grenal and the Abri Moula, France (630, 1319), and broadly contemporaneous, cut-marked and burned near-modern human fragments have been recovered from the MSA layers of Klasies River Mouth, South Africa (609, 2454). Cannibalism may also explain the especially large quantity of fragmentary Neanderthal bones found at Hortus Cave, southern France (approximately one hundred fragments from at least twenty individuals) (1408), and the even larger number recovered at Krapina Cave, Croatia (almost nine hundred pieces from fourteen or more individuals) (1424, 1724, 2231, 2479). At least nineteen of the Krapina bones were also scored by stone tools (546). However, rockfalls and postdepositional crushing could account for bone fragmentation at Krapina (1854, 2219) and perhaps also at Hortus (2327). In addition, the form and anatomical disposition of the Krapina cut marks might reflect the defleshing of partially decomposed bodies before burial rather than butchery (1853, 2327). Arguably, Krapina was a specialized burial site, since it contains relatively limited evidence for actual occupation. The approximately nine hundred Neanderthal bones are almost as numerous as Mousterian artifacts (1955).

Cannibalism was very rare among historic humans (72), and in most sites where broken and dispersed Neanderthal bones occur, carnivore consumption of bodies that were buried in shallow graves or that were left on the surface is a more likely explanation. Partly empty graves such as those at Teshik-Tash and Kiik-Koba provide circumstantial evidence for carnivore disturbance. The spotted hyena is known to exhume human bodies in Africa today, and spotted hyenas were widespread throughout both Africa and Eurasia in the Neanderthal time range. Spotted hyenas apparently occupied Krapina at least occasionally, and tooth marks of hyenas or other carnivores occur on some of the Krapina Neanderthal bones (1853, 1854). Carnivore damage has also been reported on fossil human bones from other sites, including the partial skull from the Florisbad MSA site, South Africa (503, 2138), the limb bones of the Neanderthal child from Teshik-Tash (1590), and the famous Circeo 1 Neanderthal skull from Grotta Guattari near Rome, Italy (809, 2464).

Neanderthal graves present the best case for Neanderthal spirituality or religion (196, 629), but, more prosaically, they may have been dug simply to remove corpses from habitation areas. In sixteen of twenty well-documented Mousterian graves in Europe and western Asia, the bodies were tightly flexed (in near fetal position) (969), which could imply a burial ritual or simply a desire to dig the smallest possible burial trench. Ritual has been inferred from well-made artifacts or once meaty animal bones found in at least fourteen of thirty-three Mousterian graves for which information is available (969), but there are no Mousterian burials in which the grave goods differ significantly from the artifacts and bones in the surrounding deposit, and in virtually all cases accidental incorporation in the grave fill is thus a distinct possibility. In their lack of truly unusual or distinctive items, Mousterian graves contrast sharply with many later, Upper Paleolithic burials (452, 969).

Perhaps the best case for ceremonial treatment of Neanderthal bones comes from Grotta Guattari, where construction work in 1939 exposed a cave that had been sealed by rock debris for perhaps 50 ky (241, 452, 2081). Inside, on the cave floor, an irregular ring of rocks reportedly surrounded a Neanderthal skull, whose base had been broken away as if to extract the brains. Unfortunately the original position of the skull inside the ring is open to question, since it was moved and replaced before it was first seen by scientists. In addition, the skull exhibits no evidence of human manipulation, and the only diagnostic antemortem damage is from carnivore teeth (2464).

Among other possible evidence for Neanderthal ritual, there are the scratched bones at Krapina, if one assumes the scratches were produced by methodical defleshing before burial, and the items associated with Neanderthal burials at Teshik-Tash (1590) and Shanidar Cave (2015, 2016). At Teshik-Tash the skull of an eight- to nine-year-old boy in a shallow grave was surrounded by five or six pairs of mountain goat (ibex) horns. Goat horns were common throughout the Teshik-Tash deposit, however, and no plot of their overall distribution has ever been published to demonstrate the special character of the horn circle. At Shanidar, the grave fill associated with skeleton 4 contained numerous clumps of flower pollen. But the fill was heavily disturbed by rodent burrows, and the pollen may have been intrusive (452). The Shanidar flower burial will probably remain problematic as long as it is unique.

In sum, there is little or no evidence that the Neanderthals or their near-modern contemporaries practiced ritual or ceremony in disposing of the dead. Both kinds of people clearly employed burial, at least sometimes, but the motivation need not have been religious, and the graves tended to be far simpler than those of their fully modern successors.

Population Numbers

There are no practical or theoretical grounds for estimating the absolute population densities of the Neanderthals or their contemporaries, though there is good evidence that climatic change forced population fluctuations. Thus the rarity or absence of Mousterian sites in deposits implying peak cold in northern France (2244) and Poland (716) almost certainly reflects population shrinkage under adverse conditions, and comparable site paucity during the later Mousterian (Aterian)/MSA in northern and southern Africa suggests a similar population crash, probably owing to extreme aridity during the corresponding (middle) part of the Last Glaciation. Continuing mid-glacial aridity probably accounts for the rarity of early LSA/Upper Paleolithic sites in both northern and southern Africa.

Together with an obvious preservation bias against older sites, the possible effects of climatic change must be considered before one uses site numbers or densities to draw inferences about the relative numbers of Mousterian/MSA or later people. Still, including only sites that are about equally rich in occupational debris and that were probably occupied about the same length of time under broadly similar climatic conditions, it appears that, per unit time, Mousterian sites were much less abundant in Europe than were Upper Paleolithic ones. In southwestern France and neighboring northwestern Spain, there is probably less than one Mousterian cave for every five Upper Paleolithic ones (478, 1535, 1536, 2057). The implication is that the Mousterians were less numerous than their successors, probably because they exploited the available animal and plant resources less efficiently. In addition, the bodily robusticity of the Neanderthals, which was probably linked to their relatively unsophisticated technology, meant they required more calories per individual than their successors, which in turn meant they could support fewer individuals with a given number of calories. Thus, even if Neanderthals extracted as much energy from nature as their successors, their population would have been smaller.

Smaller site numbers could also be used to infer smaller Mousterian/MSA populations in Africa relative to Upper Paleolithic/LSA ones, but the data are more problematic because of a smaller total site sample and because of the known complication introduced by climatic effects. However, in southern Africa smaller MSA populations are probably implied by the relatively large size of limpets and tortoises in MSA sites compared with those in LSA sites occupied under broadly similar climatic conditions (figs. 6.41, 6.42) (1243). Both limpets and tortoises grow more or less continuously through life, and larger average size in MSA sites implies limited human predation pressure, probably because MSA populations themselves were relatively small.

Demography and Disease

The Neanderthals and their contemporaries are represented by both sexes and by both children and adults. Yet it is impossible to estimate most Neanderthal demographic statistics, because the total fossil sample is relatively small and because it is a composite of many subsamples. In the terminology used to describe large ungulate mortality in the section above on Mousterian/MSA ecology, individual Neanderthal subsamples could reflect "attritional" mortality, "catastrophic" mortality, or some mix of the two. There is also the problem of pre- and postdepositional bias, particularly against younger individuals, whose bones are much less durable than those of adults and are therefore far more likely to have been fragmented or destroyed by carnivore feeding, profile compaction, and leaching. In short, the available sample is unlikely to reflect the fundamental, time-averaged pattern of attritional mortality among

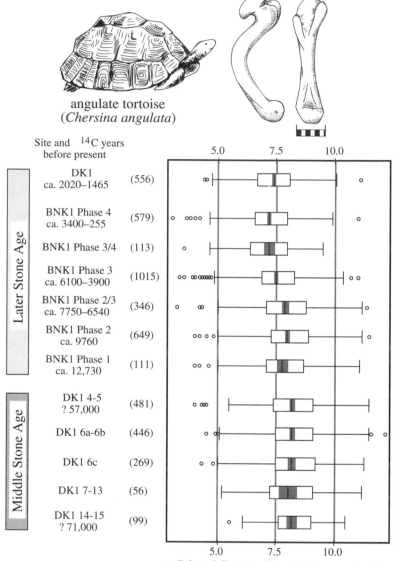

angulate tortoise
(*Chersina angulata*)

Site and ¹⁴C years before present		

Figure 6.41. Mediolateral diameters (breadths) of angulate tortoise (*Chersina angulata*) distal humeri from the MSA and LSA deposits of Die Kelders Cave 1 and the LSA deposits of nearby Byneskranskop Cave 1, South Africa. The boxplot format is described in the caption to figure 6.40. The most essential elements are the median, indicated by the vertical line near the center of each plot; the 95% confidence limits for the median, indicated by the shaded rectangle around each median; and the number of measured humeri, shown in parentheses. Medians whose 95% confidence limits do not overlap are distinct in the conventional statistical sense. The plots show that MSA tortoises were significantly larger than LSA specimens. The difference suggests that MSA people collected tortoises less intensively, probably because MSA human populations were smaller. (DK = Die Kelders; BNK = Byneskranskop.)

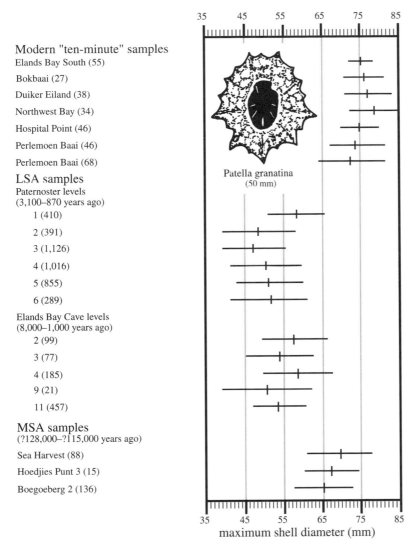

Figure 6.42. Maximum diameter of limpet (*Patella granatina*) shells in modern samples, Later Stone Age (LSA) samples from the Paternoster Midden and Elands Bay Cave, and Middle Stone Age (MSA) samples from the Sea Harvest Midden site, Hoedjies Punt 3, and Boegoeberg 2, South Africa. In each case the vertical line indicates the mean, the horizontal line indicates one standard deviation from the mean, and the figure in parentheses after the sample name is the number of specimens measured. The modern samples were collected by students, at ten-minute intervals, from intertidal rocks that are not being exploited today. The data suggest that LSA people exploited limpets more intensively than did their MSA predecessors, probably because the LSA people were more numerous. (The modern and LSA data are from 382; the MSA data are original here.)

Neanderthals, and the only vital statistic that it may reasonably reveal is maximum life expectancy.

If we assume that Neanderthals and living people shared the same basic pattern of skeletal development, the ages of adult Neanderthals at death can be estimated from macroscopic features such as the morphology of the pubic symphysis and the sacroiliac joint, the degree of cranial suture obliteration, and the extent of dental wear. Among these, dental wear is the least reliable in absolute terms, because on average the Neanderthals plainly wore their teeth more rapidly than most living people do. Ultimately, the most accurate age estimates for Neanderthal adults may come from microscopic examination of fossilized bone cells or osteons in long bone shafts (2180). At the moment, macroscopic and osteon aging of the oldest Neanderthals suggests they were only in their late thirties to middle forties (2221, 2231, 2239). They may have been even younger, if the rate of aging in Neanderthal adults matched the relatively rapid rate of development that has been suggested from the microscopic growth increments (perikymata) on the incisor enamel of a Neanderthal child from Devil's Tower, Gibraltar. These increments accumulate at the rate of about one a week, and their number implies that the Devil's Tower child died at about four years of age, although its state of dental eruption previously suggested an age closer to five (619, 2091, 2092). Neanderthals and earlier humans also obtained their second and third molars (M2s and M3s) at an earlier age than living people do (2204).

As discussed in the next chapter, in contrast to the Neanderthals, their early modern successors had a maximum life expectancy exceeding fifty years. One important implication is that early modern populations probably contained many more individuals beyond reproductive age. The knowledge and experience that older people could provide, especially in times of crisis, probably compensated for the economic burden they may have placed on younger people.

If Neanderthal life expectancy was short by later standards, then Neanderthal skeletons suggest that the proximate cause was great physical stress (194, 2231). Thus, in a sample of 669 carefully examined teeth from roughly 165 Neanderthals, hypoplastic enamel defects (pitting or grooving indicating periods of arrested enamel growth) appeared on 36%, representing 57% of the individuals (1630). The frequency of hypoplastic defects rarely reaches this level in living humans, and it was much lower in Upper Paleolithic people (350). The implication is that young Neanderthals were much more commonly subjected to food shortage, trauma, or disease (1611). In addition, the skeletons of older Neanderthals often exhibit healed fractures of the skull or limbs, degeneration of the joints, advanced osteoarthritis of the vertebral column, and periodontal disease. The most famous case is certainly the "old man" of La Chapelle-aux-Saints, who at age forty (or less) suffered from severe periodontal disease that caused substantial antemortem tooth loss and from joint degeneration or arthritis that are manifest in the articulation of the mandible to the skull and in the spinal column, hip, and foot (587, 2220). Other Neanderthals with obvious pathologies or injuries include Kebara 2, Tabun 1, Shanidar individuals 1 and 3–5, La Ferrassie 1 and 2, La Quina 5, the original (Feldhofer) Neanderthal from Germany, as many as five of the Krapina Cave Neanderthals, and the Sala 1 individual (Czech Republic) (194, 2211, 2215).

Kebara 2 had healed fractures of the fifth thoracic vertebra and the left second metacarpal. Shanidar 1 was afflicted by numerous partly interrelated traumatic and degenerative lesions, including especially a crushing fracture of the left orbit and cheekbone (fig. 6.43), a withered right upper arm from which the forearm had been lost before death, and degenerative or posttraumatic deformities of both legs that probably caused a limp (566). Shanidar 3 suffered from debilitating arthritis of both the right ankle and adjacent foot joints and probably died of a stab wound that pierced the lung, leaving an ugly scar on the left ninth rib (752, 2049). Shanidar 4 had a healed rib fracture (also seen in the individual from La Chapelle-aux-Saints), and Shanidar 5 had a large scar, caused by a sharp blow to the head, on the left frontal. One of the Krapina Neanderthals had a broadly similar scar on one parietal. Like other aged Neanderthals, La Ferrassie 1 endured periodontal disease. He also had a healed fracture on the right femur and bony lesions on both femurs, tibiae, and fibulae that reflect either systemic infection or carcinoma (723). Tabun 1 and La Ferrassie 2 had damaged fibulae. La Quina 5 suffered a withered left arm. The original Neanderthal individual and one of the Krapina Neanderthals had injured ulnae that deformed their forearms. One or more of the Krapina people had a broken clavicle (collarbone) and a concussed parietal. The Sala 1 person had a healed lesion on the right supraorbital torus.

Antemortem trauma has also been observed in adult skeletons of the earliest fully modern people, but much less frequently (2225). In fact, in the frequency of trauma and in its anatomical patterning, the Neanderthals closely recall modern rodeo riders (194), although the activities that produced the

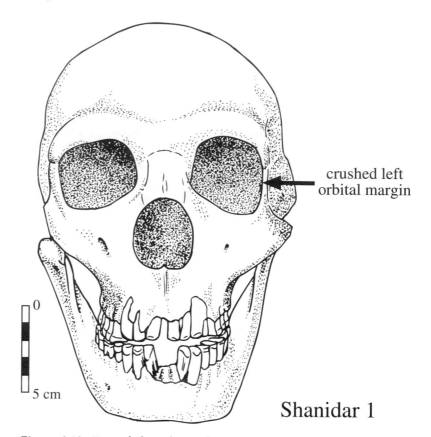

crushed left
orbital margin

Shanidar 1

Figure 6.43. Face of Shanidar 1, showing the crushed outer margin of the left orbit (drawn by Kathryn Cruz-Uribe from a photograph; © 1999 by Kathryn Cruz-Uribe). The crushing was not fatal and the orbit healed, but associated brain damage may have partially paralyzed the man's right side, accounting for a withered right arm. Like other injured or aged Neanderthals, he survived only with the help of other members of his group.

Neanderthal injuries were surely more mundane. The most fundamental implication is that Neanderthal technology (or culture in general) was relatively ineffective at reducing wear and tear on Neanderthal bodies. However, the same skeletal pathologies and injuries that show the Neanderthals lived risky lives and aged early also reveal a strikingly human feature of their social life. The La Chapelle-aux-Saints and Shanidar 1 individuals, for example, must have been severely incapacitated and would have died even earlier without substantial help and care from their comrades. The implicit group concern for the old and sick may have permitted Neanderthals to live longer than any of their predecessors, and it is the most recognizably human, nonmaterial aspect of their behavior that can be directly inferred from the archeological record.

Conclusion: The Fate of the Neanderthals

It is not difficult to understand why the Neanderthals failed to survive after behaviorally modern humans appeared. The archeological record shows that in virtually every detectable aspect—artifacts, site modification, ability to adapt to extreme environments, subsistence, and so forth—the Neanderthals were behaviorally inferior to their modern successors, and to judge from their distinctive morphology, this behavioral inferiority may have been rooted in their biological makeup.

A far more difficult question is how the Neanderthals succumbed. Were they physically replaced by anatomically modern intruders, or did they evolve into modern people? Increasingly the answer seems to be that they were physically usurped, but the issue remains controversial. It is best addressed by considering the evidence from various regions. Figure 6.44 encapsulates some of the key time-space relationships.

Western Europe

Physical displacement seems almost certain for western Europe, now that the Saint-Césaire skeleton, the Arcy temporal fragment and (taurodont) teeth, and the Zafarraya mandible show that Neanderthals survived in France and Spain to 40 ky ago and perhaps to as recently as 30 ky ago. Assuming that Neanderthals were the sole makers of Mousterian artifacts in western Europe, their presence to roughly 40 ky ago is reconfirmed by TL dates from Le Moustier in France (2272) and U-series dates from Abric Romaní in Spain (228, 229). The geologic antiquity of the oldest anatomically modern fossils in western Europe is not well fixed, but (early Aurignacian) artifact associations suggest ages between 40 and 30 ky ago for fully modern mandibles from Isturitz and Les Rois, France (776). The much more complete early modern skeletons from the famous Cro-Magnon shelter are probably about 30 ky old, based on their presumed (later Aurignacian) artifact associations (777, 1592). If the oldest known west European Aurignacian artifacts signal modern human presence even when human fossils are absent, then ^{14}C dates from Spain and Belgium show that modern humans appeared in western Europe about 40 ky ago (2066).

The Saint-Césaire and Arcy Neanderthal remains might be used to argue for continuity between Neanderthals and later people, since the accompanying artifacts are assigned to the Châtelperronian Industry. This is commonly regarded as the earliest Upper Paleolithic manifestation in the limited area of

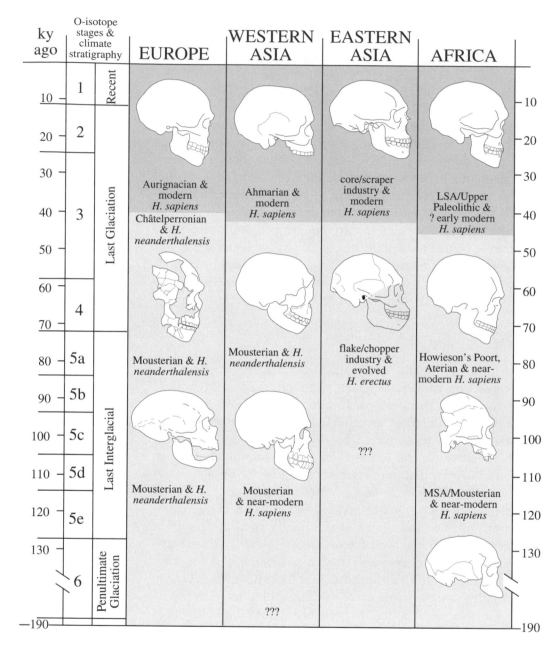

Figure 6.44. Approximate chronological arrangement of various late Pleistocene cultural units and fossil humans types in Africa and Eurasia (modified after 1242, p. 6).

northern Spain and western and central France (west of the Rhône River) where it occurs (fig. 6.45) (970–972, 1362), and the Upper Paleolithic, in turn, is usually equated with anatomically modern people. However, in terms of stone artifacts the Châtelperronian resembles the Mousterian of Acheulean Tradition Type B (with numerous backed knives), and its distinctive Upper Paleolithic traits may reflect the diffusion of

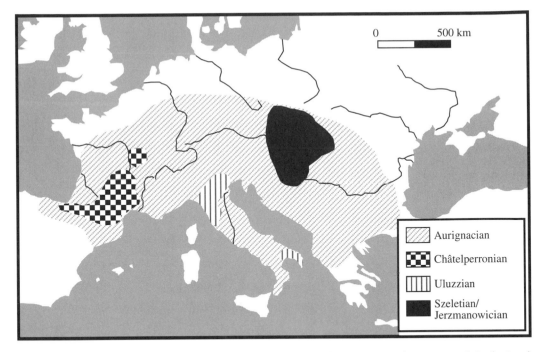

Figure 6.45. The distribution of the Aurignacian, Châtelperronian, Uluzzian, and Szeletian/Jerzmanowician Industries in Europe (redrawn after 1538, fig. 1).

Upper Paleolithic elements into a local Mousterian context (646, 970, 972, 1232). These elements include well-made end-scrapers and burins as well as formal bone artifacts and pierced animal teeth from the remarkable Châtelperronian layers of the Grotte du Renne at Arcy-sur-Cure (fig. 6.46) (707–709, 1358, 1593).

At the Grotte du Renne, the Châtelperronians also modified their living area to an extent that is common only in the Upper Paleolithic. The Grotte du Renne Châtelperronian layers contain traces of several "hut emplacements," of which the best preserved is a rough circle of eleven postholes enclosing an area 3–4 m across that was partially paved with limestone plaques (fig. 6.47). Pollen analysis indicates that wood was rare nearby, and the postholes probably supported mammoth tusks, which are more abundant in the Grotte du Renne than in any other known Paleolithic cave.

The most probable diffusionary source for the Châtelperronian was the Upper Paleolithic Aurignacian Industry, which clearly intruded into western Europe. The interstratification of Châtelperronian and early Aurignacian layers at the Roc du Combe and Le Piage shelters (southwestern France) and at El Pendo Cave (northern Spain) plainly demonstrates chronological overlap. Intersite correlations based on paleoclimatic

Figure 6.46. Stone and bone artifacts from the Châtelperronian levels of the Grotte du Renne, Arcy-sur-Cure, north-central France (redrawn after 1593, figs. 3–8). Numerous well-made burins, bone artifacts, and items of personal adornment justify assigning the Châtelperronian to the Upper Paleolithic, but the morphology of the backed knives suggests links to an immediately preceding variant of the Mousterian (Middle Paleolithic), and the possibility exists that the Châtelperronian resulted from the diffusion of Upper Paleolithic traits into a basically Mousterian context. Châtelperronian people, insofar as they are known, were Neanderthals.

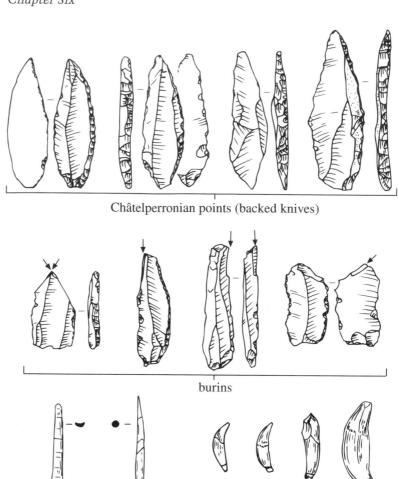

Châtelperronian points (backed knives)

burins

modified animal teeth ("pendants")

0   5 cm

Grotte du Renne,
Arcy-sur-Cure

bone artifacts

inferences from pollen and sediments suggest that Aurignacian people may have spread progressively northward within western France (1360). This interpretation is tentatively supported by previously cited ^{14}C dates that place the earliest Aurignacian near 40 ky ago in eastern and northern Spain, but only 33 ky ago near the Grotte du Renne (Paris Basin), which is the northernmost known Châtelperronian site. However, the Aurignacian had also penetrated Belgium by 40 ky ago (2066), and fresh ^{14}C dates may show that it appeared equally early throughout

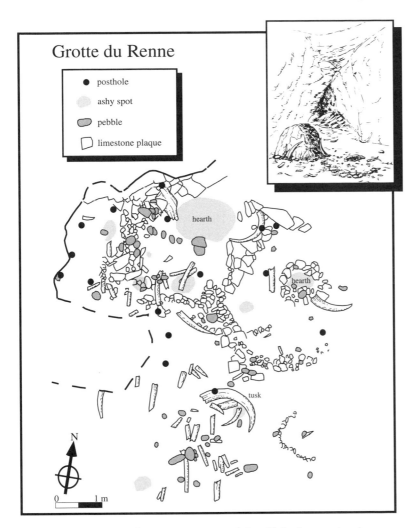

Figure 6.47. Plan and reconstruction of the Châtelperronian hut emplacement in level XI of the Grotte du Renne, Arcy-sur-Cure, north-central France (redrawn after 707, fig. 5). The arrangement of postholes, plaques, hearths, and large bones suggests a small covered hut that was immediately outside the mouth of the cave. The apparent patterning on the living floor is typical only for the Upper Paleolithic.

France. If so, it perhaps overlapped the Châtelperronian for only a millennium or two about 40 ky ago. A southward spread of the Aurignacian about the same time may explain the Uluzzian Industry in Italy, which resembles the Châtelperronian in its mix of Middle and Upper Paleolithic stone tool types (211, 840, 841, 1646).

Central Europe

In central Europe the skeletal evidence bearing on the relation between Neanderthals and early modern people is more ambiguous, partly because the Neanderthal sample consists almost entirely of fragments rather than of complete or nearly complete skulls and skeletons. Based on geological context and artifact or faunal associations, the central European Neanderthals have been divided into an early group (from Krapina in Croatia, Gánovce and Ochoz in the Czech Republic, and Subalyuk Cave in Hungary), which perhaps dates from the Last Interglacial and the earliest part of the Last Glaciation, and a later group (from Vindija Cave in Croatia and Kulna Cave, Sipka Cave, and perhaps Sala in the Czech Republic), which may date from the middle of the Last Glaciation (1998, 2000). Most specimens come from Krapina (about nine hundred pieces) (1724, 1955) and Vindija (about forty-five pieces) (2002, 2003, 2490). Compared with earlier central European Neanderthals, the later ones had smaller, thinner, and less projecting browridges, foreheads that were perhaps less receding, smaller and somewhat less prognathous faces, mandibles with a greater tendency for chin development, and perhaps smaller anterior teeth (1998, 2000, 2003, 2487). They may thus have been evolving toward early modern central Europeans, such as those from the Mladec (Lautsch) Caves and Predmostí in the Czech Republic, who perhaps approached the late Neanderthals in browridge size and general cranial ruggedness.

However, the argument for a direct link between central European Neanderthals and their earliest fully modern successors is problematic, not only because the Neanderthal fossils tend to be so fragmentary but also because the sample of truly diagnostic specimens is relatively small, especially with regard to the supposed later Neanderthals. In addition, most of the specimens are weakly dated, and their context and associations are often poorly established or specified. Finally, the trends that have been postulated from the earlier to the later Neanderthals plainly involve reductions in craniofacial size or ruggedness more than they do changes in morphology (1085, 2075). The early modern (Aurignacian and Pavlovian) inhabitants of central Europe, especially the males, certainly had rugged skulls, often with large browridges, prominent muscle markings, and occipitals that exhibit some posterior projection (a hemibun) and extensive nuchal planes. In overall morphological pattern, however, both cranially and postcranially they are unmistakably modern, with no true Neanderthal features (fig. 6.48).

Parallel to the skeletal remains, central European artifact assemblages may imply continuity between the Mousterian and the Upper Paleolithic, but the case is not persuasive. The problem is partly the bewildering array of industries that central European archeologists have identified. Thus, apart from

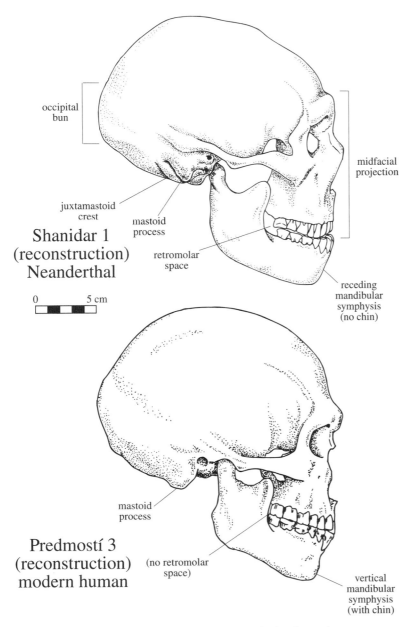

occipital
bun

midfacial
projection

juxtamastoid
crest

mastoid
process

retromolar
space

receding
mandibular
symphysis
(no chin)

Shanidar 1
(reconstruction)
Neanderthal

0   5 cm

mastoid
process

Predmostí 3
(reconstruction)
modern human

(no retromolar
space)

vertical
mandibular
symphysis
(with chin)

Figure 6.48. Reconstructed skulls of individual 1 from the Mousterian levels of Shanidar Cave, Iraq, and of individual 3 from the early Upper Paleolithic site of Predmostí, Czech Republic (drawn by Kathryn Cruz-Uribe from photographs and casts; © 1999 by Kathryn Cruz-Uribe). The Shanidar skull is from a typical Neanderthal, whereas the Predmostí one comes from a robust, early anatomically modern European. Like other Neanderthal skulls, the one from Shanidar differs from modern skulls in its relatively longer, lower braincase and long, forwardly projecting face, as well as in more detailed features such as its large juxtamastoid crest. The striking protrusion of the Shanidar face is perhaps most obvious in the forward placement of the nose and front teeth relative to the orbits. Like other Neanderthal mandibles, the Shanidar specimen has a large retromolar space, reflecting the combination of a long mandible, a short postcanine (premolar and molar) row, and a relatively narrow ascending ramus.

assemblages ascribed to the Mousterian, broadly similar ones have been assigned to the Altmühlian, Micoquian (or Micoquian-Prondnikian), Bohunician, Jankovichian, Jerzmanowician, and Szeletian Industries (25, 26, 2116, 2119, 2120, 2280, 2281). The distinctions between the Bohunician, Jankovichian, Jerzmanowician, and Szeletian are particularly vague, and for present purposes they can all be lumped into the Szeletian, which was described first (25, 26). Even after lumping, however, there are still few Szeletian assemblages that comprise more than a few dozen retouched pieces, and the implication is that Szeletian populations were generally very sparse. At the type site of Szeleta Cave (Hungary) and at other key localities, cave bears seem to have been present more often than people.

The Szeletian is unified by the presence of bifacially worked pieces, mainly (but not always) leaf-shaped (foliate) points (fig. 6.49). In other respects Szeletian assemblages are remarkably diverse. Most are dominated by Mousterian sidescrapers, denticulates, and so forth, sometimes on Levallois flakes, but some are relatively rich in endscrapers and burins, and it is these that suggest continuity from the Mousterian to the Upper Paleolithic. Three isolated human teeth (two mandibular incisors and a mandibular canine) associated with Szeletian (or Jankovichian) artifacts at Upper Remete Cave (Máriaremete-Felsö, Hungary) have been described as Neanderthal (774), and this could mean that Neanderthal Szeletians were directly ancestral to early modern central Europeans.

There is the problem, however, that some important Szeletian assemblages come from surface sites and others were excavated long ago with very weak stratigraphic control. In both cases an apparent mix of Mousterian and Upper Paleolithic stone artifact types could result from the inadvertent amalgamation of initially separate Mousterian and Upper Paleolithic components. Moreover, even if an Upper Paleolithic Szeletian truly existed, it need not imply a local origin for the Upper Paleolithic. Instead, like the Châtelperronian and the Uluzzian, it could signal the diffusion of Aurignacian traits into a Mousterian context (25, 26, 2280, 2281, 2283). Radiocarbon dating suggests that the Szeletian antedates 30 ky ago, and it may date mainly from between 45 and 40 ky ago when the early Aurignacian appeared locally.

As in western Europe, so in central Europe, the Aurignacian represents the earliest unequivocal Upper Paleolithic, and it was unquestionably intrusive (26, 1265, 1539, 2280, 2281). TL dates from Temnata in Bulgaria and ^{14}C dates from Istállóskö in Hungary, Bacho Kiro in Bulgaria, and Willendorf in Austria show that the Aurignacian encroached on central Europe before

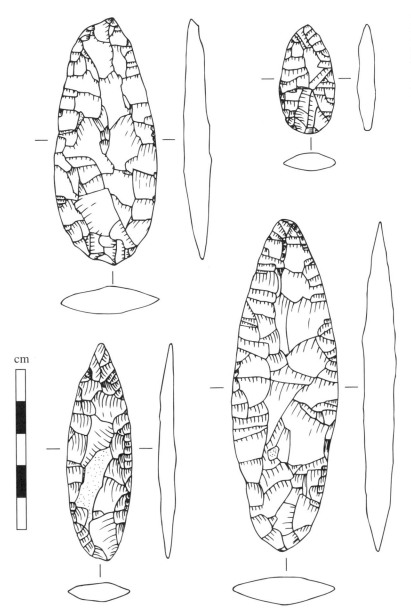

Figure 6.49. Bifacial, leaf-shaped points from the upper deposits of Szeleta Cave, Hungary (redrawn after 25, fig. 18). Such points typify the Szeletian variant of the Middle Paleolithic in central Europe.

40 ky ago (26, 1262, 1267, 1538, 1539, 2066, 2119). The available dates tentatively suggest that the Aurignacian intruded on central Europe a few millennia before it appeared in western Europe, but many fresh dates are needed for confirmation. A difference in time may in fact be too small to detect with existing dating methods.

The physical identity of the earliest central European Aurignacians is not firmly established, and it has been suggested that they were Neanderthals based on the discovery of an Aurignacian split-base bone point in the topmost layer (G1) with Neanderthal fossils at Vindija Cave (1183, 2002). The association is not compelling, however, partly because cryoturbation (frost heaving) is said to have heavily disturbed the Vindija deposits and partly because published reports on the stratigraphy are inconsistent (25). A human parietal fragment, two very incomplete mandibles, and five isolated teeth from Bacho Kiro Cave hint that the earliest Aurignacians were anatomically modern (1262). Somewhat later Aurignacians, like those whose remains have been putatively dated between 40 and 30 ky ago at the Mladec Caves, were unquestionably modern (756, 776, 1093, 2004).

Eastern Europe

Until recently it could be claimed that both Neanderthals and modern humans produced Mousterian artifacts in eastern Europe, and from this that Neanderthals evolved locally into modern humans. The evidence came mainly from the Crimea, where Mousterian artifacts were accompanied by Neanderthal remains at Kiik-Koba and Zaskal'naya VI and by fully modern remains at Starosel'e Cave (1260, 2269, 2330). However, the Starosel'e association was problematic from the time it was found, since the excavator could not distinguish any stratigraphy in a 4 m thick Mousterian profile, and chemical analysis suggested that the modern human bones postdated accompanying animal bones (1229). A meticulous reexcavation of Starosel'e has now uncovered two additional modern skeletons at roughly the same depth in the Mousterian deposit as the original skeleton (1449). However, the two new skeletons clearly lay in intrusive seventeenth-century or later Tatar graves. One of them was laid out exactly like the original one, and since the positioning accords with known Tatar custom, the previously proposed Mousterian/early modern human association at Starosel'e should be discarded. As in western and central Europe, so in eastern Europe, it still appears that only Neanderthals made Mousterian artifacts.

There is no compelling evidence for continuity between the Mousterian and the Upper Paleolithic in eastern Europe, but the Upper Paleolithic was established locally by 40 ky ago (60, 1029, 1033), as early as or earlier than in central and western Europe. The earliest east European Upper Paleolithic industries are remarkably diverse, and with minor and dubious exceptions, they exhibit little or no similarity to the Aurignacian. Thus, if the Upper Paleolithic generally signals the arrival of modern humans, their migration to eastern Europe clearly followed a distinct course. The reason may be that, unlike central and western Europe, eastern Europe was inhabited by Neanderthals only along its western and southern peripheries. Over most of eastern Europe, Upper Paleolithic people colonized environments where no one had lived before.

Western Asia

Western Asia arguably provides the best case for an evolution of Neanderthals into modern humans— from Neanderthals like those represented at Tabun, Amud, Kebara, and Dederiyeh Caves into near-modern people like those at Skhul and Qafzeh Caves (fig. 6.50). Such a succession is contradicted, however, by associated faunas and artifacts, and above all by TL and ESR dates showing that the Neanderthals mainly or entirely postdate the Skhul/Qafzeh people. With this in mind, it is notable that the near-modern people are known only from Israel, on the immediate southwest Asian periphery of Africa, and that they were associated with typical African mammals. The Neanderthals in contrast have been found throughout western Asia, and they were associated with exclusively Eurasian faunal elements. The sum suggests that the Skhul/Qafzeh people were simply near-modern Africans who extended their range slightly to the northeast during the relatively mild and moist conditions of the Last Interglacial, between 127 and 71 ky ago. The Neanderthals may have been migrants from the north or east who displaced the Israeli near-moderns when climatic conditions deteriorated at the beginning of the Last Glaciation, roughly 71 ky ago (116, 1069, 2154, 2155, 2304, 2305).

The (Tabun C–type) artifact assemblages that accompany the Skhul/Qafzeh people have no compelling northeast African antecedents, but the (Tabun B–type) assemblages that accompany the Neanderthals at Tabun, Kebara, Amud, and Dederiyeh Caves could derive from earlier assemblages farther north and east in Europe or western Asia (1448), and they could thus support a Neanderthal incursion. The Neanderthals may have accomplished the replacement, because they had relatively thick trunks and short limb segments that specially adapted them to cooler climatic conditions. There is nothing in the cultural debris of the Israeli near-moderns (or of their putative near-modern African ancestors and contemporaries) to suggest that they had a cultural (behavioral) advantage over the Neanderthals. In fact, from a strictly archeological (behavioral) perspective, there is nothing to differentiate the Qafzeh/Skhul people (or their African contemporaries) from the Neanderthals in either western Asia or Europe.

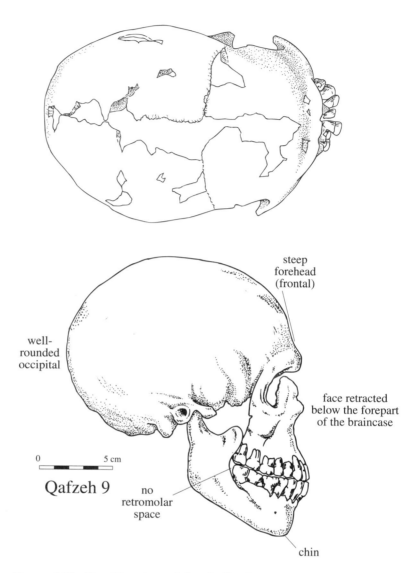

steep
forehead
(frontal)

well-
rounded
occipital

face retracted
below the forepart
of the braincase

0   5 cm

Qafzeh 9   no
retromolar
space

chin

Figure 6.50. *Top:* Top view of the skull 9 from Qafzeh Cave, Israel, partially reconstructed. *Bottom:* Lateral view of the same skull, more fully reconstructed (adapted from 2304, figs. 9, 10, 43). The Qafzeh fossils are dated to roughly 90 ky ago, when Neanderthals were the sole inhabitants of Europe, and like the Neanderthals, the Qafzeh people produced Mousterian artifacts. Unlike the Neanderthals, however, they closely resembled modern people in key morphological respects, including a relatively high, short braincase that overlapped the face below. Retraction of the face beneath the braincase is reflected in the absence of a retromolar space and in the presence of a chin. The descendants of the Qafzeh people may have been displaced by Neanderthals, who spread to Israel after 80 ky ago.

In western Asia as in Europe, the disappearance of the Neanderthals is closely tied to the appearance of the Upper Paleolithic Industrial Complex. This is distinguished from the preceding Mousterian by essentially the same features as in Europe (820, 821), including the manufacture of formal bone artifacts and art objects or items of personal adornment (126, 1042). Upper Paleolithic human remains are scarce in western Asia (126, 1017, 1956), but where they occur they are uniformly modern. This is true not only of later Upper Paleolithic people, postdating 20 ky ago, but also of earlier ones, including especially the remains of a child from an early Upper Paleolithic grave at Ksar 'Akil, Lebanon. The associated (Ahmarian) artifacts and ^{14}C dates from overlying layers suggest the child died at least 37 ky ago (195), and the implication is that west Asian Upper Paleolithic people were modern from the very beginning.

The best available dates place the Upper Paleolithic in western Asia by 40 ky ago and perhaps by 45 ky ago (125, 126, 1042, 1447, 1677, 2014). If a time approaching 45 ky ago is corroborated in the future, it might mean the European Upper Paleolithic originated in western Asia. The earliest west Asian Upper Paleolithic industry, the Ahmarian, has no known European counterpart, but it was followed by a variant of the Aurignacian from which the European early Aurignacian might derive (1532). As mentioned previously, a possible west Asian origin for the Upper Paleolithic is strongly suggested at the Israeli site of Boker Tachtit, where a progressive change from Mousterian to Upper Paleolithic core technology and a concomitant increase in Upper Paleolithic tool types has been fixed between 47 and 38 ky ago (1447).

Eastern Asia (Including Australasia)

Eastern Asia (the eastern portion of the Indian subcontinent and the area yet farther east that Europeans call the Far East) has provided less firm evidence for modern human origins than any other major region, and this situation must be reversed for a conclusive resolution of the problem. At the moment, however, for the time period between 200 and 50 ky ago that is of principal concern, both the fossil and archeological records suggest that eastern Asia continued to follow a unique course that was set long before. Thus, before 50–40 ky ago, when western Asia was variably occupied by Neanderthals of possible European origin or by near-modern people of possible African origin, eastern Asia seems to have been occupied by two or more distinctive human types that were neither Neanderthal nor modern and that probably evolved locally. In Java there were the Ngandong (or Solo) people, whose cranial morphology suggests direct descent from classic Javan *Homo erectus*. In China there were the populations represented by the Dali, Xujiayao, Maba, and Yinkou skulls, which might be called "archaic" *Homo sapiens* (or *Homo heidelbergensis*) if they had been found in Europe or Africa, but which are more likely to represent an evolved end product of Chinese *Homo erectus*.

The east Asian archeological record is equally distinctive. To begin with, there is no evidence that a typically Upper Paleolithic blade technology was ever introduced to southern China and adjacent Southeast Asia, and it can be argued that a relatively crude flake-and-chopper technology continued there, largely unchanged, from the time of classic *Homo erectus*, before 250 ky ago, until 12–10 ky ago or even later (1700). Even more important, there is no evidence for the relatively abrupt appearance of formal bone artifacts, art, sophisticated graves, or other innovations that signal a behavioral metamorphosis in the west between 50 and 40 ky ago. However, the impression of remarkable continuity and conservatism in the Far East is based on only a small number of excavated sites, often poorly dated and unevenly described (185, 459, 1021, 1145, 1160, 1636), and it may not be sustained as research expands. One hint of what may actually have happened comes from Sri Lanka to the southwest, where excavations in Batadomba Iena Cave have uncovered sophisticated stone tools ("geometric microliths"), formal bone artifacts, and modern human remains in a layer dated to roughly 28.5 ky ago (1211).

The pattern that may eventually emerge is also suggested by the relatively well known archeological records of northern Asia (Siberia) to the north and, even more strongly, by the record of Australasia (Australia, New Guinea, and Tasmania) to the southeast, both of which are detailed in the next chapter. The harsh climatic conditions of northern Asia seem to have largely precluded human occupation until about 40 ky ago, when people making Upper Paleolithic–like stone tools, formal bone artifacts, and art objects spread across the southern, more temperate margin of Siberia (2310). Sometime between 35 and 20 ky ago, they extended their range northward into arctic and subarctic latitudes, and by 14–13 ky ago they were at the Bering gateway to North America. Substantial houses and fireplaces like those of their western contemporaries help explain how these people could live where no one had before. They also buried their dead in graves that resembled western ones in complexity. Stone artifact assemblages imply that the same broadly "Upper Paleolithic" complex penetrated southward to Mongolia, northern China, Korea, and Japan (459, 1720, 1756). It could in fact be represented at the famous Upper Cave at Zhou-koudian, with its elaborate burials of fully modern people. Radiocarbon dates on animal bones that probably accompanied some of these burials range between 30 and 10 ky ago (1174).

The situation in Australasia is even more directly relevant, for the first Australasians surely came from Southeast Asia. The timing of initial colonization is debatable, but people were certainly present by 40 ky ago and perhaps by 60 ky ago (1162, 1791). If the date is only 40 ky ago, a fully modern skull from a layer radiocarbon dated to about 40 ky ago at Niah Cave, Borneo, could represent the parent population. The physical appearance of the first Australasians is unknown, but those after 30 ky ago were fully modern, if highly variable in morphology. They made nondescript stone artifacts, but at least those after 30 ky ago produced formal bone artifacts, art objects, and relatively elaborate graves that broadly recall those of their Upper Paleolithic/LSA contemporaries in western Asia, Europe, and Africa.

The initial colonization of Australasia might itself be regarded as an index of behavioral modernity, since it required the invention of watercraft that could cross 90 km of open ocean. In this light, the timing of the first crossing becomes critical. If it occurred 60 ky ago, the colonists would arguably be the oldest known behaviorally modern people in the world. They would antedate comparably modern Europeans by roughly 20 ky, and the Out of Africa theory of modern human origins espoused in the next chapter would have to be modified to allow for at least two modern human dispersals, an earlier one eastward to southern Asia and Australasia and later one northward and westward to western Asia and Europe. Depending on future research on the Asian mainland, the theory might even have to be abandoned in favor of one that accords a more central role to the Far East. The bottom line is that speculation surrounding a possible 60-ky-old colonization of Australia serves to underscore the murkiness of the east Asian archeological and fossil records bearing on modern human origins.

Africa

Among all the places from which skeletal evidence is available, Africa appears to present the best case for a local or regional evolution of anatomically modern people. This is suggested not only by the fossil evidence presented in this chapter, but also by genetic evidence summarized in the next. An African origin for modern humans could explain not only what happened to the Neanderthals but also why, unlike them, their fully modern successors had relatively long distal limb segments (forearms and lower legs), suggesting an ancestral adaptation to warmer climes.

If modern humans originated in Africa, as the fossil record implies, a puzzling question is why they failed to spread more widely before 50 ky ago. A partial answer, supported by archeological findings summarized in this chapter and reiterated in the next, is that it was only about 50 ky ago that they acquired an obvious competitive advantage over their nonmodern, Eurasian contemporaries. This advantage is signaled by the appearance of the Later Stone Age (Africa) and Upper Paleolithic (western Asia and Europe), and it arguably followed on a mutation that promoted the development of the fully modern brain. A neurological explanation for the Upper Paleolithic/LSA and its nonbiological alternative explanations are explored further in the next chapter. In the present context it is more pertinent that a behaviorally based spread from Africa requires that Africa contain the earliest evidence for the behavioral advance. Much of Africa seems to have been depopulated during the period between 60 and 40 ky ago when the behavioral advance occurred, but Enkapune Ya Muto, Kenya, shows that the LSA had begun in eastern Africa by at least 46 ky ago (31). This is 6 ky before the Upper Paleolithic had appeared in Europe and perhaps 3 ky before it appeared in western Asia. Enkapune Ya Muto and other early LSA sites in eastern Africa suggest that eastern Africa is the place to seek not only the earliest evidence of fully modern behavior, but also additional human fossils that could conclusively demonstrate an African origin for modern humans. Only eastern Asia remains contradictory, if the colonization of Australasia required the kind of behavior manifested in the Upper Paleolithic/LSA and if Australasia was actually colonized 60 ky ago.

Overview

In sum, both the question of what happened to the Neanderthals and the closely related question of where and when modern people originated remain incompletely resolved. On present evidence, however, it is certainly reasonable to conclude that modern morphology appeared first in Africa, perhaps including its immediate southwest Asian margin. Initially, the behavioral capabilities of early modern or near-modern Africans differed little from those of the Neanderthals, but eventually, perhaps because of a neurological change, they developed a capacity for culture that gave them a clear adaptive advantage over the Neanderthals and all other nonmodern people. The result was that they spread throughout the world, physically replacing all nonmoderns, of whom the last to succumb were perhaps the Neanderthals of western Europe. This hypothesis is developed further in the next chapter, devoted specifically to the earliest fully modern humans.

ANATOMICALLY MODERN HUMANS 7

The limited extent of modern human genetic diversity implies that all living humans share a very recent common ancestor. The mutation rates that probably underlie present diversity place the common ancestor near 200 thousand years (ky) ago, give or take a few tens of thousands of years. A combination of fossil and genetic evidence locates the ancestral population in Africa, and archeological discoveries imply an initial dispersal out of Africa about 50 ky ago. A more conclusive, more precise statement will require a much denser fossil record and either new dating methods or significant refinements to existing ones that permit dates between 200 and 50 ky ago. Much new research is also needed to determine whether there were multiple dispersals from Africa, perhaps at slightly different times and in different directions, and to establish whether African migrants exchanged genes and behavioral traits with some of the (nonmodern) Eurasians they encountered.

But even if important details remain to be fixed, the significance of modern human origins cannot be overstated. Before the emergence of modern people, the human form and human behavior evolved together slowly, hand in hand. Afterward, fundamental evolutionary change in body form ceased, while behavioral (cultural) evolution accelerated dramatically. The most likely explanation is that the modern human form—or more precisely the modern human brain—permitted the full development of culture in the modern sense and that culture then became the primary means by which people responded to natural selective pressures. As an adaptive mechanism, not only is culture far more malleable than the body, but cultural innovations can accumulate far more rapidly than genetic ones, and this explains how, in a remarkably short time, the human species has transformed itself from a relatively rare, even insignificant large mammal to the dominant life form on the planet. This chapter summarizes what is known about modern human origins, emphasizing especially the very significant behavioral differences between even very early modern people and their nonmodern predecessors.

History of Discovery

Until the mid-nineteenth century, human origins were a subject mainly for theologians, not scientists (681, 822, 881, 882). Medieval clerics summed the ages of post-Adamite generations in the Bible and concluded that the present world was created about 4,000 B.C. This estimate itself became gospel, and most eighteenth- and early nineteenth-century geologists and paleontologists accepted it. To explain extinct species they postulated a series of earlier, imperfect creations, each terminated by a great flood or some other global catastrophe. This interpretation, commonly termed "creationism" or "catastrophism," was widely questioned only after 1859, when Darwin's *Origin of Species* showed how natural selection could explain the evolution of living species from earlier ones.

Creationism clearly implied that people had not coexisted with long-extinct species. During the first half of the nineteenth century, however, fossil hunters searching in European caves repeatedly found bones of anatomically modern humans alongside those of extinct animals (881, 1626). Excavation methods were crude, and in some cases the associations were probably erroneous, created, for example, when excavators failed to recognize a relatively recent grave dug into much older deposits. In other cases, however, the associations were undoubtedly valid and implied that modern people had lived in Europe long before popular theology allowed. Since the theology was largely unquestioned, the associations were generally discounted, though some have been vindicated in the present century. Perhaps the most famous example was from Goat's Hole (Cave) at Paviland, South Wales, where the Rev. William Buckland, professor of geology at Oxford, excavated a skeleton in 1822–1823 (1571). It was covered by a layer of red ocher and was accompanied by numerous mammoth ivory artifacts, but Buckland concluded it belonged to a Welsh woman of the Roman era whose kinfolk worked tusks they found in the cave. In fact, direct radiocarbon dating now shows the skeleton is about 18.5 ky old, and it thus represents a person (actually a young male) who was a contemporary of the mammoth (1628).

Probably the best-known discovery after Paviland occurred in 1852, when a road repairman pulled a human bone from a rabbit hole in a hillside near Aurignac, southern France (1626). He dug a trench into the hillside and found a cave whose mouth was blocked by a collapsed limestone slab. Behind the slab lay the skeletons of seventeen people, while backdirt from the trench contained bones from extinct animals, some of them engraved. The skeletons were reburied in a local Christian cemetery, but the site attracted the attention of the pioneer French prehistorian Edouard Lartet, who excavated the cave floor systematically in 1860. He found some isolated human bones apparently associated with those of extinct mammals, but the association was never conclusively established. It is now thought the skeletons were of Neolithic (Holocene) age, postdating the extinct animals.

The first widely accepted discovery of clearly very ancient but anatomically modern human remains occurred only in 1868, when railway workers exposed deposits with human bones in the Cro-Magnon rockshelter in the town of Les Eyzies in the Dordogne region of southwestern France. An excavation by Edouard Lartet's son Louis showed that the bones came from a layer that also contained bones of mammoths, lions, and reindeer, along with numerous artifacts (1308). These latter included stone tools (of the kind now called Evolved Aurignacian), as well as artificially perforated seashells and animal teeth. The human bones represented five to eight people, including minimally a middle-aged male (subsequently called the "Old Man of Cro-Magnon"), two younger adult males, a young adult female, and a very young infant (776, 1675). From their well-established associations, it was clear that the human remains were very ancient, and radiocarbon dates associated with similar artifacts excavated in the 1950s at the nearby Abri Pataud now suggest an age of about 30 ky ago (1592). Ironically, for a brief period the Cro-Magnon skeletons could be cited against the concept of human evolution, for it could be argued that the Cro-Magnon people were in Europe as early as or earlier than nonmodern kinds of people, particularly the Neanderthals, who were discovered at broadly the same time. However, even before the turn of the century it became apparent that, though the Cro-Magnons were indeed very ancient, they still postdated the Neanderthals.

Not long after the discovery at the Cro-Magnon shelter, several other European sites provided modern human remains clearly associated with animal bones or artifacts indicating great antiquity. Some of the most important early discoveries were made at Lautsch (Mladec) (1881–1904), Brno (Francouzská Street) (1891), and Predmostí (1894), all in Moravia, Czech Republic (1998, 2000); at the Chancelade (Raymonden) Shelter (1888) and Combe-Capelle Cave (1909) in France; and at the

Figure 7.1. Approximate locations of major western and central European Upper Paleolithic sites.

Grimaldi Caves (Grotte des Enfants, Grotta del Caviglione, Barma Grande, and Bausso da Torre) in Italy (1874–1901) (1675, 2095) (fig. 7.1). Together these finds and those from many other, more recently excavated sites show that anatomically modern people have occupied Europe for at least the past 35 ky. The exact period may vary from place to place, as discussed below.

Remains of ancient, anatomically modern people have also been found in other parts of the world, though in smaller numbers than in Europe. The first were the Wajak skulls, found in central Java in 1890 (676). Among subsequent anatomically modern human fossils, some of the best known are those found in the "Upper Cave" at Zhoukoudian, northern China in 1933–34 (1143, 2401); at Ksar ʿAkil Cave, Lebanon, in 1938 (195, 553); at Niah Cave, Borneo, in 1958 (367, 1210); at Lukenya Hill, Kenya, in 1971 (879, 1762); at Lake Mungo, New South Wales, Australia, between 1968 and 1972 (308, 309); at Nahal Ein Gev 1, Israel, in 1976 (73); at Nazlet Khater, Egypt, in 1980 (2315); at Wadi Kubbaniya, Egypt, in 1982 (2418); and at Ohalo II, Israel, in 1991 (1017). Colloquially, all early modern people are sometimes called Cro-Magnons, after the early and historically significant French discovery. Technically, however, only people who closely resembled the original Cro-Magnons should be known by this term, and here all early anatomically modern people will be referred to simply as early moderns, meaning fossil people whose skeletons are not meaningfully distinguished from those of living people. Together with living people, early moderns are commonly assigned to the subspecies *Homo sapiens sapiens*. If it is accepted that all modern humans, living and fossil, share a very recent African ancestor, they may be separated from other humans simply as *Homo sapiens*.

Morphology

Any osteological diagnosis of modern humans must encompass a far greater range of variation than is known for earlier non-modern groups, each of which is represented by far fewer specimens. A succinct diagnosis is thus almost bound to fail, since marginal exceptions will always exist. Nonetheless, the following is a reasonable working description of modern humans as they are distinguished osteologically from other hominids:

Cranium

Average endocranial capacity variable from population to population, but generally greater than 1,350 cc; frontal bone (forehead) relatively vertical (fig. 7.2); vault (braincase) relatively high, more or less parallel-sided, usually with some outward bulging (bossing) in the parietal region; occipital contour relatively rounded and lacking a prominent transverse torus; brow-ridge development generally greater in males than in females and variable among populations, rarely forming a continuous bar (supraorbital torus) across the top of the orbits but instead usually consisting of two parts—a variably bulging lateral (superciliary) arch at the upper outer corner of each orbit separated

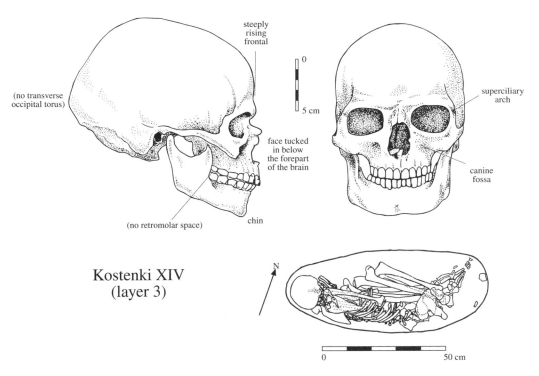

steeply
rising
frontal

(no transverse
occipital torus)

superciliary
arch

face tucked
in below
the forepart
of the brain

canine
fossa

(no retromolar space)   chin

**Kostenki XIV
(layer 3)**

0   50 cm

Figure 7.2. *Top:* Skull of an anatomically modern adult male buried more than 25 ky ago at the Upper Paleolithic site of Kostenki XIV (layer 3) on the Don River in European Russia (redrawn after 628, p.46). *Bottom:* the grave that contained the skull and associated skeleton. The skeleton was tightly flexed, suggesting the body had been firmly bound before burial, and both the skull and skeleton were covered with red mineral pigment (ocher) (redrawn after 1812, fig. 2). The skull exhibits slightly more alveolar prognathism (forward protrusion of the jaws) than do most Upper Paleolithic skulls, but it is otherwise typical of early modern Europeans. Its distinctively modern (as opposed to Neanderthal) features include the relatively high, short braincase with a nearly vertical frontal (forehead), the relatively short face with a distinct hollowing (canine fossa) between the nasal cavity and the anterior root of the zygomatic arch, the pronounced chin (vertical mandibular symphysis), and the absence of a retromolar space.

by a (supraorbital) notch or groove from an elongated (supraorbital) swelling along the upper inner margin (fig. 7.3; these swellings meet in a V between the orbits to form the so-called supraorbital trigone); face relatively flat and tucked in beneath the anterior portion of the braincase; a distinct hollowing of the bone (canine fossa) below each orbit, between the nasal cavity and the cheek bone (zygomatic arch); mandible variably robust, partly in keeping with significant interpopulational variation in tooth size, but almost always with a distinct chin; usually no gap (retromolar space) between the third molar and the ascending ramus when viewed from the side, reflecting the retraction of the face under the skull; substantial variability in the expression of many features (as discussed, for example, in references

Figure 7.3. Skull of an anatomically modern fifteen- to eighteen-year-old woman buried roughly 22 ky ago in the Pataud Rockshelter, southwestern France (drawn by Kathryn Cruz-Uribe from a slide; © 1999 by Kathryn Cruz-Uribe). The drawing has been deliberately oriented and shaded to emphasize the characteristically modern, two-part structure of the browridge. The two distinct parts are the supraorbital swelling and superciliary arch, separated by the supraorbital notch.

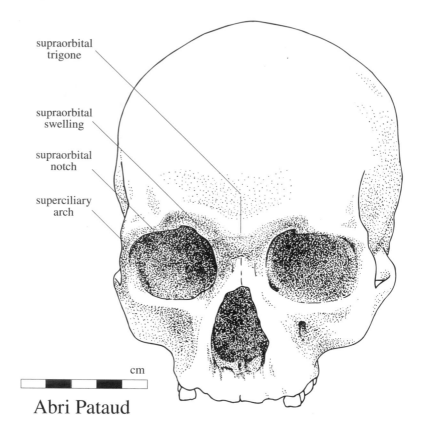

supraorbital trigone

supraorbital swelling

supraorbital notch

superciliary arch

cm

Abri Pataud

1082 and 1083), but never sufficient to incorporate earlier forms of people.

Postcranium

Limb bones sometimes very robust, with pronounced muscle markings, particularly in the earliest modern people, but still significantly less robust than in earlier kinds of people. Distinctions from the Neanderthals are especially clear (2215), thanks to many available Neanderthal postcranial bones. Besides decreased overall robusticity, they include axillary margin of the scapula sometimes bisulcate, but most commonly unisulcate, with the sulcus or groove on the ventral surface rather than on the dorsal surface as in Neanderthals; distal phalanx of the thumb usually about two-thirds the length of the proximal one, rather than being nearly equal in length as in the Neanderthals; apical tufts or tuberosities on the terminal phalanges of the hand relatively small and elongated, versus large and round in Neanderthals; cortical bone of the femur and tibia much thinner than in the Neanderthals; limbs long relative to the trunk, and distal limb segments (lower arm and lower leg) usually longer relative to the entire limb than in the Neanderthals; and pubis significantly shorter and thicker than in the Neanderthals.

The probable significance of these features was considered in chapter 6. In brief, the relatively flat, retracted face, together with some aspects of skull shape, may reflect decreased use of the anterior teeth as tools, at least by comparison with the Neanderthals. The distinctive shape of the skull may also partly mirror underlying structural changes in the brain that permit fully modern human behavior, including the unique innovative capability that Neanderthals and other nonmodern people apparently lacked. Alternatively, craniofacial differences from the Neanderthals may partly or largely reflect stochastic processes, in which Neanderthals drifted one way and ancestral moderns drifted another. The decline in overall postcranial robusticity and muscularity, coupled with the cited morphological changes in the scapula, hand, femur, and tibia, probably signal a shift toward relying less on bodily strength and more on culture (especially technology) in performing everyday tasks (2215, 2222). The selective advantage of reduced robusticity was that it reduced individual food requirements and permitted a given quantity of food to support more people. At least in Europe, where skeletal data are most abundant, during the 20 ky after modern people appeared, bodily dimensions and robusticity continued to decline, probably in relation to ongoing cultural advances (755).

Relatively long limbs and short trunks and long distal limb segments may reflect an ancestral adaptation to relatively warm climates and thus an African origin for all modern people, as discussed below. In keeping with an especially recent African origin, the earliest fully modern (early Upper Paleolithic) Europeans had even longer limbs relative to trunks than their later prehistoric and historic successors (1043). The shorter, thicker pubis might reflect a smaller average birth canal (pelvic inlet), perhaps because cultural advances enabled a shorter gestation period. As a result, infants would have been exposed to critical environmental stimuli when they were neurologically especially responsive. A shorter gestation period would also allow a reduction in birth spacing, increasing the potential for population growth. As discussed in chapter 6, however, it is perhaps more probable that the differences between Neanderthal and modern human pelvises are due to slight, if consistent, differences in habitual posture and locomotion as opposed to differences in reproductive physiology.

Much has been written on the racial affinities of early modern people, and it has sometimes been said that "racial types" occurred in areas where they were absent historically—as, for example, black Africans in Europe (Grimaldi in northern Italy and Kostenki XIV in Russia), Eskimos in Europe (Chancelade in France), and Ainu (or possibly Europeans), Melanesians, and Eskimos in northern China (Zhoukoudian Upper Cave). However, most early modern skulls do not exhibit unequivocal characteristics of any present-day race (1083, 1086, 1089), and it seems increasingly likely that the modern races formed mainly in the Holocene, after 12–10 ky ago. This is perhaps particularly clear for eastern Asia (the present-day hearth of the "Mongoloids"), but it also applies to Europe (the homeland of the "Caucasoids"). Early modern human remains are more abundant in Europe than anywhere else, and in a broad morphological sense the people did anticipate living Europeans. But as a group they tended to have larger, higher, and broader skulls with broader faces, more robust browridges, and larger mastoid processes (2301), and they should probably not be lumped with living Europeans in a "Caucasoid" race. In the present context, the more important point is that the skeletal differences that separate early modern Europeans from living humans are trivial compared with the differences that distinguish them and all living humans from the Neanderthals.

Modern Human Origins

In the previous chapter it was shown that the human form came to differ markedly between Europe and Africa after 200 ky ago, and that by 100 ky ago Europe was occupied exclusively by the highly distinctive Neanderthals, whereas Africa was inhabited by people whose looked far more modern. Much older European fossils from sites such as Swanscombe in England, Atapuerca Sima de Los Huesos in Spain, Biache-Saint-Vaast in France, and Ehringsdorf in Germany, all probably between 300 and 150 ky old, already anticipate the Neanderthals and imply that they were an autochthonous European development (77, 80, 82, 1096, 2073).

The more modern African contemporaries of the Neanderthals are represented at Klasies River Mouth, Border Cave, and Die Kelders Cave in South Africa, Mumba Shelter in Tanzania, Omo-Kibish in Ethiopia, and Dar es Soltan in Morocco (336, 1773). The famous and especially complete modern or near-modern human remains from Skhul and Jebel Qafzeh Caves in Israel probably also belong on this list, since associated "Ethiopian" mammalian species imply that they date to a time when Israel lay within a slightly expanded Africa (2154, 2155). Thermoluminescence dates on associated flints and electron spin resonance dates on animal teeth fix this time between roughly 110 and 80 ky ago (1548, 1888). As a group, the key African fossils reveal people with relatively short, high braincases overhanging the face in front, in stark contrast to the long, low braincases and forwardly mounted faces of the Neanderthals. This difference provides the main fossil support for the now famous "Out of Africa" theory, according to which modern humans spread from Africa to replace the Neanderthals and other equally archaic humans in Eurasia.

The "Out of Africa" Hypothesis and Its Multiregional Alternative

The "Out of Africa" hypothesis for modern human origins might be better called Out of Africa 2 (2073) to distinguish it from Out of Africa 1, the widely accepted original human dispersal from Africa at or before 1 my ago. In effect, Out of Africa 2 posits that Out of Africa 1 was followed by a tendency for human populations to follow different evolutionary trajectories on different continents, culminating by 100 ky ago in the emergence of at least three continentally distinct human populations. In Africa there were early modern or near-modern people, in Europe there were the Neanderthals, and in eastern Asia there were equally nonmodern people who could represent evolved end products of the *H. erectus* lineage. It its most extreme form, Out of Africa 2 posits that modern people expanded from Africa beginning 60–50 ky ago and then replaced the Neanderthals and equally archaic east Asians without gene exchange (or interbreeding). In its less extreme form, Out of Africa 2 allows for some gene flow between expanding moderns and resident archaic populations (336, 2001).

The principal alternative to Out of Africa 2 is the theory of multiregional evolution, which postulates that modern humans originated essentially everywhere that nonmodern humans had lived previously—in Africa, but also in Europe and Asia. Proponents of the multiregional model agree that widely dispersed human populations tended to diverge morphologically immediately following Out of Africa 1, but they argue that continuous gene flow ensured the rapid spread of highly adaptive novelties (like larger brains) and thereby kept all human populations on the same fundamental evolutionary track toward modern people (759, 760, 2487, 2492).

An obvious objection to multiregionalism is that it postulates substantial gene flow among small populations that were thinly scattered across three continents. In this light the multi-regional model is not so much a theory as it is an after the fact explanation for proposed morphological resemblances between nonmodern and modern populations in Asia and Europe. Multi-regionalists argue, for example, (1) that the skulls of the living Chinese share relatively nonprotruding (nonprognathous) jaws, upper facial flatness, a tendency for the development of a (sagittal) keel or torus along the top of the skull, extrasutural ("Inca") bones between the main bones of the skull, upper incisors shaped like coal shovels, and other features with fossils that have been traditionally assigned to Chinese *Homo erectus;* (2) that historic Australian Aborigines share large, sometimes shelflike brow ridges, long, flat, receding frontal bones ("foreheads"), a ridge of bone (an occipital or nuchal torus) around the back of the skull for attachment of the neck muscles, forwardly protruding (prognathous) jaws, large teeth, and other characters with fossils assigned to Indonesian *Homo erectus;* and (3) that early modern Europeans share large, prominent noses, a tendency for backward projection ("bunning") of the occiput (rear of the skull), and a propensity for a "horizontal-oval" mandibular foramen (a natural perforation on the inner surface of the mandible) with the Neanderthals. Ironically, multiregionalists do not cite comparable indications of continuity between archaic and modern Africans, perhaps because the most conspicuous similarities are ones that modern Africans share more broadly with other modern people. These include "a high, convex frontal positioned directly above a vertical face, a chin, a rounded occiput, and a short, flexed basicranium" (1372, p. 177), whose early appearance in Africa in fact comprises vital support for Out of Africa 2.

The multiregional theory has been questioned, because most key features are not simply present or absent but vary in frequency among far-flung human populations, both fossil and living. Many are actually most common in recent populations where the multiregional theory supposes them to be rare (1296, 1298). In addition, some features that do prevail where multiregionalists specify, such as large browridges and occipital tori, are primitive characters that may simply have been conserved more in some populations than in others, while other apparent regional characters, such as large noses or especially flat faces, may have evolved repeatedly (convergently) in successive archaic and modern populations (2089). Recurrent, independent evolution would be particularly likely for traits that represented adaptive responses to regional conditions, and any trait that truly indicated long-term continuity would almost have to be regionally adaptive. Otherwise its regional character would have been obscured by the interregional gene flow that multiregionalism requires (2052).

Yet other supposed indicators of regional continuity, such as occipital bunning or sagittal keeling, may not be developmentally homologous between archaic and modern humans, or they may be mechanically forced by partially shared cranial dimensions that themselves are not homologous (that is, that do not reflect close shared descent) (1372). Finally, apparent evidence for multiregional continuity may be inevitable so long as regional fossil samples remain small compared with the number of anatomical features among which multiregionalists can search for similarities (954, 1813).

In sum, the multiregional model cannot be dismissed, but at the moment the fossil evidence more strongly favors Out of Africa 2. This is particularly true with respect to Europe and western Asia, where both fossils and archeology indicate that modern humans quickly replaced the Neanderthals. The trunk and limb proportions of the earliest modern Europeans imply a tropical or subtropical origin, and the African tropics or subtropics are the obvious choice. From a strictly fossil perspective, the biggest obstacle to Out of Africa 2 is the murkiness—multiregionalists would say contrariness—of the Far Eastern fossil record. Most relevant Chinese fossils, including those from Maba (Tianshuigou), Dali, Yunxian, Yinkou (Jinniushan), Xujiayao, Yanshan (Chaoxian), and Changyang (Walongdong), remain poorly described or weakly dated, and it is still conceivable that modern humans appeared as early in China as in Africa or that the living Chinese originated from gene exchange between invading Africans and resident archaics. Regional continuity or interbreeding between invaders and residents might similarly explain the origins of the modern Australian Aborigines, but this possibility is complicated by the remarkable morphological variability of Australian fossils, only some of which recall archaic Javan *Homo* in any meaningful sense. This variability is discussed briefly below and is a puzzle that no existing theory of modern human origins parsimoniously accommodates.

Genetic Evidence for Modern Human Origins

In theory, the pattern of genetic diversity in living humans affords an independent means of deciding between the Out of Africa 2 and multiregional theories of modern human origins. A now famous study by Cann, Stoneking, and Wilson (420) illustrates both the potential and the pitfalls. Cann and her colleagues analyzed the variability of mitochondrial (nonnuclear or cytoplasmic) DNA (mtDNA) from 147 individuals characterized as sub-Saharan Africans (20), Asians (34), Caucasians (Europeans, north Africans, and Near Easterners) (46), Aboriginal Australians (21), and New Guineans (26). They chose mitochondrial (as opposed to nuclear) DNA because it diversifies more rapidly and because it is inherited more simply. (Individuals receive their mtDNA exclusively from their mothers, in contrast to nuclear DNA, which they receive in equal measures from each parent. A vital corollary is that, unlike nuclear DNA, mtDNA cannot undergo recombination during meiosis and, barring mutation, is thus identical in mother and offspring.) The 147 subjects possessed 133 different types of mtDNA, all presumed to originate ultimately from a single type by mutation—or, more precisely, by a series of mutations. Assuming that the smallest number of possible mutations is closest to the actual number that created the observed DNA diversity, Cann and her colleagues computed a genealogical diagram or tree linking the observed DNA types by their degree of similarity. The tree turned out to have two main branches—one comprising exclusively Africans and the other some Africans and everyone else. Since the individual branches that required the most mutations to explain (and that had thus probably existed longest) were African, Cann and her colleagues concluded that the most plausible root for the tree was in Africa.

By itself the mtDNA tree did not imply that the common mitochondrial ancestor was already modern, but the estimated mutation rate behind the mtDNA variability indicated that the common ancestor had lived in Africa about 200 ky ago. Since this was a time when archaic Eurasian lineages, including most notably the Neanderthals, were already distinct, it seemed clear that archaic Eurasians had contributed little, if any, mtDNA to living people. The sum implied an exclusively African origin for living humans, and the female bearer of the original African mtDNA type was popularly dubbed "African Eve." A subsequent, more comprehensive and methodologically more sophisticated analysis of mtDNA variation in living humans (2321) reconfirmed the structure of the original tree and continued to suggest that the last shared mitochondrial ancestor of all living humans had lived in Africa roughly 200 ky ago.

Human paleontologists had begun to postulate a largely African origin for living humans before Cann and her colleagues published their landmark study (2084), but the mtDNA tree appeared to be far less equivocal than the relatively scanty fossil evidence, and it convinced many observers that Out of Africa 2 was correct. Unfortunately for those who thought the issue had been largely settled, the mtDNA results proved to be at least as ambiguous as the fossil record. The fundamental problem was that Cann and her coworkers inadvertently misapplied the statistical (parsimony) analysis behind their mtDNA tree. Remedial analyses of the same data, in fact, produce a plethora of trees, of which some root in Africa and some elsewhere (999, 1389, 1417, 2160). There are no statistical grounds for preferring African trees to their non-African alternatives, and even a substantial increase in the database will not solve the problem. In short, for now the detailed mtDNA results that appeared to support Out of Africa 2 so conclusively must be placed aside.

However, this is not to say that genetics has become irrelevant to resolving modern human origins. To begin with, there is the spectacular recovery of a small segment of mtDNA from the humerus of the type (original) Feldhofer Cave Neanderthal specimen (1169, 1275, 1381, 2386). On average, the sequence of this segment differs in twenty-seven respects (twenty-seven substitutions) from homologous segments of living people. In contrast, segments from a wide variety of living humans differ among themselves in only eight respects on average. This implies that the last shared ancestor of Neanderthals and living humans lived long before the last shared ancestor of all living humans. If we assume that the differences between human and chimpanzee mtDNA accumulated over the past 5–4 my (the approximate antiquity of the last shared human-and-chimp ancestor), then the lineage providing the original Neanderthal specimen split from the one that produced living humans between 690 and 550 ky ago. It will be difficult to check the Feldhofer result in other relevant fossils, because DNA will rarely be suitably preserved and the recovery process damages or destroys precious fossils (554). However, if mtDNA from other Neanderthals proves to be equally divergent while mtDNA from geographically dispersed fossil modern humans is much less so, Neanderthals would have to be assigned to an extinct side branch of humanity that contributed few if any genes to modern populations.

In addition, the rough estimate of 200 ky ago for the last common mtDNA ancestor of all living humans remains unquestioned (1427, 1747, 2052), and it is supported by similar estimates for the last common ancestor of human Y-chromosomal genes (672, 946, 1644). These are in a sense the male counterpart of mtDNA, in that they are inherited only through males and for the most part are not subject to recombination. The coalescence of present mtDNA variability to a single type near 200 ky ago implies that the contemporaneous population comprised fewer than 10,000 breeding females (955), and such a small population would surely have become extinct if it were scattered throughout Africa and Eurasia, as multiregionalism requires. Survival was obviously far more likely if the population was concentrated in a limited geographic region, as Out of Africa 2 postulates. In addition, there are at least three other lines of genetic information that bear on modern human origins: between-population variation in the frequencies of nuclear DNA genes; the total amount of human genetic variability and geographic patterning in variability; and mtDNA sequence differences between individuals that reflect changes in ancient population size.

The analysis of nuclear gene frequency variation has been under way longest, and it has become more refined as technology has increased the number of genetic loci that can be examined. The focus is on loci that are polymorphic, that is, that can be occupied by different alleles, or alternative forms of a gene. An example is the locus for the ABO blood group, which can be occupied by the allele for the A, B, or O blood type. The first step is to calculate the frequencies of different alleles within various populations. The frequencies can then be compared to construct trees of genetic similarity that broadly resemble phylogenetic trees in form, though not necessarily in meaning.

Figure 7.4 (top) illustrates the procedure for a very simple case in which the data are frequencies of the ABO alleles in three populations. The resultant tree first links together populations 1 and 2 because their ABO frequencies are more similar to each other than they are to the frequencies of population 3. One possible reason for this is that populations 1 and 2 more commonly exchange genes with each other more than they do with population 3, but another possibility is that populations 1 and 2 share a more recent common ancestor than either shares with population 3. Mathematically more elaborate analyses involving a wide range of nuclear DNA polymorphisms all tend to produce trees in which non-African branches join together before any of them join to African branches (fig. 7.4, bottom) (303, 436, 1583, 1584, 1609, 1610). This could mean that non-African populations have recently exchanged genes far more with each other than they have with Africans (954), or it could mean that non-African populations all share a more recent common ancestor with each other than they do with Africans. The second interpretation is patently compatible only with Out of Africa 2, and it is supported by observations on alleles

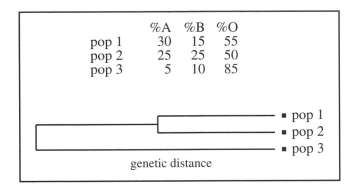

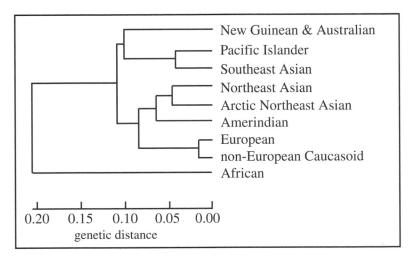

Figure 7.4. *Top:* The frequencies of the A, B, and O blood group alleles in three hypothetical populations, and a tree (or cluster) diagram linking populations based on their frequency similarity. *Bottom:* A tree diagram based on allele frequencies in forty-four blood group, protein and enzyme, and human lymphocyte antigen polymorphisms in nine geographically distinct populations (bottom redrawn after 1584, fig. 1). The tree diagram suggests either that Africans split from the other populations before they split from each other or that the other populations have exchanged genes with each other far more than they have with Africans.

that chimpanzees and people share. These alleles were presumably present in the common ancestor of chimpanzees and people, and their frequencies peak in African populations (152, 303, 1584). This is the expected result if African human populations, among all others, have the longest evolutionary history.

Two points may be made regarding the overall degree of genetic variability and its geographic patterning. First, living humans exhibit far less mtDNA and nuclear DNA variability than do chimpanzees, suggesting that living humans share a much more recent common ancestor with each other than they do with chimpanzees (419, 1815, 1858). If we assume that the rate of genetic divergence has been more or less constant and that the common ancestor of chimpanzees and people lived about 7 my ago, it would follow that the last common ancestor of all living humans existed between 175 and 125 ky ago (419). Similar computations based on a single locus (alpha-globin 2 *Alu 1*) that is especially invariable in living people place their last shared ancestor at 45 or 30 ky ago, depending on whether the date of chimpanzee/human divergence is taken as 7 or 4.7 my ago (1254). Estimates of this order obviously support Out of Africa 2 far more strongly than its multiregional alternative.

Second, analyses of mtDNA and of nuclear DNA microsatellites (simple, noncoding sequence repeats in the DNA molecule) show that genetic diversity is greater in Africa than on any other continent (303, 419, 1747, 2190). Africans are also the most diverse people in dental morphology (1111) and in cranial form (1747, 1748). Since dental form is known to be highly heritable and trees of craniometric similarity broadly resemble gene trees, greater dental and cranial heterogeneity supplement the probability of greater genetic diversity in Africa. Genetic diversity is a function of both population size and time, and greater diversity could mean simply that on average there have usually been more people in Africa than on other continents (954, 1747, 1749). As discussed in the previous chapter, however, archeological evidence suggests that Africa was lightly populated during much of the time when modern humans were emerging. Arguably then, greater genetic variability in Africa does not reflect larger population size but implies instead that African populations have been evolving longer. A related inference is that all non-Africans derive from an African subpopulation that existed after the time of the last common mtDNA ancestor, about 200 ky ago. The bottom line again supports only Out of Africa 2.

The reconstruction of past population size change is grounded in pairwise comparisons of mtDNA sequences. The number of differences or mismatches between successive pairs is scored and then plotted on a histogram (955, 1814). Thus a pair of sequences that differs at two sites would be scored as 2, a pair that differs at four sites as 4, and so forth. In a population whose size has been stable for a long time, the mode of such a histogram will be zero. In a population that has recently expanded, the mode or peak of pairwise differences will be positive, and the greater its value, the longer the probable time since the expansion began. The time itself can be estimated using the mutation rate that produced the pairwise differences.

This is fundamentally the same rate that has been used to estimate an age of roughly 200 ky ago for the last shared mtDNA
ancestor of all living humans, as discussed above.

Mismatch distributions in twenty-three of twenty-five examined populations suggest expansions that began on average
about 50 ky ago (1747, 1931). This estimate coincides closely
with the apparent time when anatomically modern invaders replaced the Neanderthals, but it does not provide unqualified
support for Out of Africa 2, because the mismatch distributions
actually imply multiple expansions in populations that were
already isolated from one another; that is, they imply that population divergence occurred before population expansion (1815).
This could mean that there were multiple, temporally successive dispersals of modern humans from different parts of Africa
to different parts of Eurasia (1300). In particular, a relatively
early dispersal eastward, perhaps across the Bab el Mandeb strait
separating the Horn of Africa from the southwestern Arabian
Peninsula at the southern end of the Red Sea, could explain
both the possible presence of people in Australia by 60 ky ago
and the remarkable cranial robusticity of some Aboriginal Australians. A somewhat later dispersal northeastward through the
Sinai Desert of Egypt could explain the replacement of the west
Asian and European Neanderthals beginning perhaps 45 ky ago.
More generally, taken at face value, the mismatch distributions
undermine the strict Out of Africa version that postulates global
replacement from a single African source, and they breathe fresh
life into variants that postulate a more complicated spread that
may have occurred in fits and starts and may sometimes have
involved interbreeding between invaders and residents.

In sum, on balance, genetic evidence provides strong if
qualified support for the Out of Africa model of modern human
origins. Pending the successful extraction of DNA from a wide
range of fossil bones, it is unlikely that genetics will support
Out of Africa more strongly, and for now closure must depend
mainly on fossil form and on dating. The genetic evidence hints
that a more complete and better-dated fossil record, perhaps
above all in eastern Asia, may one day support a more complex
version of Out of Africa in which modern Africans spread to
different parts of Eurasia at different times and only some Eurasian populations were completely replaced.

Archeology and Modern Human Origins

The remaining sections of this chapter focus mainly on the
archeology of the earliest modern humans. The present sub-
section anticipates information presented further on, and it also

draws on material in the previous chapter to show how archeology illuminates modern human origins.

A potentially vital objection to Out of Africa 2 concerns the failure of near-modern people to expand from Africa immediately after they appeared there, at least 100 ky ago. Instead, they seem to have been confined to Africa until roughly 50 ky ago, and it is even possible that they were replaced by Neanderthals on the southwest Asian margin of Africa (in what is now Israel) about 80 ky ago (2154, 2155). Archeology provides a partial solution to the apparent dilemma. As summarized in chapter 6, the people who inhabited Africa between 100 and 60 ky ago may have been near-modern in form, but in behavior they closely resembled the Neanderthals and other archaic humans (121, 1141, 1243). Admittedly, they had far more than an elementary grasp of stone flaking; they often collected naturally occurring iron and manganese compounds that they could have used as pigments; they apparently built fires at will; they buried their dead, at least on occasion; and they routinely acquired large animals as food. In all these respects and perhaps others, they may have been relatively advanced over earlier, archaic people. However, in common with earlier people and with their Neanderthal contemporaries, they manufactured a relatively small range of recognizable stone artifact types; their artifact assemblages varied remarkably little through time and space (in spite of notable environmental variation); they obtained stone raw materials overwhelmingly from local (vs. far distant) sources (suggesting relatively small home ranges or very simple social networks); they rarely if ever utilized bone, ivory, or shell to produce formal artifacts; they left little or no evidence for structures or for any other formal modification of their campsites; they were relatively ineffectual hunter-gatherers who lacked, for example, the ability to fish; their populations were apparently very sparse, even by historic hunter-gatherer standards; and they left no indisputable evidence for art or decoration.

Based on what early near-modern Africans did and did not do, it seems reasonable to conclude that they were cognitively human, but not cognitively modern in the sense that all living people are. It was only when they became cognitively modern, with the fully modern capacity for culture, that they obtained an adaptive advantage over their archaic Eurasian contemporaries. If Out of Africa 2 is correct, we would expect Africa to contain the oldest secure evidence for art and for other indicators of modern mental abilities. In fact, the oldest reliable dates for decorative items come from Enkapune Ya Muto Cave in the central Rift Valley of Kenya, Mumba Cave in northern Tanzania, and Border Cave in KwaZulu/Natal, South Africa, while the oldest secure dates for well-made bone artifacts come from Blombos Cave on the south coast of South Africa.

At Enkapune Ya Muto, Mumba, and Border Caves, the decorative objects are ostrich eggshell beads, broadly similar to ones that some Africans still made historically. At Enkapune Ya Muto, radiocarbon shows that the beads are at least 39.9 ky old, and the degree of chemical change (hydration) in associated obsidian artifacts suggests they may be several thousand years older (31). At Mumba, amino-acid racemization brackets the beads between 45 and 40 ky old (361, 950), and at Border Cave concordant racemization and radiocarbon dates place them near 38 ky ago (1558). Radiocarbon indicates that beads from Kisese II Rockshelter, Tanzania, may be only slightly younger (611, 1109). At each site the beads are important not just as decorative items, but because they may signal an exchange system like the one recently observed among !Kung-San hunter-gatherers in the Kalahari Desert, Botswana (31). !Kung groups in somewhat different environments regularly exchange bead-work, and the result is a network of relationships that enhances group survival in times of environmental stress (2467).

At Blombos Cave, the bone artifacts comprise points and other formal objects in layers that have been radiocarbon-dated to more than 40 ky ago (1016). The associated stone artifacts include numerous small, highly standardized bifacial "Still Bay" stone points that are an additional marker of behavioral change. The immediately underlying layers contain typical Middle Stone Age (MSA) artifacts that include neither formal bone tools nor comparably standardized stone implements. Until recently, Still Bay points were known almost entirely as undated surface finds, and they were commonly assigned to the MSA rather than to the succeeding Later Stone Age (LSA) because the associated artifacts did not include very small ("micro-lithic") scrapers and backed pieces like those that mark many well-known LSA assemblages after 20 ky ago. However, Blombos Cave is one of the very few southern African sites that records even a portion of the period between 50 and 20 ky ago, and the early LSA thus remains all but unknown. The typological difference between the Still Bay assemblage and much later LSA assemblages need imply only that the early LSA, like the early European Upper Paleolithic, differed significantly from later variants. Even the later LSA, after 20 ky ago, includes variants that lack microlithic pieces and that resemble other LSA assemblages mainly in the abundance of bone artifacts and art objects (612). Like the Upper Paleolithic, the LSA is united not so much by specific artifact types as by more broadly defined behavioral markers, including the use of bone to make formal artifacts, the production of unquestionable art objects, and parallel tendencies for stone tools to become more standardized and for artifact assemblages to vary more through time and space.

Enkapune Ya Muto and Blombos Caves indicate that formal bone and shell artifacts and art objects probably appeared in Africa several thousand years before analogous markers of modern behavior appeared in Europe. At least arguably, when modern human behavioral traits appear in Europe, they occur earlier in southeastern Europe (before 40 ky ago) than in western Europe (at about 40 ky ago) (1538, 1539). An African origin is further implied by the appearance of broadly similar modern behavioral markers in southwestern Asia in the interval between their earliest appearances in Africa and Europe (125).

In sum, the relevant archeological evidence both supports and supplements the fossil evidence for the African origin of modern humans.

The Relation between Biological and Cultural Change

The archeological evidence bearing on modern human origins is uneven and incomplete, but wherever it is full enough for judgment, it implies that a metamorphosis in human behavior occurred about 50 ky ago. Before this time, morphology and behavior appear to have evolved very slowly, more or less in tandem, but after this time morphology remained relatively stable while behavioral (cultural) change accelerated rapidly. Arguably, barring only the development of those typically human traits that produced the oldest known archeological sites between 2.5 and 2 million years ago, the behavioral transformation that occurred 50 ky ago represents the most dramatic behavioral shift that archeologists will ever detect, and it demands explanation. There are two fundamental possibilities: that humans had long been capable of fully modern behavior but expressed this capacity only after some momentous social or demographic change; or that the development of fully modern behavior depended intimately on biological (neural) change.

Social change independent of biological change could have involved, for example, the initial development of the nuclear family as the fundamental productive unit and of modern notions of kinship and descent to relate individuals and groups (2009, 2011). Such a metamorphosis could have stimulated rapid population growth, and larger, denser populations could in turn explain the accelerating pace of technological innovation, the proliferation of symbols (art), and other novelties that mark the archeological record after 50 ky ago. The obvious problem with this kind of explanation is that it offers no hypothesis for why social relations changed when they did or for why they changed at all.

In these circumstances it seems far more economic to tie the basic behavioral shift, which may have included major changes in social organization, to a neurological change that launched the fully modern human ability to manipulate culture as an adaptive mechanism. A neural change in turn requires no special explanation, since it would follow from the kind of fortuitous but highly advantageous mutation that must underlie all macroevolutionary change. The neural hypothesis in fact follows logically from the notion that selection for larger and presumably more sophisticated brains was a vital aspect of human evolution for a very long time before the origin of modern humans and from the observation that earlier advances in human behavior, from the Oldowan to the Acheulean to the Middle Paleolithic, corresponded broadly to changes in brain size and probably also in brain organization. Fossil and genetic evidence suggest that the last crucial neural change occurred in Africa, but an African origin is not central to the argument. The underlying mutation could have occurred in Europe, in which case the author and readers of this book would be Neanderthals contemplating the strange people who used to live in Africa.

Assuming for argument's sake that a neural change explains the appearance of fully modern behavior, it is surely reasonable to suppose that the change promoted the fully modern capacity for rapidly spoken phonemic speech, that is, for "fully vocal language—phonemicized, syntactical, and infinitely open and productive" (1565, p. 579). This inference follows not simply from the way living humans use modern phonemic language to communicate, but even more from the way they use it cognitively to model both nature and culture (209). Perhaps most important in the present context is that without modern language, modern humans could probably not ask the hypothetical or "what if" questions that underlie the modern capacity for innovation. Archeological evidence for this capacity becomes manifest only after 50 ky ago.

However, fully modern language requires not only fully modern neural organization, but also a fully modern vocal apparatus. The primitive condition, which is retained in the apes and in human infants, is for the larynx (voice box) to reside high in the neck. Such a placement sharply reduces the likelihood of choking, since it allows food and air to descend along separate digestive and respiratory tracts (1301). Newborn humans run little risk of choking while they feed, but the danger is greatly enhanced between one and a half and two years of age when the larynx begins to drop in the neck and the digestive and respiratory tracts partially merge. This potentially dangerous arrangement is explicable only if it confers a strong selective counteradvantage, and the most plausible one is that it enlarges the supralaryngeal space and thus allows humans to produce a wider range of sounds (1378–1380). Among these sounds are the vowels and consonants that separately or together characterize all known languages and that are essential for the production and decoding of rapidly articulated speech. The larynx does not fossilize, but its adult position corresponds to the flexion or upward arching of the adult cranial base (1301, 1303). Modern human adults have highly flexed cranial bases that restrict the space available for the larynx, forcing it downward in the neck. In contrast, on average the Neanderthals had much flatter cranial bases, which may mean that they had a more highly placed larynx and a more limited ability to produce crucial sounds. If this is accepted, then a mutation that enhanced the neural capacity for spoken language would have benefited only near-modern people like those represented at Skhul and Qafzeh Caves, Israel, whose upward arched basicraniums imply a fully modern laryngeal position.

A recent evolution for modern linguistic ability is supported by lexical similarities (that is, cognates or homologous words), suggesting that all modern languages share a common linguistic ancestor or "mother tongue" (1847–1849). The age of this ancestor cannot be precisely estimated, but it would probably be undetectable if it existed before 50 ky ago. An argument can even be made for language origins in Africa, since linguistic diversity is arguably greater there than anywhere else (1847–1849). But linking modern human origins to a linguistic advance is problematic for at least three reasons. First, there is the possibility that the position of the Neanderthal larynx has been misestimated. There are only three Neanderthal skulls (from La Chapelle-aux-Saints, Saccopastore, and Guattari [Circeo]) on which basicranial flexion can be reliably assessed (751). In each case flexion falls outside the modern range, but a larger sample might reveal some overlap. Second, and more important, the influence of basicranial flexion on the position of the larynx remains arguable (1064), and the modern shape of a tongue (hyoid) bone preserved with the Neanderthal skeleton from Kebara, Israel, may imply an essentially modern vocal tract (75, 76).

The third and perhaps most serious problem is the existence of fossil skulls that are as old as or older than the Neanderthal specimens and that likewise antedate the putative neural change, but that exhibit an essentially modern degree of basicranial flexion (751). This is true not only for the well-known near-modern skull 5 from Skhul Cave, Israel, but also for the much older fossil skulls from Petralona, Greece, and Broken Hill (Kabwe), Zambia, discussed in chapter 5. Advocates of a link between linguistic change and modern human origins must thus explain why the vocal tract would have assumed a potentially dangerous configuration among people who presumably lacked the neural basis for modern speech. So far there is no compelling answer.

In sum, the greatest strength of the neural hypothesis is that it parsimoniously explains the burst of behavioral innovation that ushers in fully modern humans and that surely explains their spread. Its most obvious weakness is that on present knowledge it cannot be explored independently in fossils, since the putative change was in brain organization, not size, and fossil skulls provide little or no secure evidence for brain structure. Neanderthal skulls, for example, differ dramatically in shape from modern ones, but they were as large or larger, and the difference in skull shape does not imply a significant difference in brain function (1054). No one doubts an essential link between the human brain and uniquely human behavior, but paleoneurologists disagree on many key details of brain evolution (614, 615, 705, 706, 1055), and the details are in any case not detectable in the fossil record. The bottom line is that the neural hypothesis for modern human origins is not a proper scientific hypothesis, since it is not currently testable apart from the circumstantial archeological evidence on which it is founded.

In addition to this objection, there are at least two others that are more strictly archeological. Both are founded in the rich archeological record of Europe, and in essence both concern the correlation between Neanderthals and Mousterian (Middle Paleolithic) artifacts on the one hand and between anatomically modern people and Upper Paleolithic artifacts on the other. The first objection goes to the heart of the assumption that Neanderthals were biologically precluded from behaving in a fully modern way, and it is the more difficult objection to dismiss. The second concerns the possibility that fully modern behavior did not appear abruptly with the advent of the Upper Paleolithic roughly 40 ky ago but evolved gradually afterward.

Were Neanderthals Fundamentally Incapable of Fully Modern Behavior?

As outlined here, Out of Africa 2 postulates that the Neanderthals were replaced because they could not compete culturally with their modern human successors. The argument is bolstered over most of Europe by the relatively abrupt nature of the replacement. At many sites, Cro-Magnon/Upper Paleolithic occupations overlie Neanderthal/Mousterian layers with no evidence for a major break in time or for any transition between the two, suggesting the replacement took only decades, or at most centuries. Demographic modeling shows that only a 1% or 2% rise in Neanderthal mortality would have extinguished Neanderthal populations within a thousand years (2536), and the Cro-Magnons might have induced such a rise simply by excluding the Neanderthals from essential resources. To accept this possibility, however, we must assume there was little or no gene flow or cultural trait diffusion between Neanderthal residents and Cro-Magnon invaders. Except to some multiregionalists, the fossil record provides no compelling evidence for gene flow, and even if this were possible, it might have been precluded by the biologically grounded behavioral gulf between Neanderthals and Cro-Magnons. In this sense the Cro-Magnon invasion of Europe would have differed fundamentally from the historic European invasion of the Americas or Australia, where the indigenes and invaders clearly had the same biological capacity for culture and where interbreeding was rampant.

However, there is a significant problem with the idea that the Neanderthals could not behave like moderns. This is the occasional discovery of artifact assemblages that comprise a blend of Neanderthal/Mousterian and Cro-Magnon/Upper Paleolithic artifact types. At some sites such "mixed" assemblages may have been created when excavators inadvertently merged the contents of adjacent Mousterian and Upper Paleolithic layers, but this is surely not the case at several "Châtelperronian" sites in central and western France and adjacent northern Spain. As discussed in the previous chapter, Châtelperronian stone artifact assemblages generally combine typical Mousterian sidescrapers, denticulates, and backed knives with numerous characteristic Upper Paleolithic endscrapers and burins (970–972). At one site, the singular Grotte du Renne at Arcy-sur-Cure in the Paris Basin, typical Châtelperronian stone artifacts are unequivocally accompanied by carefully shaped bone artifacts and by bone beads and pendants (fig. 6.46) (103, 707–709, 1358, 2131). The stone and bone artifacts were recovered from occupation floors with patterned arrangements of postholes, mammoth tusks, limestone plaques, and hearths that probably mark the positions of ancient huts (fig. 6.47). By themselves the stone artifacts might be ambiguous, but the bone artifacts, ornaments, and highly structured floors point unequivocally to the Upper Paleolithic. Remarkably, as indicated in chapter 6, an associated human temporal fragment and three teeth all exhibit Neanderthal (as opposed to Cro-Magnon) features (1098). Neanderthal authorship of the Châtelperronian is implied even more conspicuously at La Roche à Pierrot Rock-shelter, Saint-Césaire, west-central France, where a partial Neanderthal skeleton was directly associated with typical Châtelperronian stone tools (1361).

Châtelperronian layers have been dated variously by radiocarbon, by thermoluminescence, and by correlation to regional or global climate stratigraphy. The results are somewhat inconsistent, but as discussed in chapter 6, a reasonable inference now is that the Châtelperronian existed for a few centuries or a millennium about 40 ky ago. It was during this period that the earliest undeniable Upper Paleolithic culture or culture complex, known as the Aurignacian, appeared widely in southeastern, central, and western Europe (1538, 1539, 2066). As summarized below, the Aurignacian is marked by a multiplicity of highly formalized, distinctive Upper Paleolithic stone and bone artifact types and by a variety of art objects, including human and animal representations (105, 2443). At most sites where the Aurignacian and the Mousterian occur together, the Aurignacian abruptly overlies the Mousterian, and in the version of Out of Africa 2 favored here, the Aurignacian is a plausible artifactual manifestation of the Cro-Magnon invasion. Physically the makers of the early Aurignacian are poorly known, but sparse, fragmentary fossils from France and more numerous and complete ones from Moravia, Czech Republic, suggest they were fully modern (rather than Neanderthal) (756, 776, 1093, 2004).

How, then, to explain the Châtelperronian? Where Mousterian, Châtelperronian, and Aurignacian layers are superimposed, the Châtelperronian layers almost always lie in the middle. The exceptions are three sites where late Châtelperronian and early Aurignacian layers appear to interfinger, suggesting a brief period of culture contact. In this light, it is tempting to conclude that the Châtelperronian reflects cultural diffusion from Cro-Magnon Aurignacians to Mousterian Neanderthals, before the Neanderthals fully succumbed. But even if this appears credible, it begs one fundamental question: If the Upper Paleolithic way of doing things was clearly superior and Neanderthals could imitate it (that is, they were not biologically precluded from doing so), why did they not acculturate more widely, with the result that they or their genes would have persisted much more conspicuously into Upper Paleolithic times (after 40 ky ago)? There is no compelling answer, and the Châtelperronian remains a major puzzle whose solution is important for closure on Out of Africa 2.

Does the Upper Paleolithic Truly Represent an Abrupt Departure?

The Upper Paleolithic contrasts with the Mousterian in many ways, of which the most often cited is the widespread Upper Paleolithic emphasis on stone flakes whose length was at least twice their width. Archeologists distinguish such elongated flakes as "blades." Most Mousterian people produced very few blades, and at Mousterian sites where blades are common, they were made mainly by a variant of the Levallois technique that many Mousterian knappers also used to produce flakes (1265, 1539). In contrast, Upper Paleolithic people developed much more sophisticated techniques to produce blades regularly and consistently. In general, compared with Mousterian flake technology, Upper Paleolithic blade production provided more cutting edge from a given stone core (269), and among the very earliest Upper Paleolithic people it may have helped to conserve scarce raw material. Later Upper Paleolithic people probably produced blades mainly for historical reasons, however, as a part of their cultural heritage.

Besides emphasizing blades, most Upper Paleolithic people routinely manufactured stone tool types that are generally rare and crudely made ("atypical") in Mousterian assemblages. The most commonly cited Upper Paleolithic types are probably endscrapers (elongated flakes or blades with smooth, continuous retouch on the edge opposite the striking platform) and burins (flakes or blades from which a second smaller flake or blade [a burin spall] was struck along one edge, leaving a scar at an abrupt angle to the ventral surface of the parent) (fig. 7.5). Burins and endscrapers do characterize most Upper Paleolithic assemblages, but as truly diagnostic Upper Paleolithic types, they are probably surpassed by carefully made leaf-shaped points with flat invasive bifacial retouch; backed or truncated pieces on which a lateral edge or end has been methodically dulled or blunted; and carinate and nosed scrapers on which the presumed working edge has been formed by removing numerous small thin blades ("bladelets") (1265, 2025–2028) (fig. 7.5). Singly or in combination, finely made leaf-shaped points, backed or truncated elements, and carinate or nosed scrapers distinguish many Upper Paleolithic industries, and they are correspondingly rare in Mousterian ones.

Further, unlike Mousterians, Upper Paleolithic people commonly cut, carved, polished, or otherwise shaped bone, ivory, and antler into a wide variety of formal artifact types. These include not only pieces that were probably projectile points, awls, punches, needles, and so forth, but also nonutili-

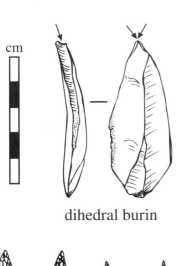

cm

dihedral burin

endscraper on a blade

Figure 7.5. *Top:* A burin and an end-scraper on a blade, tool types that are conventionally said to distinguish the Upper Paleo-lithic from the pre-ceding Mousterian. *Bottom:* Bifacial points, backed blades, and carinate (or keeled) end-scrapers, types that singly or together especially distin-guish the Upper Paleolithic.

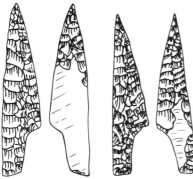

leaf-shaped points with
flat, invasive, bifacial retouch

backed blades

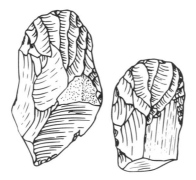

carinate endscrapers

tarian items that are clearly interpretable as art objects or items of personal adornment. As noted in the previous chapter, un-equivocal art and decorative items are essentially unknown in a Mousterian context.

In general, Upper Paleolithic people made a much wider variety of readily distinguishable artifact types than did their predecessors, and Upper Paleolithic industries or "cultures" varied far more in time and space. Finally, as detailed below, in virtually every conceivable category of recoverable items— graves, house ruins, fireplaces, and such—Upper Paleolithic examples are commonly far more elaborate than Mousterian ones. In all the ways that the Upper Paleolithic differs from the Mousterian, the Mousterian differs much less from earlier cultures, and it thus seems reasonable to conclude that the Upper Paleolithic represents a quantum change from everything that went before.

It has sometimes been suggested that an "early Upper Paleolithic" antedating 25–20 ky ago should be distinguished from a "late Upper Paleolithic" afterward and that only the late Upper Paleolithic was significantly different from the Mousterian (476, 1382, 2067). In this view the early Upper Paleolithic was transitional from the Mousterian and did not depart radically from it. The case is perhaps clearest for Cantabrian Spain (2057, 2064, 2065, 2071), where there was a dramatic increase in the number of sites after about 20 ky ago, accompanied by an artistic florescence and by important changes in the location of settlements and in hunting patterns. However, direct radiocarbon dating at Chauvet and Cosquer Caves now shows that early Upper Paleolithic people were producing spectacular wall art in nearby France more than 27 ky ago (106). Moreover, even if northern Spain and neighboring parts of western Europe appear quiescent before 20 ky ago, parts of central and eastern Europe were not. The cultural elaboration indicated by sites like Vogelherd, Hohlenstein-Stadel, and Geissenklösterle in southwestern Germany (942, 943), Stratzing/Krems-Rehberg (Galgenberg) and Willendorf in Austria (722, 1612), Dolní Vestonice, Pavlov, and Predmostí in the Czech Republic (1252, 2012, 2117, 2118), or Sungir' in Russia (101), all antedating 20 or even 30 ky ago, rivals any later developments in Spain or France. Moreover, it could be argued that the "late" Upper Paleolithic of central Europe began with a cultural decline, since site density and richness declined significantly about 20 ky ago and recovered to early Upper Paleolithic standards only five or six millennia later (940, 1266, 1269, 1641, 1807, 2115, 2118, 2427).

For the most part, the shifting locus of especially abundant, rich, or elaborate Upper Paleolithic sites probably reflects the vicissitudes of Last Glaciation climatic change (fig. 7.6), which sometimes favored one area, sometimes another (933). Significantly, the time at or about 20 ky ago was a major climatic inflection point, corresponding to the maximum development of the Last Glaciation ice sheets. Following this time, the Scandinavian ice sheet on the north and the Alpine ice sheet on the

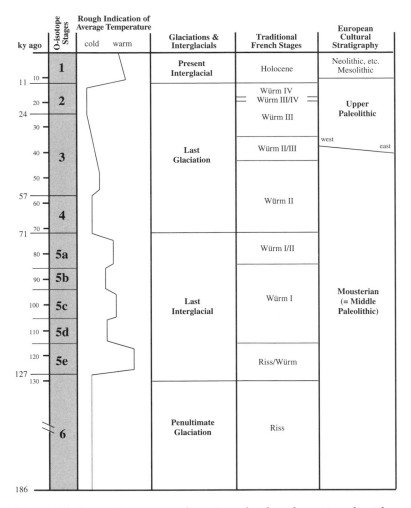

Figure 7.6. Later Quaternary climatic and cultural stratigraphy. The period referred to in the text as "late Paleolithic" is the later part of the Last Glaciation, between 40–35 and 10 ky ago. This period corresponds to the time span of the European Upper Paleolithic.

south advanced to within 600 km (360 miles) of each other across central Europe. The ice-free area in between became a frigid near-desert, and this surely explains why much of central Europe was so sparsely populated between roughly 20 and 14 ky ago. Paradoxically, however, from a human perspective the Last Glacial Maximum probably had its most dramatic effect in Africa, where people were able to reoccupy areas they had largely abandoned before 30 or 40 ky ago, both in the north (416, 2419) and in the south (610, 1243). The reason was a significant increase in regional precipitation. Together with temperature variation, similar, if less dramatic long-term moisture fluctuations probably also contributed to local long-term population and cultural fluctuations in Last Glaciation Europe.

This is not to say that all the known Upper Paleolithic traits were present from the very beginning or that all subsequent elaboration (or decline) was directly due to climate. Some of the specific cultural innovations discussed below unquestionably were developed after the Upper Paleolithic began, and they diffused more or less widely depending on geographic constraints, their local utility, and so forth. Undoubtedly some did promote local or regional population increases or cultural florescences, but none were as fundamental as those that distinguish the Upper Paleolithic from everything that went before. In fact, to the extent that an "early" Upper Paleolithic can be distinguished from a "late," the apparent differences are less than those between many historic hunter-gatherer cultures, and nowhere is the early Upper Paleolithic truly a link or evolutionary transition between the Mousterian and the late Upper Paleolithic.

In sum, from the characteristics that have been listed and that will be discussed further below, one can argue that the Upper Paleolithic signals the most fundamental change in human behavior that the archeological record may ever reveal, barring only the primeval development of those uniquely human behaviors that made archeology possible. Excepting the puzzle posed by the association of Neanderthals with the Châtelperronian, the strong correlation between Upper Paleolithic artifacts and modern human remains clearly suggests that it was the modern human physical type that allowed the Upper Paleolithic (and all subsequent cultural developments).

Cultural Variability

Previous chapters have emphasized that before 40 ky ago—that is, before the Upper Paleolithic and comparable cultural manifestations had completely supplanted earlier ones—vast areas were characterized by remarkably uniform artifact assemblages that differed from one another mainly in the relative abundance of the same basic artifact types. In addition, artifactual change through time was painfully slow: basic assemblage types lasted tens or even hundreds of thousands of years. After 40 ky ago, however, the general pattern changed radically. Like-aged artifact assemblages from neighboring regions often differed qualitatively, and within single regions the pace of artifactual change accelerated dramatically. The greatly enhanced tendency to spatial variability is dramatically illustrated by the divergence between the Upper Paleolithic of Europe, western Asia, and northern Africa on the one hand and the contemporaneous Later Stone Age (LSA) of sub-Saharan Africa on the other. Unlike their Middle Paleolithic and Middle Stone Age antecedents, whose partition reflects scholarly tradition rather than artifactual content, the Upper Paleolithic and the LSA differed artifactually from their very beginnings. The punched blades and varied burins that are a hallmark of the Upper Paleolithic never seem to have been an important element in the LSA. Instead, LSA people more commonly produced a range of scrapers and other retouched forms on flakes and flake blades (612, 2354). Some LSA retouched artifacts broadly resemble MSA ones, but they are commonly much smaller, and at least in the better-known later LSA industries (after 20 ky ago), they are more standardized. In fact, to the extent that the LSA recalls the Upper Paleolithic, it is in relatively abstract features like a greater artifact standardization, greater assemblage variation through space, accelerated assemblage change through time, and of course the routine production of formal bone artifacts and of art objects or items of personal decoration. The sum suggests a common mind-set that differed qualitatively from the mind-set of earlier peoples.

Within Africa, the Upper Paleolithic tendency to cultural diversity is particularly conspicuous in the Nile Valley, where differences in stone artifact types through time and space reveal no fewer than six cultures between 40–35 and 17 ky ago (1677, 2414). These include the Shuwikhat Industry and the Fakhurian, Kubbaniyan, Idfuan, Halfan, and Gemaian complexes, some of which existed in different parts of the Nile Valley at the same time. Within Europe, the highly diverse nature of Upper Paleolithic cultures is amply demonstrated by the numerous Upper Paleolithic sites of the east European plain, many of which have provided artifact assemblages that are unique and that cannot be assigned to an industry or culture represented at other sites, even in the same region (60, 1029, 1031, 1228, 1232, 2006, 2007, 2010).

The dramatic acceleration in change through time is perhaps best exemplified by the classic Upper Paleolithic sequence of southwestern France (1313, 1761, 2023, 2024). This was established long ago by careful excavations in caves such as La Ferrassie and Laugérie-Haute and has been repeatedly confirmed by meticulous modern excavations, including above all those at the Abri Pataud (1540, 1594, 1595). The sum shows that in the period between 40–35 and 11 ky ago, France harbored a remarkable succession of Upper Paleolithic industries, comprising most fundamentally (from older to younger) the Aurignacian, Gravettian (Perigordian), Solutrean, and Magdalenian. Each industry (or culture) was characterized by specific artifact types that are rare or unknown in the others. Thus the

Figure 7.7. Typical Aurignacian split-base bone points, "pendants," and chipped stone artifacts. The split-base points (redrawn after 203, p. 98) come from the cave of El Castillo, northern Spain. The remaining artifacts (redrawn after 855, pp. 36, 39, 153) come from the nearby cave of El Pendo.

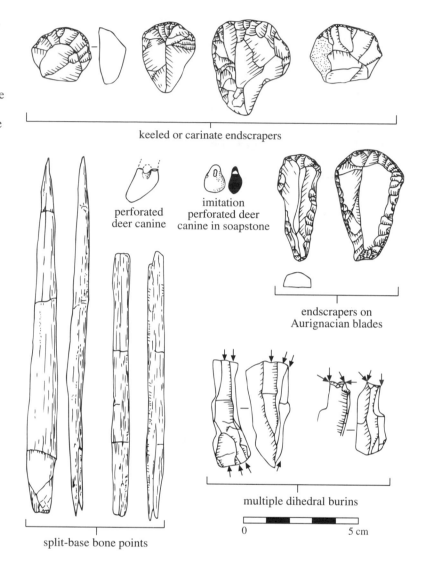

keeled or carinate endscrapers

perforated deer canine

imitation perforated deer canine in soapstone

endscrapers on Aurignacian blades

multiple dihedral burins

0   5 cm

split-base bone points

Aurignacian, which probably intruded into France from the east about 40 ky ago and was present until perhaps 28 ky ago, was distinguished by large blades with invasive, scalar retouch on the lateral edges, "beaked" burins, nosed and keeled (carinate) endscrapers, and special kinds of bone points, the most famous being those with split bases (fig. 7.7).

The Gravettian, spanning the interval between about 28 and 21 ky ago and extending eastward to European Russia and southward to Italy and Spain, was marked especially by numerous small, narrow, parallel-edged, often pointed, steeply backed blades (fig. 7.8). In western Europe later Gravettian people made characteristic tanged or stemmed points, while their central and eastern European contemporaries produced shouldered

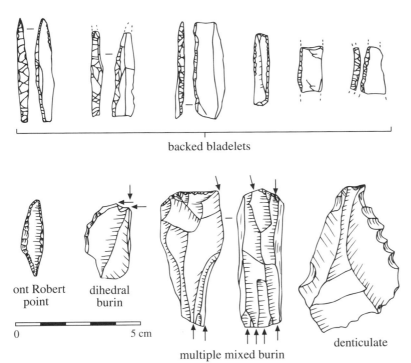

backed bladelets

ont Robert point   dihedral burin

0 ———————— 5 cm

multiple mixed burin

denticulate

Figure 7.8. Range of Gravettian (Perigordian) stone artifacts. Backed bladelets are especially characteristic. The ones here (redrawn after 203, p. 145) come from the Gravettian deposits of Cueva Morín, northern Spain. The remaining pieces (redrawn after 855, p. 113) come from Gravettian levels in the nearby cave of El Pendo.

(or highly asymmetric stemmed) forms (1641). Aurignacian-type bone points were absent throughout, and the principal bone artifacts were "awls," "punches," and other presumably domestic implements, accompanied by well-made bone art objects and items of personal adornment.

The Solutrean, present between about 21 and 16.5 ky ago and essentially confined to France and Spain, was characterized above all by finely made foliate (leaf-shaped) stone points of various shapes and sizes (fig. 7.9). Some forms are regionally restricted (2064), implying that local Solutrean groups developed their own particular styles. Finally, the Magdalenian, between about 16.5 and 11 ky ago, restricted initially to France and found later also in northern Spain, Switzerland, Germany, Belgium, and southern Britain, was distinguished primarily by a sophisticated bone and antler technology that produced points, harpoons, and other implements and weapons (fig. 7.10), sometimes elaborately incised or decorated. As in the Solutrean, regional differences in artifact form can be used to define Magdalenian subcultures. Some German Magdalenian sites may provide the oldest known evidence for domestic dogs (2426).

The late Upper Paleolithic of Europe seems to have been particularly diverse, and though various kinds of Solutreans and Magdalenians occupied the north and west, people who

Figure 7.9.
Solutrean artifacts
from La Riera Cave,
northern Spain
(redrawn after 2070,
pp. 99, 103, 107).
The Solutrean is
distinguished espe-
cially by various
kinds of well-made
unifacial and bi-
facial points.

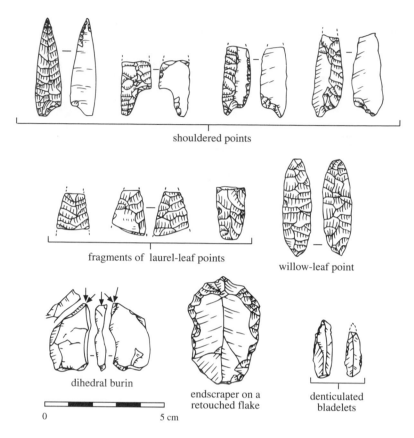

shoulded points

fragments of laurel-leaf points

willow-leaf point

dihedral burin

endscraper on a
retouched flake

denticulated
bladelets

0   5 cm

continued to emphasize Gravettian-like backed bladelets lived
to the south and east in Italy, Hungary, Bulgaria, Greece, and
the former Yugoslavia (778, 1272, 1572, 1602, 1850). The artifact assemblages these people produced are often lumped as
"Epi-Gravettian."

Within any given region, not only did each major Upper
Paleolithic industry tend to replace its predecessor far more
quickly than earlier industries did, but artifactual change within
each industry was also far more conspicuous. Thus archeologists can readily recognize distinctive subdivisions within
the Aurignacian, Gravettian, Solutrean, and Magdalenian. The
number and content of the subdivisions varies from region to
region, thereby complicating formal definition (2060, 2068), but
this only underlines the remarkable internal diversity of the
Upper Paleolithic.

Some of the extensive spatial and temporal variability that
characterized artifact assemblages after 40–35 ky ago was undoubtedly functional, reflecting the fact that some people began
to manufacture new or distinctive artifacts because they had a
new or distinctive purpose for them. Much of the variability was
probably stylistic, however, reflecting culturally different ways

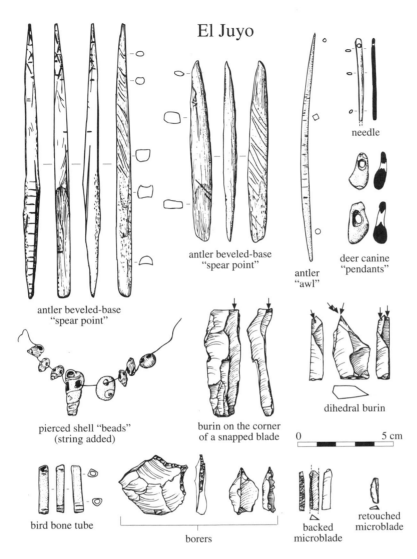

El Juyo

antler beveled-base "spear point"

antler beveled-base "spear point"

needle

antler "awl"

deer canine "pendants"

pierced shell "beads" (string added)

burin on the corner of a snapped blade

dihedral burin

0 5 cm

bird bone tube

borers

backed microblade

retouched microblade

Figure 7.10. Magdalenian bone and stone artifacts from the cave of El Juyo, northern Spain. The Magdalenian is distinguished above all by a wide variety of well-made bone and antler artifacts, including many decorated or artistic pieces.

of doing the same thing. Although it may seem peculiar to us that material items should suggest so little cultural differentiation before 40–35 ky ago, the amount that characterizes the succeeding period is reminiscent of more recent history. The extensive cultural variability of the past 40–35 ky almost certainly required the existence of modern people, with their seemingly infinite capacity for innovation.

The Late Paleolithic

This section describes various aspects of culture in the period between 40–35 and 12–10 ky ago, when early anatomically modern people had completely replaced their predecessors. This period encompasses the later part of the Last Glaciation, and it

corresponds to the time span of the European Upper Paleolithic, as it is conventionally defined (fig. 7.6). For simplicity's
sake, the people who lived in this period will be referred to here
as "late Paleolithic" to include both Upper Paleolithic people
in Europe and their early modern contemporaries elsewhere
in the world, particularly the contemporaneous LSA people of
Africa.

Economy

Although the emphasis here is on the differences between late
Paleolithic (early modern) people and their predecessors, economically they were broadly similar. All Paleolithic people
lived entirely by hunting and gathering wild resources. Only in
the transition from the Pleistocene to the Holocene, between
roughly 11 and 9 ky ago, is there sound evidence for a significant change, when some people, particularly in the Near East,
began to domesticate animals, plants, or both (128). Even these
people, however, probably continued to depend on wild resources for centuries, if not millennia.

Ethnographic observations on historic hunter-gatherers
show that the gathered, mostly vegetal component of the diet
commonly outweighs the hunted component. Undoubtedly,
late Paleolithic people also relied heavily on plants, but plant
residues are seldom preserved in Paleolithic sites. The water-
logged deposits of Ohalo II, Israel, provide a rare exception. Here
the unusual anaerobic conditions preserved fragments of acorns
and numerous seeds of wild wheat and barley that were collected by Upper Paleolithic people roughly 19 ky ago (1604).
At a handful of other sites the use of water, sometimes chemi-
cally charged, to "float" off small, light objects from sediment
samples has occasionally revealed large numbers of ancient
seeds, for example in the 22–10 ky old (and later) levels of
Franchthi Cave in Greece (948) and in the 14–13 ky old layers
of El Juyo Cave in Spain (766). At Franchthi, the seeds from deposits older than 13 ky come from species whose cultural utility is unclear, and they may in fact have been introduced by
birds or rodents. However, the seeds in deposits younger than
13 ky come mainly from wild lentils, vetches, pistachios, and
almonds that were probably eaten by people, and many of the
El Juyo seeds also come from possible food plants. Still, Ohalo II,
Franchthi, and El Juyo remain exceptional. In this light it is
understandable that archeologists have probably overstressed
the importance of hunting in Paleolithic economies. The over-
emphasis is probably most misleading for low-latitude sites
where gatherable plant foods were probably always more abun-

dant than game. It is less misleading for middle- and upper-latitude sites where, relatively speaking, game animals were much more numerous, particularly during glacial periods.

Game was especially abundant at middle latitudes in late Paleolithic Eurasia, because glacial cold and aridity favored the formation of grassy steppes over vast areas where forests prevailed during interglacials, including the Present or Holocene Interglacial. The principal species were gregarious ungulates such as the woolly mammoth, reindeer, bison, horse, and saiga antelope that could never have prospered in forested environments (935, 936). Reindeer reached the far south of France, and some even penetrated the Pyrenees to northern Spain. In France, in keeping with independent evidence for especially cold climate, reindeer were particularly common during the early Aurignacian (before 30 ky ago) and again during the Solutrean and Magdalenian (between roughly 20 and 11 ky ago) (315). Their bones dominate many French Magdalenian sites, including such well-known ones as Verberie, Pincevent, Laugérie-Haute, La Madeleine, and Isturitz (91, 92, 104, 295, 636–638). Because of regional environmental differences, other species prevail in contemporaneous sites elsewhere—for example, red deer at La Riera, El Juyo, Tito Bustillo, and other sites in northern Spain (28, 29, 478, 2065, 2070) and bison at Amvrosievka, Bol'shaya Akkarzha, Zolotovka I, and other sites on the south Russian plain (1029, 1035, 1232, 2010)—but everywhere there was a common emphasis on large gregarious herbivores.

In contrast to large herbivores, large carnivores such as lions, hyenas, and bears rarely occur in late Paleolithic archeological sites, though one or more large carnivore species overlapped people everywhere in Eurasia and Africa. At Istállóskö Cave in Hungary, Bacho Kiro Cave in Bulgaria, and other central or southeastern European early Upper Paleolithic (Aurignacian) cave sites, cave bear bones abound, but these are spatially or stratigraphically separate from the human occupational debris, and they probably result from natural deaths during hibernation (26, 1262, 1264). Human population growth and more intensive human use of caves may have extinguished the cave bear over most of its range in the later Upper Paleolithic (probably before 25 ky ago) (35), but in general Upper Paleolithic people and large carnivores probably avoided one another, since each had more to lose than to gain from confrontation.

Large herbivores not only were a vital source of food but also provided critical raw materials, including hides, sinew, and most obviously, bone, antler, or ivory. Late Paleolithic people not only used bone, antler, and ivory to fashion a wide variety of implements and art objects but also used larger pieces as weights, supports, and other structural elements in their dwellings. Patterned arrangements of large bones constitute key evidence for dwellings at the Gravettian sites of Krakow-Spadzista Street B in Poland (1263, 1273), Pavlov I and Milovice in the Czech Republic (1252, 1633), and at Mezhirich, Yudinovo, Mezin, and other Upper Paleolithic sites in Ukraine and Russia (845, 1029, 1228, 1232, 2006, 2007). In parts of central and eastern Europe where trees were especially scarce, the people even used fresh bone for fuel, judging by the large amount of bone ash and charred bone fragments in their fireplaces. Like some historic Indians on the American Great Plains, they probably also burned the dried dung of large herbivores.

Although it is probable that late Paleolithic people exploited local resources more effectively than did earlier people, firm evidence remains relatively limited. In western Europe perhaps the best indication is that, per unit time, Upper Paleolithic sites are much more numerous than Mousterian ones (478, 1535, 1536, 2057), implying that Upper Paleolithic populations were larger and denser. The principal alternative—that Upper Paleolithic people simply moved camp more often— seems unlikely, since Upper Paleolithic sites tend, if anything, to be richer and more extensive than Mousterian ones. Since Upper Paleolithic and Mousterian people lived under broadly the same conditions, Upper Paleolithic people could have been more numerous only if they used local resources more efficiently. Theoretically, this should be reflected in contrasts between Upper Paleolithic and Mousterian food refuse, but studies to check this are being undertaken only now (448, 450, 2051). For the moment, the most obvious difference is that bird and fish bones are more common in Upper Paleolithic sites, and it may be that active fishing and fowling were unique to the Upper Paleolithic. Increased reliance on fish may especially account for Upper Paleolithic florescence in parts of northern Spain, southwestern France, and southern Russia, where annual salmon runs probably provided knowledgeable hunter-gatherers with an unusually rich and reliable resource.

It has also been suggested that some Upper Paleolithic people were unique or advanced in their tendency to specialize on just one or two herbivorous species, such as reindeer in southwestern France or red deer in northern Spain (fig. 7.11) (1530, 1535, 2057, 2441). However, Mousterian levels in southwestern France are sometimes dominated by a single species, for example, by reindeer at Combe-Grenal (273, 448, 450), by aurochs at La Borde (1135), and by bison at Mauran (710, 711). One or two species also strongly dominate some Mousterian

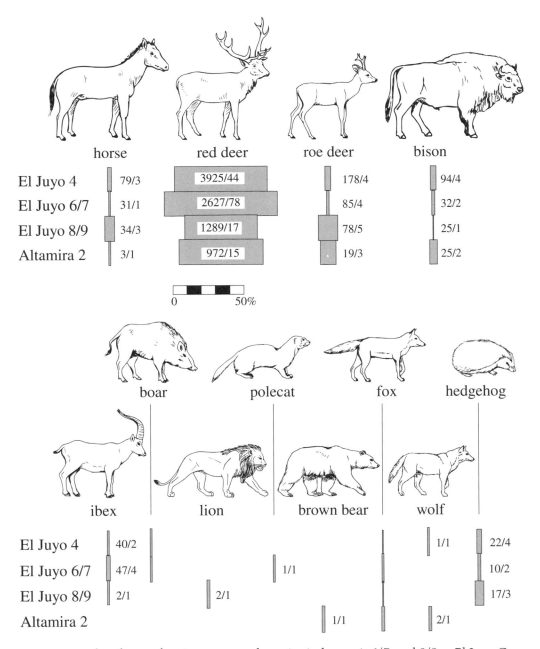

Figure 7.11. Abundance of various mammal species in layers 4, 6/7, and 8/9 at El Juyo Cave and in layer 2 at nearby Altamira Cave, northern Spain. The associated artifacts belong to the local "Lower" Magdalenian, bracketed between roughly 15 and 13 ky ago. The bars represent the percentages of each species in each layer, as measured by the minimum number of individuals (MNI) necessary to account for the bones of each species. The numbers with the bars are the number of specimens assigned to each species divided by the MNI. The chart illustrates the abundance of red deer and the rarity of large carnivores, which are characteristic aspects of most Magdalenian sites in northern Spain.

sites elsewhere, for example, bison at Wallertheim (785, 786) and reindeer at Salzgitter-Lebenstedt (393, 2203) in Germany; horses at Zwolen in Poland (1879); wild asses at Starosel'e, Kabazi II, and other sites in the Crimea (Ukraine) (441, 1229); and bison at Rozhok I, Sukhaya Mechetka, and Il'skaya on the south Russian plain (1030, 1035).

More fundamentally, in virtually all cases where bones of a single species predominate at an Upper Paleolithic or earlier site, the species could also have dominated the ancient environs of the site, and the bones therefore need not demonstrate human hunting specialization or preference. The extraordinary abundance of reindeer at late Upper Paleolithic sites in southwestern France is a case in point, since it almost certainly reflects climatic deterioration that all but eliminated other herbivores (1761). Finally, it is not necessarily true that people hunted or even ate the species whose bones are most common at a site. This is most obvious with respect to the woolly mammoth, whose bones predominate in many Mousterian and Upper Paleolithic sites in central and eastern Europe. In many cases mammoth bones appear to have been used extensively as building material, as fuel, or as raw material for artifact manufacture, and they may simply have been scavenged from long-dead animals (1232, 2006, 2010). This possibility is directly implied at the artifactually unique late Upper Paleolithic site of Mezin (Ukraine), where chemical analysis shows that the mammoth bones composing a "ruin" came from individuals that probably died decades apart (1685). It is also suggested at the Eastern Gravettian site of Krakow-Spadzista Street B (Poland), where many of the mammoth bones used to build three structures were conspicuously gnawed by carnivores (1271). At many other sites it may be indicated by differences in superficial bone weathering, although differential exposure of bones after site abandonment might produce the same result.

As discussed in chapter 6, the extreme southern tip of Africa has provided perhaps the best evidence for a late Paleolithic advance in resource exploitation (1243, 1247). In brief summary, unlike their MSA predecessors, local LSA people living under similar environmental conditions actively fished and fowled, and they killed dangerous game such as buffalo and wild pigs much more often. In addition, the significantly smaller size of shellfish and tortoises in LSA sites probably reflects larger, denser LSA populations. LSA advances in resource exploitation were probably due mainly to technological innovations, including the development of fishing and fowling gear and also of snares or weapons that reduced the hunters' exposure to danger. The contrasts are stark, but unfortunately it is not yet clear that they characterize the very earliest LSA people, whose sites are locally rare, probably because they date from a time (between 60 and 30 ky ago) when hyperaridity had greatly reduced human numbers.

Technology

Late Paleolithic people were far more inventive than their predecessors and made technological innovations at an unprecedented pace. It is these innovations that are particularly stressed here. Many remain imprecisely dated, and even when dates are available it seems unlikely that the earliest known occurrences were actually the first. For this reason, and because specific dates are not crucial to the main point, the discussion will treat the entire late Paleolithic period, 40–10 ky ago, as a single unit. The basic information for this summary is drawn from broad syntheses (in 269, 278, 500, 521, 2024) as well as from more specialized sources cited below.

In general, late Paleolithic artifact assemblages contain a much wider range of recognizable artifact types than do earlier ones, suggesting that late Paleolithic people were engaged in a wider range of activities. These clearly included shaping formal bone, ivory, and antler artifacts, and they probably also involved manufacturing more tools to create other tools rather than for immediate use as hide scrapers, projectile points, and so on. To judge by the small size or the shape of many late Paleolithic stone artifacts and by the form of many bone pieces, the people probably also manufactured many more composite tools, that is, implements combining separate pieces of stone, bone, or other materials. Unfortunately, because the composite tools were held together mainly by perishable glues, leather thongs, and so forth, few have survived to the present.

Although Mousterians and even earlier people had adapted to glacial climates in central and western Europe, late Paleolithic people were the first to inhabit the harsh environments of easternmost Europe and northern Asia (Siberia) (fig. 7.12), where the winters were exceptionally long and cold even during interglacials. Upper Paleolithic innovations for dealing with intense cold probably included both better clothing and better housing. Winter clothing almost certainly incorporated fur, and east European Upper Paleolithic sites including, for example, Mezin, Mezhirich, Eliseevichi I, Avdeevo, and Kostenki XIV have provided the oldest known evidence for systematic fur trapping (1029, 1232, 2006). The remains of wolves or arctic foxes are extraordinarily abundant in these sites, and they tend to occur either as whole or nearly whole skeletons lacking the paws

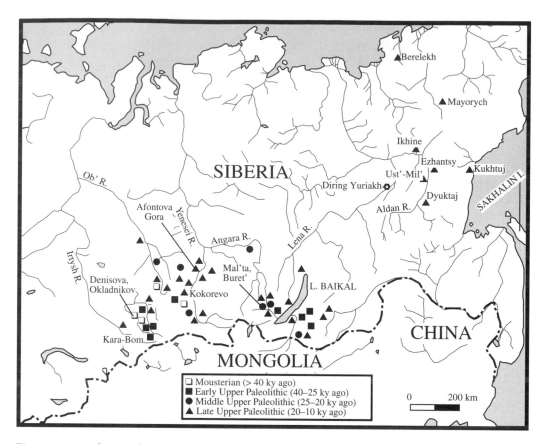

Figure 7.12. Siberia, showing the approximate locations of the Mousterian and Upper Paleolithic sites mentioned in the text (adapted from 2310, fig. 1). Mousterian and early/middle Upper Paleolithic sites occur only in southern Siberia. Late Upper Paleolithic sites have also been found in the central and northern sections.

or as articulated paw skeletons, occurring separately. The implication is that the people removed the feet with the skins, as modern trappers often do, and then discarded the skinned carcasses. The "awls," "punches," and other pointed bone objects that appear in even the earliest Upper Paleolithic sites could have been used to sew skins together, and eyed bone needles that would have made the process much easier were certainly in use by 19–18 ky ago (1881, 2053, 2064). The existence of sophisticated, well-tailored clothing is directly documented in some Upper Paleolithic burials, especially those found at the site of Sungir', 210 km northeast of Moscow (101). Here discolored soil and strings of ivory beads surrounding, girdling, and paralleling the skeletons of three people buried more than 22 ky ago even suggest the details of fur or leather garments, comprising a cap, a shirt, a jacket, trousers, and moccasins. The beads and other objects found with the skeletons were apparently sewn on the clothing as decorations or fasteners.

Like their predecessors, late Paleolithic people often occupied rockshelters or cave mouths in areas where these were available, such as southwestern France and northern Spain. Meticulous modern excavations like those at Cueva Morín (765) and El Juyo (135, 766) in northern Spain show that they sometimes built walls or otherwise modified natural shelters to make them more habitable. Over much of their range, however, late Paleolithic people did not have access to caves, and in such situations they commonly constructed dwellings that were far more substantial than any known or inferred for earlier people. The superstructures were made mainly of wooden poles, hides, and other materials that long ago vanished, but the foundations have been exposed in careful excavations. The most spectacular examples come from the harsh open plains of central and eastern Europe, where they variously comprise large, artificial depressions, regular arrangements of postholes, patterned concentrations of large bones or stone blocks (serving as construction material), sharply defined concentrations of cultural debris, or a combination of these features (figs. 7.13–7.15) (1029, 1229, 1232, 1266, 1270, 1271, 1273, 2006, 2007, 2116, 2117). The "ruins" usually border, encircle, or cover patches of ash and charcoal that mark ancient fireplaces for heating and cooking. Some of the fireplaces are more complex than any earlier ones and have intentionally corrugated floors or small ditches leading out from the central ash accumulation. Modern experiments show that these modifications increase oxygen flow and thus produce a hotter flame.

For the most part, we do not know precisely what late Paleolithic people did with their individual artifacts, but it is safe to assume that they used many to obtain or process foods and raw materials. Bone shaping and incising, for example, almost certainly involved stone burins, named for the modern metal engraving tools they broadly resemble. Hide processing was probably often accomplished by using stone endscrapers to remove hair and fat and then using spatulate bone "burnishers" to smooth the skin or spread softening agents or pigment. Enigmatic bone objects found at many sites may have been trips or other parts of compound traps for snaring foxes, hares, and other valuable fur-bearing animals. More persuasively, many pointed stone and bone artifacts must have been used to tip spears or arrows whose perishable wooden shafts have long since disappeared. Proof can be found in points still stuck in animal bones—for example, a flint point in a wolf skull at the East Gravettian (or "Pavlovian") site of Dolní Vestonice in the Czech Republic (1252), another in a reindeer vertebra at the late Magdalenian–like "Hamburgian" site of Stellmoor in northern

Figure 7.13.
Top: Roughly circular concentration of large bones believed to mark the base of a ruined hut at the Upper Paleolithic site of Mezin (complex 1) in Ukraine. *Below:* Hypothetical reconstruction of the hut, showing the bones used as weights to hold down skins stretched over a wooden framework (redrawn after 283, fig. 1).

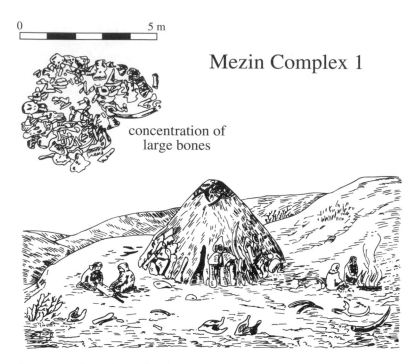

0   5 m

Mezin Complex 1

concentration of
large bones

Germany (1878), and a bone point in a bison scapula at the Siberian late Paleolithic site of Kokorevo I (2).

Like other formal bone artifacts, bone points were essentially a late Paleolithic innovation, and their variety constitutes a further indication of late Paleolithic inventiveness. Some, including the very earliest examples from Europe, have split or beveled bases (fig. 7.7) that would have made hafting easier. Others that are especially common in Siberia bear longitudinal grooves or slots into which sharp stone bits were probably inserted to produce a more ragged wound (fig. 7.16). Points with the inserts still in place have been found at several sites, including Kokorevo I mentioned above. On some points the grooves were too broad and shallow to accept inserts, and they may have been runnels to promote bleeding in a wounded animal. Finally, among all the known point types, among the most striking are those with backward-pointing barbs. Almost certainly these were harpoon heads used for spearing fish or perhaps reindeer or other mammals crossing a stream. Probable harpoons appeared in various parts of Eurasia by 14 ky ago or a little after. Excepting the barbed points from the Katanda sites, Zaire, whose age is controversial (see the section on nonstone artifacts in the chapter 6), barbed points also appear in the Sahara and eastern Africa in the late Paleolithic, mainly after 12 ky ago.

Bone points are only one of many late Paleolithic novelties that enhanced hunting efficiency. For acquiring large, mobile

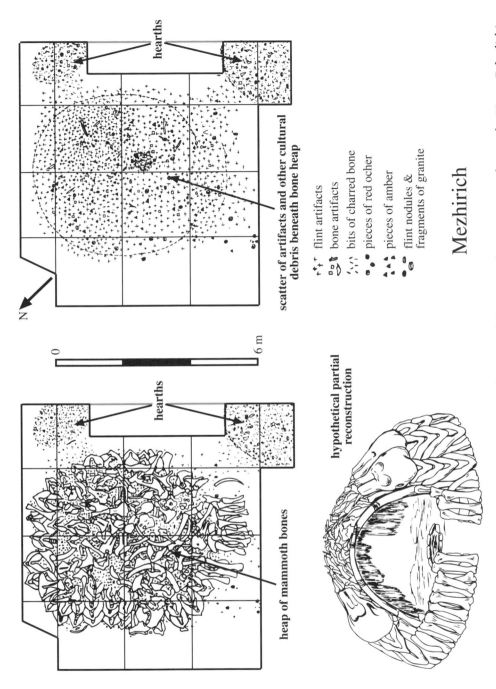

hearths

scatter of artifacts and other cultural debris beneath bone heap

N

0   6 m

↑⁺⁺   flint artifacts
ᴅ◦↕   bone artifacts
`‚`‚`‚`   bits of charred bone
◦●   pieces of red ocher
◢◣▲   pieces of amber
⬮◉   flint nodules &
      fragments of granite

Mezhirich

hearths

heap of mammoth bones

hypothetical partial reconstruction

Figure 7.14. *Top left:* Plan view of a circular heap of mammoth bones used to construct a hut at the Upper Paleolithic site of Mezhirich in Ukraine about 15 ky ago. *Top right:* Distribution of artifacts, fragmentary animal bones, and other debris beneath the bone heap. *Bottom left:* Conjectural reconstruction of the hut (adapted from 845, p. 165).

Figure 7.15. *Top:* Plan of a roughly rectangular, 30 cm deep depression dug by the Upper Paleolithic inhabitants of Pushkari I in Ukraine, showing the internal distribution of hearths and large bones. *Below:* Reconstruction of a hypothetical three-part structure that may have covered the depression (redrawn after 283, fig. 3).

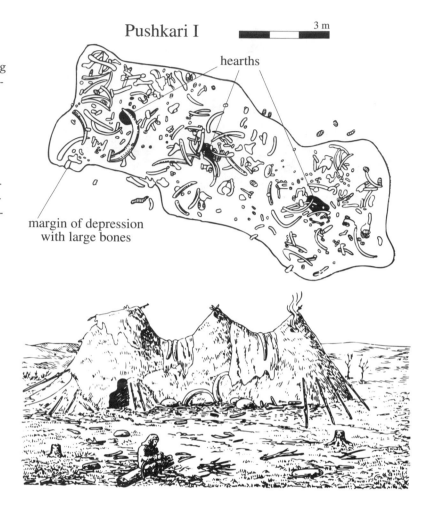

game, the most important new items were surely projectile weapons, including particularly the spear-thrower and the bow and arrow. The spear-thrower is a bone or wooden rod hooked at one end to accommodate the dimpled or notched nonpointed end of a spear shaft (fig. 7.17). With the shaft resting against the rod, the rod extends a person's arm so that the spear can be thrown harder and farther. Bone spear-throwers were in use by at least 14 ky ago.

The antiquity of the bow and arrow is more difficult to establish because the most diagnostic parts were made of perishable materials. However, small sharp stone bits, well-fashioned stone points, or small backed bladelets very similar to ones that tipped arrows later on occur from at least 20 ky ago in various parts of Eurasia and Africa. Once the bow and arrow were invented, they would obviously have diffused very rapidly, and a sharp increase in the abundance of tiny backed bladelets in

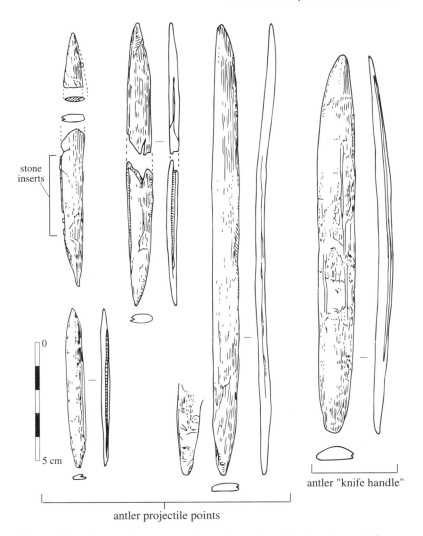

stone
inserts

0

5 cm

antler projectile points

antler "knife handle"

Figure 7.16. Grooved antler artifacts from the Siberian Upper Paleolithic site of Kokorevo I, dated to approximately 13–14 ky ago (redrawn after 4, fig. 1). The grooves were probably designed to receive
microblades, some of which were still present in one piece (left center). Most of the artifacts were probably projectile points, and the broken piece in the upper left hand corner was found stuck in a bison
shoulder blade. The piece on the far right is less pointed than the
others and may have been a knife handle. The grooved antler and
microblade technology is well documented in Siberia only after 18 ky
ago, but it may have been present earlier.

both Africa and Eurasia about 21–20 ky ago may signal their
spread. More concretely, blunt-ended bone rods strikingly similar to historic or protohistoric linkshafts or foreshafts circumstantially place the bow and arrow in southern Africa by 20 ky
ago or a little later (612). However, the oldest conclusive evidence, consisting of fragmentary wooden bows or arrows, comes

Figure 7.17. *Top:* Diagrammatic indication of how a spear-thrower "lengthens" the arm, permitting a spear to be thrown harder and farther (adapted from 943, p. 102). *Below:* Decorated "the fawn and the bird" spear-thrower from the Magdalenian deposits of Mas d'Azil, southern France (redrawn after 2024, p. 55).

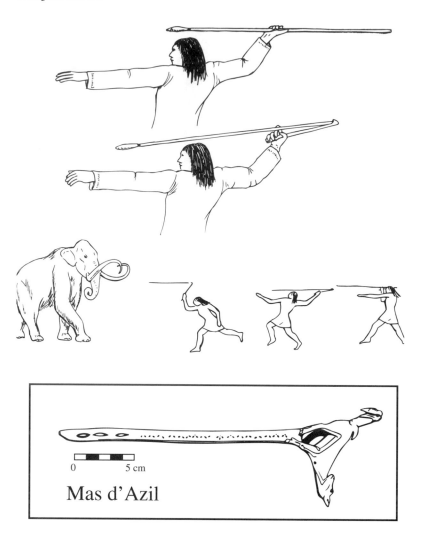

Mas d'Azil

only from late Magdalenian or related sites, dating between 12 and 10 ky ago in France and northern Germany (2267).

Late Paleolithic sites in Europe have also provided the oldest known fishhooks, dated to roughly 14 ky ago, while objects believed to be "fish gorges"—shaped bone slivers resembling double-pointed toothpicks (fig. 7.18)—occur in South African sites that are only slightly younger. Fishing with spears, traps, and so on had probably been under way for some time before 14 ky ago, to judge by fish bones found in earlier sites.

In sum, late Paleolithic people were obviously very creative, and it seems likely that by 12–10 ky ago they had devised the entire range of technology observed among historic hunter-gatherers. The Eastern Gravettian (or Pavlovian) people who occupied the Czech sites of Dolní Vestonice, Pavlov, Predmostí, and Petrkovice 28–27 ky ago even discovered that soft clay,

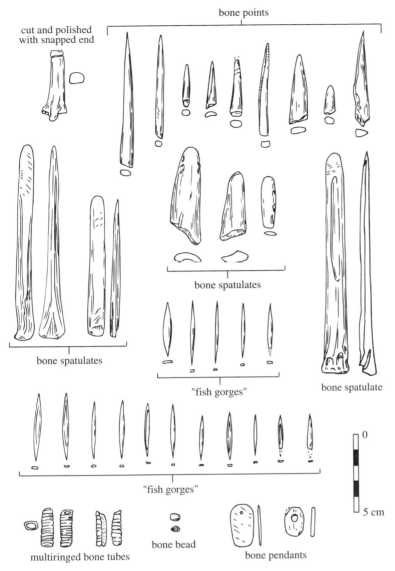

cut and polished
with snapped end

bone points

bone spatulates

bone spatulates

"fish gorges"

bone spatulate

"fish gorges"

multiringed bone tubes

bone bead

bone pendants

0

5 cm

Figure 7.18. Range of formal bone artifacts from the LSA layers of Nelson Bay Cave, South Africa (after 612, p. 176). Formal bone artifacts are rare or absent in MSA sites, and where they occur they could easily represent undetected intrusions from LSA levels. By ethnographic analogy, the Nelson Bay LSA "fish gorges" were probably baited and attached to lines to catch many of the fish represented in the same layers. In preceding MSA sites, objects interpretable as fishing gear are absent and fish bones are correspondingly rare.

properly mixed with other materials (temper) and heated to a high temperature, hardens into a much more durable material (2309). The sites have provided more than 10,000 fire-hardened clay objects, including more than 3,700 fragments of animal and human figurines. A handful of amorphous fragments preserve unique impressions from woven fabrics or nets that provide additional insight into late Paleolithic ingenuity (1717). The greatest number of hardened pieces (more than 6,700) occurred at Dolní Vestonice 1, which also preserved two walled structures that probably represent kilns. Roughly 15 ky later, about 13–12 ky ago, some very late Paleolithic occupants of Japan independently rediscovered fired clay technology, which they used to manufacture the world's oldest known ceramic vessels (1755, 2162).

Social Organization

The archeological record provides little direct indication of late Paleolithic social organization, except perhaps for greater contact or interchange between groups than ever before. In Europe and northern Asia, "luxury" items such as amber and seashells occur at Upper Paleolithic sites that are scores or even hundreds of kilometers from where the items originated (971, 1603, 2426, 2427). Even the stone used to make tools was sometimes transported over great distances. Perhaps the most spectacular examples come from very late Upper Paleolithic (terminal Pleistocene) sites on the great plain of north-central Europe, which demonstrate that particularly desirable types of flint were routinely transported 100–200 km and even more (1878). There are also earlier examples, however, such as those from the "Spitsyn" set of Upper Paleolithic sites near Kostenki-on-the-Don, in European Russia (1029, 1228, 2006, 2007); from the very early Aurignacian levels of Bacho Kiro, Bulgaria (1265); and from the Eastern Gravettian occupations at Pavlov and Dolní Vestonice, Czech Republic (2117). The Spitsyn people who lived near Kostenki before 32 ky ago imported almost all their flint from 150–300 km away; the early Aurignacians who occupied Bacho Kiro by 40 ky ago carried more than 50% of their flint at least 120 km; and the Eastern Gravettians who inhabited Pavlov and Dolní Vestonice roughly 27 ky ago imported more than 90% of their flint over a distance of 120 km. Not all late Paleolithic people routinely moved "luxury" or "basic" goods over such great distances, but before the late Paleolithic no one did. The wider intergroup contacts that such long-distance movements may imply could have been both cause and effect of the enhanced cognitive and communication abilities that are probably implied by late Paleolithic art.

In the absence of direct evidence for social organization, the physical and material cultural similarities between late Paleolithic people and many historic hunter-gatherers suggest that late Paleolithic populations were similarly organized or, perhaps more precisely, that they enjoyed broadly the same range of social structures as their historic counterparts. Some late Paleolithic people occupying more marginal environments in which relatively limited resources supported only sparse populations probably lived in small egalitarian bands of related families, like many of the historic hunter-gatherers of Australia and of the dry interiors of western North America and southern Africa.

At the other extreme, some late Paleolithic people inhabiting very rich settings that supported much denser populations may have lived in complex, "ranked" societies like those of the historic hunter-gatherers of the American Pacific Northwest. In this instance a cadre of hereditary chiefs may have coordinated many activities, including food acquisition and distribution, rituals and ceremonies, trade, and even warfare. Relatively complex social organization has been posited particularly for the very late Paleolithic (ca. 16–11 ky old) Magdalenian people of southwestern France and northern Spain, whose sites are especially numerous, rich, and closely packed. It may also have characterized some of the (ca. 26–14 ky old) Gravettian and "Epi-Gravettian" peoples of Ukraine and Russia, whose sites contain elaborate structural "ruins" and impressive food storage pits (2006, 2008). Whereas the simpler societies of the late Paleolithic may have differed little in basic organization from Middle Paleolithic and even earlier ones, the more complex societies that characterized some late Paleolithic people probably had no earlier counterparts.

"Ideology": Art and Graves

The thoughts, ideas, beliefs, and values of late Paleolithic people are not preserved in the archeological record, but their art and their graves provide the first clear evidence for ideological systems like those of historic people. The art has been summarized or analyzed many times (for example, in 3, 106, 108, 352, 533–535, 767, 883, 1268, 1359, 1945, 2268, 2442), and it may be divided between two basic categories—wall art, comprising paintings and engravings on rock surfaces, and portable or home art, comprising items that occur alongside other artifacts in the ground.

Paintings and engravings weather off rock surfaces that are exposed to the elements, and late Paleolithic wall art is therefore confined almost entirely to caves. The first examples were discovered in 1879 by Marcelino Sanz de Sautuola in Altamira Cave in the Cantabrian region of northern Spain, but for many years their great antiquity and authenticity were disputed (108, 2065). Full acceptance came only in the early 1900s, after important additional examples were found nearby in southwestern France. More than 150 caves with late Paleolithic paintings or engravings have now been identified in Franco-Cantabria (fig. 7.19), making it by far the richest region in the

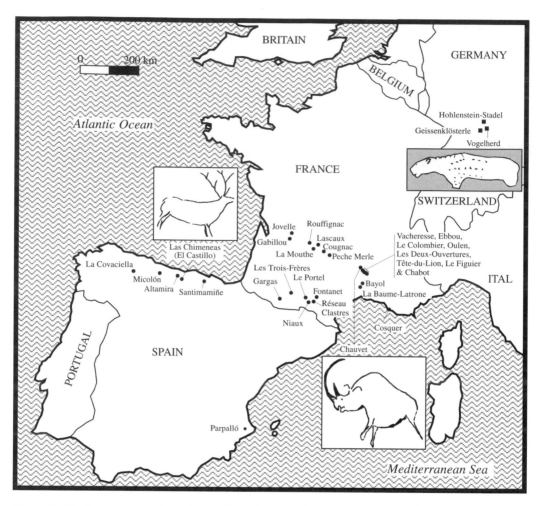

Figure 7.19. Approximate locations of key decorated caves in France and Spain and of caves in Germany that have provided some of the oldest known ivory figurines. The overwhelming majority of decorated caves occur in northern (Cantabrian) Spain and southern France, which together comprise the Franco-Cantabrian region. Exploration continues to reveal new paintings or engravings in previously known caves, and a previously unknown decorated cave is found every year or two. The insets show a lion figurine from the early Aurignacian deposits of Vogelherd Cave, dated to roughly 32 ky ago, a painted rhinoceros from Chauvet Cave that is probably about 31 ky old, and a red deer from Las Chimeneas Cave (El Castillo) that is probably 14–13 ky old. (Base map and rhinoceros modified after 453, p. 13; Vogelherd figurine after 942, fig. 3; and Las Chimeneas deer after 2068, fig. 9.)

world. In other regions with apparently suitable caves, late Paleolithic wall art is either rare or absent (120), suggesting it was not an aspect of local culture. But it is also possible that, like diverse historic or late prehistoric hunter-gatherers throughout the world, many late Paleolithic peoples produced wall art on exposed rock surfaces where it no longer survives. It is both puzzling and fortunate that, almost uniquely, the late Paleolithic artists of Franco-Cantabria often chose cave walls as a basic medium.

For more than a hundred years after Franco-Cantabrian cave art was first discovered, it was assigned to the Pleistocene largely because it depicted mammoths, bison, reindeer, wild horses, and other locally or wholly extinct animals. Stylistic comparisons with animal figures carved from or engraved in bone, antler, or ivory suggested that the final Pleistocene (16–11 ky old) Magdalenians were the main artists at famous caves like Lascaux, Les Trois-Frèrcs, Niaux, and Altamira, but until recently direct dating was impossible. This was true even though the black pigment used in many paintings was based on charcoal that is amenable to the radiocarbon method. The problem was that the amount of carbon necessary for a conventional radiocarbon date would require the destruction of whole paintings or even of multiple paintings. Now, with the advent of the accelerator radiocarbon method, it is possible to date charcoal fragments the size of a pinprick (half a milligram), whose removal does little or no damage, and this shows that the Magdalenians indeed produced most if not all the remarkable art at Niaux, Altamira, El Castillo Cave, and other Franco-Cantabrian sites (105, 106, 453, 1392, 2271).

In addition, however, accelerator dating has also shown that some French cave art substantially antedates the Magdalenian. The most spectacular examples are at Cougnac Cave (southwestern France), where paintings of the extinct Irish elk have been bracketed between 25 and 20 ky ago (1391, 1392); at Cosquer Cave near Marseilles, where stenciled or negative handprints (produced by blowing paint over a human hand) and naturalistic animal paintings have been fixed at roughly 27 and 18.5 ky ago respectively (515–517); and at Chauvet Cave in the Ardèche Valley (south-central France), where naturalistic paintings have been dated to about 31 ky ago (453, 514). Cosquer Cave contains numerous finger tracings in the once-soft cave wall that are thought to be coeval with the stenciled hands, which they sometimes overlie, and both Cosquer and Chauvet caves contain engraved animals whose species identity and style imply broad contemporaneity with the painted ones. At Chauvet the artists were presumably Aurignacians, and at Cosquer the earliest ones could have been Aurignacians or Gravettians and the later ones were probably Solutreans. At both sites, subject matter and style further support a pre-Magdalenian age for the art. This is particularly true at Chauvet, where rhinoceroses, lions, mammoths, and bears occur far more often than in well-known Magdalenian art, in which horses, bison, deer, and ibex generally dominate (513). The focus at Chauvet recalls the prominence of rhinoceroses, bears, mammoths, and lions among the seventeen Aurignacian ivory figurines found in the caves of Vogelherd, Geissenklösterle, and Hohlenstein-Stadel, southwestern Germany (fig. 7.19). The ivory statuettes are as old as the Chauvet paintings, or older, and as already indicated, they join Chauvet in underscoring the gulf between even the early Upper Paleolithic and the Mousterian.

Scanning electron microscopy, X-ray diffraction, and other advances in technology that can reveal the physicochemical composition of infinitesimal pigment samples have now also shown that the artists often mixed iron and manganese oxide or charcoal pigments with minerals that served as paint extenders or with plant oils that served as binders (512, 518, 1392, 1393). The identification of distinct pigment recipes has illuminated the probable order in which paintings were executed, and together with radiocarbon dating, pigment variation among paintings suggests that successive generations of artists may have visited a single cave for centuries or even millennia. Replication experiments have shown further that the artists could have applied paint in a variety of ways, including not only brushing but also spitting in the manner of some historic Australian Aborigines (1390).

If the techniques of the artists have become clearer, however, the meaning or purpose of their art remains mysterious. Perhaps the most secure inferences can be drawn from historic hunter-gatherers, who rarely produced art for its own sake. Instead they embedded their art in other aspects of culture, where it variously functioned to enhance hunting success, to ensure the bounty of nature, to illustrate sacred beliefs and traditions (perhaps on ritual occasions), or to mark the territorial boundaries of an identity-conscious group. Conceivably much Paleolithic wall art symbolizes or encodes the social structure or worldview of its makers, and like the much more recent rock art of southern Africa, some could register the visions of shamans or medicine men in the trance state (1366–1369). The deeply cultural (vs. strictly artistic) meaning of much Franco-Cantabrian art is probably reflected in its location not only in caves but sometimes deep within them, in chambers or passages that were difficult to reach. To penetrate the darkness, the artists used wooden torches or burned animal fat or oil on limestone and sandstone slabs. Splotches of possible torch charcoal from attempts to expose unburned wood underneath occur on the walls of Chauvet Cave, and shallow sculpted rock bowls that broadly resemble historic Inuit (Eskimo) oil lamps have been found in several French Magdalenian sites (164). Some even retain chemical traces of vegetal wicks and of the animal fats that served as fuel.

By its mode of occurrence, portable or home art, including items of personal adornment, is much easier to date. In Europe, perforated animal teeth that were probably pendants or beads, carefully shaped ivory or soft stone beads, and other carved, incised, or engraved bone objects date from the very dawn of the Upper Paleolithic (778, 941, 942, 1029, 1262, 1593, 2024, 2443), and in Africa ostrich eggshell beads remarkably similar to those produced by historic natives occur in even the earliest LSA, dating between 45 and 38 ky ago at Enkapune Ya Muto in central Kenya, Mumba Cave in northern Tanzania, and Border Cave in Kwa-Zulu/Natal, South Africa (31, 160, 162, 361). Africa has also provided some of the oldest known paintings—animal figures on rock slabs from deposits dated to at least 19 and perhaps as much as 27.5 ky ago at Apollo 11 Rockshelter in Namibia (2421).

Like the stone artifacts that accompany them, art objects and personal ornaments vary significantly in form through time and space. For example, in Europe well-produced naturalistic engravings of animals are concentrated in Magdalenian (16–11 ky old) sites, whereas human figurines occur most commonly in Gravettian (28–20 ky old) levels (fig. 7.20). Like the wall art, the portable art is impossible to interpret precisely, but little of it probably was done for artistic purposes alone. Some enigmatic engraved or incised objects may have been gaming pieces, and others were perhaps counting or recording devices, even lunar calendars (570, 1453, 1454, 1458). Many animal figurines could be the totemic symbols of kinship groups, and the human figurines obviously could represent deities or spirits. Most are highly stylized, lacking facial features or details of the hands and feet. Many, known popularly as "Venus figurines" (figs. 7.20, 7.21), have exaggerated buttocks and breasts, leading to speculation that they were fertility symbols or depictions of earth mother goddesses. Whatever the case, with the rest of the art, they clearly imply that late Paleolithic people not only were physically identical to their living descendants but possessed basically the same cognitive and communication faculties, including languages that were as complex as any historic ones. Enhanced cognition and communication were surely crucial to many late Paleolithic activities, particularly to ones that depended upon within- and between-group cooperation, and the art therefore provides a vital clue to why other aspects of late Paleolithic culture appear so advanced.

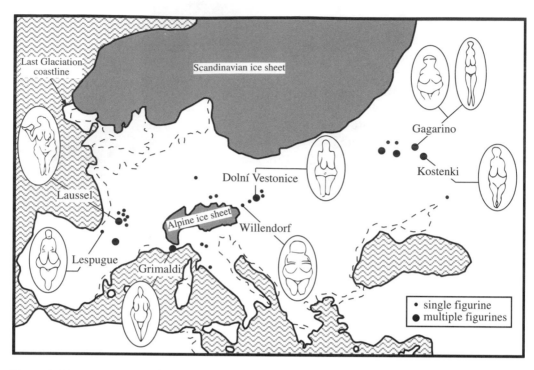

Figure 7.20. Approximate locations of the European Upper Paleolithic sites that have provided female figurines or engravings (redrawn after 445, fig. 3.19). At most sites the figurines or engravings were certainly or probably associated with the "Gravettian" culture complex, dated to between roughly 28 and 21 ky ago.

Figure 7.21. The famous "Venus of Lespugue," from a probable Gravettian layer at the site of Lespugue, southwestern France (drawn by Kathryn Cruz-Uribe from a cast; © 1999 by Kathryn Cruz-Uribe).

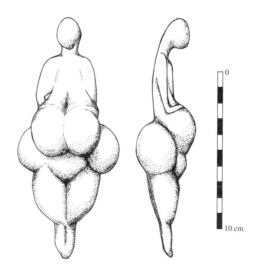

Graves are of course known from Mousterian sites, and late Paleolithic people were thus not the first to bury their dead. However, late Paleolithic sites are the earliest to contain undoubted multiple burials. The famous Aurignacian Cro-Magnon skeletons (1308) probably came from a communal grave, and some truly spectacular examples have been found at Pavlovian (Eastern Gravettian) open-air sites in the Czech Republic, radiocarbon-dated to 27–26 ky ago. These Pavlovian cases include a young woman and two young men, possibly her brothers, buried together at Dolní Vestonice (27, 1253) and a veritable graveyard at Predmostí (5, 1998, 2000). Here excavations near the end of the previous century uncovered a large (4 m × 2.5 m) oval pit that was covered by limestone slabs and mammoth bones and that contained the skeletons of eighteen individuals of various ages and both sexes. Even richer cemeteries, containing dozens of individuals, have been found at terminal Pleistocene (ca. 15–10 ky old) sites such as Taforalt and Afalou-Bou-Rhummel in northwestern Africa (fig. 7.1), where they are associated with artifacts of the Upper Paleolithic Iberomaurusian Industry (416).

Equally important, recall from the previous chapter that Mousterian graves tend to be very simple, with no clear indication of a burial ritual or of the inclusion of valued items or "grave goods." Many Upper Paleolithic graves are also relatively simple (180, 2063, 2121), but others are much more elaborate, and individuals were often buried with special bone, shell, or stone artifacts (fig. 7.22) (969). Clusters of perforated seashells occurred with the skeletons from the early Aurignacian layers of the famous Cro-Magnon Shelter in southwestern France (1308), and clusters of pierced shells or animal teeth, dense concentrations of ocher, or both characterize several other Upper Paleolithic graves, including most notably the multiple examples in the Aurignacian and Gravettian layers of the Grimaldi Caves (Italian Riviera) (1602, 1675).

An equally old or older double grave at Sungir', Russia, mentioned above for its evidence of clothing, illustrates the great difference between some Upper Paleolithic burials and all Mousterian ones. The Sungir' grave was dug into permanently frozen subsoil (permafrost) more than 22 ky ago and contained the extended skeletons of two children, one arguably male and the other female, placed head-to-head (2445). The putative male was covered with 4,903 beads whose arrangement suggests they were attached to closely fitting clothing. In addition, there were 250 perforated arctic fox canines placed as if they had been attached to a belt at the waist, an ivory animal pendant on the chest, an ivory pin (perhaps a fastener) near the throat, a large sculpted ivory mammoth under the left shoulder, a highly polished human femur shaft packed with red ocher on the left side, and a 2.4 m long ivory "lance" on the right side. The putative female was covered and surrounded by 5,374 beads or bead fragments that were also probably attached to clothing,

Figure 7.22.
Skeleton of a child
from a grave at the
Siberian Upper
Paleolithic site of
Mal'ta (redrawn
after 807, fig. 36).
Radiocarbon indi-
cates the grave
dates from about
21 ky ago (1294).
Like many other
late Paleolithic
graves, the one at
Mal'ta contained
special artifacts
that may represent
funeral offerings,
the deceased's per-
sonal belongings,
or both.

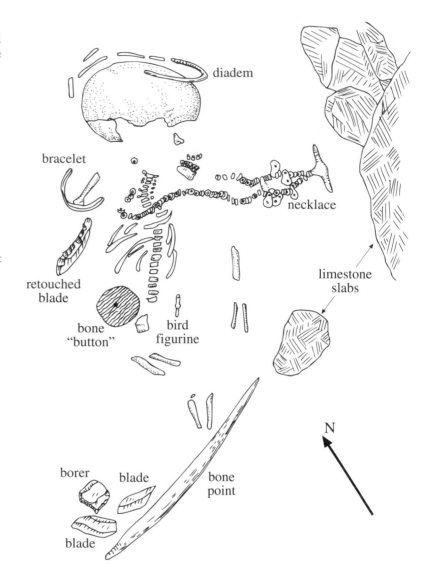

diadem

bracelet

necklace

limestone
slabs

retouched
blade

bone
"button"

bird
figurine

N

borer   blade   bone
point

blade

and there was an ivory pin at the throat, several small ivory
"lances" on both sides, and on one side, two perforated antler
"wands," one of them decorated with rows of shallow drilled
holes. Experimentation suggests that the beads alone required
thousands of hours to manufacture (2445). The nearby burial of
an adult male was equally elaborate, and the sum suggests a
burial ritual and perhaps notions of an afterlife similar to those
recorded among historic hunter-gatherers.

   In contrast to Mousterian graves, some Upper Paleolithic
ones, such as that at Předmostí, were covered by large rocks or
bones, perhaps to complete a ritual or to prevent wolves or hyenas from exhuming the bodies. Like earlier sites, however, late
Paleolithic ones often contain isolated scraps of human bone,

and carnivore exhumation of burials could be responsible. Alternatively, disarticulated bones might sometimes reflect complex burial practices that deliberately dissociated the skull from the postcranium or separated various postcranial elements from each other. The most persuasive example is probably from the 22 ky old late Perigordian (or Protomagdalenian) layer of the Abri Pataud, France, where the skull of a fifteen- to eighteen-year-old woman was positioned among three stones about 4 m away from a cluster of postcranial bones that may include hers and those of her newborn child (1594, 2327). In yet other instances, including the polished human femur in the double grave at Sungir' and intentionally shaped or perforated human bones and teeth from Isturitz, Saint-Germain-la-Rivière, and other French Upper Paleolithic sites, isolated elements may represent trophies or heirlooms (2024, 2327).

Finally, the possibility exists that some disarticulated bones represent food debris, although it seems unlikely that cannibalism was routine or widespread among late Paleolithic people. Most probably lived in small hunting-gathering bands that had more to lose than to gain from hunting their neighbors. Cannibalism was never unambiguously recorded among historic hunter-gatherers (72), and compelling archeological evidence for it is limited to post-Pleistocene agriculturalist sites like Fontbrégoua Cave, a Neolithic site in southeastern France, dated to about 6 ky ago (2327, 2329), and various Anasazi Pueblo sites in the American Southwest, dated between roughly 1100 and 700 years ago (2456, 2457). Unlike any known Paleolithic site, Fontbrégoua and the Anasazi sites each contain large numbers of disarticulated human bones that were demonstrably broken, burned, cut, or otherwise treated like the bones of other animals and that cannot be easily attributed to mortuary practices, carnivore exhumation, or postdepositional disturbance.

Mortality and Disease

The vital statistics of late Paleolithic people can be only crudely estimated, because of the relatively small number of skeletons whose age and sex can be established. However, a composite sample of seventy-six Eurasian Upper Paleolithic skeletons from various times and places, supported by a remarkable series of 163 skeletons from the terminal Pleistocene Ibero-maurusian levels of Taforalt in Morocco, suggest that the common mortality pattern resembled that of most later prehistoric and historic hunter-gatherers (2278). Child mortality was high, and women apparently had a lower life expectancy than men, probably because of the risks of childbearing. Women probably rarely reached forty years of age, and men probably rarely reached sixty or even fifty. Significantly, however, maximum life expectancy probably exceeded that of the Neanderthals, perhaps by as much as 20%. As a consequence, late Paleolithic human groups probably contained more older people, whose presence could enhance survival. They could obviously serve as repositories of vital knowledge, particularly about how to respond to very occasional, unusual crises, and they might provide direct economic support. Among Hadza hunter-gatherers in northern Tanzania today, young women are often freed to have additional children much sooner because their mothers or aunts, beyond childbearing age themselves, contribute heavily to children's diets (976). Arguably, it was such economically advantageous "grandmothering" that initially selected for the long reproductive life spans that now distinguish humans from other apes (977), and the availability of economically active older women in the late Paleolithic could help explain why late Paleolithic populations were so much larger (denser) than earlier ones.

Unlike Neanderthal skeletons, late Paleolithic ones rarely show evidence of serious accidents or disease, suggesting that late Paleolithic culture provided a far more effective shield against environmentally induced trauma. Skeletal anomalies that may reveal cause of death are particularly rare but include conspicuous lesions, perhaps caused by a severe fungal infection (578), on the skull, mandible, pelvis, and femur of the famous Old Man of Cro-Magnon (actually a person in his late forties) and dental abscessing that may have produced a fatal septicemia (blood infection) in the fifteen- to eighteen-year-old woman buried about 22 ky ago at the Abri Pataud, very near Cro-Magnon (579, 1356). Her dental infection was probably related to partial destruction of her normal right upper molars when aberrant, supernumerary teeth erupted alongside. This abnormality has not been observed in any other Upper Paleolithic people, most of whom had relatively healthy teeth, probably because their diets included few foods that encouraged caries or plaque formation.

Additional late Paleolithic skeletal anomalies that probably reflect cause of death have been reported from the cave sites of Rochereil in France and Romito in Italy. At Rochereil, a child in a Magdalenian grave had a bulging forehead and other features that suggest hydrocephaly (a normally fatal excess of cerebrospinal fluid in the skull) (579, 2279). An artificial perforation that was probably intended to provide relief may have been the immediate cause of death. At Romito, an adolescent male in a late "Epi-Gravettian" or "Romanellian" grave apparently suffered from a kind of genetic dwarfism (acromesomelic dysplasia) that certainly reduced life expectancy under Paleolithic conditions (757). The Romito dwarf is dated to 12–11 ky ago and antedates the next oldest instance by at least 5 ky. It is the only case of dwarfism yet recorded in a prehistoric Stone Age context.

Other known late Paleolithic skeletal abnormalities, reported mainly from Czech and French sites, were variably debilitating but probably not fatal. Like the ones already cited, none indicate epidemic diseases, which were probably rare until the greater population densities permitted by the development of food production, beginning 12–10 ky ago. The Czech examples include bone deformation of the left temporomandibular joint, suggesting partial facial paralysis, in one woman buried at Dolní Vestonice (1251, 1252) and a shortened, deformed right leg accompanied by pronounced spinal curvature to the left (scoliosis) in a second (1253). The French cases include fused cervical vertebrae (cervicoarthrosis) in skeletons from Cro-Magnon, Chancelade, and Combe-Capelle; a healed skull fracture, bone lesions or degeneration implying a permanently dislocated left shoulder, and a laterally deviated right big toe (*hallux valgus*) in the skeleton from Chancelade; and an asymmetric sacrum, reflecting lateral curvature of the spine, in the one skeleton from Combe-Capelle (579). By modern analogy, the occurrence of cervical fusion may mean that some older Upper Paleolithic people were relatively sedentary, literally remaining seated much of the time. Similarly, if modern people are guides, the laterally deviated big toe of the Chancelade skeleton may reflect poorly fitting footwear, though other explanations are possible.

Skeletal evidence for deliberate injury is also rare (1821), probably because, like most ethnographically recorded hunter-gatherers, late Paleolithic ones rarely engaged in warfare or interpersonal violence. In some instances, such as the healed fractures on male skulls from Chancelade (579) and Dolní Vestonice (2121), the cause could have been accidental, and in many others bone fracturing or crushing probably occurred after death, in the ground. Prominent examples of damage that was once believed to be antemortem but that was probably postdepositional include the fractured female skull (individual 2) from the Cro-Magnon site (578) and the four fractured or crushed skulls from the Upper Cave at Zhoukoudian (1669). But even if evidence for violence is rare, it does exist—for example, at the Grimaldi Caves in northern Italy, where a (Aurignacian?) child was buried with a projectile point embedded in its spinal column (579); at Wadi Kubbaniya near Aswan in

Egypt, where a young adult male, buried perhaps 25–20 ky ago, had a healed parry fracture on the right ulna, a stone chip embedded in the left humerus, and two blades (projectile armatures?) in the abdominal cavity (2418); and above all in an extraordinary terminal Pleistocene (ca. 14–12 ky old) cemetery near Jebel Sahaba in Sudanese Nubia (36, 2407). Nearly half the fifty-nine individuals exhumed there either had unhealed antemortem skeletal injuries or had stone artifacts lodged in or near their bones. They provide a remarkably graphic exception to the stated generalization that violence is limited among hunter-gatherers.

Both the Old Man of Cro-Magnon and the man buried at Chancelade were probably too disabled to fend for themselves, and their survival shows that Upper Paleolithic people, like Neanderthals before them and historic people later, cared for their old and sick. Such care need not have been entirely philanthropic or unselfish, since older people in hunter-gatherer societies commonly possess vital knowledge and experience. But care in a more emotional, abstract sense not known for the Neanderthals may be indicated for the Dolní Vestonice woman with the deformed temporomandibular joint. Both the face engraved on an ivory fragment and the face of a sculpted clay head found nearby droop on the left side, just as hers probably did. They could represent the oldest known portraits.

Late Paleolithic Population Expansion

In previous sections it was noted that, in Europe and southern Africa, late Paleolithic peoples were probably much more abundant than their predecessors. At least equally important, late Paleolithic people greatly extended the geographic range of humankind by colonizing the easternmost part of Europe (in what are today Ukraine, Belarus, and European Russia), central and northern Siberia (Asiatic Russia), the Americas, and Australasia (Australia and neighboring islands).

Easternmost Europe

Mousterian and earlier sites have been found only on the western and southern margins of easternmost Europe, in spite of extensive archeological reconnaissance and the kind of intensive commercial activity that leads to site discovery. Therefore, as each year passes it seems increasingly likely that Mousterian and earlier people simply could not occupy most of eastern Europe because of the harsh continental climate that prevailed even during interglacials. Upper Paleolithic people obviously

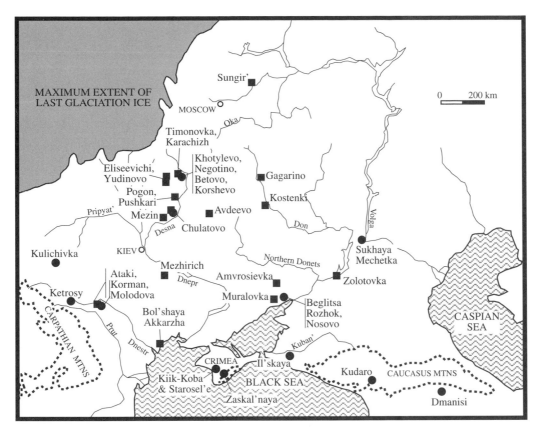

Figure 7.23. Approximate locations of major Paleolithic sites in eastern Europe (the western part of the former Soviet Union) (modified after 1030, fig. 1). Squares mark key Upper Paleolithic sites; circles mark earlier sites. Note the position of the Sungir' Upper Paleolithic site north and east of Moscow. Upper Paleolithic people were the first to live so far north and east, during either a glaciation or an interglacial.

did not find the climate an insuperable problem, and the Sungir' site is spectacular proof that even early Upper Paleolithic people could inhabit areas of permafrost (where the subsurface never thawed) north and east of present-day Moscow (fig. 7.23). On the far northeast (at the longitude of the Ural Mountains, which separate European Russia from Siberia), early Upper Paleolithic people even settled near the Arctic Circle (60, 1031).

A striking feature of the earliest east European Upper Paleolithic sites is that they vary remarkably in artifactual content, and with rare, arguable exceptions, none can be assigned to the early Aurignacian Culture (60, 1031). Remember that the early Aurignacian represents the oldest undeniable manifestation of the Upper Paleolithic over most of central and western Europe. If, as seems increasingly likely, the early east European Upper Paleolithic approaches or even exceeds the 40 ky age of the earliest Aurignacian, the implication could be for a yet earlier Upper Paleolithic that gave rise to both. The reason for such early divergence between east and west is obscure, but it may relate to the fact that over most of eastern Europe, early Upper Paleolithic people were the first humans, whereas in central and western Europe early Upper Paleolithic (Aurignacian) people occupied areas that the Neanderthals (Mousterians) had occupied before. The conspicuous internal diversity of the earliest European Upper Paleolithic is a further reminder of how greatly even the earliest Upper Paleolithic differed from the much more monotonous Mousterian that preceded it.

Siberia

The southwestern corner of Siberia enjoys a cool temperate climate like that of adjacent west-central Asia, and it was similarly occupied by Mousterians before 40 ky ago (2310). Seven isolated teeth from Denisova and Okladnikov Caves suggest that the southwest Siberian Mousterians were Neanderthals living near the northeastern edge of their range (2256). Radiocarbon dates from Kara-Bom indicate that the local Mousterian was replaced by a typical blade-and-burin Upper Paleolithic industry at least 40 ky ago (853). Broadly similar Upper Paleolithic assemblages that may be equally old occur across southern Siberia to Lake Baikal and beyond (808, 852, 1294, 2310) (fig. 7.12). Occupation of southeastern Siberia by 35–30 ky ago is probably implied by the coeval colonization of Japan (1754, 1755). Most likely this occurred via a land bridge from Siberia to Sakhalin Island, which was the northern end of a long peninsula that also comprised the four main Japanese Islands (Hokkaido, Honshu, Shikokou, and Kyushu) during much of the Last Glaciation.

The central and northern parts of Siberia experience subarctic and arctic climates as harsh as or harsher than those of easternmost Europe, and in contrast to southern Siberia, they may not have been occupied before 20 ky ago (851). The most stunning claim to the contrary comes from the Diring Yuriakh open-air site in the Lena Basin (fig. 7.12), where flaked stones are covered by eolian (windblown) sands that have been dated by thermoluminescence to more than 260 ky ago (2396). The dating is intriguing, but the flaked pieces are not clearly artifactual (1294). The next oldest artifacts may come from alluvial deposits at Ezhantsy, Ikhine I and II, and Ust'-Mil' II, also in the Lena Basin. In each case radiocarbon dates on wood fragments or stratigraphic correlations to radiocarbon-dated deposits elsewhere bracket the artifacts between 35 and 20 ky ago (671,

1293, 1295, 1306, 1488, 1576). The dating is questionable, however (1061, 1294, 2528), partly because the artifact samples tend to be very small and may have been displaced from higher levels, partly because some of the dated wood fragments could have been reworked from older sediments (this could explain some stratigraphic inversions in the dates), and partly because there are inconsistencies in the published stratigraphic descriptions. Human occupation of central and northern Siberia between 20 and 10 ky ago is uncontroversial and has been well documented at sites like Dyuktaj Cave, Verkhnetroitskaya, Kukhtuj, Berelekh, Mayorych, and Ushki Lake (figs. 7.12 and 7.24) (671, 1230, 1305, 1306).

Siberian late Paleolithic sites postdating 20 ky ago tend to share an abundance of microblades struck from polyhedral "wedge-shaped" (Gobi) cores, burins, bifacial lanceolate (leaf-shaped) stone points, and grooved or slotted antler points that together suggest a distinctive "Siberian Upper Paleolithic." This assemblage differed in detail from its west Asian and European counterparts, but it shared with them the routine manufacture of bone, ivory, and antler artifacts; the presence of readily identifiable art objects and decorative items; the construction of substantial dwellings; relatively elaborate burial of the dead; significant stylistic variation through time and space; and so forth. By and large, Siberian Upper Paleolithic people seem to have lived very much as did their European contemporaries, subsisting largely on gregarious herbivores that had become especially numerous and widespread under glacial climatic conditions.

The Americas

Genes and phenotypic traits both show unequivocally that Native Americans originated in northeastern Asia (890, 2130, 2205, 2377, 2530). The oldest known Native American skeletal remains, dated to between 10 and 8 ky ago, support the same conclusion (2041). Native Americans and northeast Asians particularly share striking dental features that include strong tendencies for the upper incisors to have shovel-shaped crowns, for the upper third molars to be very small, for the lower first molars to have an extra (third) root, and for the lower second molars to exhibit five (versus four) cusps. These characters and others have been grouped in a pattern known as "Sinodonty," in contrast to "Sundadonty," a more generalized dental pattern that characterizes southeast Asians and their Polynesian relatives, all of whom tend to have nonshoveled upper incisors, large upper third molars, and so forth (2252–2255, 2257). The Sinodont/

Sundadont split occurred only after 30 ky ago, and this may place a lower limit on the time when the Americas were first colonized. Determining this time, however, depends mainly on archeological evidence in northeastern Siberia, in the Americas, and in the region of the Bering Strait, where Siberia and North America approach each other most closely.

The growth of glaciers requires moisture as well as cold, and during late Paleolithic times Siberia was mainly too dry for large glaciers to form. However, the huge ice sheets of Europe and especially North America locked up so much water that sea level fell by up to 140 m, exposing wide swaths of land at the margins of the continents. A dry land bridge up to 1,000 km wide formed across the Bering Strait (fig. 7.24), and Alaska and the adjacent ice-free areas of northwestern Canada became an extension of Siberia, largely separated from the rest of North America by the thick ice sheets that covered most of Canada. The last glacial ice sheets began to melt about 14 ky ago, but the land connection was not fully severed until about 10 ky ago. Most archeologists agree that the first Americans were simply Siberian Upper Paleolithic people who, like the saiga antelope, the yak, and other north Asian species, naturally extended their range eastward across the land bridge. The timing of the extension is not established, but after many years of exploration and commercial activity on both sides of the land bridge, the oldest unequivocal sites all postdate 13–12 ky ago (679, 1036). Older sites may be absent because the land bridge was sparsely vegetated between 25 and 14 ky ago and supported few large herbivores (522, 523). Perhaps even more important (935), wood (for fuel) was absent until about 13 ky ago, when climatic amelioration allowed trees to recolonize sheltered valleys.

The earliest known Alaskans left behind endscrapers, sidescrapers, distinctive small teardrop-shaped and triangular bifacial points, and other artifacts that have been assigned to the Nenana Complex (854). This was first identified at the Walker Road and Dry Creek sites in the Nenana Valley of central Alaska, and it has been radiocarbon-dated to between 12 and 11 ky ago (1032, 1036, 1715). It has no obvious northeast Asian counterpart, but it was succeeded between 11 and 10 ky ago by the Denali Complex, which includes wedge-shaped cores, lanceolate points, and other items that closely recall Siberian artifacts postdating 20–15 ky ago (2430).

The Alaskan archeology may imply two early west-to-east migrations, perhaps corresponding to two of the movements that linguists and geneticists have postulated (888–890, 2205, 2206). More specifically, Nenana and Denali could represent

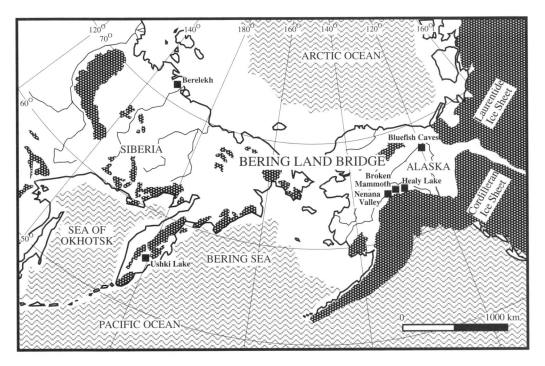

Figure 7.24. Northeastern Siberia and northwestern North America, showing the extent of the Bering land bridge in relation to the modern land configuration and the maximum extent of the Last Glaciation ice sheets. Some important archeological sites that may bear on the initial colonization of the Americas are also shown. Arguably, the oldest are the Bluefish Caves (Yukon) (6, 1576), where putative bone artifacts have been radiocarbon-dated to nearly 25 ky ago. Less controversial are Berelekh (Yakutia, Siberia) (1570) and Ushki Lake (Kamchatka Peninsula, Siberia) (659), which were occupied by 13 ky ago, and the Nenana Valley sites, Broken Mammoth, and Healy Lake (central Alaska), which were occupied between 12 and 11 ky ago (2429). The first human penetration of the land bridge may have occurred only about 12 ky ago, when trees that could supply fuel reappeared after an absence of 20 ky or more. (Base map modified after 2130, fig. 1.)

respectively the original speakers of the two oldest American linguistic stocks: Amerind, widespread throughout the Americas, and Na-Dene, confined largely to Alaska and western Canada with outliers in the southwestern United States. Speakers of the third major stock, Eskaleut (or Eskimo-Aleut), probably arrived more recently and settled almost exclusively in the Arctic. At the moment, the most serious objection to this scenario centers on the integrity of the Amerind stock. Many authorities believe this is an artificial conglomeration of languages with no demonstrable relationships (1614) and that such great linguistic diversity would require tens of thousands of years to develop, assuming it stemmed from a single source (814). Alternatively, the heterogeneity of the putative Amerind phylum might imply multiple west-to-east migrations after

12 ky ago. Cranial and dental morphology could be cited in support, since American and northeast Asian skulls dated between 10 and 8 ky ago resemble each other more closely than they resemble most later skulls in both regions (1613). Only Native American and northeast Asian skulls postdating 7–8 ky ago exhibit typical, derived "Mongoloid" features such as especially broad, short braincases, broad, flat faces with high, frontally directed cheekbones, narrow noses, and high frequencies of Sinodont dental traits. Thus, if Amerind speakers all originated from a common ancestor 12 ky ago, subsequent populations in the Americas and in northeastern Asia exhibited remarkable parallelism in cranial and dental change. A more plausible alternative is that Amerind speakers originated variously from a migration and from one or more additional migrations that occurred much later (1297, 1299).

The antiquity of human occupation south of Alaska has been hotly debated for decades (349, 1543). Many authorities see little or no compelling evidence for human presence before 12–11 ky ago, while others point to sites where stone artifacts or other humanly modified objects may be much older. However, most such sites are highly problematic, and the claimed antiquity variously depends on absolute dates that are inconsistent with each other or with other stratigraphic evidence; on the crude nature of the artifacts; on artifacts whose artifactual quality is often arguable or whose stratigraphic context is not really clear; on charred wood, earth, or bones that could have been burned naturally; or on other data that can be readily challenged (667, 1642). But fresh claims continue to surface, including important recent ones from the rockshelter of Toca do Boqueirão da Pedra Furada in northeastern Brazil (930–932) and from the streamside site of Monte Verde in south-central Chile (661–663, 665, 1541, 1544, 1546, 1577).

At Pedra Furada, a stratigraphically consistent series of radiocarbon dates implies human occupation from before 48 ky ago until roughly 6.1 ky ago. The cultural origin of the flaked stones and of well-delineated hearths dated between 10.4 and 6.1 ky ago is not disputed, but the evidence for older occupation is more problematic because it is based primarily on crudely flaked quartzite cobbles and on dispersed charcoal. The cobbles originated from a cemented layer (or conglomerate) in the cliff face approximately 100 m above the shelter, and the flaked examples might thus be "geofacts," created when cobbles weathered out naturally and struck the hard ground below (1545). Such a process is unlikely to have mimicked human flaking very often, but the flaked specimens were selected from a vastly larger number of unflaked ones. Occasional natural brushfires nearby could similarly explain the dispersed charcoal, and Pedra Furada may soon join the long list of dubious claims referred to above.

At Monte Verde, eight battered or crudely flaked stone artifacts, three naturally fractured pebbles that show traces of use, and fifteen naturally fractured pebbles that were apparently carried to the locality but not necessarily used were found deeply buried in riverine sands. Charcoal from the same level nearby has been radiocarbon-dated to more than 33 ky ago. Basically similar but more numerous naturally fractured or crudely flaked artifacts also occur at a much higher level, dated to about 12.5 ky ago, where they are associated with gomphothere (mastodon) bones and hide fragments, a wishbone-shaped mound of sand, gravel, and animal fat, parts of many different edible or medicinal plants drawn from a wide variety of environments, and a mass of logs and branches, all preserved in dense, boggy deposits (660, 662, 666). The spatial arrangement of the logs and branches suggests they could have anchored very small pole-frame huts.

The Monte Verde ensemble implies some unusual, even unique human behavior, and it might be dismissed if mastodon activity or some other nonhuman agency could more obviously explain it. The 12.5 ky old occupation is as old as or older than any widely accepted human habitation site in the Americas, and if both this date and the claims for yet older occupations at Monte Verde and Pedra Furada are sustained, their practical and theoretical implications are profound. Scholars will be hard pressed to explain either why Native Americans seem to have been so rare in the millennia preceding 12 ky, in contrast to contemporaneous Eurasians and Africans, or conversely, why, unlike the earliest Americans, their Eurasian and African contemporaries left hundreds of rich, unmistakable sites with well-made artifacts and an abundance of other cultural debris.

There is also the problem that Last Glacial migrants from Siberia to Alaska would have found it difficult or impossible to spread farther south in North America. Canada was almost completely covered by two great ice sheets, the Cordilleran on the west and the Laurentide on the east, and if people arrived in Alaska before 25 ky ago, they could have moved south only through a relatively narrow, inhospitable ice-free corridor in western Canada. Alternatively, if they arrived only between 25 and 13 ky ago, during the peak of Last Glacial cold, dispersal southward would probably have been blocked by continuous ice (1125, 1855, 2509). Southward movement would have become

practical only after 13 ky ago, when the ice sheets began to retreat. Together with environmental changes in the Beringian
source region, this could help explain why the oldest firm and
universally accepted archeological sites south of the Cana-
dian/American border are only about 11.2 ky old (989). These
sites are commonly assigned to the Clovis Complex of the
Paleo-Indian Tradition (985–987, 1542, 1543), which extended
more or less throughout North America as far south as Panama.
It is especially well known at well-excavated sites in the western United States (fig. 7.25).

The most distinctive Clovis artifacts are finely crafted, bifacial, concave-based, lanceolate projectile points with one or
more basal flutes (elongated flake scars) proceeding from the
base on both faces. Accompanying artifacts include endscrapers, sidescrapers, large bifacial implements, occasional burins
(known as "gravers" in North America), and bone or ivory implements that recall late Paleolithic examples in Eurasia. Conceivably the Clovis Complex stemmed from the Nenana Complex of Alaska (854, 1036), which may be a century or two older
and contains bifacial points that differ from Clovis ones mainly
in smaller average size and lack of fluting.

Clovis people did not penetrate South America, but other
groups reached the very tip of the continent by 11 ky ago (664,
1543). The earliest South Americans made a range of regionally
diverse projectile points that imply either a long-standing pre-
Clovis occupation (990, 1820) or a very rapid differentiation
from a Nenana- or Clovis-like base. At present both alternatives may seem improbable, but a rapid differentiation is less
so, and it would imply that people colonized South America
not in a smooth wave, but in a series of skips and jumps that
quickly isolated populations in pockets of especially favorable
habitat.

The ecological shock of human arrival, perhaps combined
with dramatic climatic change in the transition to the present
interglacial, may explain why North America lost thirty-three
large-mammal genera, more than 70% of its total, between perhaps 12 and 10 ky ago (1292, 1465). An even larger number apparently disappeared from South America at about the same
time (1465). The vanished "megafauna" included mammoths,
horses, and native camels, whose brief association with Paleo-
Indians of the Clovis Complex is now established.

The possibility that the Paleo-Indians caused or contributed to large-mammal extinctions is a potent reminder of the
level of hunting-gathering competence that late Paleolithic
peoples may have achieved. Similar, though less extensive, extinctions occurred at or near the end of the Pleistocene in

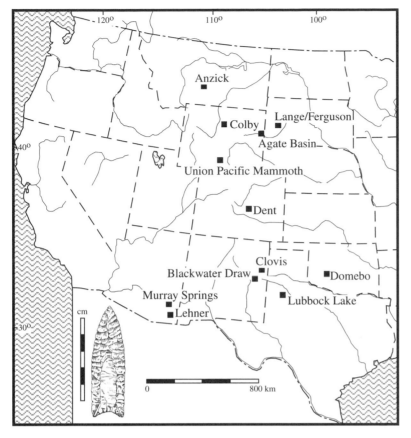

Figure 7.25. The main in situ occurrences of Clovis Paleo-Indian artifacts in the western United States (adapted from 268; point redrawn after 2506, fig. 68). The most distinctive Clovis artifacts are finely crafted bifacial, concave-based projectile points with one or more basal flutes (elongated flake scars) proceeding from the base on both faces. Bracketed by radiocarbon dating to between 12 and 11 ky ago, the Clovis tradition constitutes the oldest universally accepted evidence for human presence in the region. At Clovis sites where animal bones are preserved, some usually come from extinct species, including mammoth, camel, and horse.

Eurasia (2313) and Africa (1239), where people perhaps delivered the coup de grâce to species whose numbers and distribution had already been reduced by environmental change. Environmental change alone is an inadequate explanation for the American, Eurasian, and African extinctions, since the extinct species had survived the similar change that occurred at the end of the Penultimate Glaciation, roughly 127 ky ago. What differentiated the end of the Last Glaciation most clearly was the presence of more advanced hunter-gatherers, whose behavioral innovations and capabilities have been emphasized here.

Australasia

Unlike the Americas, Australia has not been connected to another continent since the late Cretaceous, roughly 70 my ago. Lower sea levels dramatically enlarged the Malay Peninsula, fusing it with Sumatra, Java, Borneo, Bali, and smaller Southeast Asian islands to form a subcontinent paleogeographers call "Sunda Land," while Tasmania and New Guinea were similarly joined to Australia to produce a supercontinent called "Sahul Land." However, Sunda and Sahul always remained separate (fig. 7.26). Travelers between the two could have island hopped, but some open sea travel remained unavoidable, including several voyages of roughly 30 km and at least one of 90 km (226). Such distances required the invention of boats that could maintain buoyancy for several days, and arguably only late Paleolithic people, with the fully modern ability to innovate, could have made the crossing. The boats they used probably perished long ago, and in any case the sites where remnants might persist are now inaccessible on the drowned continental shelves of Sunda and Sahul.

Until the mid-1960s one could argue that Sahul was not colonized before 10 ky ago, but it is now clear that people were present by 40–35 ky ago (1159–1162, 1394, 1622, 2182, 2438–2440) and possibly by 60 ky ago. An antiquity between 60 and 50 ky ago is strongly implied by thermoluminescence (TL) dates, optically stimulated luminescence (OSL) dates, or both on unburned quartz sands enclosing and overlying artifacts at the rockshelters of Malakunanja II (1790) and Nauwalabila I (1791), Northern Territory, Australia. OSL and calibrated ^{14}C dates on the upper levels of Nauwalabila I agree closely, but the 60–50 ky old TL and OSL determinations are still controversial. Both methods in effect date the last time that sand grains were exposed to sunlight. The energy of the sun releases electrons that were trapped in defects within the quartz structure, and these defects begin to fill again only after burial. The rate at which they fill depends on readily measurable background radioactivity in the surrounding sediments.

Objections to the Malakunanja II and Nauwalabila I dates center on the possibility that the dated quartz grains, the artifacts, or both have been displaced, and that they could thus substantially antedate the artifacts (21, 1618). At both sites, the artifacts dated to greater than 40 ky ago occur less than a meter below levels dated to less than 40 ky ago, and termite tunneling is known to selectively move relatively heavy items vertically over such a distance. In addition, no Australian archeological site has yet provided a radiocarbon date of more than

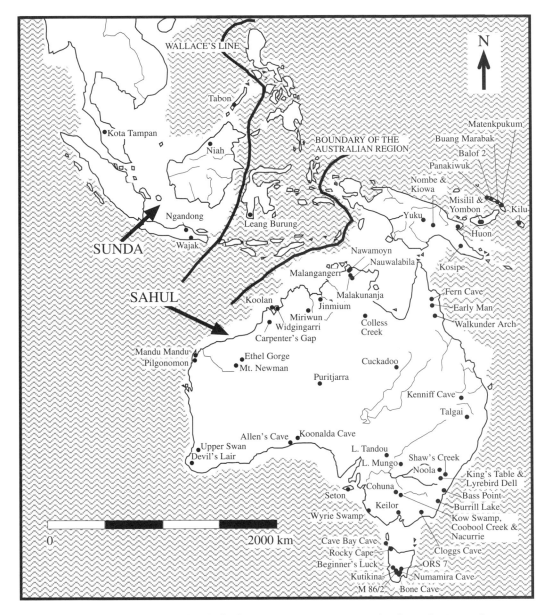

Figure 7.26. Map of Sunda and Sahul, showing major zoogeographic boundaries and some important late Pleistocene archeological or human fossil sites dating between 60 and 10 ky ago (adapted from 2439, fig. 1). Present-day landmasses are outlined in black. Additional land that would be exposed by a 200 m drop in sea level is shown in white. Even with a drop of this magnitude, substantial stretches of open sea would separate Sahul from Sunda, indicating that the earliest human inhabitants of Sahul arrived by boat.

40 ky, though strictly geologic localities have done so (23). However, as indicated previously, if a 60 ky old colonization of Sahul is ultimately established, it would have profound implications for Out of Africa 2, since it would imply at least two separate modern human dispersals from Africa: a first, eastward along

the south Asian coast, perhaps partly by boat, that resulted in the modern human colonization of Sahul roughly 60 ky ago, and a second, northward and then westward around the Mediterranean Basin that resulted in the modern human colonization of Europe roughly 20 ky later.

If a 60 ky old movement from Africa occurred, it should eventually be detectable on the south Asian mainland. Alternatively, of course, the earliest Australians may have been neither recent African immigrants nor fully modern. An ancient Asian origin may be suggested by TL dates that fix putative artifacts at Jinmium Rockshelter, Northern Territory, between 176 and 116 ky ago (772). One wall of the shelter is festooned with approximately 3,500 small humanly pecked cup-shaped pits, and an exfoliated fragment with two similar pits occurs in a Jinmium layer that is TL-dated to between 75 and 58 ky ago. If this age is accepted, art in Australia would precede confirmed art in Europe by 20 ky or more. However, the artifactual origin of Jinmium objects older than 60 ky is debatable, and even strong advocates of very early art are skeptical of the Jinmium evidence (107). The problem is that the dated sands may not have been totally "zeroed" (bleached of their TL signal by sunlight) before burial or that they may have originated from sandstone fragments that decomposed in the ground long after they were last exposed to sunlight (812, 1618). This might be checked by OSL, which is more sensitive to the possibility that traps in the quartz crystal structure were not fully emptied or zeroed before burial. Malakunanja II may provide a preview of what an OSL check is likely to show. Sand accumulation here did not involve the in situ decomposition of sandstone fragments, and TL shows that it began nearly 110 ky ago. However, artifacts occur only in sandy layers that postdate 60 ky ago (1790).

Discounting the Jinmium dates, then, the best estimate for the colonization of Sahul is between 60 and 40 ky ago. The seafaring ability of the early colonists remains uncontested, and between 35 and 30 ky ago some had even reached the oceanic islands of New Britain and New Ireland to the east of New Guinea (21, 22, 1160, 1664). Minimally this required two 30 km hops from New Guinea to New Britain and a third voyage of similar length from New Britain to New Ireland. The early colonists did not stop at New Ireland, and by 28 ky ago they had reached Buka, the northernmost island in the Solomon chain (2466). Travel to Buka from New Ireland required either a single direct voyage of 175 km or several smaller, island-hopping trips of up to 50 km each. Fish bones figure prominently in the most ancient sites on both New Ireland and Buka, and fishing may thus be added to seafaring as an marker of local behavioral modernity.

Assuming that Sahul was first colonized by boat from Sunda, the most likely point of entry was in the northwest, where coastal environments closely resembled those of Sunda. To begin with the settlers may have spread mainly along the coasts (304, 306), but they had occupied much of the Australian interior by 25–20 ky ago. Their artifacts, their hunting-gathering way of life, and, on occasion, even their rock art are now well documented at more than fifty sites dating between 40 and 10 ky ago, in various parts of New Guinea, Australia, and Tasmania (307, 561, 583, 1159, 1160, 1162, 2438, 2440). The earliest Australian flaked stone artifacts are remarkably informal, and they have been assigned to a loosely defined Core Tool and Scraper Tradition that persisted basically unchanged until roughly 4 ky ago. Similar stone artifacts occur widely in Southeast Asia in the late Pleistocene and early Holocene, and where they are found alone the behavioral modernity of the makers can be questioned. But at several Australian sites the flaked stones are accompanied by such modern behavioral markers as formal bone artifacts, art, complex burials, or a combination of these. A layer that has been radiocarbon-dated to roughly 32 ky ago at Mandu Mandu Rockshelter, Western Australia, contains artificially modified cone shells that are among the oldest known beads anywhere (1579), and a rock painting dated to about 25 ky ago at Laura South, northern Australia, rivals European Upper Paleolithic paintings in its antiquity (2395).

In addition, a stone artifact assemblage that is at least 40 ky old at the Bobongara site on the Huon Peninsula, northeastern Papua New Guinea, includes distinctive large ax-shaped flakes that were intentionally grooved or "waisted" around the middle (909). Similar tools are known from other New Guinea sites dated to roughly 25 ky ago, and smaller, morphologically similar flakes with systematically ground edges occur in layers dated between 24 and 20 ky ago in northern Australia. The grooving or waisting could have simplified mounting on wooden handles, and the various pieces might then have been used to chop or to work wood. The watercraft by which Sahul was colonized indirectly reflect sophisticated working of wood, bamboo, or related materials, but just as on other continents, on Sahul prehistoric wooden artifacts have rarely survived. A prominent exception occurs at Wyrie Swamp, South Australia, where a fortuitous preservation demonstrates the invention of the boomerang by 10 ky ago.

Among the best-documented and certainly most informative early Australian sites are those at Lake Mungo in the Willandra Lakes district of western New South Wales (182, 308–310, 1643). Here parts of three anatomically modern human skeletons labeled Mungo 1–3 have been dated directly or indirectly by radiocarbon and thermoluminescence. Mungo 1, a twenty- to twenty-five-year-old female dated to about 26 ky ago, had been partially cremated before her bones were intentionally fragmented and placed in a small, shallow pit. Mungo 2, found with Mungo 1, was too fragmentary and too poorly represented for analysis. Mungo 3, an older adult male dated to about 30 ky ago, had been laid out in a shallow grave and liberally sprinkled with red ocher before burial. Artifacts and faunal remains found nearby show that the Mungo people belonged to groups who exploited shellfish, fish, emus, marsupials, and probably plants around the Willandra Lakes between roughly 33 and 15 ky ago, when climatic change caused the lakes to shrink and eventually disappear.

Curiously, compared with the historic Australian Aborigines, the Lake Mungo people, together with broadly contemporaneous or somewhat younger ones from Keilor, Victoria, and Lake Tandou, New South Wales possessed relatively high-vaulted, thin-walled, smooth-browed, spherical skulls with relatively flat faces, whereas the people who lived at Kow Swamp, Coobool Creek, and Nacurrie, northern Victoria, between perhaps 14 and 9 ky ago and at the probably contemporaneous sites of Cohuna, Victoria, and Talgai, Queensland, had exceptionally rugged skulls with relatively low vaults, thick walls, flat and receding foreheads, strong browridges, and projecting faces. A similar, arguably even more rugged morphology characterizes a cranial vault known as Willandra Lakes Hominid 50 that eroded from ancient lake deposits just north of Lake Mungo and that may be closer in age to the gracile Mungo specimens (2183). The range of variation is extraordinary and may indicate that Australia was colonized more than once, by very different people (1162, 2181, 2182). A very late incursion, perhaps about 3–4 ky ago, is suggested indirectly by a major turnover in stone artifact types across the Australian continent and more directly by the coeval introduction of the dingo, a semidomesticated dog widely associated with the historic Aborigines (305, 1160).

The Lake Mungo people may derive from a population represented by broadly contemporaneous skeletal remains from Wajak on Java, Niah Cave on Borneo, Tabon on Palawan Island in the Philippines, and farther afield, from Zhoukoudian

(Upper Cave), Liujiang, and Ziyang in China. The origin of people like those represented at Kow Swamp, Coobool Creek, and Nacurrie is much less clear, and it has been suggested they inherited their rugged skulls from Southeast Asian *Homo erectus* (937, 938, 2184, 2483, 2493). In this view, they could descend via Willandra Lakes Hominid 50 from the Ngandong (Solo) people of Sunda (Java). Remember from previous chapters that bovid teeth associated with the Ngandong fossils have been bracketed between 53 and 27 ky ago by the ESR and mass-spectrometric U-series methods (2122).

The possibility of a line from Ngandong to Kow Swamp–Coobool Creek–Nacurrie cannot be discounted, but there are at least three arguments against it. First, on present knowledge, the Kow Swamp–Coobool Creek–Nacurrie morphology apparently postdates the more gracile morphology of Lake Mungo, perhaps by a substantial interval. Second, unlike the Kow Swamp skulls, the postcranial bones are morphologically modern in every determinable respect. This is particularly clear for the Kow Swamp femurs, which lack the thick cortical bone and peculiarities of shaft shape that mark the femurs of archaic *Homo* (1207). Third, sophisticated morphometric analysis employing several key cranial, facial, and dental measurements does not in fact differentiate the Kow Swamp skulls strongly from their Lake Mungo predecessors (376), and their difference from later prehistoric and historic Aboriginal skulls could simply reflect their greater average size and robusticity (377). If the smaller size and reduced robusticity of later people reflect relaxed selection under the mild conditions of the Holocene, the especially robust Kow Swamp morphology might reflect a local response to the especially harsh conditions of the millennia surrounding the Last Glacial Maximum, 20–18 ky ago, a time of great aridity that dramatically reduced human populations throughout Australia. The origin of the Kow Swamp morphology and, by extension, of its less robust historic successor might be resolved by the recovery of fossils firmly dated between those of Lake Mungo and Kow Swamp, but Aboriginal opposition to the study of prehistoric human remains may preclude this (1599).

Like the peopling of the Americas, the peopling of Sahul may be linked to a wave of extinctions involving the disappearance of about fifty species of large vertebrates, mostly herbivorous marsupials, from Sahul between roughly 40 and 15 ky ago (1063, 1600). A time before 30 ky ago is perhaps implied at archeological sites in Tasmania, which contain the largest well-dated, regional late Pleistocene faunal assemblages. At dates up to 30 ky ago, these assemblages include marsupials that no longer occur locally, but none of the large, totally extinct species that have been found in more poorly dated contexts elsewhere (561).

Unlike the American extinctions, the Australian ones cannot be attributed to terminal Pleistocene climatic change, which they clearly antedate, and a human role may be even more strongly implied. There were remarkably few predator species in Australia before human arrival, and the indigenous herbivores may have been particularly vulnerable to the sudden appearance of an unusually potent one. Alternatively, the extinctions could have resulted from early Aboriginal firing of the landscape and consequent environmental modification (1160). The historic Aborigines often set brushfires to aid movement or to concentrate game, and stark pollen changes, concentrations of charcoal, or both in ancient sediments suggest the practice has deep roots. Human involvement in extinctions remains as debatable in Sahul as it is in the Americas, however. A stronger case will require both more precise estimates of when the extinctions occurred and firmer evidence that the extinct species were still present when Sahul was colonized (2437).

Conclusion

Together, the fossil and archeological records suggest that the modern physical form evolved before the modern capacity for culture. From a behavioral (archeological) perspective, the earliest anatomically modern or near-modern people were not significantly different from their nonmodern predecessors and contemporaries, and this probably explains why they were confined to Africa for thousands or even tens of thousands of years. It was only sometime between 60 and 40 ky ago, when anatomically modern people developed the fully modern capacity for culture, that they were able to spread widely through Eurasia. The importance of their unique behavioral (cultural) capabilities is underlined by the apparent rapidity of their spread as well as by their subsequent evolution. Although the basic human form did not change significantly in the ensuing 40 ky, cultural evolution accelerated dramatically. Plainly it was culture and not body form that propelled the human species from a relatively rare and insignificant large mammal 40 ky ago to a geologic force today, impinging on all other species as an agent of natural selection.

However, if the broad outline of modern human origins is established, many details remain obscure. A rapid replacement of nonmodern humans is well established only in Europe, particularly western Europe, but even here the evidence is mainly artifactual, not physical. Additional skeleton remains of the earliest modern Europeans are sorely needed to demonstrate their African origin beyond a shadow of a doubt. In eastern Asia the rapidity and even the dating of the replacement is much less clear, and only a much enlarged fossil and archeological record can show whether modern invaders essentially extinguished nonmodern residents, as in Europe, or interbred with them much more extensively. Finally, if we accept that modern human behavior provided the competitive advantage that allowed modern humans to spread from Africa, it remains uncertain what promoted the behavioral advance. Did it follow strictly on social, economic, or technological change, or as may seem more likely a priori, was it sparked by a neurological change that perhaps fostered the development of fully modern language? The outstanding problems will not be resolved quickly, but the relative recency of modern human origins and the relative richness of the record mean that the likelihood for conclusive answers is far greater than for any other important event in human evolution.

CONCLUSION: ANATOMY, BEHAVIOR, AND MODERN HUMAN ORIGINS 8

The previous chapters have emphasized four fundamental stages in human evolution:

1. The initial divergence of protohumans from protochimpanzees between 7 and 5 million years (my) ago. The earliest people were distinguished from apes primarily in their lower limbs, which permitted efficient bipedal locomotion. They were remarkably apelike in brain size and in many aspects of their dentition, trunk, and upper limbs, and if they had survived to the present, they would probably be called bipedal apes. More formally, they are commonly referred to as the australopithecines, from *Australopithecus*, the first genus to be described.

2. The emergence of the genus *Homo* and the appearance of the oldest archeological sites roughly 2.5 my ago. Earliest *Homo* was distinguished from the australopithecines by brain expansion and by a reduction in the size of the premolars and molars. The oldest archeological sites are clusters of crudely flaked stones and fragmentary animal bones that imply an increased reliance on technology and carnivory. Brain enlargement, artifact manufacture, and carnivory were probably codependent, and increased tool use and greater reliance on meat probably explain the decrease in cheek tooth size.

3. The emergence of *H. ergaster* and of the Acheulean Industrial Tradition roughly 1.7–1.6 my ago. In distinction from all previous hominids, *H. ergaster* approximated living people in its nearly exclusive reliance on bipedalism (as opposed to a mix of bipedalism and tree climbing), in its body size and proportions, and in its reduced level of sexual dimorphism. It was the first hominid species to invade truly arid, highly seasonal environments, and sometime before 1 my ago, it became the first to colonize Eurasia.

4. The emergence and spread of fully modern humans about 50 thousand years (ky) ago.

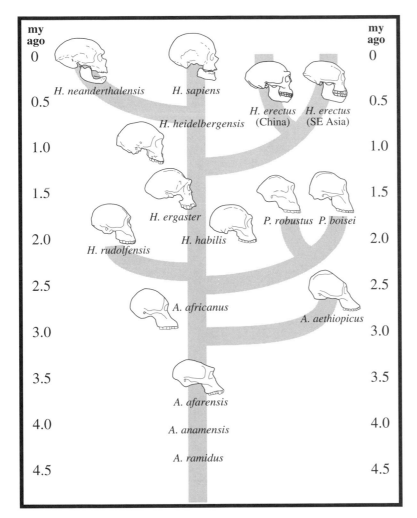

my ago

Figure 8.1. A working phylogeny linking the human species discussed in the text. (*A.* = *Australopithecus; P.* = *Paranthropus; H.* = *Homo*). The postulated branching events within *Homo* after 2–1.5 million years ago are particularly controversial.

The australopithecines, the genus *Homo,* and *H. ergaster* probably all appeared first in equatorial Africa, and each may have evolved abruptly, in a punctuational speciation event sparked by global climatic change (647, 648, 2350, 2352). Fully modern humans also appeared first in Africa, and their emergence may have been equally sudden, but it was not clearly linked to climate change. This chapter summarizes the evidence behind each stage or event, with special emphasis on the emergence and spread of modern humans. Arguably this was the most significant event that paleoanthropology will ever reveal, and its recency means it has left an extraordinary wealth of fossil and archeological evidence.

Figure 8.1 presents the working phylogeny that underlies this summary, and figure 8.2 illustrates the basic artifactual changes it refers to.

Figure 8.2. Some common artifact types in the main culture-stratigraphic units discussed in the text. The individual drawings are not to scale.

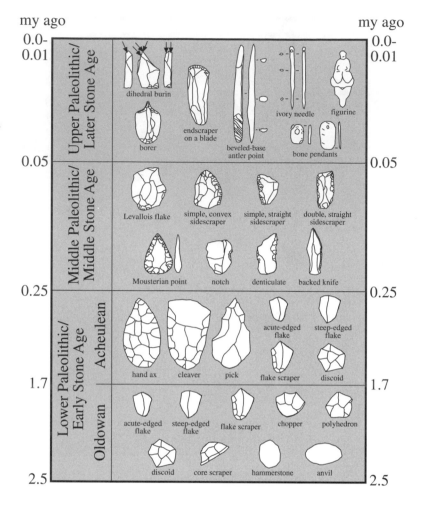

The Australopithecines

The fossil record stubbornly refuses to illuminate the divergence of people and chimpanzees, and the time of 7–5 my ago is founded on their degree of genetic differentiation (111). The oldest known human (hominid) fossils come from Aramis in the Middle Awash Valley of Ethiopia, where they date to about 4.4 my ago. They have been assigned to the species *Ardipithecus* (or *Australopithecus*) *ramidus* (2460, 2461), and they reveal creatures that were strikingly like chimpanzees in many aspects of their dentition, including their relatively thin dental enamel. Their arms were strongly muscled, with elbows that could have locked to aid in tree climbing. As now described, *A. ramidus* is linked to later hominids mainly by the incisorlike form of its canines and by the forward position of the foramen magnum on the basioccipital. Leg bones that await description will show that it was bipedal, but perhaps less completely so than its less apelike successors. It is unquestionably the most primitive hominid yet discovered, and it demonstrates unequivocally that humans descend from apes. Associated animal remains indicate that it preferred relatively moist, wooded environments, and it probably resembled chimpanzees in diet and in many aspects of its behavior and social organization.

Arguably the human family tree was bushy from the very beginning, and *A. ramidus* may represent a side branch that produced no descendants. Alternatively, about 4.2–4.1 my ago it evolved into *Australopithecus anamensis,* whose fossils have been found at Kanapoi and Allia Bay in the Lake Turkana Basin of northern Kenya (1340, 2384). *A. anamensis* resembled both chimpanzees and *A. ramidus* in its relatively large canines and in other aspects of its dentition and skull. Unlike *A. ramidus,* it anticipated later australopithecines in its relatively thick dental enamel and broadened molars. It is the oldest known hominid for which leg bones unequivocally demonstrate bipedalism. Like all the australopithecines, however, it retained powerful apelike arms, and it probably mixed bipedalism with a significant amount of tree climbing.

About 3.9–3.8 my ago, *Australopithecus anamensis* was succeeded by the much more fully known species, *Australopithecus afarensis. A. afarensis* has been found at seven or eight sites from north-central Ethiopia southward to northern Tanzania, but the most significant fossils come from Laetoli, Tanzania, and especially from Hadar, Ethiopia (900, 1151, 1152, 1220). The oldest specimens differ little from those of *A. anamensis,* and additional finds may suggest that *A. anamensis* should be sunk into *A. afarensis.* Like *A. anamensis, A. afarensis* was remarkably apelike in its skull, dentition, torso form, and arms. It is the oldest hominid species for which a complete skull is known, and this exhibits many chimpanzeelike features, including strong forward projection of the face and jaws, a very broad and flat cranial base, and a braincase that was small both absolutely and in relation to the size of the face and body. A partial skeleton from Hadar, various isolated lower limb bones, and some spectacular footprints preserved at Laetoli show that *A. afarensis* walked bipedally in broadly the same manner as living people, but its upper limb skeleton shows that it also retained an apelike ability to climb trees. It was as sexually dimorphic as the common chimpanzee, which it may have resembled in its social organization.

A. afarensis is usually considered ancestral to all later human species, and most authorities agree that it founded two distinct hominid lineages after 3 my ago. In the view that is

tentatively accepted here (fig. 8.1), it split into the still poorly
known *Australopithecus aethiopicus* and the much more completely known *Australopithecus africanus* (1512). *A. aethiopicus* was exclusively east African, whereas *A. africanus* is firmly
known only from southern Africa. Both species had significantly larger cheek teeth (premolars and molars) than *A. afarensis*, but the cheek teeth of *A. aethiopicus* were particularly massive and were mounted in a skull that had a sagittal crest, a
dish-shaped face, and other craniofacial structures that allowed
powerful grinding between the upper and lower cheek tooth
rows. *A. aethiopicus* soon became extinct without issue, but
A. africanus or its immediate descendant gave rise near 2.5 my
ago to two further lineages. The first included the "robust" australopithecines in the genus *Paranthropus*, and the second was
the stem for the genus *Homo*.

A highly defensible alternative position is that shortly after 3 my ago *A. afarensis* split between *A. africanus* and a so far
unknown second lineage; *A. africanus* subsequently became extinct without issue near 2.5 my ago; the second (undiscovered)
lineage split between *Homo* and *A. aethiopicus* near 2.5 my ago;
and *A. aethiopicus* was then directly ancestral to *Paranthropus*
(2055). In this case *A. aethiopicus* would become *P. aethiopicus*. The two alternatives stem from a frequent conundrum in
paleontology—disagreement about which derived similarities
between taxa reflect descent from a closely shared common ancestor and which ones may have evolved independently (convergently) as the taxa adapted to similar environmental conditions. The relatively close linkage of *Paranthropus* to *Homo*
(through *A. africanus*) accepted here depends on the argument
that *Paranthropus* and *Homo* shared brain expansion, a more
flexed cranial base, a reduction in subnasal facial projection,
and numerous other derived features that neither one shared
with *A. aethiopicus*, and that they probably inherited these
similarities from a common (*A. africanus*-like) ancestor. By extension, the massive cheek teeth and associated craniofacial
specializations that mark both *A. aethiopicus* and *Paranthropus* must be parallelisms that developed independently
as each taxon adopted a diet that included numerous hard or
grit-encrusted items. Conversely, the alternative view—that
A. aethiopicus was ancestral to *Paranthropus*—depends on the
argument that *Paranthropus* inherited its massive cheek teeth
and associated jaw musculature directly from *A. aethiopicus*
and that its unique similarities to *Homo* evolved in parallel. A
final choice between the alternatives will depend on a much
enlarged fossil record, particularly for the time between 2.9 and
2.3 my ago in eastern Africa.

The precise origins of *Paranthropus* aside, its coexistence with *Homo* after 2.5 my ago is not in dispute. By 2 my ago, *Paranthropus* had itself differentiated between a "hyperrobust" form, *P. boisei*, in eastern Africa and a somewhat less robust form, *P. robustus*, in southern Africa. Arguably *P. boisei* and *P. robustus* were simply geographic variants of a single widespread species. Both were "robust" strictly in their massive cheek teeth and in the associated craniofacial structures that reflect the power of the jaw muscles. In average body size they were no larger than *A. afarensis*, and they were significantly smaller than living humans (1509, 1511). They retained the basic australopithecine structural plan in which small braincases and a somewhat apelike upper body were mounted on legs designed for habitual bipedalism. They may have eaten no more meat than chimpanzees do, and they probably did not flake stone, though they may have used unflaked rocks, sticks, or even bones in the relatively unsophisticated manner that marks chimpanzees.

Fossils of *Paranthropus* actually outnumber those of *Homo* in deposits that antedate 1.5 my ago, but they are sparser in later deposits, and *Paranthropus* probably became extinct near 1 my ago. Its demise may have resulted from global climatic change that produced drier, more seasonally variable conditions throughout its range, from unsuccessful competition with the evolving genus *Homo*, or from the interaction of both factors.

Homo habilis and the Oldest Stone Tools

Sparse fossils now show that *Homo* had emerged in eastern Africa by 2.5–2.4 my ago, although it is well represented only in deposits that postdate 2 my ago. Most authorities assign all fossils of very early *Homo* to a single species, *H. habilis*, which differed from the australopithecines primarily in its larger brain and in its reduced cheek teeth (2199) (fig. 8.3). The increase in brain size was particularly significant, since it occurred without any significant increase in body size (1509). There is the complication, however, that both endocranial volume and cheek tooth size varied greatly within *H. habilis*. Some individuals had large skulls and large, australopithecine-sized teeth, while others had much smaller, australopithecine-sized skulls and relatively small cheek teeth, similar in size to those of the succeeding species, *Homo ergaster* (early African *H. erectus*). This variability might imply extraordinary differences between the sexes (sexual dimorphism), or it might mean that *H. habilis* was actually two species: *H. habilis* (in the narrow sense) for

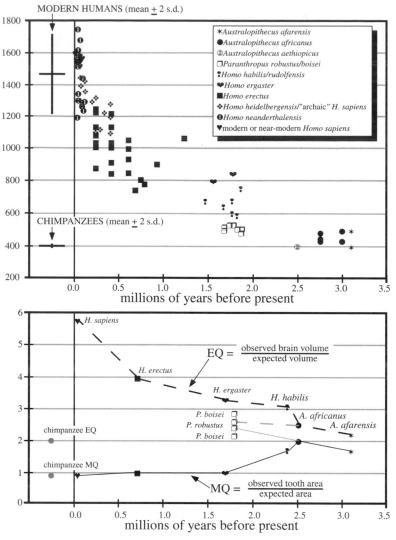

Figure 8.3. *Top:* Endocranial volume versus time, illustrating the remarkable increase in volume that characterized the evolution of *Homo* (data from 11, pp. 188–189). *Bottom:* Relative brain size (endocranial volume) (EQ) and relative cheek tooth size (MQ) versus time, showing that long-term increase in endocranial volume in *Homo* was not due simply to increased body size and that it was accompanied by a less dramatic, but still conspicuous decrease in the relative size of the cheek teeth (premolars and molars). The data are from 1511, which explains how relative endocranial volume and relative cheek tooth size were calculated.

the individuals with smaller brains and teeth, and *H. rudolfensis* for those with larger brains and teeth (2499, 2501, 2502). In the phylogeny that is tentatively accepted here (fig. 8.1), *H. habilis* narrowly understood is taken as the stem species for later *Homo*, and *H. rudolfensis* is placed to the side. Arguably, however, neither *H. habilis* nor *H. rudolfensis* constitutes a morphologically suitable ancestor for later *Homo*, and there may have been yet a third contemporaneous species of very early *Homo*. The most likely candidate for this third species would be *H. ergaster*, which is unambiguously documented in the fossil record only after 1.8–1.7 my ago.

The brain expansion that marks *H. habilis* (or the variants into which it may ultimately be split) is notable not only in itself but also because it coincided closely with the appearance of the oldest known archeological sites (1912). These clusters of stone artifacts and fragmentary animal bones provide the first nonanatomical evidence for human behavior. The artifacts are usually grouped in the Oldowan Industrial Complex, named for Olduvai Gorge, where Oldowan artifact assemblages were first thoroughly described (1326, 1327). In general, Oldowan tools include a range of sharp flakes and the cores or core ("pebble") tools from which they were struck (fig. 8.2) (1876). Oldowan stoneworking technology was primitive by later standards, and individual pieces are notoriously difficult to assign to discrete types. Still, the Oldowan Complex reflects an ability to flake stone that living chimpanzees probably cannot acquire (2210), and the artifacts and associated fragmentary animal bones demonstrate a commitment to artifact manufacture and to carnivory beyond anything known in apes. They show further that by 2.5 my ago at least one species of early *Homo* had developed the uniquely human habit of accumulating garbage at favored spots on the landscape.

A causal link between brain expansion, cheek tooth reduction, and stone tool manufacture plainly depends on the assumption that early *Homo* produced most if not all of the earliest stone tools, and it would be weakened if it could be shown that *Paranthropus* manufactured a significant number. This possibility cannot be evaluated directly, but it has been suggested by a strong similarity in thumb form between *Paranthropus robustus* and *Homo*, including *H. habilis* (2104). Thumb form in *Homo* enables the precision grasp, while thumb form in chimpanzees promotes the power grasp. Thumb bones of *A. afarensis* indicate that the earliest australopithecines had a chimpanzeelike power grasp.

If natural selection for dexterity in stone knapping drove the development of the human thumb, then it is reasonable to argue that *P. robustus* made at least some early stone tools. However, *P. robustus/P. boisei* and early *Homo* coexisted for perhaps a million years, from before 2 my ago until perhaps 1.2 my ago (fig. 8.1). During this long period, stone artifacts changed significantly, but evidence exists for only one evolving tradition, not two. Additionally, there is no obvious rupture in the archeological record at the time that *P. robustus/P. boisei* became extinct near 1.2 my ago. The same kinds of artifacts were made before and after, when only *Homo* survived. The most economical conclusion, then, is that *Homo* produced most if not all of the earliest stone tools. Conceivably *P. robustus* applied its precision grasp to food processing or to some other tool-using activity besides stone knapping. Alternatively, the thumb bone that suggests the precision grasp in *P. robustus* may actually have come from early *Homo*, which is represented in the same deposit (at Swartkrans Cave, South Africa), but by many fewer diagnostic craniodental fragments (2103).

Homo ergaster, Hand Axes, and the Initial Colonization of Eurasia

In most current textbooks, *Homo habilis* (or one of its variants) is said to have evolved into *Homo erectus* approximately 1.8–1.7 my ago. The oldest specimens of *H. erectus* include a nearly complete skull, a second partial skull, several lower jaws, and some postcranial bones from East Turkana in northern Kenya (1774, 1775, 2499, 2501), a skull and an associated partial skeleton from West Turkana (2360, 2370), and a skull from Swartkrans Cave in South Africa (507). The skulls differ from those of *H. habilis* in various features, including especially the presence of a conspicuous supraorbital torus (browridge) across the top of the orbits, thicker skull walls, and a yet larger brain (greater endocranial volume). Much of the increase in brain size, however, may relate to a dramatic concomitant increase in body size, especially in females (1509, 1511).

The type specimens of *H. erectus* come from China and Java, and these arguably differ from early African *H. erectus* in having somewhat larger, lower, flatter, and more angular braincases, yet thicker skull walls, thicker brow ridges from top to bottom, and ear and nasal specializations that together justify placing the African fossils in a separate, more primitive species, for which the name *Homo ergaster* is available (911, 912, 2499). The distinction of *H. ergaster* from *H. erectus* is admittedly subtle, and it will require careful, continuous reevaluation as the fossil record grows. However, *H. ergaster* is provisionally accepted here as the stem species for all later humans, including *H. erectus* narrowly understood (in the Far East) and other forms of primitive *Homo* in Africa and Europe (fig. 8.1).

Limb bones and inner-ear structure show that *Homo ergaster* (or "early African *H. erectus*," if preferred) was the first human species to rely almost exclusively on bipedalism, in contrast to the australopithecines and even *H. habilis*, which depended on a mix of bipedalism and tree climbing (2031). *H. ergaster* was also the first human species in which body weight and stature increased to approximately their modern values, and it appears to have been the first in which males and females differed in size no more than in living people (1510). Both the australopithecines and *H. habilis* were much more dimorphic, resembling chimpanzees in this regard. In extant higher primates, strong sexual dimorphism tends to reflect polygynous mating systems in which males compete vigorously for females, while reduced sexual dimorphism is associated with monogamous mating systems in which males and females pair for long periods. Decreased dimorphism in *H. ergaster* may thus mark the beginnings of a distinctively human pattern of sharing and cooperation between the sexes, prefiguring the social organization of historic human hunter-gatherers.

The appearance of *H. ergaster* in eastern Africa coincided with the appearance of more sophisticated stone artifacts 1.7–1.6 my ago (1797), and a causal connection is likely. The novel artifacts include the more extensively flaked bifacial tools known as hand axes (fig. 8.2). These are the hallmark of the Acheulean Industrial Complex, named for St. Acheul, in northern France, where the complex was first identified in the nineteenth century. More sophisticated technology, essentially modern body form, and implicit changes in cognition and social organization enabled *H. ergaster* to expand its range into previously unoccupied environments. It became the first human species to invade truly arid, highly seasonal environments in Africa, and it was also the first to disperse from Africa to Eurasia.

The timing of human dispersal to Eurasia is controversial, and new claims appear regularly. The ancient lakeside site of 'Ubeidiya in the Jordan Rift Valley of Israel provides the oldest unequivocal evidence. 'Ubeidiya has not furnished diagnostic human remains, but it does contain numerous hand axes and other stone tools like those that *H. ergaster* manufactured in Africa (122). Associated animal species imply a date near 1.4 my ago (2155). Arguably, however, 'Ubeidiya does not represent an Out of Africa event so much as a slight ecological expansion of Africa to incorporate its immediate southwest Asian periphery. In this case large, well-excavated artifact assemblages from the Nihewan Basin, 150 km west of Beijing, China, probably constitute the oldest unambiguous indication of people in Eurasia. The magnetism of the enclosing sediments places the oldest Nihewan assemblages near 1 my ago (1875).

Yet older Eurasian sites may exist, and two current candidates deserve special mention. These are Dmanisi, Georgia, where a human mandible and associated artifacts have been dated to roughly 1.8 my ago by paleomagnetism and by radiopotassium analysis of an underlying volcanic tuff (122, 294, 680, 775), and two localities in Java where fossils of classic *Homo erectus* have been dated to 1.8–1.6 my ago by radiopotassium (2123). The problem at Dmanisi is that the stratigraphic context and faunal associations require additional study, and a fuller evaluation may produce an age near 1–0.9 my ago (123). The advanced morphology of the human mandible itself suggests a date of this order (345, 617). The problem in Java is that the stratigraphic relation between the dated samples and *H. erectus* fossils is not firmly established (603). If the Javan dates are eventually confirmed, however, their implications would be revolutionary. They could mean that *H. ergaster* and *H. erectus* shared a more ancient common ancestor (*H. habilis?*) or even that *H. erectus* was ancestral to *H. ergaster*. Since the Javan dates precede the emergence of the Acheulean in Africa, they could also explain an otherwise puzzling aspect of Far Eastern artifact assemblages: these always retained an Oldowan cast, and they never included Acheulean hand axes (1875).

It has usually been assumed that Europe and Asia were occupied more or less simultaneously, but many putative European sites older than 500 ky are problematic, either because the dates are insecure or because the evidence for human presence depends on flaked stones that may be natural in origin. Until recently the oldest known European human fossils were believed to date to about 500 ky ago, and some specialists have argued that Europe was first colonized only about this time (651, 1811). If we accept this, the oldest known European fossils—including, for example, a mandible from Mauer in Germany, a skull from Petralona in Greece, a partial skull from Arago in France, and a tibia shaft from Boxgrove in England—would in fact represent the first Europeans (2086), and the oldest artifacts would be well-made Acheulean hand axes and associated pieces from Boxgrove (1786) and from broadly contemporaneous, well-documented Acheulean sites like Hoxne in England (1979), Cagny-la-Garenne in France (300, 2246, 2326), Fontana Ranuccio in Italy (1907), and Torralba and Ambrona in Spain (764, 1072, 1076).

An initial colonization of Europe only about 500 ky ago could explain why the Petralona skull is notably similar to its presumed African contemporaries, including above all the famous Broken Hill (Kabwe) skull from Zambia (Rhodesian man). Unlike like-aged skulls of Far Eastern *H. erectus* that have relatively small, angular braincases with barlike brow-ridges that flare out laterally, the skulls from Petralona and Broken Hill have somewhat more rounded and expanded brain-cases with browridges that arch upward over each orbit and thin toward the sides (912). In most textbooks Petralona, Broken Hill, and other broadly contemporaneous African and European fossils are referred to as "archaic *Homo sapiens,*" though some authorities prefer the term *Homo heidelbergensis* (2073). This reflects the relatively early discovery of the Mauer mandible and the proximity of Mauer to Heidelberg.

Conceivably, humans could not colonize Europe perma-nently until it provided ample scavenging opportunities, and the supply of carcasses may have been inadequate until 500 ky ago, when two species of large hyenas disappeared from Europe (2251). However, it is at least equally plausible that a human influx precipitated the hyenas' extinction, and it is difficult to imagine how carnivore turnover or other environmental factors could have excluded people from Europe for at least 500 ky af-ter they had reached northern China. Perhaps people did reach Europe before 500 ky ago but were present only sporadically and only along its southern margin. Human fossils and artifacts recently dated to near 800 ky ago at the Trinchera (Gran) Dolina site, Atapuerca, Spain (200, 424), strongly support an earlier, if perhaps transitory colonization event.

All supposed European artifact assemblages that antedate 500 ky ago lack hand axes and other bifacial tools, whereas most sites that shortly postdate 500 ky ago have them. Un-equivocal sites dated to near 500 ky ago and after are also far more abundant, and the sum suggests that the similarity of 500 ky old African and European skulls could reflect a wave of new African immigrants who introduced the Acheulean Indus-trial Tradition. Strikingly Africanlike Acheulean artifacts at the Israeli site of Gesher Benot Ya'aqov may mark an early stop in a major African exodus about 600–500 ky ago (122, 871).

Artifact assemblages remained similar between Africa and Europe after 500 ky ago, whereas the Far East continued on its own distinctive course. About 250–200 ky ago, Acheulean (Early Stone Age/Lower Paleolithic) assemblages were widely replaced in both Africa and Europe by Middle Stone Age/Middle Paleolithic assemblages emphasizing a refined flake technol-ogy, usually without hand axes (fig. 8.2). After 250 ky ago, how-ever, the human form came to differ markedly between the two continents, and by 100 ky ago Europe was occupied exclusively by the highly distinctive Neanderthals, whereas Africa was in-habited by people whose appearance was far more modern. Much older European fossils from sites like Swanscombe in

England, the Sima de los Huesos, Atapuerca, in Spain, Biache-Saint-Vaast in France, and Ehringsdorf in Germany, all probably between 300 and 150 ky old, already anticipate the Neanderthals and imply that they were an autochthonous European development (79, 82, 2073, 2086).

The more modern African contemporaries of the Neanderthals are represented at Klasies River Mouth, Border Cave, and Die Kelders Cave in South Africa, Mumba Shelter in Tanzania, Omo Kibish in Ethiopia, and Dar es Soltan in Morocco (336, 1773). The famous modern or near-modern human remains from Skhul and Jebel Qafzeh Caves in Israel probably also belong on this list, since associated "Ethiopian" mammal species imply that they date to a time when Israel lay within a slightly expanded Africa (2154, 2155). Thermoluminescence dates on associated flints and electron spin resonance dates on animal teeth fix this time between roughly 110 and 80 ky ago (1548, 1888). As a group, the key African fossils reveal people with relatively short, high braincases overhanging the face in front, in contrast to the long, low braincases and forwardly mounted faces of the Neanderthals. It is this fossil difference that most strongly supports the now-famous "Out of Africa" theory, according to which modern humans spread from Africa to replace the Neanderthals and other equally archaic humans in Eurasia.

The "Out of Africa" Hypothesis and the Evolution of Human Behavior

The Out of Africa hypothesis for modern human origins is more appropriately called Out of Africa 2 (2073), since it really concerns the pattern of human evolution after Out of Africa 1, the widely accepted original human dispersal from Africa at or before 1 my ago. At base, Out of Africa 2 is grounded in the fossil record, which shows that human populations diverged morphologically between Africa, Europe, and the Far East, especially after 500 ky ago. From this time onward, there were at least three evolving human lineages (fig. 8.1): *Homo sapiens* in Africa, *H. neanderthalensis* in Europe, and *H. erectus* in the Far East. The European lineage is the best documented, and it is marked by the progressive accumulation of Neanderthal features, culminating in the classic Neanderthals by 100 ky ago. Between 500 and 100 ky ago, Europe was generally much cooler than it has been historically, and some conspicuous Neanderthal distinctions, including massive trunks and short limbs, were probably physiological adaptations to cold. Other key distinctions—including, for example, the strong forward projection of the face along the midline, the unique configuration of the mastoid region and the occipital, and some peculiarities of the postcranium—may have resulted mainly from gene drift in small, isolated populations.

The pertinent African fossil record is much less complete, but it contains no specimens that anticipate the Neanderthals, and it shows that modern or near-modern people were widespread in Africa by 100 ky ago, when the Neanderthals were alone in Europe. The Far Eastern record is the most sketchy, and it may actually comprise two distinct evolutionary trajectories: one in Indonesia (Southeast Asia) that suggests continuity within Javan *Homo erectus* from before 500 ky ago until perhaps 50 ky ago (2122), and a second in China that may indicate evolution from classic Chinese *H. erectus* before 500 ky ago to populations that, by 100 ky ago, retained few distinctive *H. erectus* features and that approached *H. sapiens* in braincase size and form.

As revealed at present, the Chinese fossil record is perhaps the biggest impediment to unqualified acceptance of Out of Africa 2, but it is actually not so much contradictory as poorly known. China has provided only a handful of fossils that probably postdate classic *H. erectus,* and none of these is very precisely dated. In this light, the real problem with the Chinese fossil record is that its bearing on Out of Africa 2 is uncertain. A fuller record may suggest that, like European *H. neanderthalensis,* archaic Chinese populations (advanced *H. erectus?*) were replaced by immigrant African *H. sapiens* 50–40 ky ago, or it may suggest that local populations and African immigrants interbred. In the latter instance it might even reveal a dispersal from Africa to the Far East long before 50 ky ago. If interbreeding (gene flow) was demonstrated, the species distinction between invaders and locals could not be maintained. It is the absence of evidence for interbreeding and the abruptness of the replacement in Europe that most strongly support a species distinction between *H. sapiens* and *H. neanderthalensis.*

Out of Africa 2 is clearly bolstered by genetic evidence that all living humans shared a common ancestor at or after 200 ky ago and that this ancestor probably lived in Africa (303, 419, 436, 813, 947, 1584, 2052, 2496). It has been further strengthened by the sequencing of a mitochondrial DNA fragment from a humerus of the type Neanderthal fossil (1275). The difference between this sequence and homologous sequences in living humans supports fossil evidence that Neanderthals and living people last shared a common ancestor roughly 600–500 ky ago. The extraction of DNA from a wide range of human fossils would probably allow the most compelling check on Out of

Africa 2. Yet even if fossil DNA eventually provides the strongest possible support, it would not demonstrate the separate species postulated here. In fact, genetic divergence may never have preceded to the point that interbreeding was impossible, and behavioral separation (not genetic incompatibility) is adequate to explain the apparent absence of interbreeding between *H. sapiens* and *H. neanderthalensis.*

An obvious objection to Out of Africa 2 is the failure of modern or near-modern humans to expand from Africa immediately after they appeared, by 100 ky ago or before. Instead they seem to have been confined to Africa until roughly 60–50 ky ago, and it is even possible that they were replaced by Neanderthals on the southwest Asian margin of Africa (in what is now Israel) roughly 80 ky ago (2154, 2155). Archeology provides a partial answer to the apparent dilemma. The people who inhabited Africa between 100 and 60 ky ago may have been physically modern or near-modern, but they were behaviorally very similar to the Neanderthals and other nonmodern humans (121, 1141, 1243). The relatively full African and European archeological records show a distinct rupture 50–40 ky ago, when the Middle Stone Age in Africa and the broadly similar Middle Paleolithic in Europe gave way to the Later Stone Age and Upper Paleolithic, respectively. It is only the Later Stone Age and the Upper Paleolithic, after 50–40 ky ago, that reveal the full range of behavioral traits that mark many later prehistoric hunter-gatherer sites (table 8.1). There are few reliable dates for the beginning of the Later Stone Age and the Upper Paleolithic, but provisionally the available dates indicate that modern human behavioral markers appeared first in Africa, probably between 50 and 45 ky ago (31), that they spread to western Asia and eastern Europe by 43–40 ky ago, and finally that they reached western Europe about 40 ky ago (125, 1538, 1539). The geographic sequence is clearly consistent with an expansion of modern humans from Africa beginning 50–45 ky ago.

The modern human behavioral traits that appear after 50 ky ago imply the fully modern capacity for innovation that underlies culture in the narrow anthropological sense. The next section briefly explores the difficult problem of explaining why modern innovative ability developed when it did, but the point here is that it provided the adaptive advantage that allowed fully modern humans to replace their nonmodern contemporaries. The time when modern behavior appeared in the Far East remains debatable, and some have argued that the Far Eastern archeological record fails to reveal it until 10 ky ago or later (2492). In fact, however, like the companion fossil record, the

Table 8.1. Some Attributes of Fully Modern Human Behavior Detectable in the Archaeological Record Beginning 50–40 Ky Ago

---

Substantial growth in the diversity and standardization of artifact types

Rapid increase in the rate of artifactual change through time and in the degree of artifact diversity through space

First shaping of bone, ivory, shell, and related materials into formal artifacts ("points," "awls," "needles," "pins," etc.)

Earliest appearance of incontrovertible art

Oldest undeniable evidence for spatial organization of camp floors, including elaborate hearths and the oldest indisputable structural "ruins"

Oldest evidence for the transport of large quantities of highly desirable stone raw material over scores or even hundreds of kilometers

Earliest secure evidence for ceremony or ritual, expressed both in art and in relatively elaborate graves

First evidence for human ability to live in the coldest, most continental parts of Eurasia (northeastern Europe and northern Asia)

First evidence for human population densities approaching those of historic hunter-gatherers in similar environments

First evidence for fishing and for other significant advances in human ability to acquire energy

---

Note: For elaboration, see chapters 6 and 7.

Far Eastern archeological record is marked more by an absence of evidence than by evidence of an absence, and much fresh archeological research will be necessary to determine whether the pattern in the Far East was truly distinctive.

Some Problems with "Out of Africa 2"

Out of Africa 2 is the most plausible and parsimonious explanation of the available fossil and archeological data, but previous chapters provided some contrary observations, and these cannot simply be ignored. Some specialists also believe that proponents of Out of Africa 2 have unwittingly imposed their intellectual preconceptions on highly contrary data, and this leads them to reject Out of Africa 2 a priori (475–477). Carried to its logical extreme, however, this perspective precludes any decision on Out of Africa 2, barring the unlikely development of data collection procedures that do not require advance assumptions or expectations. New intellectual frameworks or paradigms may yet prove helpful, but the aim of this section is to reiterate some problems with Out of Africa 2 that are more evidentiary than epistemological.

    1. What explains the relatively abrupt appearance of modern human behavior (the modern capacity for culture) 50 ky

ago? The simplest answer is probably that it stemmed from a fortuitous mutation that promoted the fully modern brain. But this argument relies primarily on two circumstantial observations: that natural selection for more effective brains largely drove the earlier phases of human evolution, and that the relation between morphological and behavioral change shifted abruptly about 50 ky ago. Before this time morphology and behavior appear to have evolved more or less in tandem, very slowly, but after this time morphology remained relatively stable while behavioral (cultural) change accelerated rapidly. What could explain this better than a neural change that promoted the extraordinary modern human ability to innovate? This is not to say that the Neanderthals and their nonmodern contemporaries had apelike brains or that they were as biologically and behaviorally primitive as yet earlier humans. It is only to suggest that an acknowledged genetic link between morphology and behavior in yet earlier people persisted until the emergence of fully modern ones and that the postulated genetic change 50 ky ago fostered the uniquely modern ability to adapt to a remarkable range of natural and social circumstances with little or no physiological change.

The central problem with the neural hypothesis is that it is currently untestable from fossils. Earlier on in human evolution, a link between behavioral and neural change can be inferred from conspicuous increases in brain size, but humans virtually everywhere had achieved modern or near-modern brain size by 200 ky ago. Any neural change that occurred 50 ky ago would thus have to be in organization, and fossil skulls provide only speculative evidence for brain structure. Neanderthal skulls, for example, differ dramatically in shape from modern ones, but they were as large or larger, and on present evidence it is not clear that the difference in form implies a significant difference in function.

In these circumstances it might seem equally reasonable to argue that fully modern behavior originated among people who had long had the neural capacity for it but who expressed their modern potential only after some biologically irrelevant technological or social change (984, 2009). This kind of explanation is more circular than the neural (biological) alternative, however, since it does not explain why social organization or technology changed so suddenly and fundamentally. Surely it is more economical to invoke a selectively advantageous neural change like those that must partly underlie earlier behavioral advances. Arguably the last key neural change promoted the modern capacity for rapidly spoken phonemic speech, that is, for "a fully vocal language, phonemicized, syntactical, and infinitely open and productive" (1566, p. 321). This suggestion follows logically from the obvious dependence of modern culture on modern language, but the evidence for it is admittedly circumstantial and fragmentary.

2. Were Neanderthals fundamentally incapable of fully modern behavior? As outlined here, Out of Africa 2 postulates that the Neanderthals were replaced because they could not compete culturally with their modern human successors. The argument is bolstered over most of Europe by the relatively abrupt nature of the replacement. At most sites, Cro-Magnon/Upper Paleolithic occupations overlie Neanderthal/Middle Paleolithic layers with no evidence for a major break in time or for any transition between the two, suggesting that the replacement took only decades, or at most centuries. However, there is the occasional discovery of artifact assemblages that constitute a blend of Neanderthal/Middle Paleolithic and Cro-Magnon/Upper Paleolithic artifact types. The most compelling examples come from western France and northern Spain, where they have been assigned to the Châtelperronian Industry (972). The Châtelperronian probably existed for a few centuries or perhaps a millennium about 40 ky ago, and human remains from La Roche à Pierrot Rockshelter, Saint-Césaire (1361), and the Grotte du Renne, Arcy-sur-Cure (1098), indicate that the makers were Neanderthals.

At Arcy, the Neanderthals not only produced a mix of typical Middle and Upper Paleolithic stone artifacts, they also manufactured typical Upper Paleolithic bone tools and personal ornaments, and they modified their living space in a characteristically Upper Paleolithic fashion (708, 709, 1358). One can argue that the Châtelperronian Neanderthals borrowed Upper Paleolithic cultural traits from fully Upper Paleolithic (Aurignacian) neighbors whose ultimate origin was in Africa. But even if this is accepted, it begs one fundamental question: If Upper Paleolithic culture was clearly superior and the Neanderthals could imitate it (that is, they were not biologically precluded from behaving in an Upper Paleolithic way), why didn't they acculturate more widely, with the result that they or their genes would have persisted much more conspicuously into Upper Paleolithic times (after 40–35 ky ago)? There is no compelling answer, and the Châtelperronian is a major puzzle whose solution is important for closure on Out of Africa 2.

3. Why weren't the earliest modern humans as heavily built as the Neanderthals? Neanderthal limb bones are remarkably robust, with very conspicuous muscle markings, implying

that Neanderthals of both sexes were exceptionally powerful people. In spite of this, they often broke their bones, they commonly developed arthritis or other senile pathologies in their twenties or thirties, and they seldom survived beyond age forty (194, 350, 2231, 2237). The sum suggests that they led extraordinarily stressful lives. In contrast, their fully modern, Cro-Magnon successors were much less heavily built, they broke their bones much less often, and their maximum life expectancy was significantly greater. Since Neanderthals were culturally (artifactually) much less sophisticated, a reasonable explanation for the difference is that Neanderthals often accomplished physically what later people accomplished culturally (technologically).

The downside of Neanderthal robusticity was that it required a great deal of energy to sustain, and it is presumably this that favored reduced robusticity in modern humans—the same number of calories could now support larger populations, and larger populations are a measure of evolutionary success. However, the most completely known early modern contemporaries of the Neanderthals, from Qafzeh Cave in Israel, were much less heavily built than the Neanderthals, even though they made similar, relatively unsophisticated artifacts (132, 2227, 2229). When the smaller body mass of the Qafzeh people is considered, they may have equaled the Neanderthals in the robusticity of their legs, but their arms were significantly less powerful (2236). To the extent that this casts doubt on the adaptive superiority of fully modern technology, it presents a problem for the version of Out of Africa 2 preferred here.

4. What kind of people first occupied the Americas and Australia? A probable corollary of Out of Africa 2 is that they were fully modern. With regard to the Americas, this follows from the likelihood that first entry was across a land bridge that linked northeastern Asia to Alaska during glacial periods (periods of lowered sea level) and from archeological evidence that northeastern Asia was itself first colonized only after 35 ky ago (851, 1293), when fully modern humans developed the housing, clothing, and other cultural wherewithal to survive in very harsh, continental climates. The colonization of the Americas by fully modern humans is fully consistent with the archeology of North and South America, neither of which was indisputably occupied before the closing phases of the Last Glaciation, after 14 ky ago (988, 1036, 1542, 1543). At least south of Alaska, human occupation before this time was probably precluded by an ice sheet that extended more or less continuously across Canada. In short, the American record presents no problem for Out of Africa 2.

The same may not be true for Australia, where the argument for an initial colonization by fully modern humans follows from the need to cross 80–100 km of open water, the minimum distance separating Australia from Southeast Asia, even during periods of lowered sea level. It would be hard to deny an essentially modern capacity for culture to people who could produce sufficiently seaworthy watercraft. Until recently, it appeared that the first Australians were in fact fully modern people who arrived between 40 and 30 ky ago, bringing with them complex burial practices, fishing technology, art, and probably other modern behavioral markers (1161, 1162). An entry at about 40 ky ago could itself be regarded as an indicator of the modern human ability to innovate, this time with respect to water transport.

It now appears possible, however, that Australia was occupied much earlier, by 60 ky ago or before. The relevant dates were obtained by the thermoluminescence and optically stimulated luminescence methods on unburned quartz sands enclosing and overlying artifacts at the Malakunanja II and Nauwalabila I sites in northern Australia (1791). Both methods are experimental, but the results are particularly compelling at Nauwalabila I, where optical and calibrated ^{14}C dates on the upper layers agree closely. At Nauwalabila I, a layer dated by the optical method to approximately 53 ky ago contains ground hematite fragments that the excavators believe were used for painting. If so, this could mean that the first Australians were behaviorally advanced over their European and African contemporaries. However, the hematite might have been used for hide processing or some equally mundane purpose, and similar hematite fragments are common in Middle Paleolithic/Neanderthal sites, with no other evidence for art (270, 528). They also occur without apparent art at many (Middle Stone Age) sites antedating 50 ky ago in southern Africa (2165). Modified hematite fragments are actually more abundant in such sites than they are in much younger (Later Stone Age) ones, including some whose occupants surely painted on nearby rock faces.

For proponents of Out of Africa 2, then, the problem is not that 60 ky old Australian dates imply an especially early, non-African emergence of art, but that they raise two other fundamental questions. Is it possible that modern humans left Africa as much as 60 ky ago? And assuming they did, how is it that they reached the far east (Australasia) 20 ky before they reached the far west (France and Spain)? In this context it is important to note that the Middle Paleolithic/Upper Paleolithic interface in France and Spain cannot be much older than 40 ky. This estimate is based not on radiocarbon dates, which provide only minimal ages in the 40 ky range, but on thermoluminescence dates from Le Moustier, France (2272) and on uranium-series dates from Abric Romaní, Spain (228, 229), which show that the Middle Paleolithic survived in western Europe until roughly 40 ky ago.

5. Is it really true that modern behavioral markers appear widely only about 50–40 ky ago? With regard to art, for example, virtually all specialists agree that it becomes commonplace only after this time and that earlier examples are both rare and crude. But authorities disagree sharply on what this combination of rarity and simplicity implies. To some (for example, 166, 984, 1457) it means that modern cognitive abilities were present but were weakly expressed before 50 ky ago, while to others (for example, 447, 452, 582) and to me it suggests that the fully modern capacity for culture may have appeared only about this time.

Some of the very rare art objects that antedate 50 ky ago are probably younger intrusions that even the most careful excavation cannot detect, while others are probably the result of human or natural actions that will inevitably, on rare occasions, mimic crude human attempts at art. In this regard, credible claims for art or other modern human behavioral markers before 50 ky ago must involve relatively large numbers of highly patterned objects from well-documented contexts. Using this criterion, perhaps the most serious obstacle to the Out of Africa scenario favored here is the discovery of carefully shaped barbed points and accompanying evidence for fishing at the Katanda sites in the Semliki Valley of Zaire/Congo (363, 2526). Electron spin resonance dates on associated hippopotamus teeth and thermoluminescence dates on covering sands suggest an age between 155 and 90 ky ago. If this estimate is valid, it implies that modern behavioral traits and modern morphology may have appeared together, at or before 100 ky ago, and we will be forced to find a nonbehavioral explanation for why modern or near-modern humans were confined to Africa until roughly 50 ky ago. We will also have the difficult task of explaining why even larger, well-excavated artifact and bone assemblages from apparently contemporaneous African sites completely lack evidence either for formal bone working or for fishing.

The Katanda dates illustrate a recurrent problem in archeology—whether one or two startling observations are sufficient to reverse an inference based on broad patterning in many other observations. The discrepancy in this case might be resolved by applying radiocarbon dating directly to the Katanda bone points.

Conclusion

Since the 1910s, fossil and archeological discoveries have suggested that fully modern (Cro-Magnon) immigrants replaced the Neanderthals in Europe. Fossil and archeological support for an abrupt replacement grew stronger in succeeding decades, but it became particularly compelling after the middle 1980s, when new dating methods showed that near-modern or modern humans inhabited Africa by 100 ky ago; new fossils (mainly from the Sima de los Huesos, Atapuerca, Spain) documented the evolution of the Neanderthals in Europe; and increasingly sophisticated genetic analyses showed that the last shared ancestor of living humans postdated the emergence of the Neanderthal lineage. Some of the new (and old) evidence is ambiguous, circumstantial, or even contradictory, but if the study of human origins were a jury trial, the verdict would surely be that modern humans, originating in Africa, exterminated the Neanderthals. The jury might hang on the question whether modern humans expanding eastward also extinguished non-modern people in the Far East, and they might ask the judge to postpone that aspect of the trial pending the acquisition of much more evidence.

In fact, of course, human origins research differs from a jury trial in that no verdict need ever be final, and new evidence and new jury members are always welcome. I predict that the verdict on the extinction of the Neanderthals will not be reversed, but future juries, faced with new evidence, may decide differently on a range of other controversial issues, including, for example, the relation between climatic change and speciation early on in human evolution and the number and timing of early human dispersals from Africa. These matters are inherently more difficult to resolve than modern human origins, because they center on much older time periods for which the record is far more sparse. They may never be resolved as completely as modern human origins, but uncertainty stimulates data collection, and it is fresh data—expanding genetic studies and new fossils and artifacts from well-documented contexts—that have eliminated all reasonable doubt in the century-old controversy over the fate of the Neanderthals.

REFERENCES

1. Abbate, E., Albianelli, A., Azzaroli, A., Benvenuti, M., Tesfama-riam, B., Bruni, P., Cipriani, N., Clarke, R. J., Piccarelli, G., Macchia-relli, R., Napoleone, G., Papini, M., Rook, L., Sagri, M., Tecle, T. M., Toree, D., and Villa, I. 1998. A one-million-year old *Homo* cranium from the Danakil (Afar). *Nature* 393:458–460.
2. Abramova, Z. A. 1964. On the question of hunting in the Upper Paleolithic (in Russian). *Sovetskaya Arkheologiya* 4:177–180.
3. Abramova, Z. A. 1967. Palaeolithic art in the USSR. *Arctic Anthropology* 4:1–179.
4. Abramova, Z. A. 1982. Zur Jagd im Jungpaläolithikum. *Archäo-logisches Korrespondenzblatt* 12:1–9.
5. Absolon, K., and Klima, B. 1977. *Predmostí: Ein Mammutjäger-platz in Mähren.* Prague: Academia.
6. Ackerman, R. E. 1996. Bluefish Caves. In *American beginnings: The prehistory and palaeoecology of Beringia,* ed. F. H. West, pp. 511–513. Chicago: University of Chicago Press.
7. Adam, K. D. 1985. The chronological and systematic position of the Steinheim skull. In *Ancestors: The hard evidence,* ed. E. Delson, pp. 272–276. New York: Alan R. Liss.
8. Aguirre, E., and de Lumley, M.-A. 1977. Fossil men from Ata-puerca, Spain: Their bearing on human evolution in the Middle Pleistocene. *Journal of Human Evolution* 6:681–738.
9. Aguirre, E., and Rosas, A. 1985. Fossil man from Cueva Mayor, Ibeas, Spain: New findings and taxonomic discussion. In *Hominid evolution: Past, present and future,* ed. P. V. Tobias, pp. 319–328. New York: Alan R. Liss.
10. Aiello, L. C. 1986. The relationships of the Tarsiiformes: A review of the case for the Haplorhini. In *Major topics in primate and hu-man evolution,* ed. B. Wood, L. Martin, and P. Andrews, pp. 47–65. Cambridge: Cambridge University Press.
11. Aiello, L. C., and Dunbar, R. J. M. 1993. Neocortex size, group size, and the evolution of language. *Current Anthropology* 34:184–192.
12. Aiello, L. C., and Wheeler, P. 1995. The expensive-tissue hypothe-sis: The brain and digestive system in human and primate evolu-tion. *Current Anthropology* 36:199–221.
13. Aigner, J. S. 1978. Important archaeological remains from North China. In *Early Paleolithic in south and east Asia,* ed. F. Ikawa-Smith, pp. 163–232. The Hague: Mouton.
14. Aigner, J. S. 1978. Pleistocene faunal and cultural stations in South China. In *Early Paleolithic in south and east Asia,* ed. F. Ikawa-Smith, pp. 129–160. The Hague: Mouton.

15. Aigner, J. S. 1986. The age of Zhoukoudian Locality 1: The newly proposed O^{18} correspondences. *Anthropos* (Brno) 23:157–173.

16. Aitken, M. J. 1994. Optical dating: A non-specialist review. *Quaternary Science Reviews* 13:503–508.

17. Aitken, M. J., Huxtable, J., and Debenham, N. C. 1986. Thermoluminescence dating in the Palaeolithic: Burned flint, stalagmitic calcite and sediment. In *Chronostratigraphie et faciés culturels du Paléolithique inférieur et moyen dans l'Europe du Nord-Ouest*, ed. A. Tuffreau and J. Sommé, pp. 7–14. Paris: Supplément au Bulletin de l'Association Française pour l'Étude du Quaternaire.

18. Aitken, M. J., and Valladas, H. 1993. Luminescence dating relevant to human origins. In *The origin of modern humans and the impact of chronometric dating*, ed. M. J. Aitken, C. B. Stringer, and P. A. Mellars, pp. 27–39. Princeton: Princeton University Press.

19. Akazawa, T., Muhesen, S., Dodo, Y., Kondon, O., and Mizoguchi, Y. 1995. Neanderthal infant burial. *Nature* 377:585–586.

20. Alimen, H. M. 1975. Les "isthmes" hispano-marocain et siculo-tunisien aux temps acheuléens. *Anthropologie* 79:399–436.

21. Allen, J. 1994. Radiocarbon determinations, luminescence dating and Australian archaeology. *Antiquity* 68:339–343.

22. Allen, J., Gosden, C., Jones, R., and White, J. P. 1988. Pleistocene dates for the human occupation of New Ireland, northern Melanesia. *Nature* 331:707–709.

23. Allen, J., and Holdoway, S. 1995. The contamination of Pleistocene radiocarbon determinations in Australia. *Antiquity* 69:101–112.

24. Allsworth-Jones, P. 1986. Middle Stone Age and Middle Palaeolithic: The evidence from Nigeria and Cameroun. In *Stone Age prehistory: Studies in memory of Charles McBurney*, ed. G. N. Bailey and P. Callow, pp. 153–168. Cambridge: Cambridge University Press.

25. Allsworth-Jones, P. 1986. *The Szeletian and the transition from the Middle to Upper Palaeolithic in central Europe*. Oxford: Oxford University Press.

26. Allsworth-Jones, P. 1990. The Szeletian and the stratigraphic succession in central Europe and adjacent areas: Main trends, recent results, and problems for resolution. In *The emergence of modern humans: An archaeological perspective*, ed. P. Mellars, pp. 160–242. New York: Cornell University Press.

27. Alt, K. W., Pichler, S., Vach, W., Klíma, B., Vlcek, E., and Sedlmeier, J. 1997. Twenty-five-thousand-year-old triple burial from Dolní Vestonice: An Ice-Age family? *American Journal of Physical Anthropology* 102:123–131.

28. Altuna, J. 1972. Fauna de mamíferos de los yacimientos prehistóricos de Guipuzcoa. *Munibe* 24:1–464.

29. Altuna, J. 1992. El medio ambiente durante el Pleistoceno superior en la región Cantábrica con referencia especial a sus faunas de mamíferos. *Munibe* 44:13–29.

30. Amani, F., and Geraads, D. 1993. Le gisement moustérien du Djebel Irhoud, Maroc: Précisions sur la fauna et la biochronologie, et description d'un nouveau reste humain. *Comptes Rendus de l'Académie des Sciences, Paris*, ser. 2, 316:847–852.

31. Ambrose, S. H. 1998. Chronology of the Later Stone Age and food production in east Africa. *Journal of Archaeological Science* 25:377–392.

32. Ambrose, S. H., and Lorenz, K. G. 1990. Social and ecological models for the Middle Stone Age in southern Africa. In *The emergence of modern humans: An archaeological perspective*, ed. P. Mellars, pp. 3–33. Ithaca: Cornell University Press.

33. Ameloot-Van der Heijden, A., and Tuffreau, A. 1993. Les industries lithiques de Riencourt-les-Bapaume dans le contexte de l'Europe du Nord-Ouest. In *Riencourt-les-Bapaume (Pas-de-Calais): Un gisement du Paléolithique moyen*, ed. A. Tuffreau, pp. 107–111. Paris: Maison des Sciences de l'Homme.

34. Anconetani, P., Giusberti, G., and Peretto, C. 1994. Considérations taphonomiques à propos des os du *Bison schoetensacki* Freudenberg du gisement paléolithique de Isernia la Pineta (Molise—Italie). In *Outillage peu élaboré en os et bois des cervidés*, ed. M. Patou-Mathis, pp. 173–178. Treignes, Belgium: Éditions du Centre d'Études et de Documentation Archéologique.

35. Anderson, E. 1984. Who's who in the Pleistocene: A mammalian bestiary. In *Quaternary extinctions: A prehistoric revolution*, ed. P. S. Martin and R. G. Klein, pp. 40–89. Tucson: University of Arizona Press.

36. Anderson, J. E. 1968. Late Paleolithic skeletal remains from Nubia. In *The prehistory of Nubia*, ed. F. Wendorf, 2:996-1040. Dallas: Southern Methodist University Press.

37. Anderson-Gerfaud, P. 1990. Aspects of behaviour in the Middle Palaeolithic: Functional analysis of stone tools from southwest France. In *The emergence of modern humans*, ed. P. Mellars, pp. 389–418. Ithaca: Cornell University Press.

38. Andrews, P. 1980. Ecological adaptations of the smaller fossil apes. *Zeitschrift für Morphologie und Anthropologie* 71:164–173.

39. Andrews, P. 1981. Hominoid habitats of the Miocene. *Nature* 289:749.

40. Andrews, P. 1981. A short history of Miocene field palaeontology in western Kenya. *Journal of Human Evolution* 10:3–9.

41. Andrews, P. 1981. Species diversity and diet in monkeys and apes during the Miocene. In *Aspects of human evolution*, ed. C. B. Stringer, pp. 25–61. London: Taylor and Francis.

42. Andrews, P. 1983. The natural history of *Sivapithecus*. In *New interpretations of ape and human ancestry*, ed. R. L. Ciochon and R. S. Corruccini, pp. 441–463. New York: Plenum Press.

43. Andrews, P. 1983. Small mammal faunal diversity at Olduvai Gorge, Tanzania. *British Archaeological Reports* 163:77–85.

44. Andrews, P. 1984. On the characters that define *Homo erectus*. *Courier Forschungsinstitut Senckenberg* 69:167–178.

45. Andrews, P. 1985. Family group systematics and evolution among catarrhine primates. In *Ancestors: The hard evidence*, ed. E. Delson, pp. 14–22. New York: Alan R. Liss.

46. Andrews, P. 1985. Improved timing of hominoid evolution with a DNA clock. *Nature* 314:498–499.

47. Andrews, P. 1986. Aspects of hominoid phylogeny. In *Molecules and morphology in evolution: Conflict or compromise*, ed. C. Patterson, pp. 21–53. Cambridge: Cambridge University Press.

48. Andrews, P. 1986. Fossil evidence on human origins and dispersal. *Cold Spring Harbor Symposia on Quantitative Biology* 51:419–428.

49. Andrews, P. 1988. Review of *Comparative primate biology. Cladistics* 4:297–304.

50. Andrews, P. 1992. Evolution and environment in the Hominoidea. *Nature* 360:641–646.

51. Andrews, P., and Cook, J. 1985. Natural modification to bones in a temperate setting. *Man,*n.s., 20:675–691.

52. Andrews, P., and Cronin, J. E. 1982. The relationships of *Sivapithecus* and *Ramapithecus* and the evolution of the orang-utan. *Nature* 297:545–546.

53. Andrews, P., and Fernandez Jalvo, Y. 1997. Surface modifications of the Sima de los Huesos fossil remains. *Journal of Human Evolution* 33:191–217.

54. Andrews, P., Harrison, T., Delson, E., Bernor, R. L., and Martin, L. 1996. Distribution and biochronology of European and southwest Asian Miocene catarrhines. In *The evolution of western Eurasian Neogene mammal faunas*, ed. R. L. Bernor, V. Fahlbusch, and H.-W. Mittman, pp. 168–207. New York: Columbia University Press.

55. Andrews, P., and Martin, L. 1987. Cladistic relationships of extant and fossil hominoids. *Journal of Human Evolution* 16:101–118.

56. Andrews, P., and Martin, L. 1987. The phyletic position of the Ad Dabtiyah hominoid. *Bulletin of the British Museum of Natural History (Geology)* 41:383–393.

57. Andrews, P., Martin, L., and Whybrow, P. 1987. Earliest known member of the great ape and human clade. *American Journal of Physical Anthropology* 72:174–175.

58. Andrews, P., and Tekkaya, I. 1980. A revision of the Turkish Miocene hominoid *Sivapithecus meteai. Palaeontology* 23:85–95.

59. Andrews, P., and van Couvering, J. A. H. 1975. Palaeoenvironments in the east African Miocene. *Contributions to Primatology* 5:62–103.

60. Anikovich, M. 1992. Early Upper Paleolithic industries of eastern Europe. *Journal of World Prehistory* 6:205–245.

61. Anthony, B. W. 1972. The Still Bay question. In *Actes de VIe Congrès Panafricain de Préhistoire, Dakar 1967*, ed. H. J. Hugot, pp. 80–82. Chambéry: Imprimeries Réunies de Chambéry.

62. Antón, S. 1997. Developmental age and taxonomic affinity of the Modjokerto Child, Java, Indonesia. *American Journal of Physical Anthropology* 102:497–514.

63. Antón, S. C. 1994. Mechanical and other perspectives on Neandertal craniofacial morphology. In *Integrative paths to the past*, ed. R. S. Corruccini and R. L. Ciochon, pp. 677–695. Englewood Cliffs, N.J.: Prentice Hall.

64. Anzidei, A. P. 1995. La Polledrara, a Middle Pleistocene site near Rome (Italy): Its bone and stone industry. In *Congreso Internacional de Paleontologia Humana (Orce, September 1995), 3a Circular*, p. 100. Orce, Spain.

65. Arambourg, C. 1955. A recent discovery in human paleontology: *Atlanthropus* of Ternifine (Algeria). *American Journal of Physical Anthropology* 13:191–202.

66. Arambourg, C. 1963. L'*Atlanthropus mauritanicus. Mémoires et Archives de l'Institut de Paléontologie Humaine* 32:37–190.

67. Arambourg, C., and Biberson, P. 1956. The fossil human remains

from the Paleolithic site of Sidi Abderrahman (Morocco). *American Journal of Physical Anthropology* 13:191–202.

68. Arambourg, C., and Coppens, Y. 1967. Sur la découverte, dans le Pléistocène inférieur de la vallée de l'Omo (Éthiopie), d'une mandibule d'australopithécien. *Comptes Rendus de l'Académie des Sciences, Paris,* ser. D, 265:589–590.

69. Arambourg, C., and Coppens, Y. 1968. Découverte d'un australopithécien nouveau dans les gisements de l'Omo (Éthiopie). *South African Journal of Science* 64:58–59.

70. Ardrey, R. 1961. *African genesis.* New York: Dell.

71. Ardrey, R. 1976. *The hunting hypothesis.* London: Collins.

72. Arens, W. 1979. *The man-eating myth: Anthropology and anthropophagy.* New York: Oxford University Press.

73. Arensburg, B. 1977. New Upper Paleolithic human remains from Israel. *Eretz-Israel* 13:208–215.

74. Arensburg, B., Bar-Yosef, O., Chech, M., Goldberg, P., Laville, H., Meignen, L., Rak, Y., Tchernov, E., Tillier, A.-M., and Vandermeersch, B. 1985. Une sépulture néanderthalien dans la grotte de Kebara (Israel). *Comptes Rendus de l'Académie des Sciences, Paris,* ser. 2, 300:227–230.

75. Arensburg, B., Schepartz, L. A., Tillier, A. M., Vandermeersch, B., and Rak, Y. 1990. Reappraisal of the anatomical basis for speech in Middle Palaeolithic hominids. *American Journal of Physical Anthropology* 83:137–146.

76. Arensburg, B., Tillier, A. M., Vandermeersch, V., Duday, H., Schepartz, L. A., and Rak, Y. 1989. A Middle Palaeolithic human hyoid bone. *Nature* 338:758–760.

77. Arsuaga, J. L., Bermúdez de Castro, J. M., and Carbonell, E. 1994. La Sierra de Atapuerca: Los homínidos y sus actividades. *Revista de Arqueología* 15:12–25.

78. Arsuaga, J. L., Carretero, J. M., Martinez, I., and Gracia, A. 1991. Cranial remains and long bones from Atapuerca/Ibeas (Spain). *Journal of Human Evolution* 20:191–230.

79. Arsuaga, J. L., Gracia, A., Martínez, I., and Lorenzo, C. 1996. The Sima de los Huesos (Sierra de Atapuerca, Spain) cranial evidence and the origin of the Neanderthals. In *The last Neandertals, the first anatomically modern humans,* ed. E. Carbonell and M. Vaquero, pp. 39–49. Tarragona, Spain: Fundacio Catalana per a la Recerca.

80. Arsuaga, J. L., Martínez, I., Gracia, A., Carretero, J. M., and Carbonell, E. 1993. Three new human skulls from the Sima de los Huesos Middle Pleistocene site in Sierra de Atapuerca, Spain. *Nature* 362:534–537.

81. Arsuaga, J. L., Martínez, I., Gracia, A., Carretero, J. M., Lorenzo, C., García, N., and Ortega, A. I. 1997. Sima de los Huesos (Sierra de Atapuerca, Spain). The site. *Journal of Human Evolution* 33:109–127.

82. Arsuaga, J. L., Martínez, I., Gracia, A., and Lorenzo, C. 1997. The Sima de los Huesos crania (Sierra de Atapuerca, Spain): A comparative study. *Journal of Human Evolution* 33:219–281.

83. Ascenzi, A., Biddittu, I., Cassoli, P. F., Segre, A. G., and Naldini, E. S. 1995. Middle Pleistocene human calvaria, central Italy. In *Congreso Internacional de Paleontologia Humana (Orce, September 1995), 3a Circular,* p. 83. Orce, Spain.

84. Ascenzi, A., Biddutu, I., Cassoli, P. F., Segre, A. G., and Segre-Naldini, E. 1996. A calvarium of late *Homo erectus* from Ceprano, Italy. *Journal of Human Evolution* 31:409–423.

85. Asfaw, B. 1983. A new hominid parietal from Bodo, Middle Awash Valley, Ethiopia. *American Journal of Physical Anthropology* 61: 367–371.

86. Asfaw, B. 1987. The Belohdelie frontal: New evidence of early hominid cranial morphology from the Afar of Ethiopia. *Journal of Human Evolution* 16:611–624.

87. Asfaw, B., Beyene, Y., Semaw, S., Suwa, G., White, T., and Wolde-Gabriel, G. 1991. Fejej: A new paleoanthropological research area in Ethiopia. *Journal of Human Evolution* 21:137–143.

88. Asfaw, B., Beyene, Y., Semaw, S., Suwa, G., White, T., and WoldeGabriel, G. 1993. Tephra from Fejej, Ethiopia—a reply. *Journal of Human Evolution* 25:519–521.

89. Asfaw, B., Beyene, Y., Suwa, G., Walter, R. C., White, T. D., WoldeGabriel, G., and Yemane, T. 1992. The earliest Acheulean from Konso-Gardula. *Nature* 360:732–735.

90. Asfaw, B., Suwa, G., and Beyene, Y. 1995. Progress in paleoanthropology at Konso Gardula. *American Journal of Physical Anthropology*, suppl., 20:60.

91. Audouze, F. 1987. The Paris Basin in Magdalenian times. In *The Pleistocene Old World: Regional perspectives*, ed. O. Soffer, pp. 183–200. New York: Plenum Press.

92. Audouze, F., David, F., and Enloe, J. G. 1989. Habitats magdaléniens: Les apports des modèles ethno-archéologiques. *Le Courrier du CNRS: Dossiers Scientifiques* 73:12–14.

93. Avery, G. 1990. Avian fauna, palaeoenvironments and palaeoecology in the late Quaternary of the western and southern Cape, South Africa. Ph.D. thesis, University of Cape Town.

94. Avery, G., Cruz-Uribe, K., Goldberg, P., Grine, F. E., Klein, R. G., Lenardi, M. J., Marean, C. W., Rink, W. J., Schwarcz, H. P., Thackeray, A. I., and Wilson, M. L. 1997. The 1992–1993 excavations at the Die Kelders Middle and Later Stone Age Cave Site, South Africa. *Journal of Field Archaeology* 24:263–291.

95. Azzaroli, A. 1985. Historical, chronological and paleoenvironmental background to the study of *Oreopithecus bambolii*. *American Journal of Physical Anthropology* 65:142.

96. Azzaroli, A., Boccaletti, M., Delson, E., Moratti, G., and Torre, D. 1986. Chronological and paleogeographical background to the study of *Oreopithecus bambolii*. *Journal of Human Evolution* 15: 533–540.

97. Azzaroli, A., De Giuli, C., Ficcarelli, G., and Torre, D. 1988. Late Pliocene to early Mid-Pleistocene mammals in Eurasia: Faunal succession and dispersal events. *Palaeogeography, Palaeoclimatology, Palaeoecology* 66:77–100.

98. Baba, H., and Aziz, F. 1992. Human tibial fragment from Sambungmachan, Java. In *The evolution and dispersal of modern humans in Asia*, ed. T. Akazawa, K. Aoki, and T. Kimura, pp. 349–361. Tokyo: Hokusen-Sha.

99. Baba, M., Darga, L., and Goodman, M. 1980. Biochemical evidence on the phylogeny of Anthropoidea. In *Evolutionary biology of the New World monkeys and continental drift*, ed. R. L. Ciochon and A. B. Chiarelli, pp. 423–443. New York: Plenum Press.

100. Bada, J. L. 1985. Amino acid racemization dating of fossil bones. *Annual Review of Earth and Planetary Science* 13:241–268.

101. Bader, O. N. 1978. *The Sungir' Upper Paleolithic site* (in Russian). Moscow: Nauka.

102. Badgley, C., Qi, G., Chen, W., and Han, D. 1988. Paleoecology of a Miocene, tropical, upland fauna: Lufeng, China. *National Geographic Research* 4:178–195.

103. Baffier, D., and Julien, M. 1990. L'outillage en os des niveaux châtelperroniens d'Arcy-sur-Cure (Yonne). In *Paléolithique moyen récent et Paléolithique supérieur ancien en Europe*, ed. C. Farizy, 3:329–334. Nemours: Mémoires du Musée de Préhistoire d'Île de France.

104. Bahn, P. G. 1983. Late Pleistocene economies of the French Pyrenees. In *Hunter-gatherer economy in prehistory*, ed. G. Bailey, pp. 168–186. Cambridge: Cambridge University Press.

105. Bahn, P. G. 1994. New advances in the field of Ice Age art. In *Origins of anatomically modern humans*, ed. M. H. Nitecki and D. V. Nitecki, pp. 121–132. New York: Plenum Press.

106. Bahn, P. G. 1995–1996. New developments in Pleistocene art. *Evolutionary Anthropology* 4:204–215.

107. Bahn, P. G. 1996. Further back down under. *Nature* 383:577–578.

108. Bahn, P. G., and Vertut, J. 1988. *Images of the Ice Age*. New York: Facts on File.

109. Bailey, W. J. 1993. Hominoid trichotomy: A molecular overview. *Evolutionary Anthropology* 2:100–108.

110. Bailey, W. J., Fitch, D. H. A., Tagle, D. A., Czelusniak, J., Slightom, J. L., and Goodman, M. 1991. Molecular evolution of the psi-eta-globin gene locus: Gibbon phylogeny and the hominoid slowdown. *Molecular Biology and Evolution* 8:155–184.

111. Bailey, W. J., Hayasaka, K., Skinner, C. G., Kehoe, S., Sieu, L. C., Slightom, J. L., and Goodman, M. 1992. Reexamination of the African hominoid trichotomy with additional sequences from the primate beta-globin gene cluster. *Molecular Phylogenetics and Evolution* 1:97–135.

112. Baksi, A. K., Hsu, V., McWilliams, M. O., and Ferrar, E. 1992. ^{40}Ar/^{39}Ar dating of the Brunhes-Matuyama geomagnetic field reversal. *Science* 256:356–357.

113. Balout, L. 1955. *Préhistoire de l'Afrique du Nord*. Paris: Arts et Métiers Graphiques.

114. Balout, L., Biberson, P., and Tixier, J. 1967. L'Acheuléen de Ternifine (Algérie): Gisement de l'Atlanthrope. *Anthropologie* 71:217–238.

115. Bar-Yosef, O. 1980. The prehistory of the Levant. *Annual Review of Anthropology* 9:101–133.

116. Bar-Yosef, O. 1987. Pleistocene connections between Africa and southwest Asia: An archaeological perspective. *African Archaeological Review* 5:29–38.

117. Bar-Yosef, O. 1989. Geochronology of the Levantine Middle Paleolithic. In *The human revolution: Behavioural and biological perspectives on the origins of modern humans*, ed. P. Mellars and C. B. Stringer, pp. 589–610. Edinburgh: University of Edinburgh Press.

118. Bar-Yosef, O. 1992. Middle Paleolithic human adaptations in the Mediterranean Levant. In *The evolution and dispersal of modern*

humans in Asia, ed. T. Akazawa, K. Aoki, and T. Kimura, pp. 189–216. Tokyo: Hokusen-Sha.

119. Bar-Yosef, O. 1993. The role of western Asia in modern human origins. In *The origin of modern humans and the impact of chronometric dating*, ed. M. J. Aitken, C. B. Stringer, and P. A. Mellars, pp. 132–147. Princeton: Princeton University Press.

120. Bar-Yosef, O. 1994. Beyond stone tools. *Cambridge Archaeological Journal* 4:107–109.

121. Bar-Yosef, O. 1994. The contributions of southwest Asia to the study of the origin of modern humans. In *Origins of anatomically modern humans*, ed. M. H. Nitecki and D. V. Nitecki, pp. 23–66. New York: Plenum Press.

122. Bar-Yosef, O. 1994. The Lower Paleolithic of the Near East. *Journal of World Prehistory* 8:211–265.

123. Bar-Yosef, O. 1995. The Lower and Middle Paleolithic in the Mediterranean Levant: Chronology, and cultural entities. In *Man and environment in the Palaeolithic*, ed. H. Ullrich, pp. 246–263. Liège: Université de Liège.

124. Bar-Yosef, O. 1995. The role of climate in the interpretation of human movements and cultural transformations. In *Paleoclimate and evolution with emphasis on human origins*, ed. E. S. Vrba, G. H. Denton, T. C. Partridge, and L. H. Burckle, pp. 507–523. New Haven: Yale University Press.

125. Bar-Yosef, O., Arnold, M., Mercier, N., Belfer-Cohen, A., Goldberg, P., Housley, R., Laville, H., Meignen, L., Vogel, J. C., and Vandermeersch, B. 1996. The dating of the Upper Paleolithic layers in Kebara Cave, Mt. Carmel. *Journal of Archaeological Science* 23:297–306.

126. Bar-Yosef, O., and Belfer-Cohen, A. 1988. The early Upper Paleolithic in Levantine caves. *British Archaeological Reports International Series* 437:23–41.

127. Bar-Yosef, O., and Goren-Inbar, N. 1993. The lithic assemblages of 'Ubeidiya: A Lower Paleolithic site in the Jordan Valley. *Oedem* 45:1–266.

128. Bar-Yosef, O., and Meadow, R. H. 1995. The origins of agriculture in the Near East. In *Last hunters, first farmers: New perspectives on the prehistoric transition to agriculture*, ed. T. D. Price and A. B. Gebauer, pp. 39–77. Santa Fe: School of American Research.

129. Bar-Yosef, O., and Meignen, L. 1992. Insights into Middle Paleolithic cultural variability. In *The Middle Paleolithic: Adaptation, behavior, and variability*, ed. H. L. Dibble and P. Mellars, pp. 163–182. Philadelphia: University of Pennsylvania Museum.

130. Bar-Yosef, O., and Vandermeersch, B. 1981. Notes concerning the possible age of the Mousterian layers in Qafzeh Cave. In *Préhistoire du Levant*, ed. J. Cauvin and P. Sanlaville, pp. 281–285. Paris: Centre National de la Recherche Scientifique.

131. Bar-Yosef, O., and Vandermeersch, B., eds. 1991. *Le squelette moustérien de Kébara 2*. Paris: Centre National de la Recherche Scientifique.

132. Bar-Yosef, O., and Vandermeersch, B. 1993. Modern humans in the Levant. *Scientific American* 267 (4): 94–100.

133. Bar-Yosef, O., Vandermeersch, B., Arensburg, B., Belfer-Cohen, A., Goldberg, P., Laville, H., Meignen, L., Rak, Y., Speth, J. D.,

Tchernov, E., Tillier, A. M., and Weiner, S. 1992. The excavations in Kebara Cave, Mt. Carmel. *Current Anthropology* 33:497–550.

134. Bar-Yosef, O., Vandermeersch, B., Arensburg, B., Goldberg, P., Laville, H., Meignen, L., Rak, Y., Tchernov, E., and Tillier, A.-M. 1986. New data on the origin of modern man in the Levant. *Current Anthropology* 27:63–64.

135. Barandiarán, I., Freeman, L. G., González Echegaray, J., and Klein, R. G. 1985. Excavaciones en la cueva del Juyo. *Monografiyas del Centro de Investigacion y Museo de Altamira* 14:1–224.

136. Bard, E., Arnold, M., Fairbanks, R. G., and Hamelin, B. 1993. ^{230}Th/^{234}U and ^{14}C ages obtained by mass spectrometry on corals. *Radiocarbon* 35:191–199.

137. Bard, E., Hamelin, B., Fairbanks, R. G., and Zindler, A. 1990. Calibration of the ^{14}C timescale over the past 30,000 years using mass spectrometric U-Th ages from Barbados corals. *Nature* 345:405–410.

138. Barham, L. S. 1995. The Mumbwa Caves project, Zambia, 1993–94. *Nyame Akuma* 43:66–71.

139. Barham, L. S., and Smart, P. L. 1996. An early date for the Middle Stone Age of central Zambia. *Journal of Human Evolution* 30:287–290.

140. Barry, J., C., Lindsay, E. H., and Jacobs, L. L. 1982. A biostratigraphic zonation of the middle and upper Siwaliks of the Potwar Plateau of north Pakistan. *Paleogeography, Paleoclimatology and Paleoecology* 37:95–130.

141. Barry, J. C., Jacobs, L. L., and Kelley, J. 1986. An early Middle Miocene catarrhine from Pakistan with comments on the dispersal of catarrhines into Eurasia. *Journal of Human Evolution* 15:501–508.

142. Bartlein, P. J., Edwards, M. E., Shafer, S. L., and Barker, E. D. 1995. Calibration of radiocarbon ages and the interpretation of paleoenvironmental records. *Quaternary Research* 44:417–424.

143. Barton, C. M., and Clark, G. A. 1993. Cultural and natural formation processes in late Quaternary cave and rockshelter sites of western Europe and the Near East. *Monographs in World Archaeology* 17:33–52.

144. Bartstra, G.-J. 1983. The fauna from Trinil, type locality of *Homo erectus*: A reinterpretation. *Geologie en Mijnbouw* 62:329–336.

145. Bartstra, G.-J. 1984. Dating the Pacitanian: Some thoughts. *Courier Forschungsinstitut Senckenberg* 69:253–258.

146. Bartstra, G.-J. 1985. Sangiran, the stone implements of Ngebung, and the Paleolithic of Java. *Modern Quaternary Research in Southeast Asia* 9:99–113.

147. Bartstra, G. J. 1982. *Homo erectus erectus:* The search for his artifacts. *Current Anthropology* 23:318–320.

148. Bartstra, G. J. 1982. The river-laid strata near Trinil, site of *Homo erectus erectus*, Java, Indonesia. *Modern Quaternary Research in Southeast Asia* 7:97–130.

149. Baryshnikov, G., and Hoffecker, J. F. 1994. Mousterian hunters of the NW Caucasus: Preliminary results from recent investigations. *Journal of Field Archaeology* 21:1–14.

150. Bassinot, F. C., Labeyrie, L. D., Vincent, E., Quidelleur, X., Shackleton, N. J., and Lancelot, Y. 1994. The astronomical theory

of climate and the age of the Brunhes-Matuyama magnetic reversal. *Earth and Planetary Science Letters* 126:91–108.

151. Bate, D. M. A. 1951. The mammals from Singa and Abu Hugar. *Fossil Mammals of Africa* 2:1–28.

152. Batzer, M. A., Stoneking, M., Alegria-Hartman, M., Bazan, H., Kass, D. H., Shaikh, T. H., Novick, G. E., Ioannou, P. A., Scheer, W. D., Herrera, R. J., and Deininger, P. L. 1994. African origin of human-specific polymorphic *Alu* insertions. *Proceedings of the National Academy of Sciences* 91:12288–12292.

153. Baumler, M. F., and Speth, J. D. 1993. A Middle Paleolithic assemblage from Kunji Cave, Iran. In *The Paleolithic prehistory of the Zagros-Taurus*, ed. D. I. Olszewski and H. L. Dibble. University Museum Symposium Series 5 (University Museum Monograph 83), pp. 1–73. Philadelphia: University Museum, University of Pennsylvania.

154. Beard, K. C. 1990. Do we need a newly proposed order Proprimates? *Journal of Human Evolution* 19:817–820.

155. Beard, K. C., Dagosto, M., Gebo, D. L., and Godinot, M. 1988. Interrelationships among primate higher taxa. *Nature* 331:712–714.

156. Beard, K. C., Krishtalka, L., and Stucky, R. K. 1991. First skulls of the early Eocene primate *Shoshonius cooperi* and the anthropoid-tarsier dichotomy. *Nature* 349:64–66.

157. Beard, K. C., Tao Qi, Dawson, M. R., Wang, B., and Li, C. 1994. A diverse new primate fauna from middle Eocene fissure-fillings in southeastern China. *Nature* 368:604–609.

158. Beard, K. C., and Wang, B. 1991. Phylogenetic and biogeographic significance of the tarsiiform primate *Asiomomys changbaicus* from the Eocene of Jilin Province, People's Republic of China. *American Journal of Physical Anthropology* 85:159–166.

159. Beaulieu, J.-L. de, and Reille, M. 1992. The last climatic cycle at La Grande Pile (Vosges, France). A new pollen profile. *Quaternary Science Reviews* 11:431–438.

160. Beaumont, P. B. 1980. On the age of Border Cave hominids 1–5. *Palaeontologia Africana* 23:21–33.

161. Beaumont, P. B. 1982. Aspects of the Northern Cape Pleistocene Project. *Palaeoecology of Africa* 15:41–44.

162. Beaumont, P. B., De Villiers, H., and Vogel, J. C. 1978. Modern man in sub-Saharan Africa prior to 49,000 years B.P.: A review and evaluation with particular reference to Border Cave. *South African Journal of Science* 74:409–419.

163. Beaumont, P. B., van Zinderen Bakker, E. M., and Vogel, J. C. 1984. Environmental changes since 32 KYRS B.P. at Kathu Pan, Northern Cape, South Africa. In *Late Cenozoic Palaeoclimates of the Southern Hemisphere*, ed. J. C. Vogel, pp. 324–338. Rotterdam: A. A. Balkema.

164. Beaune, S. A. de, and White, R. 1993. Ice Age lamps. *Scientific American* 269 (3): 108–113.

165. Beden, M. 1979. Données récentes sur l'évolution des Proboscidiens pendant le Plio-Pléistocène en Afrique orientale. *Bulletin de la Société Géologique de France* (Paris) 21:271–276.

166. Bednarik, R. G. 1992. Palaeoart and archaeological myths. *Cambridge Archaeological Journal* 2:27–42.

167. Bednarik, R. G. 1995. Concept-mediated marking in the Lower Palaeolithic. *Current Anthropology* 36:605–634.

168. Begun, D., and Walker, A. 1993. The endocast. In *The Nariokotome* Homo erectus *skeleton,* ed. A. Walker and R. Leakey, pp. 326–358. Cambridge: Harvard University Press.

169. Begun, D. R. 1991. European Miocene catarrhine diversity. *Journal of Human Evolution* 20:521–536.

170. Begun, D. R. 1992. *Dryopithecus crusafonti* sp. nov., a new Miocene hominoid species from Can Ponsic (northeastern Spain). *American Journal of Physical Anthropology* 67:291–310.

171. Begun, D. R. 1992. Miocene fossil hominids and the chimp-human clade. *Science* 257:1929–1933.

172. Begun, D. R. 1994. Relations among the great apes and humans: New interpretations based on the fossil great ape *Dryopithecus. Yearbook of Physical Anthropology* 37:11–64.

173. Begun, R. D. 1989. A large pliopithecine molar from Germany and some notes on the Pliopithecinae. *Folia Primatologia* 52: 156–166.

174. Behm-Blancke, G. 1960. Altsteinzeitliche Restplätze in Travertingebeit von Taubach, Weimar, Ehringsdorf. *Alt Thüringen* 4:1–246.

175. Behrensmeyer, A. K. 1976. Lothagam Hill, Kanapoi, and Ekora: A general summary of stratigraphy and faunas. In *Earliest man and environments in the Lake Rudolf Basin,* ed. Y. Coppens, F. C. Howell, G. L. Isaac, and R. E. F. Leakey, pp. 163–170. Chicago: University of Chicago Press.

176. Behrensmeyer, A. K. 1986. Comment on "Systematic butchery by Plio/Pleistocene hominids at Olduvai Gorge, Tanzania" by H. T. Bunn and E. Kroll. *Current Anthropology* 5:443–444.

177. Behrensmeyer, A. K., Gordon, K. D., and Yanagi, G. T. 1986. Trampling as a cause of bone surface damage and pseudo cut marks. *Nature* 319:768–771.

178. Behrensmeyer, A. K., Potts, R., Plummer, T. W., Tauxe, L., Opdyke, N., and Jorstad, T. 1995. The Pleistocene locality of Kanjera, Western Kenya: Stratigraphy, chronology and paleoenvironments. *Journal of Human Evolution* 29:247–274.

179. Behrensmeyer, A. K., Todd, N. E., Potts, R., and McBrinn, G. E. 1997. Late Pliocene faunal turnover in the Turkana Basin, Kenya and Ethiopia. *Science* 278:1589–1594.

180. Belfer-Cohen, A., and Hovers, E. 1992. In the eye of the beholder: Mousterian and Natufian burials in the Levant. *Current Anthropology* 33:463–471.

181. Belitsky, S., Goren-Inbar, N., and Werker, E. 1991. A Middle Pleistocene wooden plank with man-made polish. *Journal of Human Evolution* 20:349–353.

182. Bell, W. T. 1991. Thermoluminescence dates for the Lake Mungo aboriginal fireplaces and the implications for the radiocarbon time scale. *Archaeometry* 33:43–50.

183. Belli, G., Bulluomni, G., Cassoli, P. F., Cecchi, S., Cucarzi, M., Delitala, L., Fornaciari, G., Mallegni, F., Piperno, M., Segre, A. G., and Segre-Naldini, E. 1991. Découverte d'un fémur humain Acheuléen à Notarchirico (Venosa, Basilicate). *Anthropologie* 95: 47–88.

184. Bellomo, R. V. 1994. Methods of determining early hominid behavioral activities associated with the controlled use of fire at FxJj 20 Main, Koobi Fora, Kenya. *Journal of Human Evolution* 27:173–195.

185. Bellwood, P. 1990. From late Pleistocene to early Holocene in Sundaland. In *The World at 18 000 BP*, vol. 2, *Low latitudes,* ed. C. Gamble and O. Soffer, pp. 255–263. London: Unwin Hyman.

186. Benefit, B. R., and McCrossin, M. L. 1997. Earliest known Old World monkey skull. *Nature* 388:368–371.

187. Benefit, B. R., and Pickford, M. 1986. Miocene fossil cercopithecoids from Kenya. *American Journal of Physical Anthropology* 69:441–464.

188. Berger, A. L. 1992. Astronomical theory of paleoclimates and the Last Glacial–Interglacial cycle. *Quaternary Science Reviews* 11:571–581.

189. Berger, L. R. 1993. A preliminary estimate of the age of the Gladysvale australopithecine site. *Paleontologia Africana* 30:51–55.

190. Berger, L. R., and Clarke, R. J. 1995. Eagle involvement in the accumulation of the Taung child fauna. *Journal of Human Evolution* 29:275–200.

191. Berger, L. R., Keyser, A. W., and Tobias, P. V. 1993. Gladysvale: First early hominid site discovered in South Africa since 1948. *American Journal of Physical Anthropology* 92:107–111.

192. Berger, L. R., and Parkington, J. E. 1995. A new Pleistocene hominid-bearing locality at Hoedjiespunt, South Africa. *American Journal of Physical Anthropology* 98:601–609.

193. Berger, R. 1979. Radiocarbon dating with accelerators. *Journal of Archaeological Science* 6:101–104.

194. Berger, T. D., and Trinkaus, E. 1995. Patterns of trauma among the Neandertals. *Journal of Archaeological Science* 22:841–852.

195. Bergman, C. A., and Stringer, C. B. 1989. Fifty years after: Egbert, an early Upper Palaeolithic juvenile from Ksar Akil, Lebanon. *Paléorient* 15:99–111.

196. Bergounioux, F. M. 1958. "Spiritualité" de l'homme de Néandertal. In *Hundert Jahre Neanderthaler,* ed. G. H. R. von Koenigswald, pp. 151–166. Utrecht: Kemink.

197. Bermúdez de Castro, J. M. 1988. Dental remains from Atapuerca/Ibeas (Spain) II. Morphology. *Journal of Human Evolution* 17:279–304.

198. Bermúdez de Castro, J. M. 1993. The Atapuerca dental remains: New evidence (1987–1991 excavations) and interpretations. *Journal of Human Evolution* 24:339–371.

199. Bermúdez de Castro, J. M. 1996. European Middle Pleistocene mortality patterns: The case of the Atapuerca-SH hominids. In *The last Neandertals, the first anatomically modern humans,* ed. E. Carbonell and M. Vaquero, pp. 21–38. Tarragona, Spain: Fundacio Catalana per a la Recerca.

200. Bermúdez de Castro, J. M., Arsuaga, J. L., Carbonell, E., Rosas, A., Martínez, I., and Mosquera, M. 1997. A hominid from the Lower Pleistocene of Atapuerca, Spain: Possible ancestor to Neandertals and modern humans. *Science* 276:1392–1395.

201. Bermúdez de Castro, J. M., and Nicolás, M. E. 1995. Posterior dental size reduction in hominids: The Atapuerca hominids. *American Journal of Physical Anthropology* 96:335–356.

202. Bermúdez de Castro, J. M., and Nicolás, M. E. 1997. Palaeo-demography of the Atapuerca-SH Middle Pleistocene hominid sample. *Journal of Human Evolution* 33:333–355.

203. Bernaldo de Quiros Guidotti, F. 1982. Los inicios del Paleolitico superior Cantabrico. *Centro de Investigacion y Museo de Altamira, Monografiyas* 8:1–347.

204. Beynon, A. D., and Wood, B. A. 1987. Patterns and rates of enamel growth in the molar teeth of early hominids. *Nature* 326:493–496.

205. Beyries, S. 1986. Approche fonctionelle de l'outillage provenant d'un site Paléolithique moyen du Nord de la France. In *Chronostratigraphie et faciés culturels du Paléolithique inférieur et moyen dans l'Europe du Nord-Ouest*, ed. A. Tuffreau and J. Sommé, pp. 219–224. Paris: Supplément au Bulletin de l'Association Française pour l'Étude du Quaternaire.

206. Beyries, S. 1987. *Variabilité de l'industrie lithique au Moustérien: Approche fonctionelle sur quelques gisements.* International Series. Oxford: British Archaeological Reports.

207. Beyries, S. 1988. Functional variability of lithic sets in the Middle Paleolithic. In *Upper Pleistocene prehistory of western Eurasia*, ed. H. L. Dibble and A. Montet-White, pp. 213–224. Monograph 54. Philadelphia: University of Pennsylvania Museum.

208. Biberson, P. 1964. La place des hommes du Paléolithique marocain dans la chronologie du Pléistocène atlantique. *Anthropologie* 68:475–526.

209. Bickerton, D. 1990. *Language and species.* Chicago: University of Chicago Press.

210. Bietti, A. 1985. A late Rissian deposit in Rome: Rebibbia-Casal de'Pazzi. In *Ancestors: The hard evidence*, ed. E. Delson, pp. 271–282. New York: Alan R. Liss.

211. Bietti, A. 1997. The transition to anatomically modern humans: The case of peninsular Italy. In *Conceptual issues in modern human origins research*, ed. G. A. Clark and C. Willermet, pp. 132–147. New York: Aldine de Gruyter.

212. Billy, G. 1982. Les dents humains de la grotte du Coupe-Gorge à Montmaurin. *Bulletins et Mémoires de la Société d'Anthropologie de Paris* 9:211–225.

213. Billy, G., and Vallois, H. V. 1977. La mandibule pré-Rissienne de Montmaurin. *Anthropologie* 81:273–312.

214. Bilsborough, A. 1986. Diversity, evolution and adaptation in early hominids. In *Stone Age prehistory: Studies in memory of Charles McBurney*, ed. G. N. Bailey and P. Callow, pp. 197–220. Cambridge: Cambridge University Press.

215. Binford, L. R. 1973. Interassemblage variability—the Mousterian and the "functional" argument. In *The exploration of culture change*, ed. C. Renfrew, pp. 227–254. Pittsburgh: University of Pittsburgh Press.

216. Binford, L. R. 1981. *Bones: Ancient men and modern myths.* New York: Academic Press.

217. Binford, L. R. 1982. Comment on "Rethinking the Middle/Upper Paleolithic transition" by R. White. *Current Anthropology* 23:177–183.

218. Binford, L. R. 1984. *Faunal remains from Klasies River Mouth.* Orlando, Fla.: Academic Press.

219. Binford, L. R. 1985. Human ancestors: Changing views of their behavior. *Journal of Anthropological Archaeology* 4:292–327.

220. Binford, L. R. 1989. Isolating the transition to cultural adaptations: An organizational approach. In *The emergence of modern humans*, ed. E. Trinkaus, pp. 18–41. Cambridge: Cambridge University Press.

221. Binford, L. R., and Binford, S. R. 1966. A preliminary analysis of functional variability in the Mousterian of Levallois Facies. *American Anthropologist* 68 (2): 238–295.

222. Binford, L. R., and Ho, C. K. 1985. Taphonomy at a distance: Zhoukoudian, "the cave home of Beijing Man"? *Current Anthropology* 26:413–442.

223. Binford, L. R., Mills, M. G. L., and Stone, N. 1988. Hyena scavenging behavior and its implications for the interpretation of faunal assemblages from FLK 22 (the Zinj floor) at Olduvai Gorge. *Journal of Anthropological Archaeology* 7:99–135.

224. Binford, S. R. 1968. A structural comparison of disposal of the dead in the Mousterian and Upper Paleolithic. *Southwestern Journal of Anthropology* 24:139–151.

225. Binford, S. R., and Binford, L. R. 1969. Stone tools and human behavior. *Scientific American* 220 (4): 70–84.

226. Birdsell, J. B. 1977. The recalibration of a paradigm for the first peopling of Greater Australia. In *Sunda and Sahul: Prehistoric studies in Southeast Asia, Melanesia, and Australia*, ed. J. Allen, J. Golson, and R. Jones, pp. 11–167. London: Academic Press.

227. Bischoff, J. L., Fitzpatrick, J. A., León, L., Arsuaga, J. L., Falguéres, C., Bahain, J. J., and Bullen, T. 1997. Geology and preliminary dating of the hominid-bearing sedimentary fill of the Sima de los Huesos Chamber, Cueva Mayor of the Sierra de Atapuerca, Burgos, Spain. *Journal of Human Evolution* 33:129–154.

228. Bischoff, J. L., Julia, R., and Mora, R. 1988. Uranium-series dating of the Mousterian occupation at Abric Romaní, Spain. *Nature* 332:668–670.

229. Bischoff, J. L., Ludwig, K., Garcia, J. F., Carbonell, E., Vaquero, M., Stafford, T. W. J., and Jull, A. J. T. 1994. Dating of the Basal Aurignacian Sandwich at Abric Romaní (Catalunya, Spain) by radiocarbon and uranium-series. *Journal of Archaeological Science* 21:541–551.

230. Bischoff, J. L., Menking, K. M., Fitts, J. P., and Fitzpatrick, J. A. 1997. Climatic oscillations 10,000–155,000 yr B.P. at Owens Lake, California reflected in glacial rock flour abundance and lake salinity in Core OL-92. *Quaternary Research* 48:313–325.

231. Bischoff, J. L., and Rosenbauer, R. J. 1981. Uranium series dating of human skeletal remains from the Del Mar and Sunnyvale sites, California. *Science* 213:1003–1005.

232. Bischoff, J. L., Soler, N., Marot, J., and Julia, R. 1989. Abrupt Mousterian/Aurignacian boundary at c. 40 ka bp: Accelerator 14C dates from L'Arbreda Cave (Catalunya, Spain). *Journal of Archaeological Science* 16:563–576.

233. Bishop, M. J., and Friday, A. E. 1986. Molecular sequences and hominoid phylogeny. In *Major topics in primate and human evolution*, ed. B. Wood, L. Martin, and P. Andrews, pp. 150–156. Cambridge: University of Cambridge Press.

234. Bishop, W. W. 1971. The late Cenozoic history of east Africa in relation to hominoid evolution. In *Late Cenozoic glacial ages*, ed. K. K. Turekian, pp. 493–527. New Haven: Yale University Press.

235. Bishop, W. W. 1978. The Lake Baringo Basin, Kenya. In *Geological background to fossil man*, ed. W. W. Bishop, pp. 207–373. Toronto: University of Toronto Press.

236. Bishop, W. W., Chapman, G. R., Hill, A., and Miller, J. A. 1971. Succession of Cainozoic vertebrate assemblages from the northern Kenya Rift. *Nature* 233:389–394.

237. Bishop, W. W., Hill, A., and Pickford, M. 1978. Chesowanja: A revised geological interpretation. In *Geological background to fossil man*, ed. W. W. Bishop, pp. 309–336. Toronto: University of Toronto Press.

238. Bishop, W. W., Pickford, M., and Hill, A. 1975. New evidence regarding the Quaternary geology, archeology and hominids of Chesowanja, Kenya. *Nature* 258:204–208.

239. Blackwell, B., and Schwarcz, H. P. 1986. U-series analyses of the lower travertine at Ehringsdorf, DDR. *Quaternary Research* 25:215–222.

240. Blackwell, B., Schwarcz, H. P., and Debénath, A. 1983. Absolute dating of hominids and Palaeolithic artifacts of the cave of La Chaise-de-Vouthon (Charente), France. *Journal of Archaeological Science* 10:493–513.

241. Blanc, A. C. 1958. Torre in Pietra, Saccopastore, Monte Circeo: On the position of the Mousterian in the Pleistocene sequence of the Rome area. In *Hundert Jahre Neanderthaler*, ed. G. H. R. von Koenigswald, pp. 167–174. Utrecht: Kemink.

242. Blumenschine, R. J. 1986. *Early hominid scavenging opportunities: Implications of carcass availability in the Serengeti and Ngorongoro ecosystems*. International Series 283. Oxford: British Archaeological Reports.

243. Blumenschine, R. J. 1987. Characteristics of an early hominid scavenging niche. *Current Anthropology* 38:383–407.

244. Blumenschine, R. J. 1995. Percussion marks, tooth marks, and experimental determinations of the timing of hominid and carnivore access to long bones at FLK *Zinjanthropus*, Olduvai Gorge, Tanzania. *Journal of Human Evolution* 29:21–51.

245. Blumenschine, R. J., and Cavallo, J. A. 1992. Scavenging and human evolution. *Scientific American* 268 (4): 90–96.

246. Blumenschine, R. J., and Masao, F. T. 1991. Living sites at Olduvai Gorge, Tanzania? Preliminary landscape archaeology results in the basal Bed II lake margin zone. *Journal of Human Evolution* 21:451–462.

247. Boaz, N. T., Bernor, R. L., Brooks, A. S., Cooke, H. B. S., de Heinzelin, J., Dechamps, R., Delson, E., Gentry, A. W., Harris, J. W. K., Meylan, P., Pavlakis, P. P., Sanders, W. J., Stewart, K. M., Verniers, J., Williamson, P. G., and Winkler, A. J. 1992. A new evaluation of the significance of the late Neogene Lusso Beds, Upper Semliki Valley, Zaire. *Journal of Human Evolution* 22:505–517.

248. Boaz, N. T., El-Arnauti, A., Gaziry, A. W., de Heinzelin, J., and Boaz, D. D., eds. 1987. *Neogene paleontology and geology of Sahabi*. New York: Alan R. Liss.

249. Boëda, E. 1988. Le concept Levallois et évaluation de son champ

d'application. *Études et Recherches Archéologiques de l'Université de Liège* 31:13–26.

250. Boëda, E. 1991. Approche de la variabilité des systèmes de production lithique des industries du Paléolithique inférieur et moyen: Chronique d'une variabilité attendue. *Techniques et Culture* 17–18:37–79.

251. Boëda, E. 1993. *Le concept Levallois: Variabilité des méthodes.* Paris: Centre National de la Recherche Scientifique.

252. Boëda, E., Geneste, J. M., and Meignen, L. 1990. Identification de chaînes opératoires lithiques du Paléolithique ancien et moyen. *Paléo* 2:43–80.

253. Boesch, C., and Boesch, H. 1989. Hunting behavior of wild chimpanzees in the Taï National Park. *American Journal of Physical Anthropology* 78:547–573.

254. Boesch-Achermann, H., and Boesch, C. 1994. Hominization in the rainforest: The chimpanzee's piece of the puzzle. *Evolutionary Anthropology* 3:9–16.

255. Bonifay, E. 1991. Les premiers industries du Sud-Est de la France et du Massif-Central. In *Les premiers Européens,* ed. E. Bonifay and B. Vandermeersch, pp. 63–80. Paris: Éditions du Comité des Travaux Historiques et Scientifiques.

256. Bonifay, E., Bonifay, M.-F., Panattoni, R., and Tiercelin, J.-J. 1976. Soleihac (Blanzac, Haute-Loire): Nouveau site préhistorique du début du Pléistocène moyen. *Bulletin de la Société Préhistorique Française* 73:293–304.

257. Bonifay, E., and Tiercelin, J.-J. 1977. Existence d'une activité volcanique et tectonique au début du Pléistocène moyen dans le bassin du Puy (Haute-Loire). *Comptes Rendus de l'Académie des Sciences, Paris,* ser. D, 284:2455–2457.

258. Bonis, L. de, Bouvrain, G., and Koufos, G. 1986. Succession and dating of the late Miocene primates of Macedonia. In *Primate evolution,* ed. J. G. Else and P. C. Lee, pp. 107–114. Cambridge: Cambridge University Press.

259. Bonis, L. de, Geraads, D., Guérin, G., Haga, A., Jaeger, J.-J., and Sen, S. 1984. Découverte d'un hominidé fossile dans le Pléistocène de la République de Djibouti. *Comptes Rendus de l'Académie des Sciences, Paris,* ser. D, 299:1097–1100.

260. Bonis, L. de, Geraads, D., Jaeger, J.-J., and Sen, S. 1988. Vertébrés du Pléistocène de Djibouti. *Bulletin de la Société Géologique de France* 4:323–334.

261. Bonis, L. de, Jaeger, J.-J., Coiffait, B., and Coiffait, P.-E. 1988. Découverte du plus ancien primate catarrhinien connu dans l'Éocène supérieur d'Afrique du Nord. *Comptes Rendus de l'Académie des Sciences, Paris,* ser. 2, 306:929–934.

262. Bonis, L. de, Johanson, D., Melentis, J., and White, T. 1981. Variations métriques de la denture chez les hominidés primitifs: Comparaison entre *Australopithecus afarensis* et *Ouranopithecus macedoniensis. Comptes Rendus de l'Académie des Sciences, Paris,* ser. 2, 292:373–376.

263. Bonis, L. de, and Melentis, J. 1984. La position phylétique d'*Ouranopithecus. Courier Forschungsinstitut Senckenberg* 69:13–23.

264. Bonis, L. de, and Melentis, J. 1991. Âge et position phylétique du crâne de Petralona (Grèce). In *Les premiers européens,* ed.

E. Bonifay and B. Vandermeersch, pp. 285–289. Paris: Éditions du Comité des Travaux Historiques et Scientifiques.

265. Bonnefille, R. 1984. Palynological research at Olduvai Gorge. *National Geographic Society Research Reports* 17:227–243.

266. Bonnefille, R. 1994. Palynology and paleoenvironment of east African hominid sites. In *Integrative paths to the past: Paleoanthropological advances in honor of F. Clark Howell*, ed. R. S. Corruccini and R. L. Ciochon, pp. 415–427. Englewood Cliffs, N.J.: Prentice Hall.

267. Bonnefille, R. 1995. A reassessment of the Plio-Pleistocene pollen record of east Africa. In *Paleoclimate and evolution with emphasis on human origins*, ed. E. S. Vrba, G. H. Denton, T. C. Partridge, and L. H. Burckle, pp. 299–310. New Haven: Yale University Press.

268. Bonnichsen, R. D., Stanford, and D., Fastook, J. L. 1987. Environmental change and developmental history of human adaptive patterns: The Paleoindian case. In *The geology of North America*, vol. K-3, ed. W. F. Ruddiman and H. E. Wright Jr., pp. 403–424. Boulder, Colo.: Geological Society of America.

269. Bordaz, J. 1970. *Tools of the Old and New Stone Age*. New York: Natural History Press.

270. Bordes, F. 1952. Sur l'usage probable de la peinture corporelle dans certaines Moustériennes. *Bulletin de la Société Préhistorique Française* 49:169–171.

271. Bordes, F. 1976–1977. Moustérien et Atérien. *Quaternaria* 19: 19–34.

272. Bordes, F. 1981. Comment on "Le Néandertalien de Saint-Césaire." *Recherche* 12:644–645.

273. Bordes, F., and Prat, F. 1965. Observations sur les faunes du Riss et du Würm I en Dordogne. *Anthropologie* 69:31–45.

274. Bordes, F. H. 1947. Étude comparative des différentes techniques de taille du silex et des roches dures. *Anthropologie* 51:1–29.

275. Bordes, F. H. 1953. Essai de classification des industries moustériennes. *Bulletin de la Société Préhistorique Française* 50:457–466.

276. Bordes, F. H. 1961. Mousterian cultures in France. *Science* 134: 803–810.

277. Bordes, F. H. 1961. Typologie du Paléolithique ancien et moyen. *Institut de Préhistoire de l'Université de Bordeaux Mémoire* 1:1–86.

278. Bordes, F. H. 1968. *The Old Stone Age*. New York: McGraw-Hill.

279. Bordes, F. H. 1972. *A tale of two caves*. New York: Harper and Row.

280. Bordes, F. H. 1978. Foreword. In *Early Palaeolithic in south and east Asia*, ed. F. Ikawa-Smith, pp. ix–x. The Hague: Mouton.

281. Bordes, F. H. 1981. Vingt-cinq ans après: Le complexe Moustérien révisité. *Bulletin de la Société Préhistorique Française* 78:77–87.

282. Bordes, F. H., and Lafille, J. 1962. Découverte d'une squelette d'enfant moustérien dans le gisement du Roc de Marsal, commune de Campagne-du-Bugue (Dordogne). *Comptes Rendus de l'Académie des Sciences, Paris*, ser. D, 254:714–715.

283. Boriskovskij, P. I. 1958. The study of Paleolithic dwellings in the USSR (in Russian). *Sovetskaya Arkheologiya* 1:3–19.

284. Boschetto, H. B., Brown, F. H., and McDougall, I. 1992. Stratigraphy of the Lothidok Range, northern Kenya, and K/Ar ages of its Miocene primates. *Journal of Human Evolution* 22:47–71.

285. Boschian, G., Mallegni, F., and Tozzi, C. 1995. The *Homo erectus* site of Visogliano Shelter (Trieste, NE Italy). In *Congreso Internacional de Paleontologia Humana (Orce, September 1995), 3a Circular*, p. 99. Orce, Spain.

286. Boschian, G., and Radmilli, A. M. 1995. Castel di Guido: A Middle Pleistocene butchering site near Rome (Italy). In *Congreso Internacional de Paleontologia Humana (Orce, September 1995), 3a Circular*, p. 100. Orce, Spain.

287. Bosinski, G. 1967. *Die Mittelpaläolithischen Funde im westlichen Mitteleuropa*. Cologne: Fundamenta Reihe A/4.

288. Bosinski, G. 1982. The transition Lower/Middle Paleolithic in northwestern Germany. *British Archaeological Reports International Series* 151:166–175.

289. Bosinski, G. 1986. Chronostratigraphie du Paléolithique inférieur et moyen en Rhénanie. In *Chronostratigraphie et faciés culturels du Paléolithique inférieur et moyen dans l'Europe du Nord-Ouest*, ed. A. Tuffreau and J. Sommé, pp. 15–34. Paris: Supplément au Bulletin de l'Association Française pour l'Étude du Quaternaire.

290. Bosinski, G. 1995. The earliest occupation of Europe: Western central Europe. In *The earliest occupation of Europe*, ed. W. Roebroeks and T. van Kolfschoten, pp. 103–128. Leiden: University of Leiden.

291. Bosinski, G. 1995. Stone artefacts of the European Lower Palaeolithic: A short note. In *The earliest occupation of Europe*, ed. W. Roebroeks and T. van Kolfschoten, pp. 263–268. Leiden: University of Leiden.

292. Bosinski, G., Brunnacker, K., and Turner, E. 1983. Ein Siedlungsbefund des frühen Mittelpaläolithikums von Ariendorf, kr. Neuwied. *Archäologisches Korrespondenzblatt* 13:157–169.

293. Bosinski, G., Gabunia, L., Justus, A., and Vekua, A. 1995. Le site de Dmanisi (Georgie, Caucase). In *Congreso Internacional de Paleontologia Humana (Orce, September 1995), 3a Circular*, p. 96. Orce, Spain.

294. Bosinski, G., and Justus, A. 1996. Die Ausgrabungen in Dmanisi (Georgien, Kaukasus). In Homo erectus heidelbergensis *von Mauer*, ed. K. W. Beinhauer, R. Kraatz, and G. A. Wagner, pp. 93–98. Sigmaringen: Jan Thorbecke.

295. Bouchud, J. 1959. Essai sur le renne et la climatologie du Paléolithique moyen et supérieur. Doctoral thesis, Université de Paris.

296. Boule, M. 1911–1913. L'homme fossile de la Chapelle-aux-Saints. *Annales de Paléontologie* 7:21–56, 85–192.

297. Boule, M. 1911–1913. L'homme fossile de la Chapelle-aux-Saints. *Annales de Paléontologie* 6:11–172.

298. Boule, M. 1911–1913. L'homme fossile de la Chapelle-aux-Saints. *Annales de Paléontologie* 8:1–70.

299. Boule, M., and Vallois, H. V. 1957. *Fossil men*. New York: Dryden Press.

300. Bourdier, F. 1976. Les industries paléolithiques anté-wurmiennes dans le Nord-Ouest. In *La préhistoire française*, ed. H. de Lumley, 1:956–963. Paris: Centre National de la Recherche Scientifique.

301. Bouysonnie, A., Bouysonnie, J., and Bardon, L. 1913. La station moustérienne de la "Bouffia" Bonneval à la Chapelle-aux-Saints. *Anthropologie* 24:609–636.

302. Bouyssonie, J. 1954. Les sépultures moustériennes. *Quaternaria* 1:107–115.

303. Bowcock, A. M., Ruiz-Linares, A., Tomfohrde, J., Minch, E., Kidd, K. R., and Cavalli-Sforza, L. L. 1994. High resolution of human evolutionary trees with polymorphic microsatellites. *Nature* 368:455–457.

304. Bowdler, S. 1977. The coastal colonization of Australia. In *Sunda and Sahul: Prehistoric studies in Southeast Asia, Melanesia, and Australia,* ed. J. Allen, J. Golson, and R. Jones, pp. 205–246. London: Academic Press.

305. Bowdler, S. 1989. The archaeology of aboriginal society. In *The growing scope of human biology,* ed. L. H. Schmitt, L. Freedman, and N. W. Bruce, pp. 179–186. Nedlands, Australia: Centre for Human Biology, University of Western Australia.

306. Bowdler, S. 1990. Peopling Australasia: The "coastal colonization" hypothesis re-examined. In *The emergence of modern humans: An archaeological perspective,* ed. P. Mellars, pp. 327–343. Ithaca: Cornell University Press.

307. Bowdler, S. 1992. *Homo sapiens* in Southeast Asia and the Antipodes: Archaeological versus biological interpretation. In *The evolution and dispersal of modern humans in Asia,* ed. T. Akazawa, K. Aoki, and T. Kimura, pp. 559–589. Tokyo: Hokusen-Sha.

308. Bowler, J. M., Jones, R., Allen, H., and Thorne, A. G. 1970. Pleistocene human remains from Australia: A living site and human cremation from Lake Mungo, western New South Wales. *World Archaeology* 2:39–60.

309. Bowler, J. M., and Thorne, A. G. 1976. Human remains from Lake Mungo: Discovery and excavation of Lake Mungo III. In *The origin of the Australians,* ed. R. L. Kirk and A. G. Thorne, pp. 95–112. Canberra: Australian Institute of Aboriginal Studies.

310. Bowler, J. M., Thorne, A. G., and Pollach, H. A. 1972. Pleistocene man in Australia: Age and significance of the Mungo skeleton. *Nature* 240:48–50.

311. Bown, T. M., and Kraus, M. J. 1988. Geology and paleoenvironment of the Oligocene Jebel Qatrani Formation and adjacent rocks. *United States Geological Survey Professional Paper* 1452:1–64.

312. Bown, T. M., and Larriestra, C. N. 1990. Sedimentary paleoenvironments of fossil platyrrhine localities, Miocene Pinturas Formation, Santa Cruz Province, Argentina. *Journal of Human Evolution* 19:87–119.

313. Bown, T. M., M. J. Kraus, M. J., Wing, S. L., Fleagle, J. G., Tiffney, B. H., Simons, E. L., and Vondra, C. F. 1982. The Fayum primate forest revisited. *Journal of Human Evolution* 11:603–632.

314. Bown, T. M., and Rose, K. D. 1991. Evolutionary relationships of a new genus and three new species of Omomyid primates (Willwood Formation, Lower Eocene, Bighorn Basin, Wyoming). *Journal of Human Evolution* 20:465–480.

315. Boyle, K. V. 1990. *Upper Palaeolithic faunas from south-west France.* Oxford: British Archaeological Reports.

316. Brace, C. L. 1982. Comment on "Upper Pleistocene hominid

evolution in south-central Europe: A review of the evidence and analysis of trends." *Current Anthropology* 23:687–688.

317. Brace, C. L., Ryan, A. S., and Smith, B. D. 1981. Comment on "Tooth wear in La Ferrassie man." *Current Anthropology* 22:426–430.

318. Brain, C. K. 1972. An attempt to reconstruct the behaviour of australopithecines: The evidence for interpersonal violence. *Zoologica Africana* 7:379–401.

319. Brain, C. K. 1976. A reinterpretation of the Swartkrans site and its remains. *South African Journal of Science* 72:141–146.

320. Brain, C. K. 1981. *The hunters or the hunted? An introduction to African cave taphonomy.* Chicago: University of Chicago Press.

321. Brain, C. K. 1982. The Swartkrans site: Stratigraphy of the fossil hominids and a reconstruction of the environment of early *Homo*. In *L'*Homo erectus *et la place de l'homme de Tautavel parmi les hominidés fossiles,* ed. M. A. de Lumley, pp. 676–706. Nice: Premier Congrès International de Paléontologie Humaine.

322. Brain, C. K. 1984. The Terminal Miocene Event: A critical environmental and evolutionary episode. In *Late Cainozoic palaeoclimates of the Southern Hemisphere,* ed. J. C. Vogel, pp. 491–498. Rotterdam: A. A. Balkema.

323. Brain, C. K. 1985. Cultural and taphonomic comparisons of hominids from Swartkrans and Sterkfontein. In *Ancestors: The hard evidence,* ed. E. Delson, pp. 72–75. New York: Alan R. Liss.

324. Brain, C. K. 1985. Interpreting early hominid death assemblages: The rise of taphonomy since 1925. In *Hominid evolution: Past, present and future,* ed. P. V. Tobias, pp. 41–46. New York: Alan R. Liss.

325. Brain, C. K. 1988. New information about the Swartkrans Cave of relevance to "robust" australopithecines. In *Evolutionary history of the "robust" australopithecines,* ed. F. E. Grine, pp. 311–316. New York: Aldine de Gruyter.

326. Brain, C. K. 1993. The occurrence of burnt bones at Swartkrans and their implications for the control of fire by early hominids. *Transvaal Museum Monograph* 8:229–242.

327. Brain, C. K. 1993. Introduction. *Transvaal Museum Monograph* 8:1–5.

328. Brain, C. K. 1993. Structure and stratigraphy of the Swartkrans Cave in the light of the new excavations. *Transvaal Museum Monograph* 8:23–33.

329. Brain, C. K. 1993. A taphonomic overview of the Swartkrans fossil assemblages. *Transvaal Museum Monograph* 8:257–264.

330. Brain, C. K., Churcher, C. S., Clark, J. D., Grine, F. E., Shipman, P., Susman, R. L., Turner, A., and Watson, V. 1988. New evidence of early hominids, their culture and environment from the Swartkrans Cave, South Africa. *South African Journal of Science* 84:828–835.

331. Brain, C. K., and Shipman, P. 1993. The Swartkrans bone tools. *Transvaal Museum Monograph* 8:195–215.

332. Brain, C. K., and Sillen, A. 1988. Evidence from the Swartkrans Cave for the earliest use of fire. *Nature* 336:464–466.

333. Bräuer, G. 1984. The "Afro-European *sapiens* hypothesis," and hominid evolution in east Asia during the Middle and Upper

Pleistocene. *Courier Forschungsinstitut Senckenberg* 69:145–165.

334. Bräuer, G. 1984. A craniological approach to the origin of anatomically modern *Homo sapiens* in Africa and implications for the appearance of modern Europeans. In *The origins of modern humans: A world survey of the fossil evidence*, ed. F. H. Smith and F. Spencer, pp. 327–410. New York: Alan R. Liss.

335. Bräuer, G. 1989. The evolution of modern humans: Recent evidence from southwest Asia. In *The human revolution: Behavioural and biological perspectives in the origins of modern humans*, ed. P. Mellars and C. B. Stringer, pp. 123–154. Edinburgh: Edinburgh University Press.

336. Bräuer, G. 1992. Africa's place in the evolution of *Homo sapiens*. In *Continuity or replacement: Controversies in* Homo sapiens *evolution*, ed. G. Bräuer and F. H. Smith, pp. 83–98. Rotterdam: A. A. Balkema.

337. Bräuer, G. 1994. How different are Asian and African *Homo erectus?* Courier Forschungsinstitut Senckenberg* 171:301–318.

338. Bräuer, G. 1995. The Dmanisi mandible and its affinities to African and Asian hominids. In *Congreso Internacional de Paleontologia Humana (Orce, September 1995), 3a Circular*, p. 96. Orce, Spain.

339. Bräuer, G., Deacon, H. J., and Zipfel, F. 1992. Comment on the new maxillary finds from Klasies River, South Africa. *Journal of Human Evolution* 22:419–422.

340. Bräuer, G., and Leakey, R. E. 1986. The ES-1693 cranium from Eliye Springs, West Turkana, Kenya. *Journal of Human Evolution* 15:289–312.

341. Bräuer, G., Leakey, R. E., and Mbua, E. 1992. A first report on the ER-3884 cranial remains from Ileret/East Turkana, Kenya. In *Continuity or replacement: Controversies in* Homo sapiens *evolution*, ed. G. Bräuer and F. H. Smith, pp. 111–119. Rotterdam: A. A. Balkema.

342. Bräuer, G., and Mbua, E. 1992. *Homo erectus* features used in cladistics and their variability in Asian and African hominids. *Journal of Human Evolution* 22:79–108.

343. Bräuer, G., and Mehlman, M. J. 1988. Hominid molars from a Middle Stone Age level at Mumba Rock Shelter, Tanzania. *American Journal of Physical Anthropology* 75:69–76.

344. Bräuer, G., and Rimbach, K. W. 1990. Late archaic and modern *Homo sapiens* from Europe, Africa, and southwest Asia: Craniometric comparisons and phylogenetic implications. *Journal of Human Evolution* 19:789–807.

345. Bräuer, G., and Schultz, M. 1996. The morphological affinities of the Plio-Pleistocene mandible from Dmanisi, Georgia. *Journal of Human Evolution* 30:445–481.

346. Bräuer, G., and Singer, R. 1996. The Klasies zygomatic bone: Archaic or modern? *Journal of Human Evolution* 30:161–165.

347. Bräuer, G., and Singer, R. 1996. Not outside the modern range. *Journal of Human Evolution* 30:173–174.

348. Bräuer, G., Yokoyama, Y., Falguères, C., and Mbua, E. 1997. Modern human origins backdated. *Nature* 386:337–338.

349. Bray, W. 1988. The Palaeoindian debate. *Nature* 382:107.

350. Brennan, M. U. 1991. Health and disease in the Middle and Upper Paleolithic of southwestern France: A bioarcheological study. Ph.D. diss., New York University.

351. Breuil, H. 1939. Le vrai niveau de l'industrie abbevillienne de la Porte du Bois (Abbeville). *Anthropologie* 41:13–34.

352. Breuil, H. 1952. *Four hundred centuries of cave art*. Montignac: Centre des Études et de Documentation Préhistorique.

353. Bricker, H. M. 1989. Comment on "Grave shortcomings: The evidence for Neandertal burial" by Robert H. Gargett. *Current Anthropology* 30:177–178.

354. Brink, J. S. 1987. The archaeozoology of Florisbad, Orange Free State. *Memoirs van die Nasionale Museum Bloemfontein* 24:1–151.

355. Brink, J. S. 1988. The taphonomy and palaeoecology of the Florisbad spring fauna. *Palaeoecology of Africa* 19:169–179.

356. Brink, J. S., and Deacon, H. J. 1982. A study of a last interglacial shell midden and bone accumulation at Herolds Bay, Cape Province, South Africa. *Palaeoecology of Africa* 15:31–40.

357. Brock, A., McFadden, P. L., and Partridge, T. C. 1977. Preliminary palaeomagnetic results from Makapansgat and Swartkrans. *Nature* 266:249–250.

358. Bromage, T. G. 1987. The biological and chronological maturation of early hominids. *Journal of Human Evolution* 16:257–272.

359. Bromage, T. G., and Dean, M. C. 1985. Re-evaluation of the age at death of immature fossil hominids. *Nature* 317:525–527.

360. Bromage, T. G., Schrenk, F., and Zonneveld, F. W. 1995. Paleoanthropology of the Malawi Rift: An early hominid mandible from the Chiwondo Beds, northern Malawi. *Journal of Human Evolution* 28:71–108.

361. Brooks, A. S. 1996. Behavior and human evolution. In *Contemporary issues in human evolution*, ed. W. E. Meickle, F. C. Howell, and N. G. Jablonski, pp. 135–166. San Francisco: California Academy of Sciences.

362. Brooks, A. S., Hare, P. E., Kokis, J. E., Miller, G. H., Ernst, R. D., and Wendorf, F. 1990. Dating Pleistocene archeological sites by protein diagenesis in ostrich eggshell. *Science* 248:60–64.

363. Brooks, A. S., Helgren, D. M., Cramer, J. S., Franklin, A., Hornyak, W., Keating, J. M., Klein, R. G., Rink, W. J., Schwarcz, H. P., Leith Smith, J. N., Stewart, K., Todd, N. E., Verniers, J., and Yellen, J. E. 1995. Dating and context of three Middle Stone Age sites with bone points in the upper Semliki Valley, Zaire. *Science* 268:548–553.

364. Broom, R. 1938. The Pleistocene anthropoid apes of South Africa. *Nature* 142:377–379.

365. Broom, R. 1949. Another new type of fossil ape-man. *Nature* 163:57.

366. Broom, R., and Schepers, G. W. H. 1946. *The South African fossil ape-men: The Australopithecinae*. Memoir 2. Pretoria: Transvaal Museum.

367. Brothwell, D. R. 1960. Upper Pleistocene human skull from Niah Caves. *Sarawak Museum Journal* 9:323–349.

368. Brown, B., Walker, A., Ward, C. V., and Leakey, R. E. 1993. New *Australopithecus boisei* calvaria from east Lake Turkana, Kenya. *American Journal of Physical Anthropology* 91:137–159.

369. Brown, F., and Feibel, C. S. 1991. Stratigraphy, depositional environments and palaeogeography of the Koobi Fora Formation. In *Koobi Fora research project*, vol. 3, *The fossil ungulates: Geology, fossil artiodactyls, and palaeoenvironments*, ed. J. M. Harris, pp. 1–30. Oxford: Clarendon Press.

370. Brown, F. H. 1994. Development of Pliocene and Pleistocene chronology of the Turkana Basin, east Africa, and its relation to other sites. In *Integrative paths to the past: Paleoanthropological advances in honor of F. Clark Howell*, ed. R. S. Corruccini and R. L. Ciochon, pp. 285–312. Englewood Cliffs, N.J.: Prentice Hall.

371. Brown, F. H. 1995. The potential of the Turkana Basin for paleoclimatic reconstruction in east Africa. In *Paleoclimate and evolution with emphasis on human origins*, ed. E. S. Vrba, G. H. Denton, T. C. Partridge, and L. H. Burkle, pp. 319–330. New Haven: Yale University Press.

372. Brown, F. H., and Feibel, C. S. 1986. Revision of lithostratigraphic nomenclature in the Koobi Fora region, Kenya. *Journal of the Geological Society, London* 143:297–310.

373. Brown, F. H., Harris, J., Leakey, R., and Walker, A. 1985. Early *Homo erectus* skeleton from west Lake Turkana, Kenya. *Nature* 316:788–792.

374. Brown, F. H., Howell, F. C., and Eck, G. G. 1978. Observations on problems of correlation of late Cenozoic hominid-bearing formations in the north Lake Turkana Basin. In *Geological background to fossil man*, ed. W. W. Bishop, pp. 473–498. Toronto: University of Toronto Press.

375. Brown, F. H., McDougall, I., Davies, T., and Maier, R. 1985. An integrated chronology for the Turkana Basin. In *Ancestors: The hard evidence*, ed. E. Delson, pp. 82–90. New York: Alan R. Liss.

376. Brown, P. 1987. Pleistocene homogeneity and Holocene size reduction: The Australian human skeletal evidence. *Archaeology in Oceania* 22:41–67.

377. Brown, P. 1993. Recent human evolution in east Asia and Australasia. In *The origin of modern humans and the impact of chronometric dating*, ed. M. J. Aitken, C. B. Stringer, and P. A. Mellars, pp. 217–233. Princeton: Princeton University Press.

378. Brugal, J.-P. 1986. Un nouveau site Acheuléen au Kenya: Isenya (Kajiado-District). Contribution à la connaissance des paléoenvironments des hominidés du Pléistocène moyen. In *Changements globaux en Afrique durant le Quaternaire*, ed. H. Faure, L. Faure, and E. S. Diop, pp. 53–56. Paris: Éditions de l'Orstom.

379. Brunet, M., Beauvilain, A., Coppens, Y., Heintz, E., Moutaye, A. H. E., and Pilbeam, D. 1995. The first australopithecine 2,500 kilometres west of the Rift Valley (Chad). *Nature* 378:273–275.

380. Brunet, M., Beauvilain, A., Coppens, Y., Heintz, E., Moutaye, A. H. E., and Pilbeam, D. 1996. *Australopithecus bahrelghazali*, une nouvelle espèce d'hominidé ancien de la région de Koro Toro (Tchad). *Comptes Rendus de l'Académie des Sciences, Paris*, ser. 2A, 322:907–913.

381. Brunnacker, K. 1975. The mid-Pleistocene of the Rhine Basin. In *After the australopithecines*, ed. K. W. Butzer and G. L. Isaac, pp. 189–224. The Hague: Mouton.

382. Buchanan, W. F., Hall, S. L., Henderson, J., Olivier, A., Pettigrew, J. M., Parkington, J. E., and Robertshaw, P. T. 1978. Coastal shell

middens in the Paternoster area, southwestern Cape. *South African Archaeological Bulletin* 33:89–93.

383. Bunn, H. T. 1981. Archaeological evidence for meat-eating by Plio-Pleistocene hominids from Koobi Fora and Olduvai Gorge. *Nature* 291:574–577.

384. Bunn, H. T. 1983. Evidence on the diet and subsistence patterns of Plio-Pleistocene hominids at Koobi Fora, Kenya and at Olduvai Gorge, Tanzania. *British Archaeological Reports* 163:21–30.

385. Bunn, H. T. 1986. Patterns of skeletal representation and hominid subsistence activities at Olduvai Gorge, Tanzania and Koobi Fora, Kenya. *Journal of Human Evolution* 15:673–690.

386. Bunn, H. T. 1991. A taphonomic perspective in the archaeology of human origins. *Annual Review of Anthropology* 20:433–467.

387. Bunn, H. T., and Ezzo, J. A. 1993. Hunting and scavenging by Plio-Pleistocene hominids: Nutritional constraints, archaeological patterns, and behavioural implications. *Journal of Archaeological Science* 20:365–398.

388. Bunn, H. T., Harris, J. W. K., Isaac, G. L., Kaufulu, Z., Kroll, E., Schick, K., Toth, N., and Behrensmeyer, A. K. 1980. FxJj 50: An early Pleistocene site in northern Kenya. *World Archaeology* 12:109–136.

389. Bunn, H. T., and Kroll, E. M. 1986. Systematic butchery by Plio/Pleistocene hominids at Olduvai Gorge, Tanzania. *Current Anthropology* 5:431–452.

390. Burney, D. A., and MacPhee, R. D. E. 1988. Mysterious island: What killed Madagascar's large native animals? *Natural History* 97(7):46–55.

391. Butler, P. M. 1986. Problems of dental evolution in the higher primates. In *Major topics in primate and human evolution,* ed. B. Wood, L. Martin, and P. Andrews, pp. 90–106. Cambridge: University of Cambridge Press.

392. Butzer, K. W. 1965. Acheulian occupation sites at Torralba and Ambrona, Spain: Their geology. *Science* 150:1718–1722.

393. Butzer, K. W. 1971. *Environment and archeology: An ecological approach to prehistory.* Chicago: Aldine-Atherton.

394. Butzer, K. W. 1973. A provisional interpretation of the sedimentary sequence from Montagu Cave (Cape Province), South Africa. *University of California Anthropological Records* 28:89–92.

395. Butzer, K. W. 1974. Paleoecology of South African australopithecines: Taung revisited. *Current Anthropology* 15:367–382.

396. Butzer, K. W. 1976. Lithostratigraphy of the Swartkrans Formation. *South African Journal of Science* 72:136–141.

397. Butzer, K. W. 1976. The Mursi, Nkalabong and Kibish Formations, lower Omo Basin, Ethiopia. In *Earliest man and environments in the Lake Rudolf Basin: Stratigraphy, paleoecology, and evolution,* ed. Y. Coppens, F. C. Howell, G. L. Isaac, and R. E. F. Leakey, pp. 12–23. Chicago: University of Chicago Press.

398. Butzer, K. W. 1978. Climate patterns in an un-glaciated continent. *Geographical Journal* 51:201–208.

399. Butzer, K. W. 1980. The Taung australopithecine: Contextual evidence. *Palaeontologia Africana* 23:201–208.

400. Butzer, K. W. 1981. Cave sediments, Upper Pleistocene stratigraphy and Mousterian facies in Cantabrian Spain. *Journal of Archaeological Science* 8:133–183.

401. Butzer, K. W. 1984. Archeogeology and Quaternary environment in the interior of southern Africa. In *Southern African prehistory and paleoenvironments*, ed. R. G. Klein, pp. 1–64. Rotterdam: A. A. Balkema.

402. Butzer, K. W. 1986. Paleolithic adaptations and settlement in Cantabrian Spain. *Advances in World Archaeology* 6:201–252.

403. Butzer, K. W. 1988. Sedimentological interpretation of the Florisbad spring deposits, South Africa. *Palaeoecology of Africa* 19: 181–189.

404. Butzer, K. W., Beaumont, P. B., and Vogel, J. C. 1978. Lithostratigraphy of Border Cave, KwaZulu, South Africa: A Middle Stone Age sequence beginning c. 195,000 B.P. *Journal of Archaeological Science* 5:317–341.

405. Butzer, K. W., Brown, F. H., and Thurber, D. L. 1969. Horizontal sediments of the lower Omo Valley: Kibish Formation. *Quaternaria* 11:15–29.

406. Butzer, K. W., Stuckenrath, R., Bruzewicz, A. J., and Helgren, D. M. 1978. Late Cenozoic paleoclimates of the Gaap Escarpment, Kalahari Margin, South Africa. *Journal of Archaeological Science* 5: 317–341.

407. Bye, B. A., Brown, F. H., Cerling, T. E., and McDougall, I. 1987. Increased age estimate for the Lower Paleolithic hominid site at Olorgesailie, Kenya. *Nature* 329:237–239.

408. Cabrera Valdés, V., and Bischoff, J. L. 1989. Accelerator 14C dates for early Upper Paleolithic (Basal Aurignacian) at El Castillo Cave (Spain). *Journal of Archaeological Science* 16:577–584.

409. Caccone, A., and Powell, J. R. 1989. DNA divergence among hominoids. *Evolution* 43:925–942.

410. Cahen, D., and Michel, J. 1986. Le site Paléolithique moyen ancien de Mesvin IV (Hainaut, Belgique). In *Chronostratigraphie et faciés culturels du Paléolithique inférieur et moyen dans l'Europe du Nord-Ouest*, ed. A. Tuffreau and J. Sommé, pp. 89–102. Paris: Supplement au Bulletin de l'Association Française pour l'Étude du Quaternaire.

411. Cain, A. J. 1960. *Animal species and their evolution.* New York: Harper and Row.

412. Callow, P. 1986. The flint tools. In *La Cotte de St. Brelade 1961–1978: Excavations by C. B. M. McBurney*, ed. P. Callow and J. M. Cornford, pp. 252–314. Norwich, England: Geo Books.

413. Callow, P., and Cornford, J. M., eds. 1986. *La Cotte de St. Brelade, Jersey: Excavations by C. B. M. McBurney, 1961–1978.* Norwich, England: Geo Books.

414. Callow, P., Walton, D., and Shell, C. A. 1986. The use of fire at La Cotte de St. Brelade. In *La Cotte de St. Brelade, Jersey: Excavations by C. B. M. McBurney, 1961–1978*, pp. 193–195. Norwich, England: Geo Books.

415. Campbell, B. 1966. *Human evolution.* Chicago: Aldine.

416. Camps, G. 1974. *Les civilisations préhistoriques de l'Afrique du Nord et du Sahara.* Paris: Doin.

417. Camps, G. 1975. The prehistoric cultures of north Africa: Radiocarbon chronology. In *Problems in prehistory: North Africa and the Levant*, ed. F. Wendorf and A. E. Marks, pp. 181–192. Dallas: Southern Methodist University Press.

418. Cande, S. C., and Kent, D. V. 1992. A new geomagnetic polarity

time scale for the Late Cretaceous and Cenozoic. *Journal of Geophysical Research* 97:13917–13951.

419. Cann, R. L., Rickards, O., and Lum, J. K. 1994. Mitochondrial DNA and human evolution: Our one Lucky Mother. In *Origins of anatomically modern humans,* ed. M. H. Nitecki and D. V. Nitecki, pp. 135–148. New York: Plenum Press.

420. Cann, R. L., Stoneking, M., and Wilson, A. C. 1987. Mitochondrial DNA and human evolution. *Nature* 329:111–112.

421. Capaldo, S. D. 1997. Experimental determinations of carcass processing by Plio-Pleistocene hominids and carnivores at FLK 22 (*Zinjanthropus*), Olduvai Gorge, Tanzania. *Journal of Human Evolution* 33:555–597.

422. Capaldo, S. D. 1998. Simulating the formation of dual-patterned archaeofaunal assemblages with experimental control samples. *Journal of Archaeological Science* 25:311–330.

423. Capitan, L., and Peyrony, D. 1911. Un nouveau squellete humain fossile. *Revue Anthropologique* 21:148–150.

424. Carbonell, E., Bermúdez de Castro, J. M., Arsuaga, J. L., Diez, J. C., Rosas, A., Cuenca-Bescós, G., Salar, R., Mosquera, M., and Rodríguez, X. P. 1995. Lower Pleistocene hominids and artifacts from Atapuerca-TD6 (Spain). *Science* 269:826–832.

425. Carbonell, E., and Castro-Curel, Z. 1992. Palaeolithic wooden artefacts from the Abric Romaní (Capellades, Barcelona, Spain). *Journal of Archaeological Science* 19:707–719.

426. Carbonell, E., and Rodríguez, X. P. 1994. Early Middle Pleistocene deposits and artefacts in the Gran Dolina site (TD4) of the "Sierra de Atapuerca" (Burgos, Spain). *Journal of Human Evolution* 26:291–311.

427. Carney, J., Hill, A., Miller, J. A., and Walker, A. 1971. Late australopithecine from Baringo District, Kenya. *Nature* 329:111–112.

428. Carretero, J. M., Arsuaga, J. L., and Lorenzo, C. 1997. Clavicles, scapulae and humeri from the Sima de los Huesos site (Sierra de Atapuerca, Spain). *Journal of Human Evolution* 33:357–408.

429. Cartmill, M. 1974. Rethinking primate origins. *Science* 184:436–443.

430. Cartmill, M. 1975. *Primate origins.* Minneapolis: Burgess.

431. Cartmill, M. 1982. Basic primatology and prosimian evolution. In *A history of American physical anthropology, 1930–1980,* ed. F. Spencer, pp. 147–186. New York: Academic Press.

432. Cartmill, M. 1992. New views on primate origins. *Evolutionary Anthropology* 1:105–111.

433. Castro-Curel, Z., and Carbonell, E. 1992. Wood pseudomorphs from Level I at Abric Romaní, Barcelona, Spain. *Journal of Field Archaeology* 22:376–384.

434. Cattani, L., Cremaschi, M., Ferraris, M. R., Mallegni, F., Masini, F., Scola, V., and Tozzi, C. 1991. Le gisement du Pléistocène moyen de Visogliano (Trieste): Restes humains, industries, environnement. *Anthropologie* 95:9–36.

435. Causse, C., Conrad, G., Fontes, J.-C., Gasse, F., Gibert, E., and Kassir, A. 1988. Le dernier "humide" pléistocène du Sahara nord-occidental daterait de 80–100,000 ans. *Comptes Rendus de l'Académie des Sciences, Paris,* ser. 2, 306:1459–1464.

436. Cavalli-Sforza, L. L., Menozzi, P., and Piazza, A. 1994. *The his-*

tory and geography of human genes. Princeton: Princeton University Press.

437. Cerling, T. E. 1992. Development of grasslands and savannas in east Africa during the Neogene. *Palaeogeography, Palaeoclimatology, Palaeoecology* 97:241–247.

438. Cerling, T. E., Harris, J. M., McFadden, B. J., Leakey, M. G., Quade, J., Eisenmann, V., and Ehleringer, J. R. 1997. Global vegetation change through the Miocene/Pliocene boundary. *Nature* 389:153–158.

439. Cerling, T. E., and Hay, R. L. 1986. An isotopic study of paleosol carbonates from Olduvai Gorge. *Quaternary Research* 25:63–78.

440. Cerling, T. E., Quade, J., Wang, Y., Morgan, M. E., Kingston, J. D., and Marino, B. D. 1994. Expansion and emergence of C4 plants. *Nature* 371:112–113.

441. Chabai, V., Marks, A. E., and Yevtushenko, A. 1995. Views of the Crimean Middle Paleolithic: Past and present. *Préhistoire Europeéne* 7:59–80.

442. Chaline, J. 1976. Les rongeurs. In *La préhistoire française*, ed. H. de Lumley, 1:420–424. Paris: Centre National de la Recherche Scientifique.

443. Chaline, J., and Laurin, J. 1986. Phyletic gradualism in a European Plio-Pleistocene *Mimomys* lineage (Arvicolidae, Rodentia). *Paleobiology* 12:203–216.

444. Chamberlain, A. T. 1989. Variations within *Homo habilis.* In *Hominidae: Proceedings of the Second International Congress of Human Paleontology*, ed. G. Giacobini, pp. 175–181. Milan: Jaca Book.

445. Champion, T. C., Gamble, C., Shennan, C., and Whittle, A. 1984. *Prehistoric Europe.* London: Academic Press.

446. Chaplin, G., Jablonski, N. G., and Cable, N. T. 1994. Physiology, thermoregulation and bipedalism. *Journal of Human Evolution* 27:497–510.

447. Chase, P., and Dibble, H. L. 1992. Scientific archaeology and the origins of symbolism: A review of current evidence and interpretations. *Cambridge Archaeological Journal* 2:43–51.

448. Chase, P. G. 1986. *The hunters of Combe Grenal: Approaches to Middle Paleolithic subsistence in Europe.* International Series 286. Oxford: British Archaeological Reports.

449. Chase, P. G. 1988. Scavenging and hunting in the Middle Paleolithic: The evidence from Europe. In *Upper Pleistocene prehistory of western Asia*, ed. H. L. Dibble and A. Montet-White, pp. 225–232. Philadelphia: University Museum, University of Pennsylvania.

450. Chase, P. G. 1989. How different was Middle Palaeolithic subsistence? A zooarchaeological perspective on the Middle to Upper Palaeolithic transition. In *The human revolution: Behavioural and biological perspectives on the origins of modern humans*, ed. P. Mellars and C. Stringer, pp. 321–337. Edinburgh: Edinburgh University Press.

451. Chase, P. G. 1990. Sifflets du Paléolithique moyen(?): Les implications d'un coprolithe de coyote actuel. *Bulletin de la Société Préhistorique Française* 87:165–167.

452. Chase, P. G., and Dibble, H. L. 1987. Middle Paleolithic symbol-

ism: A review of current evidence and interpretations. *Journal of Anthropological Archaeology* 6:263–296.

453. Chauvet, J.-M., Brunel Deschampes, É., and Hillaire, C. 1995. *Dawn of art: The Chauvet Cave.* New York: Harry N. Abrams.

454. Chavaillon, J. 1976. Mission archéologique Franco-Éthiopienne de Melka-Kunturé: Rapport préliminaire. *L'Éthiopie avant Histoire* 1:1–11.

455. Chavaillon, J. 1979. Stratigraphie du site archéologique de Melka-Kunturé (Éthiopie). *Bulletin de la Société Géologique de la France* 21:227–232.

456. Chavaillon, J. 1982. Position chronologique des hominidés fossiles d'Éthiopie. In *L'*Homo erectus *et la place de l'homme de Tautavel parmi les hominidés fossiles,* ed. M. A. de Lumley, pp. 766–797. Nice: Premier Congrès International de Paléontologie Humaine.

457. Chavaillon, J., Brahimi, C., and Coppens, Y. 1974. Première découverte d'hominidé dans l'un des sites acheuléens de Melka-Kunturé (Éthiopie). *Comptes Rendus de l'Académie des Sciences, Paris,* ser. D, 278:3299–3302.

458. Chavaillon, J., Chavaillon, N., Hours, F., and Piperno, M. 1979. From the Oldowan to the Middle Stone Age at Melka Kunturé (Ethiopia): Understanding cultural changes. *Quaternaria* 21:87–114.

459. Chen, C., and Olsen, J. W. 1990. China at the Last Glacial Maximum. In *The world at 18 000 BP,* vol. 1, *High latitudes,* ed. O. Soffer and C. Gamble, pp. 276–295. London: Unwin Hyman.

460. Chen, T., Yang, Q., and Wu, E. 1994. Antiquity of *Homo sapiens* in China. *Nature* 368:55–56.

461. Chen, T., and Yuan, S. 1988. Uranium-series dating of bones and teeth from Chinese Palaeolithic sites. *Archaeometry* 30:59–76.

462. Chen, T., Yuan, S., and Gao, S. 1984. The study of uranium-series dating of fossil bones and an absolute age sequence for the main Paleolithic sites of north China. *Acta Anthropologica Sinica* 3:268–269.

463. Chen, T.-M., Yang, Q., Hu, Y.-Q., Bao, W.-B., and Li, T.-Y. 1997. ESR dating of tooth enamel from Yunxian *Homo erectus* site, China. *Quaternary Science Reviews* 16:455–458.

464. Chernysh, A. P., ed. 1982. *Molodova I: A unique Mousterian settlement on the Middle Dniestr* (in Russian). Moscow: Nauka.

465. Chia, L.-p. 1975. *The cave home of Peking man.* Peking: Foreign Languages Press.

466. Chippindale, C. 1990. Piltdown: Who dunit? Who Cares? *Science* 250:1162–1163.

467. Churchill, S. E., Pearson, O. M., Grine, F. E., Trinkaus, E., and Holliday, T. W. 1996. Morphological affinities of the proximal ulna from Klasies River Main site: Archaic or modern? *Journal of Human Evolution* 31:213–237.

468. Churchill, S. E., and Trinkaus, E. 1990. Neandertal scapular glenoid morphology. *American Journal of Physical Anthropology* 83:147–160.

469. Ciochon, R., Long, V. T., Larick, R., Gonzalez, L., Grün, R., de Vos, J., Yonge, C., Taylor, L., Yoshida, H., and Reagan, M. 1996. Dated co-occurrence of *Homo erectus* and *Gigantopithecus* from Tham Khuyen Cave, Vietnam. *Proceedings of the National Academy of Sciences* 93:3016–3020.

470. Ciochon, R. L. 1983. Hominoid cladistics and the ancestry of modern apes and humans: A summary statement. In *New interpretations of ape and human ancestry*, ed. R. L. Ciochon and R. S. Corruccini, pp. 783–843. New York: Plenum Press.

471. Ciochon, R. L. 1985. Fossil ancestors of Burma. *Natural History* 94 (10):26–37.

472. Ciochon, R. L. 1986. Paleoanthropological and archaeological research in the Socialist Republic of Vietnam. *Journal of Human Evolution* 15:623–633.

473. Ciochon, R. L., and Chiarelli, A. B. 1980. Paleobiogeographic perspectives on the origin of the Platyrrhini. In *Evolutionary biology of the New World monkeys and continental drift*, ed. R. L. Ciochon and A. B. Chiarelli, pp. 459–493. New York: Plenum Press.

474. Ciochon, R. L., Savage, D. E., Tint, T., and Maw, B. 1985. Anthropoid origins in Asia? New discovery of *Amphipithecus* from the Eocene of Burma. *Science* 229:756–759.

475. Clark, G. A. 1992. Continuity or replacement? Putting modern human origins in evolutionary context. In *The Middle Palaeolithic: Adaptation, behavior, and variability*, ed. H. L. Dibble and P. A. Mellars, pp. 626–676. Philadelphia: University of Pennsylvania Museum.

476. Clark, G. A. 1997. The Middle-Upper Paleolithic transition in Europe: An American perspective. *Norwegian Archaeological Review* 30:25–53.

477. Clark, G. A., and Lindly, J. 1991. On paradigmatic biases and Paleolithic research traditions. *Current Anthropology* 32:577–587.

478. Clark, G. A., and Straus, L. G. 1983. Late Pleistocene hunter-gatherer adaptations in Cantabrian Spain. In *Hunter-gatherer economy in prehistory: A European perspective*, ed. G. Bailey, pp. 131–148. Cambridge: Cambridge University Press.

479. Clark, J. D. 1955. A note on a wooden implement from the level of Peat I at Florisbad, Orange Free State. *Navorsinge van die Nasionale Museum (Bloemfontein)* 1:135–140.

480. Clark, J. D. 1958. The natural fracture of pebbles from the Batoka Gorge, northern Rhodesia, and its bearing on the Kafuan Industries of Africa. *Proceedings of the Prehistoric Society* 24:64–77.

481. Clark, J. D. 1959. Further excavations at Broken Hill, northern Rhodesia. *Journal of the Royal Anthropological Institute* 89:201–231.

482. Clark, J. D. 1967. The Middle Acheulian occupation site at Latamne, northern Syria, I. *Quaternaria* 9:1–68.

483. Clark, J. D. 1968. *Further palaeo-anthropological studies in northern Lunda*. Publicaçoes culturais 78. Diamang, Angola: Museu do Dundo.

484. Clark, J. D. 1968. The Middle Acheulian occupation site at Latamne, northern Syria, II. Further excavations (1965): General results, definition, and interpretation. *Quaternaria* 10:1–71.

485. Clark, J. D. 1969. *Kalambo Falls prehistoric site.* Vol. 1. Cambridge: Cambridge University Press.

486. Clark, J. D. 1975. A comparison of the Late Acheulian industries of Africa and the Middle East. In *After the australopithecines*, ed. K. W. Butzer and G. L. Isaac, pp. 605–659. The Hague: Mouton.

487. Clark, J. D. 1982. The cultures of the Middle Palaeolithic/

Middle Stone Age. In *The Cambridge history of Africa*, ed. J. D. Clark, 1:248–341. Cambridge: Cambridge University Press.

488. Clark, J. D. 1987. Transitions: *Homo erectus* and the Acheulian—the Ethiopian sites of Gadeb and the Middle Awash. *Journal of Human Evolution* 16:809–826.

489. Clark, J. D. 1988. The Middle Stone Age of east Africa and the beginnings of regional identity. *Journal of World Prehistory* 2:235–305.

490. Clark, J. D. 1993. Stone artefact assemblages from Members 1–3, Swartkrans Cave. *Transvaal Museum Monograph* 8:167–194.

491. Clark, J. D. 1994. The Acheulian Industrial Complex in Africa and elsewhere. In *Integrative paths to the past: Paleoanthropological advances in honor of F. Clark Howell*, ed. R. S. Corruccini and R. L. Ciochon, pp. 451–469. Englewood Cliffs, N.J.: Prentice Hall.

492. Clark, J. D., Asfaw, B., Assefa, G., Harris, J. W. K., Kurashina, H., Walter, R. C., White, T. D., and Williams, M. A. J. 1984. Palaeoanthropological discoveries in the Middle Awash Valley, Ethiopia. *Nature* 307:423–428.

493. Clark, J. D., Brothwell, D. R., Powers, R., and Oakley, K. P. 1968. Rhodesian man: Notes on a new femur fragment. *Man* 3:107–111.

494. Clark, J. D., de Heinzelin, J., Schick, K. D., Hart, W. K., White, T. D., WoldeGabriel, G., Walter, R. C., Suwa, G., Asfaw, B., Vrba, E., and H-Selassie, Y. 1994. African *Homo erectus*: Old radiometric ages and young Oldowan assemblages in the Middle Awash Valley, Ethiopia. *Nature* 264:1907–1910.

495. Clark, J. D., and Harris, J. W. K. 1985. Fire and its roles in early hominid lifeways. *African Archaeological Review* 3:3–27.

496. Clark, J. D., and Haynes, C. V. 1970. An elephant butchery site at Mwanganda's village, Karonga, Malawi and its relevance for Palaeolithic archaeology. *World Archaeology* 1:390–411.

497. Clark, J. D., and Kurashina, H. 1979. Hominid occupation of the east-central highlands of Ethiopia in the Plio-Pleistocene. *Nature* 282:33–39.

498. Clark, J. D., Oakley, K. P., Wells, L. H., and McClelland, J. A. C. 1950. New studies on Rhodesian man. *Journal of the Royal Anthropological Institute* 77:7–32.

499. Clark, J. D., Williamson, K. D., Michels, J. W., and Marean, C. W. 1984. A Middle Stone Age occupation site at Porc Epic Cave, Dire Dawa (east-central Ethiopia). *African Archaeological Review* 2:37–71.

500. Clark, J. G. D. 1967. *The Stone Age hunters*. New York: McGraw-Hill.

501. Clarke, R. J. 1976. New cranium of *Homo erectus* from Lake Ndutu, Tanzania. *Nature* 262:485–487.

502. Clarke, R. J. 1977. A juvenile cranium and some adult teeth of early *Homo* from Swartkrans, Transvaal. *South African Journal of Science* 73:46–49.

503. Clarke, R. J. 1985. A new reconstruction of the Florisbad cranium, with notes on the site. In *Ancestors: The hard evidence*, ed. E. Delson, pp. 301–305. New York: Alan R. Liss.

504. Clarke, R. J. 1988. Habiline handaxes and paranthropine pedigree at Sterkfontein. *World Archaeology* 20:1–12.

505. Clarke, R. J. 1990. The Ndutu cranium and the origin of *Homo sapiens*. *Journal of Human Evolution* 19:699–736.

506. Clarke, R. J. 1994. On some new interpretations of Sterkfontein stratigraphy. *South African Journal of Science* 90:211–214.

507. Clarke, R. J. 1994. The significance of the Swartkrans *Homo* to the *Homo erectus* problem. *Courier Forschungsinstitut Senckenberg* 171:185–193.

508. Clarke, R. J., Howell, F. C., and Brain, C. K. 1970. More evidence of an advanced hominid at Swartkrans. *Nature* 225:1219–1222.

509. Clarke, R. J., and Tobias, P. V. 1995. Sterkfontein Member 2 foot bones of the oldest South African hominid. *Science* 269:521–524.

510. Close, A. E. 1991. On the validity of Middle Paleolithic tool types: A test case from the eastern Sahara. *Journal of Field Archaeology* 18:256–263.

511. Close, A. E., Wendorf, F., and Schild, R. 1990. Patterned use of a Middle Palaeolithic landscape: Bir Tarfawi and Bir Sahara East, eastern Sahara. *Sahara* 3:21–34.

512. Clottes, J. 1993. Paint analyses from several Magdalenian caves in the Ariège region of France. *Journal of Archaeological Science* 20:223–235.

513. Clottes, J. 1996. Thematic changes in Upper Palaeolithic art: A view from the Grotte Chauvet. *Antiquity* 70:276–288.

514. Clottes, J., Chauvet, J.-M., Brunel-Deschamps, E., Hillaire, C., Daugas, J.-P., Arnold, M., Cachier, H., Evin, J., Fortin, P., Oberlin, C., Tisnerat, N., and Valladas, H. 1995. Les peintures paléolithiques de la grotte Chauvet-Pont d'Arc, à Vallon-Pont-d'Arc (Ardèche, France): Datations directes et indirectes par la méthode du radiocarbone. *Comptes Rendus de l'Académie des Science, Paris,* ser. 2a, 320:1133–1140.

515. Clottes, J., and Courtin, J. 1993. Neptune's Ice Age gallery. *Natural History* 102:64–71.

516. Clottes, J., and Courtin, J. 1995. *The cave beneath the sea: Palaeolithic images at Cosquer.* New York: Harry N. Abrams.

517. Clottes, J., Courtin, J., and Valladas, H. 1992. A well-dated Paleolithic cave: The Cosquer Cave at Marseilles. *Rock Art Research* 9:122–129.

518. Clottes, J., Menu, M., and Walter, P. 1990. New light on the Niaux paintings. *Rock Art Research* 7:21–26.

519. Coffing, K., Feibel, C., Leakey, M., and Walker, A. 1994. Four-million-year-old hominids from east Lake Turkana, Kenya. *American Journal of Physical Anthropology* 93:55–65.

520. Cole, G. H. 1967. The later Acheulian and Sangoan of southern Uganda. In *Background to evolution in Africa*, ed. W. W. Bishop and J. D. Clark, pp. 481–528. Chicago: University of Chicago Press.

521. Coles, J. M., and Higgs, E. S. 1969. *The archaeology of early man.* London: Faber and Faber.

522. Colinvaux, P., and West, F. H. 1984. The Beringian ecosystem. *Quarterly Review of Archaeology* 5 (3): 10–16.

523. Colinvaux, P. A. 1996. Reconstructing the environment. In *American beginnings: The prehistory and palaeoecology of Beringia*, ed. F. H. West, pp. 13–19. Chicago: University of Chicago Press.

524. Collins, D. 1969. Culture traditions and environment of early man. *Current Anthropology* 10:267–316.

525. Coltorti, M., Cremaschi, M., Delitala, C., Esu, D., Fornasari, M., McPherron, A., Nicoletti, M., van Otterloo, R., Peretto, C., Sala, B., Schmidt, V., and Sevink, J. 1982. Reversed magnetic polarity at an early Lower Palaeolithic site in central Italy. *Nature* 300: 173–176.

526. Coltorti, M., van Otterloo, R., Sevink, J., Cremaschi, M., Esu, D., Delitala, M. C., Fornasari, M., Nicoletti, M., McPherron, A., Schmidt, V., Sala, B., Peretto, C., Giusberti, G., Guerreschi, A., Bubellini, A., Lombardini, G., Russo, P., and Martinelli, G. 1983. *Isernia la Pineta: Un accampamento più antico di 700.000 anni.* Bologna: Calderini Editore.

527. Combier, J. 1971. Le gisement pré-moustérien et acheuléen d'Orgnac. *Études Préhistoriques* 1:24–26.

528. Combier, J. 1988. Témoins moustériens d'activités volontaires. In *De Néandertal à Cro-Magnon,* ed. J.-B. Roy and A.-S. LeClerc, pp. 69–72. Nemours: Musée de Préhistoire d'Île de France.

529. Commont, V. 1908. Les industries de l'ancien Saint-Acheul. *Anthropologie* 19:527–572.

530. Conard, N. J. 1990. Laminar lithic assemblages from the last interglacial complex in northwestern Europe. *Journal of Anthropological Research* 46:243–262.

531. Conard, N. J. 1992. *Tönchesberg and its position in the Paleolithic prehistory of northern Europe.* Bonn: Dr. Rudolf Habelt.

532. Condemi, S. 1991. Les plus anciens fossiles humains de la péninsule italique. In *Les premiers européens,* ed. E. Bonifay and B. Vandermeersch, pp. 299–306. Paris: Éditions du Comité des Travaux Historiques et Scientifiques.

533. Conkey, M. W. 1981. A century of Palaeolithic cave art. *Archaeology* 34 (4): 11–28.

534. Conkey, M. W. 1983. On the origins of Paleolithic art: A review and some critical thoughts. *British Archaeological Reports International Series* 164:201–227.

535. Conkey, M. W. 1987. New approaches in the search for meaning? A review of research in "Paleolithic art." *Journal of Field Archaeology* 14:412–430.

536. Conroy, G. C. 1990. *Primate evolution.* New York: Norton.

537. Conroy, G. C. 1994. *Otavipithecus,* or How to build a better hominid—not. *Journal of Human Evolution* 27:373–383.

538. Conroy, G. C. 1997. *Reconstructing human origins: A modern synthesis.* New York: Norton.

539. Conroy, G. C., Jolly, C. J., Cramer, D., and Kalb, J. E. 1978. Newly discovered fossil hominid skull from the Afar Depression, Ethiopia. *Nature* 275:67–70.

540. Conroy, G. C., Pickford, M., Senut, B., and Mein, P. 1993. Additional Miocene primates from the Otavi Mountains, Namibia. *Comptes Rendus de l'Académie des Sciences, Paris,* ser. 2, 317: 987–990.

541. Conroy, G. C., Pickford, M., Senut, B., and Mein, P. 1993. Diamonds in the desert: The discovery of *Otavipithecus namibiensis. Evolutionary Anthropology* 2:46–52.

542. Conroy, G. C., Pickford, M., Senut, B., Van Couvering, J., and

Mein, P. 1992. *Otavipithecus namibiensis,* first Miocene hominoid from southern Africa. *Nature* 356:144–148.

543. Conroy, G. C., Senut, B., Gommery, D., Pickford, M., and Mein, P. 1996. New primate remains from the Miocene of Namibia, southern Africa. *American Journal of Physical Anthropology* 99:487–493.

544. Conroy, G. C., and Vannier, M. W. 1987. Dental development of the Taung skull from computerized tomography. *Nature* 329:625–627.

545. Conway, B., McNabb, J., and Ashton, N., eds. 1996. *Excavations at Barnfield Pit, Swanscombe, 1968–72.* London: British Museum.

546. Cook, J. 1991. Comment on "The question of ritual cannibalism at Grotta Guattari." *Current Anthropology* 32:126–127.

547. Cook, J., and Ashton, N. 1991. High Lodge, Mildenhall. *Current Archaeology* 11:133–138.

548. Cook, J., Stringer, C. B., Currant, A. P., Schwarcz, H. P., and Wintle, A. G. 1982. A review of the chronology of the European Middle Pleistocene hominid record. *Yearbook of Physical Anthropology* 25:19–65.

549. Cooke, H. B. S. 1978. Africa: The physical setting. In *Evolution of African mammals,* ed. V. J. Maglio and H. B. S. Cooke, pp. 17–45. Cambridge: Harvard University Press.

550. Cooke, H. B. S. 1984. Horses, elephants and pigs as clues in the African later Cenozoic. In *Late Cainozoic palaeoclimates of the Southern Hemisphere,* ed. J. C. Vogel, pp. 473–482. Rotterdam: A. A. Balkema.

551. Cooke, H. B. S., Malan, B. D., and Wells, L. H. 1943. Fossil man in the Lebombo Mountains, South Africa: The "Border Cave," Ingwavuma District, Zululand. *Man* 45:6–13.

552. Cooke, H. B. S., and Wilkinson, A. F. 1978. Suidae and Tayassuidae. In *Evolution of African mammals,* ed. V. J. Maglio and H. B. S. Cooke, pp. 435–482. Cambridge: Harvard University Press.

553. Coon, C. S. 1962. *The origin of races.* New York: Alfred A. Knopf.

554. Cooper, A., Poinar, H. N., Pääbo, S., Radovcic, J., Debénath, A., Caparros, M., Barroso-Ruiz, C., Bertranpetit, J., Nielsen-Marsh, C., Hedges, R. E. M., and Sykes, B. 1997. Neanderthal genetics. *Science* 177:1021–1023.

555. Copeland, L. 1975. The Middle and Upper Paleolithic of Lebanon and Syria in the light of recent research. In *Problems in prehistory: North Africa and the Levant,* ed. F. Wendorf and A. E. Marks, pp. 317–350. Dallas: Southern Methodist University Press.

556. Coppens, Y., Maglio, V. J., Madden, C. T., and Beden, M. 1978. Proboscidea. In *Evolution of African mammals,* ed. V. J. Maglio and H. B. S. Cooke, pp. 336–367. Cambridge: Harvard University Press.

557. Cornelissen, E., Boven, A., Dabi, A., Hus, J., Ju Yong, K., Keppens, E., Langohr, R., Moeyersons, J., Pasteels, P., Pieters, M., Uytterschaut, H., van Noten, F., and Workineh, H. 1990. The Kapthurin Formation revisited. *African Archaeological Review* 8:23–75.

558. Corruccini, R. S. 1992. Metrical reconsideration of the Skhul IV and IX and Border Cave crania in the context of modern human origins. *American Journal of Physical Anthropology* 87:433–445.

559. Corvinus, G. 1975. Palaeolithic remains at the Hadar in the Afar region. *Nature* 256:468–471.

560. Corvinus, G. 1976. Prehistoric exploration at Hadar, Ethiopia. *Nature* 256:571–572.

561. Cosgrove, R., Allen, J., and Marshall, B. 1990. Palaeo-ecology and Pleistocene human occupation in south central Tasmania. *Antiquity* 64:59–78.

562. Covert, H. H. 1986. Biology of the early Cenozoic primates. In *Comparative primate biology,* ed. D. R. Swindler and J. Erwin, 1:335–359. New York: Alan R. Liss.

563. Cox, A. 1969. Geomagnetic reversals. *Science* 163:217–245.

564. Cox, A. 1972. Geomagnetic reversals—their frequency, their origin and some problems of correlation. In *Calibration of hominoid evolution,* ed. W. W. Bishop and J. A. Miller, pp. 93–105. Edinburgh: Scottish University Press.

565. Cronin, J. E., Sarich, V. M., and Ryder, O. 1984. Molecular evolution and speciation in the lesser apes. In *The lesser apes,* ed. D. J. Chivers, H. Preuschoft, W. Y. Brockelman, and N. Creel, pp. 467–485. Edinburgh: Edinburgh University Press.

566. Crubézy, E., and Trinkaus, E. 1992. Shanidar 1: A case of hyperostotic disease (DISH) in the Middle Paleolithic. *American Journal of Physical Anthropology* 89:411–420.

567. Cuenca-Bescós, G., Conesa, C. L., Canudo, J. I., and Arsuaga, J. L. 1997. Small mammals from Sima de los Huesos. *Journal of Human Evolution* 33:175–190.

568. Czarnetski, A. 1991. Nouvelle découverte d'un fragment de crâne d'un hominidé archaïque dans le sud-ouest de l'Allemagne (rapport préliminaire). *Anthropologie* 95:103–112.

569. D'Errico, F. 1995. Comment on "Concept-mediated marking in the Lower Palaeolithic." *Current Anthropology* 36:618–620.

570. D'Errico, F., and Cacho, C. 1994. Notation versus decoration in the Upper Palaeolithic: A case-study from Tossal de la Roca, Alicante, Spain. *Journal of Archaeological Science* 21:185–200.

571. D'Errico, F., and Villa, P. 1997. Holes and grooves: The contribution of microscopy and taphonomy to the problem of art origins. *Journal of Human Evolution* 33:1–31.

572. Dalrymple, G. B., and Lanphere, M. A. 1969. *Potassium-argon dating: Principles, techniques, and applications to geochronology.* San Francisco: W. H. Freeman.

573. Dart, R. 1949. The predatory implemental technique of *Australopithecus. American Journal of Physical Anthropology* 7:1–38.

574. Dart, R. A. 1925. *Australopithecus africanus:* The man-ape of South Africa. *Nature* 115:195–199.

575. Dart, R. A. 1957. The osteodontokeratic culture of *Australopithecus africanus. Memoirs of the Transvaal Museum* 10:1–105.

576. Dart, R. A., and Craig, D. 1959. *Adventures with the missing link.* New York: Viking Press.

577. Darwin, C. 1871. *The descent of man and selection in relation to sex.* London: John Murray.

578. Dastugue, J. 1982. Les maladies des nos ancêtres. *Recherche* 13:980–988.

579. Dastugue, J., and de Lumley, M.-A. 1976. Les maladies des homme préhistoriques du Paléolithique et du Mésolithique. In *La*

préhistoire française, ed. H. de Lumley, 1:612–622. Paris: Centre National de la Recherche Scientifique.

580. David, J. H. M. 1989. Seals. In *Oceans of life off southern Africa,* ed. A. I. L. Payne and R. J. H. Crawford, pp. 288–302. Cape Town: Vlaeberg.

581. Davidson, I. 1990. Bilzingsleben and early marking. *Rock Art Research* 7:52–56.

582. Davidson, I., and Noble, W. 1989. The archaeology of perception: Traces of depiction and language. *Current Anthropology* 30:125–155.

583. Davidson, I., and Noble, W. 1992. Why the first colonisation of the Australian region is the earliest evidence of modern human behaviour. *Archaeology in Oceania* 27:135–142.

584. Davis, R., Ranov, V. A., and Dodonov, A. E. 1980. Early man in Soviet central Asia. *Scientific American* 243 (6): 130–137.

585. Davis, S. J. M., Rabinovich, R., and Goren-Inbar, N. 1988. Quaternary extinctions and population increase in western Asia: The animal bones from Biq'at Quneitra. *Paléorient* 14:95–105.

586. Dawkins, R. 1987. *The blind watchmaker.* New York: W. W. Norton.

587. Dawson, J. E., and Trinkaus, E. 1997. Vertebral osteoarthritis of the La Chapelle-aux-Saints 1 Neanderthal. *Journal of Archaeological Science* 24:1015–1021.

588. Day, M. H. 1971. Postcranial remains of *Homo erectus* from Bed IV, Olduvai Gorge, Tanzania. *Nature* 232:383–387.

589. Day, M. H. 1972. The Omo human skeletal remains. In *The origin of* Homo sapiens, ed. F. H. Bordes, pp. 31–35. Paris: UNESCO.

590. Day, M. H. 1982. The *Homo erectus* pelvis: Punctuation or gradualism? In *L'*Homo erectus *et la place de l'homme de Tautavel parmi les hominidés fossiles,* ed. M. A. de Lumley, pp. 411–421. Nice: Premier Congrès International de Paléontologie Humaine.

591. Day, M. H. 1984. The postcranial remains of *Homo erectus* from Africa, Asia, and possibly Europe. *Courier Forschungsinstitut Senckenberg* 69:113–121.

592. Day, M. H. 1985. Hominid locomotion—from Taung to the Laetoli footprints. In *Hominid evolution: Past, present and future,* ed. P. V. Tobias, pp. 115–127. New York: Alan R. Liss.

593. Day, M. H. 1986. Bipedalism: Pressures, origins and modes. In *Major topics in primate and human evolution,* ed. B. A. Wood, L. Martin, and P. Andrews, pp. 188–202. Cambridge: Cambridge University Press.

594. Day, M. H. 1986. *Guide to fossil man.* Chicago: University of Chicago Press.

595. Day, M. H. 1995. Continuity and discontinuity in the postcranial remains of *Homo erectus. Études et Recherches Archéologiques de l'Université de Liège* 62:181–190.

596. Day, M. H., Leakey, M. D., and Magori, C. 1980. A new hominid fossil skull (L.H. 18) from the Ngaloba Beds, Laetoli, northern Tanzania. *Nature* 284:55–56.

597. Day, M. H., and Stringer, C. B. 1982. A reconsideration of the Omo-Kibish remains and the *erectus-sapiens* transition. In *L'*Homo erectus *et la place de l'homme de Tautavel parmi les hominidés fossiles,* ed. M. A. de Lumley, pp. 814–846. Nice: Centre National de la Recherche Scientifique.

598. Day, M. H., and Stringer, C. B. 1991. Les restes craniens d'Omo-Kibish et leur classification à l'intérieur du genre *Homo. Anthropologie* 95:373–394.

599. de Jong, W. W., and Goodman, M. 1988. Anthropoid affinities of *Tarsius* supported by lens alpha A-crystallin sequences. *Journal of Human Evolution* 17:575–582.

600. de Mortillet, G. 1883. *Le préhistorique: Antiquité de l'homme.* Paris: C. Reinwald.

601. De Villiers, H. 1973. Human skeletal remains from Border Cave, Ingwavuma District, KwaZulu, South Africa. *Annals of the Transvaal Museum* 28:229–256.

602. De Villiers, H. 1976. A second adult human mandible from Border Cave, Ingwavuma District, KwaZulu, South Africa. *South African Journal of Science* 72:212–215.

603. de Vos, J., Sondaar, P., and Swisher, C. C. 1994. Dating hominid sites in Indonesia. *Science* 266:1726–1727.

604. Deacon, H. J. 1966. The dating of the Nahoon footprints. *South African Journal of Science* 62:111–113.

605. Deacon, H. J. 1970. The Acheulian occupation at Amanzi Springs, Uitenhage District, Cape Province. *Annals of the Cape Provincial Museums (Natural History)* 8:89–189.

606. Deacon, H. J. 1989. Late Pleistocene paleoecology and archaeology in the southern Cape, South Africa. In *The human revolution: Behavioural and biological perspectives on the origins of modern humans,* ed. P. A. Mellars and C. B. Stringer, pp. 547–564. Edinburgh: Edinburgh University Press.

607. Deacon, H. J. 1995. Two late Pleistocene-Holocene archaeological depositories from the southern Cape, South Africa. *South African Archaeological Bulletin* 50:121–131.

608. Deacon, H. J., and Geleijnse, V. B. 1988. The stratigraphy and sedimentology of the main site sequence, Klasies River, South Africa. *South African Archaeological Bulletin* 43:5–14.

609. Deacon, H. J., and Shuurman, R. 1992. The origins of modern people: The evidence from Klasies River. In *Continuity or replacement: Controversies in* Homo sapiens *evolution,* ed. G. Bräuer and F. H. Smith, pp. 121–129. Rotterdam: A. A. Balkema.

610. Deacon, H. J., and Thackeray, J. F. 1984. Late Pleistocene environmental changes and implications for the archaeological record in southern Africa. In *Late Cainozoic palaeoclimates of the Northern Hemisphere,* ed. J. C. Vogel, pp. 375–390. Rotterdam: A. A. Balkema.

611. Deacon, J. 1966. An annotated list of radiocarbon dates for sub-Saharan Africa. *Annals of the Cape Provincial Museums* 5:5–84.

612. Deacon, J. 1984. Later Stone Age people and their descendants in southern Africa. In *Southern African prehistory and paleoenvironments,* ed. R. G. Klein, pp. 221–328. Rotterdam: A. A. Balkema.

613. Deacon, J. 1995. An unsolved mystery at the Howieson's Poort name site. *South African Archaeological Bulletin* 50:110–120.

614. Deacon, T. W. 1992. Brain-language coevolution. In *The evolution of human languages,* ed. J. A. Hawkins and M. Gell-Mann, pp. 49–83. Redwood City, Calif.: Addison-Wesley.

615. Deacon, T. W. 1998. *The symbolic species: The co-evolution of language and the brain.* New York: W. W. Norton.

616. Dean, D., and Delson, E. 1992. Second gorilla or third chimp? *Nature* 359:676–677.

617. Dean, D., and Delson, E. 1995. *Homo* at the gates of Europe. *Nature* 373:472–473.

618. Dean, D., Hublin, J.-J., Ziegler, R., and Holloway, R. 1994. The Middle Pleistocene pre-Neanderthal partial skull from Reilingen (Germany). *American Journal of Physical Anthropology*, suppl., 18:77.

619. Dean, M. C., Stringer, C. B., and Bromage, T. C. 1986. Age at death of the Neanderthal child from Devil's Tower, Gibraltar and the implications for studies of general growth and development in Neanderthals. *American Journal of Physical Anthropology* 70:301–310.

620. Debénath, A. 1976. Les civilisations du Paléolithique inférieur en Charente. In *La préhistoire française*, ed. H. de Lumley, 1 (2):929–935. Paris: Centre National de la Recherche Scientifique.

621. Debénath, A. 1977. The latest finds of ante-Würmian human remains in Charente (France). *Journal of Human Evolution* 6:297–302.

622. Debénath, A. 1980. Nouveaux restes humains atériens du Maroc. *Comptes Rendus de l'Académie des Sciences, Paris*, ser. D, 290:851–852.

623. Debénath, A., Raynal, J.-P., and Texier, J.-P. 1982. Position stratigraphique des restes humains paléolithiques marocains sur la base des travaux récents. *Comptes Rendus de l'Académie des Sciences, Paris*, ser. D, 294:972–976.

624. Debénath, A. 1988. Recent thoughts on the Riss and early Würm assemblages of la Chaise de Vouthon (Charente, France). In *Upper Pleistocene prehistory of western France*, ed. H. L. Dibble and A. Montet-White, pp. 85–93. Philadelphia: University of Pennsylvania Museum.

625. Debénath, A. 1994. L'Atérien du nord de l'Afrique et du Sahara. *Sahara* 6:21–30.

626. Debénath, A., Raynal, J.-P., Roche, J., Texier, J.-P., and Ferembach, D. 1986. Stratigraphie, habitat, typologie et devenir de l'Atérien marocain: Données récentes. *Anthropologie* 90:233–246.

627. Debénath, A., and Sbihi-Alaoui, F. 1979. Découverte de deux nouveaux gisements préhistoriques près de Rabat (Maroc). *Bulletin de la Société Préhistorique Française* 76:11–12.

628. Debets, G. F. 1955. Paleoanthropological finds at Kostenki (in Russian). *Sovetskaya Arkheologiya* 1:43–53.

629. Defleur, A. 1993. *Les sépultures moustériennes*. Paris: CNRS Editions.

630. Defleur, A., Dutour, O., Valladas, H., and Vandermeersch, V. 1993. Cannibals among the Neanderthals. *Nature* 362:214.

631. Deino, A., and Potts, R. 1990. Single-crystal ^{40}Ar/^{39}Ar dating of the Olorgesailie Formation, southern Kenya Rift. *Journal of Geophysical Research* 95:8453–8470.

632. Deino, A. L., Renne, P. R., and Swisher, C. C. I. 1998. ^{40}Ar/^{39}Ar dating in paleoanthropology and archaeology. *Evolutionary Anthropology* 6:63–75.

633. Delfino, V. P., and Vacca, E. 1993. An archaic human skeleton discovered at Altamura (Bari, Italy). *Rivista di Antropologia, Roma* 71:249–257.

634. Delfino, V. P., and Vacca, E. 1996. The cave and the human fossil skeleton at Contrada Lamalunga, Altamura. *Bulletin of the Thirteenth Congress of the International Union of Prehistoric and Protohistoric Sciences—Forlì—Italy* 5:101–105.

635. Deloison, Y. 1986. Description d'un calcanéum fossile de primate et sa comparaison avec des calcanéums de pongidés, d'australopithèques et d'*Homo*. *Comptes Rendus de l'Académie des Sciences, Paris*, ser. 3, 302:257–262.

636. Delpech, F. 1975. Les faunes du Paléolithique supérieur dans le sud-ouest de la France. Doctoral thesis, Université de Bordeaux.

637. Delpech, F. 1983. *Les faunes du Paléolithique supérieur dans le sud-ouest de la France*. Paris: Centre National de la Recherche Scientifique.

638. Delpech, F. 1989. L'environnement animal des Magdaléniens. In *Le Magdalénien en Europe: La structuration du Magdalénien*, ed. M. Otte, pp. 5–30. Liège: Université de Liège.

639. Delson, E. 1979. *Prohylobates* (Primates) from the early Miocene of Libya: A new species and its implications for cercopithecid origins. *Geobios* 12:725–733.

640. Delson, E. 1984. Cercopithecoid biochronology of the African Plio-Pleistocene: Correlation among eastern and southern hominid-bearing localities. *Courier Forschungsinstitut Senckenberg* 69:199–218.

641. Delson, E. 1985. Catarrhine evolution. In *Ancestors: The hard evidence*, ed. E. Delson, pp. 9–13. New York: Alan R. Liss.

642. Delson, E. 1988. Chronology of South African australopith site units. In *Evolutionary history of the "robust" australopithecines*, ed. F. E. Grine, pp. 317–324. New York: Aldine de Gruyter.

643. Delson, E. 1997. One skull does not a species make. *Nature* 389:446–447.

644. Delson, E., and Andrews, P. 1975. Evolution and interrelationships of the catarrhine primates. In *Phylogeny of the primates*, ed. W. P. Luckett and F. S. Szalay, pp. 405–446. New York: Plenum Press.

645. Delson, E., and Rosenberger, A. L. 1980. Phyletic perspectives on platyrrhine origins and anthropoid relationships. In *Evolutionary biology of the New World monkeys and continental drift*, ed. R. L. Ciochon and A. B. Chiarelli, pp. 445–458. New York: Plenum Press.

646. Demars, P. Y., and Hublin, J.-J. 1989. La transition néandertaliens/hommes de type modern en Europe occidentale: Aspects paléontologiques et culturels. *Etudes et Recherches Archéologiques de l'Université de Liège* 34:23–27.

647. deMenocal, P. B. 1995. Plio-Pleistocene African climate. *Science* 270:53–59.

648. deMenocal, P. B., and Bloemendal, J. 1995. Plio-Pleistocene climatic variability in subtropical Africa and the paleoenvironment of hominid evolution: A combined data-model approach. In *Paleoclimate and evolution with emphasis on human origins*, ed. E. S. Vrba, G. H. Denton, T. C. Partridge, and L. H. Burckle, pp. 262–288. New Haven: Yale University Press.

649. Dennell, R. 1983. *European economic prehistory: A new approach*. London: Academic Press.

650. Dennell, R. 1997. The world's oldest spears. *Nature* 385:787–788.

651. Dennell, R., and Roebroeks, W. 1996. The earliest colonization of Europe: The short chronology revisited. *Antiquity* 70:535–542.

652. Dennell, R. W. 1983. A new chronology for the Mousterian. *Nature* 301:199–200.

653. Dennell, R. W., Rendell, H., and Hailwood, E. 1988. Late Pliocene artefacts from northern Pakistan. *Current Anthropology* 29:495–498.

654. Dewar, R. E. 1984. Extinctions in Madagascar. In *Quaternary extinctions: A prehistoric revolution*, ed. P. S. Martin and R. G. Klein, pp. 574–593. Tucson: University of Arizona Press.

655. Dibble, H. H. 1988. Typological aspects of reduction and intensity of utilization of lithic resources in the French Mousterian. In *Upper Pleistocene prehistory of western Eurasia*, ed. H. L. Dibble and A. Montet-White, pp. 181–197. Philadelphia: University of Pennsylvania Museum.

656. Dibble, H. L. 1988. The interpretation of Middle Paleolithic scraper reduction patterns. *Études et Recherches Archéologiques de l'Université de Liège* 31:49–58.

657. Dibble, H. L., and Holdaway, S. J. 1993. The Middle Paleolithic industries of Warwasi. In *The Paleolithic prehistory of the Zagros-Taurus*, ed. D. I. Olszewski and H. L. Dibble, pp. 75–99. Philadelphia: University Museum, University of Pennsylvania.

658. Dibble, H. L., and Rolland, N. 1992. On assemblage variability in the Middle Paleolithic of western Europe: History, perspectives, and a new synthesis. In *The Middle Paleolithic: Adaptation, behavior, and variability*, ed. H. L. Dibble and P. A. Mellars, pp. 1–28. Monograph 72. Philadelphia: University of Pennsylvania Museum.

659. Dikov, N. N. 1996. The Ushki sites, Kamchatka Peninsula. In *American beginnings: The prehistory and palaeoecology of Beringia*, ed. F. H. West, pp. 244–250. Chicago: University of Chicago Press.

660. Dillehay, T. D. 1984. A late Ice-Age settlement in southern Chile. *Scientific American* 251 (4): 106–117.

661. Dillehay, T. D. 1987. By the banks of the Chinchilhuapi. *Natural History* 96 (4): 8–12.

662. Dillehay, T. D. 1989. *Monte Verde: A late Pleistocene settlement in Chile.* Vol. 1. *Palaeoenvironment and site context.* Washington, D.C.: Smithsonian Institution Press.

663. Dillehay, T. D., ed. 1997. *Monte Verde: A Late Pleistocene settlement in Chile.* Vol. 2. *The archaeological context and interpretation.* Washington, D.C.: Smithsonian Institution Press.

664. Dillehay, T. D., Ardila Calderón, G., Politis, G., and Conceicão de Moraes Coutinho Beltrão, M. da. 1992. Earliest hunters and gatherers of South America. *Journal of World Prehistory* 6:145–204.

665. Dillehay, T. D., and Collins, M. B. 1988. Early cultural evidence from Monte Verde in Chile. *Nature* 332:150–152.

666. Dillehay, T. D., Pino, M., Valastro, S., Varela, A. G., and Casamiquela, R. 1982. Monte Verde: Radiocarbon dates from an early man site in south-central Chile. *Journal of Field Archaeology* 9:547–550.

667. Dincauze, D. F. 1984. An archaeo-logical evaluation of the case for pre-Clovis occupations. *Advances in World Archaeology* 3:275–323.

668. Djian, P., and Green, H. 1989. Vectorial expansion of the involu-crin gene and the relatedness of the hominoids. *Proceedings of the National Academy of Sciences* 88:7401–7404.

669. Dobosi, V. T. 1988. Le site paléolithique inférieur de Vértesszölös, Hongrie. *Anthropologie* 92:1041–1050.

670. Dobzhansky, T. 1962. *Mankind evolving.* New Haven: Yale University Press.

671. Dolitsky, A. B. 1985. Siberian Paleolithic archaeology: Approaches and analytic methods. *Current Anthropology* 26:361–378.

672. Dorit, R. L., Akashi, H., and Gilbert, W. 1995. Absence of poly-morphism at the SFY locus on the human Y chromosome. *Science* 268:1183–1185.

673. Drake, R., and Curtis, G. H. 1987. K-Ar geochronology of the Laetoli fossil localities. In *Laetoli: A Pliocene site in northern Tanzania,* ed. M. D. Leakey and J. D. Harris, pp. 48–51. Oxford: Clarendon Press.

674. Drennan, M. R. 1953. A preliminary note on the Saldanha skull. *South African Journal of Science* 50:7–11.

675. Dreyer, T. F. 1935. A human skull from Florisbad, Orange Free State, with a note on the endocranial cast, by C. U. Ariëns Kappers. *Koninklijke Akademie van Wetenschappen te Amsterdam, Proceedings* 38:3–12.

676. Dubois, E. 1922. The proto-Australian fossil man of Wadjak. *Koninklijke Akademie van Wetenschappen te Amsterdam,* ser. B, 23:1013–1051.

677. Dubois, E. 1994. Paleontological investigations on Java (transla-tion of "Palaeontologische onderzoekingen op Java" [1892]). In *Naming our ancestors: An anthology of hominid taxonomy,* ed. W. E. Meikle and S. T. Parker, pp. 37–40. Prospect Heights, Ill.: Waveland.

678. Dugard, J. 1995. Palaeontologist Ron Clarke and the discovery of "Little Foot": A contemporary history. *South African Journal of Science* 91:563–566.

679. Dumond, D. E. 1982. The archaeology of Alaska and the peopling of America. *Science* 209:984–991.

680. Dzaparidze, V., Bosinski, G., Bugianisvili, T., Gabunia, L., Justus, A., Kloptovskaja, N., Kvavadze, E., Lordkipanidze, D., Majsuradze, G., Mgeladze, N., Nioradze, M., Pavlenisvili, E., Schmincke, H.-U., Sologasvili, D., Tusabramisvili, D., Tval-crelidze, M., and Vekua, A. 1992. Der altpaläolithische Fundplatz Dmanisi in Georgien (Kaukasus). *Jahrbuch des Römisch-Germanischen Zentralmuseums Mainz* 36 (1989): 67–116.

681. Eiseley, L. 1961. *Darwin's century: Evolution and the men who discovered it.* New York: Doubleday.

682. Eldredge, N. 1991. *Fossils: The evolution and extinction of spe-cies.* New York: Harry N. Abrams.

683. Eldredge, N. 1995. *Reinventing Darwin: The great debate at the high table of evolutionary theory.* New York: John Wiley.

684. Eldredge, N., and Cracraft, J. 1980. *Phylogenetic patterns and the evolutionary process.* New York: Columbia University Press.

685. Eldredge, N., and Gould, S. J. 1972. Punctuated equilibrium: An alternative to phyletic gradualism. In *Models in paleobiology,* ed. T. Schopf, pp. 82–115. San Francisco: W. H. Freeman.

686. Elster, H., Gil-Av, E., and Weiner, S. 1991. Amino acid racemization of fossil bone. *Journal of Archaeological Science* 18:605–617.

687. Emiliani, C. 1955. Pleistocene temperatures. *Journal of Geology* 3:538–578.

688. Emiliani, C. 1969. The significance of deep-sea cores. In *Science in archeology*, ed. D. Brothwell and E. Higgs, pp. 109–117. London: Thames and Hudson.

689. Ennouchi, E. 1962. Un Néandertalien: L'homme du Jebel Irhoud (Maroc). *Anthropologie* 66:279–299.

690. Ennouchi, E. 1963. Les Néanderthaliens du Jebel Irhoud (Maroc). *Comptes Rendus de l'Académie des Sciences, Paris* 256:2459–2460.

691. Ennouchi, E. 1968. Le deuxième crâne de l'homme d'Irhoud. *Annales de Paléontologie* 55:117–128.

692. Ennouchi, E. 1969. Présence d'un enfant néanderthalien au Jebel Irhoud (Maroc). *Annales de Paléontologie* 55:251–265.

693. Eschassoux, A. 1995. Les grands mammifères de la grotte du Vallonnet, Roquebrune-Cap-Martin, Alpes Maritimes, France: Étude paléontologique, paléoécologique et taphonomique. In *Congreso Internacional de Paleontologia Humana (Orce, September 1995), 3a Circular*, p. 75. Orce, Spain.

694. Estes, R. D. 1992. *The behaviour guide to African mammals.* Berkeley: University of California Press.

695. Etler, D. A. 1996. The fossil evidence for human evolution in Asia. *Annual Review of Anthropology* 25:275–301.

696. Evernden, J. F., and Curtis, G. H. 1965. The potassium-argon dating of Late Cenozoic rocks in east Africa and Italy. *Current Anthropology* 6:343–385.

697. Evin, J. 1990. Validity of the radiocarbon dates beyond 35,000 years B.P. *Palaeogeography, Palaeoclimatology, Palaeoecology* 80:71–78.

698. Falguères, C., Lumley, H. de, and Bischoff, J. L. 1992. U-series dates for stalagmitic flowstone E (Riss/Würm Interglaciation) at Grotte du Lazaret, Nice, France. *Quaternary Research* 38:227–233.

699. Falk, D. 1983. Cerebral cortices of east African early hominids. *Science* 221:1072–1074.

700. Falk, D. 1983. A reconsideration of the endocast of *Proconsul africanus:* Implications for primate brain evolution. In *New interpretations of ape and human ancestry*, ed. R. L. Ciochon and R. S. Corrucini, pp. 239–248. New York: Plenum Press.

701. Falk, D. 1985. Hadar AL-162-28 endocast as evidence that brain enlargement preceded cortical reorganization in hominid evolution. *Nature* 313:45–47.

702. Falk, D. 1986. Endocast morphology of the Hadar hominid AL 162–28. *Nature* 321:536–537.

703. Falk, D. 1987. Hominid paleoneurology. *Annual Review of Anthropology* 16:13–30.

704. Falk, D. 1990. Brain evolution in *Homo*—the radiator theory. *Behavioral and Brain Sciences* 13:333–343.

705. Falk, D. 1992. *Braindance: New discoveries about human origins and brain evolution.* New York: Henry Holt.

706. Falk, D. 1992. *Evolution of the brain and cognition in hominids.* New York: American Museum of Natural History.

707. Farizy, C. 1990. Du Moustérien au Châtelperronien à Arcy-sur-Cure: Un état de la question. In *Paléolithique moyen récent et Paléolithique supérieur ancien en Europe,* ed. C. Farizy, 3:281–289. Nemours: Mémoires du Musée de Préhistoire d'Île de France.

708. Farizy, C. 1990. The transition from Middle to Upper Palaeolithic at Arcy-sur-Cure (Yonne, France): Technological, economic and social aspects. In *The emergence of modern humans: An archaeological perspective,* ed. P. Mellars, pp. 303–326. Ithaca: Cornell University Press.

709. Farizy, C. 1994. Behavioral and cultural changes at the Middle to Upper Paleolithic transition in western Europe. In *Origins of anatomically modern humans,* ed. M. H. Nitecki and D. V. Nitecki, pp. 93–100. New York: Plenum Press.

710. Farizy, C., and David, F. 1992. Subsistence and behavioral patterns of some Middle Paleolithic local groups. In *The Middle Paleolithic: Adaptation, behavior, and variability,* ed. H. L. Dibble and P. Mellars, pp. 87–96. Philadelphia: University Museum, University of Pennsylvania.

711. Farizy, C., David, J., Jaubert, J., Eisenmann, V., Girard, M., Grün, R., Krier, V., Leclerc, J., Miskovsky, J.-C., and Simonnet, R. 1994. *Hommes et bisons du Paléolithique moyen à Mauran (Haute-Garonne).* Paris: CNRS Editions.

712. Farrand, W. R. 1979. Chronology and palaeoenvironment of Levantine prehistoric sites as seen from sediment studies. *Journal of Archaeological Science* 6:369–392.

713. Farrand, W. R. 1982. Environmental conditions during the Lower/Middle Paleolithic transition in the Near East and the Balkans. *British Archaeological Reports International Series* 151:105–112.

714. Farrand, W. R. 1994. Confrontation of geological stratigraphy and radiometric dates from Upper Pleistocene sites in the Levant. In *Late Quaternary chronology and paleoclimates of the eastern Mediterranean,* ed. O. Bar-Yosef and R. S. Kra, pp. 33–53. Cambridge, Mass.: American School of Prehistoric Research.

715. Feathers, J. K. 1996. Luminescence dating and modern human origins. *Evolutionary Anthropology* 5:25–36.

716. Féblot-Augustins, J. 1993. Mobility strategies in the late Middle Palaeolithic of central Europe and western Europe: Elements of stability and variability. *Journal of Anthropological Archaeology* 12:211–265.

717. Feibel, C. S., Brown, F. H., and McDougall, I. 1989. Stratigraphic context of fossil hominids from the Omo Group deposits: Northern Turkana Basin, Kenya and Ethiopia. *American Journal of Physical Anthropology* 78:595–622.

718. Feibel, C. S., Falk, D., Baker, E., Hill, A., Ward, S., Deino, A., Curtis, G., and Drake, R. 1992. Earliest *Homo* debate. *Nature* 358:289–290.

719. Fejfar, O. 1976. Plio-Pleistocene mammal sequences. In *Quaternary glaciations in the Northern Hemisphere, report no. 3,* ed. D. J. Easterbook and V. Sibrava, pp. 351–366. Project 73/1/24. Bellingham, Wash.: IGCP.

720. Fejfar, O. 1976. Recent research at Prezletice. *Current Anthropology* 17:343–344.

721. Feldesman, M. R., Lundy, J. K., and Kleckner, J. G. 1989. The

femur-stature ratio and estimates of stature in mid-to-late-Pleistocene fossil hominids. *American Journal of Physical Anthropology* 78:219–220.

722. Felgenhauer, F., ed. 1959. *Willendorf im der Wachau.* Vienna: R. M. Rohrer.

723. Fennell, K. J., and Trinkaus, E. 1997. Bilateral femoral and tibial periostitis in the La Ferrassie 1 Neanderthal. *Journal of Archaeological Science* 24:985–995.

724. Ferembach, D. 1976. Les restes humains atériens de Témara (1975). *Bulletins et Mémoires de la Société d'Anthropologie de Paris 3*, ser. 13, 2:175–180.

725. Ferembach, D. 1976. Les restes humains de la grotte de Dar es Soltane II (Maroc), campagne 1975. *Bulletins et Mémoires de la Société d'Anthropologie de Paris 3*, ser. 13, 2:183–193.

726. Ferring, C. R. 1975. The Aterian in north African prehistory. In *Problems in prehistory: North Africa and the Levant,* ed. F. Wendorf and A. E. Marks, pp. 113–126. Dallas: Southern Methodist University Press.

727. Fleagle, J. G. 1978. Size distributions of living and fossil primate faunas. *Paleobiology* 4:67–76.

728. Fleagle, J. G. 1983. Locomotor adaptations of Oligocene and Miocene hominoids and their phyletic implications. In *New interpretations of ape and human ancestry,* ed. R. L. Ciochon and R. S. Corruccini, pp. 301–324. New York: Plenum Press.

729. Fleagle, J. G. 1984. Are there any fossil gibbons? In *The lesser apes,* ed. H. Preuschoft, D. J. Chivers, W. Y. Brockleman, and N. Creel, pp. 432–447. Edinburgh: Edinburgh University Press.

730. Fleagle, J. G. 1986. Early anthropoid evolution in Africa and South America. In *Primate evolution,* ed. P. G. Else and P. C. Lee, pp. 133–142. Cambridge: Cambridge University Press.

731. Fleagle, J. G. 1986. The fossil record of early catarrhine evolution. In *Major topics in primate and human evolution,* ed. B. Wood, L. Martin, and P. Andrews, pp. 130–149. Cambridge: University of Cambridge Press.

732. Fleagle, J. G. 1988. *Primate adaptation and evolution.* San Diego: Academic Press.

733. Fleagle, J. G. 1990. New fossil platyrrhines from the Pintura Formation, southern Argentina. *Journal of Human Evolution* 19:61–85.

734. Fleagle, J. G. 1994. Anthropoid origins. In *Integrative paths to the past: Paleoanthropological advances in honor of F. Clark Howell,* ed. R. S. Corruccini and R. L. Ciochon, pp. 17–35. Englewood Cliffs, N.J.: Prentice Hall.

735. Fleagle, J. G., Bown, T. M., Obradovich, J. D., and Simons, E. L. 1986. Age of the earliest African anthropoids. *Science* 234:1247–1249.

736. Fleagle, J. G., Bown, T. M., Obradovich, J. D., and Simons, E. L. 1986. How old are the Fayum primates? In *Primate evolution,* ed. P. G. Else and P. C. Lee, pp. 3–17. Cambridge: Cambridge University Press.

737. Fleagle, J. G., and Jungers, W. L. 1982. Fifty years of higher primate phylogeny. In *A history of American physical anthropology, 1930–1980,* ed. F. Spencer, pp. 187–230. New York: Academic Press.

738. Fleagle, J. G., and Kay, R. F. 1983. New interpretations of the phyletic position of Oligocene hominoids. In *New interpretations of ape and human ancestry,* ed. R. L. Ciochon and R. S. Corruccini, pp. 181–210. New York: Plenum Press.

739. Fleagle, J. G., and Kay, R. F. 1985. The paleobiology of catarrhines. In *Ancestors: The hard evidence,* ed. E. Delson, pp. 23–36. New York: Alan R. Liss.

740. Fleagle, J. G., and Kay, R. F. 1987. The phyletic position of the Parapithecidae. *Journal of Human Evolution* 16:483–532.

741. Fleagle, J. G., Rasmussen, D. T., Yirga, S., Bown, T. M., and Grine, F. E. 1991. New hominid fossils from Fejej, southern Ethiopia. *Journal of Human Evolution* 21:145–152.

742. Fleagle, J. G., and Simons, E. L. 1978. *Micropithecus clarki,* a small ape from the Miocene of Uganda. *American Journal of Physical Anthropology* 49:427–440.

743. Fleischer, R. L., Leakey, L. S. B., Price, P. B., and Walker, R. M. 1965. Fission track dating of Bed I, Olduvai Gorge. *Science* 148:72–74.

744. Fleischer, R. L., Price, P. B., and Walker, R. M. 1969. Quaternary dating by the fission-track technique. In *Science in archaeology,* ed. D. Brothwell and E. Higgs, pp. 58–61. London: Thames and Hudson.

745. Foley, R. A., and Lee, P. C. 1989. Finite social space, evolutionary pathways, and reconstructing hominid behavior. *Science* 243:901–906.

746. Fontes, J. C., and Gasse, F. 1989. On the ages of humid and late Pleistocene phases in north Africa—remarks on "Late Quaternary climatic reconstructions for the Maghreb (north Africa)" by P. Rognon. *Palaeogeography, Palaeoclimatology, Palaeoecology* 70:393–398.

747. Fosse, P., and Bonifay, M.-F. 1991. Les vestiges osseux de Soleilhac: Approche taphonomique. In *Les premiers Européens,* ed. E. Bonifay and B. Vandermeersch, pp. 115–133. Paris: Éditions du Comité des Travaux Historiques et Scientifiques.

748. Fossey, D. 1983. *Gorillas in the mist.* Boston: Houghton-Mifflin.

749. Fox, R. C. 1993. The primitive dental formula of the Carpolestidae (Plesiadapiformes, Mammalia) and its phylogenetic implications. *Journal of Vertebrate Palaeontology* 13:516–524.

750. Fraipont, C. 1936. Les hommes fossiles d'Engis. *Archives de l'Institut de Paléontologie Humaine* 16:1–52.

751. Franciscus, R. G. 1995. Later Pleistocene nasofacial variation in western Eurasia and Africa and modern human origins. Ph.D. diss., University of New Mexico, Albuquerque.

752. Franciscus, R. G., and Churchill, S. E. 1999. The costal skeleton of Shanidar 3 and a reappraisal of Neandertal thoracic morphology. *Journal of Human Evolution.* In press.

753. Franciscus, R. G., and Trinkaus, E. 1988. Nasal morphology and the emergence of *Homo erectus. American Journal of Physical Anthropology* 75:517–527.

754. Franciscus, R. G., and Trinkaus, E. 1995. Determinants of retromolar space presence in Pleistocene *Homo* mandibles. *Journal of Human Evolution* 28:577–595.

755. Frayer, D. W. 1984. Biological and cultural change in the European late Pleistocene and early Holocene. In *The origin of modern hu-*

mans: A world survey of the fossil evidence, ed. F. H. Smith and F. Spencer, pp. 211–250. New York: Alan R. Liss.

756. Frayer, D. W. 1986. Cranial variation at Mladec and the relationship between Mousterian and Upper Paleolithic hominids. *Anthropos* (Brno) 23:243–256.

757. Frayer, D. W., Macchiarelli, R., and Mussi, M. 1988. A case of chondrodystrophic dwarfism in the Italian late Upper Paleolithic. *American Journal of Physical Anthropology* 75:549–565.

758. Frayer, D. W., and Montet-White, A. 1989. Comment on "Grave shortcomings: The evidence for Neandertal burial" by Robert H. Gargett. *Current Anthropology* 30:180–181.

759. Frayer, D. W., Wolpoff, M. H., Thorne, A. G., Smith, F. H., and Pope, G. G. 1993. Theories of modern human origins: The paleontological test. *American Anthropologist* 95:14–50.

760. Frayer, D. W., Wolpoff, M. H., Thorne, A. G., Smith, F. H., and Pope, G. G. 1994. Getting it straight. *American Anthropologist* 96:424–428.

761. Freeman, L. G. 1975. Acheulean sites and stratigraphy in Iberia and the Maghreb. In *After the australopithecines,* ed. K. W. Butzer and G. L. Isaac, pp. 661–743. The Hague: Mouton.

762. Freeman, L. G. 1980. Occupaciones musterienses. In *El yacimiento de la Cueva de "El Pendo" (Excavaciones 1953–57),* ed. J. González-Echegaray, pp. 29–74. Madrid: Consejo Superior de Investigaciones Cientificas.

763. Freeman, L. G. 1983. More on the Mousterian: Flaked bone from Cueva Morín. *Current Anthropology* 24:366–372.

764. Freeman, L. G. 1994. Torralba and Ambrona: A review of discoveries. In *Integrative paths to the past: Paleoanthropological advances in honor of F. Clark Howell,* ed. R. S. Corruccini and R. L. Ciochon, pp. 597–637. Englewood Cliffs, N.J.: Prentice Hall.

765. Freeman, L. G., and González Echegaray, J. 1970. Aurignacian structural features and burials at Cueva Morín (Santander, Spain). *Nature* 226:722–726.

766. Freeman, L. G., González Echegaray, J., Klein, R. G., and Crowe, W. T. 1988. Dimensions of research at El Juyo: An earlier Magdalenian site in Cantabrian Spain. In *Upper Pleistocene prehistory of western Eurasia,* ed. H. L. Dibble and A. Montet-White, pp. 3–39. Philadelphia: University Museum, University of Pennsylvania.

767. Freeman, L. G., González-Echegaray, J., Bernaldo de Quiros, F., and Ogden, J. 1987. *Altamira revisited and other essays on early art.* Santander: Instituto de Investigaciones Prehistoricas.

768. Fridrich, J. 1976. The first industries from eastern and southeastern central Europe. In *Les premières industries de l'Europe (Colloque VIII),* ed. K. Valoch, pp. 8–23. Nice: Union International des Sciences Préhistoriques et Protohistoriques.

769. Fridrich, J. 1989. *Prezletice: A lower Paleolithic site in central Bohemia (Excavations 1969–1985).* Prague: Museum Nationale Pragae.

770. Fridrich, J. 1991. Les premiers peuplements humains en Bohème (Tchécoslovaquie). In *Les premiers Européens,* ed. E. Bonifay and B. Vandermeersch, pp. 195–201. Paris: Éditions du Comité des Travaux Historiques et Scientifiques.

771. Fuji, N. 1988. Palaeovegetation and palaeoclimate changes around Lake Biwa, Japan during the last ca. 3 million years. *Quaternary Science Reviews* 7:21–28.

772. Fullagar, R. L. K., Prince, D. M., and Head, L. M. 1996. Early human occupation of northern Australia: Archaeology and thermoluminescence dating of Jinmium Rock-Shelter, Northern Territory. *Antiquity* 70:751–773.

773. Gábori-Czánk, V. 1968. *La station du Paléolithique moyen d'Érd—Hongrie.* Budapest: Akadémiai Kiadó.

774. Gábori-Czánk, V. 1983. La grotte Remete "Felsö" (supérieure) et le "Szeletian" de Transdanubie. *Acta Archaeologica Academiae Scientiarum Hungaricae* 35:239–248.

775. Gabunia, L., and Vekua, A. 1995. A Plio-Pleistocene hominid from Dmanisi, east Georgia, Caucasus. *Nature* 373:509–512.

776. Gambier, D. 1989. Fossil hominids from the early Upper Paleolithic (Aurignacian) of France. In *The human revolution: Behavioural and biological perspectives in the origins of modern humans,* ed. P. Mellars and C. Stringer, pp. 194–211. Edinburgh: Edinburgh University Press.

777. Gambier, D. 1997. Modern humans at the beginning of the Upper Paleolithic in France. In *Conceptual issues in modern human origins research,* ed. G. A. Clark and C. M. Willermet, pp. 117–131. New York: Aldine de Gruyter.

778. Gamble, C. 1986. *The Palaeolithic settlement of Europe.* Cambridge: Cambridge University Press.

779. Gamble, C. 1987. Man the shoveler: Alternative models for Middle Pleistocene colonization and occupation in northern latitudes. In *The Pleistocene Old World: Regional perspectives,* ed. O. Soffer, pp. 81–98. New York: Plenum Press.

780. Gannon, P. J., Holloway, R. L., Broadfield, D. C., and Braun, A. R. 1998. Asymmetry of chimpanzee planum temporale: Humanlike pattern of Wernicke's brain language area homolog. *Science* 279:220–222.

781. García, N., Arsuaga, J. L., and Torres, T. 1997. The carnivore remains from the Sima de los Huesos Middle Pleistocene site (Sierra de Atapuerca, Spain). *Journal of Human Evolution* 33:155–174.

782. Gargett, R. H. 1989. Grave shortcomings: The evidence for Neandertal burial. *Current Anthropology* 30:157–177.

783. Garrod, D. A. E., and Bate, D. M. 1937. *The Stone Age of Mount Carmel.* Oxford: Oxford University Press.

784. Garrod, D. A. E., Buxton, L. H. D., Smith, G. E., and Bate, D. M. A. 1928. Excavation of a Mousterian rock-shelter at Devil's Tower, Gibraltar. *Journal of the Royal Anthropological Institute of Great Britain and Ireland* 58:33–113.

785. Gaudzinski, S. 1992. Wisentjäger in Wallertheim: Zur Taphonomie einer mittelpaläolithschen Freilandfundstelle in Rheinhessen. *Jahrbuch des Römisch-Germanischen Zentralmuseums Mainz* 39:246–423.

786. Gaudzinski, S. 1995. Wallertheim revisited: A re-analysis of the fauna from the Middle Palaeolithic site of Wallertheim (Rheinessen/Germany). *Journal of Archaeological Science* 22:51–66.

787. Gaudzinski, S., Bittmann, F., Boenigk, W., Frechen, M., and Kolfschoten, T. van. 1996. Palaeoecology and archaeology of the

Kärlich-Seeufer open-air site (Middle Pleistocene) in the central Rhineland, Germany. *Quaternary Research* 46:319–334.

788. Gaudzinski, S., and Turner, E. 1996. The role of early humans in the accumulation of European Lower and Middle Palaeolithic bone assemblages. *Current Anthropology* 37:153–156.

789. Gautier, A. 1986. Une histoire de dents: Les soi-disant incisives travaillées du Paléolithique moyen de Scalyn. *Hellinium* 26:177–181.

790. Gebo, D. L. 1986. Anthropoid origins—the foot evidence. *Journal of Human Evolution* 15:421–430.

791. Gebo, D. L. 1989. Locomotor and phylogenetic considerations in anthropoid evolution. *Journal of Human Evolution* 18:201–233.

792. Gebo, D. L., MacLatchy, L., Kityo, R., Deino, A., Kingston, J., and Pilbeam, D. 1997. A hominoid genus from the early Miocene of Uganda. *Science* 276:401–404.

793. Gebo, D. L., and Simons, E. L. 1987. Morphology and locomotor adaptations of the foot in early Oligocene anthropoids. *American Journal of Physical Anthropology* 74:83–101.

794. Geneste, J.-M. 1988. Économie des ressources lithiques dans le Moustérien du sud-ouest de la France. *Études et Recherches Archéologiques de l'Université de Liège* 33:75–97.

795. Geneste, J.-M. 1988. Systèmes d'approvisonnement en matières premières au paléolithique moyen et au paléolithique supérieur en Aquitaine. *Études et Recherches Archéologiques de l'Université de Liège* 35:61–70.

796. Geneste, J.-M. 1990. Développement des systèmes de production lithique au cours du paléolithique moyen en Aquitaine septentrionale. In *Paléolithique moyen récent et paléolithique supérieur ancien en Europe*, ed. C. Farizy, 3:205–207. Nemours: Mémoires du Musée de Préhistoire d'Île de France.

797. Gentner, W., and Lippolt, H. J. 1969. The potassium-argon dating of Upper Tertiary and Pleistocene deposits. In *Science in archaeology*, ed. D. Brothwell and E. Higgs, pp. 88–100. London: Thames and Hudson.

798. Geraads, D. 1980. La faune des sites à "*Homo erectus*" des carrières Thomas (Casablanca, Maroc). *Quaternaria* 22:65–94.

799. Geraads, D. 1993. *Kolpochoerus phacochoeroides* (Thomas, 1884) (Suidae, Mammalia), du Pliocène supérieur de Ahl Al Oughlam (Casablanca, Maroc). *Geobios* 26:731–743.

800. Geraads, D. 1995. Rongeurs et insectivores (Mammalia) du Pliocène final de Ahl Al Oughlam (Casablanca, Maroc). *Geobios* 1:99–115.

801. Geraads, D. 1996. Le *Sivatherium* (Giraffidae, Mammalia) du Pliocène final d'Ahl al Oughlam (Casablanca, Maroc), et l'évolution du genre en Afrique. *Paläontologische Zeitschrift* 70:623–629.

802. Geraads, D. 1997. Carnivores du Pliocène terminal de Ahl al Oughlam (Casablanca, Maroc). *Geobios* 30:127–164.

803. Geraads, D., Amani, F., and Hublin, J.-J. 1992. Le gisement pléistocène moyen de l'Aïn Maarouf près de El Hajeb, Maroc: Présence d'un hominidé. *Comptes Rendus de l'Académie des Sciences, Paris*,ser. 2, 314:319–323.

804. Geraads, D., Berrio, P., and Roche, H. 1980. La faune et l'industrie des sites à *Homo erectus* des carrières Thomas (Maroc): Préci-

sions sur l'âge de ces hominidés. *Comptes Rendus de l'Académie des Sciences, Paris,* ser. D, 291:195–198.

805. Geraads, D., Hublin, J.-J., Jaeger, J.-J., Tong, H., Sen, S., and Tourbeau, P. 1986. The Pleistocene hominid site of Ternifine, Algeria: New results on the environment, age, and human industries. *Quaternary Research* 25:380–386.

806. Geraads, D., and Tchernov, E. 1983. Fémurs humains du Pléistocène moyen de Gesher Benot Ya'acov (Israël). *Anthropologie* 87: 138–141.

807. Gerasimov, M. M. 1935. Excavations of the paleolithic site in the village of Mal'ta (preliminary report on the 1928–32 work) (in Russian). *Izvestiya Gosudarstvennoj Akademii Istorii Material'noj Kul'tury* 118:78–124.

808. Germonpré, M., and Lbova, L. 1996. Mammalian remains from the Upper Palaeolithic site of Kamenka, Buryatia (Siberia). *Journal of Archaeological Science* 23:35–57.

809. Giacobini, G. 1991. Comment on "The question of ritual cannibalism at Grotta Guattari." *Current Anthropology* 32:130–131.

810. Gibbons, A. 1996. *Homo erectus* in Java: A 250,000-year anachronism. *Science* 274:1841–1842.

811. Gibbons, A. 1997. Archaeologists rediscover cannibals. *Science* 277:635–637.

812. Gibbons, A. 1997. Doubts over spectacular dates. *Science* 278: 220–221.

813. Gibbons, A. 1997. Y chromosome shows that Adam was an African. *Science* 278:804–805.

814. Gibbons, A. 1998. Mother tongues trace steps of earliest Americans. *Science* 279:1306–1307.

815. Gibert, J., Arribas, A., Martínez Navarro, B., Albadalejo, S., Gaete, R., Gibert, L., Oms, O., Peñas, C., and Torrico, R. 1994. Biostratigraphie et magnétostratigraphie des gisements à présence humaine et action anthropique du Pléistocène inférieur de la région d'Orce (Grenada, Espagne). *Comptes Rendus de l'Académie des Sciences, Paris,* ser. 2, 318:1277–1282.

816. Gibert, J., Campillo, D., Arqués, J. M., Garcia-Olivares, E., Borja, C., and Lowenstein, J. 1998. Hominid status of the Orce cranial fragment reasserted. *Journal of Human Evolution* 34:203–217.

817. Gibert, J., Campillo, D., Martinez, B., Sanchez, F., Caporicci, R., Jimenez, C., Ferrandez, C., and Ribot, F. 1991. Nouveaux restes d'hominidés dans les gisements d'Orce et de Cueva Victoria (Espagne). In *Les premiers Européens,* ed. E. Bonifay and B. Vandermeersch, pp. 273–282. Paris: Éditions du Comité des Travaux Historiques et Scientifiques.

818. Gibert, J., Ribot, F., Fernandez, C., Martínez, B., Caporicci, R., and Campillo, D. 1989. Anatomical study: Comparison of the cranial fragment from Venta Micena (Orce; Spain) with fossil and extant mammals. *Human Evolution* 4:283–305.

819. Gibert, J., Sánchez, F., Malgosa, A., and Martínez Navarro, B. 1994. Découvertes de restes humains dans les gisements d'Orce (Granada, Espagne). *Comptes Rendus de l'Académie des Sciences, Paris,* ser. 2, 319:963–968.

820. Gilead, I. 1981. Upper Palaeolithic tool assemblages from the Negev and Sinai. In *Préhistoire du Levant,* ed. J. Cauvin and

P. Sanlaville, pp. 331–342. Paris: Centre National de la Recherche Scientifique.

821. Gilead, I. 1991. The Upper Paleolithic period in the Levant. *Journal of World Prehistory* 5:105–154.

822. Gillespie, C. C. 1951. *Genesis and geology.* New York: Harper.

823. Gillespie, R., Hedges, R. E. M., and Wand, J. O. 1984. Radiocarbon dating of bone by accelerator mass spectrometry. *Journal of Archaeological Science* 11:165–170.

824. Gingerich, P. D. 1973. First record of the Palaeocene primate *Chiromyoides* from North America. *Nature* 244:517–518.

825. Gingerich, P. D. 1977. Radiation of Eocene Adapidae in Europe. *Geobios, Mémoire Spéciale* 1:165–182.

826. Gingerich, P. D. 1980. Eocene Adapidae, paleobiogeography, and the origin of South American Platyrrhini. In *Evolutionary biology of the New World monkeys and continental drift,* ed. R. L. Ciochon and A. B. Chiarelli, pp. 123–138. New York: Plenum Press.

827. Gingerich, P. D. 1984. Paleobiology of tarsiiform primates. In *Biology of tarsiers,* ed. C. Niemitz, pp. 34–44. Stuttgart: Gustav Fischer.

828. Gingerich, P. D. 1984. Primate evolution. *University of Tennessee Studies in Geology* 8:167–184.

829. Gingerich, P. D. 1984. Primate evolution: Evidence from the fossil record, comparative morphology, and molecular biology. *Yearbook of Physical Anthropology* 27:57–72.

830. Gingerich, P. D. 1986. Early Eocene *Cantius torresi*—oldest primate of modern aspect from North America. *Nature* 319:319–321.

831. Gingerich, P. D. 1986. *Plesiadapis* and the delineation of the order Primates. In *Major topics in primate and human evolution,* ed. B. Wood, L. Martin, and P. Andrews, pp. 32–46. Cambridge: University of Cambridge Press.

832. Gingerich, P. D. 1986. Temporal scaling of molecular evolution in primates and other mammals. *Molecular biology and evolution* 3:205–221.

833. Gingerich, P. D. 1989. New earliest Wasatchian mammalian fauna from the Eocene of northwestern Wyoming: Composition and diversity in a rarely sampled high-floodplain assemblage. *University of Michigan Papers on Paleontology* 28:1–97.

834. Gingerich, P. D. 1990. African dawn for primates. *Nature* 346:411.

835. Gingerich, P. D. 1990. Mammalian order Proprimates—response to Beard. *Journal of Human Evolution* 19:821–822.

836. Gingerich, P. D. 1993. Oligocene age of the Gebel Qatrani Formation, Fayum, Egypt. *Journal of Human Evolution* 24:207–218.

837. Gingerich, P. D., and Sahni, A. 1984. Dentition of *Sivaladapis nagrii* (Adapidae) from the late Miocene of India. *International Journal of Primatology* 5:63–79.

838. Gingerich, P. D., and Schoeninger, M. 1977. The fossil record and primate phylogeny. *Journal of Human Evolution* 6:483–505.

839. Ginsburg, L. 1986. Chronology of the European pliopithecids. In *Primate evolution,* ed. J. G. Else and P. C. Lee, pp. 46–57. Cambridge: Cambridge University Press.

840. Gioia, P. 1988. Problems related to the origins of Italian Upper

Palaeolithic: Uluzzian and Aurignacian. *Études et Recherches Archéologiques de l'Université de Liège* 35:71–108.

841. Gioia, P. 1990. La transition Paléolithique moyen/Paléolithique supérieur en Italie et la question d'Uluzzien. In *Paléolithique moyen récent et Paléolithique supérieur ancien en Europe,* ed. C. Farizy, pp. 241–250. Mémoires, no. 3. Nemours: Musée de Préhistoire d'Île de France.

842. Girard, M., Miskovsky, J.-C., and Evin, J. 1990. La fin du Würm moyen et le début du Würm supérieur à Arcy-sur-Cure (Yonne). In *Paléolithique moyen récent et Paléolithique supérieur ancien en Europe,* ed. C. Farizy, pp. 295–303. Mémoires, no. 3. Nemours: Musée de Préhistoire d'Île de France.

843. Gisis, I., and Bar-Yosef, O. 1974. New excavations in Zuttiyeh Cave. *Paléorient* 2:175–180.

844. Gladiline, V., and Sitlivy, V. 1991. Les premières industries en Subcarpatie. In *Les premiers Européens,* ed. E. Bonifay and B. Vandermeersch, pp. 217–231. Paris: Éditions du Comité des Travaux Historiques et Scientifiques.

845. Gladkih, M. I., Kornietz, N. L., and Soffer, O. 1984. Mammoth bone dwellings on the Russian plain. *Scientific American* 251 (5): 164–175.

846. Gleadow, A. J. W. 1980. Fission track age of the KBS Tuff and associated hominids in northern Kenya. *Nature* 284:225–230.

847. Glen, W. 1990. What killed the dinosaurs? *American Scientist* 78:354–370.

848. Gobert, E. G. 1950. Le gisement paléolithique de Sidi Zin, avec une notice sur la faune de Sidi Zin de R. Vaufrey. *Karthago* 1:1–64.

849. Godinot, M., and Mahboubi, M. 1992. Earliest known simian primate found in Algeria. *Nature* 357:324–326.

850. Godinot, M., and Mahboubi, M. 1994. Les petits primates simiiformes de Glib Zegdou (Éocène inférieur à moyen d'Algérie). *Comptes Rendus de l'Académie des Sciences, Paris,* ser. 2, 319: 357–364.

851. Goebel, T. 1995. The record of human occupation of the Russian Subarctic and Arctic. *Byrd Polar Research Center Miscellaneous Series* M-335:41–46.

852. Goebel, T., and Aksenov, M. 1995. Accelerator radiocarbon dating of the initial Upper Palaeolithic in southeast Siberia. *Antiquity* 69:349–357.

853. Goebel, T., Derevianko, A. P., and Petrin, V. T. 1993. Dating the Middle-to-Upper-Paleolithic transition at Kara-Bom. *Current Anthropology* 34:452–458.

854. Goebel, T., Powers, R., and Bigelow, N. 1991. The Nenana Complex of Alaska and Clovis origins. In *Clovis origins and adaptations,* ed. R. Bonnichsen and K. Turnmire, pp. 49–79. Corvallis, Ore.: Center for the Study of the First Americans, Oregon State University.

855. González Echegaray, J. 1980. El yacimiento de la Cueva de "El Pendo" (Excavaciones 1953–57). *Bibliotheca Praehistorica Hispana* 17:1–270.

856. Goodall, J. 1986. *The chimpanzees of Gombe.* Cambridge: Harvard University Press.

857. Goodman, M., Koop, B. F., Czelusniak, J., Fitch, D. H. A., Tagle,

D. A., and Slightom, J. L. 1989. Molecular phylogeny of the family of apes and humans. *Genome* 31:316–335.

858. Goodman, M., Tagle, D. A., Fitch, D. H. A., Bailey, W., Czelusniak, J., Koop, B. F., Benson, P., and Slightom, J. L. 1990. Primate evolution at the DNA level and a classification of the hominoids. *Journal of Molecular Evolution* 30:260–266.

859. Goodwin, A. J. H. 1928. An introduction to the Middle Stone Age in South Africa. *South African Journal of Science* 25:410–418.

860. Goodwin, A. J. H. 1929. The Middle Stone Age. *Annals of the South African Museum* 29:95–145.

861. Gore, R. 1997. The dawn of humans: Tracking the first of our kind. *National Geographic* 192 (3): 92–99.

862. Goren, N. 1981. The lithic assemblages of the site of Ubeidiya, Jordan Valley. Ph.D. diss., Hebrew University of Jerusalem.

863. Goren-Inbar, N. 1985. The lithic assemblage of the Berekhat Ram Acheulian site, Golan Heights. *Paléorient* 11:7–28.

864. Goren-Inbar, N. 1986. A figurine from the Acheulian site of Berekhat Ram. *M'tekufat Ha'even* 19:71–12.

865. Goren-Inbar, N., ed. 1990. *Quneitra: A Mousterian site on the Golan Heights*. Jerusalem: Hebrew University of Jerusalem.

866. Goren-Inbar, N. 1992. The Acheulian site of Gesher Benot Ya'aqov—an Asian or African entity? In *The evolution and dispersal of modern humans in Asia*, ed. T. Akazawa, K. Aoki, and T. Kimura, pp. 67–82. Tokyo: Hokusen-sha.

867. Goren-Inbar, N. 1994. The Lower Paleolithic of Israel. In *The archaeology of society in the Holy Land*, ed. T. E. Levy, pp. 93–109. London: Leicester University Press.

868. Goren-Inbar, N., Belitsky, S., Goren, Y., Rabinovich, R., and Saragusti, I. 1992. Gesher Benot Ya'aqov—the "Bar": an Acheulean assemblage. *Geoarchaeology* 7:27–40.

869. Goren-Inbar, N., and Belitzky, S. 1989. Structural position of the Pleistocene Gesher Benot Ya'aqov site in the Dead Sea Rift Zone. *Quaternary Research* 31:371–376.

870. Goren-Inbar, N., and Peltz, S. 1995. Additional remarks on the Berekhat Ram figure. *Rock Art Research* 12:131–132.

871. Goren-Inbar, N., and Saragusti, I. 1996. An Acheulian biface assemblage from Gesher Benot Ya'aqov, Israel: Indications of African affinities. *Journal of Field Archaeology* 23:15–30.

872. Gould, S. J., Eldredge, N. 1977. Punctuated equilibria: Tempo and mode of evolution reconsidered. *Paleobiology* 3:115–151.

873. Gowlett, J. A. J. 1978. Kilombe—an Acheulian site complex in Kenya. In *Geological background to fossil man*, ed. W. W. Bishop, pp. 337–360. Edinburgh: Scottish Academic Press.

874. Gowlett, J. A. J. 1991. Kilombe—review of an African site complex. In *Cultural beginnings: Approaches to understanding early hominid lifeways in the African savannah*, ed. J. D. Clark, pp. 129–136. Mainz: Römisch-Germanisches Zentralmuseum.

875. Gowlett, J. A. J., and Crompton, R. H. 1994. Kariandusi: Acheulean morphology and the question of allometry. *African Archaeological Review* 12:3–42.

876. Gowlett, J. A. J., Harris, J. W. K., Walton, D. A., and Wood, B. A. 1981. Early archaeological sites, further hominid remains and traces of fire from Chesowanja, Kenya. *Nature* 294:125–129.

877. Gowlett, J. A. J., and Hedges, R. E. M. 1986. Lessons of context and contamination in dating the Upper Palaeolithic. In *Archaeological results from accelerator dating*, ed. J. A. J. Gowlett and R. E. M. Hedges, pp. 63–71. Oxford: Oxford University Committee for Archaeology.

878. Grace, R. 1996. Use-wear analysis: The state of the art. *Archaeometry* 38:20–229.

879. Gramly, R. M. 1976. Upper Pleistocene archaeological occurrences at site GvJM/22, Lukenya Hill, Kenya. *Man* 11:319–344.

880. Grausz, H. M., Leakey, R. E., Walker, A. C., and Ward, C. V. 1988. Associated cranial and postcranial bones of *Australopithecus boisei*. In *Evolutionary history of the "robust" australopithecines*, ed. F. E. Grine, pp. 127–132. New York: Aldine de Gruyter.

881. Grayson, D. K. 1983. *The establishment of human antiquity.* New York: Academic Press.

882. Grayson, D. K. 1990. The provision of time depth for paleoanthropology. *Geological Society of America Special Paper* 242: 1–13.

883. Graziosi, P. 1960. *Paleolithic art.* New York: McGraw-Hill.

884. Green, H. S. 1981. The first Welshman: Excavations at Pontnewydd. *Antiquity* 55:184–195.

885. Green, H. S., Bevins, R. E., Bull, P. A., Currant, A. P., Debenham, N. C., Embleton, C., Ivanovich, M., Livingston, H., Rae, A. M., Schwarcz, H. P., and Stringer, C. B. 1989. Le site acheuléen de la grotte de Pontnewydd, Pays de Gaulle: Géomorphologie, stratigraphie, chronologie, fauna, hominidés fossiles, géologie et industrie lithique dans le contexte paléoécologique. *Anthropologie* 83:15–52.

886. Green, H. S., Stringer, C. B., Collcutt, S. N., Currant, A. P., Huxtable, J., Schwarcz, H. P., Debenham, N., Embleton, C., Bull, P., Molleson, T. I., and Bevins, R. E. 1981. Pontnewydd Cave in Wales—a new Middle Pleistocene hominid site. *Nature* 294: 707–713.

887. Green, P. 1979. Tracking down the past. *New Scientist* 84:624–626.

888. Greenberg, J. H. 1987. *Language in the Americas.* Stanford: Stanford University Press.

889. Greenberg, J. H. 1996. Beringia and New World origins: The linguistic evidence. In *American beginnings: The prehistory and palaeoecology of Beringia*, ed. F. H. West, pp. 525–536. Chicago: University of Chicago Press.

890. Greenberg, J. H., Turner, C. G., and Zegura, S. L. 1986. The settlement of the Americas: A comparison of the linguistic, dental and genetic evidence. *Current Anthropology* 27:477–497.

891. Greenwood, P. 1968. Fish remains. In *The prehistory of Nubia*, ed. F. Wendorf, 2:100–109. Dallas: Southern Methodist University Press.

892. Gregory, W. K., and Hellman, M. 1939. The dentition of the extinct South African man-ape *Australopithecus (Plesianthropus) transvaalensis* Broom: A comparative and phylogenetic study. *Annals of the Transvaal Museum* 19:339–373.

893. Gribbin, J. 1979. Making a date with radiocarbon. *New Scientist* 82:532–534.

894. Grine, F. E. 1981. Trophic differences between "gracile" and "robust" australopithecines: A scanning electron microscope analysis of occlusal events. *South African Journal of Science* 77:203–230.

895. Grine, F. E. 1982. A new juvenile hominid (Mammalia, Primates) from Member 3, Kromdraai Formation, Transvaal, South Africa. *Annals of the Transvaal Museum* 33:165–239.

896. Grine, F. E. 1985. Dental morphology and the systematic affinities of the Taung fossil hominid. In *Hominid evolution: Past, present and future*, ed. P. V. Tobias, pp. 247–253. New York: Alan R. Liss.

897. Grine, F. E. 1985. Was interspecific competition a motive force in early hominid evolution? *Transvaal Museum Memoir* 4:143–152.

898. Grine, F. E. 1986. Dental evidence for dietary differences in *Australopithecus* and *Paranthropus:* A quantitative analysis of permanent molar microwear. *Journal of Human Evolution* 15:783–822.

899. Grine, F. E. 1989. New hominid fossils from the Swartkrans Formation (1979–1986 excavations): Craniodental specimens. *American Journal of Physical Anthropology* 79:409–449.

900. Grine, F. E. 1993. Australopithecine taxonomy and phylogeny: Historical background and recent interpretation. In *The human evolution source book*, ed. R. L. Ciochon and J. G. Fleagle, pp. 198–210. Englewood Cliffs, N.J.: Prentice Hall.

901. Grine, F. E. 1993. Description and preliminary analysis of new hominid craniodental fossils from the Swartkrans Formation. *Transvaal Museum Monograph* 8:75–116.

902. Grine, F. E., Demes, B., Jungers, W. L., and Cole, T. M. 1993. Taxonomic affinity of the early *Homo* cranium from Swartkrans, South Africa. *American Journal of Physical Anthropology* 92:411–426.

903. Grine, F. E., Jungers, W. L., and Schultz, J. 1996. Phenetic affinities among early *Homo* crania from east and south Africa. *Journal of Human Evolution* 30:189–225.

904. Grine, F. E., Jungers, W. L., Tobias, P. V., Pearson, and O. M. 1995. Fossil *Homo* femur from Berg Aukas, northern Namibia. *American Journal of Physical Anthropology* 97:151–185.

905. Grine, F. E., and Klein, R. G. 1985. Pleistocene and Holocene human remains from Equus Cave, South Africa. *Anthropology* 8:55–98.

906. Grine, F. E., and Klein, R. G. 1993. Late Pleistocene human remains from the Sea Harvest site, Saldanha Bay, South Africa. *South African Journal of Science* 89:145–152.

907. Grine, F. E., Klein, R. G., and Volman, T. P. 1991. Dating, archaeology, and human fossils from the Middle Stone Age levels of Die Kelders, South Africa. *Journal of Human Evolution* 21:363–395.

908. Grine, F. E., and Susman, R. L. 1991. Radius of *Paranthropus robustus* from Member 1, Swartkrans Formation, South Africa. *American Journal of Physical Anthropology* 84:229–248.

909. Groube, L., Chappell, J., Muke, J., and Price, D. 1986. A 40,000 year old occupation site at Huon Peninsula, Papua New Guinea. *Nature* 324:453–455.

910. Groves, C. P. 1986. Systematics of the great apes. In *Comparative*

primate biology, ed. D. R. Swindler and J. Erwin, 1:187–217. New York: Alan R. Liss.

911. Groves, C. P. 1989. *A theory of human and primate evolution.* Oxford: Clarendon Press.

912. Groves, C. P. 1992. The origin of modern humans. *Interdisciplinary Science Reviews* 19:23–34.

913. Groves, C. P. 1996. Hovering on the brink: Nearly but not quite getting to Australia. *Perspectives in Human Biology* 2:83–87.

914. Groves, C. P., and Lahr, M. M. 1994. A bush not a ladder: Speciation and replacement in human evolution. *Perspectives in Human Biology* 4:1–11.

915. Groves, C. P., and Mazák, V. 1975. An approach to the taxonomy of the Hominidae: Gracile Villafranchian hominids of Africa. *Casopis pro Mineralogii Geologii* 20:225–247.

916. Gruet, M. 1954. Le gisement moustérien d'El-Guettar. *Karthago* 5:1–79.

917. Gruet, M. 1958. Le gisement d'El-Guettar et sa flore. *Libyca* 6:79–126.

918. Grün, R. 1993. Electron spin resonance dating in paleoanthropology. *Evolutionary Anthropology* 2:172–181.

919. Grün, R. 1996. A re-analysis of electron spin resonance dating results associated with the Petralona hominid. *Journal of Human Evolution* 30:227–241.

920. Grün, R., Beaumont, P. B., and Stringer, C. B. 1990. ESR dating evidence for early modern humans at Border Cave in South Africa. *Nature* 344:537–540.

921. Grün, R., Brink, J. S., Spooner, N. A., Taylor, L., Stringer, C. B., Franciscus, R. G., and Murray, A. S. 1996. Direct dating of Florisbad hominid. *Nature* 382:500–501.

922. Grün, R., Huang, P.-H., Wu, X., Stringer, C. B., Thorne, A. G., and McCulloch, M. 1997. ESR analysis of teeth from the palaeoanthropological site of Zhoukoudian, China. *Journal of Human Evolution* 32:83–91.

923. Grün, R., Schwarcz, H. P., Ford, D. C., and Hentzsch, B. 1988. ESR dating of spring deposited travertines. *Quaternary Science Reviews* 7:429–432.

924. Grün, R., and Stringer, C. B. 1991. Electron spin resonance dating and the evolution of modern humans. *Archaeometry* 33:153–199.

925. Grün, R., Stringer, C. B., and Schwarcz, H. P. 1991. ESR dating of teeth from Garrod's Tabun Cave collection. *Journal of Human Evolution* 20:231–248.

926. Grün, R., and Thorne, A. 1997. Dating the Ngandong humans. *Science* 276:1575–1576.

927. Guérin, C., Bar-Yosef, O., Debard, E., Faure, M., Shea, J., and Tchernov, E. 1996. Mission archéologique et paléontologique dans le Pléistocène ancien d'Oubédiyeh (Israël): Résultats 1992–1994. *Comptes Rendus de l'Académie des Sciences, Paris,* ser. 2a, 322:709–712.

928. Guérin, C., and Faure, M. 1989. Les grands mammifères du Pléistocène ancien d'Oubeidiyeh (Israël): Importance stratigraphique et écologique. *British Archaeological Reports International Series* 497:19–24.

929. Guichard, J. 1976. Les civilisations du Paléolithique moyen en Périgord. In *La préhistoire française*, ed. H. de Lumley, 2:1053–1069. Paris: Centre National de la Recherche Scientifique.

930. Guidon, N. 1989. On stratigraphy and chronology of Pedra Furada. *Current Anthropology* 30:641–642.

931. Guidon, N., and Arnaud, B. 1991. The chronology of the New World: Two faces of one reality. *World Archaeology* 23:167–178.

932. Guidon, N., and Delibrias, G. 1986. Carbon-14 dates point to man in the Americas 32,000 years ago. *Nature* 321:769–771.

933. Guillien, Y., and Laplace, G. 1978. Les climates et les hommes en Europe et en Afrique septentrionale de 28000 B.P. à 10000 B.P. *Bulletin de l'Association Française pour l'Étude du Quaternaire* 4:187–193.

934. Guiot, A., Pons, A., de Beaulieu, J. L., and Reille, M. 1989. A 140,000-year continental climate reconstruction from two European pollen records. *Nature* 338:309–313.

935. Guthrie, R. D. 1990. *Frozen fauna of the Mammoth Steppe*. Chicago: University of Chicago Press.

936. Guthrie, R. D. 1996. The Mammoth Steppe and the origin of Mongoloids and their dispersal. In *Prehistoric Mongoloid dispersals*, ed. T. Akazawa and E. J. E. Szathmáry, pp. 172–186. Oxford: Oxford University Press.

937. Habgood, P. J. 1985. The origin of the Australian Aborigines: An alternative approach and view. In *Hominid evolution: Past, present and future*, ed. P. V. Tobias, pp. 367–380. New York: Alan R. Liss.

938. Habgood, P. J. 1989. The origins of anatomically modern humans in Australia. In *The human revolution: Behavioural and biological perspectives in the origins of modern humans*, ed. P. Mellars and C. Stringer, pp. 245–273. Edinburgh: Edinburgh University Press.

939. Haesaerts, P., and Dupuis, C. 1986. Contribution à stratigraphie des nappes alluviales de la Somme et de l'Avre dans la région d'Amiens. In *Chronostratigraphie et faciés culturels du Paléolithique inférieur et moyen dans l'Europe du Nord-Ouest*, ed. A. Tuffreau and J. Sommé, pp. 171–186. Paris: Supplément au Bulletin de l'Association Française pour l'Étude du Quaternaire.

940. Hahn, J. 1976. Das Gravettien in Westlichen Mitteleuropa. In *Périgordien et Gravettien en Europe*, ed. B. Klíma, pp. 100–120. Nice: Neuvième Congrès de l'Union Internationale des Sciences Préhistoriques et Protohistoriques.

941. Hahn, J. 1977. Aurignacien—das altëre Jungpaläolithikum in Mittel- und Osteuropa. *Fundamenta* (Cologne) A9:1–355.

942. Hahn, J. 1993. Aurignacian art in central Europe. In *Before Lascaux: The complex record of the early Upper Paleolithic*, ed. H. Knecht, A. Pike-Tay, and R. White, pp. 229–241. Boca Raton, Fla.: CRC Press.

943. Hahn, J., Müller-Beck, H., and Taute, W. 1985. *Eiszeithöhlen im Lonetal*. Stuttgart: Konrad Theiss.

944. Hall, C. M., Walter, R. C., Westgate, J. A., and York, D. 1984. Geochronology, stratigraphy and geochemistry of the Cindery Tuff in the Pliocene hominid-bearing sediments of the Middle Awash, Ethiopia. *Nature* 308:26–31.

945. Hall, C. M., and York, D. 1984. The applicability of ^{40}Ar/^{39}Ar dating to young volcanics. In *Quaternary dating methods*, ed. W. C. Mahaney, pp. 67–74. Amsterdam: Elsevier.

946. Hammer, M. F. 1995. A recent common ancestry for human Y chromosomes. *Nature* 378:376–378.

947. Hammer, M. F., and Zegura, S. L. 1996. The role of the Y chromosome in human evolutionary studies. *Evolutionary Anthropology* 5:116–134.

948. Hansen, J., and Renfrew, J. 1978. Palaeolithic-Neolithic seed remains at Franchthi Cave, Greece. *Nature* 271:349–352.

949. Hare, P. E. 1980. Organic geochemistry of bone and its relation to the survival of bone in the natural environment. In *Fossils in the making*, ed. A. K. Behrensmeyer and A. Hill, pp. 208–219. Chicago: University of Chicago Press.

950. Hare, P. E., Goodfriend, G. A., Brooks, A. S., Kokis, J. E., and Von Endt, D. W. 1993. Chemical clocks and thermometers: Diagenetic reactions of amino acids in fossils. *Carnegie Institution of Washington Yearbook* 92:80–85.

951. Harland, W. B., Armstrong, R. L., Cox, A. V., Craig, L. E., Smith, A. G., and Smith, D. G. 1990. *A geologic time scale 1989.* Cambridge: Cambridge University Press.

952. Harmon, R. S., Glazek, J., and Nowak, K. 1980. ^{230}Th/^{234}U dating of travertine from the Bilzingsleben archaeological site. *Nature* 284:132–135.

953. Harmon, R. S., Land, L. S., Mitterer, R. M., Garrett, P., Schwarcz, H. P., and Larson, G. J. 1981. Bermuda sea level during the last interglacial. *Nature* 239:481–484.

954. Harpending, H. C. 1994. Gene frequencies, DNA sequences, and human origins. *Perspectives in Biology and Medicine* 37:384–394.

955. Harpending, H. C., Sherry, S. T., Rogers, A. R., and Stoneking, M. 1993. The genetic structure of ancient human populations. *Current Anthropology* 34:483–496.

956. Harris, J. M. 1985. Age and paleoecology of the Upper Laetolil Beds, Laetoli, Tanzania. In *Ancestors: The hard evidence*, ed. E. Delson, pp. 76–81. New York: Alan R. Liss.

957. Harris, J. M., Brown, F. H., and Leakey, M. G. 1988. Stratigraphy and paleontology of Pliocene and Pleistocene localities west of Lake Turkana, Kenya. *Contribution in Science, Natural History Museum of Los Angeles County* 399:1–128.

958. Harris, J. M., and White, T. D. 1979. Evolution of the Plio-Pleistocene African Suidae. *Transactions of the American Philosophical Society* 69:1–128.

959. Harris, J. W. K. 1983. Cultural beginnings: Plio-Pleistocene archaeological occurrences from the Afar, Ethiopia. *African Archaeological Review* 1:3–31.

960. Harris, J. W. K., and Gowlett, J. A. 1980. Evidence of early stone industries at Chesowanja, Kenya. In *Proceedings of the Eighth Panafrican Congress of Prehistory and Quaternary Studies, Nairobi 1977*, ed. R. E. F. Leakey and B. A. Ogot, pp. 208–212. Nairobi: International Louis Leakey Memorial Institute for African Prehistory.

961. Harris, J. W. K., and Harris, K. 1981. A note on the archaeology at Laetoli. *Nyame Akuma* 18:18–21.

962. Harris, J. W. K., and Isaac, G. L. 1976. The Karari Industry: Early Pleistocene archaeological evidence from the terrain east of Lake Turkana, Kenya. *Nature* 262:102–107.

963. Harris, J. W. K., and Semaw, S. 1989. Further archaeological studies at the Gona River, Hadar, Ethiopia. *Nyame Akuma* 31:19–21.

964. Harris, J. W. K., Williamson, P. G., Morris, P. J., de Heinzelin, J., Verniers, J., Helgren, D., Bellomo, R. V., Laden, G., Spang, T. W., Stewart, K., and Tappen, M. J. 1990. *Archaeology of the Lusso Beds.* Memoir 1. Martinsville: Virginia Museum of Natural History.

965. Harris, J. W. K., Williamson, P. G., Verniers, J., Tappen, M. J., Stewart, K., Helgren, D., de Heinzelin, J., Boaz, N. T., and Bellomo, R. V. 1987. Late Pliocene hominid occupation of the Senga 5A site, Zaire. *Journal of Human Evolution* 16:701–728.

966. Harrison, T. 1986. New fossil anthropoids from the Middle Miocene of east Africa and their bearing on the origin of the Oreopithecidae. *American Journal of Physical Anthropology* 71:265–284.

967. Harrison, T. 1986. A reassessment of the phylogenetic relationships of *Oreopithecus bambolii* Gervais. *Journal of Human Evolution* 15:541–583.

968. Harrison, T. 1987. The phylogenetic relationships of the early catarrhine primates: A review of the current evidence. *Journal of Human Evolution* 16:41–80.

969. Harrold, F. B. 1980. A comparative analysis of Eurasian Paleolithic burials. *World Archaeology* 12:195–211.

970. Harrold, F. B. 1983. The Chatelperronian and the Middle-Upper Paleolithic transition. *British Archaeological Reports International Series* 164:123–140.

971. Harrold, F. B. 1988. The Chatelperronian and the early Aurignacian in France. *British Archaeological Reports International Series* 437:157–191.

972. Harrold, F. B. 1989. Mousterian, Chatelperronian and Early Aurignacian in western Europe: Continuity or discontinuity? In *The human revolution: Behavioural and biological perspectives on the origins of modern humans*, ed. P. A. Mellars and C. B. Stringer, pp. 677–713. Edinburgh: Edinburgh University Press.

973. Hartwig, W. C. 1994. Patterns, puzzles, and perspectives on platyrrhine origins. In *Integrative paths to the past: Paleoanthropological advances in honor of F. Clark Howell*, ed. R. S. Corruccini and R. L. Ciochon, pp. 69–93. Englewood Cliffs, N.J.: Prentice Hall.

974. Hartwig-Scherer, S., and Martin, R. D. 1991. Was "Lucy" more human than her child? Observations on early hominid postcranial skeletons. *Journal of Human Evolution* 21:439–449.

975. Hasegawa, M., Kishino, H., and Yano, T. 1989. Estimation of branching dates among primates by molecular clocks of nuclear DNA which slowed down in Hominoidea. *Journal of Human Evolution* 18:461–476.

976. Hawkes, K., O'Connell, J. F., and Blurton Jones, N. G. 1997. Hadza women's time allocation, offspring provisioning, and the evolution of long postmenopausal life spans. *Current Anthropology* 38:551–577.

977. Hawkes, K., O'Connell, J. F., Blurton Jones, N. G., Alvarez, H., and Charnov, E. L. 1998. Grandmothering, menopause, and the evolution of human life histories. *Proceedings of the National Academy of Sciences* 95:1336–1339.

978. Hawkes, K., O'Connell, J. F., and Blurton Jones, N. 1991. Hunting income patterns among the Hadza: Big game, common goods, foraging goals and the evolution of the human diet. *Philosophical Transactions of the Royal Society* 334:243–251.

979. Hawkes, K., O'Connell, J. F., and Rogers, L. 1997. The behavioral ecology of modern hunter-gatherers, and human evolution. *Trends in Ecology and Evolution* 12:29–32.

980. Hay, R. L. 1976. *Geology of the Olduvai Gorge: A study of sedimentation in a semiarid basin.* Berkeley: University of California Press.

981. Hay, R. L. 1987. Geology of the Laetoli area. In *Laetoli: A Pliocene site in northern Tanzania*, ed. M. D. Leakey and J. M. Harris, pp. 23–47. Oxford: Oxford University Press.

982. Hay, R. L. 1990. Olduvai Gorge: A case history in the interpretation of hominid paleoenvironments in east Africa. *Geological Society of America Special Paper* 242:23–37.

983. Hay, R. L., and Leakey, M. D. 1982. The fossil footprints of Laetoli. *Scientific American* 246 (2): 50–57.

984. Hayden, B. 1993. The cultural capacities of the Neandertals: A review and re-evaluation. *Journal of Human Evolution* 24:113–146.

985. Haynes, C. V. 1980. The Clovis culture. *Canadian Journal of Anthropology* 1:115–121.

986. Haynes, C. V. 1982. Were Clovis progenitors in Beringia? In *Paleoecology of Beringia*, ed. D. M. Hopkins, J. V. Mathews, C. E. Schweger, and S. B. Young, pp. 383–398. New York: Academic Press.

987. Haynes, C. V. 1984. Stratigraphy and late Pleistocene extinction in the United States. In *Quaternary extinctions: A prehistoric revolution*, ed. P. S. Martin and R. G. Klein, pp. 345–353. Tucson: University of Arizona Press.

988. Haynes, C. V. 1992. Contributions of radiocarbon dating to the geochronology and peopling of the New World. In *Radiocarbon after four decades*, ed. R. E. Taylor, A. Long, and R. S. Kra, pp. 355–374. New York: Springer-Verlag.

989. Haynes, C. V. 1993. Clovis-Folsom geochronology and climatic change. In *From Kostenki to Clovis: Upper Paleolithic Paleo-Indian adaptations*, ed. O. Soffer and N. D. Praslov, pp. 219–236. New York: Plenum Press.

990. Haynes, C. V., Reanier, R. E., Barse, W. T., Roosevelt, A. C., Lima da Costa, M., Brown, L. J., Douglas, J. E., O'Donnell, M., Quinn, E., Kemp, J., Machado, C. L., Imazio da Silveira, M., Feathers, J., and Henderson, A. 1997. Dating a Paleoindian site in the Amazon in comparison with Clovis Culture. *Science* 275:1948–1952.

991. Haynes, G. 1988. Longitudinal studies of African elephant death and bone deposits. *Journal of Archaeological Science* 15:131–157.

992. Haynes, G. 1991. *Mammoths, mastodonts, and elephants: Biology, behavior, and the fossil record.* Cambridge: Cambridge University Press.

993. Hays, J. D., Imbrie, J., and Shackleton, N. J. 1976. Variations in the earth's orbit: Pacemaker of the ice ages. *Science* 194:1121–1132.

994. Heberer, G. 1963. Über einen neuen archanthropinen Typus aus der Oldoway-Schlucht. *Zeitschrift für Morphologie und Anthropologie* 53:171–177.

995. Hedberg, H. D. 1976. *International stratigraphic guide.* New York: John Wiley.

996. Hedges, R. E. M. 1981. Radiocarbon dating with an accelerator: Review and preview. *Archaeometry* 23:3–18.

997. Hedges, R. E. M., and Gowlett, J. A. J. 1986. Radiocarbon dating by accelerator mass spectrometry. *Scientific American* 254 (1): 100–107.

998. Hedges, R. E. M., Housley, R. A., Bronk Ramsey, C., and van Klinken, G. J. 1994. Radiocarbon dates from the Oxford AMS system: *Archaeometry* datelist 18. *Archaeometry* 36:337–374.

999. Hedges, S. B., Kumar, S., Tamura, K., and Stoneking, M. 1992. Human origins and analysis of mitochondrial DNA sequences. *Science* 255:737–739.

1000. Heim, J. L. 1976. Les hommes fossiles de la Ferrassie I. *Archives de l'Institut de Paléontologie Humaine* 35:1–331.

1001. Heinrich, B. E., Rose, M. D., Leakey, R. E., and Walker, A. C. 1993. Hominid radius from the Middle Pliocene of Lake Turkana, Kenya. *American Journal of Physical Anthropology* 92:139–148.

1002. Heinrich, W.-D. 1982. Zur Evolution und Biostratigraphie von *Arvicola* (Rodentia, Mammalia) im Pleistozän Europas. *Zeitschrift für Geologische Wissenschaften* 10:683–735.

1003. Heinrich, W.-D. 1987. Neue Ergebnisse zur Evolution und Biostratigraphie von *Arvicola* (Rodentia, Mammalia) im Quartär Europas. *Zeitschrift für Geologische Wissenschaften* 15:389–406.

1004. Heinz, E., Brunet, B., and Battail, B. 1981. A cercopithecoid primate from the late Miocene of Moloyan, Afghanistan, with remarks on *Mesopithecus*. *International Journal of Primatology* 2:273–284.

1005. Heinzelin, J. de. 1983. The Omo Group. *Musée Royale de l'Afrique Centrale, Science Géologique,* ser. 8, 85:1–365.

1006. Heinzelin, J. de. 1994. Rifting, a long-term African story, with considerations on early hominid habitats. In *Integrative paths to the past: Paleoanthropological advances in honor of F. Clark Howell,* ed. R. S. Corruccini and R. L. Ciochon, pp. 313–320. Englewood Cliffs, N.J.: Prentice Hall.

1007. Hendey, Q. B. 1974. The late Cenozoic Carnivora of the southwestern Cape Province. *Annals of the South African Museum* 63:1–369.

1008. Hendey, Q. B. 1981. Palaeoecology of the Late Tertiary fossil occurrences in "E" Quarry, Langebaanweg, South Africa, and a reinterpretation of their geological context. *Annals of the South African Museum* 84:1–104.

1009. Hendey, Q. B. 1982. *Langebaanweg: A record of past life.* Cape Town: South African Museum.

1010. Hendey, Q. B. 1984. Southern African late Quaternary vertebrates. In *Southern African prehistory and paleoenvironments,* ed. R. G. Klein, pp. 81–106. Rotterdam: A. A. Balkema.

1011. Hennig, G. J., and Grün, R. 1983. ESR dating in Quaternary geology. *Quaternary Science Reviews* 2:157–238.

1012. Hennig, G. J., Grün, R., and Brunnacker, K. 1983. Speleothems, travertines, and paleoclimates. *Quaternary Research* 20:1–29.

1013. Hennig, G. J., Herr, W., Weber, E., and Xirotiris, N. L. 1982. Petralona Cave dating controversy. *Nature* 299:281–282.

1014. Hennig, W. 1966. *Phylogenetic systematics.* Urbana: University of Illinois Press.

1015. Henri-Martin, G. 1965. La grotte de Fontéchevade. *Bulletin de la Association Française pour l'Étude du Quaternaire* 3–4:211–216.

1016. Henshilwood, C., and Sealy, J. 1997. Bone artefacts from the Middle Stone Age at Blombos Cave, southern Cape, South Africa. *Current Anthropology* 38:890–895.

1017. Hershkovitz, I., Speirs, M. S., Frayer, D., Nadel, D., Wish-Baratz, S., and Arensburg, B. 1995. Ohalo II H2: A 19,000-year-old skeleton from a water-logged site at the Sea of Galilee, Israel. *American Journal of Physical Anthropology* 96:215–234.

1018. Hester, J. J. 1987. The significance of accelerator dating in archaeological method and theory. *Journal of Field Archaeology* 14:445–451.

1019. Hewes, G. W. 1961. Food transport and the origin of hominid bipedalism. *American Anthropologist* 63:687–710.

1020. Hewes, G. W. 1964. Hominid bipedalism: Independent evidence for the food-carrying theory. *Science* 146:416–418.

1021. Higham, C. 1989. *The archaeology of mainland Southeast Asia.* Cambridge: Cambridge University Press.

1022. Hill, A. 1985. Early hominid from Baringo District, Kenya. *Nature* 315:222–224.

1023. Hill, A. 1994. Late Miocene and early Pliocene hominoids from Africa. In *Integrative paths to the past: Paleoanthropological advances in honor of F. Clark Howell,* ed. R. S. Corruccini and R. L. Ciochon, pp. 123–145. Englewood Cliffs, N.J.: Prentice Hall.

1024. Hill, A., Drake, R., Tauxe, L., Monaghan, M., Barry, J. C., Behrensmeyer, A. K., Curtis, G., Jacobs, B. F., Jacobs, L., Johnson, N., and Pilbeam, D. 1985. Neogene palaeontology and geochronology of the Baringo Basin, Kenya. *Journal of Human Evolution* 14:759–773.

1025. Hill, A., and Ward, S. 1988. Origin of the Hominidae: The record of African large hominoid evolution between 14 my and 4 my. *Yearbook of Physical Anthropology* 31:49–83.

1026. Hill, A., Ward, S., and Brown, B. 1992. Anatomy and age of the Lothagam mandible. *Journal of Human Evolution* 22:439–451.

1027. Hill, A., Ward, S., Deino, A., Curtis, G., and Drake, R. 1992. Earliest *Homo. Nature* 355:719–722.

1028. Hodell, D. A., Benson, R. H., Kent, D. V., Boersma, A., and Rakic-el Bied, K. 1994. Magnetostratigraphic, biostratigraphic and stable isotope stratigraphy of an Upper Miocene drill core from the Salé Briqueterie (northwestern Morocco): A high resolution chronology for the Messinian Stage. *Paleooceanography* 9:835–855.

1029. Hoffecker, J. F. 1986. Upper Paleolithic settlement on the Russian plain. Ph.D. diss., University of Chicago.

1030. Hoffecker, J. F. 1987. Upper Pleistocene loess stratigraphy and Paleolithic site chronology on the Russian plain. *Geoarchaeology* 2:259–284.

1031. Hoffecker, J. F. 1988. Early Upper Paleolithic sites of the European USSR. *British Archaeological Reports International Series* 437:237–272.

1032. Hoffecker, J. F. 1996. Introduction to the archaeology of Beringia. In *American beginnings: The prehistory and palaeoecology of Beringia*, ed. F. H. West, pp. 149–153. Chicago: University of Chicago Press.

1033. Hoffecker, J. F. 1999. Eastern Europe and modern human origins. *Evolutionary Anthropology.* In press.

1034. Hoffecker, J. F., and Baryshnikov, G. 1997. Neanderthal ecology: New data from the northern Caucasus. Unpublished manuscript.

1035. Hoffecker, J. F., Baryshnikov, G., and Potapova, O. 1991. Vertebrate remains from the Mousterian site of Il'skaya I (northern Caucasus, U.S.S.R.): New analysis and interpretation. *Journal of Archaeological Science* 18:113–147.

1036. Hoffecker, J. F., Powers, W. R., and Goebel, T. 1993. The colonization of Beringia and the peopling of the New World. *Science* 259:46–53.

1037. Hoffstetter, R. 1969. Un primate de l'Oligocène inférieur sud-américain: *Branisella boliviana* gen. et sp. nov. *Comptes Rendus de l'Académie des Sciences, Paris*, ser. D, 269:434–437.

1038. Hoffstetter, R. 1974. Phylogeny and geographical deployment of the primates. *Journal of Human Evolution* 3:327–350.

1039. Hoffstetter, R. 1977. Phylogénie des primates: Confrontation des résultats obtenus par les diverses voies d'approches de problème. *Bulletins et Mémoires de la Société Anthropologique de Paris*, ser. 13, 4:327–346.

1040. Hoffstetter, R. 1980. Origin and deployment of New World monkeys emphasizing the southern continents route. In *Evolutionary biology of the New World monkeys and continental drift*, ed. C. R. L. and A. B. Chiarelli, pp. 103–122. New York: Plenum Press.

1041. Hoffstetter, R. 1988. Relations phylogéniques et position systématique de *Tarsius:* Nouvelles controverses. *Comptes Rendus de l'Académie des Sciences*, ser. 2, 307:1837–1840.

1042. Hole, F., and Flannery, K. V. 1967. Prehistory of southwestern Iran: A preliminary report. *Proceedings of the Prehistoric Society* 33:147–206.

1043. Holliday, T. W. 1997. Body proportions in late Pleistocene Europe and modern human origins. *Journal of Human Evolution* 32:423–447.

1044. Holloway, R. L. 1970. Australopithecine endocast (Taung specimen, 1924): A new volume determination. *Science* 168:966–968.

1045. Holloway, R. L. 1975. Early hominid endocasts: Volumes, morphology, and significance for hominid evolution. In *Primate functional morphology and evolution*, ed. R. H. Tuttle, pp. 393–415. The Hague: Mouton.

1046. Holloway, R. L. 1980. Indonesian "Solo" (Ngandong) endocranial reconstructions: Some preliminary observations and comparisons with Neandertal and *Homo erectus* groups. *American Journal of Physical Anthropology* 53:285–295.

1047. Holloway, R. L. 1981. The Indonesian *Homo erectus* brain casts revisited. *American Journal of Physical Anthropology* 55:43–58.

1048. Holloway, R. L. 1981. Volumetric and asymmetry determinations on recent hominid endocasts: Spy I and II, Djebel Irhoud I, and the Salé *Homo erectus* specimens, with some notes on Neanderthal brain size. *American Journal of Physical Anthropology* 55:385–393.

1049. Holloway, R. L. 1982. *Homo erectus* brain endocasts: Volumetric and morphological observations with some comments on cerebral asymmetries. In *L'Homo erectus et la place de l'homme de Tautavel parmi les hominidés fossiles,* ed. M. A. de Lumley, pp. 355–369. Nice: Premier Congrès International de Paléontologie Humaine.

1050. Holloway, R. L. 1983. Cerebral brain endocast pattern of *Australopithecus afarensis* hominid. *Nature* 303:420–422.

1051. Holloway, R. L. 1983. Human brain evolution: A search for units, models and synthesis. *Canadian Journal of Anthropology* 3:215–230.

1052. Holloway, R. L. 1985. The poor brain of *Homo sapiens neanderthalensis,* see what you please. . . . In *Ancestors: The hard evidence,* ed. E. Delson, pp. 319–324. New York: Alan R. Liss.

1053. Holloway, R. L. 1988. "Robust" australopithecine brain endocasts: Some preliminary observations. In *Evolutionary history of the "robust" australopithecines,* ed. F. E. Grine, pp. 97–106. New York: Aldine de Gruyter.

1054. Holloway, R. L. 1991. The Neandertal brain: What was primitive. *American Journal of Physical Anthropology,* suppl., 12:94.

1055. Holloway, R. L. 1998. Language's source: A particularly human confluence of hard wiring and soft. *American Scientist* 86:184–186.

1056. Holloway, R. L., and de la Coste-Lareymondie, M. C. 1982. Brain endocast asymmetry in pongids and hominids: Some preliminary finds on the paleontology of cerebral dominance. *American Journal of Physical Anthropology* 58:101–110.

1057. Holloway, R. L., and Kimbel, W. H. 1986. Endocast morphology of Hadar hominid AL 162–28. *Nature* 321:536–537.

1058. Holmes, E. C., Pesole, G., and Saccone, C. 1989. Stochastic models of molecular evolution and the estimation of phylogeny and rates of nucleotide substitution in the hominoid primates. *Journal of Human Evolution* 18:775–794.

1059. Hooghiemstra, H. 1995. Environmental and paleoclimatic evolution in late Pliocene-Quaternary Colombia. In *Paleoclimate and evolution with emphasis on human origins,* ed. E. S. Vrba, G. H. Denton, T. C. Partridge, and L. H. Burckle, pp. 249–261. New Haven: Yale University Press.

1060. Hooker, P. J., and Miller, J. A. 1979. K-Ar dating of the Pleistocene fossil hominid site at Chesowanja, north Kenya. *Nature* 282:710–712.

1061. Hopkins, D. M. 1985. Comment on "Siberian Paleolithic archaeology: Approaches and analytic methods." *Current Anthropology* 26:371–372.

1062. Horai, S., Satta, Y., Hayasaka, K., Kondo, R., Inoue, T., Ishida, T., Hayashi, S., and Takahata, N. 1992. Man's place in Hominoidea revealed by mitochondrial DNA genealogy. *Journal of Molecular Evolution* 35:32–43.

1063. Horton, D. R. 1984. Red kangaroos: Last of the Australian megafauna. In *Quaternary extinctions: A prehistoric revolution*, ed. P. S. Martin and R. G. Klein, pp. 639–680. Tucson: University of Arizona Press.

1064. Houghton, P. 1993. Neandertal supralaryngeal vocal tract. *American Journal of Physical Anthropology* 90:139–146.

1065. Hovers, E., Rak, Y., and Kimbel, W. H. 1996. Neandertals of the Levant. *Archaeology* 1996:49–50.

1066. Hovers, E., Rak, Y., Lavi, R., and Kimbel, W. H. 1995. Hominid remains from Amud Cave in the context of the Levantine Middle Paleolithic. *Paléorient* 21:47–61.

1067. Howe, B. 1967. The Palaeolithic of Tangier, Morocco: Excavations at Capc Ashakar, 1939–1947. *Bulletin of the American School of Prehistoric Research* 22:1–200.

1068. Howell, F. C. 1951. The place of Neanderthal man in human evolution. *American Journal of Physical Anthropology* 9:379–416.

1069. Howell, F. C. 1957. The evolutionary significance of variation and varieties of "Neanderthal" man. *Quarterly Review of Biology* 32:330–347.

1070. Howell, F. C. 1958. Upper Pleistocene men of the southwest Asian Mousterian. In *Neanderthal centenary, 1856–1956*, ed. G. H. R. von Koenigswald, pp. 185–198. Utrecht: Kemink.

1071. Howell, F. C. 1960. European and northwest African Middle Pleistocene hominids. *Current Anthropology* 1:195–232.

1072. Howell, F. C. 1966. Observations on the earlier phases of the European Lower Paleolithic. *American Anthropologist* 68 (2/2): 88–201.

1073. Howell, F. C. 1978. Hominidae. In *Evolution of African mammals*, ed. V. J. Maglio and H. B. S. Cooke, pp. 154–248. Cambridge: Harvard University Press.

1074. Howell, F. C. 1978. Overview of the Pliocene and earlier Pleistocene of the lower Omo Basin, southern Ethiopia. In *Early hominids of Africa*, ed. C. Jolly, pp. 85–130. London: Duckworth.

1075. Howell, F. C. 1986. Variabilité chez *Homo erectus*, et problème de la présence de cette espèce en Europe. *Anthropologie* 90:447–481.

1076. Howell, F. C., Butzer, K. W., Freeman, L. G., and Klein, R. G. 1996 [1991]. Observations on the Acheulean occupation site of Ambrona (Soria Province, Spain) with particular reference to recent investigations (1980–1983) and the lower occupation. *Jahrbuch des Römisch-Germanischen Zentralmuseums Mainz* 38: 33–82.

1077. Howell, F. C., Cole, G. H., and Kleindienst, M. R. 1962. Isimila, an Acheulian occupation site in the Iringa Highlands. In *Actes du Quatrième Congrès Panafricain de Préhistoire et de l'Étude du Quaternaire*. Tervuren: Musée Royal de l'Afrique Centrale.

1078. Howell, F. C., Cole, G. H., Kleindienst, M. R., Szabo, B. J., and Oakley, K. P. 1972. Uranium series dating of bone from the Isimila prehistoric site, Tanzania. *Nature* 237:51–52.

1079. Howell, F. C., Haesaerts, P., and de Heinzelin, J. 1987. Depositional environments, archeological occurrences and hominids from Members E and F of the Shungura Formation (Omo Basin, Ethiopia). *Journal of Human Evolution* 16:665–700.

1080. Howells, W. W. 1966. *Homo erectus. Scientific American* 215 (5): 46–53.

1081. Howells, W. W. 1967. *Mankind in the making.* New York: Doubleday.

1082. Howells, W. W. 1973. Cranial variation in man: A study by multivariate analysis. *Peabody Museum Papers* 67:1–259.

1083. Howells, W. W. 1973. *The evolution of the genus* Homo. Reading, Mass.: Addison-Wesley.

1084. Howells, W. W. 1980. *Homo erectus*—who, when, and where: A survey. *Yearbook of Physical Anthropology* 23:1–23.

1085. Howells, W. W. 1982. Comment on "Upper Pleistocene hominid evolution in south-central Europe: A review of the evidence and analysis of trends." *Current Anthropology* 23:688–689.

1086. Howells, W. W. 1989. Skull shapes and the map: Craniometric analyses in the dispersion of modern *Homo. Papers of the Peabody Museum of Archaeology and Ethnology, Harvard University* 79:1–189.

1087. Howells, W. W. 1990. Review of *Eugene Dubois and the apeman from Java* (by Bert Theunissen). *American Journal of Physical Anthropology* 81:133–134.

1088. Howells, W. W. 1991. Review of *The Story of Peking man* (by Jia Lanpo and Huang Weiwen). *American Journal of Physical Anthropology* 85:237–238.

1089. Howells, W. W. 1995. Ethnic identification of crania from measurements. *Papers of the Peabody Museum of Archaeology and Ethnology, Harvard University* 82:1–108.

1090. Hublin, J.-J. 1985. Human fossils from the north African Middle Pleistocene and the origins of *Homo sapiens.* In *Ancestors: The hard evidence,* ed. E. Delson, pp. 282–288. New York: Alan R. Liss.

1091. Hublin, J.-J. 1986. Some comments on the diagnostic features of *Homo erectus. Anthropos* (Brno) 23:175–187.

1092. Hublin, J.-J. 1988. Les plus anciens représentants de la lignée prénéandertalienne. *Études et Recherches Archéologiques de l'Université de Liège* 30:81–94.

1093. Hublin, J.-J. 1990. Les peuplements paléolithiques de l'Europe: Un point de vue paléobiogéographique. *Mémoires du Musée de Préhistoire d'Île de France* 3:29–37.

1094. Hublin, J.-J. 1992. Le fémur humain pléistocène moyen de l'Aïn Maarouf (El Hajeb, Maroc). *Comptes Rendus de l'Académie des Sciences, Paris,* ser. 2, 314975–980.

1095. Hublin, J.-J. 1993. Recent human evolution in northwestern Africa. In *The origins of modern humans and the impact of chronometric dating,* ed. M. J. Aitken, C. B. Stringer, and P. A. Mellars, pp. 118–131. Princeton: Princeton University Press.

1096. Hublin, J.-J. 1996. The first Europeans. *Archaeology* 49 (1): 36–44.

1097. Hublin, J.-J., Barroso Ruiz, C., Medina Lara, P., Fontugne, M., and Reyss, J.-L. 1995. The Mousterian site of Zafarraya (Andalucia, Spain): Dating and implications on the palaeolithic peopling processes of western Europe. *Comptes Rendus de l'Académie des Sciences, Paris,* ser. 2a, 321:931–937.

1098. Hublin, J.-J., Spoor, F., Braun, M., and Zonneveld, F. 1996. A late Neanderthal associated with Upper Paleolithic artifacts. *Nature* 381:224–226.

1099. Hublin, J.-J., and Tillier, A.-M. 1988. Les enfants moustériens de Jebel Irhoud (Maroc): Comparaison avec les Néandertaliens juvéniles d'Europe. *Bulletin et Mémoire de la Société d'Anthropologie de Paris*, ser. 14, 5:237–246.

1100. Hublin, J.-J., Tillier, A.-M., and Tixier, J. 1987. L'humérus d'enfant moustérien (Homo 4) du Jebel Irhoud (Maroc) dans son contexte archéologique. *Bulletin et Mémoire de la Société d'Anthropologie de Paris*, ser. 14, 4:115–142.

1101. Huntley, D. J., Godfrey-Smith, D. I., and Thewalt, M. L. W. 1985. Optical dating of sediments. *Nature* 313:105–107.

1102. Hurford, A. J., Gleadow, A. J. W., and Naeser, C. W. 1976. Fission-track dating of pumice from the KBS tuffs. *Nature* 263:738–740.

1103. Hutterer, K. L. 1985. The Pleistocene archaeology of Southeast Asia in regional context. *Modern Quaternary Research in Southeast Asia* 9:1–23.

1104. Huxley, T. H. 1863. *Zoological evidences as to man's place in nature*. London: Williams and Norgate.

1105. Hylander, W. L., and Johnson, K. R. 1992. Strain gradients in the craniofacial region of primates. In *The biological mechanisms of tooth movement and craniofacial adaptation*, ed. Z. Davidovitch, pp. 559–569. Columbus: Ohio State University College of Dentistry.

1106. Imbrie, J., Berger, A., and Shackleton, N. J. 1993. Role of orbital forcing: A two-million-year perspective. In *Global changes and the perspective of the past*, ed. J. A. Eddy, pp. 263–277. London: John Wiley.

1107. Imbrie, J., Hays, J. D., Martinson, D. G., McIntyre, A., Mix, A. C., Morley, J. J., Pisias, N. G., Prell, W. L., and Shackleton, N. J. 1984. The orbital theory of Pleistocene climate: Support from a revised chronology of the marine delta 18O record. In *Milankovitch and climate: Understanding the response to astronomical forcing*, part 1, ed. A. L. Berger et al., pp. 169–305. Boston: Reidel.

1108. Imbrie, J., and Imbrie, K. P. 1979. *Ice Ages: Solving the mystery*. New York: Macmillan.

1109. Inskeep, R. R. 1962. The age of the Kondoa rock paintings in light of recent excavations at Kisese II rock shelter. In *Actes du Quatrième Congrès Panafricain de Préhistoire et de l'Étude du Quaternaire*, ed. G. Mortelmans and J. Nenquin, pp. 249–256. Tervuren: Musée Royal de l'Afrique Centrale.

1110. Inskeep, R. R. 1987. *Nelson Bay Cave, Cape Province, South Africa: The Holocene levels*. International Series 357. Oxford: British Archaeological Reports.

1111. Irish, J. D. 1998. Ancestral dental traits in recent sub-Saharan Africans and the origins of modern humans. *Journal of Human Evolution* 34:81–98.

1112. Isaac, G. L. 1967. The stratigraphy of the Peninj Group—early Middle Pleistocene formations west of Lake Natron, Tanzania. In *Background to evolution in Africa*, ed. W. W. Bishop and J. D. Clark, pp. 229–257. Chicago: University of Chicago Press.

1113. Isaac, G. L. 1975. Stratigraphy and cultural patterns in east

Africa during the middle ranges of Pleistocene time. In *After the australopithecines*, ed. K. W. Butzer and G. L. Isaac, pp. 543–569. The Hague: Mouton.

1114. Isaac, G. L. 1977. *Olorgesailie.* Chicago: University of Chicago Press.

1115. Isaac, G. L. 1978. The food sharing behavior of proto-human hominids. *Scientific American* 238 (4): 90–108.

1116. Isaac, G. L. 1981. Archaeological tests of alternative models of early human behavior: Excavation and experiments. *Philosophical Transactions of the Royal Society of London,* ser. B, 292: 177–188.

1117. Isaac, G. L. 1982. Early hominids and fire at Chesowanja, Kenya. *Nature* 296:870.

1118. Isaac, G. L. 1984. The archaeology of human origins: Studies of the Lower Pleistocene in east Africa: 1971–1981. In *Advances in world archaeology*, ed. F. Wendorf and A. E. Close, 3:1–87. New York: Academic Press.

1119. Isaac, G. L. 1986. Foundation stones: Early artifacts as indicators of activities and abilities. In *Stone Age prehistory*, ed. G. N. Bailey and P. Callow, pp. 221–242. Cambridge: Cambridge University Press.

1120. Isaac, G. L., and Curtis, G. H. 1974. The age of early Acheulian industries in east Africa—new evidence from the Peninj Group, Tanzania. *Nature* 249:624–627.

1121. Isaac, G. L., and Harris, J. W. K. 1978. Archaeology. In *Koobi Fora research project*, ed. M. G. Leakey and R. E. F. Leakey, 1:64–85. Oxford: Clarendon Press.

1122. Ishida, H., Pickford, M., Nakaya, H., and Nakano, Y. 1984. Fossil anthropoids from Nachola and Samburu Hills, Samburu District, Kenya. *African Studies Monographs* (Kyoto), suppl., 2:73–85.

1123. Ivanhoe, F. 1970. Was Virchow right about Neandertal? *Nature* 227:577–579.

1124. Jablonski, N. G., and Chaplin, G. 1994. Avant les premiers pas: L'origine de la bipédie. *Recherche* 261:80–81.

1125. Jackson, L. E., and Duk-Rodkin, A. 1996. Quaternary geology of the ice-free corridor: Glacial controls on the peopling of the New World. In *Prehistoric Mongoloid dispersals*, ed. T. Akazawa and E. J. E. Szathmáry, pp. 214–227. Oxford: Oxford University Press.

1126. Jacob, T. 1975. Morphology and paleoecology of early man in Java. In *Paleoanthropology, morphology and paleoecology*, ed. R. H. Tuttle, pp. 311–325. The Hague: Mouton.

1127. Jacob, T. 1978. The puzzle of Solo man. *Modern Quaternary Research in Southeast Asia* 4:31–40.

1128. Jacob, T. 1980. The *Pithecanthropus* of Indonesia: Phenotype, genetics and ecology. In *Current argument on early man: Report from a Nobel symposium*, ed. L.-K. Königsson, pp. 170–179. Oxford: Pergamon Press.

1129. Jacob, T., Soejono, R. P., Freeman, L. G., and Brown, F. H. 1978. Stone tools from mid-Pleistocene sediments in Java. *Science* 202:885–887.

1130. Jaeger, J.-J. 1975. The mammalian faunas and hominid fossils of the Middle Pleistocene in the Maghreb. In *After the australopithecines*, ed. K. W. Butzer and G. L. Isaac, pp. 375–397. The Hague: Mouton.

1131. James, S. R. 1989. Hominid use of fire in the Lower and Middle Pleistocene: A review of the evidence. *Current Anthropology* 30:1–26.

1132. Jánossy, D. 1975. Mid-Pleistocene microfaunas of continental Europe and adjoining areas. In *After the australopithecines*, ed. K. W. Butzer and G. L. Isaac, pp. 375–397. The Hague: Mouton.

1133. Jansen, E., and Sjoeholm, J. 1991. Reconstruction of glaciation over the past 6 myr from ice-borne deposits in the Norwegian Sea. *Nature* 349:600–603.

1134. Janus, C. G. 1975. The Peking man fossils: Progress of the search. In *Paleoanthropology, morphology and paleoecology*, ed. R. H. Tuttle, pp. 291–300. The Hague: Mouton.

1135. Jaubert, J., Lorblanchet, M., Laville, H., Slott-Moller, R., Turq, A., and Brugal, J.-P. 1990. *Les chaseurs d'aurochs de La Borde: Une site du Paléolithique moyen (Livernon, Lot)*. Documents d'Archéologie Française, 27. Paris: Éditions de la Maison des Sciences de l'Homme.

1136. Jelinek, A. J. 1977. The Lower Paleolithic: Current evidence and interpretations. *Annual Review of Anthropology* 6:11–32.

1137. Jelinek, A. J. 1981. The Middle Paleolithic in the southern Levant from the perspective of the Tabun Cave. In *Préhistoire de Levant*, ed. J. Cauvin and P. Sanlaville, pp. 265–280. Paris: Centre National de la Recherche Scientifique.

1138. Jelinek, A. J. 1982. The Middle Palaeolithic in the southern Levant, with comments on the appearance of modern *Homo sapiens*. *British Archaeological Reports International Series* 151:57–104.

1139. Jelinek, A. J. 1982. The Tabun Cave and Paleolithic man in the Levant. *Science* 216:1369–1375.

1140. Jelinek, A. J. 1992. The chronology of the Middle Paleolithic and the first appearance of early modern *Homo sapiens* in southwest Asia. In *The evolution and dispersal of modern humans in Asia*, ed. T. Akazawa, K. Aoki, and T. Kimura, pp. 253–275. Tokyo: Hokusen-Sha.

1141. Jelinek, A. J. 1994. Hominids, energy, environment, and behavior in the late Pleistocene. In *Origins of anatomically modern humans*, ed. M. H. Nitecki and D. V. Nitecki, pp. 67–92. New York: Plenum Press.

1142. Jellema, L. M., Latimer, B., and Walker, A. 1993. The skull. In *The Nariokotome* Homo erectus *skeleton*, ed. A. Walker and R. Leakey, pp. 294–325. Cambridge: Harvard University Press.

1143. Jia, L. 1980. *Early man in China*. Beijing: Foreign Languages Press.

1144. Jia, L. 1985. China's earliest Palaeolithic assemblages. In *Palaeoanthropology and Palaeolithic archaeology in the People's Republic of China*, ed. R. Wu and J. W. Olsen, pp. 135–145. Orlando, Fla.: Academic Press.

1145. Jia, L., and Huang, W. 1985. The late Palaeolithic of China. In *Palaeoanthropology and Palaeolithic archaeology in the People's Republic of China*, ed. R. Wu and J. W. Olsen, pp. 211–223. Orlando, Fla.: Academic Press.

1146. Jia, L., and Huang, W. 1990. *The story of Peking man: From archaeology to mystery*. Oxford: Oxford University Press.

1147. Johanson, D. C. 1989. A partial *Homo habilis* skeleton from

Olduvai Gorge, Tanzania: A summary of preliminary results. In *Hominidae: Proceedings of the Second International Congress of Human Paleontology*, ed. G. Giacobini, pp. 155–166. Milan: Jaca Book.

1148. Johanson, D. C., Edey, M. E. 1981. *Lucy: The beginnings of humankind*. New York: Simon and Schuster.

1149. Johanson, D. C., Masao, F. T., Eck, G. G., White, T. D., Walter, R. C., Kimbel, W. H., Asfaw, B., Manega, P., Ndessokia, P., and Suwa, G. 1987. New partial skeleton of *Homo habilis* from Olduvai Gorge, Tanzania. *Nature* 327:205–209.

1150. Johanson, D. C., Taieb, M., and Coppens, Y. 1982. Pliocene hominids from the Hadar Formation, Ethiopia (1973–1977): Stratigraphic, chronologic, and paleoenvironmental contexts, with notes on hominid morphology and systematics. *American Journal of Physical Anthropology* 57:373–402.

1151. Johanson, D. C., and White, T. D. 1979. A systematic assessment of early African hominids. *Science* 202:321–330.

1152. Johanson, D. C., White, T. D., and Coppens, Y. 1978. A new species of the genus *Australopithecus* (Primates: Hominidae) from the Pliocene of eastern Africa. *Kirtlandia* 28:2–14.

1153. Johnsen, S. J., Clausen, H. B., Dansgaard, W., Fuhrer, K., Gundestrup, N. S., Hammer, C. U., Iversen, P., Jouzel, J., Stauffer, B., and Steffersen, J. P. 1992. Irregular glacial interstadials recorded in a new Greenland ice core. *Nature* 359:311–313.

1154. Johnsen, S. J., Dansgaard, W., Clausen, H. B., and Langway, C. C. 1972. Oxygen isotope profiles through the Antarctic and Greenland ice sheets. *Nature* 235:429–434.

1155. Johnson, B. J., and Miller, G. H. 1997. Archaeological applications of amino acid racemization. *Archaeometry* 39:265–287.

1156. Jones, D. L., Brock, A., and McFadden, P. L. 1986. Palaeomagnetic results from the Kromdraai and Sterkfontein hominid sites. *South African Journal of Science* 82:160–163.

1157. Jones, P. R. 1980. Experimental butchery with modern stone tools and its relevance for Palaeolithic archaeology. *World Archaeology* 12:153–175.

1158. Jones, P. R. 1981. Experimental implement manufacture and use: A case study from Olduvai Gorge. *Philosophical Transactions of the Royal Society of London*, ser. B, 292:189–195.

1159. Jones, R. 1979. The fifth continent: Problems concerning the human colonization of Australia. *Annual Review of Anthropology* 8:445–466.

1160. Jones, R. 1989. East of Wallace's Line: Issues and problems in the colonization of the Australian continent. In *The human revolution: Behavioural and biological perspectives on the origins of modern humans*, ed. P. Mellars and C. Stringer, pp. 743–782. Edinburgh: Edinburgh University Press.

1161. Jones, R. 1990. From Kakadu to Kutikina: The southern continent at 18 000 years ago. In *The world at 18 000 BP*, vol. 2, *Low latitudes*, ed. C. Gamble and O. Soffer, pp. 264–295. London: Unwin Hyman.

1162. Jones, R. 1992. The human colonisation of the Australian continent. In *Continuity or replacement: Controversies in* Homo sapiens *evolution*, ed. G. Bräuer and F. H. Smith, pp. 289–301. Rotterdam: A. A. Balkema.

1163. Jouzel, J., Barkov, N. I., Barnola, J. M., Bender, M., Chappellaz, J., Genthon, C., Kotlyakov, V. M., Lipenkov, V., Lorius, C., Petit, J. R., Raynaud, D., Raisbeck, G., Ritz, C., Sowers, T., Stievenard, M., Yiou, F., and Yiou, P. 1993. Extending the Vostok ice-core record of palaeoclimate to the penultimate glacial period. *Nature* 364: 407–412.

1164. Jouzel, J., Lorius, C., Petit, J. R., Genthon, C., Barkov, N. I., Kotlyakov, V. M., and Petrov, V. M. 1987. Vostok ice core: A continuous isotope temperature record over the last climatic cycle (160,000 years). *Nature* 329:403–407.

1165. Jouzel, J., Raisbeck, G., Benoist, J. P., Yiou, F., Lorius, C., Raynaud, D., Petit, J. R., Barkov, N. I., Korotkevitch, Y. S., and Kotlyakov, V. M. 1989. A comparison of deep Antarctic ice cores and their implications for climate between 65,000 and 15,000 years ago. *Quaternary Research* 31:135–150.

1166. Juel Jensen, H. 1988. Functional analysis of prehistoric flint tools by high-power microscopy: A review of west European research. *Journal of World Prehistory* 2:53–88.

1167. Jungers, W. L. 1988. New estimates of body size in australopithecines. In *The Evolutionary history of the "robust" australopithecines*, ed. F. E. Grine, pp. 115–116. New York: Aldine de Gruyter.

1168. Kahlke, H. D. 1962. Zur relativen Chronologie ostasiatischer Mittelpleistozän-Faunen und Hominoidea-Funde. In *Evolution und Hominisation*, ed. G. Kurth, pp. 84–107. Stuttgart: Gustav Fischer.

1169. Kahn, P., and Gibbons, A. 1997. DNA from an extinct human. *Science* 277:176–178.

1170. Kalb, J. E., Jolly, C., Mebrate, A., Tebedge, S., Smart, C., Oswald, E. B., Cramer, D., Whitehead, P., Wood, C. B., Conroy, G. C., Adefris, T., Sperling, L., and Kana, B. 1982. Fossil mammals and artifacts from the Awash Group, Middle Awash Valley, Afar, Ethiopia. *Nature* 298:25–29.

1171. Kalb, J. E., Jolly, C. J., Oswald, E. B., and Whitehead, P. F. 1984. Early hominid habitation in Ethiopia. *American Scientist* 72: 168–178.

1172. Kalb, J. E., Oswald, E. B., Tebedge, S., Mebrate, A., Tola, E., and Peak, D. 1982. Geology and stratigraphy of Neogene deposits, Middle Awash Valley, Ethiopia. *Nature* 298:17–25.

1173. Kallfelz-Klemish, C. F., and Franciscus, R. G. 1997. Static bite force production in Neandertals and modern humans. *American Journal of Physical Anthropology*, suppl., 24:140.

1174. Kamminga, J. 1992. New interpretations of the Upper Cave, Zhoukoudian. In *The evolution and dispersal of modern humans in Asia*, ed. T. Akazawa, K. Aoki, and T. Kimura, pp. 379–400. Tokyo: Hokusen-Sha.

1175. Kamminga, J., and Wright, R. V. S. 1988. The Upper Cave at Zhoukoudian and the origin of the Mongoloids. *Journal of Human Evolution* 17:739–767.

1176. Kappelman, J. 1984. Plio-Pleistocene environments of Bed I and lower Bed II, Olduvai Gorge, Tanzania. *Palaeogeography, Palaeoclimatology, Palaeoecology* 48:171–196.

1177. Kappelman, J. 1986. Plio-Pleistocene marine-continental correlation using habitat indicators from Olduvai Gorge, Tanzania. *Quaternary Research* 25:141–149.

1178. Kappelman, J. 1992. The age of the Fayum primates as determined by paleomagnetic reversal stratigraphy. *Journal of Human Evolution* 22:495–503.

1179. Kappelman, J. 1993. The attraction of paleomagnetism. *Evolutionary Anthropology* 2:89–99.

1180. Kappelman, J., and Fleagle, J. G. 1995. Age of early hominids. *Nature* 376:558–559.

1181. Kappelman, J., Simons, E. L., and Swisher, C. C. 1992. New age determinations for the Eocene-Oligocene boundary sediments in the Fayum Depression, northern Egypt. *Journal of Geology* 100:647–668.

1182. Kappelman, J., Swisher, C. C., Fleagle, J. G., Yirga, S., Bown, T. M., and Feseha, M. 1996. Age of *Australopithecus afarensis* from Fejej, Ethiopia. *Journal of Human Evolution* 30:39–46.

1183. Karavanic, I. 1995. Upper Paleolithic occupation levels and late occurring Neandertal at Vindija Cave (Croatia) in the context of central Europe and the Balkans. *Journal of Anthropological Research* 51:9–35.

1184. Kashiwaya, K., Yamamoto, A., and Fukuyama, K. 1988. Statistical analysis of grain size distribution in Pleistocene sediments from Lake Biwa, Japan. *Quaternary Research* 30:12–18.

1185. Kaufulu, Z. M., and Stern, N. 1987. The first stone artefacts found *in situ* within the Plio-Pleistocene Chiwondo Beds in northern Malawi. *Journal of Human Evolution* 16:729–740.

1186. Kay, R. F. 1981. The nut-crackers: A new theory of the adaptation of the Ramapithecinae. *American Journal of Physical Anthropology* 55:141–151.

1187. Kay, R. F. 1990. The phyletic relationships of extant and fossil Pitheciinae (Platyrrhini, Anthropoidea). *Journal of Human Evolution* 19:175–208.

1188. Kay, R. F., Fleagle, J. C., and Simons, E. L. 1981. A revision of the Oligocene apes of the Fayum Province, Egypt. *American Journal of Physical Anthropology* 55:293–322.

1189. Kay, R. F., and Grine, F. E. 1988. Tooth morphology, wear and diet in *Australopithecus* and *Paranthropus* from southern Africa. In *Evolutionary history of the "robust" australopithecines*, ed. F. E. Grine, pp. 427–448. New York: Aldine de Gruyter.

1190. Kay, R. F., Madden, R. H., Plavcan, J. M., Cifelli, R. L., and Diaz, J. G. 1987. *Stirtonia victoriae*, a new species of Miocene Colombian primate. *Journal of Human Evolution* 16:173–196.

1191. Kay, R. F., and Simons, E. L. 1980. The ecology of Oligocene African Anthropoidea. *International Journal of Primatology* 1:21–37.

1192. Kay, R. F., and Simons, E. L. 1983. Dental formulae and dental eruption patterns in Parapithecidae (Primates, Anthropoidea). *American Journal of Physical Anthropology* 62:363–375.

1193. Kay, R. F., Thewissen, J. G. M., and Yoder, A. D. 1992. Cranial anatomy of *Ignacius graybullianus* and the affinities of the Plesiadapiformes. *American Journal of Physical Anthropology* 89:477–498.

1194. Kay, R. F., Thorington, R. W., and Houde, P. 1990. Eocene plesiadapiform shows affinities with flying lemurs not primates. *Nature* 345:342–344.

1195. Keeley, L. H. 1977. The functions of Paleolithic flint tools. *Scientific American* 237 (5): 108–126.

1196. Keeley, L. H. 1980. *Experimental determination of stone tool use: A microwear analysis.* Chicago: University of Chicago Press.

1197. Keeley, L. H., and Toth, N. 1981. Microwear polishes on early stone tools from Koobi Fora, Kenya. *Nature* 293:464–465.

1198. Keith, A. 1928. *The antiquity of man.* Philadelphia: Lippincott.

1199. Keller, C. M. 1973. Montagu Cave in prehistory. *University of California Anthropological Records* 28:1–150.

1200. Kelley, J. 1986. Species recognition and sexual dimorphism in *Proconsul* and *Rangwapithecus. Journal of Human Evolution* 15:461–495.

1201. Kelley, J., and Etler, D. 1989. Hominoid dental variability and species number at the late Miocene site of Lufeng, China. *American Journal of Primatology* 18:15–34.

1202. Kelley, J., and Pilbeam, D. 1986. The Dryopithecines: Taxonomy, comparative anatomy, and phylogeny of Miocene large hominoids. In *Comparative primate biology*, ed. D. R.. Swindler and J. Erwin, 1:361–411. New York: Alan R. Liss.

1203. Kelley, J., and Qinghua, X. 1991. Extreme sexual dimorphism in a Miocene hominoid. *Nature* 352:151–153.

1204. Kelly, R. L. 1983. Hunter-gatherer mobility strategies. *Journal of Anthropological Research* 393:277–306.

1205. Kelly, R. L. 1992. Mobility/sedentism: Concepts, archaeological measures, and effects. *Annual Review of Anthropology* 21:43–66.

1206. Kennedy, G. E. 1983. Some aspects of femoral morphology in *Homo erectus. Journal of Human Evolution* 12:587–616.

1207. Kennedy, G. E. 1984. Are the Kow Swamp hominids "archaic"? *American Journal of Physical Anthropology* 65:163–168.

1208. Kennedy, G. E. 1984. The emergence of *Homo sapiens:* The postcranial evidence. *Man* 19:94–110.

1209. Kennedy, K. A. R. 1975. *Neanderthal man.* Minneapolis: Burgess.

1210. Kennedy, K. A. R. 1979. The deep skull of Niah: An assessment of twenty years of speculation concerning its evolutionary significance. *Asian Perspectives* 20:32–50.

1211. Kennedy, K. A. R., and Deraniyagala, S. U. 1989. Fossil remains of 28,000 year-old hominids from Sri Lanka. *Current Anthropology* 30:394–399.

1212. Kennedy, K. A. R., Sonakia, A., Chiment, J., and Verma, K. K. 1991. Is the Narmada hominid an Indian *Homo erectus? American Journal of Physical Anthropology* 86:475–496.

1213. Kennett, J. P. 1995. A review of polar climatic evolution during the Neogene, based on the marine sediment record. In *Paleoclimate and evolution with emphasis on human origins*, ed. E. S. Vrba, G. H. Denton, T. C. Partridge, and L. H. Burckle, pp. 49–64. New Haven: Yale University Press.

1214. Kerr, R. A. 1996. New mammal data challenge evolutionary pulse theory. *Science* 273:431–432.

1215. Keyser, A. W. 1998. Drimolen: Excursion guide. In *Dual Congress excursion guide to Gladysvale and Drimolen*, ed. L. Berger and A. Keyser, p. 9–13. Pretoria: Desktop Creations.

1216. Kibunjia, M. 1994. Pliocene archaeological occurrences in the Lake Turkana Basin. *Journal of Human Evolution* 27:159–171.

1217. Kibunjia, M., Roche, H., Brown, F. H., and Leakey, R. E. 1992. Pliocene and Pleistocene archaeological sites of Lake Turkana, Kenya. *Journal of Human Evolution* 23:432–438.

1218. Kidder, J. H., Jantz, R. L., and Smith, F. H. 1992. Defining modern humans: A multivariate approach. In *Continuity or replacement: Controversies in* Homo sapiens *evolution*, ed. G. Bräuer and F. H. Smith, pp. 157–177. Rotterdam: A. A. Balkema.

1219. Kimbel, W. H. 1995. Hominid speciation and Pliocene climatic change. In *Paleoclimate and evolution with emphasis on human origins*, ed. E. S. Vrba, G. H. Denton, T. C. Partridge, and L. H. Burckle, pp. 425–437. New Haven: Yale University Press.

1220. Kimbel, W. H., Johanson, D. C., and Rak, Y. 1994. The first skull and other new discoveries of *Australopithecus afarensis* at Hadar, Ethiopia. *Nature* 368:449–451.

1221. Kimbel, W. H., and Rak, Y. 1993. The importance of species taxa in paleoanthropology and an argument for the phylogenetic concept of the species category. In *Species, species concepts, and primate evolution*, ed. W. H. Kimbel and L. B. Martin, pp. 461–484. New York: Alan R. Liss.

1222. Kimbel, W. H., Walter, R. C., Johanson, D. C., Reed, K. E., Aronson, J. L., Assefa, Z., Marean, C. W., Eck, G. G., Bobe, R., Hovers, E., Rak, Y., Vondra, C., Yemane, T., York, D., Chen, Y., Evensen, N. M., and Smith, P. E. 1996. Late Pliocene *Homo* and Oldowan tools from the Hadar Formation (Kada Hadar Member), Ethiopia. *Journal of Human Evolution* 31:549–561.

1223. Kimbel, W. H., and White, T. D. 1988. Variation, sexual dimorphism and the taxonomy of *Australopithecus*. In *Evolutionary history of the "robust" australopithecines*, ed. F. E. Grine, pp. 175–192. New York: Aldine de Gruyter.

1224. Kimbel, W. H., White, T. D., and Johanson, D. C. 1984. Cranial morphology of *Australopithecus afarensis:* A comparative study based on a composite reconstruction of the adult skull. *American Journal of Physical Anthropology* 64:337–388.

1225. Kimbel, W. H., White, T. D., and Johanson, D. C. 1985. Craniodental morphology of the hominids from Hadar and Laetoli: Evidence of *"Paranthropus"* and *Homo* in the mid-Pliocene of eastern Africa? In *Ancestors: The hard evidence*, ed. E. Delson, pp. 120–137. New York: Alan R. Liss.

1226. Kimbel, W. H., White, T. D., and Johanson, D. C. 1988. Implications of KNM-WT 17000 for the evolution of "robust" australopithecines. In *The evolutionary history of the "robust" australopithecines*, ed. F. E. Grine, pp. 259–268. New York: Aldine de Gruyter.

1227. Kitagawa, H., and van der Plicht, J. 1998. Atmospheric radiocarbon calibration to 45,000 yr B.P.: Late Glacial fluctuations and cosmogenic isotope production. *Science* 279:1187–1190.

1228. Klein, R. G. 1969. *Man and culture in the late Pleistocene: A case study.* San Francisco: Chandler.

1229. Klein, R. G. 1969. The Mousterian of European Russia. *Proceedings of the Prehistoric Society* 35:77–111.

1230. Klein, R. G. 1971. The Pleistocene prehistory of Siberia. *Quaternary Research* 1:133–161.

1231. Klein, R. G. 1973. Geological antiquity of Rhodesian man. *Nature* 244:311–312.

1232. Klein, R. G. 1973. *Ice-Age hunters of the Ukraine.* Chicago: University of Chicago Press.

1233. Klein, R. G. 1976. The mammalian fauna of the Klasies River Mouth sites, southern Cape Province, South Africa. *South African Archaeological Bulletin* 31:75–96.

1234. Klein, R. G. 1976. A preliminary report on the Duinefontein 2 "Middle Stone Age" open-air site (Melkbosstrand, south-western Cape Province, South Africa). *South African Archaeological Bulletin* 31:12–20.

1235. Klein, R. G. 1978. The fauna and overall interpretation of the "Cutting 10" Acheulean site at Elandsfontein Hopefield), south-western Cape Province, South Africa. *Quaternary Research* 10:69–83.

1236. Klein, R. G. 1979. Stone Age exploitation of animals in southern Africa. *American Scientist* 67:151–160.

1237. Klein, R. G. 1980. Environmental and ecological implications of large mammals from Upper Pleistocene and Holocene sites in southern Africa. *Annals of the South African Museum* 81:223–283.

1238. Klein, R. G. 1983. Palaeoenvironmental implications of Quaternary large mammals in the Fynbos Biome. *South African National Scientific Programmes Reports* 75:116–138.

1239. Klein, R. G. 1984. Mammalian extinctions and Stone Age people in Africa. In *Quaternary extinctions: A prehistoric revolution,* ed. P. S. Martin and R. G. Klein, pp. 553–573. Tucson: University of Arizona Press.

1240. Klein, R. G. 1987. Problems and prospects in understanding how early people exploited animals. In *The evolution of human hunting,* ed. M. H. Nitecki and D. V. Nitecki, pp. 11–45. New York: Plenum Press.

1241. Klein, R. G. 1988. The archaeological significance of animal bones from Acheulean sites in southern Africa. *African Archaeological Review* 6:3–25.

1242. Klein, R. G. 1992. The archeology of modern human origins. *Evolutionary Anthropology* 1:5–14.

1243. Klein, R. G. 1994. Southern Africa before the Iron Age. In *Integrative paths to the past: Paleoanthropological advances in honor of F. Clark Howell,* ed. R. S. Corruccini, and R. L. Ciochon, pp. 471–519. Englewood Cliffs, N.J.: Prentice Hall.

1244. Klein, R. G., and Cruz-Uribe, K. 1984. *The analysis of animal bones from archaeological sites.* Chicago: University of Chicago Press.

1245. Klein, R. G., and Cruz-Uribe, K. 1989. Faunal evidence for prehistoric herder-forager activities at Kasteelberg, Vredenburg Peninsula, western Cape Province, South Africa. *South African Archaeological Bulletin* 44:82–97.

1246. Klein, R. G., and Cruz-Uribe, K. 1991. The bovids from Elandsfontein, South Africa, and their implications for the age, palaeoenvironment, and origins of the site. *African Archaeological Review* 9:21–79.

1247. Klein, R. G., and Cruz-Uribe, K. 1996. Exploitation of large bovids and seals at middle and later Stone Age sites in South Africa. *Journal of Human Evolution* 31:315–334.

1248. Klein, R. G., Cruz-Uribe, K., and Beaumont, P. B. 1991. Environmental, ecological, and paleoanthropological implications of the late Pleistocene mammalian fauna from Equus Cave, northern Cape Province, South Africa. *Quaternary Research* 36:94–110.

1249. Kleindienst, M. R. 1961. Variability within the late Acheulean assemblage in eastern Africa. *South African Archaeological Bulletin* 16:35–51.

1250. Klicka, J., Zink, R. M. 1997. The importance of recent ice ages in speciation: A failed paradigm. *Science* 277:1666–1669.

1251. Klíma, B. 1962. The first ground-plan of an Upper Paleolithic loess settlement in middle Europe and its meaning. In *Courses towards urban life*, ed. R. J. Braidwood and G. B. Willey, pp. 193–210. Chicago: Aldine.

1252. Klíma, B. 1963. *Dolní Vestonice.* Prague: Ceskoslvenská Akademie Ved.

1253. Klíma, B. 1987. A triple burial from the Upper Paleolithic of Dolní Vestonice. *Journal of Human Evolution* 16:831–835.

1254. Knight, A., Batzer, M. A., Stoneking, M., Tiwari, H. K., Scheer, W. D., Herrera, R. J., and Deininger, P. L. 1996. DNA sequences of Alu elements indicate a recent replacement of the human autosomal genetic complement. *Proceedings of the National Academy of Sciences* 93:4360–4364.

1255. Koenigswald, G. H. R. von. 1962. *The evolution of man.* Ann Arbor: University of Michigan Press.

1256. Koenigswald, G. H. R. von. 1975. Early man in Java: Catalogue and problems. In *Paleoanthropology, morphology, and paleoecology*, ed. R. H. Tuttle, pp. 303–309. The Hague: Mouton.

1257. Koenigswald, G. H. R. von, and Weidenreich, F. 1939. The relationship between *Pithecanthropus* and *Sinanthropus. Nature* 144:926–929.

1258. Koenigswald, W. 1973. Veränderungen in der Kleinsäugerfauna von Mitteleuropa zwischsen Cromer und Eem (Pleistozän). *Eiszeitalter und Gegenwart* 23–24:159–167.

1259. Kolfschoten, T. van. 1994. Mammalian remains in a palaeolithic context. In *Archaeology, methodology and the organisation of research*, ed. S. Milliken and C. Peretto, pp. 19–35. Forli: A. B. A. C. O. Edizioni.

1260. Kolossov, Y. G., Kharitonov, V. M., and Yakimov, V. P. 1975. Paleoanthropic specimens from the site of Zaskalnaya VI in the Crimea. In *Paleoanthropology, morphology and paleoecology*, ed. R. H. Tuttle, pp. 419–428. The Hague: Mouton.

1261. Koop, B. F., Siemieniak, D., Slightom, J. L., Goodman, M., Dunbar, J., Wright, P. C., and Simons, E. L. 1989. *Tarsius* delta- and beta-globin genes: Conversions, evolution, and systematic implications. *Journal of Biological Chemistry* 264:68–79.

1262. Kozlowski, J. K., ed. 1982. *Excavation in the Bacho Kiro Cave (Bulgaria): Final report.* Warsaw: Panstwowe Wydawnictwo Naukowe.

1263. Kozlowski, J. K. 1983. Le Paléolithique supérieur en Pologne. *Anthropologie* 87:49–83.

1264. Kozlowski, J. K. 1990. Certains aspects techno-morphologiques des pointes foliacées de la fin du Paléolithique moyen au début du Paléolithique supérieur en Europe centrale. In *Paléolithique moyen récent et Paléolithique supérieur ancien en Europe*, ed. C. Farizy, 3:125–134. Mémoires 3. Nemours: Musée de Préhistoire d'Île de France.

1265. Kozlowski, J. K. 1990. A multiaspectual approach to the origins of the Upper Palaeolithic in Europe. In *The emergence of modern humans: An archaeological perspective*, ed. P. A. Mellars, pp. 419–437. Ithaca: Cornell University Press.

1266. Kozlowski, J. K. 1990. Northern central Europe c. 18 000 BP. In *The world at 18 000 BP*, vol. 1, *High latitudes*, ed. O. Soffer and C. Gamble, pp. 204–227. London: Unwin Hyman.

1267. Kozlowski, J. K. 1992. The Balkans in the Middle and Upper Paleolithic: The gateway to Europe or a cul-de-sac? *Proceedings of the Prehistoric Society* 58:1–20.

1268. Kozlowski, J. K. 1992. *L'art de la préhistoire en Europe oriental.* Paris: CNRS Éditions.

1269. Kozlowski, J. K., and Kozlowski, S. K. 1979. *Upper Paleolithic and Mesolithic in Europe: Taxonomy and paleohistory.* Warsaw: Polska Akademia Nauk.

1270. Kozlowski, J. K., and Kubiak, H. 1971. Premières huttes d'habitation du Paléolithique supérieur en os de mammouth découvertes en Pologne. *Anthropologie* 75:245–256.

1271. Kozlowski, J. K., and Kubiak, H. 1972. Late Palaeolithic dwellings made of mammoth bones in south Poland. *Nature* 237:463–464.

1272. Kozlowski, J. K., Laville, H., and Ginters, B., eds. 1992. *Temnata Cave: Excavations in the Karulkovo Karst Area, Bulgaria*, vol. 1, part 1. Cracow: Jagellonian University Press.

1273. Kozlowski, J. K., van Vliet, B., Sachse-Kozlowska, E., Kubiak, H., and Zakrzewska, G. 1974. Upper Palaeolithic site with dwellings of mammoth bones—Cracow, Spadzista Street B. *Folia Quaternaria* 44:1–110.

1274. Krause, D. W. 1991. Were paromomyids gliders? Maybe, maybe not. *Journal of Human Evolution* 21:177–188.

1275. Krings, M., Stone, A., Schmitz, R. W., Krainitzki, H., Stoneking, M., and Pääbo, S. 1997. Neanderthal DNA sequences and the origin of modern humans. *Cell* 90:19–30.

1276. Kroll, E., and Isaac, G. L. 1984. Configurations of artifacts and bones at early Pleistocene sites in east Africa. In *Intrasite spatial analysis in archaeology*, ed. H. Hietala, pp. 4–31. Cambridge: Cambridge University Press.

1277. Kroll, E. M. 1994. Behavioral implications of Plio-Pleistocene archaeological site structure. *Journal of Human Evolution* 27:107–138.

1278. Ku, T.-L. 1976. The uranium-series methods of age determination. *Annual Review of Earth and Planetary Sciences* 4:347–379.

1279. Kuhn, S. L. 1995. *Mousterian lithic technology: An ecological perspective.* Princeton: Princeton University Press.

1280. Kukla, G., An, Z. S., Melice, J. L., Gavin, J., and Xiao, J. L. 1990. Chronostratigraphy of Chinese loess. *Transactions of the Royal Society of Edinburgh: Earth Sciences* 81:263–288.

1281. Kukla, G., Heller, F., Ming, L. X., Chun, X. T., Sheng, S. T., and Sheng, A. Z. 1988. Pleistocene climates in China dated by magnetic susceptibility. *Geology* 16:811–814.

1282. Kukla, G. J. 1975. Loess stratigraphy of central Europe. In *After the australopithecines*, ed. K. W. Butzer and G. L. Isaac, pp. 99–188. The Hague: Mouton.

1283. Kukla, G. J. 1987. Loess stratigraphy in central China. *Quaternary Science Reviews* 6:191–219.

1284. Kuman, K. 1994. The archaeology of Sterkfontein—past and present. *Journal of Human Evolution* 27:471–495.

1285. Kuman, K. 1994. The archaeology of Sterkfontein: Preliminary findings on site formation and cultural change. *South African Journal of Science* 90:215–219.

1286. Kuman, K. 1996. The Oldowan Industry from Sterkfontein: Raw materials and core forms. In *Aspects of African archaeology: Papers from the Tenth Congress of the PanAfrican Association for Prehistory and Related Studies*, ed. G. Pwiti and R. Soper, pp. 139–146. Harare: University of Zimbabwe.

1287. Kuman, K., and Clarke, R. J. 1986. Florisbad—new investigation at a Middle Stone Age hominid site in South Africa. *Geoarchaeology* 1:103–125.

1288. Kuman, K., Field, A. S., and Thackeray, J. F. 1997. Discovery of new artefacts at Kromdraai. *South African Journal of Science* 93:187–193.

1289. Kuman, K., Inbar, M., and Clarke, R. J. 1999. Palaeoenvironments and cultural sequence of the Florisbad Middle Stone Age hominid site, South Africa. *Journal of Archaeological Science*. In press.

1290. Kurtén, B. 1976. *The cave bear story: Life and death of a vanished animal.* New York: Columbia University Press.

1291. Kurtén, B. 1983. Faunal sequence from Petralona Cave. *Anthropos* (Greece) 10:53–59.

1292. Kurtén, B., and Anderson, E. 1980. *Pleistocene mammals of North America.* New York: Columbia University Press.

1293. Kuzmin, Y. V. 1994. Prehistoric colonization of northeastern Siberia and migration to America: Radiocarbon evidence. *Radiocarbon* 36:367–376.

1294. Kuzmin, Y. V. 1997. Chronology of Palaeolithic Siberia and the Russian Far East. *Review of Archaeology* 18:33–39.

1295. Kuzmin, Y. V., and Tankersley, K. B. 1996. The colonization of Eastern Siberia: An evaluation of the Paleolithic age radiocarbon dates. *Journal of Archaeological Science* 23:577–585.

1296. Lahr, M. M. 1994. The multiregional model of modern human origins: A reassessment of its morphological basis. *Journal of Human Evolution* 26:23–56.

1297. Lahr, M. M. 1995. Patterns of modern human diversification: Implications for Amerindian origins. *Yearbook of Physical Anthropology* 38:163–198.

1298. Lahr, M. M. 1996. *The evolution of modern human diversity: A study of cranial variation.* Cambridge: University of Cambridge Press.

1299. Lahr, M. M. 1997. History in the bones. *Evolutionary Anthropology* 6:2–6.

1300. Lahr, M. M., and Foley, R. 1994. Multiple dispersals and modern human origins. *Evolutionary Anthropology* 3:48–60.

1301. Laitman, J. T. 1985. Evolution of the hominid upper respiratory tract: The fossil evidence. In *Hominid evolution: Past, present, and future*, ed. P. V. Tobias, pp. 281–286. New York: Alan R. Liss.

1302. Laitman, J. T., and Heimbuch, R. C. 1982. The basicranium of Plio-Pleistocene hominids as an indicator of their upper respiratory systems. *American Journal of Physical Anthropology* 59:323–344.

1303. Laitman, J. T., Heimbuch, R. C., and Crelin, C. S. 1979. The basicranium of fossil hominids as an indicator of their upper respiratory systems. *American Journal of Physical Anthropology* 51:15–34.

1304. Laporte, L. F., and Zihlman, A. L. 1983. Plates, climate and hominoid evolution. *South African Journal of Science* 79:96–110.

1305. Larichev, V., Khol'ushkin, U., and Laricheva, I. 1990. The Upper Paleolithic of northern Asia: Achievements, problems, and perspectives. 2. Central and eastern Siberia. *Journal of World Prehistory* 4:347–385.

1306. Larichev, V., Khol'ushkin, U., and Laricheva, I. 1992. The Upper Paleolithic of northern Asia: Achievements, problems, and perspectives. 3. Northeastern Siberia and the Russian Far East. *Journal of World Prehistory* 6:441–476.

1307. Larick, R., and Ciochon, R. 1996. The first Asians. *Natural History* 105 (1): 51–53.

1308. Lartet, L. 1868. Une sépulture des troglodytes du Périgord (crânes des Eyzies). *Bulletin de la Société d'Anthropologie de Paris* 3:335–349.

1309. Latham, A. G., and Schwartz, H. P. 1992. The Petralona hominid site: Uranium-series re-analysis of "Layer 10" calcite and associated palaeomagnetic analyses. *Archaeometry* 34:135–140.

1310. Latimer, B., Ohman, J. C., and Lovejoy, C. O. 1987. Talocrural joint in African hominoids: Implications for *Australopithecus afarensis*. *American Journal of Physical Anthropology* 74:155–175.

1311. Lau, B., Blackwell, B. A. B., Schwarcz, H. P., Turk, I., and Blickstein, J. I. 1997. Dating a flautist? Using ESR (electron spin resonance) in the Mousterian Cave deposits at Divje Babe I, Slovenia. *Geoarchaeology* 12:507–536.

1312. Laville, H., Raynal, J.-P., and Texier, J.-P. 1986. Le dernier interglaciare et le cycle climatique würmien dans le sud-ouest et le Massif Central français. *Bulletin de l'Association Française pour l'Étude du Quaternaire* 23:35–46.

1313. Laville, H., Rigaud, J.-P., and Sackett, J. 1980. *Rock shelters of the Perigord*. New York: Academic Press.

1314. Le Gros Clark, W. E. 1938. The endocranial cast of the Swanscombe bones. *Journal of the Royal Anthropological Institute* 68:61–67.

1315. Le Gros Clark, W. E. 1955. *The fossil evidence for human evolution*. Chicago: University of Chicago Press.

1316. Le Gros Clark, W. E. 1960. *The antecedents of man*. Chicago: Quadrangle Books.

1317. Le Gros Clark, W. E. 1964. *The fossil evidence for human evolution*. Chicago: Quadrangle Books.

1318. Le Gros Clark, W. E. 1967. *Man-apes or ape-men?* New York: Holt, Rinehart and Winston.

1319. Le Mort, F. 1989. Traces de décharnement sur les ossements néandertaliens de Combe-Grenal (Dordogne). *Bulletin de la Société Préhistorique Française* 86:79–97.

1320. Leakey, D. M., Tobias, P. V., Martyn, J. E., and Leakey, R. E. 1969. An Acheulian industry with prepared core technique and the discovery of a contemporary hominid at Lake Baringo, Kenya. *Proceedings of the Prehistoric Society* 25:48–76.

1321. Leakey, L. S. B. 1931. *The Stone Age cultures of Kenya Colony.* Cambridge: Cambridge University Press.

1322. Leakey, L. S. B. 1935. *The Stone Age races of Kenya.* London: Oxford University Press.

1323. Leakey, L. S. B. 1961. New finds at Olduvai Gorge. *Nature* 189:649–650.

1324. Leakey, L. S. B., Evernden, J. F., and Curtis, G. H. 1961. Age of Bed 1, Olduvai Gorge, Tanganyika. *Nature* 191:478.

1325. Leakey, L. S. B., Tobias, P. V., and Napier, J. R. 1964. A new species of the genus *Homo* from Olduvai Gorge, Tanzania. *Nature* 202:308–312.

1326. Leakey, M. D. 1966. A review of the Oldowan Culture from Olduvai Gorge, Tanzania. *Nature* 210:462–466.

1327. Leakey, M. D. 1971. *Olduvai Gorge: Excavations in Beds I and II, 1960–1963.* Cambridge: Cambridge University Press.

1328. Leakey, M. D. 1975. Cultural patterns in the Olduvai sequence. In *After the australopithecines,* ed. K. W. Butzer and G. L. Isaac, pp. 477–494. The Hague: Mouton.

1329. Leakey, M. D. 1977. The archaeology of the early hominids. In *A survey of the prehistory of eastern Africa,* ed. T. H. Wilson, pp. 61–79. Nairobi: Eighth Panafrican Congress of Prehistory and Quaternary Studies.

1330. Leakey, M. D. 1978. Olduvai fossil hominids: Their stratigraphic positions and associations. In *Early hominids of Africa,* ed. C. Jolly, pp. 3–16. London: Duckworth.

1331. Leakey, M. D. 1980. Early man, environment and tools. In *Current argument on early man: Report of a Nobel symposium,* ed. L. K. Königsson, pp. 114–133. Oxford: Pergamon.

1332. Leakey, M. D. 1987. Introduction. In *Laetoli: A Pliocene site in northern Tanzania,* ed. M. D. Leakey and J. M. Harris, pp. 1–22. Oxford: Clarendon Press.

1333. Leakey, M. D. 1987. Introduction (to the hominid footprints). In *Laetoli: A Pliocene site in northern Tanzania,* ed. M. D. Leakey and J. M. Harris, pp. 490–496. Oxford: Clarendon Press.

1334. Leakey, M. D. 1987. The Laetoli hominid remains. In *Laetoli: A Pliocene site in northern Tanzania,* ed. M. D. Leakey and J. M. Harris, pp. 108–117. Oxford: Clarendon Press.

1335. Leakey, M. D., and Hay, R. L. 1979. Pliocene footprints in the Laetolil Beds at Laetoli, northern Tanzania. *Nature* 278:317–323.

1336. Leakey, M. D., and Hay, R. L. 1982. The chronological position of the fossil hominids of Tanzania. In *L'Homo erectus et la place de l'homme de Tautavel parmi les hominidés fossiles,* ed. M. A. de Lumley, pp. 753–765. Nice: Premier Congrès International de Paléontologie Humaine.

1337. Leakey, M. D., Hay, R. L., Curtis, G. H., Drake, R. E., Jackes, M. K., and White, T. D. 1976. Fossil hominids from the Laetolil Beds, Tanzania. *Nature* 262:460–465.

1338. Leakey, M. D., and Roe, D., A. 1994. *Olduvai Gorge: Excavations in Beds III, IV, and the Masek Beds, 1968–1971.* Cambridge: Cambridge University Press.

1339. Leakey, M. G. 1995. The dawn of humans: The farthest horizon. *National Geographic* 190 (9): 38–51.

1340. Leakey, M. G., Feibel, C. S., McDougall, I., and Walker, A. 1995. New four-million-year-old hominid species from Kanapoi and Allia Bay, Kenya. *Nature* 376:565–571.

1341. Leakey, M. G., and Leakey, R. E., eds. 1978. *Koobi Fora research project.* Vol. 1. *The fossil hominids and an introduction to their context, 1968–1974.* Oxford: Clarendon Press.

1342. Leakey, M. G., Ungar, P. S., and Walker, A. 1995. A new genus of large primate from the Late Oligocene of Lothidok, Turkana District, Kenya. *Journal of Human Evolution* 28:519–531.

1343. Leakey, R. E., Leakey, M. G., and Walker, A. C. 1988. Morphology of *Afropithecus turkanensis* from Kenya. *American Journal of Physical Anthropology* 76:289–307.

1344. Leakey, R. E., Leakey, M. G., and Walker, A. C. 1988. Morphology of *Turkanapithecus kalakolensis* from Kenya. *American Journal of Physical Anthropology* 76:277–288.

1345. Leakey, R. E., Walker, A. C., Ward, C. V., and Grausz, H. M. 1989. A partial skeleton of a gracile hominid from the Upper Burgi Member of the Koobi Fora Formation, East Lake Turkana, Kenya. In *Hominidae: Proceedings of the Second International Congress of Human Paleontology,* ed. G. Giacobini, pp. 167–173. Milan: Jaca Book.

1346. Leakey, R. E. F., and Leakey, M. G. 1986. A new Miocene hominoid from Kenya. *Nature* 324:143–146.

1347. Leakey, R. E. F., and Leakey, M. G. 1986. A second new Miocene hominoid from Kenya. *Nature* 324:146–148.

1348. Leakey, R. E. F., and Leakey, M. G. 1987. A new Miocene small-bodied ape from Kenya. *Journal of Human Evolution* 16:369–387.

1349. Leakey, R. E. F., and Walker, A. 1976. *Australopithecus, Homo erectus,* and the single species hypothesis. *Nature* 261:572–574.

1350. Leakey, R. E. F., and Walker, A. 1983. New higher primates from the early Miocene of Buluk, Kenya. *Nature* 318:173–175.

1351. Leakey, R. E. F., and Walker, A. 1985. A fossil skeleton 1,600,000 years old: *Homo erectus* unearthed. *National Geographic* 168 (2): 625–629.

1352. Leakey, R. E. F., and Walker, A. 1985. Further hominids from the Plio-Pleistocene of Koobi Fora, Kenya. *American Journal of Physical Anthropology* 67:135–163.

1353. Leakey, R. E. F., and Walker, A. 1988. New *Australopithecus boisei* specimens from East and West Lake Turkana, Kenya. *American Journal of Physical Anthropology* 76:1–24.

1354. Lee-Thorp, J. A., and van der Merwe, N. J. 1993. Stable carbon isotope studies of Swartkrans fossils. In *Swartkrans: A cave's chronicle of early man,* ed. C. K. Brain, pp. 251–256. Pretoria: Transvaal Museum.

1355. Lee-Thorp, J. A., van der Merwe, N. J., and Brain, C. K. 1994. Diet of *Australopithecus robustus* at Swartkrans from stable carbon isotopic analysis. *Journal of Human Evolution* 27:361–372.

1356. Legoux, P. 1975. Présentation des dents des restes humains de l'Abri Pataud. *American School of Prehistoric Research Bulletin* 30:262–305.

1357. Leinders, J. J. M., Aziz, F., Sondaar, P. Y., and de Vos, J. 1985. The age of the hominid-bearing deposits of Java: State of the art. *Geologie en Mijnbouw* 64:167–173.

1358. Leroi-Gourhan, A. 1965. Le Châtelperronien: Problème ethnologique. In *Miscelanea en homenaje al Abate Henri Breuil*, ed. E. Ripoll Perello, 2:75–81. Barcelona: Diputacion Provincial de Barcelona, Instituto de Prehistória y Arqueológia.

1359. Leroi-Gourhan, A. 1965. *Treasures of prehistoric art.* New York: Abrams.

1360. Leroyer, C., and Leroi-Gourhan, A. 1983. Problèmes de chronologie: Le Castelperronien et l'Aurignacien. *Bulletin de la Société Préhistorique Française* 80:41–44.

1361. Lévêque, F., Backer, A. M., and Guilbaud, M., eds. 1993. *Context of a late Neandertal: Implications of multidisciplinary research for the transition to Upper Paleolithic adaptations at Saint-Césaire, Charente-Maritime, France.* Madison, Wis.: Prehistory Press.

1362. Lévêque, F., and Miskovsky, J.-C. 1983. Le Castelperronien dans son environnement géologique. *Anthropologie* 87:369–391.

1363. Levine, M. 1983. Mortality models and the interpretation of horse population structure. In *Hunter-gatherer economy in prehistory*, ed. G. Bailey, pp. 23–46. Cambridge: Cambridge University Press.

1364. Levinton, J. S. 1988. *Genetics, paleontology, and macroevolution.* Cambridge: Cambridge University Press.

1365. Lewin, R. 1988. Molecular clocks turn a quarter century. *Science* 239:561–563.

1366. Lewis-Williams, J. D. 1981. *Believing and seeing: Symbolic meanings in southern San rock art.* London: Academic Press.

1367. Lewis-Williams, J. D. 1982. The economic and social context of southern San rock art. *Current Anthropology* 23:429–449.

1368. Lewis-Williams, J. D., and Dowson, T. A. 1988. The signs of all times: Entoptic phenomena in Upper Palaeolithic art. *Current Anthropology* 29:201–245.

1369. Lewis-Williams, J. D., and Loubser, J. H. N. 1986. Deceptive appearances: A critique of southern African rock art studies. *Advances in World Archaeology* 5:253–289.

1370. Libby, W. F. 1955. *Radiocarbon dating.* Chicago: University of Chicago Press.

1371. Lieberman, D. E. 1993. The rise and fall of seasonal mobility among hunter-gatherers: The case of the southern Levant. *Current Anthropology* 34:599–631.

1372. Lieberman, D. E. 1995. Testing hypotheses about recent human evolution from skulls. *Current Anthropology* 36:159–197.

1373. Lieberman, D. E. 1998. Heterochrony, homology, and browridge elongation in recent human evolution. *Journal of Human Evolution.* In press.

1374. Lieberman, D. E. 1998. Neanderthal and early modern human mobility patterns: Comparing archaeological and anatomical evidence. In *Neanderthals and modern humans in west Asia*, ed. K. Aoki, T. Akazawa, and O. Bar-Yosef, pp. 263–275. New York: Plenum Press.

1375. Lieberman, D. E., Pilbeam, D. R., and Wood, B. A. 1988. A probabilistic approach to the problem of sexual dimorphism in *Homo habilis*: A comparison of KNM-ER 1470 and KNM-ER 1813. *Journal of Human Evolution* 17:503–511.

1376. Lieberman, D. E., and Shea, J. J. 1994. Behavioral differences between archaic and modern humans in the Levantine Mousterian. *American Anthropologist* 96:300–332.

1377. Lieberman, D. E., Wood, B. A., and Pilbeam, D. R. 1996. Homoplasy and early *Homo*: An analysis of the evolutionary relationships of *H. habilis* sensu stricto and *H. rudolfensis*. *Journal of Human Evolution* 30:97–120.

1378. Lieberman, P. 1991. *Uniquely human: The evolution of speech thought, and selfless behavior*. Cambridge: Harvard University Press.

1379. Lieberman, P. 1992. On Neanderthal speech and Neanderthal extinction. *Current Anthropology* 33:409–410.

1380. Lieberman, P., Laitman, J. T., Reidenberg, J. S., and Gannon, P. J. 1992. The anatomy, physiology, acoustics, and perception of speech: Essential elements in the analysis of the evolution of human speech. *Journal of Human Evolution* 23:447–467.

1381. Lindahl, T. 1997. Facts and artifacts of ancient DNA. *Cell* 90:1–3.

1382. Lindly, J. M., and Clark, G. A. 1990. Symbolism and modern human origins. *Current Anthropology* 31:233–261.

1383. Linick, T. W., Damon, P. E., Donahue, D. J., and Jull, A. J. T. 1989. Accelerator mass spectrometry: The new revolution in radiocarbon dating. *Quaternary International* 1:1–6.

1384. Lister, A. M. 1986. New results on deer from Swanscombe, and the stratigraphical significance of deer in the Middle and Upper Pleistocene of Europe. *Journal of Archaeological Science* 13:319–338.

1385. Lister, A. M. 1992. Mammalian fossils and Quaternary biostratigraphy. *Quaternary Science Reviews* 11:329–345.

1386. Lister, A. M. 1993. Evolution of mammoths and moose: The Holarctic perspective. In *Morphological change in Quaternary mammals of North America*, ed. R. A. Martin and A. D. Barnosky, pp. 178–204. New York: Cambridge University Press.

1387. Lister, A. M., and Bahn, P. 1994. *Mammoths*. New York: Macmillan.

1388. Liu, Z. 1985. Sequence of sediments at Locality 1 in Zhoukoudian and correlation with loess stratigraphy in northern China with the chronology of deep sea-cores. *Quaternary Research* 23:139–153.

1389. Long, J. C. 1993. Human molecular phylogenetics. *Annual Review of Anthropology* 22:251–272.

1390. Lorblanchet, M. 1991. Spitting images: Replicating the spotted horses of Pech-Merle. *Archaeology* 44 (6):24–31.

1391. Lorblanchet, M. 1993. From styles to dates. In *Rock art studies:*

The post-stylistic era, or Where do we go from here? ed. M. Lorblanchet and P. G. Bahn, pp. 61–76. Oxford: Oxbow Books.

1392. Lorblanchet, M. 1994. Le mode d'utilisation des sanctuaires paléolithiques. *Museo y Centro de Investigación de Altamira, Monografias* 17:235–251.

1393. Lorblanchet, M., Labeau, M., Vernet, J.-L., Fitte, P., Valladas, H., Cachier, H., and Arnold, M. 1990. Paleolithic pigments in the Quercy, France. *Rock Art Research* 7:4–20.

1394. Lourandos, H. 1987. Pleistocene Australia: Peopling of a continent. In *The Pleistocene Old World: Regional perspectives*, ed. O. Soffer, pp. 147–165. New York: Plenum Press.

1395. Lovejoy, C. O. 1979. A reconstruction of the pelvis of Al-288 (Hadar Formation, Ethiopia). Abstract. *American Journal of Physical Anthropology* 40:460.

1396. Lovejoy, C. O. 1981. The origin of man. *Science* 211:341–350.

1397. Lowe, J. J., and Walker, M. J. C. 1984. *Reconstructing Quaternary environments*. New York: Longman.

1398. Lowenstein, J. M., Molleson, T., and Washburn, S. L. 1982. Piltdown jaw confirmed as orang. *Nature* 299:294.

1399. Lumley, H. de. 1969. A Paleolithic camp at Nice. *Scientific American* 220 (5): 42–50.

1400. Lumley, H. de. 1969. Une cabane acheuléene dans la grotte du Lazaret. *Mémoires de la Société Préhistorique Française* 7:1–234.

1401. Lumley, H. de. 1975. Cultural evolution in France in its paleoecological setting during the Middle Pleistocene. In *After the australopithecines*, ed. K. W. Butzer and G. L. Isaac, pp. 745–808. The Hague: Mouton.

1402. Lumley, H. de. 1976. Les premières industries humains en Provence. In *La préhistoire française*, ed. H. de Lumley, 1:765–794. Paris: Centre National de la Recherche Scientifique.

1403. Lumley, H. de, ed. 1979. L'homme de Tautavel. *Dossiers de l'Archéologie* 36:1–273.

1404. Lumley, H. de, and Boone, Y. 1976. Les structures d'habitat au Paléolithique inférieur. In *La préhistoire française*, ed. H. de Lumley, 1:625–643. Paris: Centre National de la Recherche Scientifique.

1405. Lumley, H. de, Fournier, A., Krzepkowska, J., and Eschassoux, A. 1988. L'industrie du Pléistocène inférieur de la grotte du Vallonet, Roquebrune-Cap-Martin, Alpes-Maritimes. *Anthropologie* 92:502–614.

1406. Lumley, H. de, Fournier, A., Park, Y. C., Yokohama, Y., and Demouy, A. 1984. Stratigraphie du remplissage Pléistocène moyen de la Caune de Arago à Tautavel. *Anthropologie* 88:5–18.

1407. Lumley, H. de, and Sonakia, A. 1985. Contexte stratigraphique et archéologique de l'homme de la Narmada, Hathnora, Madhya Pradesh, Inde. *Anthropologie* 89:3–12.

1408. Lumley, M.-A. de. 1972. Les Néandertaliens de la grotte de l'Hortus (Valflaunes, Hérault). In *La grotte de l'Hortus*, ed. H. de Lumley, pp. 375–385. Marseilles: Université de Provence.

1409. Lumley, M.-A. de, and Sonakia, A. 1985. Première découverte d'un *Homo erectus* sur le continent Indien, a Hathnora, dans le Moyenne Vallée de la Narmada. *Anthropologie* 89:13–61.

1410. Lyubin, V. P. 1989. Lower Paleolithic (in Russian). In *Paleolit Kavkaza i Severnoi Azii*, pp. 9–92. Leningrad: Akademiya Nauk SSSR.

1411. Maas, M. C., Krause, D. W., and Strait, S. G. 1988. The decline and extinction of Plesiadapiformes (Mammalia: ?Primates) in North America: Displacement or replacement. *Paleobiology* 14: 410–431.

1412. MacFadden, B. J. 1985. Drifting continents, mammals, and time scales: Current developments in South America. *Journal of Vertebrate Paleontology* 5:169–174.

1413. MacFadden, B. J. 1990. Chronology of Cenozoic primate localities in South America. *Journal of Human Evolution* 19:7–21.

1414. MacFadden, B. J., Campbell, K. E. J., Cifelli, R. L., Siles, O., Johnson, N., Naeser, C. W., and Zeitler, P. K. 1985. Magnetic polarity stratigraphy and mammalian biostratigraphy of the Desedean (Late Oligocene–Early Miocene) Salla Beds of northern Bolivia. *Journal of Geology* 93:223–250.

1415. MacLarnon, A. 1993. The vertebral canal. In *The Nariokotome* Homo erectus *skeleton*, ed. A. Walker and R. Leakey, pp. 359–390. Cambridge: Harvard University Press.

1416. MacPhee, R. D. E., Cartmill, M., and Gingerich, P. D. 1983. New Palaeogene primate basicrania and the definition of the order Primates. *Nature* 301:509–511.

1417. Maddison, D. R., Ruvolo, M., and Swofford, D. L. 1992. Geographic origins of human mitochondrial DNA: Phylogenetic evidence from control region sequences. *Systematic Biology* 41: 111–124.

1418. Maglio, V. J. 1973. Origin and evolution of the Elephantidae. *Transactions of the American Philosophical Society* 6:1–149.

1419. Magori, C. C., and Day, M. H. 1983. Laetoli Hominid 18: An early *Homo sapiens* skull. *Journal of Human Evolution* 12:747–753.

1420. Maguire, J. M. 1985. Recent geological, stratigraphic and palaeontological studies at Makapansgat Limeworks. In *Hominid evolution: Past, present and future*, ed. P. V. Tobias, pp. 151–164. New York: Alan R. Liss.

1421. Maguire, J. M., Pemberton, D., and Collett, M. H. 1980. The Makapansgat Limeworks Grey Breccia: Hominids, hyaenas, hystricids or hillwash? *Palaeontologia Africana* 23:75–98.

1422. Malan, B. D., and Wells, L. H. 1943. A further report on the Wonderwerk Cave, Kuruman. *South African Journal of Science* 40:258–270.

1423. Malatesta, A., Jaccabacci, A., Nappi, G., Conato, V., Molinari Paganelli, V., van der Werff, A., Durante, S., Settepassi, F., and Biddittu, I. 1978. Torre in Pietra, Roma. *Quaternaria* 20:205–577.

1424. Malez, M. 1970. A new look at the stratigraphy of the Krapina site. In *Krapina: 1899–1969*, pp. 40–44. Zagreb: Yugoslavenska Akademija Znanosti i Umietnosti.

1425. Mallegni, F., Mariani-Constantini, R., Fornaciari, G., Longo, E. T., Giacobini, G., and Radmilli, A. M. 1983. New European fossil hominid material from an Acheulean site near Rome (Castel di Guido). *American Journal of Physical Anthropology* 62: 263–274.

1426. Mallegni, F., and Radmilli, A. M. 1988. Human temporal bone from the Lower Paleolithic site of Castel di Guido, near Rome, Italy. *American Journal of Physical Anthropology* 76:175–182.

1427. Manderscheid, E. J., and Rogers, A. R. 1996. Genetic admixture in the late Pleistocene. *American Journal of Physical Anthropology* 100:1–5.

1428. Mania, D. 1986. Die Forschungsgrabung bei Bilzingsleben. *Jahresschrift für Mitteldeutsche Vorgeschichte* 69:235–255.

1429. Mania, D. 1991. Les premiers peuplements humains dans la région de Saale-Elbe. In *Les premiers Européens,* ed. E. Bonifay and B. Vandermeersch, pp. 173–175. Paris: Éditions du Comité des Travaux Historiques et Scientifiques.

1430. Mania, D., and Mania, U. 1988. Deliberate engravings on bone artefacts of *Homo erectus. Rock Art Research* 5:91–107.

1431. Mania, D., Mania, U., and Vlcek, E. 1994. Latest finds of skull remains of *Homo erectus* from Bilzingsleben (Thuringia). *Naturwissenschaften* 81:123–127.

1432. Mania, D., and Toepfer, V. 1973. Königsaue: Gliederung, Ökologie und mittelpaläolithische Funde der letzten Eiszeit. *Veröffentlichungen des Landes Museums für Vorgeschichte in Halle (Verlag der Wissenschaften)* 26:1–164.

1433. Mania, D., and Vlcek, E. 1981. *Homo erectus* in middle Europe: The discovery from Bilzingsleben. In Homo erectus—*papers in honor of Davidson Black,* ed. B. A. Sigmon and S. Cybulski, pp. 133–151. Toronto: University of Toronto Press.

1434. Mania, U. 1995. The utilisation of large mammal bones in Bilzingsleben—a special variant of Middle Pleistocene man's relationship to his environment. *Études et Recherches Archéologiques de l'Université de Liège* 62:239–246.

1435. Mann, A. E. 1975. Paleodemographic aspects of the South African australopithecines. *University of Pennsylvania Publications in Anthropology* 1:1–171.

1436. Manzi, G., Salvadei, L., and Passarello, P. 1990. The Casal de'Pazzi archaic parietal: Comparative analysis of new fossil evidence from the late Middle Pleistocene of Rome. *Journal of Human Evolution* 19:751–759.

1437. Marder, O., Khalaily, H., Rabinovich, R., Gvirtzman, G., Wieder, M., Porat, N., Ron, H., Bankirer, R., and Saragusti, I. 1999. The Lower Paleolithic site of Revadim Quarry, preliminary finds. *Mitekufat Ha'even.* In press.

1438. Marean, C. W. 1989. Sabertooth cats and their relevance to early hominid diet and evolution. *Journal of Human Evolution* 18:559–582.

1439. Marean, C. W. 1992. Implications of late Quaternary mammalian fauna from Lukenya Hill (south-central Kenya) for paleoenvironmental change and faunal extinctions. *Quaternary Research* 37:239–255.

1440. Marean, C. W., and Kim, S. Y. 1998. Mousterian large-mammal remains from Kobeh Cave: Behavioral implications for Neanderthals and early modern humans. *Current Anthropology* 39:579-S113.

1441. Marks, A. E. 1968. The Khormusan: An Upper Pleistocene industry in Sudanese Nubia. In *The prehistory of Nubia,* ed. F. Wendorf, 1:315–391. Dallas: Southern Methodist University.

1442. Marks, A. E. 1968. The Mousterian industries of Nubia. In *The prehistory of Nubia*, ed. F. Wendorf, 1:194–314. Dallas: Southern Methodist University.

1443. Marks, A. E. 1977. Introduction: A preliminary overview of central Negev prehistory. In *Prehistory and paleoenvironments in the central Negev, Israel*, ed. A. E. Marks, 1:194–314. Dallas: Department of Anthropology, Southern Methodist University.

1444. Marks, A. E. 1981. The Middle Palaeolithic of the Negev, Israel. In *Préhistoire du Levant*, ed. J. Cauvin and P. Sanlaville, pp. 287–298. Paris: Centre National de la Recherche Scientifique.

1445. Marks, A. E. 1983. The Middle to Upper Paleolithic transition in the Levant. *Advances in World Archaeology* 2:51–98.

1446. Marks, A. E. 1988. The Middle to Upper Paleolithic transition in the southern Levant: Technological change as an adaptation to increasing mobility. *Études et Recherches Archéologiques de l'Université de Liège* 35:109–123.

1447. Marks, A. E. 1990. The Middle and Upper Palaeolithic of the Near East and the Nile Valley: The problem of cultural transformations. In *The emergence of modern humans: An archaeological perspective*, ed. P. Mellars, pp. 56–80. Ithaca: Cornell University Press.

1448. Marks, A. E. 1992. Upper Pleistocene archaeology and the origins of modern man: A view from the Levant and adjacent areas. In *The evolution and dispersal of modern humans in Asia*, ed. T. Akazawa, K. Aoki, and T. Kimura, pp. 229–251. Tokyo: Hokusen-Sha.

1449. Marks, A. E., Demidenko, Y. E., Monigal, K., Usik, V. I., Ferring, C. R., Burke, A., Rink, J., and McKinney, C. 1997. Starosele and the Starosele child: New excavations, new results. *Current Anthropology* 38:112–123.

1450. Marks, A. E., and Freidel, D. A. 1977. Prehistoric settlement patterns in the Avdat/Aqev area. In *Prehistory and paleoenvironments in the central Negev, Israel*, ed. A. E. Marks, 2:131–158. Dallas: Department of Anthropology, Southern Methodist University.

1451. Marks, J. 1992. Genetic relations among the apes and humans. *Current Opinion in Genetics and Development* 2:883–889.

1452. Marks, J. 1993. Hominoid heterochromatin: Terminal C-bands as a complex genetic trait linking chimpanzee and gorilla. *American Journal of Physical Anthropology* 90:237–246.

1453. Marshack, A. 1972. *The roots of civilization*. New York: McGraw-Hill.

1454. Marshack, A. 1972. Upper Paleolithic notation and symbol. *Science* 178:817–828.

1455. Marshack, A. 1976. Implications of the Paleolithic symbolic evidence for the origin of language. *American Scientist* 64:136–145.

1456. Marshack, A. 1989. Evolution of the human capacity: The symbolic evidence. *Yearbook of Physical Anthropology* 32:1–34.

1457. Marshack, A. 1991. A reply to Davidson on Mania and Mania. *Rock Art Research* 8:47–58.

1458. Marshack, A. 1991. The Taï plaque and calendrical notation in the Upper Paleolithic. *Cambridge Archaeological Journal* 1:25–61.

1459. Marshack, A. 1996. A Middle Paleolithic symbolic composition from the Golan Heights: The earliest known depictive image. *Current Anthropology* 37:357–365.

1460. Marshall, F. 1986. Implications of bone modification in a Neolithic faunal assemblage for the study of early hominid butchery and subsistence practices. *Journal of Human Evolution* 15:661–672.

1461. Marston, A. T. 1937. The Swanscombe skull. *Journal of the Royal Anthropological Institute* 67:339–406.

1462. Martin, L. 1985. Significance of enamel thickness in hominoid evolution. *Nature* 314:260–263.

1463. Martin, L. 1991. Teeth, sex and species. *Nature* 352:111–112.

1464. Martin, L., and Andrews, P. 1982. New ideas on the relationships of the Miocene hominoids. *Primate Eye* 18:4–7.

1465. Martin, P. S. 1984. Prehistoric overkill: The global model. In *Quaternary extinctions: A prehistoric revolution,* ed. P. S. Martin and R. G. Klein, pp. 354–403. Tucson: University of Arizona Press.

1466. Martin, R. D. 1986. Primates: A definition. In *Major topics in primate and human evolution,* ed. B. Wood, L. Martin, and P. Andrews, pp. 1–31. Cambridge: University of Cambridge Press.

1467. Martin, R. D. 1988. Several steps forward for Eocene primates. *Nature* 331:660–661.

1468. Martin, R. D. 1990. *Primate origins and evolution: A phylogenetic reconstruction.* Princeton: Princeton University Press.

1469. Martin, R. D. 1991. New fossils and primate origins. *Nature* 349:19–20.

1470. Martínez, I., and Arsuaga, J. L. 1997. The temporal bones from Sima de los Huesos Middle Pleistocene site (Sierra de Atapuerca, Spain): A phylogenetic approach. *Journal of Human Evolution* 33:283–318.

1471. Martínez Navarro, B., and Palmqvist, P. 1995. Presence of the African machairodont *Megantereon whitei* (Broom, 1937) (Felidae, Carnivora, Mammalia) in the Lower Pleistocene site of Venta Micena (Orce, Granada, Spain), with some considerations on the origin, evolution and dispersal of the genus. *Journal of Archaeological Science* 22:569–582.

1472. Martínez Navarro, B., Turq, A., Agustí Ballester, J., and Oms, O. 1997. Fuente Nueva-3 (Orce, Granada, Spain) and the first human occupation of Europe. *Journal of Human Evolution* 33:611–620.

1473. Marzke, M. W. 1997. Precision grips, hand morphology, and tools. *American Journal of Physical Anthropology* 102:91–110.

1474. Masao, F. T. 1992. The Middle Stone Age with reference to Tanzania. In *Continuity or replacement: Controversies in* Homo sapiens *evolution,* ed. G. Bräuer and F. H. Smith, pp. 99–109. Rotterdam: A. A. Balkema.

1475. Mason, R. J. 1962. *Prehistory of the Transvaal.* Johannesburg: University of the Witwatersrand Press.

1476. Mason, R. J. 1988. Cave of Hearths, Makapansgat, Transvaal. *Occasional Papers, Archaeological Research Unit, University of the Witwatersrand* 21:1–711.

1477. Matsu'ura, S. 1986. Age of the early Javanese hominids: A review. In *Primate evolution,* ed. J. G. Else and P. C. Lee, pp. 115–121. Cambridge: Cambridge University Press.

1478. Matsuda, T., Torii, M., Koyaguchi, T., Makinouchi, T., Mitsu-shio, H., and Ishida, S. 1986. Geochronology of Miocene ho-minids east of the Kenya Rift Valley. In *Primate evolution*, ed. J. G. Else and P. C. Lee, pp. 35–45. Cambridge: Cambridge University Press.

1479. Maw, B., Ciochon, R. L., and Savage, D. E. 1979. Late Eocene of Burma yields earliest anthropoid primate *Pondaungia cotteri*. *Nature* 282:65–67.

1480. Maynard Smith, J. 1989. *Evolutionary genetics*. Oxford: Oxford University Press.

1481. Mayr, E. 1950. Taxonomic categories in fossil hominids. *Cold Spring Harbor Symposia on Quantitative Biology* 15:109–118.

1482. Mayr, E. 1963. *Animal species and evolution*. Cambridge: Harvard University Press.

1483. Mazaud, A., Laj, C., Bard, E., Arnold, M., and Tric, E. 1991. Geomagnetic field control of ^{14}C production over the last 80 ky: Implications for the radiocarbon time-scale. *Geophysical Research Letters* 18:1885–1888.

1484. McBrearty, S. 1988. The Sangoan-Lupemban and Middle Stone Age sequence at the Muguruk Site, western Kenya. *World Archaeology* 19:388–420.

1485. McBrearty, S., Bishop, L., and Kingston, J. 1996. Variability in traits of Middle Pleistocene hominid behavior in the Kapthurin Formation, Baringo, Kenya. *Journal of Human Evolution* 30: 563–580.

1486. McBurney, C. B. M. 1967. *The Haua Fteah (Cyrenaica) and the Stone Age of the southeast Mediterranean*. Cambridge: Cambridge University Press.

1487. McBurney, C. B. M. 1975. Current status of the Lower and Middle Paleolithic in the entire region from the Levant through north Africa. In *Problems in prehistory: North Africa and the Levant*, ed. F. Wendorf and A. E. Marks, pp. 411–426. Dallas: Southern Methodist University Press.

1488. McBurney, C. B. M. 1976. *Early man in the Soviet Union: The implications of some recent discoveries*. Oxford: Oxford University Press.

1489. McBurney, C. B. M., and Hey, R. W. 1955. *Prehistory and Pleistocene geology in Cyrenaican Libya*. Cambridge: Cambridge University Press.

1490. McCown, T. 1937. Mugharet es-Skhul: Description and excavations. In *The Stone Age of Mount Carmel*, ed. D. A. E. Garrod and D. Bate, pp. 91–107. Oxford: Clarendon Press.

1491. McCown, T. D., and Keith, A. 1939. *The Stone Age of Mount Carmel*. Vol. 2. Oxford: Clarendon Press.

1492. McCrossin, M. L., and Benefit, B. R. 1993. Recently recovered *Kenyapithecus* mandible and its implications for great ape and human origins. *Proceedings of the National Academy of Sciences* 90:1962–1966.

1493. McCrossin, M. L., and Benefit, B. R. 1994. Maboko Island and the evolutionary history of Old World monkeys and apes. In *Integrative paths to the past: Paleoanthropological advances in honor of F. Clark Howell*, ed. R. S. Corruccini and R. L. Ciochon, pp. 95–122. Englewood Cliffs, N.J.: Prentice Hall.

1494. McDermott, F., Grün, R., Stringer, C. B., and Hawkesworth, C. J.

1993. Mass-spectrographic U-series dates for Israeli Neanderthal/early modern hominid sites. *Nature* 363:252–254.

1495. McDermott, F., Stringer, C. B., Grün, R., Williams, C. T., Din, V. K., and Hawkesworth, C. J. 1996. New Late-Pleistocene uranium-thorium and ESR dates for the Singa hominid (Sudan). *Journal of Human Evolution* 31:507–516.

1496. McDougall, I., Brown, F. H., Cerling, T. E., and Hillhouse, J. W. 1992. A reappraisal of the geomagnetic polarity time scale to 4 ma using data from the Turkana Basin, east Africa. *Geophysical Research Letters* 19:2349–2352.

1497. McDougall, I., and Harrison, T. M. 1988. *Geochronology and thermochronology by the $^{40}Ar/^{39}Ar$ method.* Oxford: Oxford University Press.

1498. McFadden, P. L. 1980. An overview of palaeomagnetic chronology with special reference to the South African hominid sites. *Palaeontologia Africana* 23:35–40.

1499. McFadden, P. L., and Brock, A. 1984. Magnetostratigraphy at Makapansgat. *South African Journal of Science* 80:482–483.

1500. McGrew, W. C. 1992. *Chimpanzee material culture: Implications for human evolution.* Cambridge: Cambridge University Press.

1501. McHenry, H. M. 1974. How large were the australopithecines? *American Journal of Physical Anthropology* 40:329–340.

1502. McHenry, H. M. 1982. The pattern of human evolution: Studies on bipedalism, mastication, and encephalization. *Annual Review of Anthropology* 11:151–173.

1503. McHenry, H. M. 1984. Relative cheek tooth size in *Australopithecus. American Journal of Physical Anthropology* 64:297–306.

1504. McHenry, H. M. 1986. The first bipeds: A comparison of the *A. afarensis* and *A. africanus* postcranium and implications for the evolution of bipedalism. *Journal of Human Evolution* 15:177–191.

1505. McHenry, H. M. 1988. New estimates of body weight in early hominids and their significance to encephalization and megadontia in "robust" australopithecines. In *The evolutionary history of the "robust" australopithecines,* ed. F. E. Grine, pp. 133–146. New York: Aldine de Gruyter.

1506. McHenry, H. M. 1991. Femoral lengths and stature in Plio-Pleistocene hominids. *American Journal of Physical Anthropology* 85:149–158.

1507. McHenry, H. M. 1991. Petite bodies of the "robust" australopithecines. *American Journal of Physical Anthropology* 86:445–454.

1508. McHenry, H. M. 1991. Sexual dimorphism in *Australopithecus afarensis. Journal of Human Evolution* 20:21–32.

1509. McHenry, H. M. 1992. How big were early hominids? *Evolutionary Anthropology* 1:15–20.

1510. McHenry, H. M. 1994. Behavioral ecological implications of early hominid body size. *Journal of Human Evolution* 27:77–87.

1511. McHenry, H. M. 1994. Tempo and mode in human evolution. *Proceedings of the National Academy of Sciences* 91:6780–6786.

1512. McHenry, H. M. 1996. Homoplasy, clades, and hominid phylogeny. In *Contemporary issues in human evolution*, ed. W. E. Meickle, F. C. Howell, and N. G. Jablonski, pp. 77–92. Memoir 21. San Francisco: California Academy of Sciences.

1513. McHenry, H. M., and Berger, L. R. 1998. Body proportions in *Australopithecus afarensis* and *A. africanus* and the origin of the genus *Homo*. *Journal of Human Evolution* 35:1–22.

1514. McHenry, H. M., and Corruccini, R. S. 1983. The wrist of *Proconsul africanus* and the origin of hominoid postcranial adaptations. In *New interpretations of ape and human ancestry*, ed. R. L. Ciochon and R. S. Corruccini, pp. 353–367. New York: Plcnum Press.

1515. McHenry, H. M., Corruccini, R. S., and Howell, F. C. 1976. Analysis of an early hominid ulna from the Omo Basin. *American Journal of Physical Anthropology* 44:295–304.

1516. McKee, J. K. 1991. Palaeo-ecology of the Sterkfontein hominids: A review and synthesis. *Palaeontologia Africana* 28:41–51.

1517. McKee, J. K. 1993. Faunal dating of the Taung hominid fossil deposit. *Journal of Human Evolution* 23:363–376.

1518. McKee, J. K. 1993. Formation and geomorphology of caves in calcareous tufas and implications for the study of the Taung fossil deposits. *Transactions of the Royal Society of South Africa* 48 (2): 307–322.

1519. McKee, J. K., Tobias, P. V., and Clarke, R. J. 1996. Faunal evidence and Sterkfontein Member 2 foot bones of early hominid. *Science* 271:1301–1302.

1520. McKenna, M. C. 1980. Early history and biogeography of South America's extinct land mammals. In *Evolutionary biology of the New World monkeys and continental drift*, ed. R. L. Ciochon and A. B. Chiarelli, pp. 43–77. New York: Plenum Press.

1521. McRae, L. E. 1990. Paleomagnetic isochrons, unsteadiness, and uniformity of sedimentation in Miocene intermontane basin sediments at Salla, eastern Andean cordillera, Bolivia. *Journal of Geology* 95:479–500.

1522. Mehlman, M. J. 1977. Excavations at Nasera Rock, Tanzania. *Azania* 12:111–118.

1523. Mehlman, M. J. 1979. Mumba-Höhle revisited: The relevance of a forgotten excavation to some current issues in east African prehistory. *World Archaeology* 11:80–94.

1524. Mehlman, M. J. 1984. Archaic *Homo sapiens* from Lake Eyasi, Tanzania: Recent misinterpretations. *Journal of Human Evolution* 13:487–501.

1525. Mehlman, M. J. 1987. Provenience, age, and associations of archaic *Homo sapiens* crania from Lake Eyasi, Tanzania. *Journal of Archaeological Science* 14:133–162.

1526. Meignen, L., and Bar-Yosef, O. 1989. Kebara et le Paléolithique moyen du Mont Carmel (Israël). *Paléorient* 14:123–130.

1527. Meignen, L., and Bar-Yosef, O. 1992. Middle Paleolithic lithic variability in Kebara Cave, Mount Carmel, Israel. In *The evolution and dispersal of modern humans in Asia*, ed. T. Akazawa, K. Aoki, and T. Kimura, pp. 129–148. Tokyo: Hokusen-Sha.

1528. Meignen, L., Bar-Yosef, O., and Goldberg, P. 1989. Les structures de combustion moustériennes de la grotte de Kebara (Mont

Carmel, Israel). In *Nature et structures des foyers préhistoriques*, ed. M. Olive and Y. Taborin, pp. 141–146. Nemours: Musée de Préhistoire de l'Île de France.

1529. Mellars, P. 1989. Major issues in the emergence of modern humans. *Current Anthropology* 30:349–385.

1530. Mellars, P. 1989. Technological changes across the Middle-Upper Paleolithic transition: Economic, social and cognitive perspectives. In *The human revolution: Behavioural and biological perspectives in the origins of modern humans*, ed. P. Mellars and C. Stringer, pp. 338–365. Edinburgh: University of Edinburgh Press.

1531. Mellars, P., and Grün, R. 1991. A comparison of the electron spin resonance and thermoluminescence dating methods: The results of ESR dating at Le Moustier (France). *Cambridge Archaeological Journal* 2:269–276.

1532. Mellars, P., and Tixier, J. 1989. Radiocarbon-accelerator dating of Ksar 'Aqil (Lebanon) and the chronology of the Upper Palaeolithic sequence in the Middle East. *Antiquity* 63:761–768.

1533. Mellars, P. A. 1965. Sequence and development of Mousterian traditions in south-western France. *Nature* 205:626–627.

1534. Mellars, P. A. 1970. The chronology of Mousterian industries in the Périgord region of south-west France. *Proceedings of the Prehistoric Society* 35:134–17.

1535. Mellars, P. A. 1973. The character of the Middle-Upper Palaeolithic transition in south-west France. In *The explanation of culture change*, ed. C. Renfrew, pp. 255–276. Pittsburgh: University of Pittsburgh Press.

1536. Mellars, P. A. 1982. On the Middle/Upper Palaeolithic transition: A reply to White. *Current Anthropology* 23:238–240.

1537. Mellars, P. A. 1986. A new chronology for the Mousterian period. *Nature* 322:410–411.

1538. Mellars, P. A. 1993. Archaeology and the population-dispersal hypothesis of modern human origins in Europe. In *The origin of modern humans and the impact of chronometric dating*, ed. M. J. Aitken, C. B. Stringer, and P. A. Mellars, pp. 196–216. Princeton: Princeton University Press.

1539. Mellars, P. A. 1996. *The Neanderthal legacy: An archaeological perspective from western Europe*. Princeton: Princeton University Press.

1540. Mellars, P. A., Bricker, H. M., Gowlett, J. A. J., and Hedges, R. E. M. 1987. Radiocarbon accelerator dating of French Upper Paleolithic sites. *Current Anthropology* 28:128–133.

1541. Meltzer, D. J. 1993. Coming to America. *Discover* 93:90–97.

1542. Meltzer, D. J. 1993. Pleistocene peopling of the Americas. *Evolutionary Anthropology* 1:157–169.

1543. Meltzer, D. J. 1995. Clocking the first Americans. *Annual Review of Anthropology* 24:21–45.

1544. Meltzer, D. J. 1997. Monte Verde and the Pleistocene peopling of the Americas. *Science* 276:754–755.

1545. Meltzer, D. J., Adovasio, J. M., and Dillehay, T. D. 1994. On a Pleistocene human occupation at Pedra Furada, Brazil. *Antiquity* 68:695–714.

1546. Meltzer, D. J., Grayson, D. K., Ardila, G., Varker, A. W.,

Dincauze, D. F., Haynes, V., Mena, F., Nuñez, L., and Stanford, D. J. 1997. On the Pleistocene antiquity of Monte Verde. *American Antiquity* 62:659–663.

1547. Mercer, J. H. 1983. Cenozoic glaciation in the Southern Hemisphere. *Annual Review of Earth and Planetary Sciences* 11:99–132.

1548. Mercier, N., and Valladas, H. 1994. Thermoluminescence dates for the Paleolithic Levant. In *Late Quaternary chronology and paleoclimates of the eastern Mediterranean*, ed. O. Bar-Yosef and R. S. Kra, pp. 13–20. Cambridge, Mass.: American School of Prehistoric Research.

1549. Mercier, N., Valladas, H., Bar-Yosef, O., Vandermeersch, B., Stringer, C., and Joron, J.-L. 1993. Thermoluminescence date for the Mousterian burial site of Es-Skhul, Mt. Carmel. *Journal of Archaeological Science* 20:169–174.

1550. Mercier, N., Valladas, H., Joron, J.-L., Reyss, J.-L., Lévêque, F., and Vandermeersch, B. 1991. Thermoluminescence dating of the late Neanderthal remains from Saint-Césaire. *Nature* 351:737–739.

1551. Mercier, N., Valladas, H., and Valladas, G. 1995. Flint thermoluminescence dates from the CFR Laboratory at Gif: Contributions to the study of the Middle Palaeolithic. *Quaternary Science Reviews* 14:351–364.

1552. Mercier, N. H., Valladas, H., Valladas, G., Reyss, J.-L., Jelinek, A., Meignen, L., and Joron, J.-L. 1995. TL dates of burnt flints from Jelinek's excavations at Tabun and their implications. *Journal of Archaeological Science* 22:495–510.

1553. Merrick, H. V., and Brown, F. H. 1984. Obsidian sources and patterns of source utilization in Kenya and northern Tanzania: Some initial findings. *African Archaeological Review* 2:129–152.

1554. Merrick, H. V., Brown, F. H., and Nash, W. P. 1994. Use and movement of obsidian in the Early and Middle Stone Ages of Kenya and northern Tanzania. In *Society, culture and technology in Africa*, ed. T. Childs, pp. 29–44. Philadelphia: MASCA/University of Pennsylvania Museum.

1555. Merrick, H. V., and Merrick, J. P. S. 1976. Archaeological occurrences of earlier Pleistocene age from the Shungura Formation. In *Earliest man and environments in the Lake Rudolf Basin*, ed. Y. Coppens, F. C. Howell, G. L. Isaac, and R. E. F. Leakey, pp. 574–584. Chicago: University of Chicago Press.

1556. Meyers, P. A., Takemura, K., and Horie, S. 1993. Reinterpretation of Late Quaternary sediment chronology of Lake Biwa, Japan, from correlation with marine glacial-interglacial cycles. *Quaternary Research* 39:154–162.

1557. Michels, J. W., Tsong, I. S. T., and Nelson, C. M. 1983. Obsidian dating and east African archeology. *Science* 219:361–366.

1558. Miller, G. H., Beaumont, P. B., Jull, A. J. T., and Johnson, B. 1993. Pleistocene geochronology and palaeothermometry from protein diagenesis in ostrich eggshells: Implications for the evolution of modern humans. In *The origin of modern humans and the impact of chronometric dating*, ed. M. J. Aitken, C. B. Stringer, and P. A. Mellars, pp. 49–68. Princeton: Princeton University Press.

1559. Miller, J. A. 1991. Does brain size variability provide evidence of multiple species in *Homo habilis? American Journal of Physical Anthropology* 84:385–398.

1560. Miller, K. G., and Fairbanks, R. G. 1985. Cainozoic δ^{18}O record of climate and sea level. *South African Journal of Science* 81:248–249.

1561. Miller, K. G., Fairbanks, R. G., and Mountain, G. S. 1987. Tertiary oxygen isotope synthesis, sea level history, and continental margin erosion. *Paleoceanography* 2:1–19.

1562. Miller, S. 1979. Lukenya Hill, GvJm 46, excavation report. *Nyame Akuma* 14:31–34.

1563. Milo, R. G. 1994. Human-animal interactions in southern African prehistory: A microscopic study of bone damage signatures. Ph.D. diss., University of Chicago.

1564. Milo, R. G. 1998. Evidence for hominid predation at Klasies River Mouth, South Africa, and its implications for the behaviour of early modern humans. *Journal of Archaeological Science* 25:99–133.

1565. Milo, R. G., and Quiatt, D. 1993. Glottogenesis and anatomically modern *Homo sapiens. Current Anthropology* 34:569–598.

1566. Milo, R. G., Quiatt, D. 1994. Language in the Middle and Late Stone Ages: Glottogenesis in anatomically modern *Homo sapiens.* In *Hominid culture in primate perspective,* ed. D. Quiatt and J. Itani, pp. 321–329. Niwot: University Press of Colorado.

1567. Minugh-Purvis, N., and Lewandowski, J. 1992. Functional anatomy, ontogeny, and behavioral implications of coronoid process morphology of Upper Pleistocene hominines. *American Journal of Physical Anthropology,* suppl., 14:124–125.

1568. Mishra, S., Venkaatesan, T. R., Rajaguru, S. N., and Somalyajulu, K. 1995. Earliest Acheulian Industry from peninsular India. *Current Anthropology* 36:847–851.

1569. Miyamoto, M. M., and Goodman, M. 1990. DNA systematics and evolution of primates. *Annual Review of Ecology and Systematics* 2:197–220.

1570. Mochanov, Y. A., and Fedoseeva, S. A. 1996. Berelyekh, Allakhovsk region. In *American beginnings: The prehistory and palaeoecology of Beringia,* ed. F. H. West, pp. 218–221. Chicago: University of Chicago Press.

1571. Molleson, T. I. 1976. Remains of Pleistocene man in Paviland and Pontnewydd Caves, Wales. *Transactions of the British Cave Research Association* 3:112–116.

1572. Montet-White, A. 1994. Alternative interpretations of the late Upper Paleolithic in central Europe. *Annual Review of Anthropology* 23:483–508.

1573. Montgomery, P. Q., Williams, H. O. L., Reading, N., and Stringer, C. B. 1994. An assessment of the temporal bone lesions of the Broken Hill cranium. *Journal of Archaeological Science* 21:331–338.

1574. Morbeck, M. E. 1983. Miocene hominoid discoveries from Rudabánya: Implications from the postcranial skeleton. In *New interpretations of ape and human ancestry,* ed. R. L. Ciochon and R. S. Corruccini, pp. 369–404. New York: Plenum Press.

1575. Morell, V. 1996. New skull turns up in northeast Africa. *Science* 271:32.

1576. Morlan, R. E. 1987. The Pleistocene archaeology of Beringia. In *The evolution of human hunting*, ed. M. H. Nitecki and D. V. Nitecki, pp. 267–307. New York: Plenum Press.

1577. Morlan, R. E. 1990. Pleistocene South Americans. *Science* 249: 937–938.

1578. Morris, A. G. 1992. Biological relationships between Upper Pleistocene and Holocene populations in southern Africa. In *Continuity or replacement: Controversies in* Homo sapiens *evolution*, ed. G. Bräuer and F. H. Smith, pp. 131–143. Rotterdam: A. A. Balkema.

1579. Morse, K. 1993. Shell beads from Mandu Mandu Creek rock-shelter, Cape Range Peninsula, Western Australia, dated before 30,000 B.P. *Antiquity* 67:877–883.

1580. Morwood, M. J., O'Sullivan, P. B., Aziz, F., and Raza, A. 1998. Fission-track ages of stone tools and fossils on the east Indonesian island of Flores. *Nature* 392:173–176.

1581. Moss, E. 1987. A review of "Investigating microwear polishes with blind tests." *Journal of Archaeological Science* 14:473–482.

1582. Mountain, E. D. 1966. Footprints in calcareous sandstone at Nahoon Point. *South African Journal of Science* 62:103–111.

1583. Mountain, J. L., and Cavalli-Sforza, L. L. 1994. Inference of human evolution through cladistic analysis of nuclear DNA restriction polymorphisms. *Proceedings of the National Academy of Sciences* 91:6515–6519.

1584. Mountain, J. L., Lin, A. A., Bowcock, A. M., and Cavalli-Sforza, L. L. 1993. Evolution of modern humans: Evidence from nuclear DNA polymorphisms. In *The origins of modern humans and the impact of chronometric dating*, ed. M. J. Aitken, C. B. Stringer, and P. A. Mellars, pp. 69–83. Princeton: Princeton University Press.

1585. Movius, H. L. 1944. Early man and Pleistocene stratigraphy in southern and eastern Asia. *Papers of the Peabody Museum* 19: 1–125.

1586. Movius, H. L. 1948. The Lower Palaeolithic cultures of southern and eastern Asia. *Transactions of the American Philosophical Society* 38:329–420.

1587. Movius, H. L. 1949. Lower Paleolithic archaeology in southern and eastern Asia. *Studies in Physical Anthropology* 1:17–81.

1588. Movius, H. L. 1950. A wooden spear of third interglacial age from lower Saxony. *Southwestern Journal of Anthropology* 6:139–142.

1589. Movius, H. L. 1953. The Mousterian cave of Teshik-Tash, southeastern Uzbekistan, central Asia. *Bulletin of the American School of Prehistoric Research* 17:11–71.

1590. Movius, H. L. 1953. Palaeolithic and Mesolithic sites in Soviet central Asia. *Proceedings of the American Philosophical Society* 97:383–421.

1591. Movius, H. L. 1955. Palaeolithic archaeology in southern and eastern Asia, exclusive of India. *Cahiers d'Histoire Mondiale* 2:257–282.

1592. Movius, H. L. 1969. The Abri de Cro-Magnon, Les Eyzies (Dordogne), and the probable age of the contained burials on the basis of the nearby Abri Pataud. *Anuario de Estudios Atlanticos* 15:323–344.

1593. Movius, H. L. 1969. The Châtelperronian in French archaeology: The evidence of Arcy-sur-Cure. *Antiquity* 43:111–123.

1594. Movius, H. L., ed. 1975. *Excavation of the Abri Pataud, Les Eyzies (Dordogne).* Cambridge, Mass.: Peabody Museum of Archaeology and Ethnology.

1595. Movius, H. L. 1977. *Excavation of the Abri Pataud, Les Eyzies (Dordogne): Stratigraphy.* Cambridge, Mass.: Peabody Museum of Archaeology and Ethnology.

1596. Moyà-Solà, S., and Köhler, M. 1996. A *Dryopithecus* skeleton and the origins of great-ape locomotion. *Nature* 379:156–159.

1597. Moyà-Solà, S., and Köhler, M. 1997. The Orce skull: Anatomy of a mistake. *Journal of Human Evolution* 33:91–97.

1598. Mturi, A. A. 1976. New hominid from Lake Ndutu, Tanzania. *Nature* 262:484–485.

1599. Mulvaney, D. J. 1991. Past regained, future lost: The Kow Swamp Pleistocene burials. *Antiquity* 65:12–21.

1600. Murray, P. 1984. Extinctions Downunder: A bestiary of extinct Australian late Pleistocene monotremes and marsupials. In *Quaternary extinctions: A prehistoric revolution,* ed. P. S. Martin and R. G. Klein, pp. 600–628. Tucson: University of Arizona Press.

1601. Musil, R., and Valoch, K. 1969. Stránska Skála: Its meaning for Pleistocene studies. *Current Anthropology* 9:534–539.

1602. Mussi, M. 1990. Continuity and change in Italy at the Last Glacial Maximum. In *The world at 18 000 BP,* vol. 1, *High latitudes,* ed. O. Soffer and C. Gamble, pp. 126–147. London: Unwin Hyman.

1603. Mussi, M., and Roebroeks, W. 1996. The big mosaic. *Current Anthropology* 37:697–699.

1604. Nadel, D., Danin, A., Werker, E., Schick, T., Kislev, M. E., and Stewart, K. 1994. 19,000-year-old twisted fibers from Ohalo II. *Current Anthropology* 35:451–458.

1605. Naeser, C. W., McKee, E. H., Johnson, N. M., and MacFadden, B. J. 1987. Confirmation of a late Oligocene–early Miocene age of the Desedean Salla Beds of Bolivia. *Journal of Geology* 95:825–828.

1606. Naeser, N. D., and Naeser, C. W. 1984. Fission-track dating. In *Quaternary dating methods,* ed. W. C. Mahaney, pp. 87–100. Amsterdam: Elsevier.

1607. Nagatoshi, K. 1987. Miocene hominoid environments of Europe and Turkey. *Palaeogeography, Palaeoclimatology, Palaeoecology* 61:145–154.

1608. Napier, J. R., and Napier, P. H. 1967. *A handbook of living primates.* New York: Academic Press.

1609. Nei, M., Livshits, G., and Ota, T. 1993. Genetic variation and evolution of human populations. In *Genetics of cellular, individual, family, and population variation,* ed. C. Sing and C. L. Hanis, pp. 239–252. New York: Oxford University Press.

1610. Nei, M., and Roychoudhury, A. K. 1993. Evolutionary relationships of human populations on a global scale. *Molecular Biology and Evolution* 10:927–943.

1611. Neiburger, E. J., Ogilvie, M. D., and Trinkaus, E. 1990. Enamel hypoplasias: Poor indicators of dietary stress. *American Journal of Physical Anthropology* 82:231–233.

1612. Neugebauer-Maresch, C. 1988. Vorbericht über die Rettungsgrabungen an der Aurignacien-Station Stratzing/Krems-Rehberg in den Jahren 1985–1988: Zum Neufund einer weiblichen Statuette. *Fundberichte österreich* 26 (1987):73–84.

1613. Neves, W., and Pucciarelli, H. 1998. The Zhoukoudian Upper Cave skull as seen from the Americas. *Journal of Human Evolution* 34:219–222.

1614. Nichols, J. 1990. Linguistic diversity and the first settlement of the New World. *Language* 66:475–521.

1615. Ninkovich, D., Burckle, L. H., and Opdyke, N. D. 1982. Palaeogeographic and geologic setting for early man in Java. In *The ocean floor*, ed. R. A. Scrutton and M. Talwani, pp. 221–225. New York: John Wiley.

1616. Novacek, M. J. 1992. Mammalian phylogeny: Shaking the tree. *Nature* 356:121–125.

1617. O'Connell, J., Hawkes, K., and Blurton Jones, N. 1988. Hadza scavenging: Implications for Plio-Pleistocene subsistence. *Current Anthropology* 29:356–363.

1618. O'Connell, J. F., and Allen, J. 1998. When did humans first arrive in Greater Australia, and why is it important to know? *Evolutionary Anthropology* 6:132–146.

1619. O'Connell, J. F., Hawkes, K., and Blurton Jones, N. 1990. Reanalysis of large mammal body part transport among the Hadza. *Journal of Archaeological Science* 17:301–316.

1620. O'Connell, J. F., Hawkes, K., and Blurton Jones, N. 1992. Patterns in the distribution, site structure and assemblage composition of Hadza kill-butchering sites. *Journal of Archaeological Science* 19:319–345.

1621. O'Connell, J. F., Hawkes, K., and Blurton Jones, N. G. 1999. Grandmothering and the evolution of *Homo erectus. Journal of Human Evolution.* In press.

1622. O'Connor, S. 1995. Carpenter's Gap Rockshelter: 40,000 years of Aboriginal occupation in the Napier Ranges, Kimberley, WA. *Australian Archaeology* 40:58–59.

1623. Oakley, K. P. 1952. Swanscombe man. *Proceedings of the Geologists' Association* 63:271–300.

1624. Oakley, K. P. 1957. Stratigraphical age of the Swanscombe skull. *American Journal of Physical Anthropology* 15:253–260.

1625. Oakley, K. P. 1959. *Man the tool-maker.* Chicago: University of Chicago Press.

1626. Oakley, K. P. 1964. The problem of man's antiquity: An historical survey. *Bulletin of the British Museum (Natural History), Geology* 9:86–155.

1627. Oakley, K. P. 1969. Analytical methods of dating bones. In *Science in archaeology*, ed. D. Brothwell and E. Higgs, pp. 35–45. London: Thames and Hudson.

1628. Oakley, K. P. 1980. Relative dating of the fossil hominids of Europe. *Bulletin of the British Museum of Natural History (Geology)* 34:1–63.

1629. Oakley, K. P., Andrews, P., Keeley, L. H., and Clark, J. D. 1977. A reappraisal of the Clacton spearpoint. *Proceedings of the Prehistoric Society* 43:13–30.

1630. Ogilvie, M. D., Curran, B. K., and Trinkaus, E. 1989. Incidence and patterning of dental enamel hypoplasia among the Neanderthals. *American Journal of Physical Anthropology* 79:25–41.

1631. Ohel, M. Y. 1977. The Clactonian: Reexamined, redefined, and reinterpreted. *Current Anthropology* 18:329–331.

1632. Ohel, M. Y. 1979. The Clactonian: An independent complex or an integral part of the Acheulian? *Current Anthropology* 20: 685–726.

1633. Oliva, M. 1988. Discovery of a Gravettian mammoth bone hut at Milovice (Moravia, Czechoslovakia). *Journal of Human Evolution* 17:787–790.

1634. Oliver, J. S. 1994. Estimates of hominid and carnivore involvement in the FLK *Zinjanthropus* fossil assemblage and some socio-ecological implications. *Journal of Human Evolution* 27: 267–294.

1635. Olsen, J. W. 1987. Recent developments in the Upper Pleistocene prehistory of China. In *The Pleistocene Old World: Regional perspectives,* ed. O. Soffer, pp. 135–146. New York: Plenum Press.

1636. Olsen, J. W., and Miller-Antonio, S. 1992. The Palaeolithic of southern China. *Asian Perspectives* 31:129–160.

1637. Olson, S. L., and Rasmussen, D. T. 1986. Paleoenvironment of the earliest hominoids: New evidence from the Oligocene avifauna of Egypt. *Science* 233:1202–1204.

1638. Olson, T. R. 1981. Basicranial morphology of the extant hominoids and Pliocene hominids: The new material from the Hadar Formation, Ethiopia, and its significance in early human evolution and taxonomy. In *Aspects of human evolution,* ed. C. B. Stringer, pp. 99–128. London: Taylor and Francis.

1639. Olson, T. R. 1985. Cranial morphology and systematics of the Hadar Formation hominids and *"Australopithecus" africanus.* In *Ancestors: The hard evidence,* ed. E. Delson, pp. 102–119. New York: Alan R. Liss.

1640. Otte, M. 1990. From the Middle to the Upper Palaeolithic: The nature of the transition. In *The emergence of modern humans: An archaeological perspective,* ed. P. Mellars, pp. 438–456. Ithaca: Cornell University Press.

1641. Otte, M. 1990. The northwestern European plain around 18 000 BP. In *The world at 18 000 BP,* vol. 1, *High latitudes,* ed. O. Soffer and C. Gamble, pp. 56–68. London: Unwin Hyman.

1642. Owen, R. C. 1984. The Americas: The case against an Ice-Age human population. In *The origins of modern humans: A world survey of the fossil evidence,* ed. F. H. Smith and F. Spencer, pp. 517–563. New York: Alan R. Liss.

1643. Oyston, B. 1996. Thermoluminescence age determinations for the Mungo III human burial, Lake Mungo, southeastern Australia. *Quaternary Science Reviews* 15:739–749.

1644. Pääbo, S. 1995. The Y chromosome and the origin of all of us (men). *Science* 268:1141–1142.

1645. Paillard, D. 1998. The timing of Pleistocene glaciations from a simple multiple-state climate model. *Nature* 391:378–381.

1646. Palma di Cesnola, A. 1980. L'Uluzzien et ses rapports avec Protoaurignacien en Italie. In *L'Aurignacien et le Gravettien (Périgordien) dans leur cadre écologique,* ed. L. Banesz and

J. K. Kozlowski, pp. 197–312. Nitra: Archeologicky Ustav Slovenskej Akadémie Vied.

1647. Palmqvist, P. 1997. A critical re-evaluation of the evidence for the presence of hominids in Lower Pleistocene times at Venta Micena, southern Spain. *Journal of Human Evolution* 33:83–89.

1648. Pan, Y. 1990. An adapid primate, *Sinoadapis*, from Lufeng, latest Miocene, Yunnan, China. *Human Evolution* 5:239–240.

1649. Papamarinopoulos, S., Readman, P. W., Maniatis, Y., and Simopoulos, A. 1987. Palaeomagnetic and mineral magnetic studies of sediments from Petralona Cave, Greece. *Archaeometry* 29:50–59.

1650. Parés, J. M., and Pérez-González, A. 1995. Paleomagnetic age for hominid fossils at Atapuerca archaeological site, Spain. *Science* 269:832–834.

1651. Parkington, J. E. 1987. Changing views of prehistoric settlement in the western Cape. In *Papers in the prehistory of the western Cape, South Africa,* ed. J. E. Parkington and M. Hall, 4–23. International Series 332. Oxford: British Archaeological Reports.

1652. Parkington, J. E. 1990. A view from the south: Southern Africa before, during, and after the Last Glacial Maximum. In *The world at 18 000 BP,* vol. 2, *Low latitudes,* ed. C. Gamble and O. Soffer, pp. 214–228. London: Unwin Hyman.

1653. Parkington, J. E., Nilssen, P., Vermuelen, C., and Henshilwood, C. 1992. Making sense of space at Dunefield Midden campsite, western Cape, South Africa. *South African Journal of Field Archaeology* 1:63–70.

1654. Partridge, T. C. 1973. Geomorphological dating of cave opening at Makapansgat, Sterkfontein, Swartkrans, and Taung. *Nature* 246:75–79.

1655. Partridge, T. C. 1978. Re-appraisal of lithostratigraphy of Sterkfontein hominid site. *Nature* 275:484–488.

1656. Partridge, T. C. 1979. Re-appraisal of lithostratigraphy of Makapansgat Limeworks hominid site. *Nature* 279:484–488.

1657. Partridge, T. C. 1982. The chronological positions of the fossil hominids of southern Africa. In *L'Homo erectus et la place de l'homme de Tautavel parmi les hominidés fossiles,* ed. M. A. de Lumley, pp. 617–675. Nice: Premier Congrès International de Paléontologie Humaine.

1658. Partridge, T. C. 1982. Some preliminary observations on the stratigraphy and sedimentology of the Kromdraai B hominid site. *Palaeoecology of Africa* 15:3–12.

1659. Partridge, T. C. 1986. Palaeoecology of the Pliocene and Lower Pleistocene hominids of southern Africa: How good is the chronological and palaeoenvironmental evidence? *South African Journal of Science* 82:80–83.

1660. Partridge, T. C., and Watt, I. B. 1991. The stratigraphy of the Sterkfontein hominid deposit and its relationship to the underground cave system. *Palaeontologia Africana* 28:35–40.

1661. Pascual, R., and Jaureguizar, E. O. 1990. Evolving climates and mammal faunas in Cenozoic South America. *Journal of Human Evolution* 19:23–60.

1662. Patterson, B., Behrensmeyer, A. K., and Sill, W. D. 1970. Geology and fauna of a new Pliocene locality in north-western Kenya. *Nature* 226:918.

1663. Patterson, B., and Howells, W. W. 1967. Hominid humeral fragment from early Pleistocene of north-western Kenya. *Science* 156:64–66.

1664. Pavlides, C., and Gosden, C. 1994. 35,000-year-old sites in the rainforests of West New Britain, Papua New Guinea. *Antiquity* 68:604–610.

1665. Peabody, F. E. 1954. Travertines and cave deposits of the Kaap Escarpment of South Africa and the type locality of *Australopithecus africanus* Dart. *Bulletin of the Geological Society of America* 65:671–706.

1666. Pearson, O. M., and Grine, F. E. 1996. Morphology of the Border Cave hominid ulna and humerus. *South African Journal of Science* 92:231–236.

1667. Pearson, O. M., and Grine, F. E. 1997. Re-analysis of the hominid radii from Cave of Hearths and Klasies River Mouth, South Africa. *Journal of Human Evolution* 32:577–592.

1668. Pei, W., and Zhang, S. 1985. A study on lithic artifacts of *Sinanthropus* (in Chinese with an English summary on pp. 259–277). *Palaeontologia Sinica*, n.s., 168:1–277.

1669. Pei, W. C. 1939. The Upper Cave Industry of Choukoutien. *Palaeontologica Sinica*, n.s. D, 9:1–41.

1670. Pelcin, A. 1994. A geological explanation for the Berkehat Ram figurine. *Current Anthropology* 35:674–675.

1671. Peretto, C. 1991. Les gisements d'Isernia la Pineta (Molise, Italie). In *Les premiers Européens*, ed. E. Bonifay and B. Vandermeersch, pp. 161–168. Paris: Éditions du Comité des Travaux Historiques et Scientifiques.

1672. Peretto, C., ed. 1994. *Le Industrie litiche del giacimento paleolitico di Isernia La Pineta*. Città di Castello: Cosmo Iannone.

1673. Pérez, P.-J., Gracia, A., Martínez, I., and Arsuaga, J. L. 1997. Paleopathological evidence of the cranial remains from the Sima de los Huesos Middle Pleistocene site (Sierra de Atapuerca, Spain): Description and preliminary inferences. *Journal of Human Evolution* 33:409–421.

1674. Perlés, C. 1976. Le feu. In *La préhistoire française*, ed. H. de Lumley, 1:679–683. Paris: Centre National de la Recherche Scientifique.

1675. Petit-Maire, N., Ferembach, D., Bouvier, J.-M., and Vandermeersch, B. 1971. France. In *Catalogue of fossil hominids*, part 2, *Europe*, ed. K. P. Oakley, B. G. Campbell, and T. I. Molleson, pp. 71–187. London: Trustees of the British Museum (Natural History).

1676. Pfeiffer, S., and Zehr, M. K. 1996. A morphological and histological study of the human humerus from Border Cave. *Journal of Human Evolution* 31:49–59.

1677. Phillips, J. L. 1994. The Upper Paleolithic chronology of the Levant and the Nile Valley. In *Late Quaternary chronology and paleoclimates of the eastern Mediterranean*, ed. O. Bar-Yosef and R. S. Kra, pp. 169–176. Tucson, Ariz.: Radiocarbon.

1678. Phillipson, D. W. 1976. *The prehistory of eastern Zambia*. Nairobi: British Institute in Eastern Africa.

1679. Phillipson, D. W. 1993. *African archaeology.* Cambridge: Cambridge University Press.

1680. Pickford, M. 1983. Sequence and environments of the Lower and Middle Miocene hominoids of western Kenya. In *New interpretations of ape and human ancestry,* ed. R. L. Ciochon and R. S. Corruccini, pp. 421–439. New York: Plenum Press.

1681. Pickford, M. 1985. *Kenyapithecus:* A review of its status based on newly discovered fossils from Kenya. In *Hominid evolution: Past, present, and future,* ed. P. V. Tobias, pp. 107–112. New York: Alan R. Liss.

1682. Pickford, M. 1986. The geochronology of Miocene higher primate faunas of east Africa. In *Primate evolution,* ed. J. G. Else and P. C. Lee, pp. 19–33. Cambridge: Cambridge University Press.

1683. Pickford, M. 1986. A reappraisal of *Kenyapithecus.* In *Primate evolution,* ed. J. G. Else and P. C. Lee, pp. 163–171. Cambridge: Cambridge University Press.

1684. Pickford, M., Johanson, D. C., Lovejoy, C. O., White, T. D., and Aronson, J. L. 1983. A hominoid humeral fragment from the Pliocene of Kenya. *American Journal of Physical Anthropology* 60:337–346.

1685. Pidoplichko, I. G. 1969. *Upper Paleolithic mammoth bone dwellings in Ukraine* (in Russian). Kiev: Naukova Dumka.

1686. Pilbeam, D. 1996. Genetic and morphological records of the Hominoidea and hominid origins: A synthesis. *Molecular Phylogenetics and Evolution* 5:155–168.

1687. Pilbeam, D., Rose, M. D., Barry, J. C., and Shah, S. M. I. 1990. New *Sivapithecus* humeri from Pakistan and the relationship of *Sivapithecus* and *Pongo. Nature* 348:237–239.

1688. Pilbeam, D. R. 1983. New hominoid skull material from the Miocene of Pakistan. *Nature* 295:232–234.

1689. Pilbeam, D. R. 1986. Hominoid evolution and hominoid origins. *American Anthropologist* 88:295–312.

1690. Pilbeam, D. R. 1988. Part 1. Human evolution. In *Human biology: An introduction to human evolution, variation, growth, and adaptability,* ed. G. A. Harrison, J. M. Tanner, D. R. Pilbeam, and P. T. Baker, pp. 3–143. Oxford: Oxford University Press.

1691. Piperno, M., Mallegni, F., and Yokohama, Y. 1990. Découverte d'un fémur humain dans les niveaux acheuléens de Notarchirico (Venosa, Basilicata, Italie). *Comptes Rendus de l'Académie des Sciences, Paris,* ser. 2, 311:1097–1102.

1692. Piveteau, J., and Condemi, S. 1988. L'os temporal Riss-Würm (BD7) provenant de la grotte de la Chaise, abri Bourgeois-Delaunay. *Études et Recherches Archéologiques de l'Université de Liège* 30:105–110.

1693. Plug, I., and Keyser, A. W. 1994. A preliminary report on the bovid species from recent excavations at Gladysvale, South Africa. *South African Journal of Science* 90:357–359.

1694. Plummer, T. W., Kinyua, A. M., and Potts, R. 1994. Provenancing of hominid and mammalian fossils from Kanjera, Kenya, using EDXRF. *Journal of Archaeological Science* 21:553–563.

1695. Plummer, T. W., and Potts, R. 1989. Excavations and new findings at Kanjera, Kenya. *Journal of Human Evolution* 18:269–276.

1696. Plummer, T. W., and Potts, R. 1995. Hominid fossil sample from Kanjera, Kenya: Description, provenance, and implications of new and earlier discoveries. *American Journal of Physical Anthropology* 96:7–23.

1697. Pollock, J. I., and Mullin, R. J. 1987. Vitamin C biosynthesis in prosimians: Evidence for the anthropoid affinity of *Tarsius*. *American Journal of Physical Anthropology* 73:65–70.

1698. Pope, G. G. 1983. Evidence on the age of the Asian Hominidae. *Proceedings of the National Academy of Sciences* 80:4988–4992.

1699. Pope, G. G. 1985. Taxonomy, dating and paleoenvironment: The paleoecology of the early Far Eastern hominids. *Modern Quaternary Research in Southeast Asia* 9:65–80.

1700. Pope, G. G. 1988. Recent advances in Far Eastern paleoanthropology. *Annual Review of Anthropology* 17:43–77.

1701. Pope, G. G. 1995. The influence of climate and geography on the biocultural evolution of the Far Eastern hominids. In *Paleoclimate and evolution with emphasis on human origins*, ed. E. S. Vrba, G. H. Denton, T. C. Partridge, and L. H. Burckle, pp. 493–506. New Haven: Yale University Press.

1702. Pope, G. G., Barr, S., Macdonald, A., and Nakabanlang, S. 1986. Earliest radiometrically dated artifacts from Southeast Asia. *Current Anthropology* 27:275–279.

1703. Pope, G. G., and Cronin, J. E. 1984. The Asian Hominidae. *Journal of Human Evolution* 13:377–396.

1704. Potts, R. 1989. Olorgesailie: New excavations and findings in Early and Middle Pleistocene contexts, southern Kenya Rift Valley. *Journal of Human Evolution* 18:477–484.

1705. Potts, R. 1996. Evolution and climate variability. *Science* 273:922–923.

1706. Potts, R. 1996. *Humanity's descent: The consequences of ecological instability.* New York: William Morrow.

1707. Potts, R., and Deino, A. 1995. Mid-Pleistocene change in large mammal faunas of east Africa. *Quaternary Research* 43:106–113.

1708. Potts, R., Shipman, P., and Ingsall, E. 1988. Taphonomy, paleoecology, and hominids at Lainyamok, Kenya. *Journal of Human Evolution* 17:597–614.

1709. Potts, R. B. 1984. Hominid hunters? Problems of identifying the earliest hunter/gatherers. In *Hominid evolution and community ecology*, ed. R. A. Foley, pp. 129–166. London: Academic Press.

1710. Potts, R. B. 1986. Temporal span of bone accumulations at Olduvai Gorge and implications for early hominid foraging behavior. *Paleobiology* 12:25–31.

1711. Potts, R. B. 1988. *Early hominid activities at Olduvai.* New York: Aldine de Gruyter.

1712. Potts, R. B. 1991. Why the Oldowan? Plio-Pleistocene toolmaking and the transport of resources. *Journal of Anthropological Research* 47:153–176.

1713. Potts, R. B., and Shipman, P. B. 1981. Cutmarks made by stone tools on bones from Olduvai Gorge, Tanzania. *Nature* 291:577–580.

1714. Poulianos, N. A. 1989. Petralona Cave within Lower-Middle

Pleistocene sites. *Palaeogeography, Palaeoclimatology, Palaeo-ecology* 73:287–294.

1715. Powers, W. R., and Hoffecker, J. F. 1989. Late Pleistocene settlement in the Nenana Valley, central Alaska. *American Antiquity* 54:263–287.

1716. Price, J. L., and Molleson, T. I. 1974. A radiographic examination of the left temporal bone of Kabwe man, Broken Hill Mine, Zambia. *Journal of Archeological Science* 1:285–289.

1717. Pringle, H. 1997. Ice Age communities may be earliest known net hunters. *Science* 277:1203–1204.

1718. Pycraft, W. P., Elliot Smith, G., Yearsley, M., Carter, J. T., Smith, R. A., Hopwood, A. T., Bate, D. M. A., and Swinton, W. E. 1928. *Rhodesian man and associated remains*. London: British Museum (Natural History).

1719. Qiu, Z. 1985. The Middle Palaeolithic of China. In *Palaeoanthropology and Palaeolithic archaeology in the People's Republic of China*, ed. R. Wu and J. Olsen, pp. 187–210. Orlando, Fla.: Academic Press.

1720. Qiu, Z. 1992. The stone industries of *Homo sapiens* from China. In *The evolution and dispersal of modern humans in Asia*, ed. T. Akazawa, K. Aoki, and T. Kimura, pp. 363–372. Tokyo: Hokusen-Sha.

1721. Quam, R. M., and Smith, F. H. 1996. Reconsideration of the Tabun C2 "Neanderthal." *American Journal of Physical Anthropology*, suppl., 22:192.

1722. Radinsky, L. B. 1975. Primate brain evolution. *American Scientist* 63:656–663.

1723. Radinsky, L. B. 1977. Early primate brains: Facts and fiction. *Journal of Human Evolution* 6:79–86.

1724. Radovcic, J., Smith, F. H., Trinkaus, E., and Wolpoff, M. H. 1988. *The Krapina hominids: An illustrated catalog of skeletal collection*. Zagreb: Mladost.

1725. Rafferty, K. L., Walker, A., Ruff, C. B., Rose, M. D., and Andrews, P. J. 1995. Postcranial estimates of body weight in *Proconsul*, with a note on a distal tibia of *P. major*. *American Journal of Physical Anthropology* 97:391–402.

1726. Rak, Y. 1983. *The australopithecine face*. New York: Academic Press.

1727. Rak, Y. 1986. The Neanderthal: A new look at an old face. *Journal of Human Evolution* 15:151–164.

1728. Rak, Y. 1998. Does any Mousterian cave present evidence of two hominid species? In *Neanderthals and modern humans in west Asia*, ed. K. Aoki, T. Akazawa, and O. Bar-Yosef, pp. 353–366. New York: Plenum Press.

1729. Rak, Y., and Arensburg, B. 1987. Kebara 2 Neanderthal pelvis: First look at a complete inlet. *American Journal of Physical Anthropology* 73:227–231.

1730. Rak, Y., Kimbel, W. H., and Hovers, E. 1994. A Neandertal infant from Amud Cave, Israel. *Journal of Human Evolution* 26:313–324.

1731. Ranov, V. A. 1991. Les sites très anciens de l'Âge de la Pierre en U.R.S.S. In *Les premiers Européens*, ed. E. Bonifay, and B. Vandermeersch, pp. 209–216. Paris: Éditions du Comité des Travaux Historiques et Scientifiques.

1732. Ranov, V. A., Carbonell, E., and Rodríguez, X. P. 1995. Kuldara: Earliest human occupation in central Asia in its Afro-Asian context. *Current Anthropology* 36:337–346.

1733. Ranov, V. A., and Davis, R. S. 1979. Toward a new outline of the Soviet central Asian Paleolithic. *Current Anthropology* 20:249–270.

1734. Rasmussen, D. T. 1986. Anthropoid origins: A possible solution to the Adapidae-Omomyidae paradox. *Journal of Human Evolution* 15:1–12.

1735. Rasmussen, D. T., Bown, T. M., and Simons, E. L. 1992. The Eocene-Oligocene transition in continental Africa. In *Eocene-Oligocene climatic and biotic evolution*, ed. D. R. Prothero and W. A. Berggren, pp. 69–82. Princeton: Princeton University Press.

1736. Rasmussen, D. T., and Simons, E. L. 1988. New specimens of *Oligopithecus savagei*, early Oligocene primate from the Fayum, Egypt. *Folia Primatologia* 51:182–208.

1737. Rasmussen, D. T., and Simons, E. L. 1991. The oldest Egyptian hyracoids (Mammalia: Pliohyracidae): New species of *Saghatherium* and *Thyrohyrax* from the Fayum. *Neues Jahrbuch für Mineralogie, Geologie und Paläontologie Abhandlungen* 182:187–209.

1738. Rasmussen, D. T., and Simons, E. L. 1992. Paleobiology of the oligopithecines, the earliest known anthropoid primates. *International Journal of Primatology* 13:477–508.

1739. Raynal, J.-P., Geraads, D., Magoga, L., El Hajraoui, A., Texier, J.-P., Lefevre, D., and Sbihi-Alaoui, F.-Z. 1993. La grotte des Rhinocéros (Carrière Oulad Hamida 1, anciennement Thomas III, Casablanca), nouveau site acheuléen du Maroc atlantique. *Comptes Rendus de l'Académie des Sciences, Paris*, ser. 2, 316:1477–1483.

1740. Raynal, J.-P., Magoga, L., Sbihi-Alaoui, F.-Z., and Geraads, D. 1995. The earliest occupation of Atlantic Morocco: The Casablanca evidence. In *The earliest occupation of Europe*, ed. W. Roebroeks and T. van Kolfschoten, pp. 255–262. Leiden: University of Leiden.

1741. Raynal, J.-P., Texier, J.-P., Geraads, D., and Sbihi-Alaoui, F.-Z. 1990. Un nouveau gisement paléontologique plio-pléistocène en Afrique du Nord: Ahl Al Oughlam (ancienne carrière Deprez) à Casablanca (Maroc). *Comptes Rendus de l'Académie des Sciences, Paris*, ser. 2, 310:315–320.

1742. Rayner, R. J., Moon, B. P., and Masters, J. C. 1993. The Makapansgat australopithecine environment. *Journal of Human Evolution* 24:219–231.

1743. Raza, S. M., Barry, J. C., Pilbeam, D., Rose, M. D., Ibrahim Shah, S. M., and Ward, S. 1983. New hominoid primates from the middle Miocene Chinji Formation, Potwar Plateau, Pakistan. *Nature* 306:52–54.

1744. Reck, H. 1914. Erste vorläufige Mittelung über den Fund eines fossilen Menschenskelets aus Zentralafrica. *Sitzungsbericht der Gesellschaft Naturforschender Freunde zu Berlin* 3:83–95.

1745. Reed, K. E. 1997. Early hominid evolution and ecological change through the African Plio-Pleistocene. *Journal of Human Evolution* 32:289–322.

1746. Reed, K. E., Kitching, J. W., Grine, F. E., Jungers, W. L., and Sokoloff, L. 1993. Proximal femur of *Australopithecus africanus* from Member 4, Makapansgat, South Africa. *American Journal of Physical Anthropology* 92:1–15.

1747. Relethford, J. H. 1995. Genetics and modern human origins. *Evolutionary Anthropology* 4:53–63.

1748. Relethford, J. H., and Harpending, H. C. 1994. Craniometric variation, genetic theory, and modern human origins. *American Journal of Physical Anthropology* 95:249–270.

1749. Relethford, J. H., and Harpending, H. C. 1995. Ancient differences in population size can mimic a recent African origin of modern humans. *Current Anthropology* 36:667–674.

1750. Rendell, H., and Dennell, R. W. 1985. Dated Lower Palaeolithic artefacts from northern Pakistan. *Current Anthropology* 26:393.

1751. Renne, P. R., Sharp, W. D., Deino, A. L., Orsi, G., and Civetta, L. 1997. ^{40}Ar/^{39}Ar dating into the historical realm: Calibration against Pliny the Younger. *Science* 277:1279–1280.

1752. Repenning, C. A. 1980. Faunal exchanges between Siberia and North America. *Canadian Journal of Anthropology* 1:37–44.

1753. Repenning, C. A. 1987. Biochronology of the microtine rodents of the United States. In *Cenozoic mammals of North America*, ed. M. O. Woodborne, pp. 236–268. Berkeley: University of California Press.

1754. Reynolds, T. E. G. 1985. The early Palaeolithic of Japan. *Antiquity* 59:93–96.

1755. Reynolds, T. E. G., and Barnes, G. L. 1984. The Japanese Palaeolithic: A review. *Proceedings of the Prehistoric Society* 50:49–63.

1756. Reynolds, T. E. G., and Kaner, S. C. 1990. Japan and Korea at 18 000 BP. In *The world at 18 000 BP*, vol. 1, *High latitudes*, ed. O. Soffer and C. Gamble, pp. 296–311. London: Unwin Hyman.

1757. Rhodes, E. J., Raynal, J.-P., Geraads, D., and Sbihi-Alaoui, F.-Z. 1994. Premières dates RPE pour l'Acheuléen du Maroc atlantique (grotte du Rhinocéros, Casablanca). *Comptes Rendus de l'Académie des Sciences, Paris*, ser. 2, 319:1109–1115.

1758. Richard, A. F. 1985. *Primates in nature.* New York: W. H. Freeman.

1759. Ridley, M. 1986. *Evolution and classification: The reformation of cladism.* London: Longman.

1760. Ridley, M. 1993. *Evolution.* Boston: Blackwell Scientific.

1761. Rigaud, J.-P., and Simek, J. 1990. The last pleniglacial in the south of France (24 000–14 000 years ago). In *The world at 18 000 BP*, vol. 1, *High latitudes*, ed. O. Soffer and C. Gamble, pp. 69–86. London: Unwin Hyman.

1762. Rightmire, G. P. 1975. Problems in the study of Later Pleistocene man in Africa. *American Anthropologist* 77:28–52.

1763. Rightmire, G. P. 1979. Cranial remains of *Homo erectus* from Beds II and IV, Olduvai Gorge, Tanzania. *American Journal of Physical Anthropology* 51:99–115.

1764. Rightmire, G. P. 1979. Implications of the Border Cave skeletal remains for later Pleistocene human evolution. *Current Anthropology* 20:23–35.

1765. Rightmire, G. P. 1980. *Homo erectus* and human evolution in the African Middle Pleistocene. In *Current argument on early man*, ed. L.-K. Königsson, pp. 70–85. Oxford: Pergamon Press.

1766. Rightmire, G. P. 1981. Patterns in the evolution of *Homo erectus*. *Paleobiology* 7:241–246.

1767. Rightmire, G. P. 1983. The Lake Ndutu cranium and early *Homo sapiens* in Africa. *American Journal of Physical Anthropology* 61:245–254.

1768. Rightmire, G. P. 1984. Comparisons of *Homo erectus* from Africa and Southeast Asia. *Courier Forschungsinstitut Senckenberg* 69:83–98.

1769. Rightmire, G. P. 1984. *Homo sapiens* in sub-Saharan Africa. In *The origin of modern humans: A world survey of the fossil evidence*, ed. F. H. Smith and F. Spencer, pp. 295–325. New York: Alan R. Liss.

1770. Rightmire, G. P. 1985. The tempo of change in the evolution of mid-Pleistocene *Homo*. In *Ancestors: The hard evidence*, ed. E. Delson, pp. 255–264. New York: Alan R. Liss.

1771. Rightmire, G. P. 1987. Africa and the origin of modern humans. In *Variation, culture and evolution in African populations: Papers in honour of Dr. Hertha de Villiers*, ed. R. Singer and J. K. Lundy, pp. 209–220. Johannesburg: Witwatersrand University Press.

1772. Rightmire, G. P. 1988. *Homo erectus* and later Middle Pleistocene humans. *Annual Review of Anthropology* 17:239–259.

1773. Rightmire, G. P. 1989. Middle Stone Age humans from eastern and southern Africa. In *The human revolution: Behavioural and biological perspectives on the origins of modern humans*, ed. P. Mellars and C. B. Stringer, pp. 109–122. Edinburgh: Edinburgh University Press.

1774. Rightmire, G. P. 1990. *The evolution of* Homo erectus: *Comparative anatomical studies of an extinct human species.* Cambridge: Cambridge University Press.

1775. Rightmire, G. P. 1992. *Homo erectus:* Ancestor or evolutionary side branch? *Evolutionary Anthropology* 1:43–49.

1776. Rightmire, G. P. 1993. Variation among early *Homo* crania from Olduvai Gorge and the Koobi Fora region. *American Journal of Physical Anthropology* 90:1–33.

1777. Rightmire, G. P. 1996. The human cranium from Bodo, Ethiopia: Evidence for speciation in the Middle Pleistocene? *Journal of Human Evolution* 31:21–39.

1778. Rightmire, G. P. 1998. Human evolution in the Middle Pleistocene: The role of *Homo heidelbergensis*. *Evolutionary Anthropology* 6:281–227.

1779. Rightmire, G. P., and Deacon, H. J. 1991. Comparative studies of late Pleistocene human remains from Klasies River Mouth, South Africa. *Journal of Human Evolution* 20:131–156.

1780. Rink, W. J., Schwarcz, H. P., Grün, R., Yalçinkaya, I., Taskiran, H., Otte, M., Valladas, H., Mercier, N., Bar-Yosef, O., and Kozlowski, J. 1994. ESR dating of the Last Interglacial Mousterian at Karaïn Cave, southern Turkey. *Journal of Archaeological Science* 21:839–849.

1781. Rink, W. J., Schwarcz, H. P., Smith, F. H., and Radovcic, J. 1995. ESR ages for Krapina hominids. *Nature* 378:24.

1782. Rink, W. J., Schwarcz, H. P., Valoch, K., Seitl, L., and Stringer, C. B. 1996. ESR dating of Micoquian industry and Neanderthal remains at Kulna Cave, Czech Republic. *Journal of Archaeological Science* 23:889–901.

1783. Robbins, L. H., Murphy, M. L., Stewart, K. M., Campbell, A. C., and Brook, G. A. 1994. Barbed bone points, paleoenvironment, and the antiquity of fish exploitation in the Kalahari Desert, Botswana. *Journal of Field Archaeology* 21:257–264.

1784. Roberts, D., and Berger, L. R. 1997. Last Interglacial (c. 117 kyr) human footprints from South Africa. *South African Journal of Science* 93:349–350.

1785. Roberts, M. B. 1986. Excavation of the Lower Palaeolithic site at Amey's Eartham Pit, Boxgrove, West Sussex: A preliminary report. *Proceedings of the Prehistoric Society* 52:215–245.

1786. Roberts, M. B., Gamble, C. S., and Bridgland, D. R. 1995. The earliest occupation of Europe: The British Isles. In *The earliest occupation of Europe*, ed. W. Roebroeks and T. van Kolfschoten, pp. 165–191. Leiden: University of Leiden.

1787. Roberts, M. B., Stringer, C. G., and Parfitt, S. A. 1994. A hominid tibia from Middle Pleistocene sediments at Boxgrove, UK. *Nature* 369:311–313.

1788. Roberts, N. 1984. Pleistocene environments in time and space. In *Hominid evolution and community ecology*, ed. R. Foley, pp. 25–53. London: Academic Press.

1789. Roberts, R., Walsh, G., Murray, A., Olley, J., Jones, R., Morwood, M., Tuniz, C., Lawson, E., Macphail, M., Bowdery, D., and Naumann, I. 1997. Luminescence dating of rock art and past environments using mud-wasp nests in northern Australia. *Nature* 387:696–699.

1790. Roberts, R. G., Jones, R., and Smith, M. A. 1990. Thermoluminescence dating of a 50,000-year-old human occupation site in northern Australia. *Nature* 345:153–156.

1791. Roberts, R. G., Jones, R., Spooner, N. A., Head, M. J., Murray, A. S., and Smith, M. A. 1994. The human colonisation of Australia: Optical dates of 53,000 and 60,000 years bracket human arrival at Deaf Adder Gorge, Northern Territory. *Quaternary Science Reviews* 13:575–586.

1792. Robertshaw, P. T. 1977. Archaeological investigations at Langebaan Lagoon, Cape Province. *Palaeoecology of Africa* 10:139–148.

1793. Robinson, J. T. 1954. Prehominid dentition and hominid evolution. *Evolution* 8:324–334.

1794. Robinson, J. T. 1963. Adaptive radiation in the australopithecines and the origin of man. In *African ecology and human evolution*, ed. F. C. Howell and F. Bourlière, pp. 385–416. Chicago: Aldine.

1795. Robinson, J. T. 1972. *Early hominid posture and locomotion.* Chicago: University of Chicago Press.

1796. Robinson, J. T., and Mason, R. J. 1957. Occurrence of stone artifacts with *Australopithecus* at Sterkfontein. *Nature* 180:521–524.

1797. Roche, H. 1995. Les industries de la limite Plio-Pléistocène et du Pléistocène ancien en Afrique. In *Congreso Internacional de Paleontologia Humana (Orce, September 1995), 3a Circular*, p. 93. Orce, Spain.

1798. Roche, H., Brugal, J.-P., Lefèvre, D., Ploux, S., and Texier, P.-J. 1988. Isenya: État des recherches sur un nouveau site acheuléen d'Afrique orientale. *African Archaeological Review* 6:27–55.

1799. Roche, H., Brugal, J.-P., Lefèvre, D., and Texier, P.-J. 1987. Premières données sur l'Acheuléen des hauts plateaux kényans: Le site d'Isenya (district de Kajiado). *Comptes Rendus de l'Académie des Sciences, Paris*, ser. 2, 305:529–532.

1800. Roche, J., and Texier, J.-P. 1976. Découverte des restes humains dans un niveau atérien supérieur de la grotte des Contrebandiers, à Témara (Maroc). *Comptes Rendus de l'Académie des Science, Paris*, ser. D, 282:45–47.

1801. Rodman, P. S., and McHenry, H. M. 1980. Energetics and the origin of hominid bipedalism. *American Journal of Physical Anthropology* 52:103–106.

1802. Roe, D. A. 1981. *The Lower and Middle Palaeolithic periods in Britain*. London: Routledge and Kegan Paul.

1803. Roe, D. A. 1995. The Orce Basin (Andalucia, Spain) and the initial Palaeolithic of Europe. *Oxford Journal of Archaeology* 14:1–12.

1804. Roebroeks, W. 1986. Archaeology and Middle Pleistocene stratigraphy: The case of Maastricht-Belvédère (NL). In *Chronostratigraphie et faciés culturels du Paléolithique inférieur et moyen dans l'Europe du Nord-Ouest*, ed. A. Tuffreau and J. Sommé, pp. 81–86. Paris: Supplément au Bulletin de l'Association Française pour l'Étude du Quaternaire.

1805. Roebroeks, W. 1988. *From find scatters to early hominid behaviour: A study of Middle Palaeolithic riverside settlements at Maastricht-Belvédère (the Netherlands)*. Leiden: [Analecta Praehistorica Leidensis].

1806. Roebroeks, W. 1996. The English Palaeolithic record: Absence of evidence, evidence of absence and the first occupation of Europe. In *The English Palaeolithic reviewed*, ed. C. S. Gamble and A. J. Lawson, pp. 57–62. Wessex: Trust for Wessex Archaeology.

1807. Roebroeks, W., Conard, N. J., and Kolfschoten, T. van. 1992. Dense forests, cold steppes, and the Palaeolithic settlement of northern Europe. *Current Anthropology* 33:551–586.

1808. Roebroeks, W., De Loecker, D., Hennekens, P., and van Ieperen, M. 1992. "A veil of stones": On the interpretation of an early Middle Palaeolithic low density scatter at Maastricht-Belvédère (the Netherlands). *Analecta Praehistorica Leidensis* 25:1–16.

1809. Roebroeks, W., De Loecker, D., Hennekens, P., and van Ieperen, M. 1993. On the archaeology of the Maastricht-Belvédère Pit. *Meddelingen Rijks Geologische Dienst* 47:69–79.

1810. Roebroeks, W., Kolen, J., and Rensink, E. 1988. Planning depth, anticipation and the organization of Middle Palaeolithic technology: The "archaic" natives meet Eve's descendants. *Helinium* 28:17–34.

1811. Roebroeks, W., and Kolfschoten, T. van. 1994. The earliest occupation of Europe: A short chronology. *Antiquity* 68:489–503.

1812. Rogachev, A. N. 1955. A burial of the Old Stone Age at the site of Kostenki XIV (Markina Gora) (in Russian). *Sovetskaya Etnografiya* 1955 (1): 29–38.

1813. Rogers, A. R. 1995. How much can fossils tell us about regional continuity? *Current Anthropology* 36:674–676.

1814. Rogers, A. R., and Harpending, H. C. 1992. Population growth makes waves in the distribution of pairwise differences. *Molecular Biology and Evolution* 9:552–569.

1815. Rogers, A. R., and Jorde, L. B. 1995. Genetic evidence on modern human origins. *Human Biology* 67:1–36.

1816. Rogers, J. 1993. The phylogenetic relationship among *Homo, Pan and Gorilla:* A population genetics perspective. *Journal of Human Evolution* 25:201–215.

1817. Rolland, N. 1981. The interpretation of Middle Palaeolithic variability. *Man* 16:15–42.

1818. Rolland, N. 1992. The Palaeolithic colonization of Europe: An archaeological and biogeographic perspective. *Trabajos de Prehistoria* 49:69–111.

1819. Rolland, N., and Dibble, H. L. 1990. A new synthesis of Middle Paleolithic variability. *American Antiquity* 55:480–499.

1820. Roosevelt, A. C., Lima da Costa, M., Lopes Machado, C., Michab, M., Mercier, N., Valladas, H., Feathers, J., Barnett, W., Imazio da Silveira, M., Henderson, A., Sliva, J., Chernoff, B., Reese, D. S., Holman, J. A., Toth, N., and Schick, K. 1996. Paleoindian cave dwellers in the Amazon: The peopling of the Americas. *Science* 272:373–384.

1821. Roper, M. K. 1969. A survey of the evidence for intrahuman killing in the Pleistocene. *Current Anthropology* 10:427–459.

1822. Rosas, A. 1995. Seventeen new mandibular specimens from the Atapuerca/Ibeas Middle Pleistocene hominids sample. *Journal of Human Evolution* 28:533–559.

1823. Rosas, A. 1997. A gradient of size and shape for the Atapuerca sample and Middle Pleistocene hominid variability. *Journal of Human Evolution* 33:319–331.

1824. Rosas, A., Bermúdez de Castro, J. M., and Aguirre, E. 1991. Mandibules et dents d'Ibeas (Espagne) dans le contexte de l'évolution humaine en Europe. *Anthropologie* 95:89–102.

1825. Rose, K. D. 1995. The earliest primates. *Evolutionary Anthropology* 3:159–173.

1826. Rose, K. D., and Bown, T. M. 1991. Additional fossil evidence on the differentiation of the earliest euprimates. *Proceedings of the National Academy of Sciences* 88:98–101.

1827. Rose, K. D., and Fleagle, J. G. 1981. The fossil history of nonhuman primates in the Americas. In *Ecology and behavior of Neotropical primates,* ed. A. F. Coimbra-Filho and R. A. Mittermeier, 1:111–167. Rio de Janeiro: Academia Brasileira de Ciências.

1828. Rose, K. D., and Walker, A. C. 1985. The skeleton of early Eocene *Cantius,* oldest lemuriform primate. *American Journal of Physical Anthropology* 66:73–89.

1829. Rose, M. D. 1983. Miocene hominoid postcranial morphology: Monkey-like, ape-like, neither or both? In *New interpretations of ape and human ancestry,* ed. R. L. Ciochon and R. S. Corruccini, pp. 405–417. New York: Plenum Press.

1830. Rose, M. D. 1984. Hominoid postcranial specimens from the Middle Miocene Chinji Formation, Pakistan. *Journal of Human Evolution* 13:503–516.

1831. Rose, M. D. 1986. Further hominoid postcranial specimens from the late Miocene Nagri Formation of Pakistan. *Journal of Human Evolution* 15:333–367.

1832. Rosenberg, K. R. 1985. Neanderthal birth canals. Abstract. *American Journal of Physical Anthropology* 66:222.

1833. Rosenberg, K. R. 1986. The functional significance of Neandertal pubic morphology. Ph D. diss., University of Michigan, Ann Arbor.

1834. Rosenberger, A. L. 1986. Platyrrhines, catarrhines and the anthropoid transition. In *Major topics in primate and human evolution,* ed. B. Wood, L. Martin, and P. Andrews, pp. 66–88. Cambridge: University of Cambridge Press.

1835. Rosenberger, A. L., and Delson, E. 1985. The dentition of *Oreopithecus bambolii:* Systematic and paleobiological implications. *American Journal of Physical Anthropology* 66:222–223.

1836. Rosenberger, A. L., Setoguchi, T., and Hartwig, W. C. 1991. *Laventiana annectens,* new genus and species: Fossil evidence for the origins of callitrichine New World monkeys. *Proceedings of the National Academy of Sciences* 88:2317–2140.

1837. Rosenberger, A. L., Setoguchi, T., and Shigehara, N. 1990. The fossil record of callitrichine primates. *Journal of Human Evolution* 19:209–236.

1838. Rosenberger, A. L., and Szalay, F. S. 1980. On the tarsiiform origins of Anthropoidea. In *Evolutionary biology of the New World monkeys and continental drift,* ed. R. L. Ciochon and A. B. Chiarelli, pp. 139–157. New York: Plenum Press.

1839. Rosenzweig, M. L. 1997. Tempo and mode of speciation. *Science* 279:1622–1623.

1840. Rothschild, B. M., Herskovitz, I., and Rothschild, C. 1995. Origin of yaws in the Pleistocene. *Nature* 378:343–344.

1841. Roubet, F.-E. 1969. Le niveau atérien dans le stratigraphie côtière à l'ouest d'Alger. *Palaeoecology of Africa* 4:124–129.

1842. Rucklidge, J. C. 1984. Radioisotope detection and dating with particle accelerators. In *Quaternary dating methods,* ed. W. C. Mahaney, pp. 17–32. Amsterdam: Elsevier.

1843. Ruff, C. B. 1993. Climatic adaptation and hominid evolution: The thermoregulatory imperative. *Evolutionary Anthropology* 2:53–60.

1844. Ruff, C. B. 1994. Morphological adaptation to climate in modern and fossil hominids. *Yearbook of Physical Anthropology* 37:65–107.

1845. Ruff, C. B., Trinkaus, E., Walker, A., and Larsen, C. S. 1993. Postcranial robusticity in *Homo* I: Temporal trends and mechanical interpretation. *American Journal of Physical Anthropology* 91:21–53.

1846. Ruff, C. B., and Walker, A. 1993. Body size and body shape. In *The Nariokotome* Homo erectus *skeleton,* ed. A. Walker and R. Leakey, pp. 234–263. Cambridge: Harvard University Press.

1847. Ruhlen, M. 1994. *On the origin of languages: Studies in linguistic taxonomy.* Stanford: Stanford University Press.

1848. Ruhlen, M. 1994. *The origin of language: Tracing the evolution of the mother tongue.* New York: John Wiley.

1849. Ruhlen, M. 1996. Multiregional evolution or "Out of Africa"? The linguistic evidence. In *Prehistoric Mongoloid dispersals,* ed. T. Akazawa and E. J. E. Szathmáry, pp. 52–65. Oxford: Oxford University Press.

1850. Runnels, C. 1995. Review of Aegean prehistory IV: The Stone Age of Greece from the Palaeolithic to the advent of the Neolithic. *American Journal of Archaeology* 99:699–728.

1851. Russell, D. E., and Gingerich, P. D. 1987. Nouveaux primates de l'Éocene du Pakistan. *Comptes Rendus de l'Académie des Sciences, Paris*, ser. D, 304:209–214.

1852. Russell, M. D. 1985. The supraorbital torus: "A most remarkable peculiarity." *Current Anthropology* 26:337–360.

1853. Russell, M. D. 1987. Bone breakage in the Krapina hominid collection. *American Journal of Physical Anthropology* 72:373–380.

1854. Russell, M. D. 1987. Mortuary practices at the Krapina Neandertal site. *American Journal of Physical Anthropology* 72:381–398.

1855. Rutter, N. W. 1980. Late Pleistocene history of the western Canadian ice-free corridor. *Canadian Journal of Anthropology* 1:1–8.

1856. Ruvolo, M., Disotell, T. R., Allard, M. W., Brown, W. M., and Honeycutt, R. L. 1991. Resolution of the African hominoid trichotomy by use of a mitochondrial gene sequence. *Proceedings of the National Academy of Sciences* 88:1570–1574.

1857. Ruvolo, M., Pan, D., Zehr, S., Goldberg, T., Disotell, T. R., and von Dornum, M. 1994. Gene trees and hominoid phylogeny. *Proceedings of the National Academy of Sciences* 91:8900–8904.

1858. Ruvolo, M., Zehr, S., von Dornum, M., Pan, D., Chang, B., and Lin, J. 1993. Mitochondrial COII sequences and modern human origins. *Molecular Biology and Evolution* 10:1115–1135.

1859. Saban, R. 1977. The place of Rabat man (Kébibat, Morocco) in human evolution. *Current Anthropology* 18:518–524.

1860. Sampson, C. G. 1974. *The Stone Age archaeology of southern Africa.* New York: Academic Press.

1861. Sanders, W. J., and Bodenbender, B. E. 1994. Morphometric analysis of lumbar vertebra UMP 67-28: Implications for spinal function and phylogeny of the Miocene Moroto hominoid. *Journal of Human Evolution* 26:203–237.

1862. Sanlaville, P., Besançon, J., Copeland, L., and Muhesen, S. 1993. *Le Paléolithique de la Vallée moyenne de l'Oronte (Syrie).* Oxford: British Archaeological Reports.

1863. Santa Luca, A. P. 1978. A re-examination of presumed Neanderthal fossils. *Journal of Human Evolution* 7:619–636.

1864. Santa Luca, A. P. 1980. The Ngandong fossil hominids. *Yale University Publications in Anthropology* 78:1–175.

1865. Santonja, M., and Villa, P. 1990. The Lower Paleolithic of Spain and Portugal. *Journal of World Prehistory* 4:45–94.

1866. Sarich, V. M. 1971. A molecular approach to the question of human origins. In *Background for man,* ed. P. Dolhinow and V. M. Sarich, pp. 60–81. Boston: Little, Brown.

1867. Sarich, V. M. 1983. Retrospective on hominoid macromolecular systematics. In *New interpretations of ape and human ancestry,* ed. R. L. Ciochon and R. S. Corruccini, pp. 137–150. New York: Plenum Press.

1868. Sarich, V. M., and Cronin, J. E. 1980. South American mammal molecular systematics, evolutionary clocks, and continental drift. In *Evolutionary biology of the New World monkeys and continental drift,* ed. R. L. Ciochon and A. B. Chiarelli, pp. 399–421. New York: Plenum Press.

1869. Sarich, V. M., and Wilson, A. C. 1967. Immunological time scale for hominid evolution. *Science* 158:1200–1203.

1870. Sartono, S. 1975. Implications arising from *Pithecanthropus* VIII. In *Paleoanthropology, morphology and paleoecology,* ed. R. H. Tuttle, pp. 327–360. The Hague: Mouton.

1871. Savage, D. E., and Russell, D. E. 1983. *Mammalian paleofaunas of the world.* Reading, Mass.: Addison Wesley.

1872. Schaller, G. B. 1963. *The mountain gorilla: Ecology and behavior.* Chicago: University of Chicago Press.

1873. Schick, K. 1994. The Movius Line reconsidered: Perspectives on the earlier Paleolithic of eastern Asia. In *Integrative paths to the past: Paleoanthropological advances in honor of F. Clark Howell,* ed. R. S. Corruccini and R. L. Ciochon, pp. 569–596. Englewood Cliffs, N.J.: Prentice Hall.

1874. Schick, K., and Toth, N. 1994. Early Stone Age technology in Africa: A review and case study into the nature and function of spheroids and subspheroids. In *Integrative paths to the past: Paleoanthropological advances in honor of F. Clark Howell,* ed. R. S. Corruccini and R. L. Ciochon, pp. 429–449. Englewood Cliffs, N.J.: Prentice Hall.

1875. Schick, K. D., and Dong, Z. 1993. Early Paleolithic of China and eastern Asia. *Evolutionary Anthropology* 2:22–35.

1876. Schick, K. D., and Toth, N. 1993. *Making silent stones speak: Human evolution and the dawn of technology.* New York: Simon and Schuster.

1877. Schick, K. D., Toth, N., Qi, W., Clark, J. D., and Etler, D. 1991. Archaeological perspectives in the Nihewan Basin, China. *Journal of Human Evolution* 21:13–26.

1878. Schild, R. 1984. Terminal Paleolithic of the north European plain: A review of lost chances, potential and hopes. *Advances in World Archaeology* 3:193–27.

1879. Schild, R., Sulgostowska, Z., Gautier, A., Bluszcz, A., Juel Jensen, H., Królik, H., and Tomaszewski, J. 1988. The Middle Paleolithic of the north European plain at Zwolen: Preliminary results. *Études et Recherches Archéologiques de l'Université de Liège* 35:149–167.

1880. Schlanger, N. 1994. Mindful technology: Unleashing the *chaîne opératoire* for an archaeology of mind. In *The ancient mind: Elements of cognitive archaeology,* ed. C. Renfrew and E. B. W. Zubrow, pp. 143–151. Cambridge: Cambridge University Press.

1881. Schmider, B. 1990. The Last Pleniglacial in the Paris Basin (22 500–17 000 BP). In *The world at 18 000 BP,* vol. 1, *High latitudes,* ed. O. Soffer and C. Gamble, pp. 41–53. London: Unwin Hyman.

1882. Schoetensack, O. 1994 [1908]. The mandible of *Homo heidelbergensis* from the Mauer Sands at Heidelberg (translation from "Der Unterkiefer des *Homo heidelbergensis* aus den Sanden von Mauer bei Heidelberg"). In *Naming our ancestors: An anthology of hominid taxonomy,* ed. W. E. Meikle and S. T. Parker, pp. 42–46. Prospect Heights, Ill.: Waveland.

1883. Schrenk, F., Bromage, T. G., Betzler, C. G., Ring, U., and Juwa-yeyi, Y. M. 1993. Oldest *Homo* and Pliocene biogeography of the Malawi Rift. *Nature* 365:833–836.

1884. Schultz, A. H. 1969. *The life of primates*. New York: Universe Books.

1885. Schwarcz, H. P. 1980. Absolute age determination of archaeological sites by uranium series dating of travertines. *Archaeometry* 22:3–24.

1886. Schwarcz, H. P. 1992. Uranium series dating in paleoanthropology. *Evolutionary Anthropology* 1:56–62.

1887. Schwarcz, H. P. 1993. Uranium-series dating and the origin of modern man. In *The origin of modern humans and the impact of chronometric dating*, ed. M. J. Aitken, C. B. Stringer, and P. A. Mellars, pp. 12–26. Princeton: Princeton University Press.

1888. Schwarcz, H. P. 1994. Chronology of modern humans in the Levant. In *Late Quaternary chronology and paleoclimates of the eastern Mediterranean*, ed. O. Bar-Yosef and R. S. Kra, pp. 21–31. Cambridge, Mass.: American School of Prehistoric Research.

1889. Schwarcz, H. P., Buhay, W. M., Grün, R., Valladas, H., Tchernov, E., Bar-Yosef, O., and Vandermeersch, B. 1989. ESR dating of the Neanderthal site, Kebara Cave. *Journal of Archaeological Science* 16:653–659.

1890. Schwarcz, H. P., and Gascoyne, M. 1984. Uranium-series dating of Quaternary sediments. In *Quaternary dating methods*, ed. W. C. Mahaney, pp. 33–51. Amsterdam: Elsevier.

1891. Schwarcz, H. P., and Grün, R. 1993. Electron spin resonance (ESR) dating of the origins of modern man. In *The origin of modern humans and the impact of chronometric dating*, ed. M. J. Aitken, C. B. Stringer, and P. A. Mellars, pp. 40–48. Princeton: Princeton University Press.

1892. Schwarcz, H. P., Grün, R., Latham, A. G., Mania, D., and Brunnacker, K. 1988. The Bilzingsleben archaeological site: New dating evidence. *Archaeometry* 30:5–17.

1893. Schwarcz, H. P., Grün, R., and Tobias, P. V. 1994. ESR dating studies of the australopithecine site of Sterkfontein, South Africa. *Journal of Human Evolution* 26:175–181.

1894. Schwarcz, H. P., Grün, R., Vandermeersch, B., Bar-Yosef, O., Valladas, H., and Tchernov, E. 1988. ESR dates for the hominid burial site of Qafzeh in Israel. *Journal of Human Evolution* 17:733–737.

1895. Schwarcz, H. P., and Latham, A. G. 1984. Uranium-series age determinations of travertines from the site of Vértesszöllös, Hungary. *Journal of Archaeological Science* 11:5–17.

1896. Schwartz, J. H. 1990. *Lufengpithecus* and its potential relationship to an orang-utan clade. *Journal of Human Evolution* 19:591–605.

1897. Schwartz, J. H. 1992. The Chinese fossil connection: Interpreting the origins of humans and apes. *Terra* 31:16–25.

1898. Schwartz, J. H., Long, V. T., Cuong, N. L., Kha, L. T., and Tattersall, I. 1994. A diverse hominoid fauna from the late Middle Pleistocene breccia cave of Tham Khuyen, Socialist Republic of Vietnam. *Anthropological Papers of the American Museum of Natural History* 73:1–11.

1899. Schwartz, J. H., and Tattersall, I. 1987. Tarsiers, adapids, and the integrity of Strepsirhini. *Journal of Human Evolution* 16:23–40.

1900. Schwartz, J. H., Tattersall, I., Huang, W., Gu, Y., Ciochon, R., Larick, R., Qiren, F., de Vos, J., Schwarcz, H., Rink, W. J., and Yonge, C. 1996. Whose teeth? *Nature* 381:201–202.

1901. Schweitzer, F. R. 1979. Excavations at Die Kelders, Cape Province, South Africa: The Holocene deposits. *Annals of the South African Museum* 78:101–233.

1902. Schweitzer, F. R., and Wilson, M. L. 1982. Byneskranskop 1, a late Quaternary living site in the southern Cape Province, South Africa. *Annals of the South African Museum* 88:1–203.

1903. Scott, K. 1980. Two hunting episodes of Middle Palaeolithic age at La Cotte de Saint-Brelade, Jersey (Channel Islands). *World Archaeology* 12:137–152.

1904. Scott, K. 1986. The bone assemblages of layers 3 and 6. In *La Cotte de St. Brelade, Jersey: Excavations by C. B. M. McBurney, 1961–1978*, pp. 159–183. Norwich, England: Geo Books.

1905. Scott, K. 1986. The large mammal fauna. In *La Cotte de St. Brelade, Jersey: Excavations by C. B. M. McBurney 1961–1978*, pp. 109–138. Norwich, England: Geo Books.

1906. Scott, K. 1989. Mammoth bones modified by humans: Evidence from La Cotte de St. Brelade, Jersey, Channel Islands. In *Bone modification*, ed. R. Bonnichsen and M. H. Sorg, pp. 335–346. Orono: Center for the Study of the First Americans, University of Maine.

1907. Segre, A., and Ascenzi, A. 1982. Fontana Ranuccio: Italy's earliest Middle Pleistocene hominid site. *Current Anthropology* 25: 230–233.

1908. Sellet, F. 1993. *Chaîne opératoire:* The concept and its application. *Lithic Technology* 18:106–112.

1909. Selvaggio, M. M. 1998. Evidence for a three-stage sequence of hominid and carnivore involvement with long bones at FLK *Zinjanthropus*, Olduvai Gorge, Tanzania. *Journal of Archaeological Science* 25:191–202.

1910. Sémah, F. 1984. The Sangiran Dome in the Javanese Plio-Pleistocene chronology. *Courier Forschungsinstitut Senckenberg* 69:245–252.

1911. Sémah, F., Sémah, A.-M., Djubitantono, T., and Simanjuntak, H. T. 1992. Did they also make stone tools? *Journal of Human Evolution* 23:439–446.

1912. Semaw, S., Renne, P., Harris, J. W. K., Feibel, C. S., Bernor, R. L., Fesseha, N., and Mowbray, K. 1997. 2.5-million-year-old stone tools from Gona, Ethiopia. *Nature* 385:333–336.

1913. Semenov, S. A. 1964. *Prehistoric technology.* New York: Barnes and Noble.

1914. Senut, B., Pickford, M., Ssemanda, I., Elepu, D., and Obwona, P. 1987. Découverte du premier Homininae (*Homo* sp.) dans le Pléistocène de Nyabusosi (Ouganda occidental). *Comptes Rendus de l'Académie des Sciences, Paris*, ser. 2, 305:819–822.

1915. Senut, B., and Tardieu, C. 1985. Functional aspects of Plio-Pleistocene hominid limb bones: Implications for taxonomy and phylogeny. In *Ancestors: The hard evidence*, ed. E. Delson, pp. 193–201. New York: Alan R. Liss.

1916. Seret, G., Guiot, J., Wansard, G., de Beaulieu, J. L., and Reille, M. 1992. Tentative paleoclimatic reconstruction linking pollen and sedimentology in La Grande Pile (Vosges, France). *Quaternary Science Reviews* 11:425–430.

1917. Seuss, H. E. 1986. Secular variations of cosmogenic 13C on earth: Their discovery and interpretation. *Radiocarbon* 28:259–265.

1918. Shackleton, N. J. 1967. Oxygen isotope analyses and Pleistocene temperatures re-assessed. *Nature* 215:259–265.

1919. Shackleton, N. J. 1975. The stratigraphic record of deep-sea cores and its implications for the assessment of glacials, interglacials, stadials, and interstadials in the Mid-Pleistocene. In *After the australopithecines*, ed. K. W. Butzer and G. L. Isaac, pp. 1–24. The Hague: Mouton.

1920. Shackleton, N. J. 1987. Oxygen isotopes, ice volumes, and sea level. *Quaternary Science Reviews* 6:183–190.

1921. Shackleton, N. J. 1995. New data on the evolution of Pliocene climatic variability. In *Paleoclimate and evolution with emphasis on human origins*, ed. E. S. Vrba, G. H. Denton, T. C. Partridge, and L. H. Burckle, pp. 242–248. New Haven: Yale University Press.

1922. Shackleton, N. J., Berger, A., and Peltier, W. R. 1990. An alternative astronomical calibration of the Lower Pleistocene time scale based on ODP site 677. *Journal of the Royal Society of Edinburgh: Earth Sciences* 81:251–261.

1923. Shackleton, N. J., Crowhurst, S., Hagelberg, T., Pisias, N. G., and Schneider, D. A. 1995. A new Late Neogene time scale: Application to leg 138 sites. *Proceedings of the Ocean Drilling Program, Scientific Results* 138:73–101.

1924. Shackleton, N. J., and Opdyke, N. 1973. Oxygen isotope and palaeomagnetic stratigraphy of equatorial Pacific core V28–238. *Quaternary Research* 3:39–55.

1925. Shapiro, H. L. 1971. The strange, unfinished saga of Peking man. *Natural History* 80 (9): 8–18, 74, 76–77.

1926. Shapiro, H. L. 1974. *Peking man.* New York: Simon and Schuster.

1927. Shea, J. J. 1988. Spear points from the Middle Paleolithic of the Levant. *Journal of Field Archaeology* 15:441–450.

1928. Shea, J. J. 1989. A functional study of the lithic industries associated with hominid fossils in the Kebara and Qafzeh Caves, Israel. In *The human revolution: Behavioural and biological perspectives in the origins of modern humans*, ed. P. A. Mellars and C. B. Stringer, pp. 611–625. Edinburgh: Edinburgh University Press.

1929. Shea, J. J. 1992. Lithic microwear analysis in archeology. *Evolutionary Anthropology* 1:143–150.

1930. Sheppard, P. J. and Kleindienst, M. R. 1996. Technological change in the Earlier and Middle Stone Age of Kalambo Falls (Zambia). *African Archaeological Review* 13:171–196.

1931. Sherry, S. T., Rogers, A. R., Harpending, H., Soodyall, H., Jenkins, T., and Stoneking, M. 1994. Mismatch distributions of mtDNA reveal recent human population expansions. *Human Biology* 66:761–775.

1932. Shiner, J. L. 1968. Miscellaneous sites. In *The prehistory of Nubia*, ed. F. Wendorf, 2:630–650. Dallas: Southern Methodist University Press.

1933. Shipman, P. 1984. The earliest tools: Re-assessing the evidence from Olduvai Gorge. *Anthroquest* 29:9–10.

1934. Shipman, P. 1984. Scavenger hunt. *Natural History* 93 (4): 20–27.

1935. Shipman, P. 1986. Scavenging or hunting in early hominids: Theoretical framework and tests. *American Anthropologist* 88: 27–43.

1936. Shipman, P. 1986. Studies of hominid-faunal interactions at Olduvai Gorge. *Journal of Human Evolution* 15:691–706.

1937. Shipman, P. 1988. Diet and subsistence strategies at Olduvai Gorge. In *Diet and subsistence: Current archaeological perspectives*, ed. B. V. Kennedy and G. M. le Moine, pp. 3–11. Calgary: Archaeological Association of the University of Calgary.

1938. Shipman, P. 1989. Altered bones from Olduvai Gorge, Tanzania: Techniques, problems, and implications of their recognition. In *Bone modification*, ed. R. Bonnichsen and M. H. Sorg, pp. 317–334. Orono: Center for the Study of the First Americans, University of Maine.

1939. Shipman, P., Bosler, W., and Davis, K. L. 1981. Butchering of giant geladas at an Acheulian site. *Current Anthropology* 22: 257–268.

1940. Shipman, P., Potts, R., and Pickford, M. 1983. Lainyamok: A new Middle Pleistocene hominid site. *Nature* 306:365–368.

1941. Shipman, P., and Rose, J. 1983. Early hominid hunting, butchering and carcass-processing behaviors: Approaches to the fossil record. *Journal of Anthropological Archaeology* 2:57–98.

1942. Sibley, C. G., and Ahlquist, J. E. 1984. The phylogeny of the hominoid primates, as indicated by DNA-DNA hybridization. *Journal of Molecular Evolution* 20:2–15.

1943. Sibley, C. G., and Ahlquist, J. E. 1987. DNA hybridization evidence of hominoid phylogeny: Results from an expanded data set. *Journal of Molecular Evolution* 26:99–121.

1944. Sibley, C. G., Comstock, J. A., and Ahlquist, J. E. 1990. DNA hybridization evidence of hominoid phylogeny: A reanalysis of the data. *Journal of Molecular Evolution* 30:202–236.

1945. Sieveking, A. 1979. *The cave artists*. London: Thames and Hudson.

1946. Sigé, B., Jaeger, J.-J., Sudre, J., and Vianey-Liaud, M. 1990. *Altiatlasius koulchii* n. gen. et sp., primate Omomyidé du Paléocène supérieur du Maroc, et les origines des euprimates. *Palaeontographica*, ser. A, 214:31–56.

1947. Sigmon, B. A. 1982. Comparative morphology of the locomotor skeleton of *Homo erectus* and the other fossil hominids, with special reference to the Tautavel innominate and femora. In *L'Homo erectus et la place de l'homme de Tautavel parmi les hominidés fossiles*, ed. M. A. de Lumley, pp. 422–446. Nice: Premier Congrès International de Paléontologie Humaine.

1948. Sillen, A. 1986. Biogenic and diagenetic Sr/Ca in Plio-Pleistocene fossils in the Omo Shungura Formation. *Paleobiology* 12:311–323.

1949. Sillen, A. 1992. Strontium-calcium ratios (Sr/Ca) of *Australopithecus robustus* and associated fauna from Swartkrans. *Journal of Human Evolution* 23:495–516.

1950. Sillen, A. 1993. Was *Australopithecus robustus* an omnivore? *South African Journal of Science* 89:71–72.

1951. Sillen, A., and Hoering, T. 1993. Chemical characterization of burnt bones from Swartkrans. *Transvaal Museum Monograph* 8:243–249.

1952. Sillen, A., and Lee-Thorp, J. A. 1993. Diet of *Australopithecus robustus. South African Journal of Science* 89:174.

1953. Sillen, A., and Morris, A. 1996. Diagenesis of bone from Border Cave: Implications for the age of the Border Cave hominids. *Journal of Human Evolution* 31:499–506.

1954. Sillen, A., and Parkington, J. E. 1996. Diagenesis of bones from Eland's Bay Cave. *Journal of Archaeological Science* 23:535–542.

1955. Simek, J. F., and Smith, F. H. 1997. Chronological changes in stone tool assemblages from Krapina (Croatia). *Journal of Human Evolution* 32:561–575.

1956. Simmons, T., Falsctti, A. B., and Smith, F. H. 1991. Frontal bone morphometrics of southwest Asian Pleistocene hominids. *Journal of Human Evolution* 20:249–269.

1957. Simons, E., and Pilbeam, D. R. 1965. Preliminary revision of the Dryopithecinae (Pongidae, Anthropoidea). *Folia Primatologia* 3:81–152.

1958. Simons, E. L. 1967. The earliest apes. *Scientific American* 217 (65): 28–35.

1959. Simons, E. L. 1972. *Primate evolution: An introduction to man's place in nature.* New York: Macmillan.

1960. Simons, E. L. 1981. Man's immediate forerunners. *Philosophical Transactions of the Royal Society of London,* ser. B, 292:21–41.

1961. Simons, E. L. 1984. Dawn ape of the Fayum. *Natural History* 93 (5): 18–20.

1962. Simons, E. L. 1985. Origins and characteristics of the first hominoids. In *Ancestors: The hard evidence,* ed. E. Delson, pp. 37–41. New York: Alan R. Liss.

1963. Simons, E. L. 1986. *Parapithecus grangeri* of the African Oligocene: An archaic catarrhine without lower incisors. *Journal of Human Evolution* 15:205–213.

1964. Simons, E. L. 1987. New faces of *Aegyptopithecus* from the Oligocene of Egypt. *Journal of Human Evolution* 16:273–289.

1965. Simons, E. L. 1989. Description of two genera and species of late Eocene Anthropoidea from Egypt. *Proceedings of the National Academy of Sciences* 86:9956–9960.

1966. Simons, E. L. 1990. Discovery of the oldest known anthropoidean skull from the Paleogene of Egypt. *Science* 247:1567–1569.

1967. Simons, E. L. 1993. Egypt's simian spring. *Natural History* 102: 58–59.

1968. Simons, E. L., Bown, T. M., and Rasmussen, D. T. 1986. Discovery of two additional prosimian primate families (Omomyidae, Lorisidae) in the African Oligocene. *Journal of Human Evolution* 15:431–437.

1969. Simons, E. L., Rasmussen, D. T., and Gebo, D. L. 1987. A new species of *Propliopithecus* from the Fayum, Egypt. *American Journal of Physical Anthropology* 73:139–148.

1970. Simons, E. L., and Rasmussen, D. T. 1989. Cranial morphology of *Aegyptopithecus* and *Tarsius* and the question of the tarsier-anthropoidean clade. *American Journal of Physical Anthropology* 79:1–23.

1971. Simons, E. L., and Rasmussen, D. T. 1991. The generic classification of Fayum Anthropoidea. *International Journal of Primatology* 12:163–178.

1972. Simons, E. L., and Rasmussen, D. T. 1994. A remarkable cranium of *Plesiopithecus terus* (Primates, Prosimii) from the Eocene of Egypt. *Proceedings of the National Academy of Sciences* 92: 9946–9950.

1973. Simons, E. L., and Rasmussen, D. T. 1994. A whole new world of ancestors: Eocene anthropoideans from Africa. *Evolutionary Anthropology* 3:128–138.

1974. Simpson, G. G. 1953. *The major features of evolution.* New York: Columbia University Press.

1975. Simpson, G. G. 1955. The Phenacolemuridae: A new family of early primates. *Bulletin of the American Museum of Natural History* 85:1–350.

1976. Simpson, G. G. 1961. *Principles of animal taxonomy.* New York: Columbia University Press.

1977. Sinclair, A. R. E., Leakey, M. D., and Norton-Griffiths, M. 1986. Migration and hominid bipedalism. *Nature* 324:307–308.

1978. Singer, R. 1958. The Rhodesian, Florisbad, and Saldanha skulls. In *Neanderthal Centenary, 1856–1956,* ed. G. H. R. von Koenigswald, pp. 53–62. Utrecht: Kemink.

1979. Singer, R., Gladfelter, B. G., and Wymer, J. J. 1993. *The Lower Paleolithic site at Hoxne, England.* Chicago: University of Chicago Press.

1980. Singer, R., and Wymer, J. J. 1976. The sequence of Acheulian industries at Hoxne, Suffolk. In *L'Évolution de l'Acheuléen en Europe (colloque X),* ed. J. Combier, pp. 14–30. Nice: Union International des Sciences Préhistoriques et Protohistoriques.

1981. Singer, R., and Wymer, J. J. 1982. *The Middle Stone Age at Klasies River Mouth in South Africa.* Chicago: University of Chicago Press.

1982. Singer, R., Wymer, J. J., Gladfelter, B. G., and Wolff, R. G. 1973. Excavation of the Clactonian Industry at the Golf Course, Clacton-on-Sea, Essex. *Proceedings of the Prehistoric Society* 39:6–74.

1983. Singer, R. S., and Wymer, J. J. 1968. Archaeological investigations at the Saldanha skull site in South Africa. *South African Archaeological Bulletin* 25:63–74.

1984. Skelton, R. R., and McHenry, H. M. 1992. Evolutionary relationships among early hominids. *Journal of Human Evolution* 23:309–349.

1985. Skelton, R. R., and McHenry, H. M. 1998. Trait list bias and a reappraisal of early hominid phylogeny. *Journal of Human Evolution* 34:109–113.

1986. Skinner, J. 1965. The flake industries of southwest Asia: A typological study. Ph.D. diss., Columbia University.

1987. Skinner, J. D., Haupt, M. A., Hoffman, M., and Dott, H. M. 1998. Bone collecting by brown hyaenas *Hyaena brunnea* in the Namib Desert: Rate of accumulation. *Journal of Archaeological Science* 25:69–71.

1988. Skinner, J. D., and Smithers, R. H. N. 1990. *The mammals of the southern African subregion.* Pretoria: University of Pretoria.

1989. Skinner, J. D., and van Aarde, R. J. 1991. Bone collecting by brown hyenas *Hyaena brunnea* in the central Namib Desert, Namibia. *Journal of Archaeological Science* 18:513–523.

1990. Skinner, J. D., van Aarde, R. J., and Goss, R. A. 1995. Space and resource use by brown hyenas *Hyaena brunnea* in the Namib Desert. *Journal of Zoology, London* 237:123–131.

1991. Skinner, M. 1991. Bee brood consumption: An alternative explanation for hypervitaminosis A in KNM-ER 1808 (*Homo erectus*) from Koobi Fora, Kenya. *Journal of Human Evolution* 20:493–503.

1992. Small, M. F. 1993. Closing the gap: New discoveries from chimpanzee populations spread across Africa reveal the roots of human culture are deeper than we thought. *Wildlife Conservation,* July–August, 17–23.

1993. Smith, A. B. 1987. Seasonal exploitation of resources on the Vredenburg Peninsula after 2000 B.P. In *Papers in the prehistory of the western Cape, South Africa,* ed. J. E. Parkington and M. Hall, pp. 393–402. International Series 332. Oxford: British Archaeological Reports.

1994. Smith, A. B. 1992. Origins and spread of pastoralism in Africa. *Annual Review of Anthropology* 21:125–141.

1995. Smith, B. H. 1986. Dental development in *Australopithecus* and early *Homo. Nature* 323:327–330.

1996. Smith, B. H. 1992. Life history and the evolution of human maturation. *Evolutionary Anthropology* 1:134–142.

1997. Smith, B. H. 1994. Patterns of dental development in *Homo, Australopithecus, Pan,* and *Gorilla. American Journal of Physical Anthropology* 94:307–326.

1998. Smith, F. H. 1982. Upper Pleistocene hominid evolution in south-central Europe. *Current Anthropology* 23:667–703.

1999. Smith, F. H. 1983. Behavioral interpretation of changes in craniofacial morphology across the archaic/modern *Homo sapiens* transition. *British Archaeological Reports International Series* 164:143–163.

2000. Smith, F. H. 1984. Fossil hominids from the Upper Pleistocene of central Europe and the origin of modern Europeans. In *The origins of modern humans: A world survey of the fossil evidence,* ed. F. H. Smith and F. Spencer, pp. 137–209. New York: Alan R. Liss.

2001. Smith, F. H. 1994. Samples, species, and populations in the study of modern human origins. In *Origins of anatomically modern humans,* ed. M. H. Nitecki and D. V. Nitecki, pp. 227–249. New York: Plenum Press.

2002. Smith, F. H., and Ahern, J. C. 1994. Additional cranial remains from Vindija Cave, Croatia. *American Journal of Physical Anthropology* 93:275–280.

2003. Smith, F. H., Boyd, D. C., and Malez, M. 1985. Additional Upper Pleistocene human remains from Vindija Cave, Croatia, Yugoslavia. *American Journal of Physical Anthropology* 68:375–383.

2004. Smith, F. H., Simek, J. F., and Harrill, M. S. 1989. Geographic variation in supraorbital torus reduction during the Later Pleistocene (c. 80 000–15 000 BP). In *The human revolution: Behavioural and biological perspectives in the origin of modern humans,* ed. P. Mellars and C. Stringer, pp. 172–193. Edinburgh: Edinburgh University Press.

2005. Smith, P., and Arensburg, B. 1977. A Mousterian skeleton from Kebara Cave. *Eretz Israel* 13:164–176.

2006. Soffer, O. 1985. *The Upper Paleolithic of the central Russian plain*. Orlando, Fla.: Academic Press.

2007. Soffer, O. 1989. The Middle to Upper Palaeolithic transition on the Russian plain. In *The human revolution: Behavioural and biological perspectives in the origins of modern humans*, ed. P. Mellars and C. Stringer, pp. 714–742. Edinburgh: Edinburgh University Press.

2008. Soffer, O. 1989. Storage, sedentism and the Eurasian Palaeolithic record. *Antiquity* 63:719–732.

2009. Soffer, O. 1990. Before Beringia: Late Pleistocene bio-social transformations and the colonization of northern Eurasia. In *Chronostratigraphy of the Paleolithic in north, central, east Asia and America*. Novosibirsk: Academy of Sciences of the USSR.

2010. Soffer, O. 1990. The Russian plain at the Last Glacial Maximum. In *The world at 18 000 BP*, vol. 1, *High latitudes*, ed. O. Soffer and C. Gamble, pp. 228–252. London: Unwin Hyman.

2011. Soffer, O. 1994. Ancestral lifeways in Eurasia—the Middle and Upper Paleolithic records. In *Origins of anatomically modern humans*, ed. M. H. Nitecki and D. V. Nitecki, pp. 101–119. New York: Plenum Press.

2012. Soffer, O., Vandiver, P., Klima, B., and Svoboda, J. 1993. The pyrotechnology of performance art: Moravian venuses and wolverines. In *Before Lascaux: The complex record of the early Upper Paleolithic*, ed. H. Knecht, A. Pike-Tay, and R. White, pp. 259–275. Boca Raton, Fla.: CRC Press.

2013. Solan, M., and Day, M. H. 1992. The Baringo (Kapthurin) ulna. *Journal of Human Evolution* 22:307–313.

2014. Solecki, R. S. 1963. Prehistory in the Shanidar Valley, northern Iraq. *Science* 139:179–193.

2015. Solecki, R. S. 1975. Shanidar IV, a Neanderthal flower burial in northern Iraq. *Science* 190:880–881.

2016. Solecki, R. S. 1989. On the evidence for Neandertal burial. *Current Anthropology* 30:324.

2017. Sommé, J., Munaut, A. V., Puisségur, J. J., and Cunat, N. 1986. Stratigraphie et signification climatique du gisement paléolithique de Biache-Saint-Vaast (Pas-de-Calais, France). In *Chronostratigraphie et faciés culturels du Paléolithique inférieur et moyen dans l'Europe du Nord-Ouest*, ed. A. Tuffreau and J. Sommé, pp. 187–195. Paris: Supplément au Bulletin de l'Association Française pour l'Étude du Quaternaire.

2018. Sonakia, A. 1985. Early *Homo* from Narmada Valley, India. In *Ancestors: The hard evidence*, ed. E. Delson, pp. 334–338. New York: Alan R. Liss.

2019. Sonakia, A. 1992. Human evolution in south Asia. In *The evolution and dispersal of modern humans in Asia*, ed. T. Akazawa, K. Aoki, and T. Kimura, pp. 337–347. Tokyo: Hokusen-Sha.

2020. Sondaar, P. Y. 1984. Faunal evolution and the mammalian biostratigraphy of Java. *Courier Forschungsinstitut Senckenberg* 69:219–235.

2021. Sondaar, P. Y., de Vos, J., and Leinders, J. J. M. 1983. Facts and fiction around the fossil mammals of Java (reply to Bartstra). *Geologie en Mijnbouw* 62:339–343.

2022. Sondaar, P. Y., van den Bergh, G. D., Mubroto, B., Aziz, F., de Vos, J., and Batu, U. L. 1994. Middle Pleistocene faunal turnover and colonization of Flores (Indonesia) by *Homo erectus*. *Comptes Rendus de l'Académie des Sciences, Paris*, ser. 2, 319:1255–1262.

2023. Sonneville-Bordes, D. de. 1963. Upper Paleolithic cultures in western Europe. *Science* 142:347–355.

2024. Sonneville-Bordes, D. de. 1973. The Upper Paleolithic: c. 33,000–10,000 B.C. In *France before the Romans*, ed. S. Piggott, G. Daniel, and C. McBurney, pp. 30–60. London: Thames and Hudson.

2025. Sonneville-Bordes, D. de, and Perrot, J. 1954. Lexique typologique du Paléolithique supérieur. *Bulletin de la Société Préhistorique Française* 51:327–335.

2026. Sonneville-Bordes, D. de, Perrot, J. 1955. Lexique typologique du Paléolithique supérieur. *Bulletin de la Société Préhistorique Française* 52:76–79.

2027. Sonneville-Bordes, D. de, and Perrot, J. 1956. Lexique typologique du Paléolithique supérieur. *Bulletin de la Société Préhistorique Française* 53:408–412.

2028. Sonneville-Bordes, D. de, and Perrot, J. 1956. Lexique typologique du Paléolithique supérieur. *Bulletin de la Société Préhistorique Française* 53:547–549.

2029. Spencer, F. 1984. The Neandertals and their evolutionary significance: A brief historical survey. In *The origins of modern humans: A world survey of the fossil evidence*, ed. F. H. Smith and F. Spencer, pp. 1–49. New York: Alan R. Liss.

2030. Spencer, F. 1990. *Piltdown: A scientific forgery*. Oxford: Oxford University Press.

2031. Spoor, F., Wood, B., and Zonneveld, F. 1994. Implications of early hominid labyrinthine morphology for evolution of human bipedal locomotion. *Nature* 369:645–648.

2032. Spoor, F., Wood, B., and Zonneveld, F. 1996. Evidence for a link between human semicircular canal size and bipedal behaviour. *Journal of Human Evolution* 30:183–187.

2033. Stafford, T. W., Hare, P. E., Currie, L., Jull, A. J. T., and Donahue, D. 1990. Accuracy of North American human skeletal ages. *Quaternary Research* 34:111–120.

2034. Stanford, C. B. 1995. Chimpanzee hunting behavior and human evolution. *American Scientist* 83:256–261.

2035. Stanford, C. B. 1996. The hunting ecology of wild chimpanzees: Implications for the evolutionary ecology of Pliocene hominids. *American Anthropologist* 98:96–113.

2036. Stanley, S. M. 1979. *Macroevolution: Pattern and process*. San Francisco: W. H. Freeman.

2037. Stanley, S. M. 1981. *The new evolutionary timetable: Fossils, genes, and the origin of species*. New York: Basic Books.

2038. Stanley, S. M. 1992. An ecological theory for the origin of *Homo*. *Paleobiology* 18:237–257.

2039. Stearns, C. E. 1984. Uranium-series dating and the history of sea level. In *Quaternary dating methods*, ed. W. C. Mahaney, pp. 53–65. Amsterdam: Elsevier.

2040. Stearns, C. E., and Thurber, D. L. 1965. Th^{230}/U^{234} dates of late

Pleistocene marine fossils from the Mediterranean and Moroccan littorals. *Quaternaria* 7:29–42.

2041. Steele, D. G., and Powell, J. F. 1993. Paleobiology of the first Americans. *Evolutionary Anthropology* 2:138–146.

2042. Stefan, V. H., and Trinkaus, E. 1998. Discrete trait and dental morphometric affinities of the Tabun 2 mandible. *Journal of Human Evolution* 34:443–468.

2043. Stepanchuk, V. N. 1993. Prolom II, a Middle Palaeolithic cave site in the eastern Crimea with non-utilitarian bone artefacts. *Proceedings of the Prehistoric Society* 19:17–37.

2044. Stern, J. T., and Susman, R. L. 1983. The locomotor anatomy of *Australopithecus afarensis. American Journal of Physical Anthropology* 60:279–317.

2045. Steudel, K. 1980. New estimates of early hominid body size. *American Journal of Physical Anthropology* 52:63–70.

2046. Steudel, K. L. 1994. Locomotor energetics and hominid evolution. *Evolutionary Anthropology* 3:42–48.

2047. Stewart, T. D. 1960. Form of the pubic bone in Neanderthal man. *Science* 131:1437–1438.

2048. Stewart, T. D. 1962. Neanderthal scapulae with special attention to the Shanidar Neanderthals from Iraq. *Anthropos* 57:781–800.

2049. Stewart, T. D. 1977. The Neanderthal skeletal remains from Shanidar Cave, Iraq: A summary of findings to date. *Proceedings of the American Philosophical Society* 121:121–165.

2050. Stiles, D. N., and Partridge, T. C. 1979. Results of recent archaeological and palaeoenvironmental studies of the Sterkfontein Extension Site. *South African Journal of Science* 75:346–352.

2051. Stiner, M. C. 1994. *Honor among thieves: A zooarchaeological study of Neandertal ecology.* Princeton: Princeton University Press.

2052. Stoneking, M. 1993. DNA and recent human evolution. *Evolutionary Anthropology* 2:60–73.

2053. Stordeur-Yedid, D. 1979. *Les aiguilles à chas au Paléolithique.* Treizième supplément à Gallia Préhistoire. Paris: Centre National de la Recherche Scientifique.

2054. Strait, D. S., and Grine, F. E. 1998. Trait list bias? A reply to Skelton and McHenry. *Journal of Human Evolution* 34:115–118.

2055. Strait, D. S., Grine, F. E., and Moniz, M. A. 1997. A reappraisal of early hominid phylogeny. *Journal of Human Evolution* 32:17–82.

2056. Strasser, E., and Delson, E. 1987. Cladistic analysis of cercopithecid relationships. *Journal of Human Evolution* 16:81–99.

2057. Straus, L. G. 1977. Of deerslayers and mountain men. In *For theory building in archaeology,* ed. L. R. Binford, pp. 41–76. New York: Academic Press.

2058. Straus, L. G. 1983. From Mousterian to Magdalenian: Cultural evolution viewed from Vasco-Cantabrian Spain and Pyrenean France. *British Archaeological Reports International Series* 164:73–111.

2059. Straus, L. G. 1985. Stone Age prehistory of northern Spain. *Science* 230:501–507.

2060. Straus, L. G. 1987. Paradigm lost: A personal view of the current state of Upper Paleolithic research. *Helinium* 27:157–171.

2061. Straus, L. G. 1989. Age of the modern Europeans. *Nature* 342: 476–477.

2062. Straus, L. G. 1989. Grave reservations: More on Paleolithic burial evidence. *Current Anthropology* 30:633–634.

2063. Straus, L. G. 1990. Comment on "Symbolism and modern human origins." *Current Anthropology* 31:248–249.

2064. Straus, L. G. 1990. The Last Glacial Maximum in Cantabrian Spain: The Solutrean. In *The world at 18 000 BP*, vol. 1, *High latitudes*, ed. O. Soffer and C. Gamble, pp. 89–108. London: Unwin Hyman.

2065. Straus, L. G. 1992. *Iberia before the Iberians: The stone age prehistory of Cantabrian Spain.* Albuquerque: University of New Mexico Press.

2066. Straus, L. G. 1993–1994. Upper Paleolithic origins and radiocarbon calibration: More new evidence from Spain. *Evolutionary Anthropology* 2:195–198.

2067. Straus, L. G. 1994. The pace of change in the Paleolithic. *American Anthropologist* 96:713–716.

2068. Straus, L. G. 1995. The Upper Paleolithic of Europe: An overview. *Evolutionary Anthropology* 4:4–16.

2069. Straus, L. G. 1997. The Iberian situation between 40,000 and 30,000 B.P., in light of European models of migration and convergence. In *Conceptual issues in modern human origins research*, ed. G. A. Clark and C. M. Willermet, pp. 235–252. New York: Aldine de Gruyter.

2070. Straus, L. G., and Clark, G. A. 1986. La Riera Cave: Stone Age hunter-gatherer adaptations in northern Spain. *Arizona State University Anthropological Research Papers* 36:1–497.

2071. Straus, L. G., and Heller, C. W. 1988. Explorations of the twilight zone: The early Upper Paleolithic of Vasco-Cantabrian Spain and Gascony. *British Archaeological Reports International Series* 437:97–133.

2072. Street, F. A. 1980. Ice Age environments. In *The Cambridge encyclopaedia of archaeology*, ed. A. Sherratt, pp. 52–56. Cambridge: Cambridge University Press.

2073. Stringer, C., and Gamble, C. 1993. *In search of the Neanderthals.* New York: Thames and Hudson.

2074. Stringer, C. B. 1979. A re-evaluation of the fossil human calvaria from Singa, Sudan. *Bulletin of the British Museum of Natural History (Geology)* 32:77–83.

2075. Stringer, C. B. 1982. Comment on "Upper Pleistocene hominid evolution in south-central Europe: A review of the evidence and analysis of trends." *Current Anthropology* 23:690–691.

2076. Stringer, C. B. 1983. Some further notes on the morphology and dating of the Petralona hominid. *Journal of Human Evolution* 12:731–742.

2077. Stringer, C. B. 1984. The definition of *Homo erectus* and the existence of the species in Africa and Europe. *Courier Forschungsinstitut Senckenberg* 69:131–144.

2078. Stringer, C. B. 1985. Middle Pleistocene hominid variability and the origin of late Pleistocene humans. In *Ancestors: The hard evidence*, ed. E. Delson, pp. 289–295. New York: Alan R. Liss.

2079. Stringer, C. B. 1986. An archaic character in the Broken Hill innominate E.719. *American Journal of Physical Anthropology* 71:115–120.

2080. Stringer, C. B. 1986. The credibility of *Homo habilis*. In *Major topics in primate and human evolution*, ed. B. Wood, L. Martin, and P. Andrews, pp. 266–294. Cambridge: Cambridge University Press.

2081. Stringer, C. B. 1986. Ice Age relation. *Geographical Magazine* 58:652–656.

2082. Stringer, C. B. 1989. The origin of early modern humans: A comparison of the European and non-European evidence. In *The human revolution: Behavioural and biological perspectives on the origins of modern humans*, ed. P. Mellars and C. B. Stringer, pp. 232–244. Edinburgh: Edinburgh University Press.

2083. Stringer, C. B. 1993. Secrets of the Pit of the Bones. *Nature* 362:501–502.

2084. Stringer, C. B. 1994. Out of Africa—a personal history. In *Origins of anatomically modern humans*, ed. M. H. Nitecki and D. V. Nitecki, pp. 150–174. New York: Plenum Press.

2085. Stringer, C. B. 1995. The evolution and distribution of Later Pleistocene human populations. In *Paleoclimate and evolution with emphasis on human origins*, ed. E. S. Vrba, G. H. Denton, T. C. Partridge, and L. H. Burckle, pp. 524–531. New Haven: Yale University Press.

2086. Stringer, C. B. 1996. The Boxgrove tibia: Britain's oldest hominid and its place in the Middle Pleistocene record. In *The English Palaeolithic reviewed*, ed. C. S. Gamble and A. J. Lawson, pp. 52–56. Wessex: Trust for Wessex Archaeology.

2087. Stringer, C. B. 1996. Current issues in modern human origins. In *Contemporary issues in human evolution*, ed. W. E. Meickle, F. C. Howell, and N. G. Jablonski, pp. 115–134. San Francisco: California Academy of Sciences.

2088. Stringer, C. B., Andrews, P., and Currant, A. B. 1996. Palaeoclimatic significance of mammalian faunas from Westbury Cave, Somerset, England. In *The early Middle Pleistocene in Europe*, ed. E. Turner, pp. 135–143. Rotterdam: Balkema.

2089. Stringer, C. B., and Bräuer, G. 1994. Methods, misreading, and bias. *American Anthropologist* 96:416–424.

2090. Stringer, C. B., Cornish, L., and Stuart-Macadam, P. 1985. Preparation and further study of the Singa skull from Sudan. *Bulletin of the British Museum of Natural History (Geology)* 38:347–358.

2091. Stringer, C. B., and Dean, M. C. 1997. Age at death of Gibraltar 2—a reply. *Journal of Human Evolution* 32:471–472.

2092. Stringer, C. B., Dean, M. C., and Martin, R. D. 1990. A comparative study of cranial and dental development within a recent British sample and among Neandertals. In *Primate life history and evolution*, ed. C. J. DeRousseau, pp. 115–152. New York: Wiley-Liss.

2093. Stringer, C. B., Grün, R., Schwarcz, H. P., and Goldberg, P. 1989. ESR dates for the hominid burial site of Es Skhul in Israel. *Nature* 339:756–758.

2094. Stringer, C. B., Howell, F. C., and Melentis, J. K. 1979. The significance of the fossil hominid skull from Petralona, Greece. *Journal of Archaeological Science* 6:235–253.

2095. Stringer, C. B., Hublin, J.-J., and Vandermeersch, B. 1984. The origin of anatomically modern humans in western Europe. In *The origins of modern humans: A world survey of the fossil evidence*, ed. F. H. Smith and F. Spencer, pp. 51–135. New York: Alan R. Liss.

2096. Stuart, A. J. 1982. *Pleistocene vertebrates in the British Isles*. London: Longman.

2097. Stuart, A. J. 1991. Mammalian extinctions in the late Pleistocene of northern Eurasia and North America. *Biological Review* 66:453–562.

2098. Stuiver, M., Brazunias, T. F., Becker, B., and Kromer, B. 1991. Climatic, solar, oceanic and geomagnetic influences on Late-Glacial and Holocene atmospheric $^{14}C/^{12}C$ change. *Quaternary Research* 35:1–24.

2099. Susman, R., and Stern, J. T. 1979. Telemetered electromyography of flexor digitorum profundus and flexor digitorum superficialis in *Pan troglodytes* and implications for the interpretation of the O.H. 7 hand. *American Journal of Physical Anthropology* 50:781–784.

2100. Susman, R. L. 1987. Pygmy chimpanzees and common chimpanzees: Models for the behavioral ecology of the earliest hominids. In *The evolution of human behavior: Primate models*, ed. W. G. Kinzey, pp. 72–86. Albany: State University of New York Press.

2101. Susman, R. L. 1988. Hand of *Paranthropus robustus* from Member 1, Swartkrans: Fossil evidence for tool behavior. *Science* 240:781–784.

2102. Susman, R. L. 1991. Who made the Oldowan tools? Fossil evidence for tool behavior in Plio-Pleistocene hominids. *Journal of Anthropological Research* 47:129–151.

2103. Susman, R. L. 1993. Hominid postcranial remains from Swartkrans. *Transvaal Museum Monograph* 8:117–136.

2104. Susman, R. L. 1994. Fossil evidence for early hominid tool use. *Science* 265:1570–1573.

2105. Susman, R. L. 1998. Hand function and tool behavior in early hominids. *Journal of Human Evolution* 35:23–46.

2106. Susman, R. L., and Stern, J. T. 1982. Functional morphology of *Homo habilis*. *Science* 217:931–934.

2107. Susman, R. L., Stern, J. T., and Jungers, W. J. 1985. Locomotor adaptations in the Hadar hominids. In *Ancestors: The hard evidence*, ed. E. Delson, pp. 184–192. New York: Alan R. Liss.

2108. Susman, R. L., Stern, J. T., and Jungers, W. L. 1984. Arboreality and bipedality in the Hadar hominids. *Folia Primatologia* 43:113–156.

2109. Sussman, R. W. 1991. Primate origins and the evolution of angiosperms. *American Journal of Primatology* 23:209–223.

2110. Suwa, G. 1988. Evolution of the "robust" australopithecines in the Omo succession: Evidence from mandibular premolar morphology. In *Evolutionary history of the "robust" australopithecines*, ed. F. E. Grine, pp. 199–222. New York: Aldine de Gruyter.

2111. Suwa, G., Asfaw, B., Beyene, Y., White, T. D., Katoh, S., Nagaoka, S., Nakaya, H., Uzawa, K., Renne, P., and Wolde-Gabriel, G. 1997. The first skull of *Australopithecus boisei*. *Nature* 389:489–492.

2112. Suwa, G., White, T. D., and Howell, F. C. 1996. Mandibular post-canine dentition from the Shungura Formation, Ethiopia: Crown morphology, taxonomic allocations, and Plio-Pleistocene hominid evolution. *American Journal of Physical Anthropology* 101: 247–282.

2113. Suzuki, H., and Takai, F., eds. 1970. *The Amud man and his cave site.* Tokyo: University of Tokyo.

2114. Svoboda, J. 1987. Lithic industries of the Arago, Bilzingsleben, and Vértesszöllös hominids: Comparison and evolutionary interpretation. *Current Anthropology* 28:219–227.

2115. Svoboda, J. 1990. Moravia during the Upper Pleniglacial. In *The world at 18 000 BP,* vol. 1, *High latitudes,* ed. O. Soffer and C. Gamble, pp. 193–203. London: Unwin Hyman.

2116. Svoboda, J. 1993. The complex origin of the Upper Paleolithic in the Czech and Slovak Republics. In *Before Lascaux: The complex record of the Early Upper Paleolithic,* ed. H. Knecht, A. Pike-Tay, and R. White, pp. 23–36. Boca Raton, Fla.: CRC Press.

2117. Svoboda, J. 1994. The Pavlov site, Czech Republic: Lithic evidence from the Upper Paleolithic. *Journal of Field Archaeology* 21:69–81.

2118. Svoboda, J., Lozek, V., Svobodová, H., and Skrdla, P. 1994. Predmostí after 110 years. *Journal of Field Archaeology* 21:457–472.

2119. Svoboda, J., and Simán, K. 1989. The Middle-Upper Paleolithic transition in southeastern central Europe (Czechoslovakia and Hungary). *Journal of World Prehistory* 3:283–322.

2120. Svoboda, J., and Svoboda, H. 1985. Les industries de type Bohunice dans leur cadre stratigraphique et écologique. *Anthropologie* 89:505–514.

2121. Svoboda, J., and Vlcek, E. 1991. La nouvelle sépulture de Dolní Vestonice (DV XVI), Tchécoslovaquie. *Anthropologie* 95:323–328.

2122. Swisher, C. C., Rink, W. J., Antón, S. C., Schwarcz, H. P., Curtis, G. H., Suprijo, A., and Widiasmoro. 1996. Latest *Homo erectus* of Java: Potential contemporaneity with *Homo sapiens* in Southeast Asia. *Science* 274:1870–1874.

2123. Swisher, C. C. I., Curtis, G. H., Jacob, T., Getty, A. G., Suprijo, A., and Widiasmoro. 1994. Age of the earliest known hominids in Java, Indonesia. *Science* 263:1118–1121.

2124. Syvanen, M. 1987. Molecular clocks and evolutionary relationships: Possible distortions due to horizontal gene flow. *Journal of Molecular Evolution* 26:16–23.

2125. Szalay, F. S. 1972. Paleobiology of the earliest primates. In *The functional biology of primates,* ed. R. H. Tuttle, pp. 3–35. Chicago: Aldine-Atherton.

2126. Szalay, F. S., and Delson, E. 1979. *Evolutionary history of the primates.* New York: Academic Press.

2127. Szalay, F. S., and Li, C. K. 1986. Middle Paleocene euprimate from southern China and the distribution of primates in the Paleogene. *Journal of Human Evolution* 15:387–397.

2128. Szalay, F. S., Rosenberger, A. L., and Dagosto, M. 1987. Diagnosis and differentiation of the order Primates. *Yearbook of Physical Anthropology* 30:75–105.

2129. Szalay, F. S., Tattersall, I., and Decker, R. L. 1975. Phylogenetic relationships of *Plesiadapis*—postcranial evidence. *Contributions to Primatology* 5:136–166.

2130. Szathmary, E. J. E. 1993. Genetics of aboriginal North Americans. *Evolutionary Anthropology* 1:202–220.

2131. Taborin, Y. 1990. Les prémices de la parure. In *Paléolithique moyen récent et Paléolithique supérieur ancien en Europe*, ed. C. Farizy, 3:335–344. Nemours: Mémoires du Musée de Préhistoire d'Île de France.

2132. Tague, R. G., and Lovejoy, C. O. 1986. The obstetric pelvis of A.L. 288-1 (Lucy). *Journal of Human Evolution* 15:237–255.

2133. Takai, M., and Anaya, F. 1996. New specimens of the oldest fossil platyrrhine, *Branisella boliviana* from Salla, Bolivia. *American Journal of Physical Anthropology* 99:301–318.

2134. Tallon, P. W. J. 1978. Geological setting of the hominid fossils and Acheulean artifacts from the Kapthurin Formation, Baringo District, Kenya. In *Geological background to fossil man*, ed. W. W. Bishop, pp. 361–373. Toronto: University of Toronto Press.

2135. Tamrat, E., Thouveny, N., Taïeb, M., and Opdyke, N. D. 1995. Revised magnetostratigraphy of the Plio-Pleistocene sedimentary sequence of the Olduvai Formation (Tanzania). *Palaeogeography, Palaeoclimatology, Palaeoecology* 114:273–283.

2136. Tankard, A. J., and Schweitzer, F. R. 1976. Textural analysis of cave sediments: Die Kelders, Cape Province, South Africa. In *Geoarchaeology*, ed. D. A. Davidson and M. L. Shackley, pp. 289–316. London: Duckworth.

2137. Tappen, M. 1995. Savanna ecology and natural bone deposition. *Current Anthropology* 36:223–260.

2138. Tappen, N. C. 1987. Circum-mortem damage to some ancient African hominid crania: A taphonomic and evolutionary essay. *African Archaeological Review* 5:39–47.

2139. Tattersall, I. 1986. Species recognition in human paleontology. *Journal of Human Evolution* 15:165–175.

2140. Tattersall, I. 1992. Species concepts and species identification in human evolution. *Journal of Human Evolution* 22:341–349.

2141. Tattersall, I. 1994. Morphology and phylogeny. *Evolutionary Anthropology* 3:40–41.

2142. Tattersall, I. 1995. *The fossil trail.* Oxford: Oxford University Press.

2143. Tattersall, I. 1996. Paleoanthropology and perception. In *Contemporary issues in human evolution*, ed. W. E. Meickle, F. C. Howell, and N. G. Jablonski, pp. 1–45. San Francisco: California Academy of Sciences.

2144. Tauber, H. 1970. The Scandinavian varve chronology and C14 dating. In *Radiocarbon variations and absolute chronology*, ed. I. U. Olsson, pp. 173–195. New York: John Wiley.

2145. Tauxe, L., Deino, A. D., Behrensmeyer, A. K., and Potts, R. 1992. Pinning down the Brunhes/Matuyama and Upper Jaramillo boundaries: A reconciliation of orbital and isotopic time scales. *Earth and Planetary Science Letters* 109:561–572.

2146. Tauxe, L., Opdyke, N. D., Pasini, G., and Elmi, C. 1983. Age of the Pliocene-Pleistocene boundary in the Vrica section, southern Italy. *Nature* 304:125–129.

2147. Taylor, R. E. 1992. Radiocarbon dating of bone: To collagen and beyond. In *Radiocarbon after four decades: An interdisciplinary perspective,* ed. R. E. Taylor, A. Long, and R. S. Kra, pp. 375–402. New York: Springer-Verlag.

2148. Taylor, R. E. 1996. Radiocarbon dating: The continuing revolution. *Evolutionary Anthropology* 4:169–181.

2149. Taylor, R. E., Donahue, D. J., Zabel, T. H., Damon, P. E., and Jull, J. T. 1984. Radiocarbon dating by particle accelerators: An archaeological perspective. *Advances in Chemistry Series* 205: 333–356.

2150. Taylor, R. E., Stuiver, M., and Reimer, P. J. 1996. Development and extension of the radiocarbon time scale: Archaeological applications. *Quaternary Science Reviews* 15:655–668.

2151. Tchernov, E. 1987. The age of the ʿUbeidiya Formation, an early Pleistocene hominid site in the Jordan Valley, Israel. *Israel Journal of Earth Science* 36:3–30.

2152. Tchernov, E. 1988. Biochronology of the Middle Paleolithic and dispersal events of hominids in the Levant. *Études et Recherches Archéologiques de l'Université de Liège* 2:153–168.

2153. Tchernov, E. 1988. La biochronologie du site de ʿUbeidiya (Vallée du Jourdain) et les plus anciens hominidés du Levant. *Anthropologie* 92:839–861.

2154. Tchernov, E. 1992. Biochronology, paleoecology, and dispersal events of hominids in the southern Levant. In *The evolution and dispersal of modern humans in Asia,* ed. T. Akazawa, K. Aoki, and T. Kimura, pp. 149–188. Tokyo: Hokusen-Sha.

2155. Tchernov, E. 1994. New comments on the biostratigraphy of the Middle and Upper Pleistocene in the southern Levant. In *Late Quaternary chronology and paleoclimates of the eastern Mediterranean,* ed. O. Bar-Yosef and R. S. Kra, pp. 333–350. Cambridge, Mass.: American School of Prehistoric Research.

2156. Tchernov, E. 1996. Rodent faunas, chronostratigraphy and paleobiogeography of the southern Levant during the Quaternary. *Acta Zoologica Cracoviana* 39:513–530.

2157. Tchernov, E., Horwitz, L. K., and Ronen, A. 1994. The faunal remains from Evron Quarry in relation to other Lower Paleolithic hominid sites in the southern Levant. *Quaternary Research* 42: 328–339.

2158. Teaford, M. F., and Walker, A. 1984. Quantitative differences in dental microwear between primate species with different diets and a comment on the presumed diet of *Sivapithecus. American Journal of Physical Anthropology* 64:191–200.

2159. Temerin, L. A., and Cant, J. G. H. 1983. The evolutionary divergence of Old World monkeys and apes. *American Naturalist* 122:335–351.

2160. Templeton, A. R. 1992. Human origins and analysis of mitochondrial DNA sequences. *Science* 255:737.

2161. Terasmae, J. 1984. Radiocarbon dating: Some problems and potential developments. In *Quaternary dating methods,* ed. W. C. Mahaney, pp. 1–15. Amsterdam: Elsevier.

2162. Teruya, E. 1986. The origins and characteristics of Jomon Ceramic Culture: A brief introduction. In *Windows on the Japanese past: Studies in archaeology and prehistory,* ed. R. J. Pearson, G. L. Barnes, and K. L. Hutterer, pp. 223–228. Ann Arbor: Center for Japanese Studies, University of Michigan.

2163. Texier, P.-J. 1995. The Oldowan assemblage from NY18 site at Nyabusosi (Toro-Uganda). *Comptes Rendus de l'Académie des Sciences, Paris*, ser. 2a, 320:647–653.

2164. Thackeray, A. I. 1989. Changing fashions in the Middle Stone Age: The stone artefact sequence from the Klasies River Main Site, South Africa. *African Archaeological Review* 7:23–57.

2165. Thackeray, A. I. 1992. The Middle Stone Age south of the Limpopo River. *Journal of World Prehistory* 6:385–440.

2166. Thackeray, J. F. 1988. Molluscan fauna from Klasies River, South Africa. *South African Archaeological Bulletin* 43:27–32.

2167. Theunissen, B. 1989. *Eugène Dubois and the ape-man from Java: The history of the first "missing link" and its discoverer.* Dordrecht: Kluwer Academic.

2168. Theunissen, B., de Vos, J., Sondaar, P. Y., and Aziz, F. 1990. The establishment of a chronological framework for the hominid-bearing deposits of Java: A historical survey. *Geological Society of America Special Paper* 242:39–54.

2169. Thieme, H. 1983. Mittelpaläolithische Siedlungsstrukturen in Rheindalen (BRD). *Ethnographisch-Archäologisches Zeitschrift* 24:362–374.

2170. Thieme, H. 1996. Altpaläolithische Wurfspeere aus Schöningen, Niedersachsen: Ein Vorbericht. *Archäologisches Korrespondenzblatt* 26:377–393.

2171. Thieme, H. 1997. Lower Palaeolithic hunting spears from Germany. *Nature* 385:807–810.

2172. Thieme, H., and Veil, S. 1985. Neue Untersuchungen zum eemzeitlichen Elefanten-Jagdplatz Lehringen, Ldkr, Verden. *Kunde*, n.s., 36:11–58.

2173. Thissen, J. 1986. Ein weiterer Fundplatz der Westwandfundschicht (B1) von Rheindahlen. *Archäologisches Korrespondenzblatt* 16:111–121.

2174. Thoma, A. 1972. On Vértesszöllös man. *Nature* 236:464–465.

2175. Thoma, A. 1978. Some notes on the Vértesszöllös occipital. *Journal of Human Evolution* 7:323–325.

2176. Thomas, H. 1979. Géologie et paléontologie du gisement acheuléen de l'erg Tihodaïne. *Mémoires du Centre de Recherches Anthropologiques, Préhistoriques et Ethnographiques* 27:1–122.

2177. Thomas, H., Roger, J., Sen, S., and Al-Sulaimani, Z. 1988. Découverte des plus anciens "Anthropoïdes" du continent arabo-africain et d'un primate tarsiiforme dans l'Oligocène du Sultanat d'Oman. *Comptes Rendus de l'Académie des Sciences, Paris*, ser. 2, 306:823–829.

2178. Thomas, H., Roger, J., Sen, S., Bourdillon-de-Grissac, C., and al-Sulaimani, Z. 1989. Découverte de vertébrés fossiles dan l'Oligocène inférieur du Dhofar (Sultanat d'Oman). *Geobios* (Lyons) 22:101–120.

2179. Thomas, H., Sen, S., Roger, J., and al-Sulaimani, Z. 1991. The discovery of *Moeripithecus markgrafi* Schlosser (Propliopithecidae, Anthropoidea, Primates), in the Ashawq Formation (Early Oligocene of Dhofar Province, Sultanate of Oman). *Journal of Human Evolution* 20:33–49.

2180. Thompson, D. D., and Trinkaus, E. 1981. Age determination for the Shanidar 3 Neanderthal. *Science* 212.

2181. Thorne, A. G. 1977. Separation or reconciliation? Biological clues to the development of Australian society. In *Sunda and Sahul: Prehistoric studies in Southeast Asia, Melanesia, and Australia,* ed. J. Allen, J. Golson, and R. Jones, pp. 197–204. London: Academic Press.

2182. Thorne, A. G. 1980. The arrival of man in Australia. In *The Cambridge encyclopaedia of archaeology,* ed. A. Sherratt, pp. 96–100. Cambridge: Cambridge University Press.

2183. Thorne, A. G. 1984. Australia's human origins—How many sources? *American Journal of Physical Anthropology* 63:227.

2184. Thorne, A. G., and Wolpoff, M. H. 1981. Regional continuity in Australasian Pleistocene hominid evolution. *American Journal of Physical Anthropology* 55:337–349.

2185. Thouveny, N., and Bonifay, E. 1984. New chronological data on European Plio-Pleistocene faunas and hominid occupation sites. *Nature* 308:355–358.

2186. Tiedemann, R., Sarnthein, M., and Shackleton, N. J. 1994. Astronomic timescale for the Pliocene Atlantic δ^{18}O and dust flux records of Ocean Drilling Program site 65f9. *Paleoceanography* 9:619–638.

2187. Tillet, T. 1984. The Aterian site of Seggedim. *Palaeoecology of Africa* 16:301–304.

2188. Tillet, T. 1985. The Palaeolithic and its environment in the northern part of the Chad Basin. *African Archaeological Review* 3:163–177.

2189. Tillier, A.-M., Arensburg, B., Vandermeersch, B., and Rak, Y. 1991. L'apport de Kébara à la palethnologie funéraire des Néanderthaliens du Proche Orient. In *Le squelette moustérien de Kébara 2,* ed. O. Bar-Yosef and B. Vandermeersch, pp. 89–95. Paris: Centre National de la Recherche Scientifique.

2190. Tishkoff, S. A., Dietzsch, E., Speed, W., Pakstis, A. J., Kidd, J. R., Cheung, K., Bonné-Tamir, B., Santachiara-Benerecetti, A. S., Moral, P., Krings, M., Pääbo, S., Watson, E., Risch, N., Jenkins, T., and Kidd, K. K. 1996. Global patterns of linkage disequilibrium at the CD4 locus and modern human origins. *Science* 271: 1380–1387.

2191. Tixier, J. D., Roe, D., Turq, A., Gibert, J., Martínez, B., Arribas, L., Gibert, R., Gaete, A., Maillo, A., and Iglesias, A. 1995. Présence d'industries lithiques dans le Pléistocène inférieur de la région d'Orce (Grenade, Espagne): Quel est l'état de la question? *Comptes Rendus de l'Académie des Sciences, Paris,* ser. 2, 321: 71–78.

2192. Tobias, P. V. 1967. *Olduvai Gorge.* Vol. 2. *The cranium and maxillary dentition* of Australopithecus (Zinjanthropus) boisei. Cambridge: Cambridge University Press.

2193. Tobias, P. V. 1967. The hominid skeletal remains of Haua Fteah. In *The Haua Fteah (Cyrenaica) and the Stone Age of the southeast Mediterranean,* ed. C. B. M. McBurney, pp. 338–352. Cambridge: Cambridge University Press.

2194. Tobias, P. V. 1971. Human skeletal remains from the Cave of Hearths, Makapansgat, northern Transvaal. *American Journal of Physical Anthropology* 34:335–368.

2195. Tobias, P. V. 1972. Progress and problems in the study of early man in sub-Saharan Africa. In *The functional and evolutionary biology of primates,* ed. R. H. Tuttle, pp. 63–93. Chicago: Aldine.

2196. Tobias, P. V. 1978. The earliest Transvaal members of the genus *Homo* with another look at some problems of hominid taxonomy and systematics. *Zeitschrift für Morphologie und Anthropologie* 69:225–265.

2197. Tobias, P. V. 1984. *Dart, Taung and the missing link*. Johannesburg: Witwatersrand University Press.

2198. Tobias, P. V. 1985. The former Taung cave system in light of contemporary reports and its bearing on the skull's provenance: Early deterrents to the acceptance of *Australopithecus*. In *Hominid evolution: Past, present and future*, ed. P. V. Tobias, pp. 25–40. New York: Alan R. Liss.

2199. Tobias, P. V. 1991. *Olduvai Gorge*. Vol. 4. *The skulls, endocasts and teeth of* Homo habilis. Cambridge: Cambridge University Press.

2200. Tobias, P. V. 1992. Piltdown: An appraisal of the case against Sir Arthur Keith. *Current Anthropology* 33:243–293.

2201. Tobias, P. V. 1993. On Piltdown: The French connection revisited. *Current Anthropology* 34:65–67.

2202. Tobias, P. V., Vogel, J. C., Oschladeus, H. D., Partridge, T. C., and McKee, J. K. 1993. New isotopic and sedimentological measurements of the Thabaseek deposits (South Africa) and the dating of the Taung hominid. *Quaternary Research* 40:360–367.

2203. Tode, A., Preul, F., Richter, A., and Kleinschmidt, A. 1953. Die Untersuchung der paläolithischen Freilandstation von Salzgitter-Lebenstedt. *Eiszeitalter und Gegenwart* 3:144–220.

2204. Tompkins, R. L. 1996. Relative dental development of Upper Pleistocene hominids compared to human population variation. *American Journal of Physical Anthropology* 99:103–118.

2205. Torroni, A., Schurr, T. G., Cabell, M. F., Brown, M. D., Neel, J. V., Larsen, M., Smith, D. G., Vullo, C. M., and Wallace, D. C. 1993. Asian affinities and continental radiation of the four founding Native American mtDNAs. *American Journal of Human Genetics* 53:563–590.

2206. Torroni, A., Schurr, T. G., Yang, C.-C., Szathmary, E. J. E., Williams, R. C., Schanfield, M. S., Troup, G. A., Knowler, W. C., Lawrence, D. N., Weiss, K. M., and Wallace, D. C. 1992. Native American mitochrondrial DNA analysis indicates that the Amerind and Nadene populations were founded by two independent migrations. *Genetics* 130:153–162.

2207. Toth, N. 1985. Archeological evidence for preferential right-handedness in the Lower and Middle Pleistocene, and its possible implications. *Journal of Human Evolution* 14:607–614.

2208. Toth, N. 1985. The Oldowan reassessed: A close look at early stone artifacts. *Journal of Archaeological Science* 12:101–120.

2209. Toth, N., and Schick, K. D. 1986. The first million years: The archaeology of protohuman culture. *Advances in Archaeological Method and Theory* 9:1–96.

2210. Toth, N., Schick, K. D., Savage-Rumbaugh, E. S., Sevick, R. A., and Rumbaugh, D. M. 1993. *Pan* the tool-maker: Investigations into the stone tool-making and tool-using capabilities of a bonobo (*Pan paniscus*). *Journal of Archaeological Science* 20:81–91.

2211. Trinkaus, E. 1978. Hard times among the Neanderthals. *Natural History* 87 (10): 58–63.

2212. Trinkaus, E. 1980. New light on the very ancient Near East. *Symbols* 1980:2–3, 11.

2213. Trinkaus, E. 1981. Neanderthal limb proportions and cold adaptation. In *Aspects of human evolution*, ed. C. B. Stringer, pp. 187–224. London: Taylor and Francis.

2214. Trinkaus, E. 1982. Evolutionary continuity among archaic *Homo sapiens*. *British Archaeological Reports International Series* 151:301–319.

2215. Trinkaus, E. 1983. Neandertal postcrania and the adaptive shift to modern humans. *British Archaeological Reports International Series* 164:165–200.

2216. Trinkaus, E. 1983. *The Shanidar Neandertals*. New York: Academic Press.

2217. Trinkaus, E. 1984. Neanderthal pubic morphology and gestation length. *Current Anthropology* 25:508–514.

2218. Trinkaus, E. 1984. Western Asia. In *The origin of modern humans: A world survey of the fossil evidence*, ed. F. H. Smith and F. Spencer, pp. 251–293. New York: Alan R. Liss.

2219. Trinkaus, E. 1985. Cannibalism and burial at Krapina. *Journal of Human Evolution* 14:203–216.

2220. Trinkaus, E. 1985. Pathology and posture of the La-Chapelle-aux-Saints Neanderthal. *American Journal of Physical Anthropology* 67:19–41.

2221. Trinkaus, E. 1986. The Neanderthals and modern human origins. *Annual Review of Anthropology* 15:193–218.

2222. Trinkaus, E. 1987. Bodies, brawn, brains and noses: Human ancestors and human predation. In *The evolution of human hunting*, ed. M. Nitecki and D. V. Nitecki, pp. 107–145. New York: Plenum Press.

2223. Trinkaus, E. 1987. The Neandertal face: Evolutionary and functional perspectives on a recent hominid face. *Journal of Human Evolution* 16:429–443.

2224. Trinkaus, E. 1989. Neandertal upper limb morphology and manipulation. In *Hominidae: Proceedings of the Second International Congress of Human Paleontology*, ed. G. Giacobini, pp. 331–337. Milan: Jaca Book.

2225. Trinkaus, E. 1989. The Upper Pleistocene transition. In *The emergence of modern humans: Biocultural adaptations in the later Pleistocene*, ed. E. Trinkaus, pp. 42–66. Cambridge: Cambridge University Press.

2226. Trinkaus, E. 1991. The evolution and dispersal of modern humans in Asia. *Current Anthropology* 32:353–355.

2227. Trinkaus, E. 1992. Morphological contrasts between the Near Eastern Qafzeh-Skhul and late archaic human samples: Grounds for a behavioral difference. In *The evolution and dispersal of modern humans in Asia*, ed. T. Akazawa, K. Aoki, and T. Kimura, pp. 278–294. Tokyo: Hokusen-Sha.

2228. Trinkaus, E. 1992. Paleontological perspectives on Neandertal behavior. *Études et Recherches Archéologiques de l'Université de Liège* 56:151–176.

2229. Trinkaus, E. 1993. Femoral neck-shaft angles of the Qafzeh-Skhul early modern humans, and activity levels among immature Near Eastern Middle Paleolithic hominids. *Journal of Human Evolution* 25:393–416.

2230. Trinkaus, E. 1993. A note on the KNM-ER 999 hominid femur. *Journal of Human Evolution* 24:493–504.

2231. Trinkaus, E. 1995. Neanderthal mortality patterns. *Journal of Archaeological Science* 22:121–142.

2232. Trinkaus, E. 1995. Near Eastern late archaic humans. *Paléorient* 21:9–23.

2233. Trinkaus, E., Churchill, S. E., Villemeur, I., Riley, K. G., Heller, J. A., and Ruff, C. B. 1991. Robusticity versus shape: The functional interpretation of Neandertal appendicular morphology. *Journal of the Anthropological Society of Nippon* 99:257–298.

2234. Trinkaus, E., and Howells, W. W. 1979. The Neanderthals. *Scientific American* 241 (6): 118–133.

2235. Trinkaus, E., and LeMay, M. 1982. Occipital bunning among later Pleistocene hominids. *American Journal of Physical Anthropology* 72:123–129.

2236. Trinkaus, E., Ruff, C. B., and Churchill, S. E. 1998. Upper limb versus lower loading patterns among Near Eastern Middle Paleolithic hominids. In *Neanderthals and modern humans in west Asia*, ed. T. Akazawa, K. Aoki, and O. Bar-Yosef, pp. 391–404. New York: Plenum Press.

2237. Trinkaus, E., and Shipman, P. 1993. *The Neandertals: Changing the image of mankind.* New York: Alfred A. Knopf.

2238. Trinkaus, E., and Shipman, P. 1993. Neandertals: Images of ourselves. *Evolutionary Anthropology* 1:194–201.

2239. Trinkaus, E., and Thompson, D. D. 1987. Femoral diaphyseal histomorphometric age determinations for the Shanidar 3, 4, 5 and 6 Neandertals and Neandertal longevity. *American Journal of Physical Anthropology* 72:123–129.

2240. Tuffreau, A. 1978. Les industries acheuléenes de Cagny-la-Garenne (Somme). *Anthropologie* 82:37–60.

2241. Tuffreau, A. 1979. Les débuts du Paléolithique moyen dans la France septentrionale. *Bulletin de la Société Préhistorique Française* 76:140–142.

2242. Tuffreau, A. 1982. On the transition Lower/Middle Palaeolithic in northern France. *British Archaeological Reports International Series* 151:137–149.

2243. Tuffreau, A. 1988. Biache-Saint-Vaast et les industries moustériennes du Pléistocène moyen récent dans la France septentrionale. In *Biache-Saint-Vaast et les industries moustériennes du Pléistocène moyen récent dans la France septentrionale,* pp. 197–207. Paris: Supplément au Bulletin de l'Association Française pour l'Étude du Quaternaire.

2244. Tuffreau, A. 1992. Middle Paleolithic settlement in northern France. In *The Middle Paleolithic: Adaptation, behavior, and variability*, ed. H. L. Dibble and P. A. Mellars, pp. 59–73. Monograph 72. Philadelphia: University of Pennsylvania Museum.

2245. Tuffreau, A., Chaline, J., Munaut, A., Piningre, J.-F., Poplin, F., Puissegur, J.-J., Sommé, J., and Vandermeersch, B. 1978. Premiers résultats de l'étude du gisement paléolithique de Biache-Saint-Vaast (Pas-de-Calais). *Comptes Rendus de l'Académie des Sciences, Paris,* ser. D, 286:457–459.

2246. Tuffreau, A., Lamotte, A., and Marcy, J.-L. 1997. Land-use and site function in Acheulean complexes of the Somme Valley. *World Archaeology* 29:225–241.

2247. Tuffreau, A., Munaut, A. V., Puisségur, J. J., and Sommé, J. 1982. Stratigraphie et environnement de la séquence archéologique de Biache-Saint-Vaast (Pas de Calais). *Bulletin de l'Association Française pour l'Étude du Quaternaire* 19:57–62.

2248. Tuffreau, A., Révillon, J., Sommé, J., Aitken, M. J., Huxtable, J., and Leroi-Gourhan, A. 1985. Le gisement paléolithique moyen de Seclin (Nord, France). *Archäeologisches Korrespondenzblatt* 15:132–138.

2249. Turk, I., Bastiani, G., Culiberg, M., Dirjec, J., Kavur, B., Krystufek, B., Ku, T.-L., Kunei, D., Nelson, D. E., Omerzel-Terlep, M., and Sercelj, A. 1997. *Mousterian "bone flute" and other finds from Divje Babe I cave site in Slovenia.* Ljubljana: Zanstvenoraziskovakni Center SAZU.

2250. Turk, I., Dirjec, J., and Kavur, B. 1995. The oldest musical instrument in Europe discovered in Slovenia? *Razprave IV. Razreda SAZU* 36:287–293.

2251. Turner, A. 1992. Large carnivores and earliest European hominids: Changing determinants of resource availability during the Lower and Middle Pleistocene. *Journal of Human Evolution* 22:109–126.

2252. Turner, C. G. 1985. The dental search for Native American origins. In *Out of Asia: Peopling of the Americas and the Pacific*, ed. R. Kirk and E. Szathmary, pp. 31–78. Canberra: Australian National University.

2253. Turner, C. G. 1987. Telltale teeth. *Natural History* 96 (1): 6–10.

2254. Turner, C. G. 1989. Teeth and prehistory in Asia. *Scientific American* 260 (2): 88–96.

2255. Turner, C. G. 1990. Major features of Sundadonty and Sinodonty, including suggestions about east Asian microevolution, population history, and late Pleistocene relationships with Australian Aboriginals. *American Journal of Physical Anthropology* 82: 295–317.

2256. Turner, C. G. 1990. Paleolithic Siberian dentition from Denisova and Okladnikov Caves, Altayskiy Kray, U.S.S.R. *Current Research in the Pleistocene* 7:65–66.

2257. Turner, C. G. 1995. Shifting continuity: Modern human origins. In *The origin and past of modern humans as viewed from DNA*, ed. S. Brenner and K. Hanihara, pp. 216–243. Singapore: World Scientific.

2258. Turner, E. 1986. The 1981–83 excavations in the Karl Schneider Quarry, Ariendorf, West Germany. In *Chronostratigraphie et faciés culturels du Paléolithique inférieur et moyen dans l'Europe du Nord-Ouest*, ed. A. Tuffreau and J. Sommé, pp. 35–42. Paris: Supplement au Bulletin de l'Association Française pour l'Étude du Quaternaire.

2259. Turner, E. 1995. The Lower Palaeolithic site at Miesenheim I. In *Congreso Internacional de Paleontologia Humana (Orce, September 1995), 3a Circular*, pp. 97–98. Orce, Spain.

2260. Turq, A. 1990. Exploitation des matières premières lithiques dans le Moustérien entre Dordogne et Lot (sud-ouest de la France). In *Le silex de sa genèse à l'outil*, ed. M.-R. Séronie-

Vivien and M. Lenoir, 2:415–427. Paris: Centre National de la Recherche Scientifique.

2261. Turq, A. 1992. Raw material and technological studies of the Quina Mousterian in Périgord. In *The Middle Paleolithic: Adaptation, behavior, and variability*, ed. H. L. Dibble and P. A. Mellars, 72:75–85. Philadelphia: University of Pennsylvania Museum Monograph.

2262. Turq, A., Martínez-Navarro, B., Palmqvist, P., Arribas, A., Agustí, J., and Rodríguez Vidal, J. 1996. Le Plio-Pleistocène de la région d'Orce, province de Grenade, Espagne: Bilan et perspectives de recherche. *Paleo* 8:161–204.

2263. Tuttle, R. H. 1981. Evolution of hominid bipedalism and prehensile capabilities. *Philosophical Transactions of the Royal Society of London*, ser. B, 292:89–94.

2264. Tuttle, R. H. 1985. Ape footprints and Laetoli impressions: A response to the SUNY claims. In *Hominid evolution: Past, present and future*, ed. P. V. Tobias, pp. 129–130. New York: Alan R. Liss.

2265. Tuttle, R. H. 1986. *Apes of the world: Their social behavior, communication, mentality and ecology.* Park Ridge, N.J.: Noyes.

2266. Tuttle, R. H. 1987. Kinesiological inferences and evolutionary implications from Laetoli bipedal trails G-1, G-2/3, and A. In *Laetoli: A Pliocene site in northern Tanzania*, ed. M. D. Leakey and J. M. Harris, pp. 503–523. Oxford: Clarendon Press.

2267. Tyldesley, J. A., Bahn, P. 1983. Use of plants in the European Palaeolithic: A review of the evidence. *Quaternary Science Reviews* 2:53–81.

2268. Ucko, P., and Rosenfeld, A. 1967. *Paleolithic cave art.* New York: McGraw-Hill.

2269. Ullrich, H. 1958. Neandertalerfunde aus der Sowjetunion. In *Neanderthal centenary, 1856–1956*, ed. G. H. R. von Koenigswald, pp. 72–106. Utrecht: Kemink.

2270. Ungar, P. S., Fennell, K. J., Gordon, K., and Trinkaus, E. 1997. Neandertal incisor bevelling. *Journal of Human Evolution* 32: 407–421.

2271. Valladas, H., Cachier, H., Maurice, P., Bernaldo de Quiros, F., Clottes, J., Cabrera Valdés, V., Uzquiano, P., and Arnold, M. 1992. Direct radiocarbon dates for prehistoric paintings at the Altamira, El Castillo and Niaux Caves. *Nature* 357:68–70.

2272. Valladas, H., Geneste, J. M., Joron, J. L., and Chadelle, J. P. 1986. Thermoluminescence dating of Le Moustier (Dordogne, France). *Nature* 322:452–454.

2273. Valladas, H., Joron, J.-L., Valladas, G., Arensburg, B., Bar-Yosef, O., Belfer-Cohen, A., Goldberg, P., Laville, H., Meignen, L., Rak, Y., Tchernov, E., Tillier, A.-M., and Vandermeersch, B. 1987. Thermoluminescence dates for the Neanderthal burial site at Kebara in Israel. *Nature* 330:159–160.

2274. Valladas, H., Reyss, J.-L., Joron, J.-L., Valladas, G., Bar-Yosef, O., and Vandermeersch, B. 1988. Thermoluminescence dating of Mousterian "Proto-Cro-Magnon" remains from Israel and the origin of modern man. *Nature* 331:614–616.

2275. Valladas, H., and Valladas, G. 1987. Thermoluminescence dating of burnt flints and quartz: Comparative results. *Archaeometry* 29:214–220.

2276. Vallois, H., and Roche, J. 1958. La mandibule acheuléene de Témara. *Comptes Rendus de l'Académie des Sciences, Paris* 246:3113–3116.

2277. Vallois, H. V. 1951. La mandibule humaine fossile de la grotte du Porc Épic près Diré Daoua (Abyssinie). *Anthropologie* 55:231–238.

2278. Vallois, H. V. 1961. The social life of early man: The evidence of skeletons. In *The social life of early man*, ed. S. L. Washburn, pp. 214–235. Chicago: Aldine.

2279. Vallois, H. V. 1971. Le crâne trépané magdalénien de Rochereil. *Bulletin de la Société Préhistorique Française* 68:485–495.

2280. Valoch, K. 1969. The beginning of the Upper Paleolithic in central Europe (in Russian). *Bulletin of the Commission for the Study of the Quaternary Period (Academy of Sciences of the USSR)* 36:63–74.

2281. Valoch, K. 1972. Rapports entre le Paléolithique moyen et le Paléolithique supérieur en Europe centrale. In *The origin of Homo sapiens*, ed. F. H. Bordes, pp. 161–171. Paris: UNESCO.

2282. Valoch, K. 1976. Aperçu des premières industries en Europe. In *Les premières industries de l'Europe (Colloque Huit)*, ed. K. Valoch, pp. 178–183. Nice: Union International des Sciences Préhistoriques et Protohistoriques.

2283. Valoch, K. 1982. Comment on "Upper-Pleistocene hominid evolution in south-central Europe: A review of the evidence and analysis of trends." *Current Anthropology* 23:692.

2284. Valoch, K. 1982. The Lower/Middle Palaeolithic transition in Czechoslovakia. *British Archaeological Reports International Series* 151:193–201.

2285. Valoch, K. 1984. Le Taubachien: Sa géochronologie, paléoecologie et paléoethnologie. *Anthropologie* 88:193–208.

2286. Valoch, K. 1986. The central European early Palaeolithic. *Anthropos* (Brno) 23:189–206.

2287. Valoch, K. 1991. Les premiers peuplements humains en Moravie (Tchécoslovaquie). In *Les premiers Européens*, ed. E. Bonifay and B. Vandermeersch, pp. 189–194. Paris: Éditions du Comité des Travaux Historiques et Scientifiques.

2288. Valoch, K. 1995. Stránská Skála I, le site cromérien près de Brno. In *Congreso Internacional de Paleontologia Humana (Orce, September 1995), 3a Circular*, pp. 102. Orce, Spain.

2289. van Andel, T. H. 1989. Late Pleistocene sea levels and the human exploitation of the shore and shelf of southern Africa. *Journal of Field Archaeology* 16:133–155.

2290. Van Couvering, J. A., and Harris, J. A. 1991. Late Eocene age of Fayum mammal faunas. *Journal of Human Evolution* 21:241–260.

2291. Van Couvering, J. H., and Van Couvering, J. A. 1976. Early Miocene mammal fossils from east Africa. In *Human origins: Louis Leakey and the east African evidence*, ed. G. L. Isaac and E. R. McCown, pp. 155–207. Menlo Park, Calif.: W. A. Benjamin.

2292. van den Bergh, G. D., de Vos, J., Aziz, F., and Sondaar, P. Y. 1995. *Homo erectus* in S.E. Asia: Time, space, migration routes— a global model. In *Congreso Internacional de Paleontologia Humana (Orce, September 1995), 3a Circular*, pp. 94–95. Orce, Spain.

2293. Van der Hammen, T. 1974. The Pleistocene changes of vegetation and climate in tropical South America. *Journal of Biogeography* 1:3–26.

2294. Van der Hammen, T., Wijmstra, A., and Zagwijn, W. H. 1971. The floral record of the late Cenozoic of Europe. In *Late Cenozoic glacial ages*, ed. K. K. Turekian, pp. 391–424. New Haven: Yale University Press.

2295. van Gijn, A. L. 1990. *The wear and tear of flint: Principles of functional analysis applied to Dutch Neolithic assemblages.* Leiden: University of Leiden.

2296. van Noten, F. 1982. *The archaeology of central Africa.* Graz: Akademische Druk- und Verlagsanstalt.

2297. van Noten, F. 1983. News from Kenya. *Antiquity* 57:139–140.

2298. van Peer, P. 1991. Interassemblage variability and Levallois styles: The case of the northern African Middle Palaeolithic. *Journal of Anthropological Archaeology* 10:107–151.

2299. van Schaik, C. P., Fox, E. A., and Sitompul, A. F. 1996. Manufacture and use of tools in wild Sumatran orangutans: Implications for human evolution. *Naturwissenschaften* 83:186–188.

2300. Van Valen, L., and Sloan, R. E. 1965. The earliest primates. *Science* 150:743–745.

2301. van Vark, G. N., Bilsborough, A., and Henke, W. 1992. Affinities of European Upper Paleolithic *Homo sapiens* and later human evolution. *Journal of Human Evolution* 23:401–417.

2302. Vandermeersch, B. 1970. Une sépulture moustérienne avec offrandes découverte dans la grotte de Qafzeh. *Comptes Rendus de l'Académie des Sciences, Paris,* ser. D, 270:280–301.

2303. Vandermeersch, B. 1976. Les sépultures néandertaliennes. In *La préhistoire française,* ed. H. de Lumley,1:725–727. Paris: Centre National de la Recherche Scientifique.

2304. Vandermeersch, B. 1981. *Les hommes fossiles de Qafzeh (Israel).* Paris: Centre National de la Recherche Scientifique.

2305. Vandermeersch, B. 1982. The first *Homo sapiens sapiens* in the Near East. *British Archaeological Reports International Series* 151:297–299.

2306. Vandermeersch, B. 1985. The origin of the Neanderthals. In *Ancestors: The hard evidence,* ed. E. Delson, pp. 306–309. New York: Alan R. Liss.

2307. Vandermeersch, B. 1989. The evolution of modern humans: Recent evidence from southwest Asia. In *Behavioural and biological perspectives in the origin of modern humans,* ed. P. A. Mellars and C. B. Stringer, pp. 155–164. Edinburgh: Edinburgh University Press.

2308. Vandermeersch, B., Tillier, A.-M., and Krukoff, S. 1976. Position chronologique des restes humains de Fontéchevade. In *Le peuplement anténeandertalien de l'Europe,* ed. A. Thoma, pp. 19–26. Nice: Union Internationale des Sciences Préhistoriques et Protohistoriques.

2309. Vandiver, P. B., Soffer, O., Klíma, B., and Svoboda, J. 1989. The origins of ceramic technology at Dolni Vestonice, Czechoslovakia. *Science* 246:1001–1008.

2310. Vasil'ev, S. A. 1993. The Upper Palaeolithic of northern Asia. *Current Anthropology* 34:82–92.

2311. Vaufrey, R. 1955. *Préhistoire de l'Afrique.* Vol. 1. *Maghreb.* Paris: Masson.

2312. Velleman, P. F. 1995. *Data Desk Version 5.0: Statistics guide.* Ithaca, N.Y.: Data Description.

2313. Vereshchagin, N. K., and Baryshnikov, G. G. 1984. Quaternary mammalian extinctions in northern Eurasia. In *Quaternary extinctions: A prehistoric revolution,* ed. P. S. Martin and R. G. Klein, pp. 483–516. Tucson: University of Arizona Press.

2314. Vermeersch, P. M., Gijselings, G., and Paulissen, E. 1984. Discovery of the Nazlet Khater man, Upper Egypt. *Journal of Human Evolution* 13:281–286.

2315. Vermeersch, P. M., Paulissen, E., Gijselings, G., Otte, M., Thoma, A., van Peer, P., and Lauwers, R. 1984. 33,000-yr old chert mining site and related *Homo* in the Egyptian Nile Valley. *Nature* 309:342–344.

2316. Vermeersch, P. M., Paulissen, E., and van Peer, P. 1990. Palaeolithic chert exploitation in the limestone stretch of the Egyptian Nile Valley. *African Archaeological Review* 8:77–102.

2317. Verosub, K., and Tchernov, E. 1991. Résultats préliminaires de l'étude magnétostratigraphique d'une séquence sédimentaire à industrie humaine en Israël. In *Les premiers Européens,* ed. E. Bonifay and B. Vandermeersch, pp. 237–242. Paris: Éditions du Comité des Travaux Historiques et Scientifiques.

2318. Vértes, L. 1964. *Tata: Eine mittelpaläolitische Travertin Siedlung in Ungarn.* Budapest: Akadémiai Kiadó.

2319. Vértes, L. 1965. Typology of the Buda industry: A pebble-tool industry from the Hungarian Lower Paleolithic. *Quaternaria* 7:185–195.

2320. Vértes, L. 1975. The Lower Palaeolithic site of Vértesszöllös, Hungary. In *Recent Archaeological Excavations in Europe,* ed. R. Bruce-Mitford, pp. 287–301. London: Routledge and Kegan Paul.

2321. Vigilant, L., Stoneking, M., Harpending, H., Hawkes, K., and Wilson, A. C. 1991. African populations and the evolution of human mitochondrial DNA. *Science* 253:1503–1507.

2322. Villa, P. 1976. Sols et niveaux d'habitat du Paléolithique inférieur en Europe et au Proche Orient. *Quaternaria* 19:107–134.

2323. Villa, P. 1983. Terra Amata and the Middle Pleistocene archaeological record of southern France. *University of California Publications in Anthropology* 13:1–303.

2324. Villa, P. 1989. On the evidence for Neandertal burial. *Current Anthropology* 30:325–326.

2325. Villa, P. 1990. Torralba and Aridos: Elephant exploitation in Middle Pleistocene Spain. *Journal of Human Evolution* 19:299–309.

2326. Villa, P. 1991. Middle Pleistocene prehistory in southwestern Europe: The state of our knowledge and ignorance. *Journal of Anthropological Research* 47:193–217.

2327. Villa, P. 1992. Cannibalism in prehistoric Europe. *Evolutionary Anthropology* 1:93–104.

2328. Villa, P. 1996. Review of "The first Italians: Le industrie litiche di giacimento paleolitico di Isernia La Pineta." *Lithic Technology* 21:71–79.

2329. Villa, P., Bouville, C., Courtin, J., Helmer, D., Mahieu, E., Shipman, P., Belluomini, G., and Branca, M. 1986. Cannibalism in the Neolithic. *Science* 233:431–437.

2330. Vlcek, E. 1975. Morphology of a Neanderthal child from Kiik-Koba in the Crimea. In *Paleoanthropology, morphology and paleoecology*, ed. R. H. Tuttle, pp. 409–418. The Hague: Mouton.

2331. Vlcek, E. 1978. A new discovery of *Homo erectus* in central Europe. *Journal of Human Evolution* 7:239–251.

2332. Vogel, J. C., Beaumont, P. B. 1972. Revised radiocarbon chronology for the Stone Age in South Africa. *Nature* 237:50–51.

2333. Vogel, J. C., and Partridge, T. 1984. Preliminary radiometric ages of the Taung tufas. In *Late Cainozoic palaeoclimates of the Southern Hemisphere*, ed. J. C. Vogel, pp. 507–514. Rotterdam: A. A. Balkema.

2334. Voigt, E. A. 1982. The molluscan fauna. In *The Middle Stone Age at Klasies River Mouth in South Africa*, ed. R. Singer and J. J. Wymer, pp. 155–186. Chicago: University of Chicago Press.

2335. Voigt, E. A. 1983. Mapungubwe: An archaeozoological interpretation of an Iron Age community. *Transvaal Museum Monograph* 11:1–203.

2336. Vollbrecht, J. 1995. Stratified Lower Palaeolithic assemblages in the Rhineland, west Germany. In *Congreso Internacional de Paleontologia Humana (Orce, September 1995), 3a Circular*, p. 97. Orce, Spain.

2337. Volman, T. P. 1978. Early archaeological evidence for shellfish collecting. *Science* 201:911–913.

2338. Volman, T. P. 1981. The Middle Stone Age in the southern Cape. Ph.D. diss., University of Chicago.

2339. Volman, T. P. 1984. Early prehistory of southern Africa. In *Southern African prehistory and paleoenvironments*, ed. R. G. Klein, pp. 169–220. Rotterdam: A. A. Balkema.

2340. Vondra, C. F., and Bowen, B. E. 1976. Plio-Pleistocene deposits and environments, East Rudolf, Kenya. In *Earliest man and environments in the Lake Rudolf Basin*, ed. Y. Coppens, F. C. Howell, G. L. Isaac, and R. E. F. Leakey, pp. 79–93. Chicago: University of Chicago Press.

2341. Vrba, E. S. 1974. Chronological and ecological implications of the fossil Bovidae at the Sterkfontein australopithecine site. *Nature* 250:19–23.

2342. Vrba, E. S. 1975. Some evidence of chronology and palaeoecology of Sterkfontein, Swartkrans and Kromdraai from the fossil Bovidae. *Nature* 254:301–304.

2343. Vrba, E. S. 1980. Morphological and environmental change: How do they relate in time? *South African Journal of Science* 76:61–84.

2344. Vrba, E. S. 1980. The significance of bovid remains as indicators of environment and predation patterns. In *Fossils in the making*, ed. A. K. Behrensmeyer and A. P. Hill, pp. 247–271. Chicago: University of Chicago Press.

2345. Vrba, E. S. 1981. The Kromdraai Australopithecine Site revisited in 1980: Recent investigations and results. *Annals of the Transvaal Museum* 33:18–60.

2346. Vrba, E. S. 1982. Biostratigraphy and chronology, based particularly on Bovidae, of southern hominid-associated assemblages: Makapansgat, Sterkfontein, Taung, Kromdraai, Swartkrans; also Elandsfontein (Saldanha), Broken Hill (now Kabwe) and Cave of Hearths. In *L'Homo erectus et la place de l'homme de Tautavel parmi les hominidés fossiles*, ed. M. A. de Lumley, pp. 707–752. Nice: Premier Congrès International de Paléontologie Humaine.

2347. Vrba, E. S. 1985. Early hominids in southern Africa: Updated observations on chronological and ecological background. In *Hominid evolution, past, present and future*, ed. P. V. Tobias, pp. 195–200. New York: Alan R. Liss.

2348. Vrba, E. S. 1985. Ecological and adaptive changes associated with early hominid evolution. In *Ancestors: The hard evidence*, ed. E. Delson, pp. 63–71. New York: Alan R. Liss.

2349. Vrba, E. S. 1988. Late Pliocene climatic events and hominid evolution. In *Evolutionary history of the "robust" australopithecines*, ed. F. E. Grine, pp. 405–426. New York: Aldine de Gruyter.

2350. Vrba, E. S. 1993. The pulse that produced us. *Natural History* 102 (5): 47–51.

2351. Vrba, E. S. 1995. The fossil record of African antelopes (Mammalia, Bovidae) in relation to human evolution and paleoclimate. In *Paleoclimate and evolution with emphasis on human origins*, ed. E. S. Vrba, G. H. Denton, T. C. Partridge, and L. H. Burckle, pp. 385–424. New Haven: Yale University Press.

2352. Vrba, E. S. 1995. On the connections between paleoclimate and evolution. In *Paleoclimate and evolution with emphasis on human origins*, ed. E. S. Vrba, G. H. Denton, T. C. Partridge, and L. H. Burckle, pp. 24–48. New Haven: Yale University Press.

2353. Vrba, E. S., and Panagos, D. C. 1982. New perspectives on taphonomy, palaeoecology and chronology of the Kromdraai apeman. *Paleoecology of Africa* 15:13–26.

2354. Wadley, L. 1993. The Pleistocene Later Stone Age south of the Limpopo River. *Journal of World Prehistory* 7:243–296.

2355. Waechter, J. D'A. 1964. The excavation of Gorham's Cave, Gibraltar, 1951–54. *Bulletin of the Institute of Archaeology* 4: 189–213.

2356. Waechter, J. D'A. 1973. The late Middle Acheulian industries of the Swanscombe area. In *Archaeological theory and practice*, ed. D. E. Strong, pp. 67–86. New York: Seminar Press.

2357. Wagner, E. 1984. Ein Jagdplatz des *Homo erectus* im mittel-pleistozänen Travertin in Stuttgart-Bad Cannstatt. *Germania* 62:229–267.

2358. Wagner, E. 1990. Ökonomie und Ökologie in den altpaläolithischen Travertinfundstellen von Bad Cannstatt. *Fundberichte aus Baden-Württemberg* 13:1–15.

2359. Wagner, G. A. 1996. Fission-track dating in paleoanthropology. *Evolutionary Anthropology* 5:165–171.

2360. Walker, A., and Leakey, R., eds. 1993. *The Nariokotome* Homo erectus *skeleton*. Cambridge: Harvard University Press.

2361. Walker, A., and Leakey, R. 1993. Perspectives on the Nariokotome discovery. In *The Nariokotome* Homo erectus *Skeleton*, ed. A. Walker and R. Leakey, pp. 411–430. Cambridge: Harvard University Press.

2362. Walker, A., and Leakey, R. 1993. The skull. In *The Nariokotome*

Homo erectus *skeleton*, ed. A. Walker and R. Leakey, pp. 63–94. Cambridge: Harvard University Press.

2363. Walker, A., Leakey, R. E., Harris, J. M., and Brown, F. H. 1986. 2.5-myr *Australopithecus boisei* from west of Lake Turkana, Kenya. *Nature* 322:517–522.

2364. Walker, A., and Ruff, C. B. 1993. The reconstruction of the pelvis. In *The Nariokotome* Homo erectus *skeleton*, ed. A. Walker and R. Leakey, pp. 221–233. Cambridge: Harvard University Press.

2365. Walker, A., Teaford, M. F., Martin, L., and Andrews, P. 1993. A new species of *Proconsul* from the early Miocene of Rusinga/Mfangano Islands, Kenya. *Journal of Human Evolution* 25: 43–56.

2366. Walker, A. C. 1978. Prosimian primates. In *Evolution of African mammals*, ed. V. J. Maglio and H. B. S. Cooke, pp. 90–99. Cambridge: Harvard University Press.

2367. Walker, A. C. 1981. Dietary hypotheses and human evolution. *Philosophical Transactions of the Royal Society of London*, ser. B, 292:57–64.

2368. Walker, A. C. 1981. The Koobi Fora hominids and their bearing on the origins of the genus *Homo*. In Homo erectus-*Papers in honor of Davidson Black*, ed. B. A. Sigmon and J. S. Cybulski, pp. 193–215. Toronto: University of Toronto Press.

2369. Walker, A. C. 1984. Extinction in hominid evolution. In *Extinctions*, ed. M. H. Nitecki, pp. 119–152. Chicago: University of Chicago Press.

2370. Walker, A. C. 1993. The origin of the genus *Homo*. In *The origin and evolution of humans and humanness*, ed. D. T. Rasmussen, pp. 29–47. Boston: Jones and Bartlett.

2371. Walker, A. C., Falk, D., Smith, R., and Pickford, M. 1983. The skull of *Proconsul africanus*: Reconstruction and cranial capacity. *Nature* 305:525–527.

2372. Walker, A. C., and Leakey, R. E. F. 1978. The hominids of East Turkana. *Scientific American* 239 (2): 54–66.

2373. Walker, A. C., and Leakey, R. E. F. 1988. The evolution of *Australopithecus boisei*. In *Evolutionary history of the "robust" australopithecines*, ed. F. E. Grine, pp. 247–258. New York: Aldine de Gruyter.

2374. Walker, A. C., and Pickford, M. 1983. New postcranial fossils of *Proconsul africanus* and *Proconsul nyanzae*. In *New interpretations of ape and human ancestry*, ed. R. L. Ciochon and R. S. Corruccini, pp. 325–351. New York: Plenum Press.

2375. Walker, A. C., Teaford, M. F., and Leakey, R. E. 1986. New information concerning the R114 *Proconsul* site, Rusinga Island, Kenya. In *Primate evolution*, ed. J. G. Else and P. C. Lee, pp. 144–149. Cambridge: Cambridge University Press.

2376. Walker, A. C., Zimmerman, M. R., and Leakey, R. E. F. 1982. A possible case of hypervitaminosis A in *Homo erectus*. *Nature* 296:248–250.

2377. Wallace, D. C., Garrison, K., and Knowler, W. C. 1985. Dramatic founder effects in Amerindian mitochondrial DNAs. *American Journal of Physical Anthropology* 68:149–155.

2378. Wallace, J. A. 1975. Dietary adaptations of *Australopithecus* and early *Homo*. In *Paleoanthropology, morphology and paleoecology*, ed. R. H. Tuttle, pp. 203–223. The Hague: Mouton.

2379. Walter, R. C., and Aronson, J. L. 1982. Revisions of the K/Ar ages for the Hadar hominid site, Ethiopia. *Nature* 295:140–142.

2380. Walter, R. C., Manega, P. C., Hay, R. L., Drake, R. E., and Curtis, G. H. 1991. Laser-fusion ^{40}Ar/^{39}Ar dating of Bed I, Olduvai Gorge, Tanzania. *Nature* 354:145–149.

2381. Wanpo, H., Ciochon, R., Yumin, G., Larick, R., Qiren, F., Schwarcz, H., Yonge, C., de Vos, J., and Rink, W. 1995. Early *Homo* and associated artefacts in Asia. *Nature* 378:275–278.

2382. Ward, C. V. 1993. Torso morphology and locomotion in *Proconsul nyanzae*. *American Journal of Physical Anthropology* 92:291–328.

2383. Ward, C. V., Ruff, C. B., and Walker, A. 1991. *Proconsul* did not have a tail. *Journal of Human Evolution* 21:215–220.

2384. Ward, C. V., Walker, A., and Leakey, M. G. 1997. New fossils of *Australopithecus anamensis* from Kanapoi and Allia Bay, Kenya. *American Journal of Physical Anthropology*, suppl., 24:235.

2385. Ward, C. V., Walker, A., Teaford, M. F., and Odhiambo, I. 1993. Partial skeleton of *Proconsul nyanzae* from Mfangano Island, Kenya. *American Journal of Physical Anthropology* 90:77–111.

2386. Ward, R., and Stringer, C. 1997. A molecular handle on the Neanderthals. *Nature* 388:225–226.

2387. Ward, S. C., and Hill, A. 1987. Pliocene hominid partial mandible from Tabarin, Baringo, Kenya. *American Journal of Physical Anthropology* 72:21–37.

2388. Ward, S. C., and Brown, B. 1986. The facial skeleton of *Sivapithecus indicus*. In *Comparative primate biology*, ed. D. R. Swindler and J. Erwin, 1:413–452. New York: Alan R. Liss.

2389. Ward, S. C., and Kimbel, W. H. 1983. Subnasal alveolar morphology and the systematic position of *Sivapithecus*. *American Journal of Physical Anthropology* 61:157–171.

2390. Ward, S. C., and Pilbeam, D. R. 1983. Maxillofacial morphology of Miocene hominoids from Africa and Indo-Pakistan. In *New interpretations of ape and human ancestry*, ed. R. L. Ciochon and R. S. Corruccini, pp. 325–351. New York: Plenum Press.

2391. Warren, S. H. 1911. Palaeolithic wooden spear from Clacton. *Quarterly Journal of the Geological Society (London)* 67:cxix.

2392. Warren, S. H. 1920. A natural "eolith" factory beneath the Thanet Sand. *Quarterly Journal of the Geological Society* 76:238–253.

2393. Washburn, S. L. 1960. Tools and human evolution. *Scientific American* 203 (9): 63–75.

2394. Washburn, S. L. 1985. Human evolution after Raymond Dart. In *Hominid evolution: Past, present and future*, ed. P. V. Tobias, pp. 3–18. New York: Alan R. Liss.

2395. Watchman, A. 1993. Evidence of a 25,000-year-old pictograph in northern Australia. *Geoarchaeology* 8:465–473.

2396. Waters, M. R., Forman, S. L., and Pierson, J. M. 1997. Diring Yuriakh: A Lower Paleolithic site in central Siberia. *Science* 275:1281–1284.

2397. Watson, V. 1993. Composition of the Swartkrans bone accumulations, in terms of skeletal parts and animals represented. *Transvaal Museum Monograph* 8:35–73.

2398. Wehmiller, J. F. 1982. A review of amino acid racemization studies in Quaternary molluscs: Stratigraphic and chronologic applications in coastal and interglacial sites. Pacific and Atlantic coasts, United States, United Kingdom, Baffin Island and tropical islands. *Quaternary Science Reviews* 1:83–120.

2399. Weidenreich, F. 1936. The mandibles of *Sinanthropus pekinensis:* A comparative study. *Palaeontologia Sinica,* ser. D, 7 (3): 1–162.

2400. Weidenreich, F. 1937. The dentition of *Sinanthropus pekinensis:* A comparative odontography of the hominids. *Palaeontologia Sinica,* n.s. D, 1:1–180.

2401. Weidenreich, F. 1939. On the earliest representatives of modern mankind recovered on the soil of east Asia. *Bulletin of the Natural History Society of Peking* 13:161–174.

2402. Weidenreich, F. 1941. The extremity bones of *Sinanthropus pekinensis:* A comparative study. *Palaeontologia Sinica,* n.s. D, 5:1–150.

2403. Weidenreich, F. 1943. The skull of *Sinanthropus pekinensis:* A comparative study on a primitive hominid skull. *Palaeontologia Sinica,* n.s. D, 10:1–485.

2404. Weidenreich, F. 1951. Morphology of Solo man. *Anthropological Papers of the American Museum of Natural History* 43:205–290.

2405. Weiner, J. S. 1955. *The Piltdown forgery.* London: Oxford University Press.

2406. Weiner, J. S., Oakley, K. P., and Le Gros Clark, W. E. 1953. The solution of the Piltdown problem. *Bulletin of the British Museum of Natural History (Geology)* 2 (3): 139–146.

2407. Wendorf, F. 1968. Site 117: A Nubian Final Paleolithic graveyard near Jebel Sahaba, Sudan. In *The prehistory of Nubia,* ed. F. Wendorf, 2:954–995. Dallas: Southern Methodist University Press.

2408. Wendorf, F., Close, A. E., and Schild, R. 1987. Recent work on the Middle Palaeolithic of the eastern Sahara. *African Archaeological Review* 5:49–63.

2409. Wendorf, F., Close, A. E., Schild, R., Gautier, A., Schwarcz, H. P., Miller, G., Kowalski, K., Królik, H., Bluszcz, A., Robins, D., and Grün, R. 1990. Le dernier interglaciaire dans le Sahara oriental. *Anthropologie* 94:361–391.

2410. Wendorf, F., Laury, E. L., Albritton, C. C., Schild, R., Haynes, C. V., Damon, P. E., Shafiqullah, M., and Scarborough, R. 1975. Dates for the Middle Stone Age of east Africa. *Science* 187:740–742.

2411. Wendorf, F., and members of the Combined Prehistoric Expedition. 1977. Late Pleistocene and Recent climatic changes in the Egyptian Sahara. *Geographical Journal* 143:211–234.

2412. Wendorf, F., and Schild, R. 1974. *A Middle Stone Age sequence from the central Rift Valley, Ethiopia.* Warsaw: Polish Academy of Sciences.

2413. Wendorf, F., and Schild, R. 1976. *Prehistory of the Nile Valley.* New York: Academic Press.

2414. Wendorf, F., and Schild, R. 1989. Summary and synthesis. In *The prehistory of Wadi Kubbaniya,* vol. 3, *Late Paleolithic archaeology,* ed. A. E. Close, pp. 768–824. Dallas: Southern Methodist University Press.

2415. Wendorf, F., and Schild, R. 1992. The Middle Paleolithic of north Africa: A status report. In *New light on the northeast African past*, ed. F. Klees and R. Kuper, pp. 41–78. Cologne: Heinrich-Barth Institut.

2416. Wendorf, F., Schild, R., Close, A. E., and Associates. 1993. *Egypt during the Last Interglacial: The Middle Paleolithic of Bir Tarfawi and Bir Sahara East.* New York: Plenum Press.

2417. Wendorf, F., Schild, R., Close, A. E., Schwarcz, H. P., Miller, G. H., Grün, R., Bluszcz, A., Stokes, S., Morawska, L., Huxtable, J., Lundberg, J., Hill, C. L., and McKinney, C. 1994. A chronology for the Middle and Late Pleistocene wet episodes in the eastern Sahara. In *Late Quaternary chronology and paleoclimates of the eastern Mediterranean*, ed. O. Bar-Yosef and R. S. Kra, pp. 147–168. Tucson: Radiocarbon.

2418. Wendorf, F., Schild, R., Close, A. E., Stewart, T. D., Angel, J. L., Kelley, J. O., Tiffany, M., and Hill, C. L. 1986. *The Wadi Kubbaniya skeleton: A late Paleolithic burial from southern Egypt.* Dallas: Southern Methodist University Press.

2419. Wendorf, F., Schild, R., and Haas, H. 1979. A new radiocarbon chronology for prehistoric sites in Nubia. *Journal of Field Archaeology* 6:219–223.

2420. Wendorf, F. E., and Schild, R. 1980. *Prehistory of the eastern Sahara.* New York: Academic Press.

2421. Wendt, W. E. 1976. "Art mobilier" from Apollo 11 Cave, south west Africa: Africa's oldest dated works of art. *South African Archaeological Bulletin* 31:5–11.

2422. Wengler, L. 1986. Position géochronologique et modalités du passage Moustérien-Atérien en Afrique du Nord: L'exemple de la grotte du Rhafas du Maroc oriental. *Comptes Rendus de l'Académie des Science, Paris*, ser. 2, 303:1153–1156.

2423. Wengler, L. 1990. Economie des matières premières et territoire dans le Moustérien et l'Atérien maghrébins: Exemples du Maroc oriental. *Anthropologie* 94:335–360.

2424. Wengler, L. 1990. Territoire et migrations humaines durant le Paléolithique moyen: Le cas du Maroc oriental. *Sahara* 3:35–44.

2425. Wengler, L. 1991. Choix des matières premières lithiques et comportement des hommes au Paléolithique moyen. In *Vingt-cinq ans d'études technologiques en préhistoire*, pp. 139–157. Juan-les-Pins, France: Editions APDCA.

2426. Weniger, G.-C. 1989. The Magdalenian of western central Europe: Settlement pattern and regionality. *Journal of World Prehistory* 3:323–372.

2427. Weniger, G.-C. 1990. Germany at 18 000 BP. In *The world at 18 000 BP*, vol. 1, *High latitudes*, ed. O. Soffer and C. Gamble, pp. 171–192. London: Unwin Hyman.

2428. Wesselman, H. B. 1995. Of mice and almost-men: Regional paleoecology and human evolution in the Turkana Basin. In *Paleoclimate and evolution with emphasis on human origins*, ed. E. S. Vrba, G. H. Denton, T. C. Partridge, and L. H. Burckle, pp. 356–368. New Haven: Yale University Press.

2429. West, F. H., ed. 1996. *American beginnings: The prehistory and palaeoecology of Beringia.* Chicago: University of Chicago Press.

2430. West, F. H. 1996. Beringia and New World origins: The archaeological evidence. In *American beginnings: The prehistory and palaeoecology of Beringia*, ed. F. H. West, pp. 537–559. Chicago: University of Chicago Press.

2431. Wheeler, P. E. 1984. The evolution of bipedality and the loss of functional body hair in hominids. *Journal of Human Evolution* 13:91–98.

2432. Wheeler, P. E. 1991. The thermoregulatory advantages of hominid bipedalism in open equatorial environments: The contribution of increased convective heat loss and cutaneous evaporative cooling. *Journal of Human Evolution* 21:107–115.

2433. Wheeler, P. E. 1992. The influence of the loss of functional body hair on the water budgets of early hominids. *Journal of Human Ecology* 23:379–388.

2434. Wheeler, P. E. 1993. The influence of stature and body form on hominid energy and water budgets: A comparison of *Australopithecus* and early *Homo* physiques. *Journal of Human Evolution* 24:13–28.

2435. White, F. J. 1996. Comparative socio-ecology of *Pan paniscus*. In *Great ape societies*, ed. W. C. McGrew, L. F. Marchant, and T. Nishida, pp. 29–41. Cambridge: Cambridge University Press.

2436. White, F. J. 1996. *Pan paniscus* 1973 to 1996: Twenty-three years of field research. *Evolutionary Anthropology* 5:11–17.

2437. White, J. P. 1996. Paleolithic colonization in Sahul Land. In *Prehistoric Mongoloid dispersals*, ed. T. Akazawa and E. J. E. Szathmáry, pp. 303–308. Oxford: Oxford University Press.

2438. White, J. P., and Habgood, P. J. 1985. La préhistoire de l'Australie. *Recherche* 16:730–737.

2439. White, J. P., and O'Connell, J. F. 1979. Australian prehistory: New aspects of antiquity. *Science* 203:21–28.

2440. White, J. P., and O'Connell, J. F. 1982. *A prehistory of Australia, New Guinea and Sahul.* New York: Academic Press.

2441. White, R. 1982. Rethinking the Middle/Upper Paleolithic transition. *Current Anthropology* 23:169–192.

2442. White, R. 1986. *Dark caves, bright visions: Life in Ice-Age Europe.* New York: W. W. Norton.

2443. White, R. 1989. Production complexity and standardization in early Aurignacian bead and pendant manufacture: Evolutionary implications. In *The human revolution: Behavioural and biological perspectives on the origins of modern humans*, ed. P. Mellars and C. Stringer, pp. 366–390. Edinburgh: Edinburgh University Press.

2444. White, R. 1990. Comment on "Symbolism and modern human origins." *Current Anthropology* 31:250–251.

2445. White, R. 1993. Technological and social dimensions of "Aurignacian-age" body ornaments across Europe. In *Before Lascaux: The complex record of the early Upper Paleolithic*, ed. H. Knecht, A. Pike-Tay, and R. White, pp. 277–299. Boca Raton, Fla.: CRC Press.

2446. White, R. 1995. Comment on "Concept-mediated marking in the Lower Palaeolithic." *Current Anthropology* 36:623–625.

2447. White, T. D. 1980. Evolutionary implications of Pliocene hominid footprints. *Science* 208:175–176.

2448. White, T. D. 1981. Primitive hominid canine from Tanzania. *Science* 213:348–349.

2449. White, T. D. 1982. Les australopithèques. *Recherche* 13:1258–1270.

2450. White, T. D. 1984. Pliocene hominids from the Middle Awash, Ethiopia. *Courier Forschungsinstitut Senckenberg* 69:57–68.

2451. White, T. D. 1985. *Acheulian man in Ethiopia's Middle Awash Valley: The implications of cut marks on the Bodo cranium.* Amsterdam: Nederlands Museum voor Anthropologie en Praehistorie.

2452. White, T. D. 1986. *Australopithecus afarensis* and the Lothagam mandible. *Anthropos* 23:79–90.

2453. White, T. D. 1986. Cut marks on the Bodo cranium: A case of prehistoric defleshing. *American Journal of Physical Anthropology* 69:503–509.

2454. White, T. D. 1987. Cannibals at Klasies? *Sagittarius* 2 (1): 6–9.

2455. White, T. D. 1987. Review of "Neogene paleontology and geology of Sahabi." *Journal of Human Evolution* 16:312–315.

2456. White, T. D. 1992. *Prehistoric cannibalism at Mancos Canyon 5MTUMR-2346.* Princeton: Princeton University Press.

2457. White, T. D., and Folkens, P. A. 1991. *Human osteology.* San Diego: Academic Press.

2458. White, T. D., and Johanson, D. C. 1989. The hominid composition of Afar Locality 333: Some preliminary observations. In *Hominidae: Proceedings of the Second International Congress of Human Paleontology*, ed. G. Giacobini, pp. 97–101. Milan: Jura Book.

2459. White, T. D., Johanson, D. C., and Kimbel, W. H. 1981. *Australopithecus africanus:* Its phyletic position reconsidered. *South African Journal of Science* 77:445–470.

2460. White, T. D., Suwa, G., and Asfaw, B. 1994. *Australopithecus ramidus*, a new species of early hominid from Aramis, Ethiopia. *Nature* 371:306–312.

2461. White, T. D., Suwa, G., and Asfaw, B. 1995. *Australopithecus ramidus*, a new species of early hominid from Aramis, Ethiopia. *Nature* 375:88.

2462. White, T. D., Suwa, G., Hart, W. K., Walter, R. C., Wolde-Gabriel, G., de Heinzelin, J., Clark, J. D., Asfaw, B., and Vrba, E. 1993. New discoveries of *Australopithecus* at Maka in Ethiopia. *Nature* 366:261–265.

2463. White, T. D., Suwa, G., Richard, G., Watters, J. P., and Barnes, L. G. 1983. "Hominid clavicle" from Sahabi is actually a fragment of cetacean rib. *American Journal of Physical Anthropology* 61:239–244.

2464. White, T. D., and Toth, N. 1991. The question of ritual cannibalism at Grotta Guattari. *Current Anthropology* 32:118–124.

2465. Wible, J. R., and Covert, H. H. 1987. Primates: Cladistic diagnosis and relationships. *Journal of Human Evolution* 16:1–22.

2466. Wickler, S., and Spriggs, M. 1988. Pleistocene human occupation of the Solomon Islands, Melanesia. *Antiquity* 62:703–706.

2467. Wiessner, P. 1982. Risk, reciprocity and social influences on !Kung San economics. In *Politics and history in band societies*, ed. E. Leacock and R. Lee, pp. 61–84. Cambridge: Cambridge University Press.

2468. Wijmstra, T. A., and van der Hammen, T. A. 1974. The Last Interglacial-Glacial cycle: State of affairs of correlation between data obtained from the land and from the ocean. *Geologie en Mijnbouw* 53:386–392.

2469. Wiley, E. O. 1981. *Phylogenetics: The theory and practice of phylogenetic systematics.* New York: John Wiley.

2470. Williams, D. F., Peck, J., Karabanov, E. B., Prokopenko, A. A., Kravchinsky, V., King, J., and Kuzmin, M. I. 1997. Lake Baikal record of continental climate response to orbital insolation during the past 5 million years. *Science* 278:1114–1117.

2471. Wintle, A. G. 1980. Thermoluminescence dating: A review of recent applications to non-pottery materials. *Archaeometry* 22: 113–122.

2472. Wintle, A. G., and Huntley, D. J. 1982. Thermoluminescence dating of sediments. *Quaternary Science Reviews* 1:31–53.

2473. Wintle, A. G., and Jacobs, J. A. 1982. A critical review of the dating evidence for Petralona Cave. *Journal of Archaeological Science* 9:39–47.

2474. Woillard, G. M. 1978. Grande Pile peat bog: A continuous pollen record for the past 140,000 years. *Quaternary Research* 9:1–21.

2475. Woillard, G. M., and Mook, W. G. 1982. Carbon-14 dates at Grande Pile: Correlation of land and sea chronologies. *Science* 215:159–161.

2476. WoldeGabriel, G., White, T. D., Suwa, G., Renne, P., de Heinzelin, J., Hart, W. K., and Helken, G. 1994. Ecological and temporal placement of early Pliocene hominids at Aramis, Ethiopia. *Nature* 371:330–333.

2477. Wolff, R. 1984. New specimens of the primate *Branisella boliviana.* *Journal of Vertebrate Paleontology* 4:570–574.

2478. Wolpoff, M. H. 1977. Some notes on the Vérteszöllös occipital. *American Journal of Physical Anthropology* 47:357–364.

2479. Wolpoff, M. H. 1979. The Krapina dental remains. *American Journal of Physical Anthropology* 50:67–114.

2480. Wolpoff, M. H. 1980. Cranial remains of the Middle Pleistocene European hominids. *Journal of Human Evolution* 9:357–364.

2481. Wolpoff, M. H. 1980. *Paleoanthropology.* New York: Alfred A. Knopf.

2482. Wolpoff, M. H. 1984. Evolution in *Homo erectus:* The question of stasis. *Paleobiology* 10:389–406.

2483. Wolpoff, M. H. 1985. Human evolution at the peripheries: The pattern at the eastern edge. In *Hominid evolution: Past, present and future,* ed. P. V. Tobias, pp. 355–365. New York: Alan R. Liss.

2484. Wolpoff, M. H. 1985. On explaining the supraorbital torus. *Current Anthropology* 26:522.

2485. Wolpoff, M. H. 1989. The place of Neandertals in human evolution. In *The emergence of modern humans: Biocultural adaptations in the later Pleistocene,* ed. E. Trinkaus, pp. 97–141. Cambridge: Cambridge University Press.

2486. Wolpoff, M. H. 1994. Time and phylogeny. *Evolutionary Anthropology* 3:38–39.

2487. Wolpoff, M. H. 1996. *Human evolution.* 1996–1997 ed. New York: McGraw-Hill.

2488. Wolpoff, M. H., and Caspari, R. 1990. On Middle Paleolithic/Middle Stone Age hominid taxonomy. *Current Anthropology* 31:394–395.

2489. Wolpoff, M. H., and Caspari, R. 1996. *Race and human evolution: A fatal attraction.* New York: Simon and Schuster.

2490. Wolpoff, M. H., Smith, F. H., Malez, M., Radovcic, J., and

Rukavina, D. 1981. Upper Pleistocene hominid remains from Vindija Cave, Croatia, Yugoslavia. *American Journal of Physical Anthropology* 54:499–546.

2491. Wolpoff, M. H., Thorne, A. G., Jelinek, J., and Zhang, Z. Y. 1994. The case for sinking *Homo erectus:* One hundred years of *Pithecanthropus* is enough! *Courier Forschungsinstitut Senckenberg* 171:341–361.

2492. Wolpoff, M. H., Thorne, A. G., Smith, F. H., Frayer, D. W., and Pope, G. G. 1994. Multiregional evolution: a World-wide source for modern human populations. In *Origins of anatomically modern humans,* ed. M. H. Nitecki and D. V. Nitecki, pp. 175–199. New York: Plenum Press.

2493. Wolpoff, M. H., Zhi, W. X., and Thorne, A. G. 1984. Modern *Homo sapiens* origins: A general theory of hominid evolution involving the fossil evidence from east Asia. In *The origins of modern humans: A world survey of the fossil evidence,* ed. F. H. Smith and F. Spencer, pp. 411–483. New York: Alan R. Liss.

2494. Wood, B. 1992. Early hominid species and speciation. *Journal of Human Evolution* 22:351–365.

2495. Wood, B. 1996. Origin and evolution of the genus *Homo.* In *Contemporary issues in human evolution,* ed. W. E. Meickle, F. C. Howell, and N. G. Jablonski, pp. 105–114. Memoir 21. San Francisco: California Academy of Sciences.

2496. Wood, B. 1997. *Ecce Homo*—behold mankind. *Nature* 390:520–521.

2497. Wood, B., and Turner, A. 1995. Out of Africa and into Asia. *Nature* 378:239–240.

2498. Wood, B. A. 1984. The origins of *Homo erectus. Courier Forschungsinstitut Senckenberg* 69:99–112.

2499. Wood, B. A. 1991. *Koobi Fora Research Project.* Vol. 4. *Hominid cranial remains.* Oxford: Clarendon Press.

2500. Wood, B. A. 1992. Old bones to match old stones. *Nature* 355:783–790.

2501. Wood, B. A. 1992. Origin and evolution of the genus *Homo. Nature* 355:783–790.

2502. Wood, B. A. 1993. Early *Homo:* How many species? In *Species, species concepts, and primate evolution,* ed. W. H. Kimbel and L. B. Martin, pp. 485–522. New York: Alan R. Liss.

2503. Wood, B. A., and van Noten, F. 1986. Preliminary observations on the BK 8518 mandible from Baringo, Kenya. *American Journal of Physical Anthropology* 69:117–127.

2504. Wood, B. A., Wood, C., and Konigsberg, L. 1994. *Paranthropus boisei:* An example of evolutionary stasis? *American Journal of Physical Anthropology* 95:117–136.

2505. Woodward, A. S. 1921. A new cave man from Rhodesia, South Africa. *Nature* 108:371–372.

2506. Wormington, H. M. 1964. *Ancient man in North America.* Denver: Denver Museum of Natural History.

2507. Wrangham, R. W. 1987. The significance of African apes for reconstructing human evolution. In *The evolution of human behavior: Primate models,* ed. W. G. Kinzey, pp. 28–47. Albany: State University of New York Press.

2508. Wrangham, R. W., Chapman, C. A., Clark-Arcadi, A. P., and

Isabirye-Basuta, G. 1996. Social ecology of Kanyawara chimpanzees: Implications for understanding the costs of great ape groups. In *Great ape societies*, ed. W. C. McGrew, L. F. Marchant, and T. Nishida, pp. 45–57. Cambridge: Cambridge University Press.

2509. Wright, H. E. 1991. Environmental conditions for Paleoindian immigration. In *The first Americans: Search and research*, ed. T. D. Dillehay and D. J. Meltzer, pp. 113–136. Boca Raton, Fla.: CRC Press.

2510. Wu, R. 1985. New Chinese *Homo erectus* and recent work at Zhoukoudian. In *Ancestors: The hard evidence*, ed. E. Delson, pp. 245–248. New York: Alan R. Liss.

2511. Wu, R., and Dong, X. 1985. *Homo erectus* in China. In *Palaeoanthropology and Palaeolithic archaeology in the People's Republic of China*, ed. R. Wu and J. W. Olsen, pp. 79–89. Orlando, Fla.: Academic Press.

2512. Wu, R., and Lin, S. 1983. Peking man. *Scientific American* 248 (6): 86–95.

2513. Wu, R., and Lin, S. 1985. Chinese palaeoanthropology: Retrospect and prospect. In *Palaeoanthropology and Palaeolithic archaeology in the People's Republic of China*, ed. R. Wu and J. W. Olsen, pp. 1–27. Orlando, Fla.: Academic Press.

2514. Wu, X. 1992. The origin and dispersal of anatomically modern humans in east and Southeast Asia. In *The evolution and dispersal of modern humans in Asia*, ed. T. Akazawa, K. Aoki, and T. Kimura, pp. 373–378. Tokyo: Hokusen-Sha.

2515. Wu, X., and Bräuer, G. 1993. Morphological comparison of archaic *Homo sapiens* crania from China and Africa. *Zeitschrift für Morphologie und Anthropologie* 79:241–259.

2516. Wu, X., and Poirier, F. E. 1995. *Human evolution in China: A metric description of the fossils and a review of the sites*. New York: Oxford University Press.

2517. Wu, X., and Wu, M. 1985. Early *Homo sapiens* in China. In *Palaeoanthropology and Palaeolithic archaeology in the People's Republic of China*, ed. R. Wu and J. W. Olsen, pp. 91–106. Orlando, Fla.: Academic Press.

2518. Würges, K. 1986. Artefakte aus den ältesten Quartärsedimenten (Schichten A–C) der Tongrube Kärlich, Kreis Mayen-Koblenz/Neuwieder Becken. *Archäologisches Korrespondenzblatt* 16: 1–6.

2519. Wymer, J. 1968. *Lower Palaeolithic archaeology in Britain*. London: John Baker.

2520. Wymer, J. J. 1955. A further fragment of the Swanscombe skull. *Nature* 176:426–427.

2521. Wymer, J. J. 1964. Excavations at Barnfield Pit, 1955–1960. In *The Swanscombe skull*, ed. C. D. Ovey, pp. 19–61. London: Royal Anthropological Institute.

2522. Wymer, J. J. 1988. Palaeolithic archaeology and the British Quaternary Sequence. *Quaternary Science Reviews* 7:79–98.

2523. Wynn, T. 1991. Tools, grammar and the archaeology of cognition. *Cambridge Archaeological Journal* 1:191–206.

2524. Wynn, T. 1995. Handaxe enigmas. *World Archaeology* 27:10–24.

2525. Yellen, J. E. 1996. Behavioural and taphonomic patterning at

Katanda 9: A Middle Stone Age site, Kivu Province, Zaire. *Journal of Archaeological Science* 23:915–932.

2526. Yellen, J. E., Brooks, A. S., Cornelissen, E., Mehlman, M. J., and Stewart, K. 1995. A Middle Stone Age worked bone industry from Katanda, Upper Semliki Valley, Zaire. *Science* 268:553–556.

2527. Yi, S., and Clark, G. A. 1983. Observations on the Lower Paleolithic of northeast Asia. *Current Anthropology* 24:181–202.

2528. Yi, S., and Clark, G. A. 1985. The "Dyuktai Culture" and New World origins. *Current Anthropology* 26:1–20.

2529. Zapfe, H. 1960. Die Primatenfunde aus der miozänen Spaltenfüllung von Neudorf an der March (Devvnská Nová Ves), Tschecoslowakei. *Schweizerische Palaeontologische Abhandlungen* 78:4–293.

2530. Zegura, S. L. 1987. Blood test. *Natural History* 96 (7): 8–11.

2531. Zhang, S. 1985. The early Palaeolithic of China. In *Palaeoanthropology and Palaeolithic archaeology in the People's Republic of China*, ed. R. Wu and J. W. Olsen, pp. 147–186. Orlando, Fla.: Academic Press.

2532. Zhang, Y. 1985. *Gigantopithecus* and "*Australopithecus*" in China. In *Palaeoanthropology and Palaeolithic archaeology in the People's Republic of China*, ed. R. Wu and J. W. Olsen, pp. 69–78. Orlando, Fla.: Academic Press.

2533. Zhen, Z. M., Kun, H. C. 1990. History of the dating of *Homo erectus* at Zhoukoudian. *Geological Society of America Special Paper* 242:69–74.

2534. Zihlman, A. L. 1982. *The human evolution coloring book*. New York: Barnes and Noble.

2535. Zihlman, A. L. 1985. *Australopithecus afarensis:* Two sexes or two species? In *Hominid evolution: Past, present and future*, ed. P. V. Tobias, pp. 213–220. New York: Alan R. Liss.

2536. Zubrow, E. 1989. The demographic modelling of Neanderthal extinction. In *The human revolution: Behavioural and biological perspectives on the origins of modern humans*, ed. P. Mellars and C. Stringer, pp. 212–231. Edinburgh: Edinburgh University Press.

2537. Zune, L. 1985. Reply to Binford and Ho. *Current Anthropology* 26:432–433.

REFERENCE INDEX

SITE INDEX

SUBJECT INDEX

Zdansky, Otto, 260
Zhoukoudianian Artifact
Industry. *See* Choukou-
tienian Artifact Industry
Zhoukoudian Locality 1 hu-
man fossils, 51, 53, 283, 299.
See also Zhoukoudian *in site
index*
Zhoukoudian Upper Cave hu-
man remains, 498, 502, 555.
See also Upper Cave *in site
index*

Zinjanthropus, 165, 213. *See
also* Olduvai Hominid 5
Zouhrah Cave human fossils,
397. *See also* Zouhrah Cave
in site index
Zuttiyeh human cranial fragment, 376, 377, 394. *See also*
Zuttiyeh *in site index*
Zygomatic arch (cheekbone),
64, 193, 199–201, 210, 212,
215, 223, 283, 288, 379